GEOLOGY OF GIANT PETROLEUM FIELDS

Memoir 14

GEOLOGY OF GIANT PETROLEUM FIELDS

a symposium

of papers on giant fields of the world including those presented at the 53rd Annual Meeting of the Association in Oklahoma City, Oklahoma, April 23-25, 1968.

Edited by

Michel T. Halbouty

Published by The American Association of Petroleum Geologists
Tulsa, Oklahoma, U.S.A., November 1970

Composed, printed and bound in the United States of America
by George Banta Company, Inc.

Contents

Foreword

When, in 1963, Oklahoma City was selected as the site for the 53rd Annual Meeting of The American Association of Petroleum Geologists, Edwin P. Kerr, Jr., Mobil Oil Corporation, was asked to assume—and he accepted—the thankless job of General Chairman. Under him, Mr. Kerr assembled—through persuasiveness and knowledge of *people*—a group of hard-driving, conscientious, and far-sighted scientists to organize the joint AAPG-SEPM program. Technical Program Coordinator was William P. Siard of the Getty Oil Company. AAPG Technical Program Chairman was Clifford B. Branan, Jr., consultant, and F. H. Hartman, Beard Oil Company, was appointed Technical Program Editor. The theme, "Geology of the Giants," was chosen by Mr. Kerr and his chairmen for the AAPG technical sessions, and invitations were sent to many well-known geologists throughout the world to present papers on the giant fields in their areas.

Mr. Kerr invited me to present the "Introduction to the Theme," a task which I gladly accepted, because this subject has been of intense interest to me. Thirty-one papers were given at these sessions on Tuesday through Thursday, April 23-25, 1968. Twenty-four of these are printed here, together with three papers from other sources, and the Introduction.

One of the papers in this volume is that by A. A. McGregor and C. A. Biggs, entitled "Bell Creek Field, an Embryonic Giant, Powder River and Carter Counties, Montana." This paper won the George C. Matson Award for the best AAPG paper presented at the Oklahoma City convention. Charles A. Biggs was given this coveted award at the Association's 54th Annual Meeting in Dallas, Texas, on April 15, 1969.

The convention committee at the outset desperately wanted the papers published in a single volume, and those on the committee were distraught when it became evident that a majority of the authors were not planning to present or submit complete papers. However, F. H. Hartman courageously took up the challenge and requested each author to give him his paper at the Oklahoma City meeting. About half

complied; this was not enough for Mr. Hartman who steadfastly pushed authors who had indicated an interest in publishing. He doubled the number of papers submitted, the last two of which were received in May 1969.

For a time, it appeared that none of these papers might be published at all for lack of funds; however, it finally and reluctantly was decided that a few would be published in the AAPG *Bulletin*. Fortunately, the AAPG Executive Committee received funds from an anonymous donor to publish the papers as a *Memoir* in a single volume. Then, because I had been intensely interested in these papers from the outset, and because I had given the "Introduction to the Theme" at Oklahoma City, the Executive Committee under President Frank B. Conselman graciously asked me in October 1968 to be Special Editor of this *Memoir*. I was more than happy to do this, because I believe strongly—then as now—that until we understand what makes oil and gas fields into true "giants" or mere single-location "bullseyes," industry never will succeed in keeping the world adequately supplied with petroleum for future generations.

I extend my thanks to the Oklahoma City Convention Committee Chairmen Kerr, Siard, Branan, and Hartman for their assistance, and I commend them for their foresight in selecting such an outstanding theme. Also, I take this opportunity to express my deepest gratitude to many other persons without whom this volume would not exist:

First: to the 52nd AAPG Executive Committee who made this volume possible: President Frank B. Conselman; Vice-President John E. Kilkenny; Secretary-Treasurer James M. Forgotson, Jr.; Editor John D. Haun; and Past-President and my successor, J. Ben Carsey.

Second: The AAPG editorial staff: A. A. Meyerhoff and Robert H. Dott, Sr., who rewrote many manuscripts, and translated some from their original languages into English; and Robert H. Dott, Sr., Ethylmae Tidwell, Emily Tompkins, and Howard A. Meyerhoff who did the final editing and proofreading.

Third: That dedicated group of Associate

viii Foreword

Editors and referees whose careful reviews improved each of these papers and helped to make them the excellent contributions that they are: W. F.-Auer, Grey H. Austin, Thomas D. Barber, James A. Barlow, Jr., Edward Barrett, Leroy E. Becker, Olin G. Bell, Robert R. Berg, Reed K. Bitter, Carl C. Branson, Jules Braunstein, A. E. H. Budwill, Angus S. Campbell, Jack O. Colle, E. J. Combs, Kenneth H. Crandall, David K. Davies, Ralph E. Edie, Ralph H. Espach, Jr., N. R. Fischbuch, George S. Garbarini, Eduardo J. Guzmán, James J. Halbouty, John C. Hazzard, Hollis D. Hedberg, Leo Hendricks, Harold N. Hickey, John M. Hills, Theodore S. Jones, Myron C. Kiess, Robert E. King, H. Douglas Klemme, Roy P. Lehman, E. M. Leavitt, Walter K. Link, Erick R. Mack, J. F. Mason, Pat McDonald, Gloria June McFarland, Donald P. McGookey, Lee H. Meltzer, John B. Miller, Peter Misch, William R. Moran, A. E. L. Morris, Edgar W. Owen, John B. Patton, R. S. Petty, Bayard D. Rea, E. C. Robinson, Amos Salvador, John R. Sandidge, O. B. Shelburne, L. L. Sloss, Walter M. Small, C. W. Spencer, J. D. Traxler, V. J. Veroda, M. A. Warner, and J. D. Weir.

Fourth: Dr. Kozo Kawai, of Tokyo, Japan, who supplied the invaluable Kudo compilation of data from mainland China, and Mrs. Nobu Ikeno Farrill, who translated the Japanese and Chinese documents assembled by Kudo and Kawai. I also thank Dr. K. M. Khudoley of The All-Union Geological Institute, Leningrad, USSR, for invaluable data on Russian fields.

Thus, in addition to the 44 authors of the 28 papers, no less than 62 additional judgments contributed to this volume. This *Memoir* is short compared with some, but the minds which produced it are many. *Appreciation* is not the only word I have; *thank you*—one and all—is the simplest way to convey my deep feelings of gratitude to all of you for your help.

MICHEL T. HALBOUTY, *Special Editor*
Houston, Texas
May 27, 1969

Geology of Giant Petroleum Fields: Introduction[1]

MICHEL T. HALBOUTY[2]

Houston, Texas 77027

PETROLEUM DOMINATES WORLD ENERGY AND WORLD POWER BALANCE

The very thought of a giant petroleum field, to most explorationists, is an exciting dream; relatively few giants have been found, so few of us ever have been associated with one, yet many more are needed to keep the world's petroleum industry alive and healthy. Therefore, the deliberate solicitation of a group of papers on giant fields to be presented at a convention of the world's largest geological society is truly a great forward step; I interpret the enthusiastic response as proof that explorationists—the oil-finders—are beginning to think positively; that they realize that continuing exploration for 1-million-bbl and 10-million-bbl "bullseyes" is *not* the way to salvage the future. Small fields are not to be ignored, I agree, but it is the giant fields which keep the nations of the world alive.

To wit: Burke and Gardner (1969) published some most informative statistics concerning only fields with more than 1 billion bbl of oil reserves; USSR-bloc nations' reserves and fields were excluded. These authors wrote that, of the then-proved 402.2 billion bbl of recoverable oil reserves originally in place, 87.9 percent were in only 71 fields. If one subtracts the 62.5 billion bbl already produced from these 71 fields, they still hold 296.7 billion bbl or 73.8 percent of the non-USSR-bloc reserves.

After Burke and Gardner's paper was published, about 69.9 billion bbl of additional reserves were added to Kuwait and Saudi Arabia, and possibly 20 billion to Alaska (Gardner, 1969). Thus the nonbloc nations' proved recoverable oil is about 386.6 billion bbl, of which 72 fields contain 80 percent, or approximately 309.3 billion bbl. Eight fields alone hold about 218.9 billion bbl, 56 percent of all proved recoverable oil outside the Soviet bloc.[3]

Similar conclusions have been reached in smaller areas, such as the United States alone, and North America alone. Halbouty (1968; this volume) shows that, of the nearly 23,000 fields in the United States, 261 oil fields and 47 gas fields will yield approximately 60 percent of the United States' ultimate production. Data of Moody *et al.* (this volume) show that 46 fields will yield 36 percent of the ultimate oil production of the North American continent.

The number of 1-billion-bbl plus fields in the Soviet bloc nations is about 20. Except for the Ploieşti complex, Rumania (and a few fields in mainland China), all fields known to be this large are in the USSR itself. Accurate oil production and reserve figures for the bloc nations are difficult to obtain. Many data which are published are suspect because of the amazing number of contradictions. In the USSR, oil-reserve data were placed in 1947 under the State Secrets Act (Campbell, 1968). Most current estimates credit the bloc nations with 56 billion bbl of proved oil reserves (Burke and Gardner, 1969), of which 35 billion are credited to the Soviet Union (Perrodon, 1966). King (1969) wrote that the proved oil reserve of the USSR probably exceeds 63 billion bbl; thus the bloc nations have a minimum proved reserve of 84 billion bbl as of the end of 1968. If extrapolation is made of Kudo's (1966) data from mainland China, proved bloc-nation reserves are between 96 and 110 billion bbl. However published mainland-China data are subject to large errors.

Perrodon (1966, p. 336) wrote that the ultimate recoverable oil from the USSR is in the order of 493.5 billion bbl. Assuming this figure to be correct, I predict that approximately 300 billion of this amount will come from less than 20 percent of the USSR's oil fields. However, the area underlain by marine sedimentary strata in the USSR is about 5.5 million mi² (14 million km²)—contrasted with 1.9 to 2.5 million mi² (4.6–6.3 million km²) in the United

[1] Manuscript received, May 25, 1969.

[2] Consulting geologist and petroleum engineer; independent producer and operator. Special Editor of this volume.

[3] These figures, based originally on Burke and Gardner (1969), have been modified by extrapolation from Gardner (1969) and other sources. More recent information subsequently has been obtained that some-

what modifies these figures. The resulting percentage changes are not significantly different from those given here. The more recent information is incorporated in the final paper of this volume.

States. Hence the estimate of 493.5 billion bbl of ultimately recoverable oil probably is low. For example, the West Siberian basin alone (750,000 mi^2 or 2,000,000 km^2) has a minimum estimated ultimate recoverable reserve of 190 billion bbl (King, 1969). This suggests that more than 1 trillion bbl of oil ultimately may be recovered from the USSR.

In 1965, according to Perrodon (1966), 45 of 570 oil *fields* (there were 1,172 oil and gas *pools* in the USSR in 1961, according to Vasil'yev and Khanin, 1963)—8 percent—held 80 percent of Russia's proved oil reserves. At least 12 of these 45 fields ultimately will produce more than 1 billion bbl.

Since 1965, at least 6 more giant oil fields have been discovered (King, 1969). In the remaining bloc countries, 5–10 fields probably hold 50 to 75 percent of the oil that will be produced, unless new fields are found. Of the bloc countries, only mainland China promises to be a major petroleum-producing nation. To date, its potential is barely tapped. Kudo (1966) has summarized what is known of mainland China's petroleum potential; there are perhaps 12 to 24 billion bbl of proved oil reserves, mainly in 3 to 6 fields.

Gas reserves apparently are distributed similarly, most gas being in a small number of fields. According to Beebe and Curtis (1968), only 82 fields in the United States have an ultimate gas recovery of more than 1 Tcf, but these 82 fields contain 45.8 percent of the remaining proved recoverable reserve. One field alone—Panhandle-Hugoton which is described in this volume—will produce about 10 percent of the gas ultimately recovered in the United States. Eleven Canadian giants produced 57.7 percent of Canada's gas in 1965, and two Mexican fields produced 53.5 percent of Mexico's gas during the same year.

In the USSR, the gas-reserve figures are not subject to the State Secrets Act. Even though the published reserve figures commonly are contradictory, the same conclusion emerges—most of Russia's gas will be produced from a few fields. The same appears to be true of mainland China, though few gas fields are developed because of the lack of a prepared market. Kudo (1966) estimated that minimum proved gas reserves as of mid-1966 were 23 Tcf.

On January 1, 1969, proved dry gas reserves of the USSR were 382.6 Tcf (Oil and Gas Jour., 1969), with about 60 Tcf of associated gas. Twenty-six gas fields in the USSR have a reserve of 309.9 Tcf, or nearly 80 percent of the nonassociated USSR total; of the 26 giants, 6 hold 60 percent of the proved national nonassociated gas reserve. These are the Urengoy (141 Tcf), Zapolyarnoye (53 Tcf), and Gubinskoye (12.5 Tcf) fields in the northeastern part of the West Siberian basin; Krasnyy Kholm (22 Tcf) in the Orenburg area, southern Volga-Ural region (King, 1969; *see* also Perrodon, 1966; Kortunov, 1967; King, 1968; World Petroleum, 1968); Gazli (17 Tcf) in Uzbekistan; and Shebelinka (14 Tcf) in the Ukraine (Oil and Gas Jour., 1969). Only Gazli and Shebelinka had produced any substantial amount of gas. King (1968) and Campbell (1968) gave proved, probable, and possible ultimate recovery as 2,100 and 2,278 Tcf, respectively, but these figures probably are low.

According to King (1969) and the January 1, 1968 report (World Petroleum, June 1968), 548 separate gas and condensate deposits exist in the USSR. Add this figure to Perrodon's (1966) 570 oil fields and the total is a minimum 1,118 fields in the Soviet Union as of January 1, 1968; this contrasts with Vasil'yev and Khanin's (1963) count of 1,172 pools in the Soviet Union on January 1, 1961. Because Vasil'yev and Khanin used the term "pool," these authors presumably included all the pools in each field to arrive at their total of 1,172. Regardless, the most significant fact from these statistics is that the USSR, with a total sedimentary-basin area of 5.5 million mi^2, had no less than 1,118 fields on January 1, 1968, contrasted with the United States' total of nearly 23,000 fields in a sedimentary-basin area of 1.9–2.5 million mi^2. *These sharply contrasting figures are a good index of the relative degree of maturity reached by the petroleum industry in the two nations.* In terms of future world power balance, these figures are most instructive and should provide some incentive for U.S. explorationists to "get with it."

Therefore, *the lesson is very plain:* giant fields keep nations in business, and it is toward such fields that we must direct our major efforts, because—contrary statements and implications notwithstanding—the day of economic production of petroleum from tar sand is only just arriving (Spragins, 1969); the day of economic production of coal and oil shale is not here, and a suitable technological "breakthrough" for economic production of such energy sources has not been developed (Halbouty, 1968; Cameron, 1969; Corcoran, 1969; Gonzalez, 1969). Petroleum still is "king."

If we do not find more petroleum—and soon —serious economic and political problems are facing several major world powers, including the United States. Some countries, notably those which are small and those which have few facilities for gaining technical knowledge, are petroleum-poor, and such nations always have had such problems. These countries—the "havenots"—become the pawns of the "haves" unless they have other large material resources. Moreover, the "haves" that are powerful nations, and that have stable governments and world-wide influence, can and *do* become "havenots." *When* they do, they are at the mercy of reserve-rich countries—some of them politically unstable, some politically hostile, and some both. The once-powerful and influential nation becomes a second-class power; it declines; and its influence wanes to nothing. This fate is a very real one and, in the United States, it is looming just over the horizon.

Our Responsibilities

For these reasons, I emphasize here our responsibilities—our need not just to be *aware* of them but also to *do* something about them and to *contribute* positively toward finding large fields—big fields—*giant* fields. Of the 27 papers which follow, 11 deal with United States fields, 14 describe foreign fields, and two (Moody *et al.;* Halbouty *et al.*) review areas outside the United States. The fields described in the first 25 papers produced 4 billion bbl of oil during 1967—30 percent of the 1967 world production figure of nearly 13.5 billion bbl.

I hope that this wide geographic distribution of papers (Fig. 1)[4] will bring home to each explorationist who reads them that geology, and large oil and gas fields, know no political boundaries; and that our scientific knowledge, thinking, and application of ideas should not be blinded or tainted by artificial barriers.

The data presented at this historic meeting of the Association will enable us to analyze what are the truly meaningful factors which constitute the anatomy of a giant. Something of each factor is in this volume: *source* of the petroleum, and its *migration* and *accumulation* in the trap; the *age of the trap,* from Cambrian through Pleistocene; the *type of trap,* anticline, anticlinal nose, monocline, faulted anticline, growth fault, reef, salt dome, stratigraphic trap, unconformity, and paleogeomorphic traps, as well as combinations

[4] Fig. 1 does not include the complete list of 269 giants summarized in the final paper (Halbouty *et al.*) for which there are separate maps.

of these—a full range of trap types: from Salt Creek to Bell Creek; from the Golden Lane to Swan Hills; from Groningen to Panhandle-Hugoton; from Ploieşti to the Asmari Limestone fields of Iran; or from Tom O'Conner to Bay Marchand–Timbalier Bay–Calliou Island; and, finally, the problems encountered, from discovery of the prospect to pumping of the petroleum through the pipeline. Engineering problems are included within the scope of many of these papers, and some indeed can and should be instructive to all explorationists. Wilmington, perhaps, is the most unusual example of a particular engineering problem.

History, too, is an important part of the volume. I refer not just to history of development of a field—how it was discovered, how it was drilled, and how it was developed—but also to history in the sense of technological achievement and advancement as, for example, at Panhandle-Hugoton and the most unusual, fascinating story of the Oklahoma City field. The sum of technological achievements during the development of the Oklahoma City field alone may have contributed more to the advancement of the petroleum industry than the advancements made during the development of any other oil field in the world.

Through reading these papers, all of us shall profit. By reading, we gain ideas. Mental retention of an *idea,* whether it be obtained from a study of Hassi er R'mel or Healdton, or from a comparison of Sarir with Ragusa and Rainbow-Zama, is the key to *new ideas.* It is these new ideas which will provide new petroleum, and which should prove once and for all that bookkeepers and accountants, lawyers and economists, who are too abundant and too powerful in the structure of exploration management (or *non*exploration management), actually contribute little to the discovery of the very petroleum which keeps these gentlemen in their jobs, from the clerk to the director level. Oil-finding is the job of explorationists—geologists and geophysicists—and as Kenneth H. Crandall (1969) has pointed out so lucidly, when the climate of exploration is bad, exploration suffers—in fact, it is *stifled.* In my dictionary, "to stifle" means ". . . kill by depriving of oxygen; to cut off (as the voice or breath); **to** withhold from circulation or expression; to become suffocated by or as if by lack of oxygen."

Nor is the nonexplorationist in management the only stifler. We stifle *ourselves*—by *not* reading, by limiting our horizons and breadth of thought, by wasting our leisure time, by

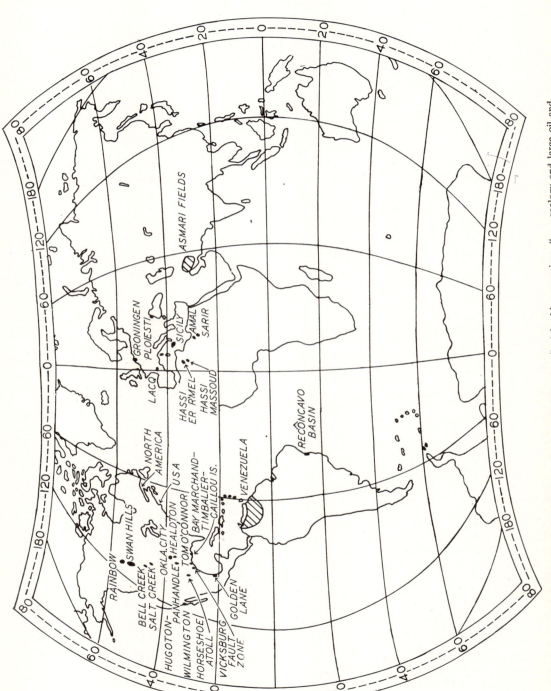

Fig. 1.—Geographic distribution of giant oil fields described in this symposium. ". . . geology and large oil and gas fields know no political boundaries. . . ."

failing to take advantage of ideas gained in other areas that could lead to petroleum discovery in our own back yards, and by failing to push our ideas, to *sell* them, because of fears of job security. We must learn, and learn well, what factors control petroleum generation, migration, and accumulation, so that we may apply our knowledge of those factors and sell our ideas to management, or to clients. If we fail this, we also shall fail to find the giants which are waiting to be found—not only in North America, but in the rest of the world.

As we make it a point to learn how these giant fields formed, we should study the modes of occurrence of the accumulations, the types of trap, how each trap formed and how it was found, the age of the reservoir and the age, or ages, of the sediments in which the petroleum generated and from which it was expelled and migrated to the trap. We should ask ourselves: first, what is usual about each of these accumulations? And second, what is *unusual?* Then we must concentrate on the unusual, for commonly it is that unusual aspect which is the key to accumulation. Prejudiced ideas should be discarded, for it is these old, ingrained, hardnosed prejudices which also stifle exploration; old prejudices must not be tolerated in our thinking of the future. The present plight of the U.S. petroleum industry is more than ample support for that last statement.

GIANT FIELDS STILL ARE TO BE FOUND

We should be encouraged and comforted in the knowledge that giant fields do exist and in the logic that, because they exist in reasonable numbers and are discovered steadily through the years, more remain to be found. They only wait for scientific knowledge and hardened determination to find them; or they wait for serendipity; or for the rank wildcatter who drills what the scientist has condemned—or where the exploration superintendent fears to tread.

The discovery of giant fields is not a "thing of the past," as so many of our industry's prophets—both within and without—like to tell us. As Kenneth Crandall (1969) has pointed out to us, and as I have stated many times, the hour has arrived to give exploration a rebirth, a new drive, and a gigantic effort. In my 1968 paper, which has been modified and updated for this volume, I show *why* there is excellent reason to believe that a respectable number of giant fields remain to be found in the United States alone—onshore as well as offshore.

WHAT CONSTITUTES A "GIANT" FIELD?

In my 1968 paper, I used the term "giant" as it is used most widely in the United States, Canada, and Mexico—a field having more than 100 million bbl of recoverable oil or more than 1 Tcf of recoverable gas. Beebe and Curtis (1968) used the term in the same way. Moody *et al.* (this volume), in contrast, define a giant oil field as one which has more than 500 million bbl of recoverable reserves. On a worldwide scale, particularly in North Africa and Middle East where 100-million-bbl and 1-Tcf fields are "a dime a dozen," it is not practical to use the limits which I used (1968; this volume) for the United States. Burke and Gardner's (1969) definition, in certain respects, is more practical—a "giant" field has more than 1 billion bbl of recoverable oil (and, I would add, for gas fields 10 Tcf, or more of recoverable gas). A "supergiant" field, in Burke and Gardner's terminology, contains more than 10 billion bbl of recoverable oil. Because only one field is now known to have more than 100 Tcf of recoverable gas, there probably are few gas fields which now can be called "supergiants," though Panhandle-Hugoton, Groningen, Hassi er R'mel, Zapolyarnoye, as well as several other gas fields nearly qualify as such. Urengoy recently has been reported to contain a gas reserve of 210 Tcf.

Therefore, instead of giving a rigid definition in terms of the absolute amount of the recoverable reserve, I prefer to define a giant in terms of its proximity to a large, accessible market. In the United States, Mexico, and much of Canada, a field with 100 million bbl or 1 Tcf of reserves still is a giant; in western Europe, such a field also is a giant—almost a supergiant. The same is true in Eastern Europe (including European USSR), southeastern Asia, Japan, China, Australia, central and southern Africa, and South America. However, in North Africa and the Middle East, as well as in Soviet Asia, a giant almost has to have 1 billion bbl or 10 Tcf of recoverable reserves to be economical. This also true of the Arctic regions of North America. Thus, in the final paper of this volume, a compromise definition is used. A giant oil field is defined as one which has 500 million bbl of reserves recoverable by present methods, and a giant gas field as one which has 3.5 Tcf of reserves recoverable by present methods.

PAPERS OF THIS VOLUME

The papers presented here—excluding the

final one (Halbouty *et al.*) in which the overall distribution of world giants is considered—include fields which have oil reserves of approximately 140 billion bbl and gas reserves of nearly 350 Tcf. This amount of oil is 3.5 times the present United States reserve, 14 times the Canadian reserve, and nearly a quarter of the world oil reserve. The gas in these fields is 20 percent more than the reserves in the United States, 8 times the Canadian gas reserve, and about a quarter of the world's gas reserves. Outside the United States, approximately 156 fields contain 80 percent of the world's reserves. Six of these foreign giants are supergiants; for example, Greater Kuwait, in a country slightly larger than Rhode Island, has 66 billion bbl of recoverable oil reserves—nearly twice the reserves of the United States. Our reserves in the United States are but 7 percent of the world total.

In the final paper by Halbouty *et al.*, all of the world's known giant oil and gas fields are summarized. *For the first time, accurate reserve figures are published for giant oil fields of the USSR and Mainland China.*

UNITED STATES RESERVE SITUATION

Of the 22,898 fields in the United States (as of January 1, 1968), including Alaska, only 308 (1.33 percent) are giant fields as defined by Halbouty (1968). Of these 308 giants, 96 were found by surface evidence and surface geology, 60 by subsurface geology, 119 by geophysics, and 33 by random drilling. Most of the U.S. giants were found by exploration directed *purposely* toward finding and exploiting structural traps.

In some areas, the search for structure located inadvertently, and certainly unintentionally, the more subtle types of traps—stratigraphic or paleogeomorphic. An important lesson to be learned from this fact is that, although only a few of these subtle traps have been found—and most of these accidentally—they are found *fairly persistently* in the discovery records through the years. From this observation, I conclude that this persistent, though "accidental," discovery history of stratigraphic-paleogeomorphic accumulations indicates that many more such traps are present in the United States, or, for that matter, in the world. If this conclusion is valid, it is not difficult to imagine what *could* be the course of future exploration if exploratory effort were directed purposely toward finding the subtle and obscure traps.

There are more East Texas-type giants to be

found in this country, and they will be found. I hope that they are found soon, and that they are found *purposely* by an enterprising company, unafraid of change, or by some independent geologist who has the courage to back his own convictions. Until several fields of this magnitude are found, we must concentrate on discovering the largest possible fields, because we need them to keep the national reserve-to-production ratio in balance; this ratio has fallen from 13 to 10 years since 1958. At the present rate of new-reserve discovery, this ratio will be down to 5 years by 1980 (Crandall, 1969).

From the preceding sections, it must be clear that the future exploration position of the United States is not the same as that of many foreign countries. Because exploration in most foreign areas has not been so extensive and thorough as in the United States (please recall the 1,118 fields of the USSR *versus* the 22,898 fields in the United States), it is apparent that further exploration in many foreign countries will result in the discovery of numerous new giants. Therefore, it is logical to assume that foreign petroleum exploration will continue to be focused on the obvious type of structural traps which can be found readily with today's exploratory tools. In contrast, the possibility of locating large structural accumulations in the United States, particularly onshore, is greatly diminished. Hence, it is imperative that we in the United States change our exploration methods and reorient exploratory thinking to search for the "unconventional" accumulations which are present. We *must* find these unconventional hidden giants to meet this nation's future needs.

UNITED STATES' NEEDS ARE EVERYONE'S NEEDS

It is evident that the use and availability of energy—whether it be oil or gas, coal, or nuclear—directly control the standard of living of people everywhere. Energy is essential to new growth, and to continuing growth, of any nation. Hence, the availability and security of energy sources are of prime importance to all industry and to all governments. Petroleum is the world's primary fuel and it will maintain that position as long as oil and gas can be found economically and in large quantities.

The petroleum industry is an *international* industry. Exploration for oil and gas is even more international than the industry itself. Moreover, this is an international symposium. I hope that every explorationist will gain some new insight, some new knowledge, and a spark of some new idea from this book, so that each of us—the finders of petroleum—can serve mankind, the world, and the petroleum industry more effectively by finding many more giant fields—everywhere.

REFERENCES CITED

Beebe, B. W., and B. F. Curtis, 1968, Natural gases of North America—a summary: Am. Assoc. Petroleum Geologists Mem. 9, v. 2, p. 2245–2355.

Burke, R. J., and F. J. Gardner, 1969, The world's monster oil fields, and how they rank: Oil and Gas Jour., v. 67, no. 2 (Jan. 13), p. 43–49.

Cameron, R. J., 1969, Outlook for oil shale, *in* Future energy outlook: a symposium given at the 53rd Annual Meeting of The American Association of Petroleum Geologists: Colorado School Mines, p. 69–77.

Campbell, R. W., 1968, The economics of Soviet oil and gas: Baltimore, Johns Hopkins Press, 279 p.

Corcoran, J., 1969, Outlook for coal, *in* Future energy outlook: a symposium given at the 53rd Annual Meeting of The American Association of Petroleum Geologists: Colorado School Mines, p. 61–67.

Crandall, K. H., 1969, Putting exploration back into focus: Am. Assoc. Petroleum Geologists Bull., v. 53, no. 10, p. 2055–2061.

Gardner, F. J., 1969, The numbers game—in Arabia: Oil and Gas Jour., v. 67, no. 25 (June 23), p. 71.

Gonzalez, R. J., 1969, Interfuel competition for future energy markets, *in* Future energy outlook: a symposium given at the 53rd Annual Meeting of The American Association of Petroleum Geologists: Colorado School Mines, p. 1–12.

Halbouty, Michel T., 1968, Giant oil and gas fields in United States: Am. Assoc. Petroleum Geologists Bull., v. 52, no. 7, p. 1115–1151.

———— Giant oil and gas fields in United States: this volume.

———— A. A. Meyerhoff, R. E. King, R. H. Dott, Sr., H. D. Klemme, and T. Shabad, 1970, World's giant oil and gas fields, geological factors affecting their formation, and basin classification: this volume.

King, R. E., 1967, Petroleum exploration and production in Europe in 1966: Am. Assoc. Petroleum Geologists Bull., v. 51, no. 8, p. 1512–1563.

———— 1968, Petroleum exploration and production in Europe in 1967: Am. Assoc. Petroleum Geologists Bull., v. 52, no. 8, p. 1439–1488.

———— 1969, Oil and gas exploration and production in the Soviet Union: London, Inst. Petroleum (Inst. Petroleum-AAPG Joint Conf.), 8 p. (Preprint.)

Kortunov, A. K., 1968, Gas in the Soviet Union—1967: World Petroleum, v. 39, no. 4, p. 28–37.

Kudo, Tadahiro, 1966, Chūgoku Tairiku no Sekiyu-shigen (Petroleum resources of continental China): Tokyo, Ajia Keizai Kenkyūjo (Inst. Asian Economic Affairs), Kaigaitōshi-Sankōshiryō (Research Material on Overseas Investment), no. 9, 42, Ichigaya-Honmurachō, Shinjuku-ku, Tokyo, 60 p. (in Japanese).

Moody, J. D., J. W. Mooney, and J. Spivak, Giant oil fields of North America: this volume.

Oil and Gas Journal, 1969, Russian gas reserves continue to soar: v. 67, no. 26 (June 30), p. 70–72.

Perrodon, A., 1966, Géologie du pétrole: Paris, Presses Univ. France, 440 p.

Spragins, F. K., 1969, Outlook for tar sands and other bitumens, *in* Future energy outlook: a symposium given at the 53rd Annual Meeting of The American Association of Petroleum Geologists: Colorado School Mines, p. 79–89.

Vasil'yev, V. G., and A. A. Khanin, 1963, Distribution of oil and gas pools in the section of the sedimentary cover of the USSR: Geologiya nefti i gaza, v. 7, no. 11; 1969, *in trans.:* Petroleum Geology, February 1969, v. 7, no. 11, p. 601–604.

World Petroleum, 1968, Siberian discoveries improve Russia's natural gas reserves: v. 39, no. 6, p. 57–60.

Giant Oil Fields of North America[1]

J. D. MOODY,[2] J. W. MOONEY,[2] and J. SPIVAK[2]
New York City, New York 10017

Abstract Forty-five oil fields with recognized ultimate recoveries larger than 500 million bbl have been found in seven provinces in North America. The earliest find was in 1865 and the latest in 1962. Recognition of more recent discoveries as giants will probably increase the population to 51 fields. Examination of reservoir age, lithology, trap and depositional environment leads to the conclusion that there are no diagnostic character-istics common to all of the giants; however, Paleozoic age, shallow-marine environment, sandstone lithology, and structural traps which have surface expression are the preferred modes. Uncommonly thick reservoir rock or unusual development of porosity and permeability are not required.

Twenty-five to twenty-eight more giant oil fields remain to be found, based upon current projections. Under present, or slightly improved, exploration economics the twenty-fifth future giant may not be found until 2060.

Introduction

There are 45 giant oil fields recognized at present in North America. The characteristics of these 45 giants are examined in order to make a prediction concerning future discoveries of fields of this size. For the purposes of this paper a giant oil field is defined as one whose ultimate recovery under known technology will equal or exceed 500 million bbl. A field is considered to be an aggregate of contiguous, overlapping or superimposed reservoirs whose combined ultimate production will satisfy this criterion. Gas, either associated or nonassociated with crude oil, has not been included.

At the moment there are about 140 giant oil fields recognized in the free world, and North America's 45 giants represent 32 percent of this total. In all of North America there are about 26,250 separate known fields, whose ultimate recovery is an estimated 132 billion bbl; the 45 giants represent only 0.2 percent of the total number of fields, but their 46 billion bbl of oil amount to 35 percent of the continent's total ultimate recovery.

[1] Read before the 53d Annual Meeting of the Association, Oklahoma City, Oklahoma, April 25, 1968. Manuscript received, May 7, 1968; accepted, July 30, 1968. Published by permission of Mobil Oil Corporation.

[2] Mobil Oil Corporation.

General Characteristics

Geographic Distribution

Figure 1 shows the geographic distribution of North America's giants, which are located in seven different producing areas. They are concentrated in the Mid-Continent area, with 13; and the Gulf Coast and California areas with 11 each.

Figure 2 shows the distribution of the giant fields by states or provinces. In this and subsequent bar-graph figures, the numbers of fields are represented by the bars and scale on the left; the adjacent bar on the right and the right-hand scale illustrate the total ultimate recoveries from that number of giants represented by the left-hand bar. The vertical scales on all of the bar graphs are the same. Texas with 14 of the giants, and California with 11, account for more than half of the total.

Historical Distribution

Figure 3 shows when the giants were found. Exploration is deemed to have started in 1860; the first giant, Bradford, was found in 1865, and the greatest number, 13, was found in the 10-year period from 1930 to 1940. The time relation of the discovery dates of the giants to the discovery dates of the first 25-million-bbl or greater field in each of the seven provinces is of interest (Fig. 4). Although 9 giants were discovered within 10 years of the discovery of a significant-size field in an area, 22, or nearly half, were found 30 years or more after a province had been demonstrated to be commercially petroliferous!

Method Leading to Discovery

It is commonly difficult to determine exactly why a particular well was drilled and, more often than not, several lines of evidence converging on a point determine the location of a wildcat. However, 23, or about half of the North American giants, are mappable and were mapped from surface indications, or are directly associated with surface seeps (Fig. 5). Another 15 were found under conditions where

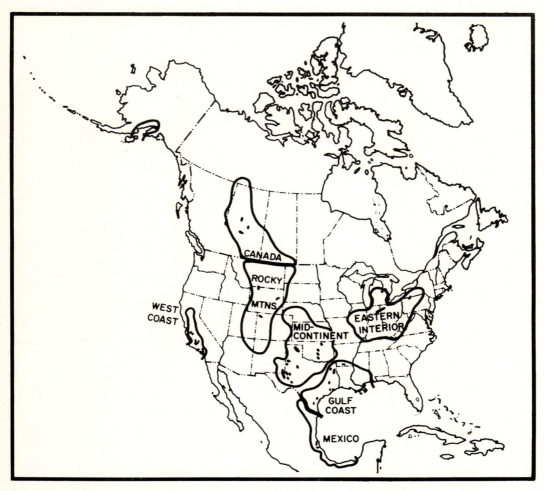

FIG. 1.—Geographic distribution of giant oil fields in North America.

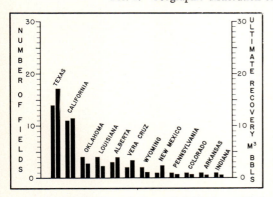

FIG. 2.—Distribution of giant oil fields by political subdivision. (*Note*: bars are paired; left-hand bar of each pair gives number of fields, right-hand bar gives ultimate recovery.)

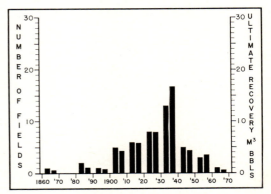

FIG. 3.—Year of discovery, plotted in decades. (*See* note for Fig. 2.)

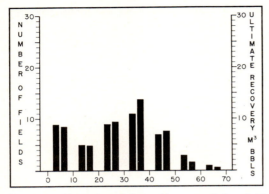

FIG. 4.—Time elapsed from discovery of first field greater than 25 million bbl to discovery of giant, plotted in 10-year periods. (*See* note for Fig. 2.)

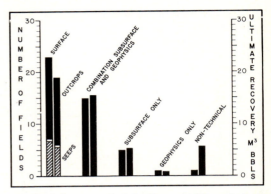

FIG. 5.—Primary method leading to drilling of giant-field discovery well. (*See* note for Fig. 2.)

it was impossible to assign the discovery specifically to either subsurface work or geophysics, and are therefore shown as combination finds. The remaining 7 are attributed to either subsurface alone (5), geophysics alone (1) or non-technical (1).

Geometric Characteristics

Ultimate recoveries.—Figure 6 shows the ultimate recoveries of the 45 currently recognized giants arranged in order of increasing size. The smallest is 500 million bbl, and the sizes increase gradually and fairly regularly to East Texas at 5,600 million bbl. Note that 13

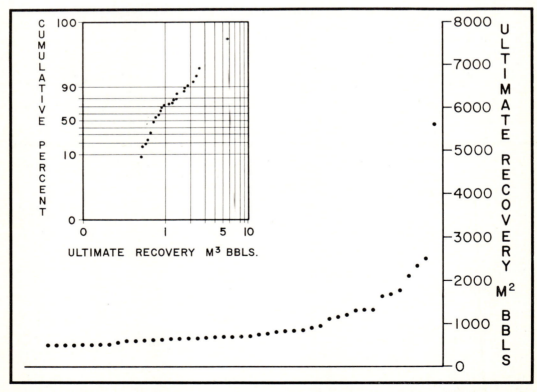

FIG. 6.—Size distribution of giant fields in order of magnitude. Inset shows data plotted on log-normal graph.

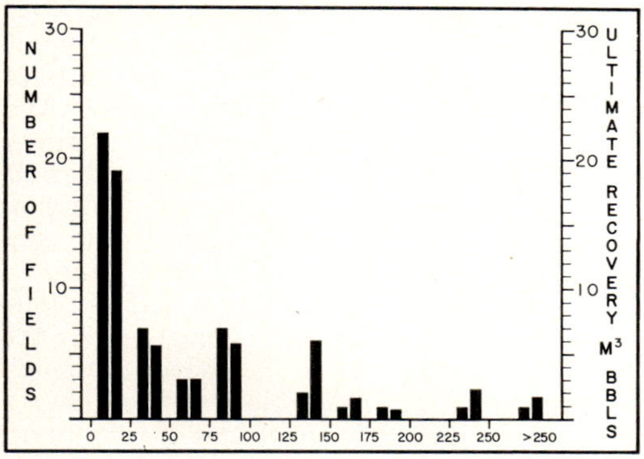

FIG. 7.—Surface productive acreage of giants, in thousands of acres. (*See* note for Fig. 2.)

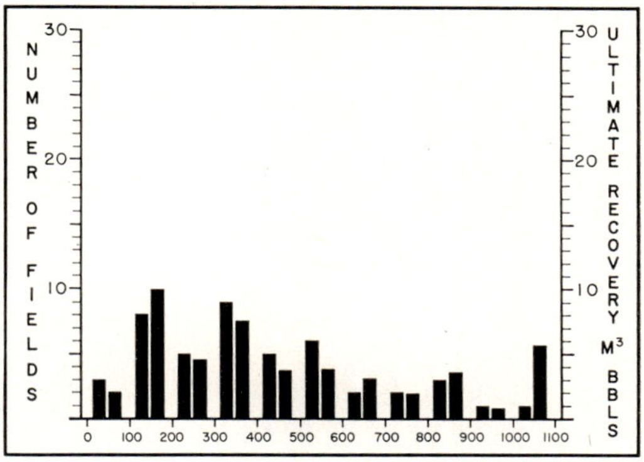

FIG. 9.—Recovery, bbl/acre-foot. (*See* note for Fig. 2.)

of the 45, or about one third, are larger than 1 billion bbl. The irregularities in the curve may represent undiscovered giants. This point will be referred to again later. The inset shows that the distribution of giant fields is approximately log normal.

Surface acreage.—Santa Fe Springs in the Los Angeles basin covers only 1,515 acres (Fig. 7). Pembina is 260 times the area of Santa Fe Springs, with 390,000 surface acres. However, almost half of the giants cover less than 25,000 surface acres.

Recovery factor.—Barrels-per-acre recovery is plotted on Figure 8. The range is from 2,000 to 558,000 (Long Beach), with two-thirds of the giants having recoveries of less than 50,000 bbl per acre. Only nine, or 20 percent of the giants, have recoveries greater than 600 bbl/acre-foot (Fig. 9). Thus, in general, giant oil fields are *not* characterized by unusually high reservoir capacity. The range is from about 50 to 1,000 bbl per acre-foot.

Average net feet of pay.—Similarly, the giant fields are not characterized by unusually thick pay zones (Fig. 10). The range is from about 30 to 1,440 ft, in terms of average net pay. Half the giants have pay thicknesses less than 100 ft, and only 9 have more than 200 ft.

Depth of pay.—The principal pay zones of two-thirds of North America's giant oil fields are above 6,000 ft, and all are above 12,000 ft (Fig. 11). The range is from 1,000 to 11,400 ft. The lower bar graph shows that practically no oil zones in the giant fields lie below 14,000 ft.

GEOLOGICAL CHARACTERISTICS

Trap

Type.—Although 19 giants are in anticlinal traps, there is a considerable spread in the trapping mechanism (Fig. 12). Fourteen are primarily stratigraphic (reef, pinchout, trunca-

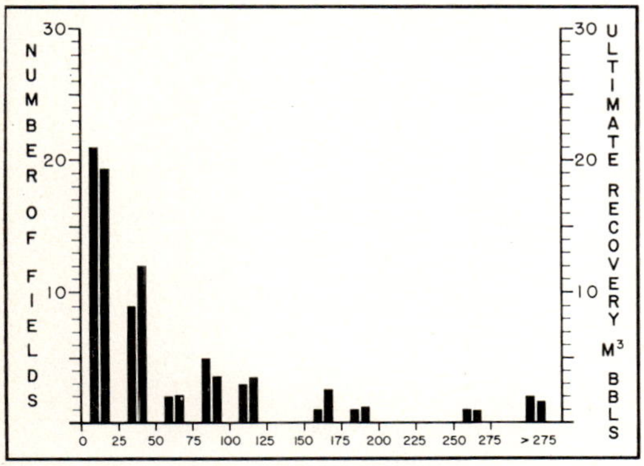

FIG. 8.—Recovery, thousands of barrels per acre. (*See* note for Fig. 2.)

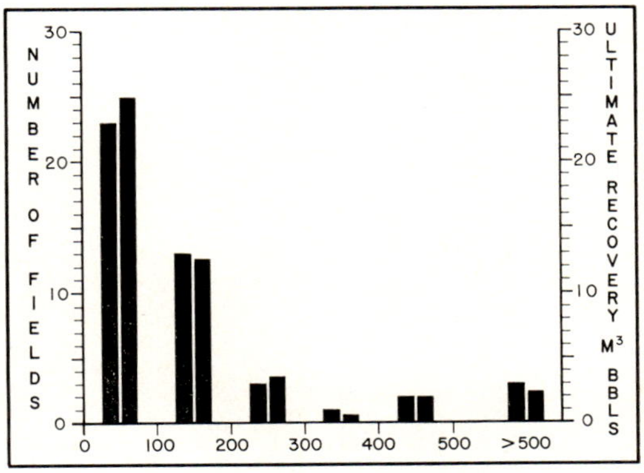

FIG. 10.—Average net feet of pay. (*See* note for Fig. 2.)

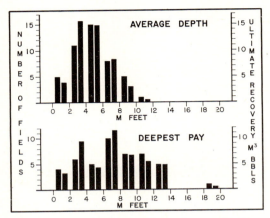

Fig. 11.—Depth of pay. Upper bar graph shows distribution of average depth of principal reservoirs. Lower bar graph shows depth to deepest oil producing horizon. (*See* note for Fig. 2.)

tion) and nine are combination structural and stratigraphic.

Basin position.—The shelf area (Fig. 13) is the preferred habitat of giant oil fields. (The term *shelf,* as used here, includes hinge lines on the shelf side of the basin.) However, the deep part and the tectonically active steep side of the basin both harbor enough giants to warrant exploratory attention.

Reservoir

Age.—Figure 14 shows the age of the reservoir rock plotted on a linear scale, with major stratigraphic subdivisions shown below. Because a number of the giants have more than one significant pay zone, the left-hand bars add up to more than 45. However, the right-hand bars still add up to 46 billion bbl. The concentration of 25 giants in Tertiary and Upper Cretaceous rocks is striking, although the giants contain reservoirs of all periods, and 23 have accumulations in late Paleozoic rocks. The greatest number of giants occur in late Tertiary rocks, but the greatest concentration of barrels falls in mid-Cretaceous, mostly in close proximity to the unconformity between Upper and Lower Cretaceous in the Gulf coastal plain.

Age and Lithology.—The principal reservoirs are reasonably evenly divided among the Cenozoic, Mesozoic, and Paleozoic; although on a reserve basis, Paleozoic reservoirs have a slight edge (Fig. 15). On a unit-of-time basis, however, Cenozoic rocks are by far the richest, as shown by the cross-hatched bars.

In terms of gross reservoir lithology, two-thirds of the giants are in sandstones, and one-third in carbonates. None of the Cenozoic giants is in carbonates, whereas 29 percent of the Mesozoic and 68 percent of the Paleozoic giants are in carbonate reservoirs.

Environment of Deposition.—Sands deposited in a shallow-marine environment have a decided edge, although winners have been found in continental, littoral, and deep-marine environments as well (Fig. 16). All the carbonate reservoirs were deposited in a shallow-marine environment, with reef or reef-complex associations predominating.

Relation to Unconformities

If there is an important unconformity in the section, a giant accumulation is more likely to be related to it than not (Fig. 17). However,

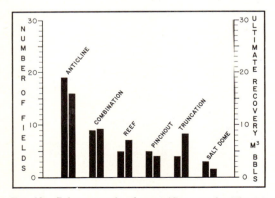

Fig. 12.—Primary mode of trap. (*See* note for Fig. 2.)

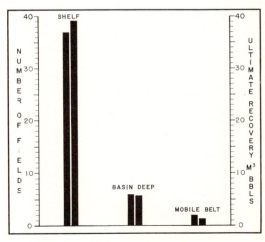

Fig. 13.—Basin position of giant fields. (*See* note for Fig. 2.)

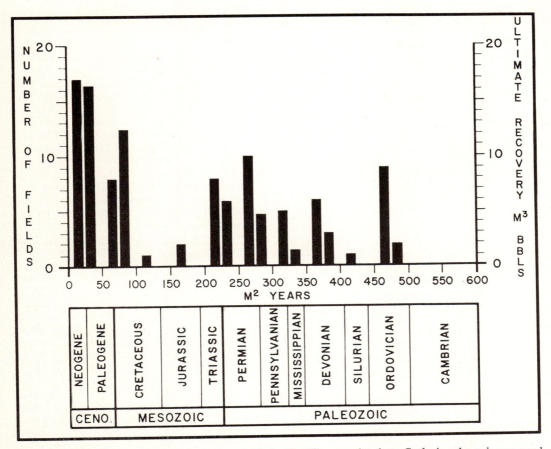

Fɪɢ. 14.—Radiometric age of reservoirs in giant fields, in 50-million year brackets. Geologic column is compared to linear scale in lower part of figure. (*See* note for Fig. 2.)

the absence of major unconformities does not rule out the occurrence of giant oil fields.

Relation to Evaporites

Twenty-six of the giants are in areas where there are evaporites (Fig. 18). Of these, 22 are related directly to the evaporite occurrence, mostly as components of facies systems containing carbonate reservoirs. Nine are found in relation to salt tectonics—three on shallow piercement salt domes and six on deep-seated domes or salt-controlled fault structures.

Conclusions

Selecting the most commonly occurring of the various parameters cited might lead to the conclusion that, to find a giant oil field, industry should look for surface indications of anti-

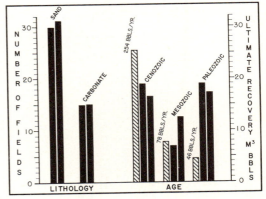

Fɪɢ. 15.—Summary of reservoir characteristics. Cross-hatched bars have been added to age breakout to show calculated number of barrels recovery per radiometric year. (*See* note for Fig. 2.)

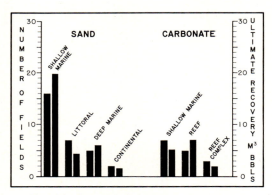

FIG. 16.—Reservoir lithology by depositional environment. (*See* note for Fig. 2.)

GIANTS OF THE FUTURE

An independent estimate indicates that about 110 billion bbl of oil remain to be found in North America. Of this total 38 billion bbl are attributed to areas where exploration has advanced to a point where it is unlikely that additional giant fields will be found. If past ratios hold for the future potential, about 25 billion bbl of the 72 billion attributed to new areas should be concentrated in giant fields. This suggests that about 25 giant oil fields remain to be found. Of the 25, eight should be in excess of 1 billion bbl. Smoothing out the size-distribution curve of ultimate recoveries (Fig. 6) by judiciously inserting the predicted 25 future giants would yield a much closer approximation to a log-normal distribution.

Discovered but as yet unrecognized giant fields probably exist. Examination of the larger, more recent, discoveries yields an estimate of six additional giant fields, which would increase the giant population from 45 to 51. On the assumption that East Texas is the largest field in a population of 51 "found" giants, a plot of the data to approximate more closely a log-normal distribution yields 28 future giants (Fig. 19).

This is a reasonable check on the previous estimate of 25, but this procedure yields 12 fields in the billion-plus category, with a hypothetical reserve of 32 billion bbl, suggesting that the estimate of 110 billion bbl yet to be discovered is conservative. There is no reason —geological, statistical, exploratory, or other— why East Texas should in fact be the largest field in North America. However, if a point were added to the curve beyond East Texas, it would be necessary to postulate the existence of a field in the 10-billion-bbl category!

clinal structure in Paleozoic sandstones on basin shelves with unconformities and evaporites, in Texas. No such combination is known to exist, much less in association with a giant oil field. There are no definable diagnostic criteria as to age, lithology, type of trap, or depositional environment—although to be sure, a shelf position in shallow-marine sediments is to be preferred. Unusually thick reservoir rock, or unusual developments of porosity and permeability are not required. The only obvious requirement is a trap of such capacity that a giant accumulation can be accommodated. The conclusion is reached that any trap sufficiently large to contain a giant accumulation above about 15,000 ft should be drilled, regardless of other parameters. The penalties of missing a giant by excluding, for example, continental or deep-marine facies, or the mobile side of the basin, or the basin deep, clearly overbalance the risks.

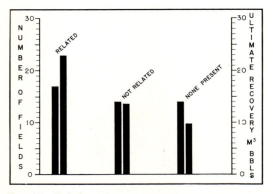

FIG. 17.—Relation of accumulations in giant fields to unconformities. (*See* note for Fig. 2.)

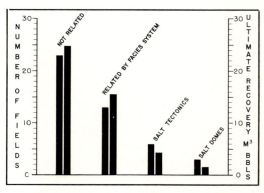

FIG. 18.—Relation of giant fields to evaporites. (*See* note for Fig. 2.)

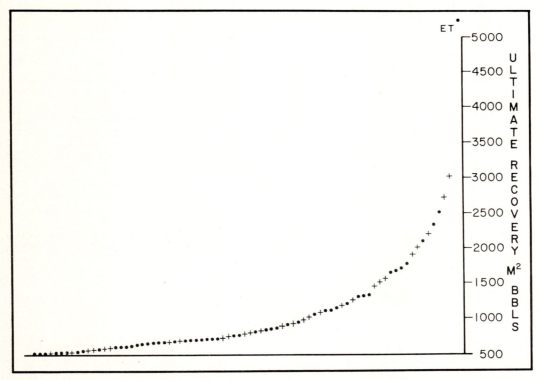

Fig. 19.—Giants in order of magnitude. Circles represent 45 known giant fields and six fields found but not yet known to be giants. Pluses represent giants of the future. **ET** = East Texas.

It is predicted, therefore, that 25 or more giant fields remain to be found in North America. They should range in size up to 3 billion bbl, with an outside chance for a 10-billion-bbl or larger field.

Discovery Rate

The solid line on Figure 20 is a cumulative time-frequency distribution curve for the discovered giant fields. The dashed part of the curve is a correction for the probability that there are giant fields already discovered but unrecognized. On the assumption of a similar time-rate of discovery for the next few decades, extrapolation of the curve along the dotted line to the limit forecast above would place discovery of the 76th giant in the year 2060.

Economic pressures may result in an improvement of this discovery rate. However, the oil industry is now operating in a deteriorating exploration environment. Industry is facing a profit squeeze resulting from relatively constant crude prices, rising costs, and a declining dis-

covery index (barrels found/foot of new-field wildcat drilled). If the exploratory environment continues to deteriorate, exploration will un-

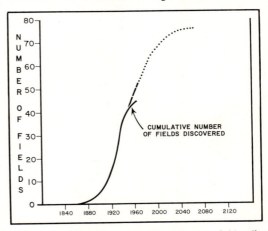

Fig. 20.—Cumulative number of giant fields discovered plotted against year of discovery. Dashed curve represents discovered but not recognized giants.

doubtedly be choked off before all, or even most, of the undiscovered giant fields are found.

Geographic Distribution

The majority of the future giants will be found in large traps, at reasonable depths, probably in sediments deposited in a shallow-marine environment, and probably on the shelf areas of depositional basins.

Just where these places might be cannot be predicted, but to quote a poet out of context:

. . . Full many a gem of purest ray serene
The dark unfathomed caves of ocean bear. . . .

Selected References

American Petroleum Institute, American Gas Association, and Canadian Petroleum Association, 1967, Reserves of crude oil, natural gas liquids, and natural gas in the United States and Canada: v. 21, July, 265 p.

Barton, D. C., and George Sawtelle, eds., 1936, Gulf Coast oil fields, a symposium: Am. Assoc. Petroleum Geologists, 1070 p.

Canadian Petroleum Association, 1966, Statistical year book: 153 p.

Cervera, Eduardo, 1967, Exploration and production in Mexico: 7th World Petroleum Conf. Proc., Mexico, 43 p.

Hubbert, M. K., 1967, Degree of advancement of petroleum exploration in United States: Am. Assoc. Petroleum Geologists Bull., v. 51, no. 11, p. 2207–2227.

Knebel, G. M., and G. Rodriguez-Eraso, 1956, Habitat of some oil: Am. Assoc. Petroleum Geologists Bull., v. 40, no. 4, p. 547–561.

Levorsen, A. I., ed., 1941, Stratigraphic type oil fields, a symposium: Am. Assoc. Petroleum Geologists, 902 p.

Oil and Gas Journal, 1967, Forecast-review: January 30, p. 136–202.

———— 1968, Forecast-review; February 5, p. 138–211.

Ver Wiebe, W. A., 1949, Oil fields in North America: Ann Arbor, Michigan, Edwards Brothers, 249 p.

APPENDIX A

North American Giant Oil Fields

(Ultimate Recovery of 500 Million Bbl or More)

Field	Location	Year Discovered	Discovery Method[1]	Cumulative Production to 1/1/67 (million bbl)	Ultimate Recovery (million bbl)
Bay Marchand, Block 2, Louisiana		1949	Gp, Sb	183	600
Bradford-Cattaraugus, Pa. and N. Y.		1865	Sp	581	658
Buena Vista, California		1909	S	575	700
Burbank, Oklahoma		1920	S	458	500
Coalinga, California		1887	Sp	568	640
Coalinga East Extension, California		1938	Sb	426	511
Caillou Island, Louisiana		1930	Gp, Sb	290	500
Conroe, Texas		1931	S, Sb, Gp, Sp	430	714
Cowden complex, Texas		1930	Sb	497	700
East Texas, Texas		1930	NT	3,712	5,600
Elk Basin, Wyoming and Montana		1915	S	317	566
Elk Hills, California		1919	S	276	1,303
Eunice—southward, New Mexico		1927	S, Gp, NT	1,635	2,334
Golden Lane, Veracruz, Mexico		1908	Sp	1,166	1,205
Golden Trend, Oklahoma		1945	Sb, Gp	310	500
Goldsmith-Andector complex, Texas		1934	Sb, Gp	517	693
Greta-Tom O'Connor complex, Texas		1934	S, Gp	394	719
Hastings, East and West, Texas		1934	Sb, Gp	365	665
Hawkins, Texas		1940	S, Gp, Sb	323	840
Huntington Beach, California		1920	S	772	1,150
Kelly-Snyder, Diamond M complex, Texas		1948	Gp	508	1,683
Kern River, California		1899	Sp	427	850
Lima Indiana, Ohio and Indiana		1885	Sp	482	514
Long Beach, California		1921	S	852	950
McElroy-Dune complex, Texas		1926	Sb, Gp	614	1,107
Midway-Sunset, California		1901	Sp	962	1,325
Oklahoma City, Oklahoma		1928	S	730	770
Panhandle, Texas		1910	Sp	1,158	1,647
Pembina, Alberta, Canada		1953	Sb, Gp	436	1,773
Poza Rica, Veracruz, Mexico		1930	S, Gp	883	2,100
Rangely, Colorado		1902	Sp	393	600
Redwater, Alberta, Canada		1948	Gp	309	817
Salt Creek, Wyoming		1906	S, Sb	442	510
Santa Fe Springs, California		1919	S, Sp	592	615
Seeligson complex, Texas		1937	Gp, Sb	411	752
Sho-Vel-Tum, Oklahoma		1914	S	743	901
Slaughter-Levelland, Texas		1936	Sb, Gp	492	691
Smackover, Arkansas		1922	Sb	489	525
South Pass Block 24, Louisiana		1950	Gp, Sb	256	500
Swan Hills and South, Alberta, Canada		1957	Sb, Gp	113	1,318
Ventura, California		1917	Sp, S	715	860
Wasson, Texas		1936	Sp, Gp	444	650
West Delta Block 73, Louisiana		1962	Gp, Sb	19	625
Wilmington, California		1936	Gp	1,098	2,500
Yates, Texas		1926	S, Sp	503	654

[1] S = Surface geology, Sb = Subsurface, Gp = Geophysics, Sp = Seepages, NT = Nontechnical.

Index map of North America showing area of study.

Geology of Middle Devonian Reefs, Rainbow Area, Alberta, Canada[1]

D. L. BARSS,[2] A. B. COPLAND,[2] and W. D. RITCHIE[2]

Calgary, Alberta

Abstract Data obtained from exploration for hydrocarbon-bearing Middle Devonian Rainbow Member reefs in northern Alberta, Canada, provides an excellent opportunity to examine the regional geological history of the Black Creek basin and the evolution of varied reef forms within the Rainbow portion of this basin.

During pre-Middle Devonian and Middle Devonian time, a cyclical sequence of redbeds (clastics and evaporites) of the Lower Elk Point subgroup were deposited in a shallow epicontinental sea. The incipient development of the Black Creek basin occurred at this time. Subsequently, negative epeirogenic movement resulted in widespread deposition of fine-grained, dark carbonate rocks of the Lower Keg River Member. Local faunal changes, as well as local changes in thickness (interpreted to be caused by "lime-mud" mounds), occur in the upper part of this unit. During the time of Upper Keg River deposition crinoidal beds were deposited in the form of a bank about 50 ft thick. The crinoidal bank extended throughout and beyond the Rainbow part of the Black Creek basin. In this bank, reef-constructing organisms flourished in several localities. The loci for concentration of reef organisms which led to rapid Rainbow Member reef growth, are believed to be the mud-mound topographic highs that were present in the underlying Lower Keg River Member. Structural control of reef growth, if it did exist, was subtle.

The reefs that grew in the Rainbow subbasin are characterized by pinnacle and atoll forms having vertical relief of up to 820 ft. Relatively rapid basin subsidence, combined with directional aspect of climate, paleography of the sea floor, and local bathymetry, controlled the external geometry, and to some extent, the internal facies of the reefs.

Detailed lithologic studies reveal 14 facies representing six depositional environments—basin, bank, fore-reef, organic reef, backreef, and lagoon. Superimposed on the original facies is a variable diagenetic history.

The growth of Shekilie barrier-reef complex across northwestern Alberta and adjacent areas of British Columbia and the Northwest Territories, and regional tectonic movements, altered depositional patterns late in Elk Point sedimentation. The barrier formed by the reef complex and structure prevented the free flow of normal marine waters southeast into the Black Creek basin. In this basin salinity of the water increased and the Black Creek Member salt was deposited, followed by the Muskeg anhydrites. The Muskeg evaporites completely infilled the Black Creek basin and covered the Rainbow Member reefs except for those present in the Shekilie barrier complex. The evaporite cover provided an effective seal for hydrocarbon entrapment. Reserves from the Rainbow Member pools in the Rainbow field and the Rainbow South field are estimated to be in excess of 1.2 billion and 165 million bbl of oil-in-place, respectively.

INTRODUCTION

The importance of Upper Devonian carbonates in providing reservoirs for hydrocarbon accumulation in Western Canada is well known. This paper reports on another Devonian carbonate reservoir—the Rainbow Member—which was discovered early in 1965. The stratigraphic relationship and distribution of significant Devonian units is shown on Figure 1. Reefs of Upper Devonian Leduc Formation occur in a basin covering about 70,000 mi[2] in central Alberta and contain in the order of 4.6 billion bbl of oil-in-place and 9.75 trillion cu ft of gas. The Swan Hills reefs of Middle-Late Devonian (Beaverhill Lake) age occur in a basin about 27,000 mi[2] in area and contain about 6.1 billion bbl of oil-in-place and 9 trillion cu ft of gas. The Rainbow Member reefs occur within the Black Creek basin, an area of about 11,000 mi[2], and contain in excess of 1.5 billion bbl of oil-in-place and 1 trillion cu ft of gas.[3] The significance of the recent discovery at Rainbow is evident from its large hydrocarbon reserves, and from the fact that a discovery of this kind leads to new cycles of exploration and discoveries in other strata. Regional studies indicate that reefs of equivalent age are present in an unexplored area exceeding 156,000 mi[2] in Western Canada.

The Rainbow Member reefs are of late Middle Devonian (Givetian) age. In the Rainbow fields area in the Black Creek basin they develop both a pinnacle and atoll form with a height up to 820 ft and an areal dimension up to approximately 6 mi[2]. In several wells the entire reef sections were cored throughout and many wells penetrated not only the reef section but also the pre-reef beds. The good quality of the data available has provided an excellent opportunity to study the lithology of the Rainbow

[1] Modified from a paper read by D. L. Barss before the 53d Annual Meeting of the Association, Giants Symposium, Oklahoma City, Oklahoma, April 24, 1968. Manuscript received, February 6, 1969.

[2] Banff Oil Ltd.

The authors thank G. E. Chin, M. E. Hriskevich, J. R. Langton, and S. Machielse for their assistance and constructive criticism. Acknowledgment is also due the Banff Oil Ltd. Drafting Department for their cooperation in preparation of the illustrations.

[3] Reserve figures from Canadian Petroleum Association 1967 Statistical Year Book, April 1968.

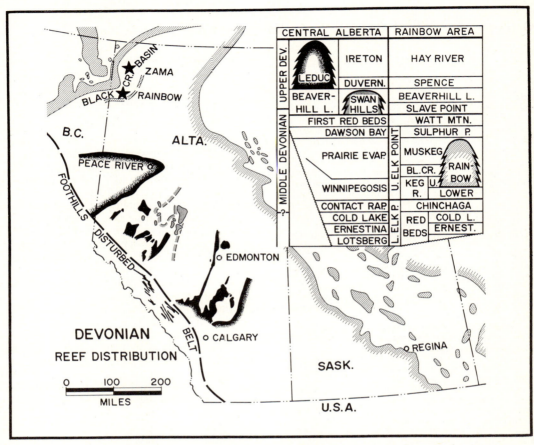

CENTRAL ALBERTA		RAINBOW AREA	
LEDUC	IRETON	HAY RIVER	
	DUVERN.	SPENCE	
BEAVER-HILL L.	SWAN HILLS	BEAVERHILL L.	
		SLAVE POINT	
FIRST RED BEDS		WATT MTN.	
DAWSON BAY		SULPHUR P.	
PRAIRIE EVAP.		MUSKEG	
		BL.CR.	RAIN-BOW
WINNIPEGOSIS		KEG R.	
			LOWER
CONTACT RAP.		CHINCHAGA	
COLD LAKE		RED BEDS	COLD L.
ERNESTINA			ERNEST.
LOTSBERG			

Fig. 1.—Devonian-reef distribution in Western Canada.

Member reefs and the pre-reef geological history.

This paper is divided into three parts—(1) regional geological history prior to, during, and after the time of Keg River deposition; (2) discussion of the inception of reef growth and forms, details of the lithofacies and the effects of post-reef salt solution, normal faults, and compaction; (3) summary of exploration highlights, reservoir factors, and reserves in the Rainbow and Rainbow South fields.

Terminology

The terms "reef," "reef complex," and "organic reef" have been defined by Klovan (1964) and appear to have been used in a consistent manner by recent authors. Klovan (1964), Lowenstam (1959), Cloud (1952), and Nelson et al. (1962), all stress the importance of the wave-resistant structure produced by reef organisms. Recently, Stanton (1967) suggested that important independent variables controlling the external form and facies of reefs are directional aspect of climate, paleography of the sea floor, and local bathymetry. He believes that these physiographic factors may exert more control on reef growth than biological factors. A study of the different external forms and facies present in the Leduc, Swan Hills, and Rainbow reef complexes—units that contain very similar organic-reef faunas—indicates that physiographic factors are significant. In addition, rate of basin subsidence may have had an important influence on the external form, and on facies.

Although physiography and subsidence were critical to the external form of the reef complex and, to some extent, the internal facies, we believe that the presence of sediment-binding and carbonate-producing organisms was of prime importance. Also, the study of the distribution of the fauna and detrital components within ancient reefs is a necessary prerequisite to reconstruction of environment of deposition and, as well, to a knowledge of reservoir characteristics. The writers have found the following definitions by Klovan (1964) to be the most satisfactory and have used them in the text.

Reef	Rigid carbonate structure with vertical dimensions significantly larger than the contemporaneous sediments, composed, at least in part, of organisms able to build and maintain the structure as a topographic feature on the sea floor and potentially in the zone of wave action.
Organic reef	That portion of the reef which is or was built directly by organisms and is responsible for the reef's wave resistant character.
Reef complex	The aggregate of reef limestone and related carbonate rocks.

Other definitions concerning reef terminology that are used in this paper are those of Langton and Chin (1968):

Pinnacle reef	Reef or reef complex as defined above, which developed without a backreef or lagoonal facies.
Atoll reef	An organic reef or reef complex which partly or completely surrounds lagoonal and backreef carbonate facies.

Although reef terminology is applied in a reasonably consistent manner by most authors, the conceptual views held by various authors of structures referred to as *banks* and *mounds* are not so clear at the present time. The definition of these terms is important in this paper as the writers examine carbonate masses of diverse shape, organic content, and possible origin.

Grabau (1913) described *banks* as structures formed by growth of shell colonies with little or no relief above the sea floor. He noted that sediments of the bank are interstratified with surrounding sediments. Cummings (1932) used a similar definition to define *shell banks* and recognized their difference from reefs. Lowenstam (1959, p. 433, 434) expanded on Cummings' concept; he differentiated well-defined low-lying structures comprised of unconsolidated banks of sponge, pelecypod, brachiopod, and crinoid material, from true reefs on the basis of the ecologic potential of the biota.

Thus, banks result from accumulation of bioclastic debris where the biota plays a passive role. Lowenstam noted that because of the passive role of the organisms, they do not grow above wave base and, above all, ". . . they lack the regenerative power to deposit carbonate in excess of loss caused by physical forces." Nelson *et al.* (1962, p. 234) followed Lowenstam's definition essentially, describing a ". . .bank as a skeletal limestone deposit formed by organisms which do not have the ecologic potential to erect a rigid, wave-resistant structure." Klement (1968) used a definition similar to that of Lowenstam. He indicated that banks are formed by organisms which do not have the ability to build a rigid, three-dimensional frame. He differentiated those banks formed by transport and accumulation of organisms from the in-place accumulations. In the latter group a three-fold subdivision is made into biogenic banks formed—by baffling action of the organisms, by binding action of organisms, and by localized growth of organisms.

Although it may be possible to recognize Klement's subdivisions in modern banks, their identification in ancient beds is not always possible. Moreover, banks may evolve into reefs, or intermittent development of reefs in predominantly bank-type structures may occur. As described by Walther and cited in Lowenstam (1959), an example of growth of sediment-binding organisms on a bank is present on the Pidgeon Bank in the Gulf of Naples. At this locality, calcareous algae are encrusted on Foraminifera-bearing sands. Lowenstam, in discussing this example, mentions that transition of banks to reefs may occur intermittently, and is dependent upon the degree of stability. At Rainbow, crinoidal banks generally underlie the reef, but in some localities stromatoporoid and coral fauna are abundant in the bank. Where this occurs, the concept of evolution from bank to organic-reef as expressed by Lowenstam, appears to be valid. The presence of facies and structures that are transitional from bank to reef does not present problems in terminology at Rainbow. The attributes of "reefs" have been examined previously. The widespread crinoidal unit at the base of the Upper Keg River Member[4] provides a good example of "bank" characteristics. The definition of this term is as follows:

---

[4] Local usage in stratigraphic nomenclature is followed in this article rather than strict application of rules of the American Commission on Stratigraphic Nomenclature.—Ed.

Bank A carbonate structure formed by organisms which do not build and maintain a structure on the sea floor in the zone of wave action. Sediments of the bank may be interstratified with sediments of either the basin or other carbonate masses and may or may not have topographic expression on the sea floor.

In the present study, several criteria are characteristic of bank sediments—massive encrusting fauna are rare or absent; typical faunas include crinoids, brachiopods, bryozoa, gastropods, branching corals, and *Stromatactis;* the accumulation of carbonate debris, particularly crinoid stems, is determined more by the natural breakdown of the organisms than by the forces of erosion; stratification with laterally equivalent nonbank beds is common and implies that the banks had little or no topographic relief.

Another type of carbonate structure is the *mud mound.* Carbonate structures that fall into this category have been described in the Mississippian of New Mexico by Pray (1958), at the base of the Niagaran reefs in Indiana by Textoris and Carozzi (1964), and in Devonian reefs of the Spanish Sahara by Dumestre and Illing (1967). The Mississippian mounds of Pray range from 25 to 350 ft in thickness and have flank dips up to 35°. The core facies comprises "lime mud"[5] and sparry calcite which passes abruptly into a flank facies of coarse crinoidal material. Fenestrate bryozoans and crinoids are the main faunal components of the core. The bryozoans occur in growth position and as fragmented debris, generally comprising less than 20 percent of the rock. Crinoids are present but in very small amount. Convincing evidence shows that the largest of these mounds had considerable topographic relief at the end of the period of growth. However, they probably did not grow into the zone of vigorous wave action.

The subreef beds of the Niagaran reefs consist of argillaceous calcisiltite which is fossiliferous near the reef structures. According to Textoris and Carozzi (1964, p. 423), the mounds started by "An apparently fortuitous bioclastic accumulation, mostly of crinoids" which "dilutes the calcite mud to form a mound of fossiliferous calcisiltite below wave base." With the addition of fistuliporid bryozoans and *Stromatactis,* the mound continued

[5] The term "lime mud," although slang, has been established in the literature, and is used herein to mean $CaCO_3$ mud.

to grow to wave base. At this stage construction of a reef frame began as a result of an increase in abundance of stromatoporoids.

The "T-bone" Middle Devonian reefs of the Spanish Sahara are thought by Dumestre and Illing (1967) to be mud mounds. These mounds are 60–90 ft thick and show depositional dips on the flanks of up to 34°. They consist essentially of "lime mud"; fauna is comprised of *Stromatactis,* auliporoid corals, finger corals, and crinoid ossicles. The mud mounds are believed to be the loci of later Middle Devonian reef growth.

The processes involved in the growth of mud mounds is not entirely clear. In Pray's (1958) view the fenestrate bryozoans in the New Mexico mounds played a significant role either by forming a current-baffle sediment trap or by forming a mesh-like fragment-constituted sediment-retaining mat. Pray suggested that the "lime mud" is indigenous and favors an algal origin for it even though distinct algal structures have not been recognized. Textoris and Carozzi (1964) favor a physio-chemical origin for the "lime mud" in the Niagaran mounds. The mounds grew simply by accumulating more bioclastic material than the surrounding areas. Binding of the "lime mud" into calcisiltite, they suggest, occurred through simple penecontemporaneous carbonate recrystallization.

A mud mound could therefore be defined as a mud-supported, fossiliferous, carbonate structure. The included organisms themselves do not have the ability to construct a rigid wave-resistant frame and for this reason mounds are believed to have grown mainly below the zone of vigorous wave action. Mud mounds have definite, though in some examples subtle, topographic relief. They are distinguished from banks in that they are mud supported, commonly have topographic relief, and have pronounced flank dips. The muds are generally somewhat argillaceous. Although the faunas present in mud mounds and banks are similar, they comprise a small percentage of mounds as compared with banks.

Local thickening of 20–50 ft in Lower Keg River strata is known beneath several reefs of the Rainbow area. Detailed log correlation has shown that in the 7–32–109–8 W6M well in the "A" pool (Fig. 11) a thickening of 35 ft occurs in the upper part of the Lower Keg River. In cores of the Lower Keg River in different wells, dips up to 15° have been observed. The faunal components include mainly crinoid material with minor bryozoans, styliolina, and gastropods, and rarely, branching corals and stromatoporoids. These faunas are present in argillaceous calcilutite and calcisiltite and rarely exceed 20 percent of the total rock volume. The evidence at hand favors the view that at least some of these Lower Keg River "thicks" are mud mounds and that they provided favorable sites for the inception of reef growth.

REGIONAL SETTING

Precambrian

Granite and hornblende gneisses, biotite granites, quartz diorites, and metasediments form the Precambrian rocks which constitute the "basement" in this region (Fig. 2). The surface on which sedimentation first began was irregular owing to the topographic expression of major tectonic trends. The three most important tectonic features are—the Tathlina high at the northern edge of the map sheet; the Peace River high in the south; and the Hay River-East Arm fault zone which crosses the area from northeast to southwest. Another feature, the Shekilie fault zone, crosses the western part of the area from northeast to southwest but its subsurface definition is poor. Other major faults are present but there is insufficient subsurface control to establish whether or not they are in trends. Burwash (1957) indicated that there are three Precambrian provinces present in northern Alberta.

Lower Elk Point Subgroup

The Lower Elk Point subgroup comprises a red, evaporitic and clastic sequence ranging from about 500 ft in the deepest part of the basin to zero where these sediments lap onto the Peace River high. Thickness of the Lower Elk Point is less than 50 ft over the Tathlina high (Fig. 3). The Lower Elk Point sediments overlie Precambrian "basement" except in the western part where they overlie Cambrian clastic rocks. The general basin configuration during the time of deposition of the Lower Elk Point persisted into Upper Elk Point deposition.

The Lower Elk Point subgroup comprises the following units in ascending order—"Basal Red Beds," Ernestina, Cold Lake, and Chinchaga (Fig. 4). The age of the Chinchaga Formation has been established by Law (1955) as Middle Devonian. The age of older units is not known, owing to lack of fossils. The Cold Lake and Ernestina Formations probably correlate with beds of the Mirage Point Formation

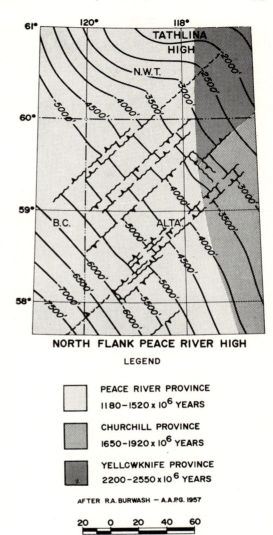

NORTH FLANK PEACE RIVER HIGH

LEGEND

PEACE RIVER PROVINCE
1180–1520 x 10⁶ YEARS

CHURCHILL PROVINCE
1650–1920 x 10⁶ YEARS

YELLOWKNIFE PROVINCE
2200–2550 x 10⁶ YEARS

AFTER R.A. BURWASH — A.A.P.G. 1957

20   0   20   40   60
MILES

FIG. 2.—Precambrian structure map.

which was considered by Norris (1965) to be Middle to Late Ordovician. More recently, Norris (personal commun.) and others have indicated that these beds may be of Middle and possibly Early Devonian age. Near the Tathlina high in the north and the Peace River high in the south, these units grade to coarse clastic deposits. Where definition of the individual units cannot be recognized, the sequence is referred to collectively as the "Elk Point sands."

The Lower Elk Point subgroup comprises several cycles of red clastic and evaporite de-

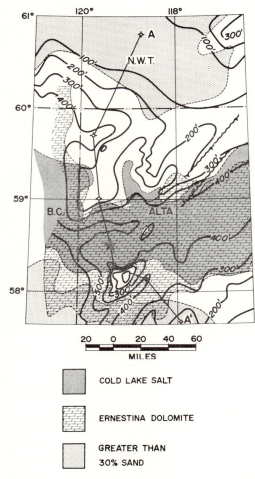

FIG. 3.—Isopach map of Lower Elk Point subgroup.
A-A' is trace of section in Figure 4. CI = 100 ft.

Map legend:
COLD LAKE SALT

ERNESTINA DOLOMITE

GREATER THAN
30% SAND

posits. These sediments were deposited in a widespread, shallow, epicontinental sea during deposition of which periodic restriction or shallowing caused development of evaporitic basins.

The first cycle consists of the "Basal Red Beds"—a red silt and shale unit with, locally, clear quartz "Granite Wash" type sandstone—overlain by red anhydrite and dense, finely crystalline dolomite of the Ernestina Formation. The maximum thickness of the Ernestina is 50 ft. Although relatively thin, it is present across a large area of northern Alberta. The distribution of this unit clearly marks the trend of the east-west Northern Alberta basin (Fig. 3); this trend persisted throughout the deposi-

tion of the rest of Middle Devonian sediments in northern Alberta.

The second cycle of the Lower Elk Point subgroup—the Cold Lake Formation—consists of a thin basal red silt and shale unit overlain by salt which ranges in thickness from 0 to 180 ft. The edge of the Cold Lake is roughly coincident with that of the underlying Ernestina Formation. There are small thickness changes in the Cold Lake Formation; to date, however, there has been no direct evidence of salt solution having taken place.

The third major cycle—the Chinchaga Formation—consists of a basal red shale and silt unit overlain by anhydrite, dolomite, and clastic rock. The anhydrite and dolomite grade into coarse clastic rock toward the Peace River high and toward the Tathlina high (Fig. 4). A clastic cycle can be recognized in the middle part of the Chinchaga and can be correlated regionally; its base is the boundary between the upper and lower members of the Chinchaga Formation as defined by Belyea and Norris (1962).

The regional deposition of the Chinchaga or equivalent strata indicates clearly that a major submergence of this northern region occurred. At the end of the Lower Elk Point deposition, the entire map area was one of very low relief. The general lack of lithologic or thickness changes of the individual units indicates little tectonic activity during deposition of the Lower Elk Point sediments; the only features of prominence were topographic highs that were present when sedimentation in this area began.

Upper Elk Point Subgroup

The Upper Elk Point subgroup contains, in ascending order—Keg River, Muskeg, Sulphur Point and Watt Mountain Formations (Fig. 5). The rocks of the Upper Elk Point subgroup are characterized by numerous facies changes; these can be related primarily to the occurrence of reef growth during the time of Keg River deposition. However, structural movement may have contributed to changes in basin configuration and depositional patterns. The Upper Elk Point subgroup ranges in thickness from about 1,100 ft in the central part of the basin to less than 100 ft over the Tathlina high (Fig. 6). Upper Elk Point strata onlap the Peace River high on the south.

Keg River Formation

The Keg River Formation is subdivided into lower and upper members which are readily recognized throughout the map area. Further

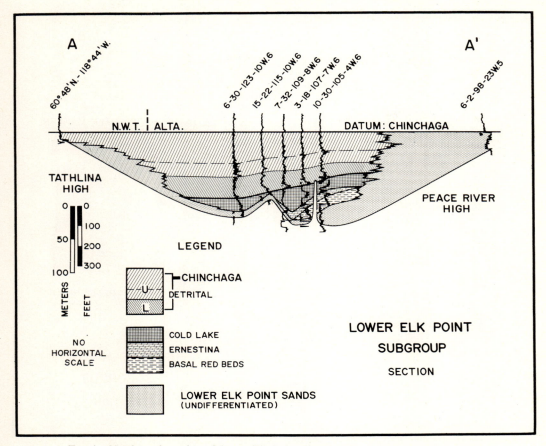

FIG. 4.—North-south section of Lower Elk Point subgroup. Trace is **A-A'** on Figure 3.

subdivision of the lower member into two units, and the upper member into three, is practical within the Rainbow subbasin.

Lower Keg River Member.[4a]—In the Rainbow subbasin the Lower Keg River Member ranges in thickness from 110 to 170 ft (Figs. 7 and 13). It has been divided into a lower and an upper unit by Hriskevich (1966). The lower unit consists of about 70 ft of dark bituminous, argillaceous, very fine-grained limestone and dolomite which contain variable amounts of crinoid debris, sporadic corals, and brachiopods; stromatoporoids are present in minor quantity. Dolomitization appears generally in the lower part of the unit. The basal contact of the lower unit of the Lower Keg River Member is generally recognized by the change from anhydrite to dolomite or limestone. However, in

some wells a thin primary dolomite is present above the anhydrite of the Chinchaga, and although the age of this dolomite is not clear, it is included with the Chinchaga Formation in this paper. The upper unit comprises beds of dark gray, black, bituminous, argillaceous, very fine-grained limestone with scattered layers of crinoids and brachiopods.

The bituminous, argillaceous limestones of the Lower Keg River or equivalent units are generally less than 150 ft thick but are present in a broad area of Western Canada sedimentary basin. The widespread nature of the carbonate rocks indicates that negative epeirogenic movement followed deposition of the Chinchaga evaporites. The bituminous content in the limestones suggest euxinic conditions—possibly produced by poor circulation on a continent-wide shelf.

Upper Keg River Member.—Within the

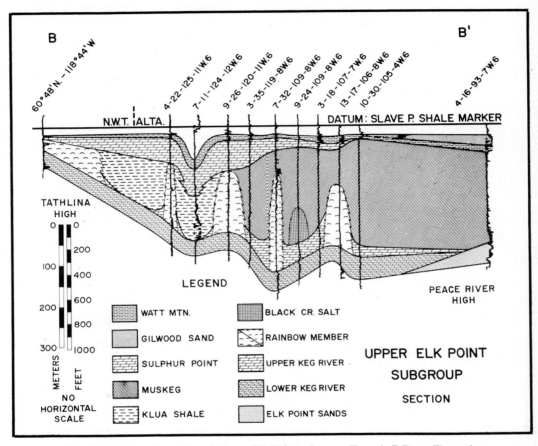

FIG. 5.—North-south section of Upper Elk Point subgroup. Trace is **B-B'** on Figure 6.

Black Creek basin, the Upper Keg River Member ranges from less than 40 ft west of the Zama area to about 150 ft in the Rainbow sub-basin. Northwest of the basin, in British Columbia and the Northwest Territories, and east of the basin in the Fort McMurray area, there is evidence that Upper Keg River beds were not deposited.

Hriskevich (1966) proposed a threefold subdivision of the Upper Keg River. Correlations of these units is readily possible throughout the Rainbow subbasin and in some other parts of the Black Creek basin. The lower unit consists of black to light-brown carbonate. It is generally dolomitized and contains a fauna comprising crinoid stems and fragments, corals, brachiopods, *Stromatactis,* and minor stromatoporoids. The lower unit ranges in thickness from zero where organic-reef facies are present, to about 50 ft in offreef sections. The middle unit

is about 50 ft thick and consists of interbedded dark, very bituminous and argillaceous, very fine-grained limestone and very fine-grained, slightly bituminous and argillaceous limestone. Tentaculites is the only significant fossil. The upper unit, also approximately 50 ft thick, is similar lithologically to the middle unit except that the bituminous and argillaceous content is less; laminated bedding is common. Fossils are sparce in the upper unit, with only a few isolated Amphipora and thin-shelled brachiopods present.

At some localities abundant massive stromatoporoids and corals became established in the lowermost unit of the Upper Keg River Member, and a wave-resistant structure was formed. These structures subsequently developed into Rainbow Member atoll and pinnacle-reef complexes. Figure 8 shows the regional distribution of reef complexes. The north-trending reef complex in the western part of the area is termed the "Shekilie barrier." This barrier-reef complex restricted circulation of water in post-reef time with the result that evaporites were deposited southeast of it. Normal marine shales were deposited on the west. The evaporites are included in the Muskeg Formation and the shales in the Klua Formation. East of the Rainbow locality another barrier-reef trend, referred to as the Hay River barrier by McCamis and Griffith (1967), is believed to be present. Definition of this barrier-reef trend is not well documented by subsurface control, but its presence is indicated by the distribution of the Black Creek salt and marked differences in anhydrite content of the lower part of the Muskeg Formation, between wells in the Rainbow area and wells southeast of the postulated barrier. The Black Creek basin, bordered by these regional barrier reefs, is the area in which pinnacle and atoll reefs grew. Figures 8 and 9 illustrate the positions of the barriers enclosing

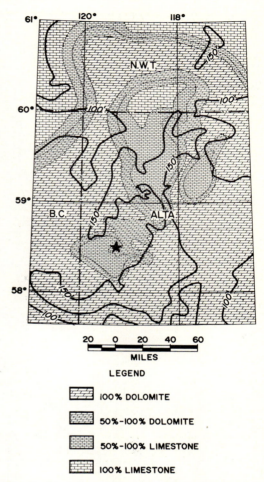

FIG. 7.—Isopach and lithofacies map of Lower Keg River Member. CI = 50 ft.

the Black Creek basin and the stable-shelf carbonates east of the Hay River barrier.

Muskeg Formation

In offreef wells in the Rainbow subbasin, the Muskeg Formation is readily subdivided into the Black Creek Member consisting of salt with minor anhydrites; and an unnamed upper member consisting of interbedded anhydrites, primary dolomites and secondary, locally porous dolomite of organic and detrital origin.

The Black Creek salt ranges locally in thickness from a maximum of 271 ft to 0, owing to solution and removal. At the base of the Black Creek salt, two thin anhydrite beds, 10–15 ft thick, are generally present and directly overlie the limestone of the Upper Keg River Member.

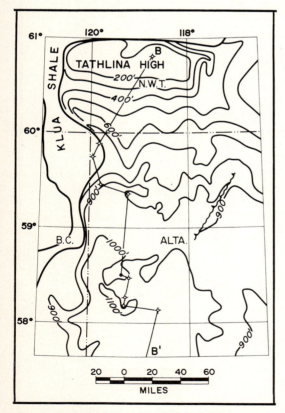

FIG. 6.—Isopach map of Upper Elk Point subgroup. B-B′ is trace of section in Figure 5. CI = 100 ft.

LEGEND

100% DOLOMITE

50%–100% DOLOMITE

50%–100% LIMESTONE

100% LIMESTONE

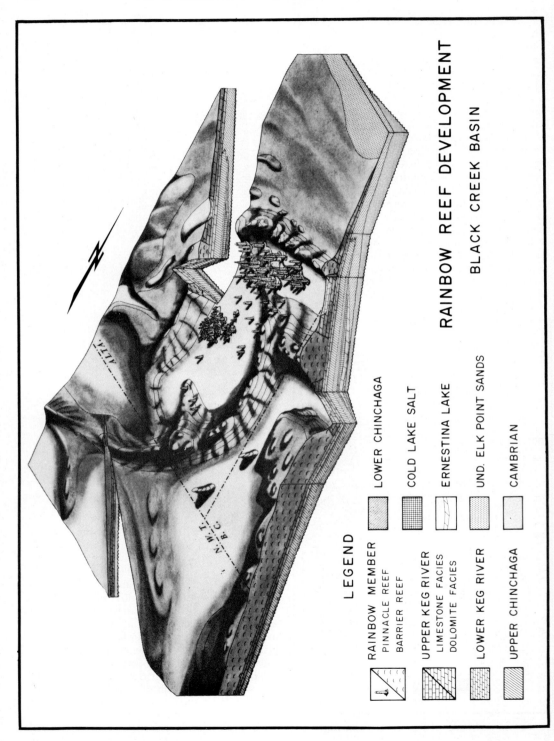

LEGEND

RAINBOW MEMBER
PINNACLE REEF
BARRIER REEF

UPPER KEG RIVER
LIMESTONE FACIES
DOLOMITE FACIES

LOWER KEG RIVER

UPPER CHINCHAGA

LOWER CHINCHAGA

COLD LAKE SALT

ERNESTINA LAKE

UND. ELK POINT SANDS

CAMBRIAN

RAINBOW REEF DEVELOPMENT

BLACK CREEK BASIN

FIG. 8.—Rainbow reef development in Black Creek basin.

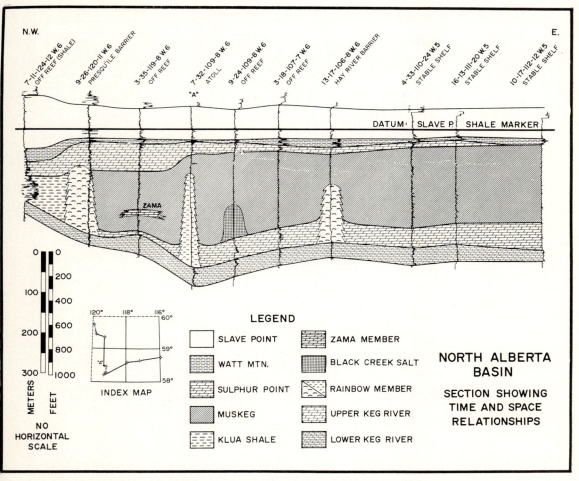

FIG. 9.—Northwest-southeast section of North Alberta basin, showing time and space relations. "A" well is in "A" pool, Figure 13.

The contact between the Black Creek Member and the overlying anhydrite is sharp.

In the Zama part of the Black Creek basin, the Muskeg sequence has been subdivided, in ascending order, into the Black Creek, Lower Anhydrite, Zama, Upper Anhydrite and Bistcho Members by McCamis and Griffith (1967). Black Creek salt, as much as 124 ft thick, is established by well control in the Zama area, and salt ranging in thickness from 0 to 170 ft is also present further to the northeast in the Steen River area. The Lower and Upper Anhydrite Members of the Muskeg Formation are separated by porous, brown, laminated, fine- to medium-crystalline saccharoidal dolomites of the Zama Member which range up to 110 ft thick. The Zama Member lies directly on the high part of reef buildups and overlies the Lower Anhydrite Member in the off-reef areas as illustrated by McCamis and Griffith

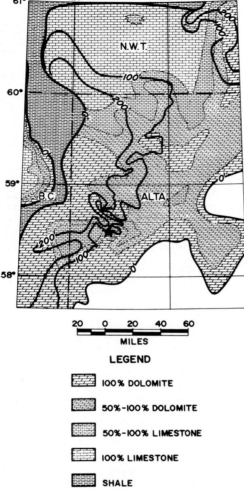

20   0   20   40   60

MILES

LEGEND

100% DOLOMITE

50%-100% DOLOMITE

50%-100% LIMESTONE

100% LIMESTONE

SHALE

FIG. 10.—Isopach and lithofacies map of Sulphur
Point Formation. CI = 100 ft.

(1967, Fig. 20). Fossils are sparse and consist
of amphiporoids, stromatoporoids, crinoids,
and brachiopods. The Zama Member is not
recognized in the Rainbow subbasin.

Sulphur Point Formation

McCamis and Griffith (1967) have introduced the name Bistcho Member to include the
beds which previously were included in the Sulphur Point and Presqu'ile Formations. The
Bistcho Member is considered by them to be a
facies variant of part of the Muskeg evaporites.

The writers favor retention of the name Sulphur Point for the following reasons—the
name and type section of the Sulphur Point

Formation was established by Norris (1965);
the name has priority and has gained wide acceptance within the industry; and the upper
and lower contacts are easily identified on
lithologic and mechanical logs in a wide area
of northern Alberta, northeast British Columbia, and the Northwest Territories. As there is
evidence of a disconformity between the Sulphur Point Formation and underlying Muskeg
Formation, we conclude that the Sulphur Point
carbonate rocks were deposited subsequent to
the Muskeg evaporites and are not a facies variant of the upper part of the Muskeg Formation.

The name Presqu'ile is not used consistently
by different authors in the area, particularly in
subsurface correlations. Generally, the name
refers to the dolomitized part of the Sulphur
Point Formation. When used in this sense, the
upper and lower contacts of the dolomites are
commonly poorly defined. For these reasons,
we recommend discontinuing use of the name
Presqu'ile.

The Sulphur Point Formation ranges in
thickness from zero along its eastern edge, to
more than 250 ft in the western part of the
map area (Fig. 10). The lower contact with
the anhydrite of the Muskeg is sharp. The presence of a disconformity is indicated by the relations of the lower contact of the Sulphur
Point with the Muskeg as shown in Figure 23.
At the top of the Muskeg Formation the presence of an extra unit in a low reef-flank position (4–9 well, Fig. 23) may indicate compaction in the underlying Muskeg beds or possibly
erosion before deposition of the Sulphur Point.
Abrupt local changes in thickness of the Sulphur Point, as seen in several wells in the Rainbow locality, also indicate the presence of a
disconformity at the base of the Sulphur Point.
It is possible, however, that these local changes
in thickness reflect normal fault movements
following Muskeg deposition or during deposition of the Sulphur Point. The upper contact
with the Watt Mountain is placed at the first
prominent occurrence of green shale. The presence of irregular and rounded limestone fragments at this contact suggests a disconformity
between the Watt Mountain and the Sulphur
Point.

The Sulphur Point Formation consists of
fine- to medium-grained limestones with biota
consisting of scattered Amphipora, calcispheres, gastropods, and algae. Although regional dolomitization trends can be mapped
(Fig. 10), dolomitization is variable in local

detail. Intergranular and vuggy porosity ranges from poor to very good. The Sulphur Point unit is a secondary, but important reservoir in northern Alberta.

Watt Mountain Formation

The Watt Mountain Formation is 10–20 ft thick locally, and ranges up to a maximum of 65 ft. It consists of apple-green and gray shale and very fine-grained, dense limestone. There is a marked increase in the relative amount of clear-quartz sandstone and siltstone in the formation as the Peace River high is approached.

RAINBOW AREA

Inception of Reef Growth

In Western Canada detailed geologic information on the incipient stage of reef growth

generally has been very meager. As a result there have been different opinions expressed on whether structure or sedimentary features, or both, have been responsible for initiation of reef growth. The entire reef section has been cored in many wells in the Rainbow field and the pre-Rainbow Member-reef beds have been drilled in many other wells. The subsurface data that have been obtained are pertinent to questions concerned with initial stages of reef growth.

In the Rainbow area there was an abundance of organisms having the ecologic potential to construct rigid wave-resistant structures that grew on the Lower Keg River surface. In the 7–32–109–8 W6 well, for example, massive stromatoporoids and corals are well developed in beds of the basal crinoidal bank of the

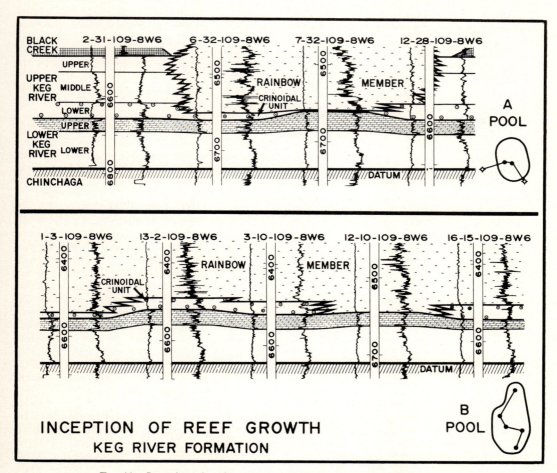

Fig. 11.—Inception of reef growth in Keg River Formation, Rainbow area.
Location of "A" and "B" pools is on Fig. 13.

D. L. Barss, A. B. Copland, and W. D. Ritchie

Upper Keg River Member (Fig. 11). Once the massive stromatoporoids and corals became established they grew upward and outward into the present form of the reef structure. The crinoidal bank contains, in addition to abundant crinoid stems and fragments, numerous branching and massive tabulate corals, even in wells not overlain by reefs. This is evident at the 2–28–109–8 W6 well about 1 mi south of the "A" pool. The presence of a fauna with the ecologic potential to construct a wave-resistant structure at this nonreef well indicates that factor(s) other than presence of reef-building organisms are critical in triggering reef growth at specific locations.

Another possible explanation for localizing reef growth is structural highs. Structural mapping of the top of the Chinchaga indicates that normal faults with displacement in the order of 20 to 80 ft are common. However, there is no coincidence between the normal fault highs and earliest reef growth. Nor is there correspondence of Lower Keg River "thicks" with residual lows on the Chinchaga surface (Fig. 12).[6] Correlation of markers within the Lower and Upper Keg River Members in off-reef wells indicates that normal faults occurred after the inception of reef growth. For example, at the "A" pool (Fig. 11), markers directly overlying the Lower Keg River–Chinchaga contact can be readily correlated. The conformity of these markers with the contact and absence of onlap of beds indicates that the basal Keg River beds were deposited on a regular surface. The evidence indicates that local changes in thickness in the Lower Keg River were not caused by faults.

As local Lower Keg River "thicks" commonly underlie the Rainbow Member reefs (Fig. 13), the possibility that they controlled inception of reef growth has been examined. In the 7–32–109–8 W6 well in the "A" Pool,

[6] Figure 12 is a modified first-order residual map on the top of the Chinchaga surface. The modification refers to adjustments that were made to the present day elevations of the Chinchaga surface. These changes were necessary in order to remove the effects of post-Devonian normal-fault movements and to isolate fault movements of post-Chinchaga to pre-Watt Mountain age. The top of the Hay River Shale is used because it is unaffected by post-Chinchaga, pre-Watt Mountain normal-fault movement and because it is not affected significantly by drape over the Rainbow Member reefs nor by solution of salt from the Black Creek Member. First order residual values obtained from the top of Hay River Shale structure-contour map were used to adjust the present-day elevations of the Chinchaga surface from which a modified first-order residual Chinchaga map was prepared.

and in other wells in the "B" Pool, thickening occurs in the upper part of the Lower Keg River Member. The top of the Lower Keg River has dips of up to 15°. Faunal evidence indicates that these features, *i.e.* the local "thicks," may be mud mounds and that they provided the locus for reef growth during Upper Keg River deposition.

Figure 11 illustrates that the earliest concentration of reef-constructing organisms occurred at the 7–32 location overlying the mud mound-like topographic high. Thus, at the "A" pool, it appears that a mud-mound high was important in controlling the inception of reef growth. As shown in Figure 13, the areal size of the mud mound and the overlying concentration of reef fauna is small compared with the reef complex that subsequently developed. In the other pools, where well control is poor compared with the "A" pool, the details of reef inception are less obvious. However, the presence of local Lower Keg River "thicks" in many of the other pools (Fig. 13) suggests a history of reef inception similar to that of the "A" pool.

Although a mud-mound control of inception of reef growth appears valid, it should be emphasized that reef growth did not start when a certain specific thickness of the Lower Keg River was present. There may have been gentle depressions and highs present on the sea floor that combined with local Lower Keg River "thicks" to initiate reef growth.

Reef Form Evolution

Once the massive stromatoporoids and corals with rigid wave-resistant characteristics became established, lateral and vertical growth proceeded rapidly. Inception of reef growth is considered to have been similar for the different reefs at Rainbow, but the size and geometric form that these reefs reached at maturity differs markedly. In the Rainbow area, two basic geometric forms are present. These are oval, characterized by the pinnacle reef, and elliptical, characterized by the atoll reef. The atoll reef may be further subdivided, on the basis of size and internal geometry, into the crescent-atoll and large-atoll forms.

The independent variables that controlled the final form and dimension of the reefs are —size of the mound or groups of mounds on which the reef growth started, rate of subsidence of the basin, and the physical environment in which the reefs grew. The presence of a massive stromatoporoid and coral fauna is of

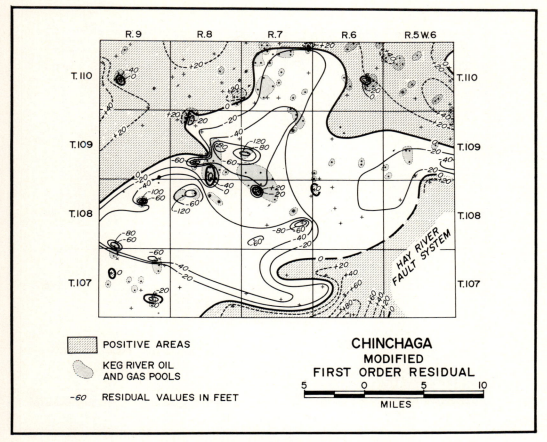

FIG. 12.—Chinchaga modified first order residual.

course important, but as similar faunas are common in other units of the Devonian, the concern here is with factors that produced reefs of the size and geometric form at Rainbow. A comparison of Rainbow Member reefs with part of the great Barrier Reef, Australia described by Fairbridge (1950) is thought to be useful. Figure 14 shows this comparison. Fairbridge illustrated the evolution of a reef from a small initial patch to a crescentic reef. The size of the patch reefs is reflected in the final shape and size of the complete reef masses. The larger ones developed into large horseshoe or crescent atolls and large atoll reefs, whereas the smaller patches developed into small horseshoe atolls or oval reefs.

At Rainbow, the size of the mound(s) on which reef growth began appears to have played an integral part in determining the size

and shape of the final reef. Relative subsidence of the basin was rapid at Rainbow, so that from a small mound the reef grew vertically to produce a small oval-shaped pinnacle reef, less than ½ mi in diameter. On larger mounds the crescent-atoll reef developed. Crescent-atoll reefs are generally more than 2 mi long and more than 1½ mi wide, and are characterized by an organic rim which is generally located on the north and east sides. Sediments typical of quiet-water deposition were deposited in a lagoonal area behind the rim. The "O" pool (Fig. 14) is a good example of the crescent atoll. Large atolls developed on a larger mound or on several small, closely spaced mounds. A more or less continuous outer organic rim developed along which growth of massive stromatoporoids and corals was prolific, owing to the presence of abundant food and oxygenated wa-

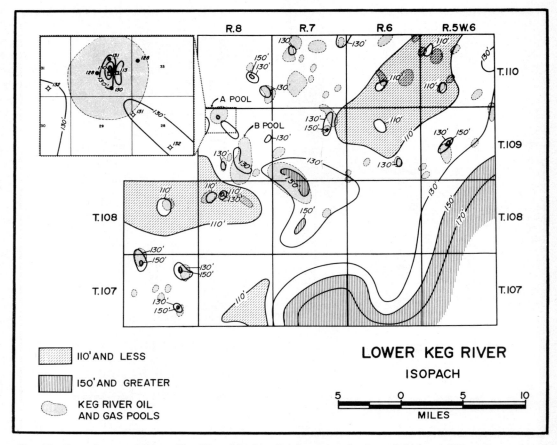

Fig. 13.—Isopach map of Lower Keg River Member, Rainbow area. Sections of "A" and "B" pools are shown in Figure 11. CI = 20 ft.

ters. A lagoonal quiet-water environment was present in the central low area. The "B" pool (Fig. 14) is typical of this form.

In addition to the influence of size of the mounds and rapid subsidence of the basin, the physical environment is believed to have exerted a strong influence on the reef forms that developed at Rainbow. Stanton (1967) indicated that three aspects of the physical environment are of major importance—directional climatic effects, particularly the wind; the paleogeographic setting (position of major shoals or land barriers, extent of deep open ocean); and local bathymetry.

The effect of wind in shaping the reef forms is illustrated in the comparison of some of the modern reefs of the Great Barrier with Rainbow Member reefs. Figure 14 illustrates the development of horseshoe-shaped reefs with con-

vexity in a windward direction. Several authors have shown this to be due to the relatively high organic growth and sediment production on the windward side. Moreover, movement of sediment by currents and waves results in a windswept crescent shape. These are convex toward the north-northeast and so indicate the prevailing wind direction at the time of reef growth.

The paleogeography of an area, that is, the position of the basin or deep water relative to the shelf and shore, determines the effectiveness of directional climatic factors and is therefore of critical significance with regard to reef forms. Because water depth is a significant factor in the consideration of paleogeography, it is inevitable that questions pertaining to the depths in which the Rainbow Member reefs grew, and relative age of the reef and offreef beds be examined also. Wave energy available for the

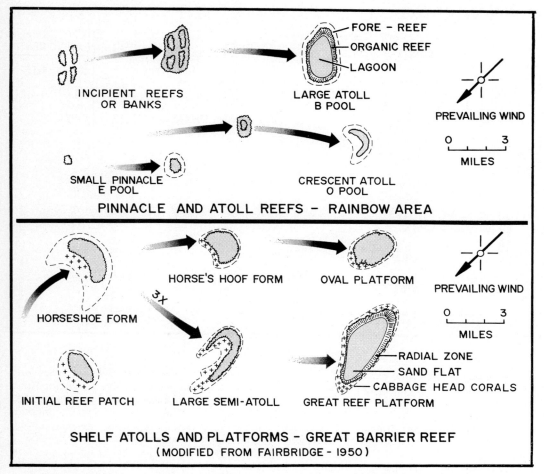

FORE – REEF
ORGANIC REEF
LAGOON

INCIPIENT REEFS
OR BANKS

LARGE ATOLL
B POOL

PREVAILING WIND

0   3
MILES

SMALL PINNACLE
E POOL

CRESCENT ATOLL
O POOL

PINNACLE AND ATOLL REEFS – RAINBOW AREA

HORSE'S HOOF FORM   OVAL PLATFORM

PREVAILING WIND

0   3
MILES

HORSESHOE FORM

3X

INITIAL REEF PATCH   LARGE SEMI-ATOLL   GREAT REEF PLATFORM

RADIAL ZONE
SAND FLAT
CABBAGE HEAD CORALS

SHELF ATOLLS AND PLATFORMS – GREAT BARRIER REEF
(MODIFIED FROM FAIRBRIDGE - 1950)

Fig. 14.—Pinnacle and atoll reefs, Rainbow area, compared with Great Barrier Reef, Australia
(modified after Fairbridge, 1950).

erosion and transport of carbonate sediments is shown to be related to the length of fetch, that is, the distance between windward barriers and the reef(s) in question. Another factor which limits wave energy is water depth. In shallow seas, the energy imparted by winds to waves is severely restricted. It is possible to estimate the maximum height of waves as a measure of wave energy. The interrelations of fetch length, water depth, and wave height is illustrated by Bretschneider (1954). Stanton (1967) discussed an example of a low-energy situation in the shallow water of the Bahama platform. Water depths of 2–3 fm over the platform effectively limit the wind energy that can be transmitted to the waves. As a result, patch reefs protected by the shallow water alone are roughly circular and are without the downwind "tails" of detritus.

The abundance of coarse clastic material and the elliptical or crescent shape of the reefs in the Rainbow area indicate that growth took place in an environment characterized by high wave energy. It is improbable that land barriers of a shallow shoaling sea were present in the windward direction. Evidence indicates that deep water surrounded the buildups. Carbonate rudites and arenites were transported to the edge of the organic reef and were deposited on the forereef slope. The lack of transport of reef-derived detritus away from the reef complex indicates low-energy deep-water conditions in the offreef area. Dips in the order of 25° in the forereef beds provide further evidence that

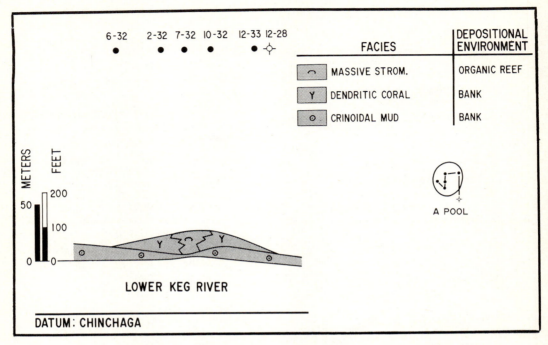

FIG. 15.—Pinnacle reef, deep-bank phase. Location of "A" pool shown on Figure 13.

strong topographic relief existed between the zone of organic production and destruction and bottom edge of the buildups.

In summary, all of the physiographic factors —directional aspect of climate, the paleogeography of the sea floor, and local bathymetry, indicate that the Rainbow Member reefs, for the most part of their growth, exhibited strong topographic relief. From this it follows that deposition of Muskeg evaporites occurred later than the laterally equivalent reef beds. However, deposition of the Black Creek salt and Muskeg anhydrites could have occurred in the offreef deeper water contemporaneously with reef growth near the surface.

Lithofacies

In the following section the lithologic and faunal constituents of Rainbow Member reefs are discussed briefly, largely on the basis of the work of Langton and Chin (1968) who examined in detail approximately 15,000 ft of core. These authors were able to recognize 14 lithofacies which they assigned to six depositional environments.

Following is a discussion of the lithofacies and depositional environments during the significant phases of reef growth at Rainbow.

Early bank phase (pinnacle).—Facies present in this phase are massive stromatoporoids, dendritic corals, and crinoidal muds (Fig. 15). The massive stromatoporoid facies containing massive corals, branching stromatoporoids, brachiopods, and gastropods, is of specific interest because of its reef-building potential. However, during the early bank phase the crinoidal mud facies was dominant and there was only minor topographic relief on the sea floor. The writers consider that the term "deep bank" most accurately describes the structure and environment at this time.

Bank to organic-reef transition phase (pinnacle).—Figure 16 illustrates the facies which are present and their distribution. The massive stromatoporoids continued to flourish and were joined by the reef-constructing massive coral facies. Coral rudite facies is present and consists of fragments derived from the massive stromatoporoids and corals. The rudites have an arenite matrix. This facies is present in the forereef position of the pinnacles. Skeletal arenites are widely distributed throughout the different environments of the reef complex, however, they are commonly abundant in this phase. In addition to the arenites and rudites, other facies present include the massive syrin-

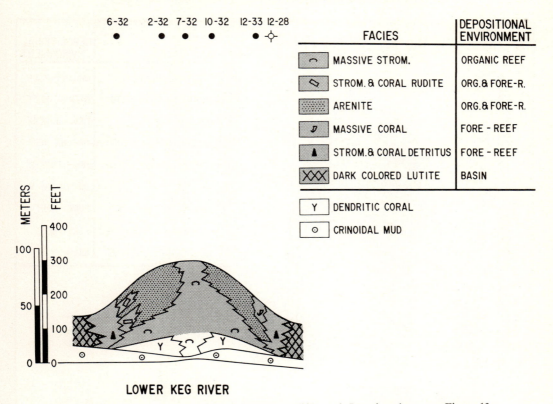

						FACIES	DEPOSITIONAL ENVIRONMENT

6-32   2-32   7-32 10-32   12-33 12-28

FACIES	DEPOSITIONAL ENVIRONMENT
MASSIVE STROM.	ORGANIC REEF
STROM. & CORAL RUDITE	ORG.& FORE-R.
ARENITE	ORG.& FORE-R.
MASSIVE CORAL	FORE - REEF
STROM.& CORAL DETRITUS	FORE - REEF
DARK COLORED LUTITE	BASIN

Y DENDRITIC CORAL	
○ CRINOIDAL MUD	

LOWER KEG RIVER

FIG. 16.—Pinnacle reef, bank/organic-deef phase, "A" pool. Location shown on Figure 13.

goporoid corals and organic-reef detritus. The organic-reef-detritus facies consists of fine fragments of stromatoporoids and branching corals in a brown, very finely crystalline dololutite matrix. The dark color, presence of fine matrix, and occurrence in a low flank position indicate deposition in relatively deep, quiet water. This facies is transitional with the dark-colored lutites. The latter consist of finely laminated bituminous calcilutites and are found in the offreef basin environment.

There was a delicate balance between reef-building ability of the organic-reef organisms and the destructive forces of the physical environment. The reef was subjected to strong turbulent water conditions, as indicated by the presence of abundant arenites and rudites.

Organic-reef phase (pinnacle).—During this phase (Fig. 17) the massive stromatoporoids and corals of the organic-reef environment became dominant and the major part of the reef

complex was constructed. The presence of rudite and arenite zones attests to the turbulent water conditions to which the organic reef was exposed. However, movement of derived detritus was restricted to the forereef slope, where deposition occurred below wave base.

Two additional facies, the light-colored lutites and the skeletal rudites, were deposited during this phase. The light-colored lutites consist of micro- to very fine-grained limestone and constitute a minor part of the organic reef; they were deposited in local, protected areas on the reef. The skeletal rudite consists of fragmented gastropods, brachiopods, colonial septate corals, and Stachyodes in the calcilutite to calcisiltite matrix. Massive stromatoporoid fragments are present sporadically.

Langton and Chin (personal commun.) have recorded the presence of massive stromatoporoid facies at the top of small pinnacle reefs and to within 50 ft of the top of some of

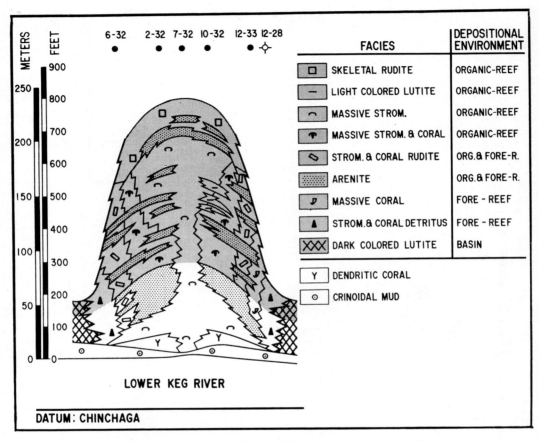

FIG. 17.—Pinnacle reef, organic-reef phase, "A" pool. Location shown on Figure 13.

the larger pinnacle and atoll reefs. However, in the uppermost part of the organic-reef phase, the areal extent of this facies became much smaller and there is an increase in the amount of skeletal rudite facies. The decrease in massive stromatoporoid facies indicates a less favorable environment—possibly an increase in salinity of the seas.

Shallow-bank phase (pinnacle).—The terminal stage of reef growth is illustrated on Figure 18. The Amphipora facies, though not extensive, is generally present in this phase. Massive stromatoporoid facies is present also, but the areal extent of this facies was reduced further from that present in the organic-reef phase. In addition to the Amphipora facies, the laminite facies is generally present in this phase. The laminites are thought to be, in part, of algal origin. There is some difficulty in separating the skeletal rudites of the organic-reef phase from those present in the shallow-bank phase. The distinction is made mainly on the basis of contained fossils and associated lithology. The skeletal rudite of the organic-reef environment contains minor massive stromatoporoids and rare amphiporoids, whereas the skeletal rudite of the shallow shoal environment contains numerous amphiporoids and rare stromatoporoids.

During this late phase, the population and variety of the fauna decreased. These changes and the presence of laminites and primary anhydrites in the upper 20–50 ft of some reefs, indicate an environmental change to shallow, highly saline waters. Minor occurrences of massive stromatoporoids probably indicate isolated topographic features where favorable salinities were maintained by introduction of fresh meteoric waters.

Atoll Reefs.—The growth of atoll reefs was

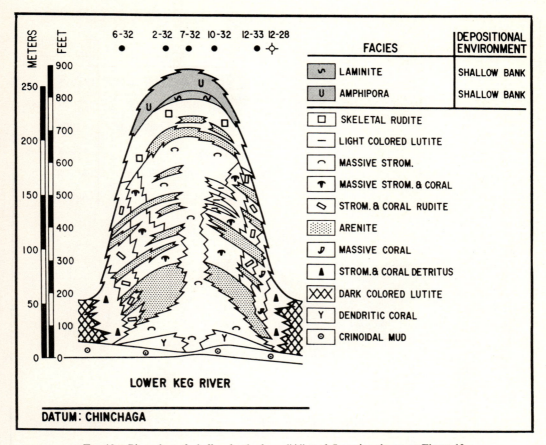

FIG. 18.—Pinnacle reef, shallow-bank phase, "A" pool. Location shown on Figure 13.

very similar to that of the pinnacles (Fig. 19). The main difference occurred during the early bank phase when the organic-reef fauna became established over a much broader area. As reef growth proceeded, the organic-reef facies grew toward the outside edge of the reef complex where nutrients were more plentiful. The upward growth of the organic-reef facies was more rapid than the other parts of the reef complex. As a result, a rim developed around the edge of the reef and resulted in the development of a relatively quiet-water interior lagoon. In the atoll reefs, in addition to the deep-bank, organic-reef, forereef, and shallow-bank environments of the pinnacle reefs, a lagoon environment is present. The facies which are present in the lagoon comprise stromatoporoid and coral rudites, arenites, nonskeletal rudites, laminites, and light-colored lutite. The arenites and stromatoporoid and coral rudites

are very similar to those found on the forereef position of the pinnacle, and represent material moved by wave action behind the rim of the reef. Nonskeletal rudites consist of fragments of lutite torn from the lagoon floor during storms and then redeposited in a quiet-water environment. The laminites and light-colored lutites are important constituents in the lagoonal environment.

Diagenesis

In addition to recrystallization of $CaCo_3$ and deposition of calcite cement, other diagenetic processes which have affected the reef are solution, dolomitization, anhydritization, and deposition of carbon. The effect of the different diagenetic processes is beyond the scope of this paper and only generalized comments on certain aspects of anhydritization, dolomitization, and solution are presented.

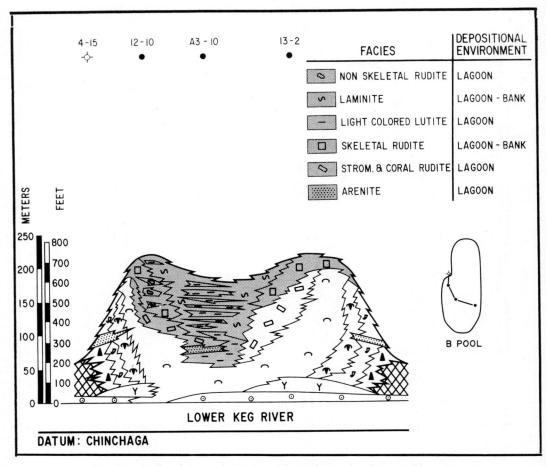

Fig. 19.—Atoll reef, lagoon-bank phase, "B" pool. Location shown on Figure 13.

Anhydritization, although present, is not appreciable except in the upper 20–50 ft of the shallow-bank phase. In addition to the presence of anhydrite in the uppermost beds of the reef, some thin beds of anhydrite have been found more than 300 ft from the top of full reef buildup. The anhydrite occurs disseminated through the carbonate, as nodules in vugs, and as a replacement mineral. It ranges from clear white to gray and microcrystalline to medium crystalline. Machielse (personal commun.) has noted that the anhydrite is associated with brown, cryptocrystalline to microcrystalline dolomite which appears to be of primary origin. Thus, although the possibility of primary origin for some of the anhydrite exists, the anhydrite which is most common is the clear-white, finely crystalline, secondary variety. Also, as this variety of anhydrite infills pores and vugs, and as it cannot be correlated from well to well, it is believed to be of secondary origin.

Three factors are important with regard to dolomitization. Langton and Chin (1968) pointed out that the degree and type of dolomitization appear to be related to reef size and its internal makeup. Also, they recognized early and late stages of dolomitization.

The relation between dolomitization and reef size is illustrated by the fact that pinnacle reefs are not dolomitized, crescent atolls are moderately to strongly dolomitized, and all large atolls are strongly dolomitized. An exception is the small pinnacle "D" pool (Fig. 24) reef which is completely dolomitized. In this case, the dolomitization may be the result of its proximity to the "B" pool atoll reef and its connate water environment. Dolomitization is characteristic of atolls and not pinnacles because, as Langton and Chin believe, the larger reefs created their own evaporitic environment. Evaporation resulted in an increase in density of the water. Movement of this relatively dense saline water through the reef resulted in dolomitization by the seepage refluxion process.

Two stages of dolomitization are clearly evident. The early stage resulted in selective replacement of the fine-grained matrix material and did not obliterate sedimentary and faunal textures to any significant extent. In its final form it resulted in a fine-grained dolomite. The color of the host rock was retained and is generally light to dark brown. The late phase of dolomitization is characterized by coarse crystallinity and light color. The white, coarsely crystalline dolomite replaces and occurs as veins in the early, brown, finely crystalline dolomite. The presence of the two kinds of dolomitization gives the rock a "marbled" appearance. In other types of late-stage dolomitization, terms such as *banded, nebulous,* and *pseudo-brecciated,* describe the texture very well. The presence of the late-stage of dolomitization is the result of late, postdepositional introduction of magnesium-rich waters, along fracture zones or zones of porosity. The magnesium-rich waters may have been displaced from the Muskeg evaporite beds, by compaction.

There has been solution, but the degree of solution was highly variable in all reefs. Large fragments of gastropods and brachiopods locally have been completely leached. Also, there was selective leaching of stromatoporoids, corals, and laminites. In addition, abundant stylolitization and the pronounced degree of compaction in atoll reefs (*e.g.* "F"-pool reef [Fig. 24]) indicate that considerable solution occurred in certain beds. The relation and relative timing of solution processes and late-stage dolomitization has not been fully established. It is evident that initial porosity was important in controlling both solution and infill processes. The porous laminates show the effects of the two processes (Langton and Chin, 1968, Pl. 3, Fig. 10). In this illustration solution occurred along the porous and permeable layers and was followed by infilling of the late-stage dolomites. Late-stage dolomitization continued inward to the less permeable parts of the rock. In some places only large remnants of the host rock are left in a matrix of coarse-grained, white dolomite. It appears that late-stage dolomitization

occurred simultaneously or immediately following solution.

Rainbow Member—Time and Space Relations

The major part of the Rainbow Member reefs are considered to be the time equivalent of the Upper Keg River Member. The facies containing the massive stromatoporoid and coral fauna and the erosional products occur from the base of each reef to the top of full reef buildups in some reefs and to within 50 ft in most other reefs. Strong topographic relief is indicated by dips on forereef beds in the order of 25°. Also, there is no evidence of intertonguing of the reef detritus with the laterally equivalent Muskeg evaporite beds; the arenites, rudites and organic-reef detritus were carried only to the reef slope where they were deposited in deep water in which the energy was too low to carry away the detritus from the reef structure. All of the factors of the physical environment indicate that deep water surrounded the Rainbow Member reefs. It follows that deposition of Muskeg evaporites occurred either subsequently to reef growth or in deep water at the same time as reef growth was taking place near the surface.

Although the age relation for the major part of Rainbow Member reef growth and Muskeg evaporite deposition seems clear, there appears to be direct evidence of contemporaneity of reef growth and evaporite deposition in the terminal, shallow-bank phase at the top of the reef. The presence of laminites and Amphipora indicate relatively quiet water. The reduction in both number and variety of fossils, and the presence of dark-brown, very finely crystalline, primary-type anhydrite, indicates a highly saline water environment. Deposition of Muskeg evaporites was probably taking place at the same time as the Rainbow Member reef growth in the terminal shallow-bank phase.

In the Zama subbasin (Figs. 9 and 20) the geologic history was different from that at Rainbow. The Upper Keg River reefs (McCamis and Griffith, 1967) are probably the age equivalent of the Rainbow Member reefs. However, the correlation between Rainbow and Zama is not clear. A distinctive stratigraphic unit—the Zama Member—overlies either reef where there is full buildup, or Muskeg evaporites in the offreef area (Fig. 20). The Zama Member consists of fine-grained dolomite with low fossil content, commonly bedded or laminated. Because it is underlain and overlain by Muskeg evaporites in offreef areas, it is

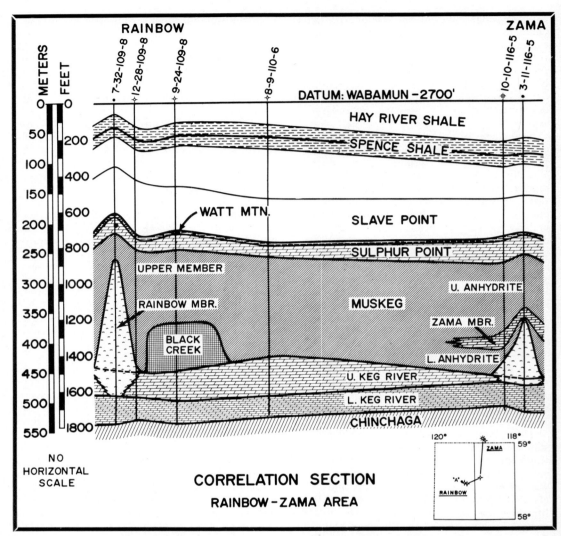

Fig. 20.—Southwest-northeast correlation section, Rainbow to Zama area.

known to be of Muskeg age. It is not recognized in the Rainbow subbasin. McCamis and Griffith (1967), however, consider it to be equivalent to the upper 150–250 ft of the Rainbow Member. We consider this correlation to be untenable. At Rainbow, growth of the organic reef occurs either to the top or to within 50 ft of the top of most of the reefs. The strong regional shoal environment which led to deposition of the Zama Member in the Zama subbasin was not present in the Rainbow subbasin.

Post-Rainbow Member Structure

The present structural attitude of the beds is due to four factors—normal fault movement, solution of the Black Creek salt, compaction of Muskeg evaporite beds, and post-Laramide regional tilting. Regional tilting is not significant in the context of this paper and is not discussed.

Normal fault movement.—In the Rainbow map area, two systems of faulting are present (Fig. 12). One system, the Hay River-East Arm, is about 10 mi south of the Rainbow

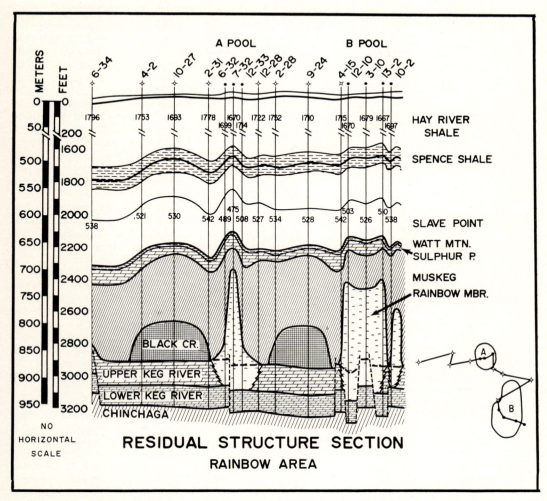

FIG. 21.—Residual structure section, showing effects of salt solution, Rainbow area. "A" and "B" pools shown in Figure 13.

field. It is composed of normal faults with throws of 100 ft or greater. The strike of the individual faults is variable, but the overall trend of the fault system is southwest-northeast. The second system of faulting is present in the general Rainbow field area. Faults in this system have throws in the order of 20–80 ft and they have no apparent trend (Figs. 12, 21).

The Hay River–East Arm fault system represents an early Precambrian movement that probably was modified later by erosion. When deposition started in this area, only prominent topographic features were present. Reactivation of normal and transcurrent faults occurred dur-

ing different periods as late as the end of Mississippian time. Other normal faults and fault systems of this nature are present in northern Alberta.

Normal faults in the second system, with present-day throws of 20–80 ft are of more direct interest as they commonly underlie the edges of the reefs in the Rainbow field (Fig. 12). Because of this relation there has been a tendency to view these faults as the loci for reef growth. As stated, the writers do not subscribe to this view. The loci for reef growth appears to have been Lower Keg River mud mounds. Detailed correlation of markers within

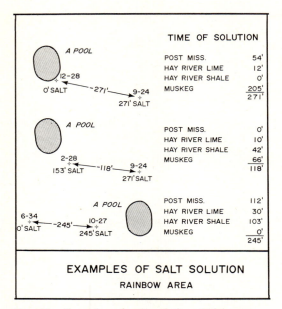

FIG. 22.—Examples of salt solution. Rainbow area.

the Lower and Upper Keg River Members (Fig. 11) indicates that normal-fault movement occurred subsequent to deposition of the lower part of the Upper Keg River Member, that is, sometime during Rainbow Member reef growth or later. As normal faults are not evident at the Watt Mountain level, the time of movement can be dated between that of the deposition of the upper Rainbow Member and the Watt Mountain Formation. Regional structure analysis supports the interpretation of normal fault movements at this time. In the area of the Tathlina high in the Northwest Territories it is clear that the Lower Keg River and probably the Upper Keg River were deposited as a normal sequence. The section between the Watt Mountain and Lower Keg River, however, is considerably thinner than that in the rest of the basin on the south. Uplift of the Tathlina high before Watt Mountain deposition—possibly during Muskeg deposition—resulted in nondeposition and possibly erosion. Along the She-kilie trend on the western edge of the Black Creek basin, the interval from the top of the Muskeg Formation to the top of Lower Keg River Formation is thin. The overlying Sulphur Point Formation, however, is regionally thick. It appears that uplift occurred toward the end of Muskeg deposition, which resulted in a thin Muskeg section, or possibly some erosion of the Muskeg beds. During Sulphur Point deposition

down-warping occurred which resulted in a relatively thick Sulphur Point section along the western part of the basin (Figs. 5, 9, and 10).

Within the Rainbow subbasin the normal faults with movement of 20–80 ft may be rebound features related to the emplacement of individual reef masses. The aggregate weight of the individual reefs and the regional barrier-reef complexes could have caused both local and broad crustal readjustments. The combination of the barrier-reef complexes and crustal readjustments along the Shekilie and Tathlina features formed a regional barrier that prevented the flow of normal marine waters southeast to the Black Creek basin.

Solution of Black Creek Member salt.—The distribution of the Black Creek Member salt is best known in the Rainbow subbasin. The salt beds have been found in nine wells, with thickness up to 271 ft. Seismic data are also useful in defining the presence of salt. Presence of the Black Creek Member in the Zama subbasin has been confirmed at the Hudson's Bay Zama No. 10–21–116–5 W6 well which penetrated 124 ft of salt. Salt is also present on the northeast in the Steen River area where six wells have penetrated up to 180 ft of the Black Creek Member. The original extent of the Black Creek salt was probably much greater than it is today.

In the Rainbow subbasin, solution of salt occurred in early Late Devonian (upper Muskeg, Hay River) and post-Mississippian times. A summary of the amounts and periods of salt solution of specific examples is shown on Figure 22 and illustrated on Figure 21. Solution during upper Muskeg deposition can be seen in many wells. For example, from the 9–24 well to the 12–28 well, 271 ft of salt was removed; 205 ft can be accounted for in the thicker Muskeg section. Salt solution in the amount of 103 ft occurred during Hay River deposition between wells 6–34 and 10–27. Between wells 2–28 and 9–24, 52 ft of thickening of the Hay River is attributed to salt solution. There was no apparent post-Mississippian salt solution between these two wells. However, 54 ft of post-Mississippian solution is indicated between the 12–28 and 9–24 wells, and 112 ft between wells 10–27 and 6–34.

Compaction.—In Figure 23, compaction of Muskeg beds is well illustrated in the 4–9, 5–9, and 15–9 wells. Detailed correlation of Muskeg units 2 to 5, inclusive, shows aggregate thickening from 260 ft over nearly full reef at well 15–9 to 330 ft over high reef flank at well

5–9. Because of the high flank position of these two wells, it is apparent that thickening of Muskeg units 2 to 5 was not influenced by salt solution. The combined thickness of units 2 to 5 on the reef flank in well 4–9 is 400 ft. Although thickening due to salt solution is possible at 4–9 because it is in a low flank position, uniformity and consistency of correlation with the other wells suggests that changes in thickness are due to differential compaction. From 15–9 to 4–9 thickening due to compaction amounts to 35 percent. In addition to the thickening described above, Figure 23 also shows 115 ft of onlap of basal Muskeg beds onto the reef between 4–9 and 5–9 and an additional 68 ft in unit 6 between wells 5–9 and 15–9, for a total of 183 ft. This fact and absence of intertonguing of reef and Muskeg beds indicate the topographic relief that existed on the reef late in Muskeg deposition.

Figure 21 illustrates compaction and thinning of beds overlying the "A" and "B" reef pools of the Rainbow field. In the interval Spence Shale "A" marker to Watt Mountain, thickening in the order of 35–55 ft occurs from a full reef buildup to completely off-reef, full Black Creek Member salt section. Also shown in Figure 21 is thickening of 15–20 ft in this same interval in the 3–10 well of the "B" atoll—a well which is located in the lagoonal area. This indicates that solution and compaction of lagoonal reef carbonates occurred during the time represented by this interval.

In the interval, top Hay River Shale to Spence "A" marker, thickening from reef to offreef is in the order of 30–50 ft. As this amount of thickening occurs where there has been no solution of the Black Creek salt, it is evident that compaction of the offreef Muskeg evaporites continued until the end of Hay River Shale deposition. However, there was only insignificant compaction following Hay River deposition. Rainbow Member reefs are not readily discernible on structural contour maps of the top of Hay River Shale.

EXPLORATION AND RESERVES—RAINBOW LAKE

Prior to the discovery of the Rainbow field by the Banff *et al.* Rainbow No. 7–32–109–8 W6 well, there had been no drilling within the outline of the current field limits (Fig. 24). Subsequent exploration and development drilling has resulted in 55 Keg River oil pools, 7 Keg River gas pools, 6 Muskeg oil pools, and 4 Sulphur Point oil pools. About 72 of the 126

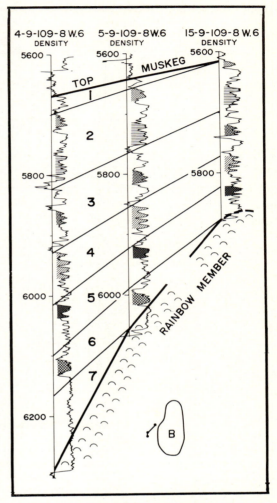

FIG. 23.—Correlation of Muskeg markers, "D" pool (wells 4–9, 5–9) and "EEE" pool (15–9), Rainbow area (Fig. 24). No horizontal scale.

exploratory wells, or approximately one well in two, have been discoveries. The exceptionally high success rate of 56 percent for exploratory wells resulted from use of modern geophysical techniques. Draping over reef structures and salt solution near reefs are identifiable seismically, particularly when "stacking" or common-depth point shooting is used. The use of seismic data has met with similar success in other areas within the Black Creek basin. In the Virgo and Zama Lake fields north of Rainbow, 110 Keg River oil pools have been found. The number of oil and gas completions in the Keg River Formation within the Black Creek

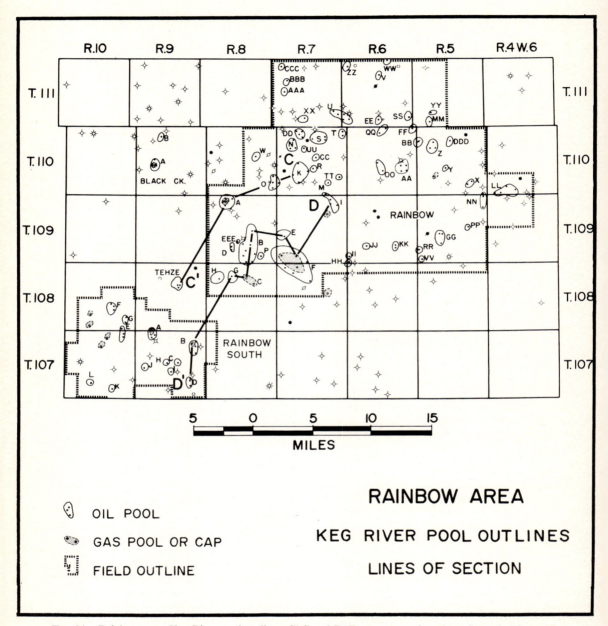

FIG. 24.—Rainbow area, Keg River pool outlines. **C′-C** and **D′-D** are traces of sections shown in Figure 25.

basin probably will increase, inasmuch as the area is still under active exploration and development.

Figure 25 illustrates two schematic sections across Rainbow and Rainbow South fields, showing reservoir parameters and recoverable reserves for 10 of the 55 Keg River oil pools.

The "C" pool on D′-D section is a gas pool and the Tehze pool is outside the current field boundaries. At the present ime, these 10 pools account for approximately 60 percent of the estimated primary recoverable reserves of the two fields.

Figure 25 also illustrates the wide range of

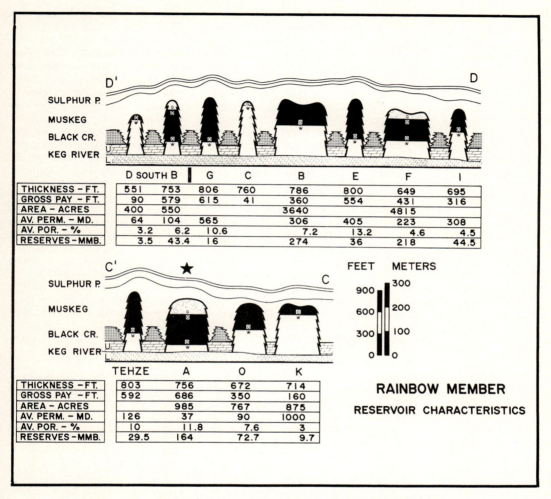

	D SOUTH B	G	C	B	E	F	I
THICKNESS – FT.	551	753	806	760	786	800	649	695
GROSS PAY – FT.	90	579	615	41	360	554	431	316
AREA – ACRES	400	550			3640		4815	
AV. PERM. – MD.	64	104	565		306	405	223	308
AV. POR. – %	3.2	6.2	10.6		7.2	13.2	4.6	4.5
RESERVES – MMB.	3.5	43.4	16		274	36	218	44.5

	TEHZE	A	O	K
THICKNESS – FT.	803	756	672	714
GROSS PAY – FT.	592	686	350	160
AREA – ACRES		985	767	875
AV. PERM. – MD.	126	37	90	1000
AV. POR. – %	10	11.8	7.6	3
RESERVES – MMB.	29.5	164	72.7	9.7

RAINBOW MEMBER

RESERVOIR CHARACTERISTICS

FIG. 25.—Reservoir characteristics and reserves, Rainbow Member, in Rainbow area.
Traces of sections **D'-D** and **C'-C** are shown in Figure 24.

net pays and reservoir factors found in Rainbow. Generally, the net pays are better in the western part of the Rainbow field, generally in the 200–500 ft range. However, there are exceptions such as the "C" pool which has only 41 ft of Keg River gas pay. Porosity and permeability values vary greatly from pool to pool, as well as within individual pools and are related to the different lithofacies and to diagenetic effects. An example of the broad relation of reservoir facies to lithofacies is illustrated in Rainbow South "A" pool and in the Rainbow "B" pool; in these, porosity is lower in the center of the atolls, averaging 6 percent, and increases significantly to about 11 percent along the flanks of the reefs. This increase is attrib-

uted to the higher percentages of arenites and rudites in the forereef position. Another example is the South "G" pool in Rainbow South field where the 4–12–108–10 W6 central well has an average of 7.4 percent porosity and the 12–1–108–10 W6 flank well has an average of 10.4 percent porosity. In addition to varying reservoir properties within a reef as a result of different lithofacies, there is variation in the same lithofacies between reefs owing to the variation in content of "lime mud" in the matrix and owing to the degree to which solution and dolomitization processes were active. Dolomitization has had a variable effect on reservoir properties. In some pools, the "F" pool, for example, dolomitization is essentially complete.

The average porosity in this pool is 4.5 percent which is lower than many of the less dolomitized reefs. The effect of solution on porosity and permeability is evident in nearly all reefs in varying degrees. An example of extreme solution is found in the small pinnacle reef at well 15–9–109–8 W6. In this well, 400 ft of dolomite and limestone was penetrated in which the porosity averaged 23 percent. Approximately 100 ft of this section had porosity values in the range of 40 to 50 percent. This type of cavernous porosity is uncommon in the Rainbow area; in the Rainbow Member reefs, porosity ranges more commonly from 5 to 11 percent. Further details on geologic and reservoir properties and their correlation were presented by Langton and Chin (1968).

With the Rainbow and South Rainbow fields, there are estimated to be 1.2 billion and 165 million bbl of oil-in-place, respectively. The reef reservoirs have good vertical relief and lend themselves to enhanced recovery schemes. Primary recovery factors range from 20 to 52 percent whereas secondary recovery methods, such as miscible flood and water or gas injection, are realistically estimated to boost the ultimate recoveries to 80–95 percent.

CONCLUSIONS

1. Rainbow Member reef growth began in the lower unit of the Upper Keg River Member. This unit contains, in addition to crinoids and brachiopods, a reef-prone fauna consisting of massive stromatoporoids, branching septate, and tabulate corals. The inception of reef growth was sensitive to "highs" on the underlying Lower Keg River sea floor. These "highs," with relief of about 20–40 ft, are thought to be mud mounds which formed in the upper part of the Lower Keg River Member. No direct evidence of structural control of reef growth exists, although there may have been gentle warping of the Lower Keg River surface which combined with mud mounds to control reef growth.

2. The major part of Rainbow Member reefs are of late Middle Devonian (Upper Keg River) age. There is the possibility that the uppermost part of the reefs, that is, the shallow-bank deposits, may be of "Muskeg age."

3. Within the Rainbow part of the Black Creek basin are Rainbow Member reefs of both pinnacle and atoll forms with vertical dimensions up to 820 ft. The size and geometry of the reefs are related to the initial size of the mound(s) on which they grew, rate of basin subsidence, directional influence of climate, paleogeography of the sea floor, and local bathymetry.

4. Numerous normal faults with 20–80 ft of throw are present. Movement occurred after Rainbow Member reef growth began and prior to deposition of the Watt Mountain sediments.

5. Salt solution and compaction features are recognized in the Black Creek basin. Solution of Black Creek salt occurred mainly in the Rainbow area late in Muskeg deposition, and in early Late Devonian, and post-Mississippian times. Compaction of the Muskeg evaporites occurred from the time of Muskeg deposition until at least the end of Hay River Shale deposition. Detailed correlation of markers in the Rainbow area indicates compaction of approximately 35 percent within the Muskeg Formation.

6. Modern seismic techniques have been used in exploration for Rainbow Member reefs. These techniques have proved to be remarkably successful as shown by a discovery of one Keg River oil or gas accumulation for every two wells drilled.

SELECTED REFERENCES

Adams, J. E., and Rhodes, M. L., 1960, Dolomitization by seepage refluxion: Am. Assoc. Petroleum Geologists Bull., v. 44, no. 12, p. 1912–1930.

Andrichuk, J. M., 1961, Stratigraphic evidence for tectonic and current control of Upper Devonian reef sedimentation, Duhamel area, Alberta, Canada: Am. Assoc. Petroleum Geologists Bull., v. 45, no. 5, p. 612–632.

Belyea, H. R., 1959, Devonian Elk Point Group, central and southern Alberta, parts of 72, 73, 82, 83: Canada Geol. Survey Paper 59–2, 14 p.

——— and D. J. McLaren, 1962, Upper Devonian formations, southern part of Northwest Territories, northeastern British Columbia, and northwestern Alberta: Canada Geol. Survey Paper 61–29, 74 p.

——— and A. W. Norris, 1962, Middle Devonian and older Palaeozoic formations of southern district of MacKenzie and adjacent areas: Canada Geol. Survey Paper 62–15, 82 p.

Bretschneider, C. L., 1954, Generation of wind waves over shallow bottom: U.S. Army Corps Engineers, Beach Erosion Board Tech. Memo. no. 51, p. 1–24.

Burwash, R. A., 1957, Reconnaissance of subsurface Precambrian of Alberta: Am. Assoc. Petroleum Geologists Bull., v. 41, no. 1, p. 70–103.

Brown, P. R., 1963, Some algae from the Swan Hills reef: Bull. Canadian Petroleum Geology, v. 11, no. 2, p. 178–182.

Cloud, P. E., 1952, Facies relationships of organic reefs: Am. Assoc. Petroleum Geologists Bull., v. 36, no. 11, p. 2125–2149.

Cummings, E. R., 1932, Reefs or bioherms?: Geol. Soc. America Bull., v. 43, no. 1, p. 331–352.

Douglas, R. J. W., 1959, Great Slave and Trout River map-areas, Northwest Territories, parts of North halves of 85 and 95: Canada Geol. Survey Paper 58–11, 57 p.

———— and D. K. Norris, 1959, Fort Liard and La Biche map-areas, Northwest Territories and Yukon, 95B and 95C: Canada Geol. Survey Paper 59–6, 23 p.

———— and ———— 1961, Camsell Bend and Root River map-areas, District of MacKenzie, Northwest Territories, 95J and K: Canada Geol. Survey Paper 61–13, 36 p.

———— and ———— 1963, Dahadinni and Wrigley map-areas, District of MacKenzie, Northwest Territories, 95N and O: Canada Geol. Survey Paper 62–33, 34 p.

Dumestre, A., and L. V. Illing, 1967, Middle Devonian Reefs in Spanish Sahara, in Internat. Symposium on the Devonian System: Calgary, Alberta Soc. Petroleum Geologists, v. 2, p. 333–350.

Fairbridge, R. W., 1950, Recent and Pleistocene coral reefs of Australia: Jour. Geology, v. 58, no. 4, p. 330–401.

Fischbuch, N. R., 1960, Stromatoporoids of the Kaybob reef, Alberta: Alberta Soc. Petroleum Geologists Jour., v. 8, no. 4, p. 113–131.

Fong, G. 1960, Geology of the Devonian Beaverhill Lake Formation, Swan Hills area, Alberta, Canada: Am. Assoc. Petroleum Geologists Bull., v. 44, no. 2, p. 195–209.

Grabau, A. W., 1913, Principles of stratigraphy: New York, A. G. Seiler, 1185 p. 2d ed., 1924, 1185 p.; Dover reprint, 1960, 2 v.

Gray, F. F., and J. R. Kassube, 1963, Geology and stratigraphy of Clarke Lake gas field, northeastern British Columbia: Am. Assoc. Petroleum Geologists, Bull., v. 47, no. 3, p. 467–483.

Grayston, L. D., D. F. Sherwin, and J. F. Alan, 1964, Middle Devonian, Chap. 5, in Geological history of Western Canada: Alberta Soc. Petroleum Geologists, p. 49–59.

Hriskevich, M. E., 1966, Stratigraphy of Middle Devonian and older rocks of Banff–Aquitaine Rainbow West 7–32 discovery well, Alberta: Bull. Canadian Petroleum Geology, v. 14, no. 2, p. 241–265.

Ingels, J. J. C., 1963, Geometry, paleontology, and petrography of Thornton reef complex, Silurian of northeastern Illinois: Am. Assoc. Petroleum Geologists Bull., v. 47, no. 3, p. 405–440.

Jenik, A. J., and J. R. Lerbekmo, 1968, Facies and geometry of Swan Hills Reef Member of Beaverhill Lake Formation (Upper Devonian), Goose River field, Alberta, Canada: Am. Assoc. Petroleum Geologists Bull., v. 52, no. 1, p. 21–56.

Klement, K., 1968, Reefs and banks; bioherms and biostromes (abs.): Oilweek, Feb. 19, p. 14, 20.

Klovan, J. E., 1964, Facies analysis of the Redwater reef complex, Alberta, Canada: Bull. Canadian Petroleum Geology, v. 12, no. 1,, p. 1–100.

Kornicker, L. S., and D. W. Boyd, 1962, Shallow-water geology and environments of Alacran reef complex, Campeche Bank, Mexico: Am. Assoc. Petroleum Geologists Bull., v. 46, no. 5, p. 640–673.

Kuenen, Ph. H., 1950, Marine geology: New York, John Wiley & Sons, 568 p.

Langton, J. R., and G. E. Chin, 1968, Rainbow Member facies and related reservoir properties, Rainbow Lake, Alberta: Bull. Canadian Petroleum Geology, v. 16, no. 1, p. 104–143; modified version, Am. Assoc. Petroleum Geologists Bull., v. 52, no. 10, p. 1925–1955.

Law, J., 1955, Geology of northwestern Alberta and adjacent areas: Am. Assoc. Petroleum Geologists Bull., v. 39, no. 10, p. 1927–1978.

Lowenstam, H. A., 1959, Niagaran reefs of the Great Lakes area: Jour. Geology, v. 58, no. 4, p. 430–487.

McCamis, J. G., and L. S. Griffith, 1967, Middle Devonian facies relationships, Zama area, Alberta: Bull. Canadian Petroleum Geology, v. 15, no. 4, p. 434–467.

Murray, J. W., 1965, Stratigraphy and carbonate petrology of the Waterways Formation, Judy Creek, Alberta, Canada: Bull. Canadian Petroleum Geology, v. 13, no. 2, p. 303–326.

Nelson, H. F., C. W. Brown, and J. H. Brineman, 1962, Skeletal limestone classification, in W. E. Ham, ed., Classification of carbonate rocks—a symposium: Am. Assoc. Petroleum Geologists Mem. 1, p. 224–252.

Norris, A. W., 1965, Stratigraphy of Middle Devonian and older Palaeozoic rocks of the Great Slave Lake region, Northwest Territories: Canada Geol. Survey Mem. 322, 180 p.

Oliver, T. A., and N. W. Cowper, 1963, Depositional environments of the Ireton Formation, central Alberta: Bull. Canadian Petroleum Geology, v. 11, no. 2, p. 183–202.

Pray, L. C., 1958, Fenestrate bryozoan core facies, Mississippian bioherms, southwestern United States: Jour. Sed. Petrology, v. 28, no. 3, p. 261–273.

Sherwin, D. F., 1962, Lower Elk Point section in east-central Alberta: Alberta Soc. Petroleum Geologists Jour., v. 10, no. 4, p. 185–191.

Sikabonyi, L. A., 1959, Paleozoic tectonics and sedimentation in the northern half of the West Canadian basin: Alberta Soc. Petroleum Geologists Jour., v. 7, no. 9, p. 193–216.

Soderman, J. W., and A. V. Carozzi, 1963, Petrography of algal bioherms in Burnt Bluff Group (Silurian), Wisconsin: Am. Assoc. Petroleum Geologists Bull., v. 47, no. 9, p. 1682–1708.

Stanton, R. J., Jr., 1967, Factors controlling shape and internal facies distribution of organic carbonate buildups: Am. Assoc. Petroleum Geologists Bull., v. 51, no. 12, p. 2462–2467.

Stout, J. L., 1964, Pore geometry as related to carbonate stratigraphic traps: Am. Assoc. Petroleum Geologists Bull., v. 48, no. 3, pt. 1, p. 329–337.

Textoris, D. A., and A. V. Carozzi, 1964, Petrography and evolution of Niagaran (Silurian) reefs, Indiana: Am. Assoc. Petroleum Geolgists Bull., v. 48, no. 4, p. 397–426.

———— and ———— 1966, Petrography of a Cayugan (Silurian) stromatolite mound and associated facies, Ohio: Am. Assoc. Petroleum Geologists Bull., v. 50, no. 7, p. 1375–1388.

Thomas, G. E., and H. S. Rhodes, 1961, Devonian limestone bank-atoll reservoirs of the Swan Hills area, Alberta: Alberta Soc. Petroleum Geologists Jour., v. 9, no. 2, p. 29–38.

Van Hees, H., 1956, Elk Point Group: Alberta Soc. Petroleum Geologists Jour., v. 4, no. 2, p. 29–37.

Geology of Beaverhill Lake Reefs, Swan Hills Area, Alberta[1]

C. R. HEMPHILL,[2] R. I. SMITH,[2] and F. SZABO[2]

Calgary, Alberta

Abstract The discovery in 1957 of oil in the remote Swan Hills region, 125 mi northwest of Edmonton, began a wave of exploration similar to that following the 1947 Leduc discovery which started the postwar oil boom in Western Canada. By the end of 1967 more than 1,800 wells had been drilled to explore and develop the Swan Hills region. Drilling has established in-place reserves of more than 5.9 billion bbl of oil and 4.5 trillion ft³ of gas.

Devonian sedimentary rocks unconformably overlie an eroded Cambrian section in the southeast part of the Swan Hills region; in the northwest part of the region, Devonian rocks lap onto the Precambrian granite of the Peace River arch.

Three positive features—the Tathlina uplift, Peace River arch, and the Western Alberta ridge—profoundly influenced Middle Devonian Upper Elk Point and Late Devonian Beaverhill Lake sedimentation. An embayment, shielded on the north by the emergent Peace River arch and on the south and west by the nearly emergent Western Alberta ridge, provided an environment conducive to reef development in the central Swan Hills region. Carbonate-bank deposition flanking the Western Alberta ridge in the south and southwestern part of the study area persisted throughout the time of Beaverhill Lake deposition. These beds merge with the overlying Woodbend reef system.

Recent changes proposed in Beaverhill Lake nomenclature include the elevation of the Beaverhill Lake to group status and the Swan Hills Member to formation status. The term Swan Hills Formation, as used herein, refers to the reef and carbonate-bank facies of the Beaverhill Lake Group, whereas the term Waterways Formation is applied to the offreef shale and limestone facies. The Swan Hills Formation is considered to be equivalent in age to the Calmut and younger members of the Waterways Formation.

The Swan Hills Formation is divided into Light Brown and Dark Brown members. Swan Hills reefs attained a thickness greater than 300 ft, whereas the carbonate-bank facies commonly exceeds 400 ft in thickness. Changing sedimentary and environmental conditions produced a complex reef facies; six major stages are postulated in the development of the undolomitized reef from which the Swan Hills field is producing. Stromatoporoids are the dominant reef-building organisms; abundant *Amphipora* characterize the restricted lagoonal facies.

Although the total impact of Swan Hills production on the provincial economy is difficult to determine, the $184 million paid by the industry to acquire Crown lands in the region during the 10-year period after the initial discovery attests to the economic importance of the Swan Hills producing region.

INTRODUCTION

Before 1957, oil production from the Swan Hills Formation of the Beaverhill Lake Group was unknown. Early in that year, the Virginia Hills field was discovered and focused the attention of industry on a remote and heavily timbered area, approximately 125 mi northwest of Edmonton, Alberta (Fig. 1). Abundant rainfall, muskeg, lack of roads, and rugged topography (1,900–4,200 ft above sea level) made exploration difficult and costly.

The exposed stratigraphic sequence consists of the Late Cretaceous Edmonton Formation and early Tertiary Paskopoo Formation, capped with unconsolidated gravel similar to the Cypress Hills Conglomerate of southern Alberta and Saskatchewan (Russel, 1967).

The Swan Hills Beaverhill Lake reef discovery was made on a large farmout block by the Home Union H. B. Virginia Hills 9–20–65–13–W5M well. On January 31, 1957, a drill-stem test of a 30-ft interval in the Upper Devonian Beaverhill Lake section flowed 40° API oil to the surface. Within 30 days another successful Beaverhill Lake well (the Home *et al.* Regent Swan Hills 8–11–68–10–W5M) was drilled 25 mi northeast of the original discovery. This was the first well of the Swan Hills field. Paradoxically, the confirmation well for the Virginia Hills discovery was a dry hole, even though it was less than 1 mi away; and the 8–11–68–10–W5M well at Swan Hills found oil in what since has proved to be one of the poorest producing areas in the entire complex. However, further drilling around Virginia Hills and Swan Hills, and the impressive number of

[1] Read before the 53rd Annual Meeting of the Association, Oklahoma City, Oklahoma, April 24, 1968. Manuscript received, May 7, 1968; accepted, October 14, 1968.

[2] Home Oil Co. Ltd.

We thank our employer, Home Oil Co. Ltd., for permitting the preparation and publication of this paper, and J. L. Carr, chief geologist, for his encouragement and helpful suggestions. We are particularly appreciative of the long hours spent by W. C. Mackenzie and his drafting department in preparing the illustrations, and also thank W. Hriskevich who compiled the necessary statistical data. R. Sears was most helpful with the parts pertaining to reservoir data, as was A. B. Van Tine with the pressure-depth relations. Finally, we are grateful to Home Oil geologists, particularly H. H. Suter, for criticisms and helpful suggestions after reading the manuscript.

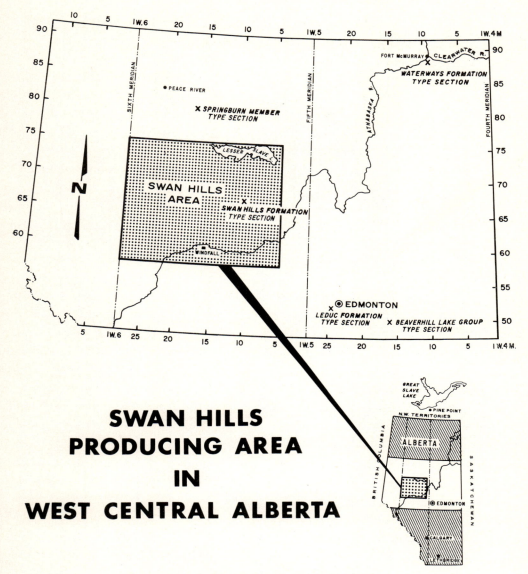

FIG. 1.—Index map, Swan Hills area, Alberta.

new-field discoveries assured continued activity through the region.

Within the map area few more than 12 wells had penetrated the Beaverhill Lake Group before the Home Oil discovery (Fig. 2). By the end of 1967, more than 1,800 wells had been drilled into or through rocks of the Beaverhill Lake Group, and resulted in the discovery of 12 oil fields and two gas fields in the Swan Hills Formation. Total initial in-place hydro-carbons as of January 1, 1968, are estimated to have been in excess of 5.932 billion bbl and 4.512 trillion ft³ of gas (Oil and Gas Conservation Board, 1967).

The Swan Hills reef complex is an excellent example of an undolomitized Devonian carbonate bank-reef development. The availability of numerous cores has facilitated detailed facies studies on several fields. The main purpose of this paper, therefore, is to relate the entire pro-

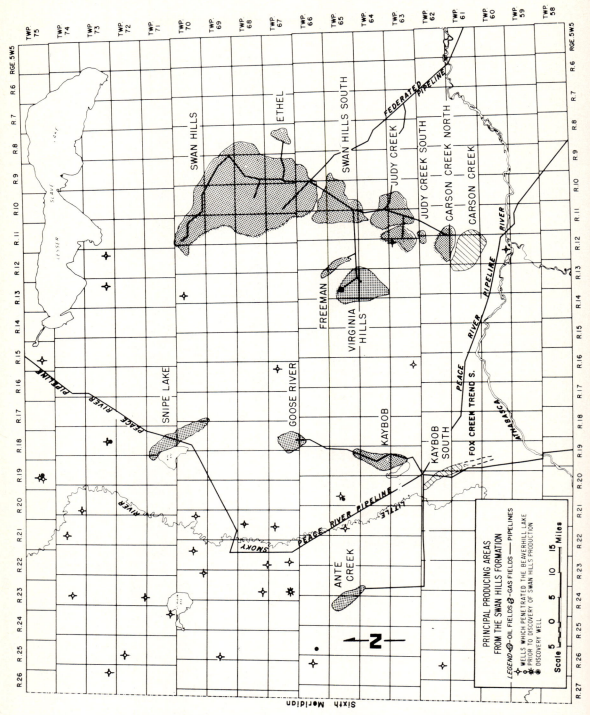

Fig. 2.—Location of Late Devonian Swan Hills Formation fields in Swan Hills region, Alberta. Well symbols are locations where rocks of Beaverhill Lake Group were penetrated before discovery. Gathering system and two major pipeline systems also are shown.

ductive area of the Swan Hills reef complexes to the regional, structural, and stratigraphic setting of the western Canadian sedimentary basin. A secondary purpose is to synthesize the published data on the Swan Hills and related beds and to propose some changes in the nomenclature to reduce confusion.

PREVIOUS WORK

Since Fong's (1959, 1960) original proposal of the type section and discussion of the geology of the Beaverhill Lake Formation, several other excellent studies have been made by Koch (1959), Carozzi (1961), and Thomas and Rhodes (1961). Edie (1961) was the first to make a detailed facies study of the Swan Hills field. Other outstanding studies include those by Fischbuch (1962), Brown (1963), Murray (1964, 1966), Jenik (1965), and Leavitt (1966).

MAP AREA

All present Swan Hills production is from a rectangular area in west-central Alberta, the dimensions of which are approximately 69 by 105 mi, or roughly 7,200 mi². The area is bounded on the east by 115° W long. and on the west by 118° W long. The south boundary is at 54° N lat and the north boundary is at 55° N lat. The area also can be described as lying between T60 and T71 and R8 and R25, W5M.

FIELDS

The Beaverhill Lake oil fields within the map area are, in order of decreasing importance, Swan Hills, Judy Creek, Swan Hills South, Virginia Hills, Kaybob, Carson Creek North, Snipe Lake, Goose River, Freeman, Ante Creek, Judy Creek South, and Ethel (Fig. 2). There are only two gas fields, Carson Creek and Kaybob South. Although most of the present fields are considered to be developed fully, some higher risk, marginal locations could be, and are being, drilled. In the Kaybob South field an important extension toward the southeast recently has been drilled at Fox Creek. Details are not available because of a highly competitive land situation.

DISCOVERY METHODS

Even though well control was extremely sparse for the deeper part of the section, several factors accounted for the gradual increase of exploration.

1. Decline in new discoveries of Leduc reefs (equivalent in age to the Woodbend) in the more accessible country, and the possibility of the presence in this area of other Leduc reef chains with large reserves.

2. Successful exploration along the Mississippian subcrop edges farther south and the possibility of similar conditions in this northern district.

3. An indicated thinning of the underlying Elk Point interval, possibly caused by the presence of a basement high which might have been the locus for reef development.

4. The numerous possibilities for the presence of Mesozoic sandstone bodies which were known to be productive in other parts of the province.

5. The presence of oölites, stromatoporoids, and *Amphipora* in the lower part of the Beaverhill Lake section in some of the older wells. (This indicated shallow-water shoaling and the potential for reef buildup.)

6. The large acreage blocks which could be assembled and the availability of additional offsetting Crown acreage through competitive bidding.

7. The discovery of oil in the "granite wash" just north of the Snipe Lake field.

Seismic studies have been of limited value in exploration for Beaverhill Lake reefs. A geophysical program was conducted on the acreage before the selection of the wellsite and, although the choice of drilling location was influenced by seismic information, the site selection in Virginia Hills was based on a somewhat nebulous feature. Similar seismic data were used to locate the first well in the Swan Hills field.

The value of seismic work in defining areas of Beaverhill Lake reefs has been argued strongly ever since. The arguments may be attributed to uncertainty in identification of the reflecting horizon, lack of velocity contrast between the reef and the enclosing rocks, and the absence of differential compaction or draping over the biohermal buildups. Both seismic structure and isochron maps have been used to select drillable locations, but generally the results have not permitted a good correlation of the well information with the seismic data.

GEOLOGIC HISTORY[3]

The western Canadian sedimentary basin is underlain by the westward continuation of Precambrian rocks of the Canadian shield. The oldest Cambrian basin on the shield rocks was restricted to the Cordilleran trough. It was a long, narrow marine trough in what is now northern British Columbia. The seaway straddled the British Columbia–Alberta boundary in the vicinity of the present Rocky Mountains.

[3] This discussion is mostly a synthesis of work by Van Hees and North (1964), Porter and Fuller (1964), Grayston *et al.* (1964), and Moyer *et al.* (1964).

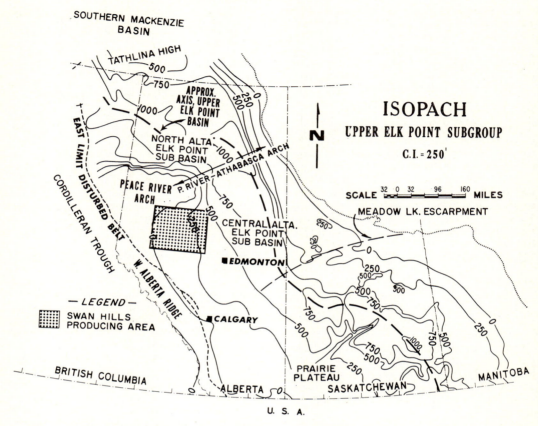

FIG. 3.—Isopach map, Upper Elk Point subgroup. Maximum thickness of Upper Elk Point strata is in northern Alberta and in Williston basin, Saskatchewan. Position of basinal axis is similar to position of Beaverhill Lake basinal axis shown in Fig. 4. Note emergent Western Alberta ridge and Peace River arch. CI = 250 ft. Redrawn from Grayston et al. (1964).

From this trough, Middle and Late Cambrian seas transgressed eastward over the cratonic shelf.

Sub-Devonian erosion removed Upper Cambrian rocks from all but the southeast part of the Swan Hills producing area. No Ordovician or Silurian rocks are present, presumably because of removal by pre-Devonian erosion. Basal Devonian strata unconformably overlie the eroded Cambrian section and in the northwest, lap onto Precambrian granite of the Peace River arch.

Thus, before the Devonian, Caledonian tectonism produced uplift and erosion which led to the widespread destruction of sedimentary beds. Three important positive areas—the Peace River–Athabasca arch, the Tathlina uplift, and the Western Alberta ridge—were present (Fig. 3), and were to exert an impor-

tant influence on sedimentation during Early, Middle, and part of Late Devonian times.

At the beginning of the Early Devonian, the Tathlina uplift and the Western Alberta ridge separated the moderately subsiding basin in Alberta from the MacKenzie basin on the north and the Cordilleran basin on the west. This interior basin, subdivided by the Peace River–Athabasca arch into northern and central Alberta Elk Point subbasins, received 600–1,300 ft of Lower Elk Point clastic and evaporite sediments. Topography controlled the lateral extent of the shallow Lower Elk Point sea.

Collapse of the eastern part of the Peace River–Athabasca arch near the end of Lower Elk Point deposition caused the northern and central Alberta subbasins to merge. The resulting basinal configuration remained essentially unaltered through the rest of the Elk Point de-

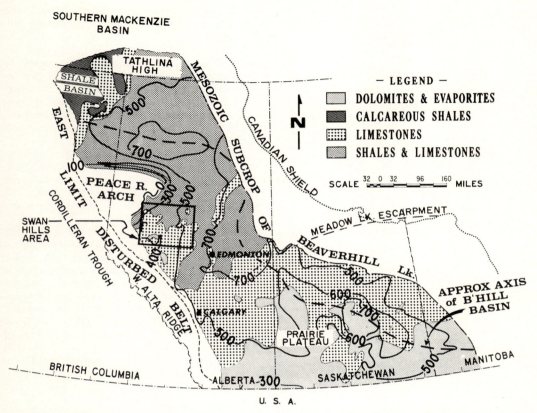

FIG. 4.—Isopach and lithofacies map, Late Devonian Beaverhill Lake Group. More than 700 ft of Beaverhill Lake present in northern and central Alberta and central Saskatchewan. Note progression of facies from outer marine shale in northeastern British Columbia to dolomite and evaporite in Southern Alberta, Saskatchewan, and Manitoba. Swan Hills producing area is southeast flank of Peace River arch. CI = 200 ft. Redrawn from Moyer (1964).

position, and for all of Beaverhill Lake deposition (Figs. 3, 4).

As subsidence and widespread incursion of more normal seawater continued, carbonate material was deposited early in Elk Point deposition. Keg River carbonate banks fringed the basin and patch reefs developed within the basin. For the first time during the Devonian the sea crossed the Meadow Lake escarpment, and sediments were deposited in southern Alberta, southern Saskatchewan, and southwest Manitoba.

The most rapid reef growth was in northeastern British Columbia; the resulting reef complex restricted the circulation of seawater into the Upper Elk Point basin. Reef growth terminated and evaporite accumulation predominated until near the end of Elk Point deposition in central Alberta. In northwestern Al-

berta fluctuating marine conditions prevailed generally near the end of Elk Point deposition. Crickmay (1957), Campbell (1950), and Law (1955) reported a hiatus between the Muskeg and Watt Mountain or Amco Formations in the Fort McMurray, Pine Point, and Steen River areas, respectively. Subaerial erosion and penecontemporaneous shallow-water deposition probably characterized the start of Watt Mountain deposition throughout the Upper Elk Point basin.

The sands of the Gilwood were derived from the Peace River arch and, according to Kramers and Lerbekmo (1967), represent a regressive-deltaic environment during Watt Mountain sedimentation. Other writers, including Guthrie (1956) and Fong (1960), believe the Gilwood Sandstone was deposited during shallow marine transgression. In the writers'

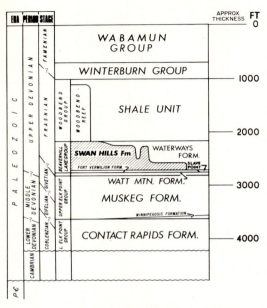

FIG. 5.—Devonian stratigraphic nomenclature, west-central Alberta. Chart shows position of Swan Hills Formation in geologic column. Relation of Swan Hills reef complex to enclosing Waterways Formation and younger Woodbend reefs is shown diagrammatically.

opinion, an oscillatory environment probably characterized Watt Mountain deposition and we agree with Griffin (1965), who wrote that the Fort Vermilion was deposited during final oscillatory conditions before the main transgressive phase which began during the time of Slave Point deposition.

The Late Devonian Beaverhill Lake transgression began with the deposition of Slave Point carbonate on a broad shelf in northeastern British Columbia, northern Alberta, and the adjacent part of the Northwest Territories. A carbonate reefoid-front facies, similar to the underlying Elk Point reefoid carbonate bank, developed in northeastern British Columbia.

Climatic changes and regional tectonism, accompanied by an influx of argillaceous material, destroyed the Slave Point biota (Griffin, 1965). Alternating limestone and shale characterize the rest of the Beaverhill Lake in northern Alberta and in the Edmonton area. However, in the Swan Hills region a shallow-water embayment, protected on the north by the emergent Peace River arch and flanked on the southwest by the Western Alberta ridge, provided a setting conducive to bank develop-

ment and subsequent reef growth. Emergence followed by deepening of water during late Beaverhill Lake deposition terminated Swan Hills reef growth. However, in places, notably the Windfall area, deposition of bank carbonate continued throughout Beaverhill Lake deposition. These carbonate beds grade into the overlying Woodbend reef. During the final deposition of the Beaverhill Lake, a fringing reef developed in the Springburn area on the south flank of the Peace River arch.

The overlying Woodbend consists of carbonate reefs and, in the offreef areas, green shale. The Winterburn consists of silty carbonate, red and green shale, and some anhydrite. The youngest Devonian unit, the Wabamun, is entirely carbonate.

The Devonian rocks within the study area and adjacent regions are subdivided in order of ascending age into Elk Point, Beaverhill Lake, Woodbend, Winterburn, and Wabamun Groups (Fig. 5). The total Devonian section along the eastern margin of the Swan Hills region is approximately 3,600 ft thick and thins markedly northwestward to 300 ft over the Peace River arch (Fig. 6). Within the east-central part of Alberta (*i.e.*, over the central part of the Elk Point and Beaverhill Lake basins) the Devonian is thicker than 4,500 ft.

ELK POINT GROUP

McGehee (1949) first defined the Elk Point Formation. Belyea (1952) elevated it to group rank and Crickmay (1954) described the type section. The writers, following Grayston *et al.* (1964), subdivide the group into the Lower and Upper Elk Point subgroups, the exact ages of which are not settled. However, we tentatively follow Grayston *et al.* (1964), Basset (1961), and Hriskevich (1966), and place the Middle-Lower Devonian boundary at the Upper-Lower Elk Point subgroup division.

Opinion is divided concerning the boundary between the Elk Point and overlying Beaverhill Lake Group. The writers prefer to include the Watt Mountain Formation in the Beaverhill Lake Group. However, we recognize the fact that the controversy is far from resolved, hence do not assign the Watt Mountain Formation to either the Elk Point Group or Beaverhill Lake Group. Murray (1964) and Thomas and Rhodes (1961) placed the boundary at the top of the Fort Vermilion anhydrite. Fong (1960), Edie (1961), and Jenik (1965) considered the boundary to be at the top of the Watt Mountain Gilwood Sandstone Member. Grayston *et al.* (1964) believed that the Watt Mountain Formation belongs with the Beaverhill Lake depositional cycle and thus favored its inclusion with the Beaverhill Lake Group.

The Lower Elk Point subgroup consists mainly of terrigenous clastic and evaporite beds and ranges in thickness from 200 to 1,300 ft.

The Upper Elk Point subgroup is subdivided into the Winnipegosis Formation below and the Muskeg Formation above. The Winnipegosis is mainly carbonate, in contrast to the salt and anhydrite which predominate in the Muskeg Formation. Total thickness of this subgroup ranges from 400 ft in the eastern part of the Swan Hills region to more than 1,000 ft in the central part of the basin.

BEAVERHILL LAKE GROUP

The Beaverhill Lake section originally was defined by the Imperial Oil Geological Staff (1950). Common industry practice in the Swan Hills area has been to include basal carbonate beds of the Cooking Lake Formation with the Beaverhill Lake Group. The proper correlation is shown in Figure 7. The writers follow Leavitt (1966) and raise the Beaverhill Lake section to group rank. As used in this paper, the Beaverhill Lake Group includes, from base to top, the Fort Vermilion, Slave Point, Swan Hills, and Waterways Formations (Fig. 5).

The Beaverhill Lake type section in the Edmonton area was defined by Imperial Oil (1950) to include the interval from 4,325 to 5,047 ft (722 ft) in Anglo Canadian Beaverhill Lake (Lsd. 11, Sec. 11, T50, R17, W4M). Fong (1960) defined the Beaverhill Lake section in the Swan Hills producing area as the interval 8,020–8,543 ft (523 ft) in Home Regent "A" Swan Hills (Lsd. 10, Sec. 10, T67, R10, W5M). The top of the interval coincides with the highest beds of dark-brown limestone below bituminous shale. Subsequent drilling between the Swan Hills region and the Beaverhill Lake type section well has indicated that the upper boundary of the Beaverhill Lake as defined in the type section has been miscorrelated with the upper boundary established by usage in the Swan Hills region. This discrepancy is shown in Figure 7.

Beaverhill Lake strata are present through much of the western Canadian sedimentary basin, but are of maximum thickness in central Alberta, where drilling has established the presence of up to 750 ft of section (Fig. 4). Figure 4 shows that the axis of the Beaverhill Lake basin closely parallels that of the Upper Elk Point basin (Fig. 3). A marked thickness reduction of the Beaverhill Lake in northeastern British Columbia, though due partly to depositional thinning, has been interpreted by Griffin (1965) as primarily the result of post-Beaverhill Lake erosion.

Beaverhill Lake beds may be divided into five main facies: the outer shale facies, carbonate-front facies, inner alternating limestone and shale facies, inner reef facies, and shelf-margin carbonate-evaporate facies.

The carbonate-front facies, extending northeastward through northern British Columbia into the Northwest Territories, formed an effective barrier to the Beaverhill Lake shelf basin, profoundly influencing sedimentation there and on the shelf margin. Seaward, the carbonate-front facies is in abrupt contact with dark, deep-water marine shale, and the central part of the Beaverhill Lake basin contains the alternating limestone and shale facies of the Waterways Formation. Southward, the limestone and shale facies grades into shelf-margin carbonate and evaporite, limestone, and primary dolomite. In southern Alberta, Saskatchewan, and Manitoba, anhydrite is a major constituent of the carbonate-evaporite facies (Fig. 4).

The inner reef facies, designated the Swan Hills Formation, is restricted to an embayment south of the Peace River arch.

McLaren and Mountjoy (1962), on faunal evidence, correlate the Moberly and Mildred —upper members of the Waterways Formation —with the Flume Formation exposed in the front ranges of the Rocky Mountains south of the Athabasca River. Lower beds are correlative with the Flume Formation north of the Athabasca River.

Agreement is lacking on the age of the Beaverhill Lake Group. Mound (1966) established a Late Devonian age from conodont studies. Clark and Ethington (1965), on the basis of conodont data, placed all but the top few feet of the Flume Formation in the Middle Devonian. Norris (1963) assigned the Waterways Formation to the early Late Devonian and the Slave Point Formation in northeastern British Columbia to the Middle Devonian. McGill (1966), on the basis of ostracods, placed the Givetian-Frasnian boundary at the base of the Waterways Formation in the Lesser Slave Lake area in the northeastern corner of the Swan Hills region. Loranger (1965) placed the entire Beaverhill Lake and the lower part

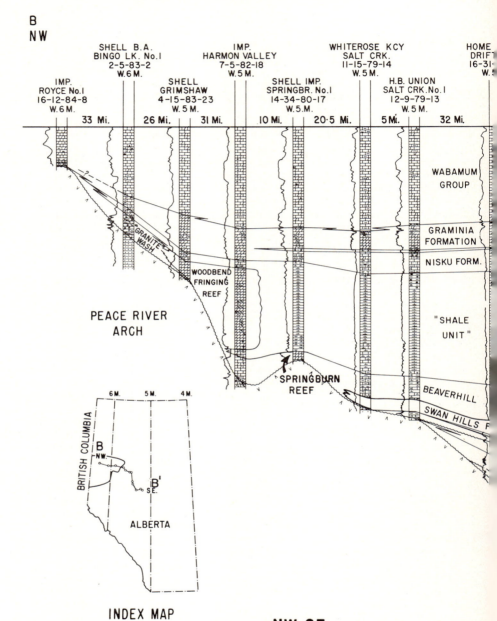

INDEX MAP

NW-SE
GENERALIZED DEVONIAN-CAMBRIAN
CROSS SECTION B-B'
OFF THE PEACE RIVER ARCH

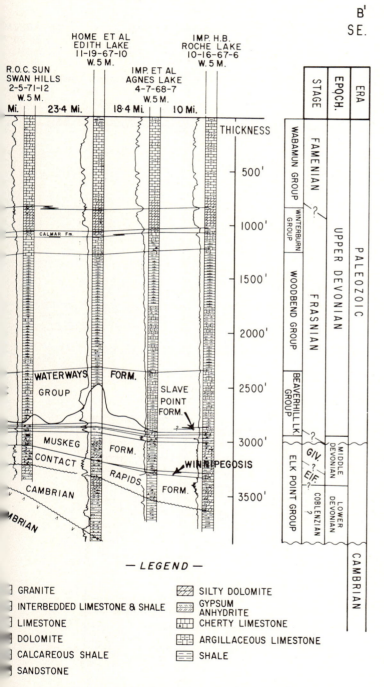

FIG. 6.—Northwest-southeast cross section **B-B'** showing thinning of Devonian over Peace River arch. Crest of arch was emergent until time of Wabamun deposition. Swan Hills Formation, deposited during early Late Devonian transgression, onlaps southeastern flank of Peace River arch. Vertical scale in feet.

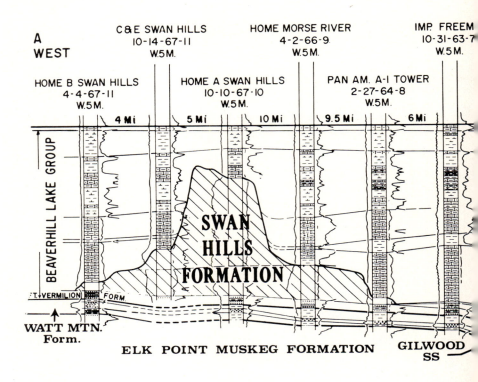

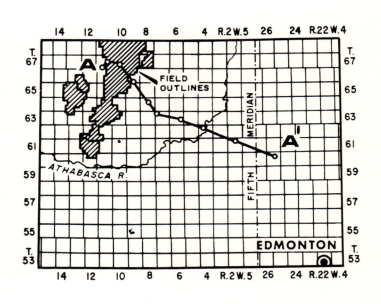

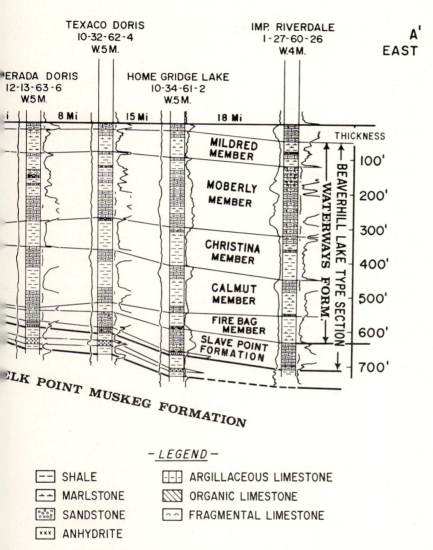

FIG. 7.—West to east cross section **A-A'** of Beaverhill Lake Group, through Swan Hills field, Alberta, into offreef basin area. Section demonstrates that Swan Hills Formation is time equivalent of Calmut and younger members of Waterways Formation, and not of Slave Point Formation as believed by some authors. Relation of upper Beaverhill Lake limestone in Swan Hills area to Beaverhill Lake type section should be noted. Thicknesses in feet.

of the overlying Woodbend Group in northeastern Alberta in the Middle Devonian.

The writers assign a Late Devonian age to the Beaverhill Lake Group in the Swan Hills region. This does not conflict with the findings of Norris, because the Slave Point Formation is the product of initial deposition of an advancing sea from the north, and thus as a transgressive deposit should become progressively younger southward.

Fort Vermilion Formation.—The name Fort Vermilion Member was introduced by Law (1955) for an evaporite unit at the base of the Slave Point Formation, which overlies the Watt Mountain Formation in the subsurface of northwestern Alberta. This anhydrite unit is only 23 ft thick in the California Standard Steen River type well (Lsd. 2, Sec. 22, T117, R5, W6M) but thickens eastward to 120 ft at the Hudson's Bay No. 1 Fort Vermilion (Sec. 32, T104, R8, W5M). Fort Vermilion anhydrite is widespread; subsequent drilling has established Fong's (1960) "Basal Beaverhill Lake" anhydrite unit as correlative with the Fort Vermilion.

Norris (1963) raised the Fort Vermilion Member to formation status. The writers concur with Norris, because the Fort Vermilion beds are widespread and because we agree with Griffin (1965), that the Fort Vermilion was deposited during final oscillatory conditions before the main Slave Point transgressive phase began.

Slave Point Formation.—Cameron (1918) assigned the name Slave Point Formation to scattered exposures of thin-bedded, medium-grained, dark-gray bituminous limestone, which he concluded were of Middle Devonian age. The limestone crops out on the north and south shores of Great Slave Lake. Cameron originally estimated the total section to be 160 ft thick, but in 1922 revised this estimate to 200 ft.

Campbell (1950) correlated a 310-ft section of limestone, shale, and dolomite penetrated in bore holes at Pine Point on the south shore of Great Slave Lake with Cameron's Slave Point section. Campbell's subsurface section includes 170 ft of fine-grained, stromatoporoidal limestone overlying 11 ft of dark-greenish-gray shale, termed the Amco Formation, which unconformably overlies 129 ft of fossiliferous limestone and dolomitic limestone. Law (1955) noted a hiatus between the Watt Mountain and Muskeg Formations penetrated in the California Standard Steen River (2–22–117–5–W6M),

in northwest Alberta, and he correlated the Watt Mountain with the Amco Formation. Law concluded that only the upper 170 ft of the carbonate rocks in the Pine Point area subsurface correlate with the Slave Point Formation of the plains (informal usage).

The Slave Point Formation of the plains attains a maximum thickness of 500 ft, in northwestern Alberta, north of the Peace River arch, northeastern British Columbia, and the District of MacKenzie. In northeastern British Columbia and the District of MacKenzie, there is an abrupt change in facies between the Slave Point carbonate facies front and the basinal Otter Park Shale. Gray and Kassube (1963) describe the facies front as being rich in stromatoporoids and *Amphipora*. Griffin (1965) defined five rock types in the vicinity of the facies front: dark stromatoporoid-bearing calcilutite, stromatoporoid biosparite, stromatoporoid biomicrite, light-colored micrite (and fossil micrite), and white and gray dolomite. Dolomitization has destroyed many of the original lithic characteristics, but the abundance of stromatoporoids suggests that the front facies is in fact a reef chain.

The Slave Point limestone thins eastward, and also south of the Peace River arch. Crickmay (1957) noted 5.5 ft of magnesium limestone between the Waterways and Muskeg Formations at the Bear Biltmore 7–11–87–17–W4M well near Fort McMurray. On the basis of fossil content, Crickmay considered this unit to be correlative with the Slave Point Formation of the outcrop area. Subsequently, Belyea (1952) correlated the unit with the basal limestone member of the Beaverhill Lake Formation type section (*i.e.,* she considered the Beaverhill Lake section in the Edmonton area to be equivalent to the Waterways Formation plus the 5.5-ft magnesium limestone section at the Waterways type locality). Norris (1963) disputed Crickmay's correlation of these beds with the Slave Point Formation and proposed the term "Livock River Formation" to include this section. Griffin (1965) confirmed Crickmay's interpretation, however, by establishing, on the basis of mechanical-log and lithologic analysis, a convincing correlation between Norris's Livock River Formation and the Slave Point Formation in northeastern British Columbia. Thus, the basal 35-ft limestone unit in the type Beaverhill Lake Formation correlates with the Slave Point Formation.

Swan Hills Formation.—Fong (1960) applied the name Swan Hills Member to the productive unit of the Beaverhill Lake in the Swan Hills area. The type section is the interval at 8,167–8,500 ft (333 ft) in the Home Regent "A" Swan Hills 10–10–67–10–W5M well. Murray (1966) proposed that the Swan Hills Member be raised to formation rank, a view also held by Leavitt (1966). The writers concur. Drilling since 1960 has established the presence of Swan Hills rocks across an area of approximately 7,000 mi². Its widespread occurrence and the economic importance support elevation to formation rank.

Fong defined the Swan Hills as an organic bioclastic limestone unit overlying and gradational with anhydrite and shale now known to be correlative with the Fort Vermilion Formation. The Swan Hills Formation is divisible into a lower Dark Brown member and an upper Light Brown member. In a typical section (Fig. 8) the Dark Brown member is 110 ft thick but may range from 80 to 160 ft, and consists of dark-brown calcarenite and calcilutite with some reef-building organisms. The Dark Brown member grades upward into the Light Brown member, which has a maximum thickness of 320 ft, and consists of calcarenite, calcilutite, and biogenic carbonate.

The Swan Hills Formation consists of organic carbonate deposits. The central part of the Swan Hills region contains organic reefs as thick as, or slightly thicker than, 300 ft. The carbonate-bank deposits which characterize the southwestern part of the region are more than 400 ft thick and grade upward into the overlying Woodbend reef deposits.

The Swan Hills is transgressive, becoming younger northwestward in the direction of the Peace River arch (Fig. 6). The unit wedges out against the lower flank of the Peace River arch. An upper Beaverhill Lake reef termed the Springburn Member is interpreted to be a local fringe reef. The Springburn is a forerunner to the extensive Woodbend fringe reef and is considered by the writers to be separate from the Swan Hills Formation.

Murray (1966) observed a very sharp contact between the reef and offreef facies in the Judy Creek area and suggested that most of the basin facies is younger than the Swan Hills Formation. Leavitt (1966) found in the Carson Creek area, in addition to a sharp reef-offreef boundary, a deeper water fauna in the adjacent offreef facies and concluded that there was considerable relief between the reef complex and the offreef basin. Apparently, this relation is not characteristic of the entire Swan Hills region, for Fischbuch (1962) reported an interfingering of Swan Hills carbonate rocks with offreef strata near the south end of the Kaybob field.

Waterways Formation.—The name Waterways originally was suggested by Warren (1933) for the limestone and shale which crop out at the confluence of the Clearwater and Athabasca Rivers in northeastern Alberta. Warren's Waterways section, as determined from outcrop and nearby salt wells, totals 405 ft. On the basis of the section penetrated in the Bear Biltmore 7–11 well, Crickmay (1957) also included beds removed by erosion in the Fort McMurray area. The type subsurface section penetrated at Bear Biltmore (Lsd. 7, Sec. 11, T87, R17, W4M), is 740 ft and was subdivided by Crickmay (1957) into five units designated in ascending order the Firebag, Calmut, Christina, Moberly, and Mildred Lake Members.

The Waterways Formation, consisting of alternating brown fragmental limestone and greenish-gray shale, is separated from the underlying Elk Point Group by the 5.5-ft Slave Point magnesian limestone bed. Griffin (1965) traced Crickmay's (1957) five-member subdivision of the Waterways Formation as far north as California Standard Mikkwa (Lsd. 12, Sec. 23, T98, R21, W4M). In addition, Griffin's correlation of the top of the Calmut Member in northeastern British Columbia provided convincing support for his suggestion that thinning of the shale-limestone sequence in northwesternmost Alberta and northeastern British Columbia is due to (1) facies equivalence of the Slave Point Formation (carbonate) with the Waterways Formation (limestone and shale) and (2) progressive northwestward erosional truncation of the Beaverhill Lake Group. These relations are illustrated in Figure 9.

AGE RELATIONS OF SWAN HILLS, SLAVE POINT, AND WATERWAYS FORMATIONS

Warren (1957) observed that ". . . the oil-bearing horizon within the Beaverhill Lake Formation [*i.e.,* the Swan Hills Formation] south of Lesser Slave Lake carries a fauna younger than that of the Firebag Member of the Waterways Formation [Fig. 7] and thus is younger than the Slave Point Formation."

According to Crickmay (1957), *Atrypa* aff. *A. independensis* is a guide fossil everywhere for the Slave Point. Koch (1959) noted the abundance of this species within the basal part of the Swan Hills Formation. Its presence sug-

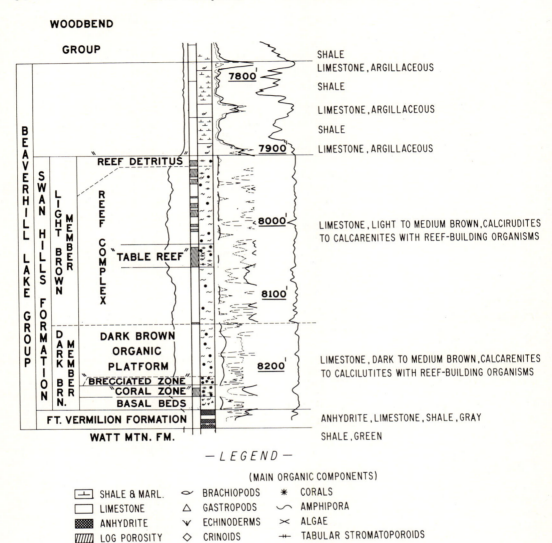

FIG. 8.—Typical Beaverhill Lake section in Swan Hills field, Alberta. Well is Home Regent Swan Hills 4-28-67-10-W5M. KB elev. = 3,322.1 ft. Well was cored almost completely through Swan Hills Formation, permitting excellent control for facies divisions shown. Depths in feet below KB.

gests that the lower beds of this formation correlate with the Slave Point Formation. Koch considered, however, the tendency for *Atrypa* of the *A. independensis* type to migrate with the nearshore carbonate facies (*i.e.,* it is a facies-controlled organism), and accordingly he preferred to correlate the basal Swan Hills with the Firebag Member.

The writers concur with Warren and believe that the basal part of the Swan Hills Formation

in the Swan Hills field correlates with the Calmut Member of the Waterways Formation. Moreover, the Slave Point Formation is known to wedge out east of the Swan Hills fields and thus was not deposited within the designated Swan Hills region. These relations are illustrated in Figure 7.

Woodbend Group.—The contact between the Beaverhill Lake and Woodbend Groups is gradational. Bioclastic limestone and shale of

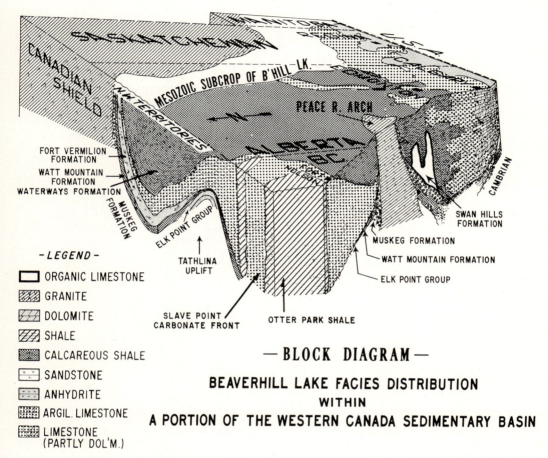

—BLOCK DIAGRAM—

BEAVERHILL LAKE FACIES DISTRIBUTION
WITHIN
A PORTION OF THE WESTERN CANADA SEDIMENTARY BASIN

FIG. 9.—Block diagram showing Beaverhill Lake facies distribution, and position of Swan Hills Formation with respect to Peace River arch and to various Beaverhill Lake facies. View is toward southeast. Basinal Otter Park shale facies is in abrupt contact with Slave Point carbonate-front facies. Calcareous shale and argillaceous limestone of Waterways Formation occupy much of Alberta and grade southeastward into primary dolomite, anhydrite, and limestone facies.

the Cooking Lake Formation compose the basal part of the Woodbend Group. The Cooking Lake Formation formed a platform for subsequent Leduc reef development in central Alberta (Fig. 10) where, in the offreef areas, it is overlain by the Duvernay bituminous shale and interbeds of dark-brown limestone. The Duvernay Formation is overlain by the Ireton Formation, a green shale section flanking and overlying the Leduc reefs.

In the Swan Hills region, the Duvernay-Cooking Lake section is represented principally by greenish-gray calcareous shale; accordingly, the Ireton-Duvernay-Cooking Lake section is referred to herein as the "shale unit." Unlike

Fong's (1960) "shale unit," it does not include the Nisku Formation of the Winterburn Group. Thus, the section in this paper is identical with the "shale unit" of Murray (1966).

Younger Devonian.—The Woodbend is overlain by the Winterburn and Wabamun Groups. The Woodbend includes shale, siltstone, anhydrite, dolomite, and silty dolomite. The Wabamun consists mostly of dolomitic limestone and some anhydrite.

STRATIGRAPHY OF SWAN HILLS FORMATION

The Swan Hills Formation is divided into two members on the basis of color and morphologic characteristics—a lower or Dark

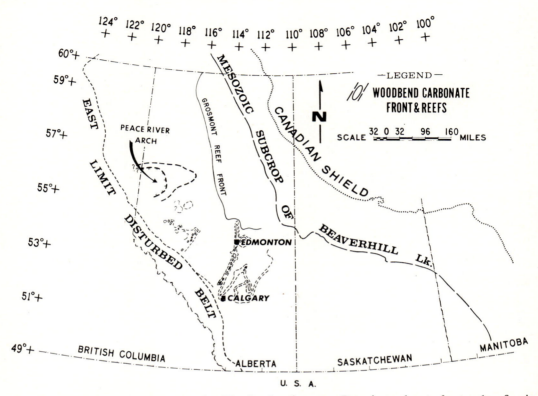

Fig. 10.—Distribution of Late Devonian Woodbend reef systems. Devonian carbonate front and reefing in Alberta reached maximum development during Late Devonian (Woodbend) time. Grosmont reef complex underlies northeastern Alberta and is followed in clockwise direction by Leduc-Meadowbrook reef chain in central Alberta. Simonette-Windfall reef system and Sturgeon Lake atoll cover large part of west-central Alberta including western and southern part of Swan Hills region. Peace River arch is flanked by broad, continuous, fringing reef system. Redrawn from Belyea (1964).

Brown member and an upper or Light Brown member (Fig. 8).

The Dark Brown member is a widespread organic platform or carbonate bank, covering an extensive area of west-central Alberta. It is fringed on the northeast, north, and northwest by stromatoporoidal limestone, which is interpreted as an organic reef (Fig. 11). This member represents the first transgressive phase of the Late Devonian sea.

Generally there is an abrupt facies change toward the east and north between the Dark Brown member and the calcareous shale and argillaceous limestone of the Waterways Formation. On the west the entire Swan Hills Formation is terminated by onlap on the Peace River arch.

The thickness of the Dark Brown member ranges from 80 ft in the Deer Mountain–

House Mountain area to approximately 160 ft in the vicinity of Carson Creek (Fig. 12). Thin-section studies show that the dark color is caused mainly by the presence of pyrobitumen and residual oil in the fine matrix rather than by a high argillaceous content.

Frame-building organisms became established at numerous localities on the top of the Dark Brown member and formed a bioherm or "reef complex," as used in this paper. This complex probably grew on topographic highs or positive areas of the underlying carbonate bank, hence its lateral extent is more restricted than that of the platform on which it rests.

The Swan Hills, Judy Creek, Carson Creek North, Kaybob, and Goose River fields are separate reeflike buildups on the platform, separated from each other by surge channels and the associated offreef shale and limestone beds

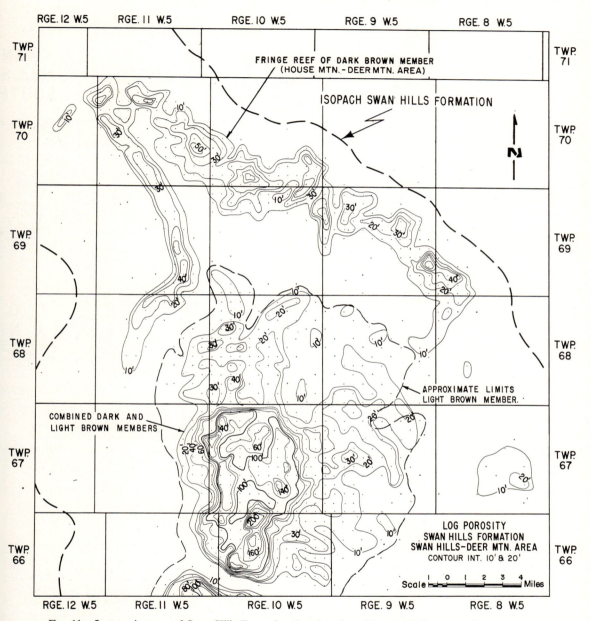

FIG. 11.—Isoporosity map of Swan Hills Formation, based on logs. Map emphasizes productive reef-run and main buildup area. Patch reefs and reef outwash are loci of porosity in intervening area. Porous trend along northern edge originally was called House Mountain–Deer Mountain area but now is included in area designated as Swan Hills field. Note line showing limits of Light Brown member. CI = 10 and 20 ft.

of the Waterways Formation (Figs. 13–15).

South and southwest of the Swan Hills region, the Light Brown member formed a massive carbonate bank with reef-building organisms around the rim of the widespread buildup. The Carson Creek, Virginia Hills, and Ante Creek fields are associated with this reef-rimmed bank (Fig. 14). Farther south and

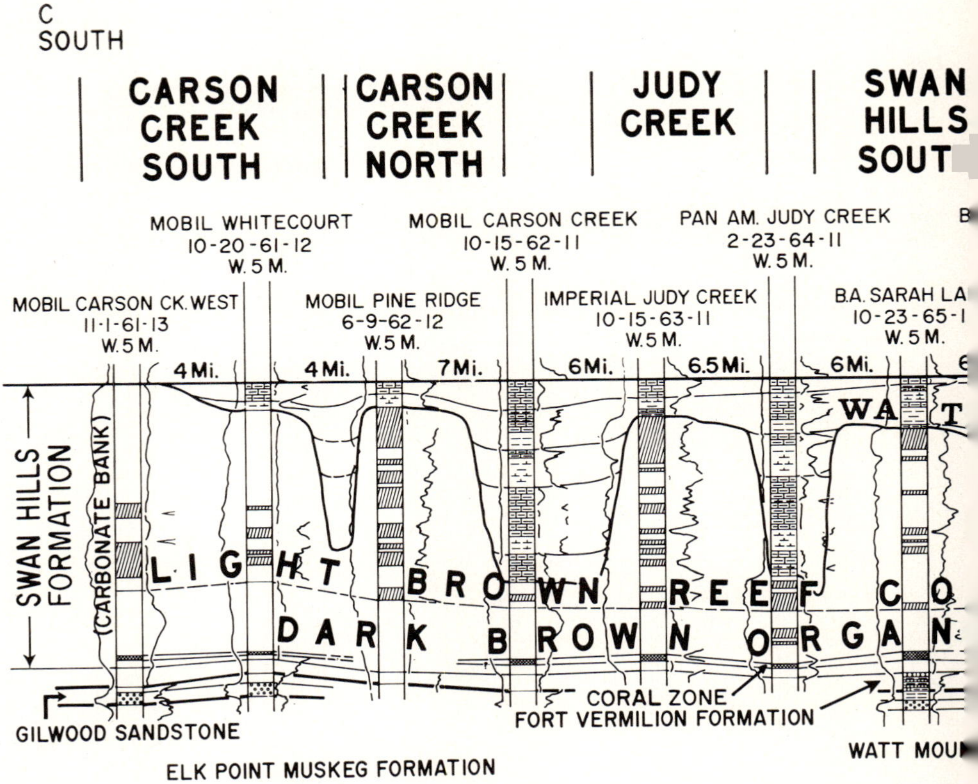

68 — C. R. Hemphill, R. I. Smith, and F. Szabo

Fig. 12.—South-north cross section C-C' of Beaverhill Lake Group, north-central Alberta. Section passes through fields of eastern reef chain. Shale and argillaceous limestone of Waterways Formation separate reef buildups. Gradual southward thickening of Dark Brown and Light Brown members, combined with thinning of gross Beaverhill Lake, results in entire Beaverhill Lake interval consisting of Swan Hills carbonates. Deer Mountain field now is northern part of Swan Hill field. Area through which section passes is shown on Figures 2 and 14; exact location shown on Figure 21.

southwest the bank formed the foundation for the Late Devonian Woodbend reefs, following deposition of the Beaverhill Lake.

Although there are facies variations from field to field in the area, several facies types predominate in all fields (Fig. 8).

Dark Brown Member

"Basal beds."—The lowermost bed of the "Basal beds" overlies the Fort Vermilion anhydrite and shale, and consists of an argillaceous, cryptograined, dark-brown to black limestone ranging in thickness from 3 to 14 ft. Anhydrite and pyrite are common in the lower part. The "Basal beds" are sparsely fossiliferous, and contain the following fauna in decreasing order of predominance: brachiopods (*Atrypa*), ostracods, echinoderms, and crinoids.

The uppermost units of the "Basal beds" are transitional with the overlying "Coral zone," and have similar organic content. These beds were deposited in aerated seawater of normal salinity. The presence of reef-building organisms in the upper part, the higher percentage

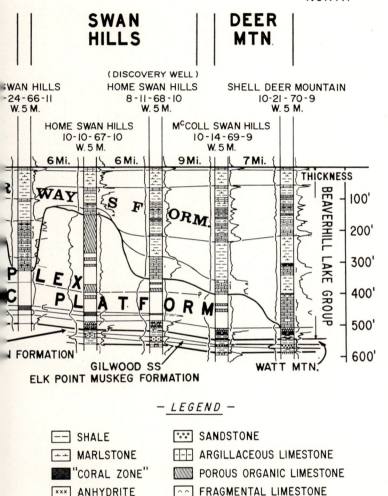

of skeletal material, and the local presence of sparry calcite cement suggest deposition in shallower water than the adjacent offreef basal beds of the Waterways Formation on the east (Fig. 7).

"Coral zone."—Overlying the "Basal beds" is light-brown, fossiliferous limestone of great lateral extent, called the "Coral zone" by Fong (1959). The thickness ranges from 4 to 20 ft. The lower and upper contacts are transitional. This *in situ* reefal biofacies contains the lowest zone of porosity in the productive areas. The

thickness, porosity, and permeability of the "Coral zone" improve where the Light Brown member also is well developed. In the marginal areas of the dark-brown organic platform this facies is not recognizable, but is replaced by stromatoporoid limestone. Organic components of the "Coral zone" in decreasing order of predominance are *Thamnopora* corals, massive and tabular stromatoporoids, *Amphipora,* and *Stachyoides.* Skeletal fragments and grains, and mud-supported carbonate rocks (calcilutite) are the main matrix components.

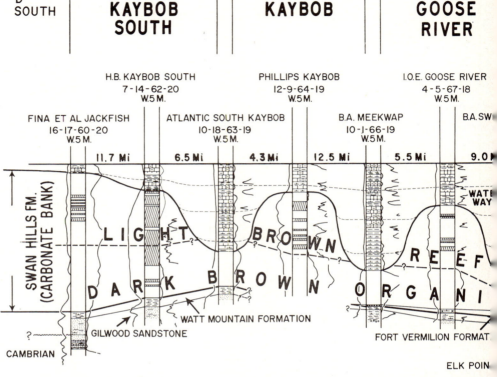

FIG. 13.—South-north cross-section **D-D′** of Beaverhill Lake Group, north-central Alberta. Section passes through western Swan Hills fields and shows relations similar to those in Fig. 12. Gradual pinchout of Fort Vermilion Formation causes basal beds of Beaverhill Lake to overlie directly Watt Mountain Formation. Gross Beaverhill Lake thins markedly between Kaybob and Kaybob South fields. Area through which sections passes is shown on Figures 14 and 21; exact location on Figure 21.

Above the "Coral zone" are dense beds which completely separate it from the Light Brown member. As a result, the "Coral zone" in the Swan Hills field is a separate producing zone of the Swan Hills Formation.

Stromatoporoid reef front.—Around the northern and northeastern margins of the Dark Brown member platform high water energy fostered the growth of a reef body with a rigid framework. The slightly different organic composition in the different reefs probably is related to local energy conditions and food supply. Along the eastern edge of the Swan Hills field, stromatoporoids predominate, but the increasing numbers of *Amphipora* and the greater amount of carbonate-mud matrix suggest a lower energy environment. Toward the central part of the platform, the reef front interfingers with the dark brown *Amphipora*

beds, and the offreef Waterways Formation is discordant. Organic components, in order of decreasing importance, are tabular and massive stromatoporoids, algae, brachiopods, cup corals, *Amphipora*, and crinoids. The matrix ranges from coarse reef detritus to fine carbonate mud and in places consists of skeletal grains, intraclasts, and some sparry calcite.

The porosity common to these beds is associated with both the organic framework and matrix. Hydrocarbon production is from the stromatoporoidal reef front at the edge of the platform north and northeast of the Swan Hills and Virginia Hills fields. Similar beds also are present in the marginal areas of the Light Brown member in the upper part of the reef complex and, where such beds are developed, they also are productive.

"Brecciated zone."—This term applies to a

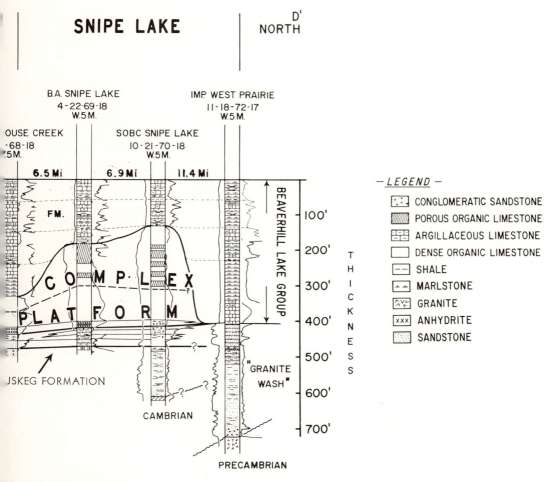

SNIPE LAKE   D'
                           NORTH

B.A. SNIPE LAKE   IMP. WEST PRAIRIE
4-22-69-18   11-18-72-17
   W.5 M.   W.5 M.

OUSE CREEK   SOBC SNIPE LAKE
-68-18   10-21-70-18
.5 M.   W.5 M.

6.5 Mi   6.9 Mi   11.4 Mi

FM.

C·O·M·P·L·E·X

P·L·A·T·F·O·R·M

JSKEG FORMATION

CAMBRIAN

PRECAMBRIAN

BEAVERHILL LAKE GROUP

100'
200'
300'
400'
500'
600'
700'

"GRANITE WASH"

T H I C K N E S S

—LEGEND—

- CONGLOMERATIC SANDSTONE
- POROUS ORGANIC LIMESTONE
- ARGILLACEOUS LIMESTONE
- DENSE ORGANIC LIMESTONE
- SHALE
- MARLSTONE
- GRANITE
- ANHYDRITE
- SANDSTONE

stromatoporoid bed overlying the "Coral zone" that has been observed in most of the Swan Hills–Ante Creek area. The name refers to the brecciated appearance of the cores, which contain numerous circular and semicircular, partly broken stromatoporoid colonies. However, examination of the cores proved that the appearance is caused by the growth pattern of the stromatoporoids. Only very limited mechanical transport is indicated, and the zone is not truly "brecciated."

The "Brecciated zone" interfingers with the stromatoporoidal reef front on the north and northeast, and with the *Amphipora* beds of the Dark Brown member on the west and southwest. The thickness ranges from 0 to 50 ft. Organic components in order of decreasing abundance include bulbous stromatoporoids, *Amphipora*, and corals. The matrix consists of fine

skeletal grains, intraclasts, and carbonate mud. Pyrobitumen also is common. The organisms of the "Brecciated zone" apparently accumulated in a lower energy environment than those of the stromatoporoidal reef front—probably in a restricted shelf lagoon. Porosity is sparingly present in association with stromatoporoids.

Dark brown Amphipora *beds.*—The dark-brown *Amphipora* beds extend throughout the Swan Hills region, and form the most common unit of the Dark Brown organic platform. The thickness ranges from 40 to 100 ft, generally increasing southward. There also are local variations in the thickness over areas where the Light Brown member is developed. The base of the unit is at the top of the "Coral zone," or at the top of the "Brecciated zone" where the latter is present. The unit interfingers with the stromatoporoidal reef front facies on the east

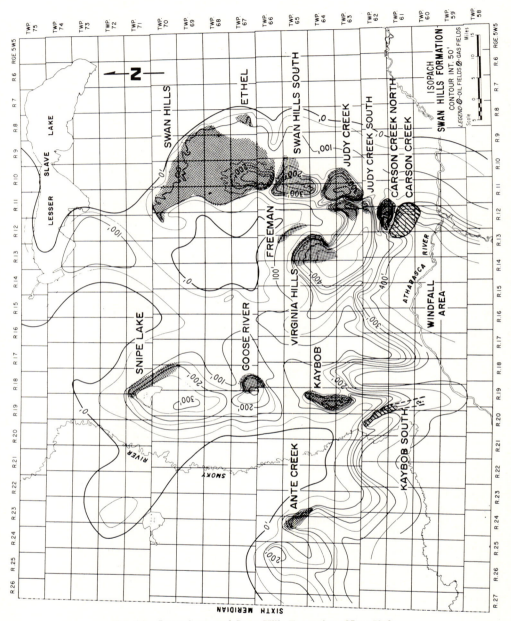

FIG. 14.—Isopach map of Swan Hills Formation. CI = 50 ft.

and northeast. The *Amphipora* content ranges from 0 to 60 percent, and the unit generally is less fossiliferous near the base. A few bulbous stromatoporoids also are present. Organic components, in order of decreasing importance, are *Amphipora,* bulbous stromatoporoids, brachiopods, and gastropods. Laminated limestone beds with sparry calcite also are common and generally are devoid of organisms. The matrix ranges from poorly sorted silt- and sand-size grains and pellets to carbonate mud, or calcilutite. Skeletal grains also are recognizable.

Porosity is developed only locally in these beds, and is associated with some *Amphipora-*

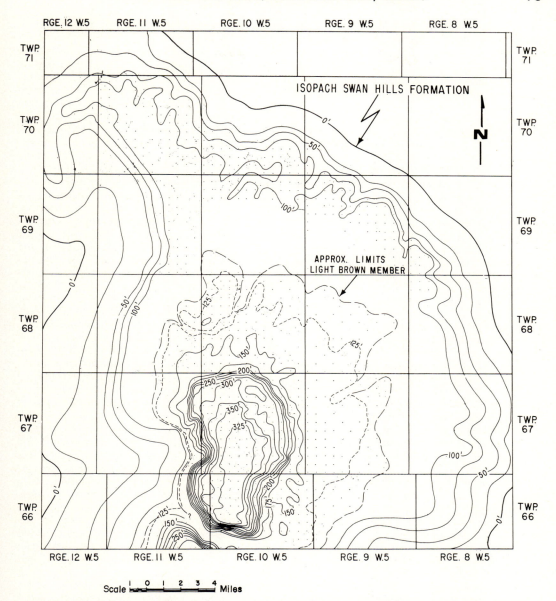

FIG. 15.—Detailed isopach map of Swan Hills Formation in Swan Hills field–Deer Mountain area. Broad plat-form is present between edge and Swan Hills main buildup. Rigid wave-resistant stromatoporoid reef wall first developed on northeast edge of platform.

rich zones. A little void space is present in a few places in granular or pelleted limestone, but for all practical purposes the dark-brown *Amphipora* beds do not contribute significantly to the pore volume of the Swan Hills Forma-tion. This unit reflects sedimentary and envi-ronmental conditions of a restricted shelf la-goon, and the laminated limestone indicates oc-casional supratidal-flat conditions.

Light Brown Member

The thickness of the Light Brown member changes more abruptly than that of the gently sloping Dark Brown platform on which it lies.

The member ranges in thickness from a few feet to approximately 320 ft south and west of Carson Creek. The alternation of higher energy conditions which produced reef buildups with brief erosional episodes and stable sea-level conditions caused considerable variation in sedimentation.

Light-brown Amphipora *beds.*—These are the most widespread beds of the Light Brown member and cover part of the Dark Brown platform. They range in thickness from 0 to approximately 200 ft, the thickest sections being over the area of the Swan Hills–Carson Creek reef complex.

The light-brown *Amphipora* beds consist of two main limestone types: (1) light-brown to medium-brown limestone, fragmental, with calcarenite matrix and *Amphipora;* and (2) light-olive-brown limestone, with *Amphipora,* calcilutite, and sparry calcite matrix. Limestone of the first type comprises the first sediments deposited on the top of the Dark Brown platform and contains only scattered stromatoporoids. The matrix is fragmental in appearance and consists of skeletal grains and intraclasts, suggesting a shallow, agitated, open-marine environment. In areas where only this fragmental *Amphipora* bed is present over the Dark Brown platform, the contact between the reef and overlying offreef Waterways Formation is sharp. In most places the organisms of the Swan Hills Formation are truncated at the contact, and only a thin zone of pyrite is present; thus very shallow postdepositional erosion is suggested.

Limestone of the second type is found most commonly in the center part of the Light Brown member, *Amphipora* being the predominant organic component. Dendroid, tabular, and bulbous stromatoporoids are present in a few areas, together with ostracods, gastropods, and calcispheres. The matrix consists of sparry calcite, carbonate mud, skeletal grains, and intraclasts. These sediments appear to have been deposited in shallow, slightly agitated water with low turbidity, typical of a restricted shelf-lagoon environment.

Porosity development commonly is associated with the matrix and organisms of the light-brown fragmental *Amphipora* beds, whereas in the light-olive-brown limestone porosity is irregular and generally poorly developed.

"Table reef" (Edie, 1961).—This easily recognizable, widespread zone of the Light Brown member overlies the light-brown fragmental *Amphipora* beds and ranges in thickness from 15 to 25 ft. Its broad lateral extent is characterized by a relatively uniform organic content. In the marginal areas north and northwest of the buildup the organisms are severely broken and reworked, forming a fragmental zone which interfingers with the *in situ* material. The presence of the broken rocks indicates short erosional periods. The principal organisms, in order of decreasing abundance, are dendroid stomatoporoids, Solenoporoid algae, brachiopods, cup corals, and *Amphipora.*

The matrix is considerably varied, but the most common constituents are the skeletal grains and debris of various sizes, with negligible amounts of carbonate mud. Porosity is common in this zone and is associated with both organisms and matrix. Interorganic vuggy porosity present between stromatoporoids suggests secondary, postdepositional leaching. The "Table reef" zone of the Swan Hills Formation probably represents an environment of shallow, agitated water. Local emergence or erosion occurred at the end of deposition, as indicated by the presence of organic debris on the slopes of the reef.

Porous calcarenite beds.—A porous zone, consisting of skeletal grains and reworked and transported organic fragments, overlies the light-brown *Amphipora* beds and, in places, the upper slopes of the Light Brown buildup. In a few areas this zone contains almost no organisms, and consists of well-sorted carbonate sandstone; in the upper slopes, most of the zone consists of well-rounded stromatoporoid and *Amphipora* fragments. The reef rubble on the top and slopes of the buildup suggests that growth terminated as a result of shallowing of the seawater rather than submergence. Porosity generally is excellent, and is intragranular. Intraorganic void space also is common, decreasing slightly toward the central part of the buildup. This unit is one of the most important reservoir rocks of the area. The porous calcarenite beds probably were formed in the most exposed parts of the organic buildup, in shal-

FIG. 16.—South-north sections showing stages I-VI in development of Swan Hills reefing, as interpreted for Swan Hills field. Similar stages can be postulated for other Beaverhill Lake fields.

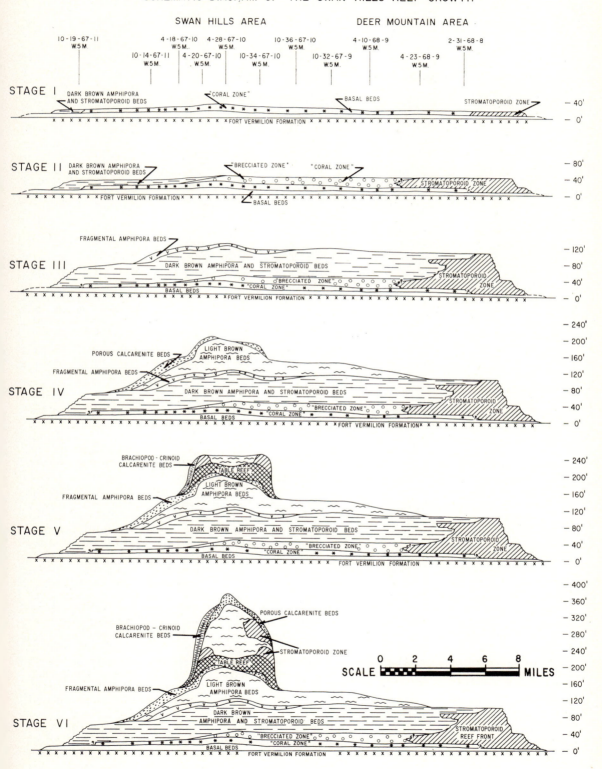

SCHEMATIC DIAGRAM OF THE SWAN HILLS REEF GROWTH

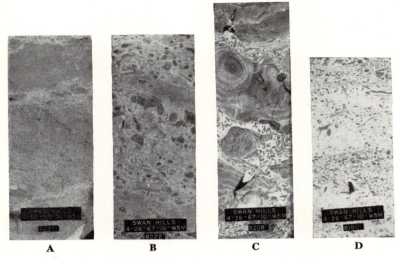

FIG. 17.—**A,** Fine-grained, dark-brown to black argillaceous beds of basal Beaverhill Lake, 3 ft below "Coral zone." Fine clasts of similar material can be seen scattered throughout, suggesting reworking of beds. **B,** *Thamnopora*-type corals in matrix of buff calcarenite, representing "Coral zone" of Swan Hills Formation. Unit forms lowermost productive bed in Swan Hills field. In this sample "Coral zone" has 8.2 percent porosity and 3.6 md permeability. **C,** Broken, bulbous and massive stromatoporoids of "Brecciated zone." Fine *Amphipora* in dark-brown micritic matrix fill interstices between larger stromatoporoid fragments. Some fractured stromatoporoids have calcite infilling. **D,** Dark-brown micritic *Amphipora* beds. Banded appearance results from long axis of *Amphipora* being deposited parallel with normal bedding, and different energy levels which prevailed in carrying *Amphipora* fragments into lagoon.

low turbulent water. The upper boundary with the offreef Waterways Formation is unconformable and appears to be an erosion surface.

Brachiopod-crinoid beds.—The brachiopod-crinoid beds are found most commonly on the west slopes of the Light Brown member, and consist of calcarenites of skeletal origin. In order of decreasing importance the organic components are brachiopods, crinoids, stromatoporoids, and *Amphipora*. The stromatoporoids and *Amphipora* gradually decrease in number, and the calcarenite grades into calcisiltite and fine carbonate mudstone farther away from the main buildup. This brachiopod-crinoid zone with fine calcarenite derived from skeletal material is typical of the reef flanks and is considered to have been deposited in open-marine, aerated water of medium energy. The beds generally are devoid of porosity and therefore have no economic significance.

DEPOSITIONAL HISTORY OF SWAN HILLS FORMATION

In the Swan Hills field area the depositional history of the Swan Hills Formation can be described in six stages, each indicating a major change in sedimentation and environment (Fig. 16). Stages I-III cover the time of Dark Brown

member deposition, and Stages IV-VI cover the time of Light Brown member deposition.

Stage I.—The Beaverhill Lake sea gradually transgressed westward from the deepest part of the basin. The area between the Peace River arch and West Alberta ridge was flooded, and the widespread shoal conditions that developed favored organic growth. The depositional environment of associated sediments was similar to that of the offreef Waterways Formation, except for the higher percentage of skeletal material and the first appearance of slightly shallow-water organisms (Fig. 17A).

After the appearance of the corals in the upper part of the "Basal beds," the area became densely colonized by frame-building organisms that created a solid base and good foothold for additional organic growth (Fig. 17B). Along the northeast rim, in areas of slightly higher wave energy, a stromatoporoid zone began to develop, whereas farther south along the platform edge in the areas of lower energy, the corals and stromatoporoids were replaced by *Amphipora*-rich beds.

Stage II.—A slight but persistent rise in sea level caused continued stromatoporoid reef growth on the northeast rim. This growth created a broad shelf lagoon behind the

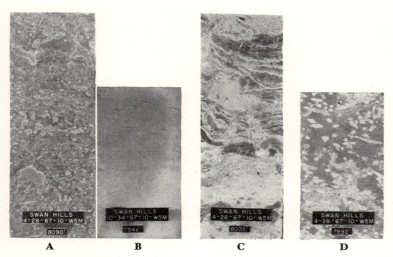

FIG. 18.—**A,** Light-brown beds of *Amphipora* bank deposits. In addition to abundant *Amphipora* there are some *Stachyoides* and very few bulbous stromatoporoids in calcarenite matrix. Because of reworking of these beds, *Amphipora* are more randomly oriented and widely scattered. **B,** Fine skeletal reef calcarenite of reworked *Amphipora* bank deposits shows no bedding or recognizable organisms, and is very homogeneous rock. Sample has 9.1 percent porosity and 18 md permeability. **C,** Almost solid framework of massive stromatoporoids bound by algal mats; forms "table reef" (Fig. 5). Calcarenite fills spaces between larger organisms. In sample shown porosity is 12.2 percent with 12 md permeability. **D,** *Amphipora* in light-buff calcarenite. This is typical lagoon deposit just behind reef rim. In sample porosity is 5.5 percent and permeability 10 md.

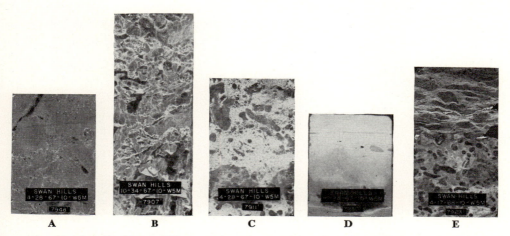

FIG. 19.—**A,** Pelletoid unfossiliferous carbonate mud deposited in quieter and deeper water of central lagoon. **B,** Stromatoporoid reef-wall material with abundant sparry calcite filling interstices. Organic porosity is 9 percent and permeability is 11 md. **C,** Bioclastic material of "Detrital zone." Reef rubble composed of large and small organic fragments enclosed in dark-brown calcarenite matrix common to top and upper slopes. Porosity is 12.6 percent and permeability 64 md. **D,** Dark-gray, cryptograined, dense, argillaceous, finely bedded calcareous shale of Waterways Formation. **E,** Contact between Swan Hills Formation and overlying Waterways Formation. Close examination of contact shows truncated organisms indicative of erosion before burial. *Boudinage* structure common to some of Waterways beds is present here

Table 1. Parameters of Swan Hills Oil Fields

Fields, Pools	Disc. Date	Name and Loc. Disc. Well	Total Wells Drld. to 12/31/67	Av. Well Depth (Ft)	Prod. Area (Acres)	Maximum Reservoir Thickness (Ft)	Av. Net Pay (Ft)	Av. Por. (%)	Av. Perm. (Md)	Water Sat. (%)	Est. Oil in Place (1,000 Bbl)	Cum. Prod. to 12/31/67 (1,000 Bbl)	Remaining Recoverable Crude Oil (1,000 Bbl)
Swan Hills													
A and B pools	3/ 2/57	Home et al. Regent 8-11-68-10-W5M	520	8,299	102,479	395	52.8	7.8	20	18.6	1,944,000	83,309	688,621
C pool	3/21/58	Texcan Mic Mac Deer Mtn. 10-14-69-9-W5M	353	7,475	55,058	120	29.8	6.2	5	10.2	553,000	17,684	141,616
Judy Creek													
A pool	2/25/59	Imperial Judy Creek 16-31-63-10-W5M	184	8,665	28,200	358	66.7	9.3	43	16	809,000	42,018	322,032
B pool	9/ 6/59	Imperial Virginia Hills 10-13-63-12-W5M	75	8,842	12,380	453	58.4	9.0	111	17	256,000	13,123	109,757
Swan Hills South	2/27/59	B.A. Pan Am. Sarah Lake 2-13-65-11-W5M	243	8,345	35,680	405	69.8	8.0	26	18.3	897,600	43,639	348,512
Virginia Hills	1/31/57	Home Union H.B. Virginia Hills 9-20-65-13-W5M	134	9,283	24,350	503	48.4	8.1	35	20.4	450,200	25,288	148,992
Kaybob	4/22/57	Phillips Kaybob 7-22-64-19-W5M	107	9,780	18,000	232	59.9	7.4	23.3	22	300,000	22,346	97,654
Carson Creek North	9/ 6/58	Mobil P.R. Carson N 6-1MU 6-1-62-12-W5M	46										
A pool—oil pool				8,632	6,916		31	8.3	56	13	66,000	4,100	24,940
—gas cap				8,580	3,720		11	8.0	15				
B pool—oil pool				8,736	12,372		51	9.1	167	21	215,000	8,939	74,911
—gas cap					377		7.8	9.9	21	21			
Snipe Lake	10/24/62	S.O.B.C. Snipe Lake 10-21-70-18-W5M	114	8,534	17,297	272	33.3	7.3	35	27	198,000	10,587	66,633
Goose River (B pool excl.)	8/28/63	B.A. Goose River 10-4-67-18-W5M	28	9,185	7,700	200	49	8.2	103	19	145,000	2,613	20,587
Freeman	10/31/62	H.B. Union Home Freeman 2-1 2-1-66-13-W5M	20	9,184	5,390	160	29	5.9	20	25	40,400	874	3,934
Ante Creek	10/15/62	Atlantic Ante Creek 4-7 4-7-65-23-W5M	24	11,270	7,510	260	25.3	6.3	6	22	34,800	1,838	3,730
Judy Creek South	3/19/60	Mobil Carson Creek 14-31 14-31-62-11-W5M	8	8,925	3,009	136	22.6	6.3	24.4	25	15,000	557	2,443
Ethel	1/28/64	Mobil Atlantic Ethel 10-11 10-11-67-8-W5M	4	7,522	1,289	87	23.6	5.7	10.2	17	8,100	19	62

wave-resistant reef front. In the quieter and more restricted waters behind the reef, the less wave-resistant bulbous stromatoporoids of the "Brecciated zone" flourished (Fig. 17C). Farther west, in the quiet, semistagnant waters of the lagoon, *Amphipora* beds were deposited in precipitated carbonate mud.

Stage III.—The rigid stromatoporoid reef wall continued to grow on the northeast side of the carbonate bank, while *Amphipora*-rich beds were deposited in the shelf lagoon behind the reef front (Fig. 17D). The writers assume that carbonate-bank growth ceased when slight eastward tilting deepened the water above the reef-front area, and at the same time part of the carbonate bank became emergent on the west. Brief exposure of the east-central part of the carbonate platform caused reworking of the upper part of the bank in the Swan Hills area and created a thin fragmental zone on this part of the carbonate platform. This marked the termination of deposition for the Dark Brown member and provided a substratum for further organic buildup on the slightly higher, emergent western side.

Stage IV.—After the submergence of the stromatoporoidal reef front, *Amphipora* grew abundantly in the quieter, slightly deeper water of the lagoon bank. The periods of quiescence were interrupted by frequent storms, which transported *Amphipora* fragments into localized carbonate-bank deposits that formed loci for the reef growth of Stage V (Fig. 18A). The presence of porous calcarenite near the top and along the flanks of these bank deposits suggests near emergence at this time, and deposition ceased (Fig. 18B).

Stage V.—Stage V is the "Table reef" which developed on the beds of Stage IV. A temporary stillstand of the sea permitted lateral growth of the organic lattice, which resulted in this widespread, fairly homogeneous buildup (Fig. 18C). Slow subsidence resulted in the formation of a circular stromatoporoidal reef atoll enclosing a central lagoon (Fig. 18D). Fluctuating water level or occasional storms caused the deposition of small amounts of fragmental material in zones within the lagoon and the reef rubble of the outer slopes. Local emergence probably terminated this stage of growth.

Stage VI.—The final stage of Swan Hills reef development was marked by a return to supratidal conditions. This caused renewed growth of *Amphipora* and the deposition of pelletoid and unfossiliferous carbonate-mud beds (Fig. 19A). Carbonate-mud accumula-

Table 2. Parameters of Swan Hills Gas Fields

Fields, Pools	Disc. Date	Name and Loc., Disc. Well	Total Wells Drilled to 12/31/67	Av. Well Depth (Ft)	Prod. Area (Acres)	Max. Reservoir Thickness (Ft)	Av. Net Pay (Ft)	Av. Por. (%)	Av. Perm. (Md)	Water Sat. (%)	Initial GIP (Bcf)	Marketable Gas Prod. (Bcf)	Remaining Marketable Gas to 12/31/67
Kaybob South	9/11/61	H.B. Union Kaybob 11-27 11-27-62-20-W5M	8	10,560	16,440	325	43	10	294.2	15	670	0	370
Carson Creek	2/26/57	Mobil Oil Whitecourt 12-13 12-13-61-12-W5M											
A pool			8	8,550	15,840	89	20	8	21	20	210	8	142
B pool			5	8,610	6,980	83	24	8	75	20	110	-15	95

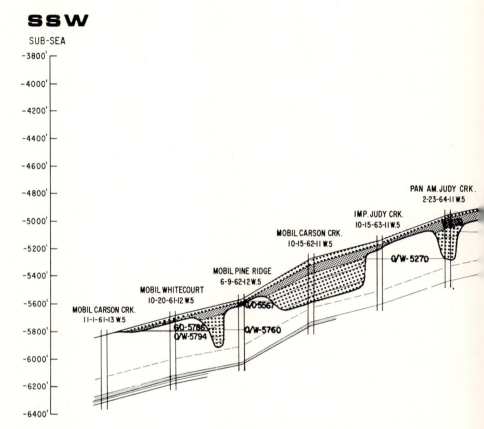

FIG. 20.—SSW-NNE structural cross section through eastern fields (Carson Creek to Swan Hills) shows lateral progression of gas-oil and oil-water interfaces. This is prime example illustrating Gussow's (1954) hypothesis of differential entrapment of hydrocarbons. Vertical scale in feet; sea-level datum. Trace is same as C-C′ (Fig. 12), shown on Figure 21.

tion alternated with the deposition of *Amphipora*-rich beds and a few thin terrigenous mud stringers. Incipient reef growth is present locally (Fig. 19B). Upward organic growth was prevented by emergence and strong erosion. The result of this activity was the formation of calcarenite beds and coarse reef rubble on the top and upper slopes of the buildup. These carbonate clastic rocks are the youngest strata of the Swan Hills Formation (Fig. 19C). Penecontemporaneous deposition of carbonate mud and clay (Waterways Formation) in the offreef areas is indicated (Fig. 19D). Sedimentation proceeded at a slightly slower rate relative to the development of the Swan Hills section.

On the east side of the buildup the contact with the stratified Waterways equivalent of the reef is sharp and unconformable (Fig. 19E). On the west side the brachiopod-crinoid beds,

which show a gradual decrease in grain size and organic content away from the reef, form a transitional zone between the two formations.

At the end of Stage VI, a sudden increase in water depth drowned the reef complex. Waterways-Woodbend clay and carbonate-mud deposits covered most of the area.

DIAGENESIS OF SWAN HILLS CARBONATES

Dolomitization.—The absence of significant dolomitization in the Swan Hills Formation makes it possible to conduct detailed facies, textural, and environmental studies of the reefs. Most of the formation is in the initial stage of compaction-current dolomitization, which involves decreased porosity and reduced pore volume. Where dolomitization has occurred, it generally plugs void space of primary organic and matrix porosity. Presumably the magne-

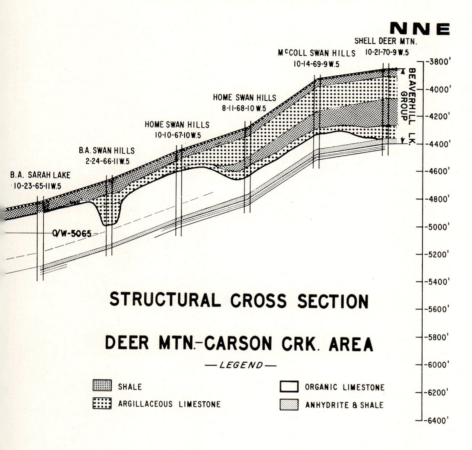

NNE

SHELL DEER MTN.

McCOLL SWAN HILLS   10-21-70-9 W.5
10-14-69-9 W.5

HOME SWAN HILLS
8-11-68-10 W.5

HOME SWAN HILLS
10-10-67-10 W.5

B.A. SWAN HILLS
2-24-66-11 W.5

B.A. SARAH LAKE
10-23-65-11 W.5

O/W-5065

BEAVERHILL LAKE GROUP

-3800'
-4000'
-4200'
-4400'
-4600'
-4800'
-5000'
-5200'
-5400'
-5600'
-5800'
-6000'
-6200'
-6400'

STRUCTURAL CROSS SECTION

DEER MTN.-CARSON CRK. AREA

—LEGEND—

SHALE   ORGANIC LIMESTONE

ARGILLACEOUS LIMESTONE   ANHYDRITE & SHALE

sium ions were brought to the site of precipitation by percolating waters after lithification.

Another form of dolomitization is associated with the organisms; *e.g.,* the axial canals in *Amphipora* commonly contain dolomite infilling in varied amounts. Generally this dolomite infilling is associated with calcite crystals. Presumably the source of magnesium was the organisms, or magnesium ions from areas of higher ion concentration which migrated to and accumulated in the fossils.

A less common form of dolomite is a scattering of perfect dolomite rhombs, about 20–50 μ in diameter, occupying space once filled by much finer calcite grains and microorganisms in a generally dense matrix.

Complete dolomitization of the Swan Hills Formation is observed south and west of the Kaybob area, and south of the Carson Creek

gas field, where it is associated mainly with the upper part of the Light Brown member. The reason for this complete dolomitization is not known to the writers, but is assumed to be related to tectonism.

Silicification.—Scarce light-gray to brown-gray dolomitic chert lenses are found in the Swan Hills Formation. Thickness ranges from 1 to 3 in. The material partly replaces skeletal and nonskeletal grains, and possibly was precipitated from chemical solutions of organic origin.

Recrystallization.—Recrystallization is here defined as the formation of new mineral grains in the limestone in the solid state, without the introduction of other elements. Recrystallization is not very important in the Swan Hills area, but it has been observed in a few places in the upper part of the Light Brown member,

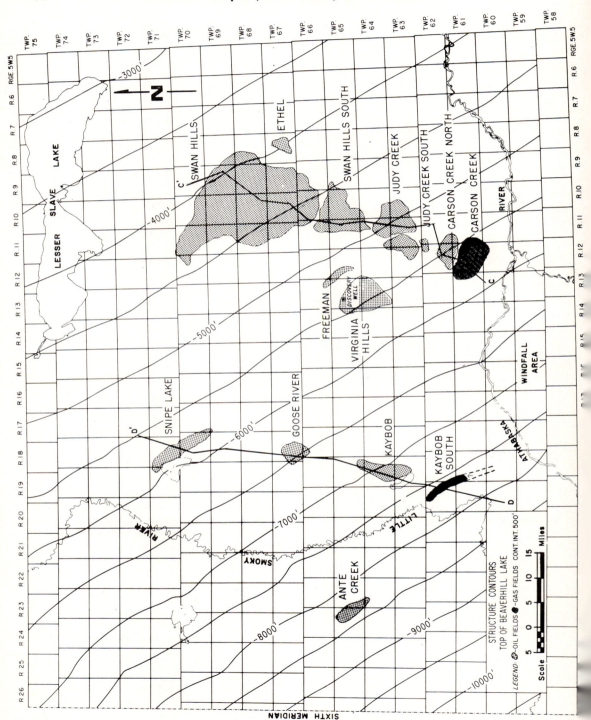

Fig. 21.—Structural contours, datum top of Beaverhill Lake, show southwesterly regional dip. Lack of differential compaction is shown by absence of contour deviation as contours pass through areas of reef buildup. CI = 500 ft. Depth subsea. Shows locations of Figures 12, 13, and 20.

generally in association with the finer grained sediments of the lagoon facies. The matrix is converted into a limestone of saccharoid texture, and the original structure of the *Amphipora* and other organisms is blurred or lost.

Fracturing.—A study was made of numerous cores from the Swan Hills field and the fractured intervals were recorded. Generally, fractures are very scarce in the Dark Brown member, but are more common in the upper part of the Light Brown member.

The matrix of the Dark Brown member consists of very fine-grained calcarenite and calcilutite, which commonly are recemented by secondary calcite. Recementation caused considerable compaction. The lithologic character of the Light Brown member reflects sedimentation in high-energy waters; the matrix ranges in lithologic character from loosely consolidated, fine-grained calcarenite to calcirudite which yielded to stress more easily by fracturing.

Detailed cross-section studies of the Beaverhill Lake Group show that this limestone interval thins slightly over areas of maximum reef buildup. The reduction in thickness probably resulted from the collapse of porous sections accompanied by the development of fractures. Most fractures are vertical or high-angle oblique and many of the fracture planes are lined with secondary calcite crystals. The importance of these fractures is that they may permit communication between otherwise isolated porous zones within the Light Brown member.

RESERVOIR CHARACTERISTICS

Reservoir parameters for each field are shown in Tables 1 and 2. Within the map area, approximately 377,000 acres is underlain by productive Swan Hills Formation. In individual wells net pay thickness is as much as 250 ft. However, the best field average is 69.8 ft and the poorest 22.6 ft. Field-weighted average-po-

Table 3. Properties of Natural Gas, Carson Creek and Kaybob South Fields

Field	Sp. Gr. Gas	Recoverable NGL Content (Bbl/MMcf)[1]	Recoverable Sulfur Content (Long Tons/MMcf Raw Gas)
Carson Ck. A	0.930	131.48	—
B	0.972	142.17	—
Kaybob South	1.0096	107	6.2

[1] Based on raw gas volume with 85 percent recovery of propane, 95 percent recovery of butane, and 100 percent recovery of heavier ends.

Table 4. Beaverhill Lake Fields, Oil Gravity and Saturation Pressures

Field	API Gravity Produced Liquid (°API)	BHT (°F)	Ps @ BHT (psig)	Ps @ 225° F (psig)
Ante Creek	46	233	4,140	4,116
Carson Creek oil leg[1]	43	198	3,767	3,848
Carson Creek gas	61	198	(3,850 DP)[2]	—
Carson Creek N. A	44	187	3.391	3,493
Carson Creek N. B[3]	44	191	3,312	3,414
Freeman	40	221	1,738	1,750
Goose River	40	233	2,052	2,028
Judy Creek A	43	205	2,290	2,350
Judy Creek B	43	206	2,940	2,997
Kaybob S.	60	238	(3,643 DP)	—
Kaybob	43	235	3,019	2,989
Snipe Lake	37	183	1,300	1,426
Swan Hills S.	42	225	2,219	2,219
Swan Hills (Main)	43	225	1,820	1,820
Swan Hills (Inverness)	42	211	1,726	1,768
Swan Hills C	42	189	1,370	1,478
Virginia Hills	38	218	1,812	1,833

[1] There is a thin (8 ft) noncommercial oil leg in this reservoir.
[2] DP denotes dew-point pressure for gas fields.
[3] Reservoir has a small gas cap but oil leg appears undersaturated.

rosity values range from a low of 5.7 percent to a high of 10 percent, whereas weighted average horizontal permeability values for the fields range from 5 to 167 md.

All of the oil pools are undersaturated and there is a general increase in API gravity toward the gas-condensate pools. An odd situation is present in the Carson Creek North field, where both the A and B pools have a gas cap but samples of the reservoir oil indicate undersaturation. Snipe Lake has the lowest gravity of 37° and Ante Creek has the highest of 46°.

Water saturation ranges from 10 to 30 percent, but generally is about 20 percent. The structural relations and the differing water tables along the eastern reef trend are depicted in Figure 20.

Oil is carried to market by the Peace River pipeline and the Federated pipeline. Posted wellhead price is approximately $2.55/bbl (Canadian).

In several fields, the presence of separate pools has been established. Four factors account for this situation:

1. Presence of oil in separate reef facies within the same reef complex (the best example is Swan Hills).
2. Vertical separation by an impermeable green shale barrier, such as is found in the Carson Creek and Carson Creek North fields.
3. Horizontal and vertical permeability barriers caused by completely "tight" reef facies; *e.g.*, the Judy Creek–Judy Creek South relation.

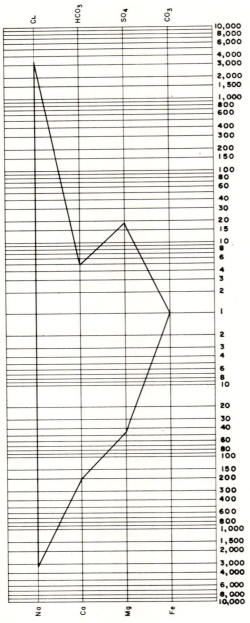

FIG. 22.—Logarithmic plot of typical water analysis, Swan Hills Formation, in milligram equivalents per unit. There is considerable uniformity in chemical composition of formation waters from Swan Hills fields throughout area. This illustration is thought to represent typical logarithmic pattern of a Swan Hills Formation water analysis.

4. Scour or surge channels between reef masses typical of the eastern reef chain, as shown on Figure 14. Dissection of this chain is apparent from the number of separate pools along the trend.

The Light Brown member of the Swan Hills Formation has the best reservoir characteristics. In fact, it is only in the Swan Hills field that the Dark Brown member and the "Coral zone" are of economic importance. There the porous beds of the Dark Brown and Light Brown members are grouped and designated as one pool for proration purposes. The "Coral zone" is separated by the impermeable carbonate of the Dark Brown member and forms the B pool of the field. Probably the principal reason for production in the Dark Brown member is the lack of definite water table, which suggests that the entire basal part of the Swan Hills Formation is above the oil-water interface as calculated from regional pressure data.

Depth of drilling is controlled primarily by the position of the well in the Alberta basin. Regional dip is southwest at approximately 40 ft/mi in the eastern part of the area and steepens to 50 ft/mi in the western part (Fig. 21). Accordingly, drilling depth ranges from 7,475 ft at the northeast end of the Swan Hills field to more than 11,000 ft in the Ante Creek area.

RESERVOIR FLUIDS

Gas.—The primary difference in gas composition between the two Beaverhill Lake gas fields (Table 2; Carson Creek and Kaybob South) is in the sulfur content. Gas from Carson Creek is sweet, whereas Kaybob South gas has 16.59 percent H_2S by volume. Both fields are rich in natural gas liquids. Table 3 gives the most important gas properties. Small gas caps are present in the Carson Creek North A and B pools.

Oil.—Oil properties differ somewhat among fields. This is to be expected because of the many reservoirs, each of which has different physical properties. Generally, the Beaverhill Lake reservoirs contain an undersaturated paraffin-base crude, with a low sulfur content ranging from a trace to 0.42 percent. The variations in API gravity are given in Table 4, together with the bottomhole temperature and saturation pressure for each field. The saturation pressures also have been adjusted to a common temperature of 225° F, to permit easier correlation of the reservoir-fluid properties.

Water.—There is an oil-water or gas-water

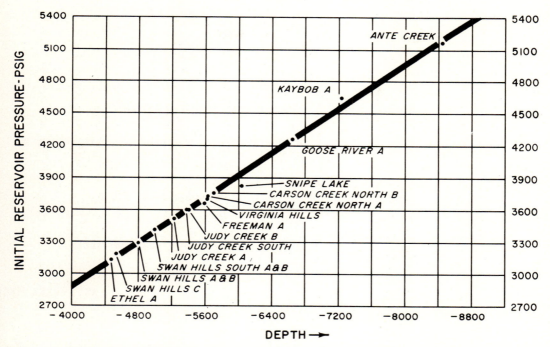

Fig. 23.—Plot of initial pressure *versus* datum (depth subsea), Beaverhill Lake Group. This pressure-depth plot of Swan Hills fields shows close relation of initial pressures, suggesting common pressure system for area. Data from Oil and Gas Conservation Board, Calgary (1967).

interface in most of the developed fields. Notable exceptions are Swan Hills, Ethel, Kaybob, and Freeman. Where a variable water table is reported, it generally can be attributed to a change in lithologic character within the interval in which the water table normally would be. A very few water occurrences which are difficult to explain have been found in the Swan Hills field. Probably rock-geometry is involved (Stout, 1964).

The chemical compositions of different formation waters can be compared by plotting the components as a logarithmic pattern. The figures used are milligram equivalents per unit, obtained by multiplying parts per million (mg/l) by the following factors: Na + K × 0.0435; Ca × 0.0499; Mg × 0.0822; SO_4 × 0.0208; Cl × 0.0282; CO_3 × 0.0333; and HCO_3 × 0.0164.

Water analyses from several Beaverhill Lake fields were plotted in this manner and a typical logarithmic pattern for Swan Hills Formation waters is shown in Figure 22.

PRESSURE RELATIONS

A pressure-depth plot has been made by the Oil and Gas Conservation Board using the original pressures for several of the Swan Hills fields. The close relation of these initial pressures suggests a common pressure system (Fig. 23).

A similar detailed plot was made by selecting the most reliable pressure data from the individual wells in the gas, oil, and water phases of the Swan Hills Formation. To it have been added pressure data from the fluid phases of the widespread Gilwood Sandstone, which is the highest reservoir rock below the Swan Hills Formation. The data indicate a connection between these two reservoirs (Fig. 24).

It is not the writers' intention to speculate on the various stages of fluid migration or to explain the reason for the Gilwood-Swan Hills relation. Many factors are involved, such as rate of compaction, thickness of overburden, regional tilting, fracturing, juxtaposition of porous beds, *etc.* In some areas the Beaverhill

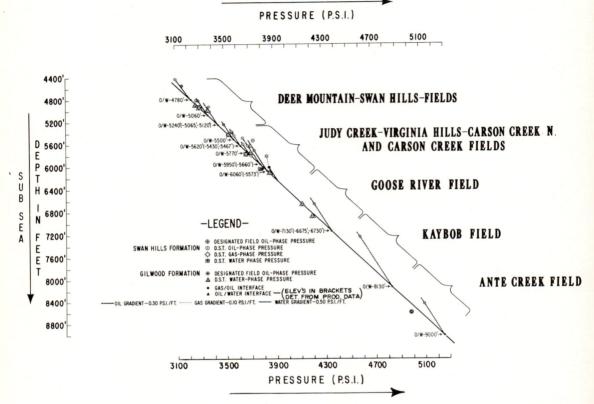

FIG. 24.—Pressure-depth plot, Beaverhill Lake–Gilwood Sandstone system, illustrates relation of Swan Hills oil and gas pools to a common reservoir system. Pressures measured within water phase in both Gilwood Sandstone and Swan Hill Formation reservoirs also relate to single system, suggesting that Swan Hills reservoir continuity is achieved through widespread Gilwood aquifer

Lake directly overlies the Gilwood Sandstone, or the "granite wash" with which the Gilwood Sandstone merges in the vicinity of the Peace River arch. Presumably these areas constitute the principle points for pressure communication.

On Figure 24 two oil-water interfaces are shown for some of the fields. One is the hypothetical interface based on the fluid-phase pressure data relation and represents the level at which the pore space is filled with 100 percent formation water. The other (in brackets) is the operating field oil-water interface, which has been established from production, drill-stem tests, and log interpretations. Field oil-water interfaces generally are established in the transition zone at a point where water-free production is obtained. This probably accounts for the differences between the calculated and reported oil-water contacts.

LOGGING

The most common logging devices used in Swan Hills fields are the induction electric log, microcaliper log, and gamma-sonic log. The microlog is an excellent tool for establishing effective net pay, whereas the sonic log has been used widely for calculating porosity in uncored wells and sections of uncored wells.

SECONDARY RECOVERY

Without exception, early production history of the field and the pools producing oil from the Swan Hills Formation presaged rapid pressure decline. None of the fields has been characterized by a strongly active water drive, and all indicated a low recovery of approximately 16 percent of the oil in place based on primary depletion. The primary-depletion recovery mechanism is rock and fluid expansion down to

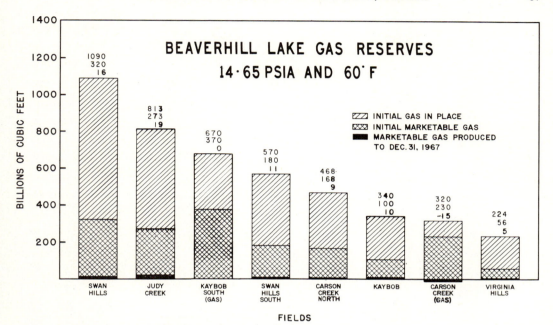

Fig. 25.—Bar diagram of Beaverhill Lake gas reserves shows initial gas in place, marketable gas, and where significant gas is produced for each of principal Swan Hills fields. Gas produced from Carson Creek North field was reinjected into Carson Creek gas field and accounts for negative value.

the bubble point, followed by a relatively inefficient solution gas drive.

Concern of the operating companies and the Alberta Oil and Gas Conservation Board resulted in exhaustive reservoir studies to find the most efficient means of maintaining field pressures above the bubble point, and thus obtain the maximum ultimate recovery of oil. The findings prompted unitization of most of the fields and, with one exception, water injection has been selected as the method of secondary recovery. For the Ante Creek field, a miscible flood is planned because of the higher initial reservoir pressure and solution gas-oil ratio resulting from greater depth of burial.

To the writers' knowledge, for every principal Swan Hills oil accumulation a secondary-recovery scheme is planned or is in effect. Secondary-recovery techniques are expected to result in recoveries of 35–60 percent of the original oil in place.

Water is obtained from surface sources, with approval of the Water Resources Branch of the Department of Agriculture, and is treated chemically for purity and bacteria control before injection. In most cases a line-drive method is adopted using downdip wells and injecting water below the field oil-water interface.

Where no field water table is present and it is necessary to inject into previously oil-producing wells, injection characteristics generally are not as good.

RESERVES

Gas.—Dissolved gas provides most gas reserves from the Swan Hills Formation. This, plus the reserves at Carson Creek and Kaybob South gas fields, accounted for a total of 4.512 trillion ft[3] initial raw gas in place to December 31, 1967. Conservation practices that require gathering most of the casinghead gas for marketing or reinjection into the formation result in an estimated initial marketable gas reserve of 1.097 trillion ft[3] from this source. Addition of the estimated marketable reserves from the two gas fields places the total for the area at 1.697 trillion ft[3].

Figure 25 shows the initial gas in place, marketable gas, and gas produced for each of the Swan Hills fields. Gas produced from the Carson Creek North field has been reinjected into the Carson Creek gas field and accounts for the negative value shown for marketable gas produced for that field.

Oil.—Total recoverable oil for all the fields producing from the Swan Hills Formation has

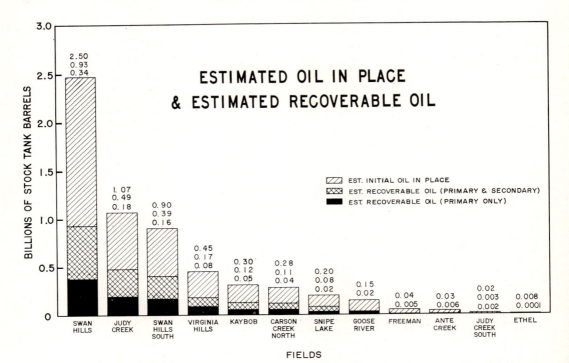

FIG. 26.—Estimated oil in place attributable to each of Swan Hills fields. Estimated recoverable oil by primary recovery and combination of primary- and secondary-recovery methods also is shown.

been estimated at 2.331 billion bbl to December 31, 1967. If only light- and medium-gravity crudes are considered, this figure amounts to 24.78 percent of the total known recoverable oil reserves for the Province of Alberta. The estimated oil in place attributable to each field, together with reserves expected to be produced by primary recovery and a combination of primary- and secondary-recovery methods, is shown in Figure 26.

The history of the petroleum industry in the Province of Alberta has been cyclical, characterized by a series of intermittent major discoveries. The effect of additional reserves from a major strike, such as that of the Swan Hills Formation, is shown in Figure 27.

ECONOMIC IMPACT

An estimate of the total funds spent within the map area after the initial discovery would require compiling figures on seismic work, road building, drilling and completion of wells, townsites, land acquisition, and pipelines. Though such a study is beyond the scope of this paper, it is interesting to examine the factor of land acquisition.

In Alberta, exclusive of early settlement areas, the Crown owns the petroleum and natural gas rights. These rights are disposed of by a system of closed bidding at periodic government sales. There were no freehold lands within the Swan Hills area. In the 2 years before the discovery, approximately $140,291 had been spent for the acquisition of petroleum and natural gas rights within the map area. In the years 1957–1967 this figure rose to $184,034,918.

SELECTED REFERENCES

Andrichuk, J. M., 1958, Stratigraphy and facies analysis of Upper Devonian reefs in Leduc, Stettler, and Redwater areas, Alberta: Am. Assoc. Petroleum Geologists Bull., v. 42, no. 1, p. 1–93.

Bassett, H. G., 1961, Devonian stratigraphy, central Mackenzie River region, Northwest Territories, Canada, in Geology of the Arctic, v. 1: Toronto, Toronto Univ. Press, p. 481–498.

——— and J. G. Stout, 1967, The Devonian stratigraphy of western Canada, in International Symposium on the Devonian System, v. 1: Alberta Soc. Petroleum Geologists, p. 717–752.

Beales, F. W., 1957, Bahamites and their significance in oil exploration: Alberta Soc. Petroleum Geologists Jour., v. 5, no. 10, p. 227–231.

——— 1958, Ancient sediments of Bahaman type:

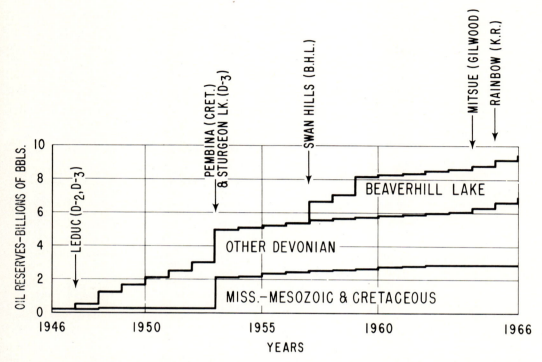

FIG. 27.—Alberta's cumulative ultimate proved oil reserves distributed according to year of discovery to Dec. 13, 1966. Share of Alberta's cumulative ultimate proved oil reserves attributable to Swan Hills fields is shown. By end of 1966 Swan Hills fields accounted for 25.87 percent of Alberta total.

Am. Assoc. Petroleum Geologists Bull., v. 42, no. 8, p. 1845–1880.
—— 1960, Limestone peels: Alberta Soc. Petroleum Geologists Jour., v. 8, no. 4, p. 132–135.
Beard, D. E., 1959, Selective solution in the Devonian Swan Hills Member: Alberta Soc. Petroleum Geologists Jour., v. 7, no. 7, p. 163–164.
Belyea, H. R., 1952, Notes on the Devonian System of the north central plains of Alberta: Canada Geol. Survey Paper 52–27, 66 p.
—— 1955, Correlations in the Devonian of southern Alberta: Alberta Soc. Petroleum Geologists Jour., v. 3, no. 9, p. 151–156.
—— 1964, Upper Devonian, pt. II, in Geological history of Western Canada: Alberta Soc. Petroleum Geologists, p. 66–81.
—— and A. W. Norris, 1962, Middle Devonian and older Paleozoic formations of southern District of Mackenzie and adjacent areas: Canada Geol. Survey Paper 62–15, 82 p.
Bonham-Carter, C. F., 1963, A study of microscopic components of the Swan Hills Devonian reef: Unpub. M.S. thesis, Toronto Univ.
Brown, P. R., 1963, Some algae from the Swan Hills reef: Bull. Canadian Petroleum Geology, v. 11, no. 2, p. 178–182.
Burwash, R. A., H. Baadsgaard, Z. E. Peterman, and G. H. Hunt, 1964, Precambrian, chap. 2, in Geological history of Western Canada: Alberta Soc. Petroleum Geologists, p. 14–19.
Cameron, A. E., 1918, Explorations in the vicinity of

Great Slave Lake: Canada Geol. Survey Summ. Rept., pt. C, 1917, p. 21–28.
—— 1922, Hay and Buffalo Rivers, Great Slave Lakes, and adjacent country: Canada Geol. Survey Summ. Rept., pt. B, 1921, p. 1–44.
Campbell, N., 1950, The Middle Devonian in the Pine Point area, N.W.T.: Geol. Assoc. Canada Proc., v. 33, p. 87–96.
Carozzi, A. V., 1961, Reef petrography in the Beaverhill Lake Formation, Upper Devonian, Swan Hills area, Alberta, Canada: Jour. Sed. Petrology, v. 31, no. 4, p. 497–513.
Century, J. R. (ed.), 1966, Oil fields of Alberta, supplement: Alberta Soc. Petroleum Geologists, 136 p.
Clark, D. L., and R. L. Ethington, 1965, Conodont biostratigraphy of part of the Devonian of the Alberta Rocky Mountains: Bull. Canadian Petroleum Geology, v. 13, no. 3, p. 382–389.
Crickmay, C. H., 1954, Paleontological correlation of Elk Point and equivalents, in Western Canada sedimentary basin: Am. Assoc. Petroleum Geologists, p. 143–158.
—— 1957, Elucidation of some Western Canada Devonian formations: Imperial Oil Ltd., unpub. rept.
De Mille, G., 1958, Pre-Mississippian history of the Peace River arch: Alberta Soc. Petroleum Geologists Jour., v. 6, no. 3, p. 61–68.
Edie, E. W., 1961, Devonian limestone reef reservoir, Swan Hills oil field, Alberta: Canadian Inst. Mining and Metallurgy Trans., v. 64, p. 278–285.
Fischbuch, N. R., 1960, Stromatoporoids of the Kay-

bob reef, Alberta: Alberta Soc. Petroleum Geologists Jour., v. 8, p. 113–131.

———— 1962, Stromatoporoid zones of the Kaybob reef, Alberta: Alberta Soc. Petroleum Geologists Jour., v. 10, no. 1, p. 62–72.

Folk, R. L., 1959, Practical petrographic classification of limestones: Am. Assoc. Petroleum Geologists Bull., v. 43, no. 1, p. 1–38.

Fong, G., 1959, Type section Swan Hills Member of the Beaverhill Lake Formation: Alberta Soc. Petroleum Geologists Jour., v. 7, no. 5, p. 95–108.

———— 1960, Geology of Devonian Beaverhill Lake Formation, Swan Hills area, Alberta, Canada: Am. Assoc. Petroleum Geologists Bull., v. 44, no. 2, p. 195–209.

Galloway, J. J., 1960, Devonian stromatoporoids from the lower Mackenzie Valley of Canada: Jour. Paleontology, v. 34, no. 4, p. 620–636.

Gray, F. F., and J. R. Kassube, 1963, Geology and stratigraphy of Clarke Lake gas field, northeastern British Columbia: Am. Assoc. Petroleum Geologists Bull., v. 47, no. 3, p. 467–483.

Grayston, L. D., D. F. Sherwin, and J. F. Allan, 1964, Middle Devonian, chap. 5, in Geological history of Western Canada: Alberta Soc. Petroleum Geologists, p. 49–59.

Griffin, D. L., 1965, The Devonian Slave Point, Beaverhill Lake, and Muskwa Formations of northeastern British Columbia and adjacent areas: British Columbia Dept. Mines and Petroleum Resoures Bull., no. 50, 90 p.

Gussow, W. C., 1954, Differential entrapment of oil and gas, a fundamental principle: Am. Assoc. Petroleum Geologists Bull., v. 38, no. 5, p. 816–853.

Guthrie, D. C., 1956, Gilwood Sandstone in the Giroux Lake area, Alberta: Alberta Soc. Petroleum Geologists Jour., v. 4, no. 10, p. 227–231.

Hriskevich, M. E., 1966, Stratigraphy of Middle Devonian and older rocks of Banff Aquitaine Rainbow West 7–32 discovery well, Alberta: Bull. Canadian Petroleum Geology, v. 14, no. 2, p. 241–265.

Illing, L. V., 1959, Deposition and diagenesis of some upper Palaeozoic carbonate sediments in Western Canada: 5th World Petroleum Cong., Sec. 1, Paper 2, p. 23–52.

Imperial Oil Limited, Geological Staff, 1950, Devonian nomenclature in the Edmonton area, Alberta, Canada: Am. Assoc. Petroleum Geologists Bull., v. 34, no. 9, p. 1807–1825.

Jenik, A. J., 1965, Facies and geometry of the Swan Hills Member, Alberta: Unpub. M.Sc. thesis, Alberta Univ.

Koch, N. G., 1959, Correlation of the Devonian Swan Hills Member, Alberta: Unpub. M.Sc. thesis, Alberta Univ.

Kramers, J. W., and J. E. Lerbekmo, 1967, Petrology and mineralogy of Watt Mountain Formation, Mitsue-Nipisi area, Alberta: Bull. Canadian Petroleum Geology, v. 15, no. 3, p. 346–378.

Law, J., 1955, Rock units of northwestern Alberta: Alberta Soc. Petroleum Geologists Jour., v. 3, no. 6, p. 81–83.

Leavitt, E. M., 1966, The petrology, paleontology and geochemistry of the Carson Creek North reef complex, Alberta: Unpub. Ph.D. thesis, Alberta Univ.

LeBlanc, R. J., and J. G. Breeding (eds.), 1957, Regional aspects of carbonate deposition: Soc. Econ. Paleontologists and Mineralogists Spec. Pub. 5, 178 p.

Loranger, D. M., 1965, Devonian paleoecology of northeastern Alberta: Jour. Sed. Petrology, v. 35,

no. 4, p. 818–838.

McGehee, J. R., 1949, Pre-Waterways Paleozoic stratigraphy of Alberta plains: Am. Assoc. Petroleum Geologists Bull., v. 33, no. 4, p. 603–613.

McGill, P., 1966, Ostracods of probable late Givetian age from Slave Point Formation, Alberta: Bull. Canadian Petroleum Geology, v. 14, no. 1, p. 104–133.

McGrossan, R. G., and R. P. Glaister (eds.), 1964, Geological history of Western Canada: Alberta Soc. Petroleum Geologists, 232 p.

McLaren, D. J., and E. W. Mountjoy 1962, Alexo equivalents in the Jasper region, Alberta, Canada: Canada Geol. Survey Paper 62–63, 36 p.

Mound, M. C., 1966, Late Devonian conodonts from Alberta subsurface (abs.): Am. Assoc. Petroleum Geologists Bull., v. 50, no. 3, p. 628.

Moyer, G. L., 1964, Upper Devonian, pt. I, chap. 6, in Geological history of Western Canada: Alberta Soc. Petroleum Geologists, p. 60–66.

Murray, J. W., 1964, Some stratigraphic and paleoenvironmental aspects of the Swan Hills and Waterways Formation, Judy Creek, Alberta, Canada: Unpub. Ph.D. thesis, Princeton Univ.

———— 1966, An oil producing reef-fringed carbonate bank in the Upper Devonian Swan Hills Member, Judy Creek, Alberta: Bull. Canadian Petroleum Geology, v. 14, no. 1, p. 1–103.

Norris, A. W., 1963, Devonian stratigraphy of northeastern Alberta and northwestern Saskatchewan: Canada Geol. Survey Mem. 313, 168 p.

Oil and Gas Conservation Board, 1967, Pressure-depth and temperature-depth relationships, Alberta crude oil pools: OGCB Rep. 67–2.

———— 1968, Reserves of crude oil, gas, natural gas liquids, and sulphur, Province of Alberta: OGCB Rept. 68–18; 175 p.

Porter, J. W., and J. G. C. M. Fuller, 1964, Ordovician-Silurian, pt. 1, chap. 4, in Geological History of Western Canada: Alberta Soc. Petroleum Geologists, p. 34–42.

Russel, L. S., 1967, Palaeontology of the Swan Hills area, north-central Alberta: Toronto Univ. Press, Life Sci. Contr. no. 71, 31 p.

Stout, J. L., 1964, Pore geometry as related to carbonate stratigraphic traps: Am. Assoc. Petroleum Geologists Bull., v. 48, no. 3, p. 329–337.

Thomas, G. E., and H. S. Rhodes, 1961, Devonian limestone bank-atoll reservoirs of the Swan Hills area, Alberta: Alberta Soc. Petroleum Geologists Jour., v. 9, no. 2, p. 29–38.

Uyeno, T. T., 1967, Conodont zonation, Waterways Formation (Upper Devonian), northeastern and central Alberta: Canada Geol. Survey Paper 67–30, 20 p.

Van Hees, H., 1958, The Meadow Lake escarpment—its regional significance to lower Paleozoic stratigraphy, in 1st Internat. Williston Basin Symposium: North Dakota Geol. Soc. and Saskatchewan Geol. Soc., p. 131–139.

———— and F. K. North, 1964, Cambrian, chap. 3, in Geological History of Western Canada: Alberta Soc. Petroleum Geologists, p. 20–33.

Walker, C. T., 1957, Correlations of Middle Devonian rocks in western Saskatchewan: Saskatchewan Dept. Mineral Resources Rept. 25, 59 p.

Warren, P. S., 1933, The age of the Devonian limestone at McMurray, Alberta: Canadian Field-Naturalist, v. 47, no. 8, p. 148–149.

———— 1957, The Slave Point Formation: Edmonton Geol. Soc. Quart., v. 1, no. 1, p. 1–2.

White, R. J. (ed.), 1960, Oil fields of Alberta: Alberta Soc. Petroleum Geologists, 272 p.

Giant Oil and Gas Fields in United States[1]

MICHEL T. HALBOUTY[2]

Houston, Texas 77027

Abstract The great importance of the major or giant oil and gas fields of the United States to the economy and strength of both the petroleum industry and the nation cannot be overemphasized. Such oil fields provided 50.7 percent of the United States production in 1966. It is estimated that they will yield almost 58 percent of the country's ultimate oil production and that they contain approximately 57 percent of the nation's crude reserves.

Accompanying this paper are 26 figures and 2 tables which give pertinent data on the 259 oil and 47 gas giants found in the United States. An important fact shown by these charts and tables is that the rate of discovery of new giant fields is declining steadily in nearly all regions of the country.

Most of the major fields were discovered by deliberate exploration for structural traps. Some structural tests found, usually by accident, other subtle stratigraphic or paleogeomorphic accumulations. Although only few such accumulations have been found—most of them accidentally—the rate of discovery of stratigraphic and paleogeomorphic traps has been relatively persistent through the years. This persistent though "accidental" discovery rate indicates that many more such accumulations must be present throughout the country.

If exploration is to succeed in finding new reserves for even the near future, geologists and management will have to place a greater emphasis on the search for the subtle traps with large accumulations. The requirements to find these large reservoirs are (1) the focusing of exploratory thinking on the geologic conditions under which such traps and accumulations form, and (2) the purposeful and deliberate drilling of more stratigraphic and paleogeomorphic wildcat tests.

INTRODUCTION

The nation's giants, those major fields which have produced or will produce at least 100 million bbl of oil or 1 trillion cu ft of gas, are of great importance to both the economy and strength of the petroleum industry and the na-

[1] Manuscript received, December 1, 1967; accepted, February 22, 1968. Reprinted from the AAPG *Bulletin,* v. 52, no. 7, p. 1115–1151, 1968.

[2] Consulting geologist and petroleum engineer; independent producer and operator.

Appreciation is extended to James J. Halbouty, geologist, who is associated with the writer, for his most valuable assistance in the preparation of this paper. The very helpful and constructive criticisms of John B. Patton and Leroy E. Becker, as well as other reviewers of the first draft of this paper are acknowledged with deep gratitude.

tion. Such oil fields provided 50.7 percent of the United States production in 1966. It is estimated that they will yield about 58 percent of the country's ultimate oil production and that they contain approximately 57 percent of the nation's remaining crude reserves.

In terms of barrels of oil, the giants, compared with all other oil fields, provided approximately 1.54 billion of 3 billion bbl produced in 1966. They will yield an ultimate production of 69.3 billion bbl out of 119.7 billion, and it is estimated that they now contain 21.45 billion bbl of the nation's total present liquid-hydrocarbon reserves of 37.54 billion bbl. These figures emphasize the great role that the giants have played during the past and will occupy during the future in the industrial development and economic strength of this country.

If the amount of United States reserves is to keep pace with anticipated consumption of petroleum products during the coming years, the nation's explorationists will be required to find many new giant fields. If not, the number of new smaller discoveries found annually will have to *increase* continuously at a much higher rate than before—much higher than during the last 20 years.

The figures accompanying this paper show the trend of giant field discoveries in the United States through the years. They show (1) when and how these fields were found, (2) the geological age of producing beds in each field, and (3) the type of trap from which each field produced or is producing. Although the information on which these charts are based is essentially correct, some of the statistics, such as type of trap, method of discovery, and the ranking of some fields as giants, may be subject to other interpretations. However, because of the scope of this survey, the many sources of information involved, and the conflicting data available for certain fields, a choice of data had to be made, and the writer tried to choose an "average" in each case where there was conflict.

HISTORY OF PETROLEUM EXPLORATION

Seepage period.—It is appropriate to review briefly the history of petroleum exploration in the United States. In the 1859–1900 period, oil seeps in California, Kansas, New York, Wyoming, Pennsylvania, Colorado, and other states attracted much attention after Drake made his famous discovery in 1859 near Titusville, Pennsylvania, on the basis of such seeps. This period of exploration possibly should be termed the "seepage years."

Altogether 10 giant fields were discovered during the seepage period. These were found as a result of surface indications (oil seeps) and random drilling. The Bradford field, Pennsylvania, discovered in 1871, is considered to be the first giant-field find. Although the McKittrick field, California, was being *mined* for oil in the early 1860s, the discovery of this giant by drilling did not occur until 1887.

Surface geology.—The discovery of the tremendous cap-rock oil reservoir at Spindletop, near Beaumont, Texas, in 1901, ushered in a new era of exploration which lasted until about 1932. Exploration during the early part of this period (1901–1910), especially in the Gulf Coast region, focused primarily on a search for surface mounds similar to that at Spindletop and for other surface indications such as paraffin dirt, oil seeps, gas seepages, sulfur water in wells, *etc*. Such surface indications and random drilling led to the discovery of several piercement-type salt domes in the Gulf Coast region.

As a result of studies which had been made to determine the relation between structure and those oil accumulations which had been discovered through surface "leads," the United States Geological Survey, at the turn of the century, demonstrated that many oil fields, especially in California, were located on surface structural anomalies. The first USGS report on these findings was issued in 1903; others were issued in 1908. These reports encouraged oil companies to employ geologists to do surface mapping. As a result, many fields were found by surface mapping during the 1910–1932 period.

Advent of geophysics and subsurface geology.—By the early 1920s nearly all areas in the Gulf Coast where surface indications occurred had been found and tested. It became apparent at that time, in view of the lack of obvious prospective areas, that new methods of exploration were needed to find subsurface structures in that region.

The new exploration methods which were developed were geophysical—the torsion balance and refraction seismograph. Both were introduced during the early 1920s. These new tools brought instantaneous and remarkably successful results to exploratory efforts.

The first salt-dome discovery by torsion balance occurred in 1923 at Nash dome, Brazoria County, Texas. In 1924, the refraction seismograph and the torsion balance were applied successfully in the finding of the Orchard dome, Fort Bend County, Texas.

From 1923 through 1929, not only geophysics, but also micropaleontology, became important bases for accurate, scientific, subsurface geology. The discovery rate increased during this period. However, this increase was made possible not only by these two new techniques—geophysics and subsurface geology—but also by a continued use of surface mapping and random drilling.

Reflection seismograph and electric log.—By the early 1930s geophysics had introduced the reflection seismograph into the exploratory effort. The phenomenal success of this instrument in locating subsurface structures gained worldwide acceptance for geophysics and relegated all other exploration methods, except subsurface geology, to minor roles in the search for petroleum.

The sagging importance of subsurface geology, which had been dependent to a great extent on paleontology and lithologic correlations, was bolstered greatly by introduction of the electric log during the early 1930s. The electric log, combined with paleontologic and lithologic correlations, greatly enhanced the accuracy and value of subsurface geology during the 1930s and has made it possible for subsurface geology to maintain its position as a required and necessary exploration method to the present.

The surge in discoveries, which was begun by the discovery of Spindletop in 1901, continued until 1952, after peaking in 1929 and again in 1937–1938. Since 1952, exploratory results, unaided by the introduction of new methods of exploration, have declined drastically.

This condensed history of petroleum exploration in the United States is reflected by Figure 1. The figure graphically indicates the "ups and

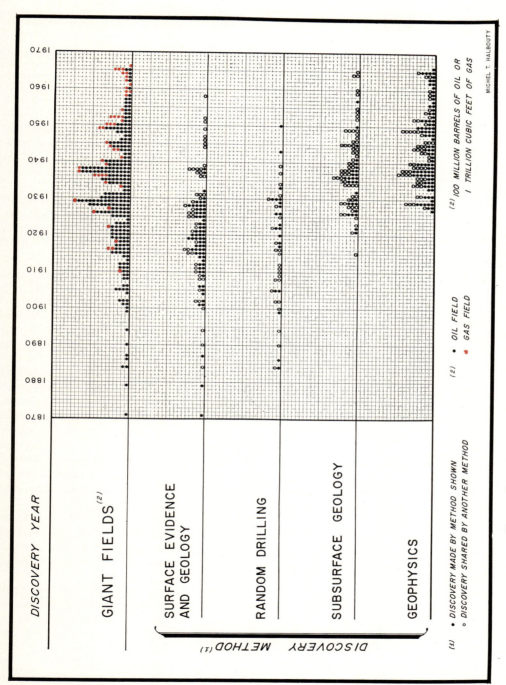

Fig. 1.—Total giant fields found in United States by year and discovery methods.

downs" of petroleum exploration in the United States and shows the effects of the various petroleum-finding methods on exploratory results to the present.

GIANT FIELDS

DISCOVERY RATES

At least 306 giant fields—259 oil and 47 gas— have been found in this country (Fig. 1). The number of giant fields credited to each of the exploratory methods are (1) surface evidence and surface geology, 94.5 fields; (2) random drilling, 33.5 fields; (3) subsurface geology, 59 fields; and (4) geophysics, 119 fields.

Most discoveries shown on Figure 1 are the results of exploration which was oriented *purposely* toward finding and exploiting the structural trap. In some places this search was successful in finding, usually *unintentionally,* the subtle stratigraphic or paleogeomorphic types of traps.

Although only a relatively few of these subtle traps have been found to date (and most of these accidentally), they appear *fairly persistently* in the discovery records throughout the years. This persistent, though "accidental," discovery history suggests that *many more such traps must occur* throughout the country. If this conclusion is valid, future exploration success lies in a search *directed purposely* toward the subtle and obscure traps. These traps must receive attention because obvious subsurface structural features are increasingly more difficult to find today simply because there are increasingly fewer to find.

AGES OF RESERVOIRS

Fields producing from Cenozoic and Paleozoic rocks far outnumber those producing from Mesozoic beds (Fig. 2). The reasons for this are not readily apparent but may be related to the seemingly greater deformation of, or structure in, Paleozoic and Cenozoic rocks in petroliferous regions. Figure 2 indicates that 134 giant fields produce from Cenozoic beds, 130 from Paleozoic rocks, and 42 from Mesozoic strata.

TYPES OF TRAPS

Concerning the difference between stratigraphic and paleogeomorphic traps Martin (1966, p. 2279) stated:

An important difference between stratigraphic and paleogeomorphologic traps is the pronounced three-dimensional aspect of the latter. The writer prefers to limit the use of the term *stratigraphic trap* to those traps which are caused by a lateral change in reservoir properties within a given stratum. Erosion surfaces, reef buildups, and other geomorphic phenomena are bounded by air or water at the time of their formation. Subsequent deposition of younger strata adjacent to but different from those that constitute such a morphological surface does not create a stratigraphic trap but a paleogeomorphic trap.

The writer agrees with Martin; however, it should be pointed out that, in this paper, stratigraphic and paleogeomorphic trap fields are grouped under the term "stratigraphic" and no attempt has been made to distinguish between the two types in either the statistical presentations or the text.

Of the 306 giant fields which have been discovered, 238 are classified as structural traps (Fig. 3). Difference of opinion may occur concerning classification of traps where production is from structural closures and from stratigraphic and other types of subtle traps in the same field. An example would be a field in which a large anticline is productive from blanket sandstone and also from closed sandstone lenses or truncated sandstone beds which lie on flanks where only monoclinal dip is found. These conditions are present in many giant fields which have been classified as structural traps. Some such fields have been considered to be combination traps.

Where structural folding has caused the migration of oil into locally developed stratigraphic traps on large structures, a structural classification has been given for this type field. Where a stratigraphic trap with a *substantial* productive column lies across a nose or anticlinal feature of *limited* closure, a combination-trap classification is used.

The 238-field total for structural giants, therefore, should be considered an approximation based on the writer's interpretation of available data. The same is true of the 35 combination-trap giant fields.

The high discovery rate for structural giants during the 1926–1940 period reflects the success of geophysical exploration and to a lesser degree, subsurface geology. From 1940 to the present, the discovery curve for structural giants has declined.

Thirty-three stratigraphic-trap giants have been found to date (Fig. 3). Although the number of this type of giant discovery is relatively small, exploration continues to discover (usually accidentally) these subtle giants.

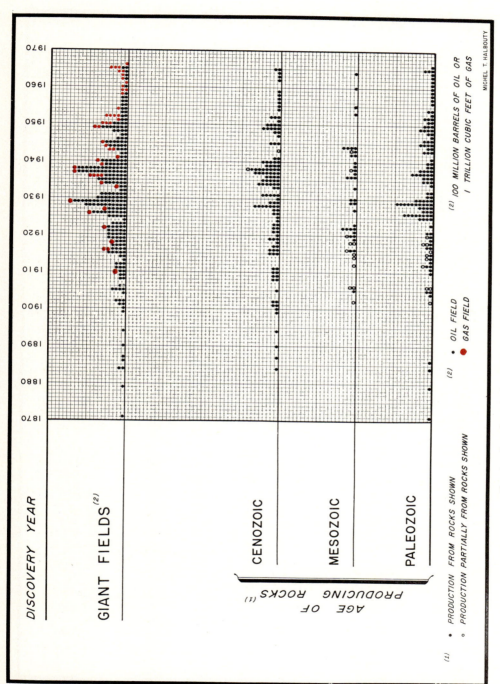

Fig. 2.—Age of producing rocks in giant fields in United States.

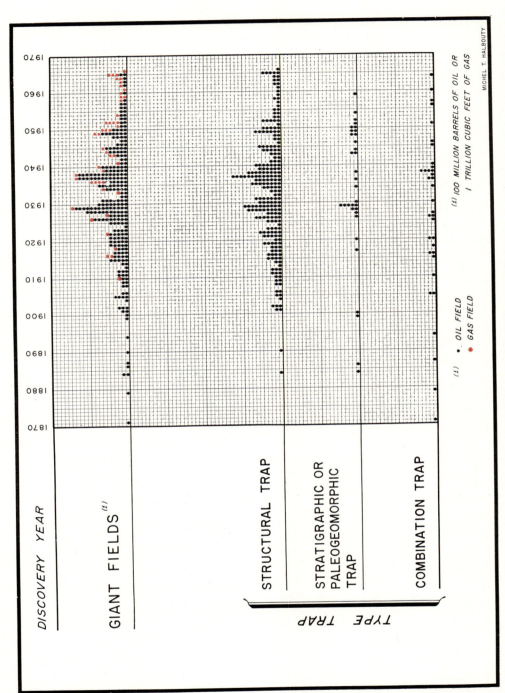

Fig. 3.—Type traps in giant fields in United States.

REGIONS OF OCCURRENCE

Figures 4–24 show giant-field discoveries in the United States by regions. Each of these figures shows the year of discovery, discovery methods, type of trap, and type of major production, whether oil or gas. These data also are listed in Table 1.

Alaska.—Six giants have been found in this new oil province (Fig. 4). The first Alaskan giant discovery, Swanson River, was made in 1957. Two discoveries which were made in 1965, Granite Point and McArthur River, already have been classified as giant oil fields. Of the 6 giants, 4 are oil fields and 2 are gas fields.

All six discoveries were based on geophysical exploration. Exploration activity in Alaska reached a new peak in 1966 and is expected to continue at a high rate.[3] Although all giants discovered to the present are structural traps, it is very probable that stratigraphic giants will be found as exploration increases. Alaska giants produce from Cenozoic rocks.

Appalachians.—This region includes New York, Pennsylvania, and West Virginia. Two giant oil fields, both found during the early "seepage period," occur in this province—Bradford in Pennsylvania and the Allegheny district in New York (Fig. 5). Both produce from traps considered to be of the combination type. Although it generally is believed that permeability and porosity variations in Devonian rocks control oil accumulation, anticlinal folding has had some influence on oil

migration in the two fields. No giant has been discovered in the Appalachian region since 1879. This province offers excellent possibilities for finding obscure and subtle traps.

Arkansas.—The four giants in this state are in southern Arkansas (Fig. 6); all produce from Mesozoic rocks (Cretaceous and Jurassic). Each is on a large anticlinal feature. Two, El Dorado and Magnolia, are considered to be partly combination traps inasmuch as important production is obtained from sandstone pinchouts on the flanks of the anticlines. No giant discovery has been made in Arkansas since 1938. Arkansas offers excellent possibilities for finding stratigraphic traps.

West Coast states.—This area includes California, Oregon, and Washington; only California, however, has provided giant field discoveries—42 of them (Fig. 7). Twenty-eight are considered to be structural traps, 4 are classified as stratigraphic traps, and 10 are regarded as combination traps. Production in all giant fields is from Cenozoic strata. Most giant discoveries in California resulted from exploration based on surface evidence, including surface geology. Possibilities for finding stratigraphic reservoirs in California are considered to be excellent. No giant field has been found in California since 1949.[4]

Illinois.—This state contains six giant fields (Fig. 8). Five are structural traps. The sixth, Old Illinois, actually includes several producing areas, two of which—Lawrence County Division and Main Consolidated—are giant producing areas.

---

[3] Alaska can claim the largest U.S. discovery of the decade—Prudhoe Bay, on the north slope—from two Triassic sandstones and from Paleozoic carbonates. A reserve of 4.8 billion bbl has been proved; a more accurate reserve is estimated to be 20 billion bbl or more.

[4] California contributed a new giant in 1968—the Dos Quadros field in the Santa Barbara Channel. Reserves of 100 million bbl have been established in lower Pliocene sandstones, with potentials in deeper zones.

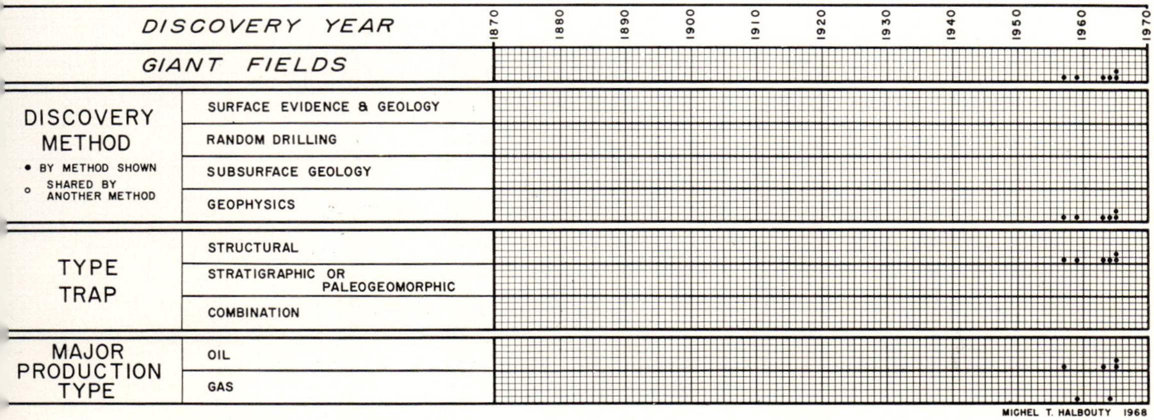

Fig. 4.—Giant fields found in Alaska; production from Cenozoic rocks.

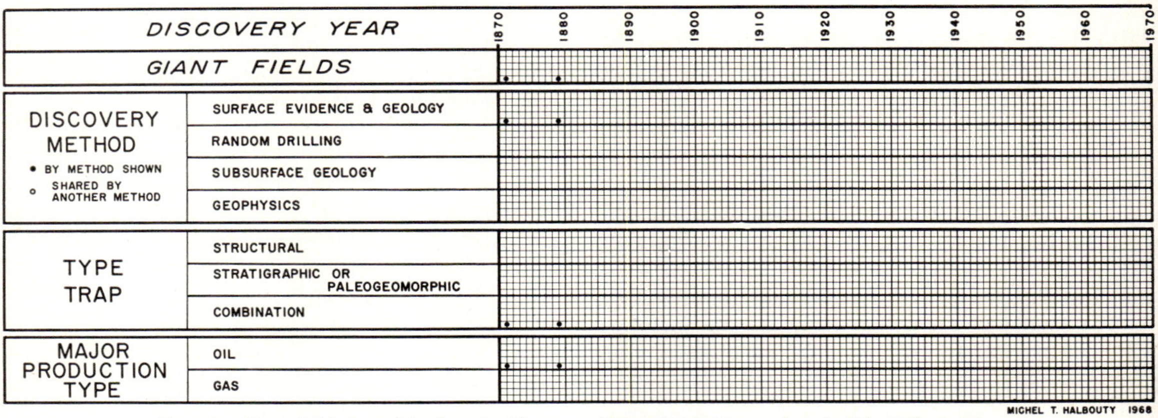

FIG. 5.—Giant fields found in Appalachian area (New York, Pennsylvania, West Virginia); production from Paleozoic rocks.

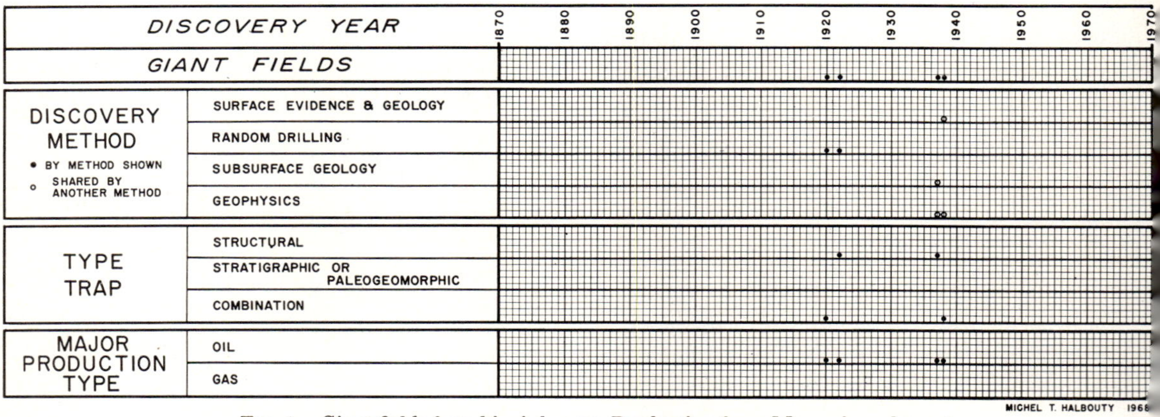

FIG. 6.—Giant fields found in Arkansas. Production from Mesozoic rocks.

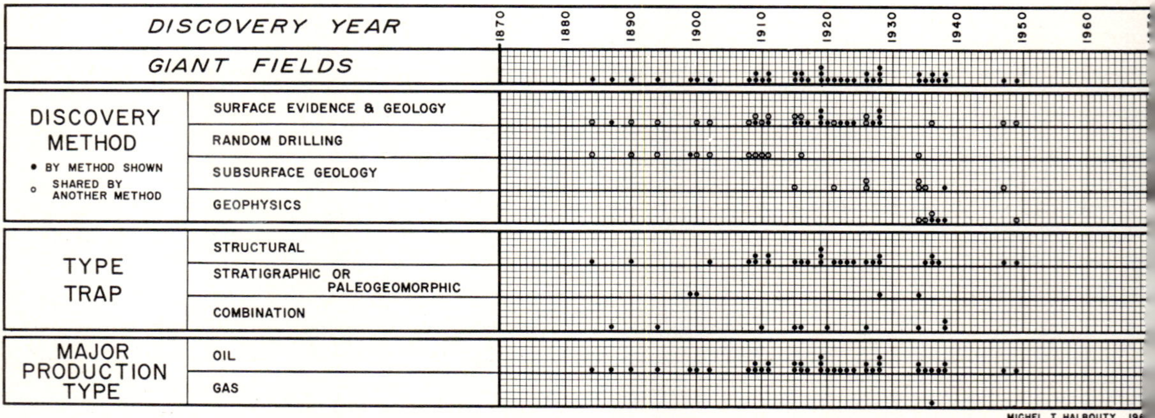

FIG. 7.—Giant fields found in West coast states (California, Oregon, Washington); production from Cenozoic rocks.

Production in the Lawrence area is structurally controlled; in Main Consolidated, production generally is from sandstone lenses in areas of monoclinal dip. Illinois giants produce from Paleozoic rocks. In Illinois there are possibilities for finding new reserves in stratigraphic traps (especially reef types). No giant discovery has been made in Illinois since 1940.

Indiana-Ohio.—Two giant fields have been found in Indiana and Ohio, one in each state (Fig. 9). Both were found during the "seepage period." In Indiana, the Trenton field gas reservoir underlies many counties. This field also has produced more than 105 million bbl of oil from areas near its northern edge. In Ohio, the Lima field is essentially a complex of several adjoining oil fields which have produced approximately 375 million bbl of oil. Both giant fields, which are on the Cincinnati arch, produce from stratigraphic traps which involve porosity and permeability variations in Paleozoic reservoir rocks. Although the two producing areas are referred to as separate fields, they are considered by many to be a single field (Lima-Indiana) which is separated arbitrarily by a state boundary. The region offers possibilities for finding subtle traps. No giant has been found in this two-state area since 1886.

Kansas.—Nine giant fields have been found in Kansas (Fig. 10). Surface geology has accounted for half of the discoveries. Five of the fields produce from structural traps. Two, Chase-Silica and Kraft-Prusa, are classified as combination traps. The Greenwood and Hugoton gas fields are considered to be stratigraphic traps. The huge Hugoton accumulation apparently is caused by a westward sealing of porosity that resulted from a facies change in the productive Paleozoic dolomites. The Hugoton field, which includes parts of Kansas, Oklahoma, and Texas, in areal extent is the largest single producing area in the United States. It includes approximately 2,500,000 acres in southwest Kansas, 925,000 acres in the Oklahoma Panhandle, and approximately 575,000 acres in the northern part of the Texas Panhandle (exclusive of the Texas Panhandle field). Kansas giants produce from Paleozoic rocks. This state is located favorably for the occurrence of stratigraphic traps. No giant field has been found in Kansas since 1951.

Kentucky.—The two Kentucky giants, Big Sinking oil field and Big Sandy gas field, are in eastern Kentucky (Fig. 11). Both apparently were found by random drilling. Big Sinking is a strati-graphic trap in an area of essentially monoclinal dip; however, some faulting and minor closure are present. The Big Sandy field, which includes approximately 650,000 acres, is classified as a combination-type trap. Included within its producing area are domes, anticlines, noses, terraces, monoclines, and synclines. The two fields produce from Paleozoic rocks. The state appears to be attractive for stratigraphic trap exploration. No giant has been found in Kentucky since 1918.

Louisiana-north.—Ten giant fields have been found in North Louisiana (Fig. 12). Of these, 7 are structural traps, 2 are combination traps, and 1 is considered to be a stratigraphic trap. The most recent giant discovery, Black Lake field, was found in 1964; it produces from a combination trap formed by porosity variations caused by biohermal reefing in Pettet limestone (Sligo Formation, Lower Cretaceous) across a structural nose. Most giant fields in North Louisiana were discovered by surface geology and random drilling. Possibilities for finding stratigraphic traps in North Louisiana are very favorable. North Louisiana giants produce from Mesozoic rocks.

Louisiana-south.—Forty-three giant fields have been found in South Louisiana—30 onshore and 13 offshore (Fig. 13). Although all are classified as structural traps associated with salt domes and anticlines, production has been from lenticular sandstones on some of these structural features. Seven giants are principally gas producers; the others are oil fields. Figure 13 emphasizes the highly important role which geophysics has played onshore and offshore in oil exploration in South Louisiana. The most recent giant discovery, West Delta Block 73, offshore, was made in 1962. Production in South Louisiana is from Cenozoic sandstone.

Gulf states (southern).—Six giant fields have been found in the southern Gulf states; this area includes Alabama and Mississippi (Fig. 14). One giant field, Citronelle, is in Alabama; 5 are in Mississippi. All are on anticlines or salt domes. Two Mississippi giants, Cranfield and Gwinville, are gas fields. Most giant field production is from Mesozoic (Cretaceous and Jurassic) beds; the Wilcox (Eocene) contains sizable reserves in the Cranfield field. Possibilities for finding stratigraphic traps are favorable in this region. The single Alabama giant, Citronelle, was discovered in 1955; no giant has been found in Mississippi since 1944.

New Mexico.—Of the 10 giants in this state, 9

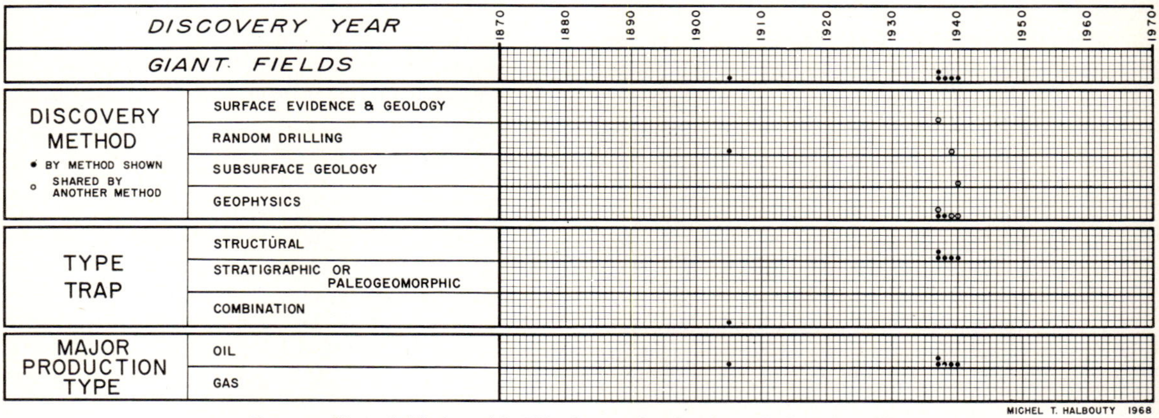

Fig. 8.—Giant fields found in Illinois; production from Paleozoic rocks.

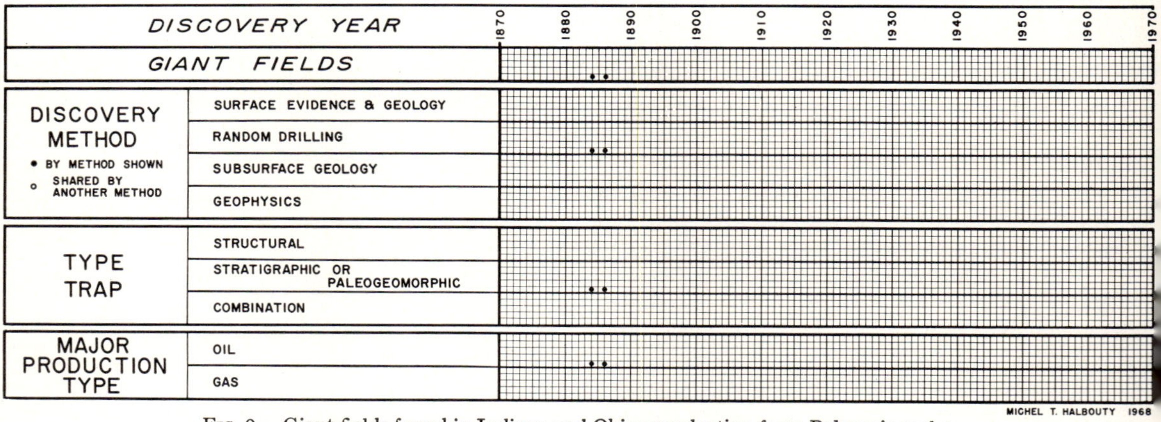

Fig. 9.—Giant fields found in Indiana and Ohio; production from Paleozoic rocks.

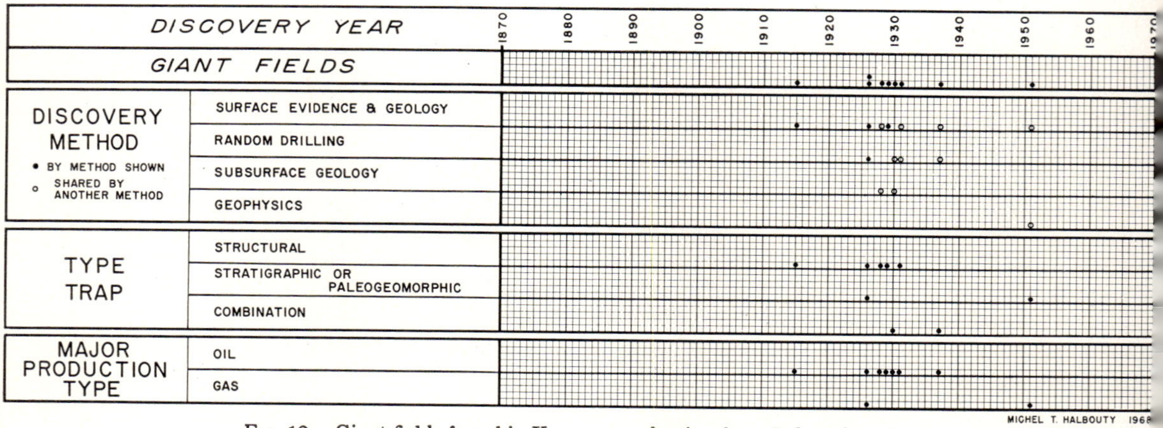

Fig. 10.—Giant fields found in Kansas; production from Paleozoic rocks.

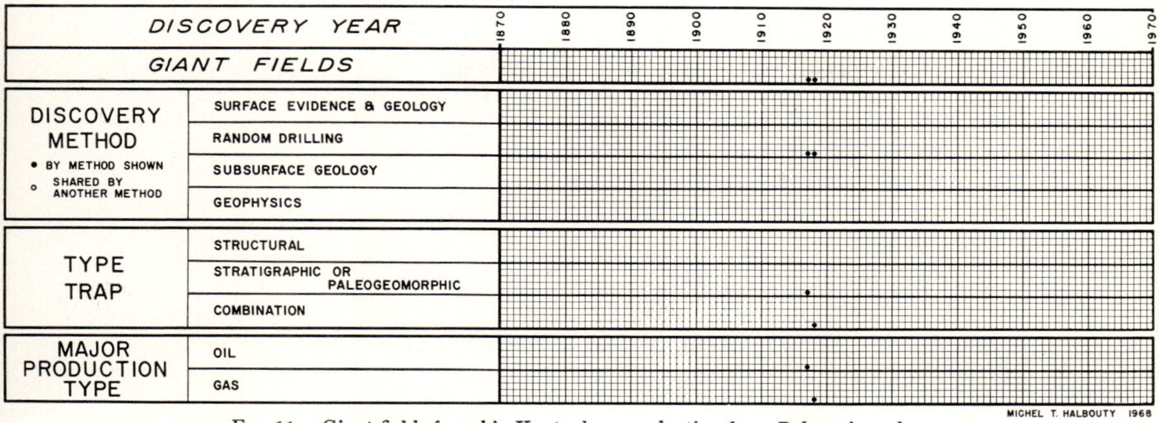

Fig. 11.—Giant fields found in Kentucky; production from Paleozoic rocks.

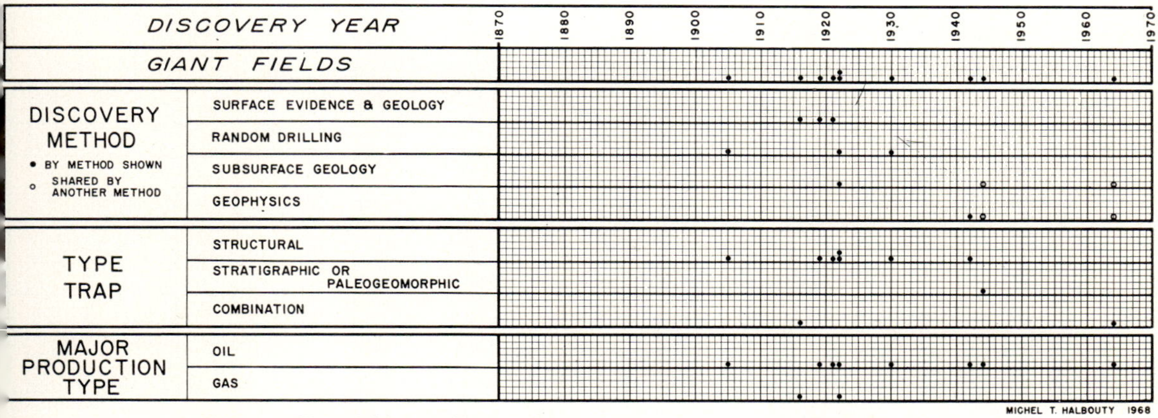

Fig. 12.—Giant fields found in North Louisiana; production from Mesozoic rocks.

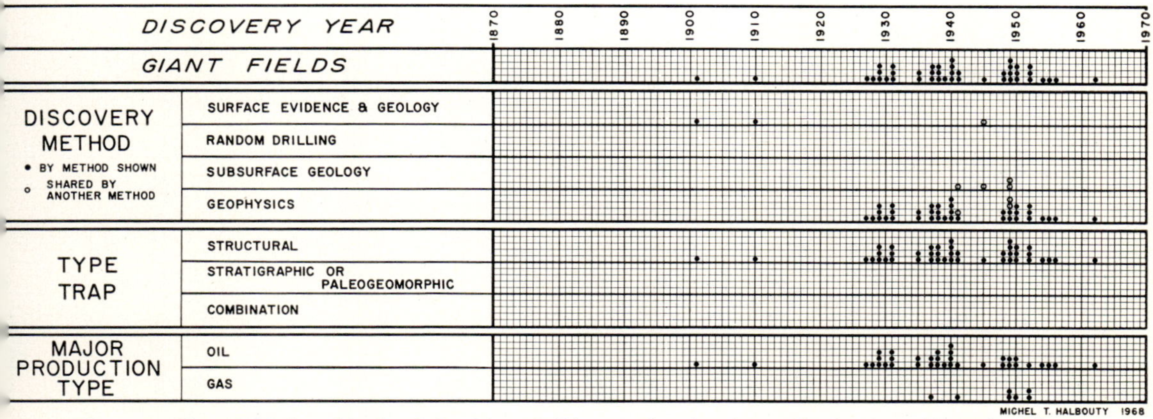

Fig. 13.—Giant fields found in South Louisiana (offshore and onshore); production from Cenozoic rocks.

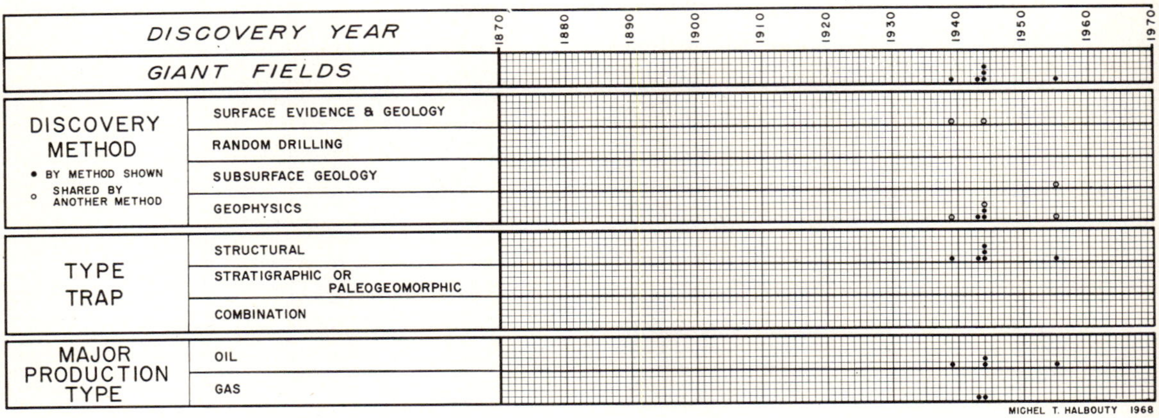

FIG. 14.—Giant fields found in southern Gulf states (Alabama, Mississippi); production from Mesozoic rocks; some from Cenozoic rocks.

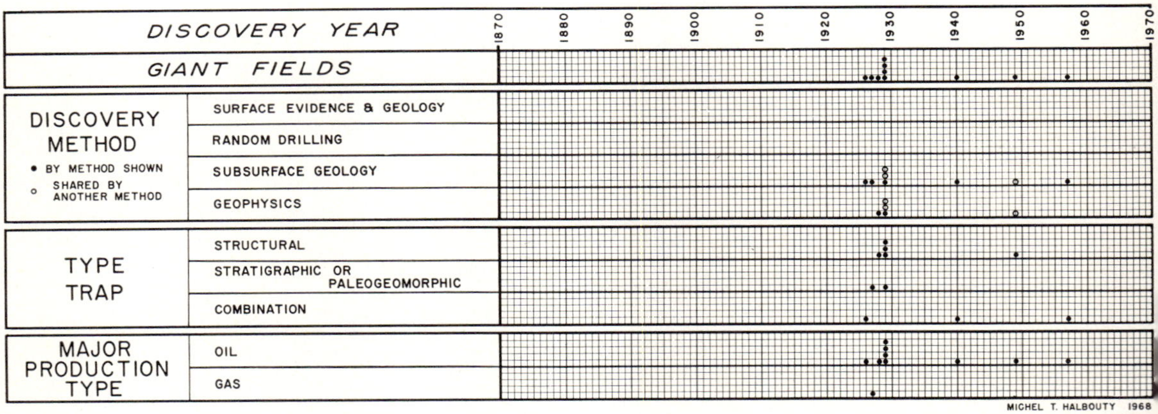

FIG. 15.—Giant fields found in New Mexico; production from Paleozoic rocks, eastern New Mexico, and from Mesozoic rocks, western New Mexico.

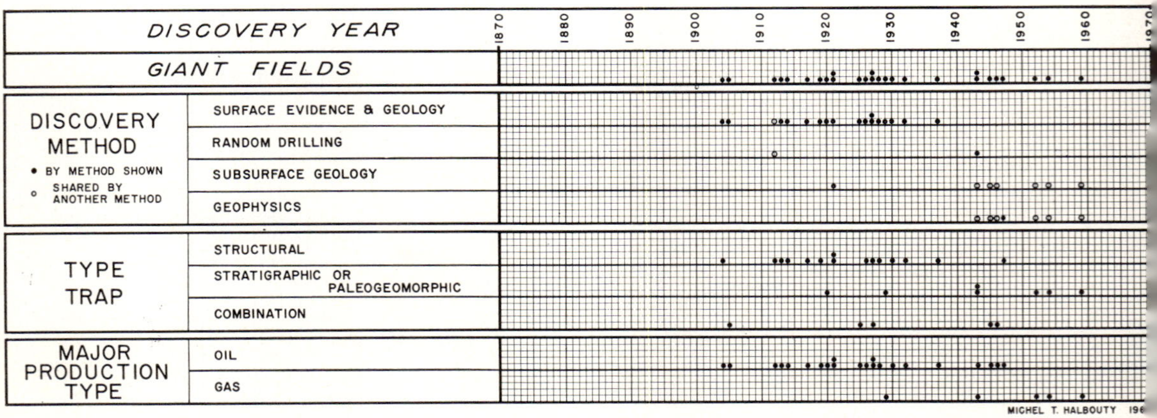

FIG. 16.—Giant fields found in Oklahoma; production from Paleozoic rocks.

are in eastern New Mexico, and 1, the Blanco Mesaverde gas field, is in the western part of the state (Fig. 15). Empire Abo, the most recent giant, was discovered in 1957. Two fields, Jalmat and Blanco Mesaverde, are stratigraphic traps. Five are considered to be structural traps, and 3 are classified as combination traps. Eastern New Mexico giants produce from Paleozoic rocks; in Blanco Mesaverde, western New Mexico, production is from Mesozoic (Cretaceous) strata. New Mexico offers good possibilities for finding stratigraphic traps.

Oklahoma.—Twenty-seven giant fields have been found in Oklahoma—22 oil fields and 5 gas pools (Fig. 16). Seven are considered to be stratigraphic traps, 5 are combination traps, and 15 are structural traps. Surface geology has led to the majority of discoveries. The most recent discovery, Putnam gas field, was in 1959. Oklahoma giants produce from Paleozoic beds which offer excellent possibilities for finding the obscure traps.

Rocky Mountains-central.—The central Rocky Mountain region includes Colorado, Nevada, Utah, and western Nebraska. Three giant fields have been found in this province—1 in Colorado and 2 in Utah (Fig. 17). The Colorado giant, Rangely field, is a structural trap. In Utah, the two giants, Aneth and Red Wash, are stratigraphic traps on structural noses. Although these fields are classified as combination traps, some geologists possibly would prefer to classify them as stratigraphic.

The original shallow Cretaceous discovery at Rangely in 1902 proved only minor reserves; major Paleozoic reserves were found in 1932. Forty-nine years elapsed between the original discovery of Rangely on the basis of surface geology and the 1951 discovery of Red Wash field through geophysical exploration. Production in the central Rocky Mountain region is from Paleozoic, Mesozoic, and Cenozoic rocks. This area offers attractive possibilities for finding stratigraphic traps. No giant discovery has been made in the province since Aneth in 1956.

Rocky Mountains-northern.—Thirteen giant fields are in the northern Rocky Mountain region; this area includes Idaho, Montana, North Dakota, South Dakota, and Wyoming (Fig. 18). One giant, the Cut Bank field, is in Montana; this is a stratigraphic trap. Two anticlinal giants

produce in North Dakota. Ten giants, one stratigraphic, the others anticlinal, have been found in Wyoming. Most fields in this region were discovered through surface geology. The most recent discovery, Tioga, in North Dakota, was in 1952; this was found by geophysical methods. No giant has been found in Wyoming since 1938 and none since 1929 in Montana.[5] Although most giant fields in the region produce primarily from Paleozoic rocks, some contain sizable reserves in Mesozoic and Cenozoic Formations.

Texas.—Two giant fields have been found in Texas Railroad Commission District 1, southwest Texas (Fig. 19; see Fig. 25 for location of Texas Railroad Commission districts). Both were discovered by surface geology. These fields produce from fault traps in areas of monoclinal dip. No giant discovery has been made in this area since 1929.

Texas Railroad Commission Districts 2 and 4 comprise the lower Texas Gulf Coast region. Nineteen giants have been discovered in this prolific area (Fig. 20). Of these, 16 are oil fields and 3 are gas producers. Except for one stratigraphic trap, the North Government Wells field, all accumulations are controlled essentially by anticlinal closure. Although substantial reserves in several of these anticlines are in sandstones which pinch out or grade laterally into shale on the flanks, these fields are considered to be structural traps because the structural folds appear to be the controlling factor for accumulation.

Stratigraphic traps in lenticular sandstone bodies in areas of monoclinal dip are numerous in the Jackson (late Eocene) trend in this region. Excellent possibilities for stratigraphic traps in lenticular lagoonal sandstone also are present in areas within the prolific Frio (Oligocene) trend. Most production in Districts 2 and 4 is from Frio sandstones. Although all exploratory methods have been used successfully in the lower Texas Gulf Coast, most giant discoveries which have been made since 1934 are based on geophysical exploration. No giant field has been found in this region since 1958.

[5] Two new stratigraphic giants were added to the Lower Cretaceous Muddy (Newcastle) Sandstone trend, Powder River basin, after this paper was written. Hilight, in Wyoming, found in 1969, is credited with 100 million bbl of reserves; and Bell Creek, in Montana (*see* next paper), opened in 1967, has 150 million.

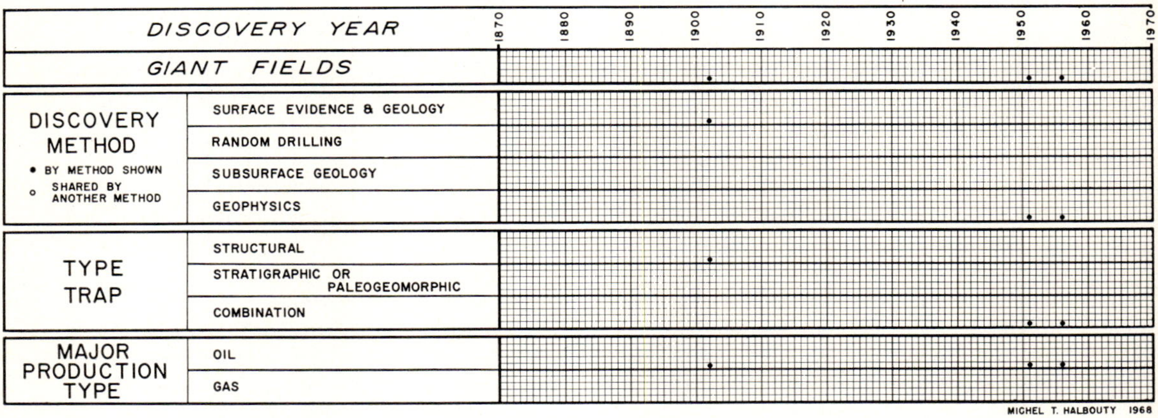

FIG. 17.—Giant fields found in central Rocky Mountain area (Colorado, west Nebraska, Nevada, and Utah); production from Paleozoic, Mesozoic, and Cenozoic rocks.

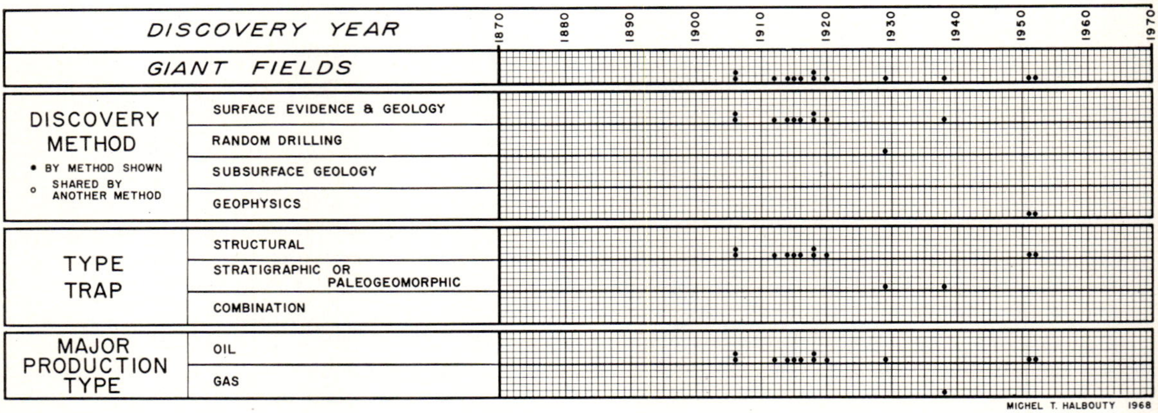

FIG. 18.—Giant fields found in northern Rocky Mountain area (Idaho, Montana, North and South Dakota, and Wyoming); production from Paleozoic, Mesozoic, and Cenozoic rocks.

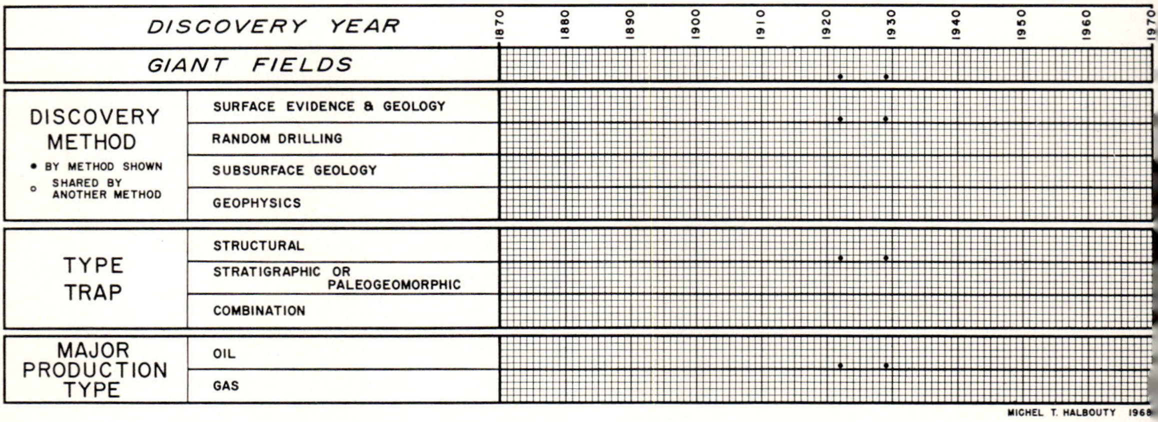

FIG. 19.—Giant fields found in Texas Railroad Commission District 1 (southwest Texas); production from Mesozoic rocks. Location of Texas Railroad Commission Districts shown on Figure 25.

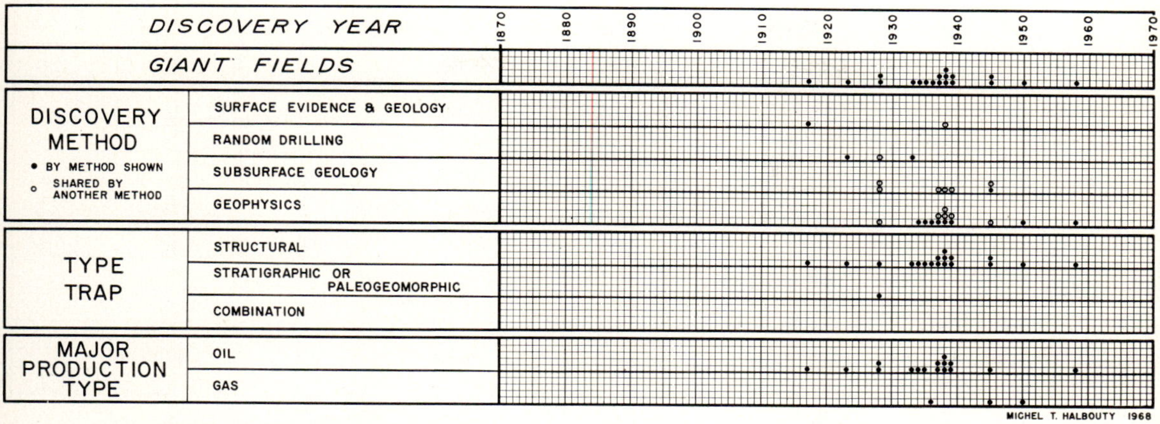

DISCOVERY YEAR

1870 1880 1890 1900 1910 1920 1930 1940 1950 1960 1970

GIANT FIELDS

DISCOVERY METHOD
• BY METHOD SHOWN
○ SHARED BY ANOTHER METHOD

SURFACE EVIDENCE & GEOLOGY
RANDOM DRILLING
SUBSURFACE GEOLOGY
GEOPHYSICS

TYPE TRAP
STRUCTURAL
STRATIGRAPHIC OR PALEOGEOMORPHIC
COMBINATION

MAJOR PRODUCTION TYPE
OIL
GAS

MICHEL T. HALBOUTY 1968

FIG. 20.—Giant fields found in Texas Railroad Commission Districts 2, 4 (lower Gulf coast); production from Cenozoic rocks. Location of Texas Railroad Commission Districts shown on Figure 25.

DISCOVERY YEAR

1870 1880 1890 1900 1910 1920 1930 1940 1950 1960 1970

GIANT FIELDS

DISCOVERY METHOD
• BY METHOD SHOWN
○ SHARED BY ANOTHER METHOD

SURFACE EVIDENCE & GEOLOGY
RANDOM DRILLING
SUBSURFACE GEOLOGY
GEOPHYSICS

TYPE TRAP
STRUCTURAL
STRATIGRAPHIC OR PALEOGEOMORPHIC
COMBINATION

MAJOR PRODUCTION TYPE
OIL
GAS

MICHEL T. HALBOUTY 1968

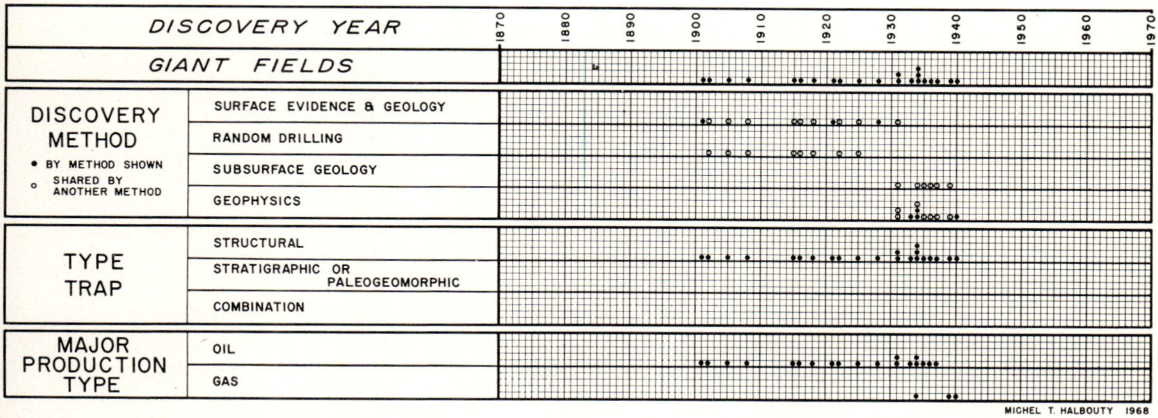

FIG. 21.—Giant fields found in Texas Railroad Commission District 3 (upper Texas Gulf Coast); production from Cenozoic rocks. Location of Texas Railroad Commission Districts shown on Figure 25.

DISCOVERY YEAR

1870 1880 1890 1900 1910 1920 1930 1940 1950 1960 1970

GIANT FIELDS

DISCOVERY METHOD
• BY METHOD SHOWN
○ SHARED BY ANOTHER METHOD

SURFACE EVIDENCE & GEOLOGY
RANDOM DRILLING
SUBSURFACE GEOLOGY
GEOPHYSICS

TYPE TRAP
STRUCTURAL
STRATIGRAPHIC OR PALEOGEOMORPHIC
COMBINATION

MAJOR PRODUCTION TYPE
OIL
GAS

MICHEL T. HALBOUTY 1968

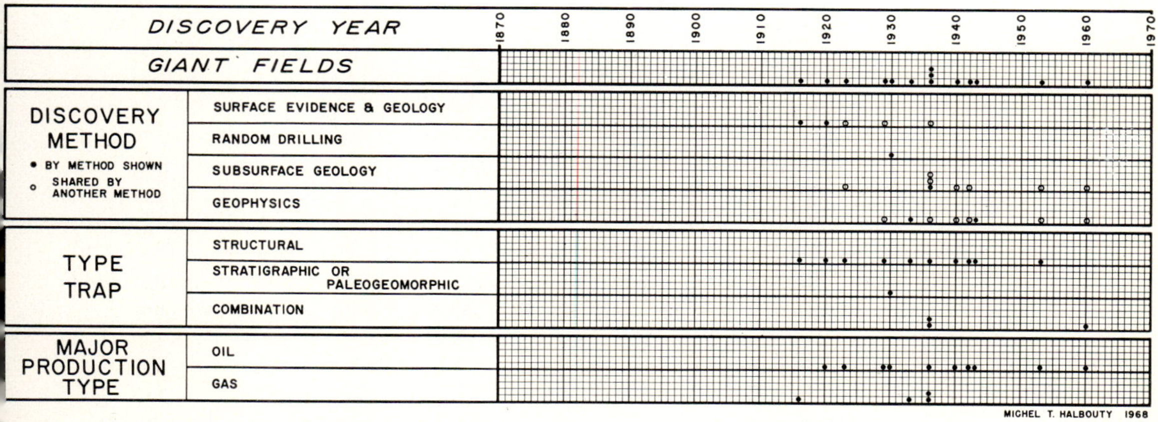

FIG. 22.—Giant fields found in Texas Railroad Commission Districts 5, 6 (East Texas); production from Mesozoic rocks. Location of Texas Railroad Commission Districts shown on Figure 25.

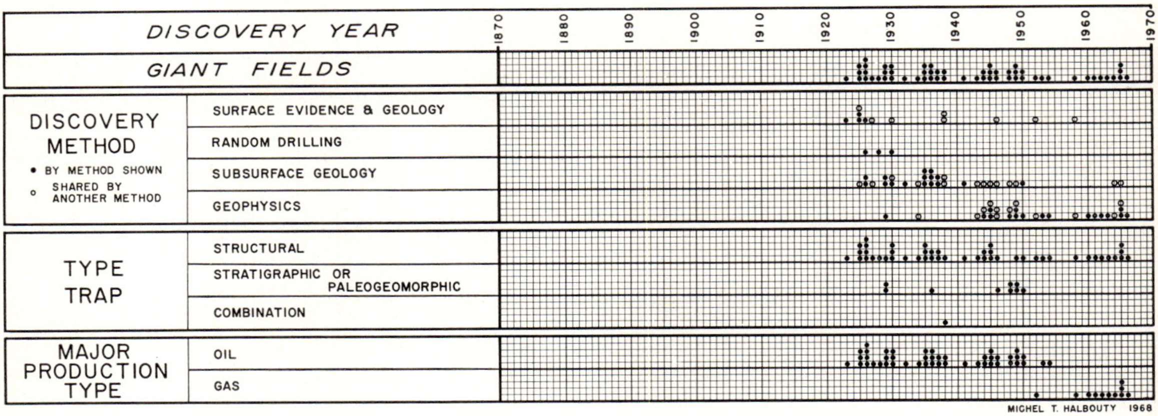

| DISCOVERY YEAR | 1870 | 1880 | 1890 | 1900 | 1910 | 1920 | 1930 | 1940 | 1950 | 1960 | 1970 |

FIG. 23.—Giant fields found in Texas Railroad Commission Districts 7-C, 8, 8-A (West Texas); production from Paleozoic rocks. Location of Texas Railroad Commission Districts shown on Figure 25.

Texas Railroad Commission District 3 consists of the upper Texas Gulf Coast. Twenty-two giant fields are located in this region (Fig. 21). Except for 3 gas-distillate fields, all are oil producers. All are on anticlines or salt domes. Although no stratigraphic trap giants have been found in the region, stratigraphic traps do occur on the flanks of some salt domes. Possibilities for finding subtle traps in District 3 are favorable in the Hackberry province of southeast Texas (and southwest Louisiana) where such traps may occur below a major Frio unconformity in buried, truncated structural features and in thick sandstone beds with limited areal extent. All exploratory methods have been responsible for giant-field discoveries in the upper Texas Gulf Coast. No giant discovery has been made in this area since 1940. Giants in the upper Gulf Coast of Texas produce from Cenozoic, mainly Frio, sandstone.

Fourteen giants have been found in East Texas, Texas Railroad Commission Districts 5 and 6 (Fig. 22). Four are gas fields, 10 are oil producers. Ten giants produce from anticlinal or fault traps. Carthage and Joaquin-Logansport gas fields, and Fairway field, which produces from a trap brought about by porosity and permeability variations in the James Limestone Member (Pearsall Formation, Lower Cretaceous), are considered to be combination traps. East Texas field, the world's most renowned stratigraphic trap, produces from truncated Woodbine Sandstone (Upper Cretaceous) on the monoclinal west flank of the Sabine uplift. Fairway field, discovered in 1960, is the most recent giant found in East Texas. East Texas giants produce principally from Mesozoic (Cretaceous) rocks. Subsurface geology and geophysics have discovered most giants in this region. Possibilities for finding stratigraphic traps in East Texas are considered good.

Fifty-seven giants have been discovered in West Texas, which includes Texas Railroad Commission Districts 7c, 8, and 8a (Fig. 23). Of these 47 are structural traps, 9 are stratigraphic and reef traps, and 1 is a combination trap. Although only one field is classified as a combination trap, several of the 47 structural giants produce to some extent from subtle traps caused by lateral and vertical permeability and porosity variations in producing strata. West Texas giants produce from Paleozoic rocks. Of much interest are the several very deep gas giants which have been found in recent years off the western edge of the Central Basin platform within the Delaware basin. These gas-producing areas are on large anticlinal features and are productive from strata between approximate depths of 14,000 and 22,000 ft. Most West Texas giants have been found by subsurface geology and geophysics. Possibilities for finding stratigraphic trap accumulations in West Texas are considered to be excellent.

Seven giants have been found in the areas included in Texas Railroad Commission District 9 (North Texas) and District 10 (Texas Panhandle)—6 in North Texas and 1, the Texas Panhandle field, in the Panhandle area (Fig. 24).

DISCOVERY YEAR		1870	1880	1890	1900	1910	1920	1930	1940	1950	1960	1970
GIANT FIELDS												
DISCOVERY METHOD — • BY METHOD SHOWN ○ SHARED BY ANOTHER METHOD	SURFACE EVIDENCE & GEOLOGY											
	RANDOM DRILLING											
	SUBSURFACE GEOLOGY											
	GEOPHYSICS											
TYPE TRAP	STRUCTURAL											
	STRATIGRAPHIC OR PALEOGEOMORPHIC											
	COMBINATION											
MAJOR PRODUCTION TYPE	OIL											
	GAS											

MICHEL T. HALBOUTY 1968

FIG. 24.—Giant fields found in Texas Railroad Commission Districts 9 (north Texas) and 10 (Texas Panhandle); production from Paleozoic rocks. Texas Railroad Commission Districts shown on Figure 25.

Most giants in this region are on anticlinal structures. Exceptions are Boonsville Bend gas field, North Texas, a huge stratigraphic trap which was found in an area of monoclinal dip by random drilling, and Hull-Silk-Sikes field, a combination trap on an anticlinal nose. Production from North Texas and Texas Panhandle giants is from Paleozoic rocks. Giant fields in this region are above and on the flanks of the buried Amarillo mountains in the Texas Panhandle area and along the crest and flanks of the Red River uplift in North Texas.

ANALYSIS OF MAJOR FIELDS

All giant fields shown in Figures 1–24, with pertinent data about each, are listed in Table 1 under regional location and in Table 2 under years of discovery. Location of each of these giants is shown on Figure 26 as a number. Identity of each field is shown by the corresponding number in Column 1, Table 1.

The important fact derived from the preceding data is *not* that 306 giant fields have been found by the petroleum industry since the first was discovered in 1871, but that the *discovery of new giant fields has almost ceased in nearly all regions of the country.* Exceptions are (a) Alaska, where exploration is relatively new, (b) West Texas, where great drilling depths have been required to find the new giants, (c) North Louisiana, and (d) East Texas.[6] Between 1960 and the present, 16 giant fields have been found in the nation; these include 4 in Alaska, 9 in West Texas, 2 in Loui-

[6] See footnotes 4 and 5.

siana, and 1 in East Texas. Several recent discoveries in offshore Louisiana probably will attain giant classification as development continues; however, to date none has been classified as such by the operators.

In no other parts of the country have any giants been discovered since 1960; in many areas it has been several decades since the last giant was discovered. Will this nation soon be running out of giants? This is possible, unless new methods of exploration are found, or unless new approaches to exploration are adopted.

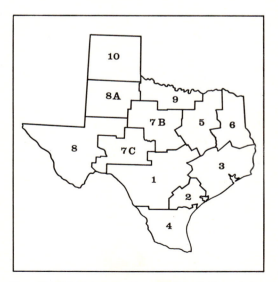

FIG. 25.—District map, Texas Railroad Commission, Oil and Gas Division, State of Texas.

TABLE 1. GIANT OIL AND GAS FIELDS IN UNITED STATES (BY REGIONS)

Field Number Fig. 26	Field	Disc. Year	Discovery Methods	Age of Reserv. Rocks	Type[1] Trap	Approx. Proved Acres	Major Prod. Type	Oil (1,000 bbl) Cum. Prod. to 1/1/67	Oil Est. Res. 1/1/67	Oil Prod. in 1966	Gas (billion cu ft) Cum. Prod. to 1/1/66	Gas Est. Res. 1/1/66	Gas Prod. in 1965	Additional Structure or Trap Data
	ALASKA (Fig. 4)													
1	Cook Inlet	1964	Geophys.	Ceno.	Struc.	4,000	Gas	—	—	—	—	—	—	Anticlinal
2	Granite Point	1965	Geophys.	Ceno.	Struc.	3,000	Oil	—	175,000	—	—	—	5.9	Anticlinal
3	Kenai	1959	Geophys.	Ceno.	Struc.	6,000	Gas	—	—	—	15.2	4,984.8	—	Anticlinal
4	McArthur River	1965	Geophys.	Ceno.	Struc.	6,000	Oil	—	160,000	—	—	—	—	Anticlinal
5	Middle Ground Shoal	1963	Geophys.	Ceno.	Struc.	6,000	Oil	2,671	197,329	2,644	—	—	—	Anticlinal
6	Swanson River	1957	Geophys.	Ceno.	Struc.	6,200	Oil	61,665	188,335	10,406	—	—	—	Anticlinal
	APPALACHIAN AREA (NEW YORK, PENNSYLVANIA, WEST VIRGINIA) (Fig. 5)													
7	Allegany (N. Y.)	1879	Surf. evidence	Paleoz.	Comb.	30,000	Oil	162,609	5,800	667	—	—	—	Anticlinal with porosity changes
8	Bradford (Pa.)	1871	Surf. evidence	Paleoz.	Comb.	85,000	Oil	634,844	22,662	3,648	—	—	—	Anticlinal with porosity changes
	ARKANSAS (Fig. 6)													
9	El Dorado, E. and S.	1920	Random drlg.	Meso.	Comb.	13,200	Oil	77,289	22,711	388	—	—	—	Anticlinal with monoclinal traps
10	Magnolia	1938	Surf. geol., Geophys.	Meso.	Comb.	4,500	Oil	141,268	53,732	5,510	—	—	—	Anticlinal with monoclinal traps
11	Schuler and East	1937	Subsurf. geol., Geophys.	Meso.	Struc.	7,900	Oil	98,810	8,190	1,320	—	—	—	Anticline, some ss. lens.
12	Smackover	1922	Random drlg.	Meso.	Struc.	23,700	Oil	488,835	36,165	3,213	—	—	—	Domal anticline, some monoclinal traps
	WEST COAST STATES (CALIFORNIA, OREGON, WASHINGTON) (Fig. 7) California, Central (San Joaquin Valley)													
13	Belridge S.	1911	Surf. evidence, Random drlg.	Ceno.	Struc.	9,200	Oil	124,728	77,078	7,662	—	—	—	Anticlinal, some truncated ss.
14	Buena Vista	1909	Surf. geol., Random drlg.	Ceno.	Struc.	14,800	Oil	539,986	75,088	7,537	887.6	—	21.7	Anticlinal, some lentic. ss.
15	Coalinga Eastside	1890	Surf. evidence, Random drlg.	Ceno.	Struc.	6,200	Oil	359,977	33,450	5,384	—	—	—	Anticlinal, some monoclinal traps
16	Coalinga Nose (E. Ext.)	1938	Subsurf. geol.	Ceno.	Comb.	4,000	Oil	389,741	93,763	8,282	—	—	—	Noses on ss. lens.
17	Coalinga Westside	1900	Surf. geol., Random drlg.	Ceno.	Strat.	9,600	Oil	208,025	28,689	4,086	—	—	—	Monoclinal with unconformity
18	Coles Levee N.	1938	Geophys.	Ceno.	Comb.	3,800	Oil	129,785	30,200	2,515	—	—	—	Anticlinal with monoclinal trap
19	Cuyama S.	1949	Surf. geol., Geophys.	Ceno.	Struc.	—	Oil	187,045	44,851	5,217	—	—	—	Faulted anticline
20	Cymric	1916	Surf. geol., Random drlg.	Ceno.	Comb.	3,300	Oil	101,035	40,577	5,630	—	—	—	Faulted monocline with unconformity
21	Edison	1934	Subsurf. geol., Random drlg.	Ceno.	Comb.	6,600	Oil	100,877	19,079	2,146	—	—	—	Monoclinal with unconformity
22	Elk Hills	1919	Surf. geol.	Ceno.	Struc.	19,600	Oil	276,095	1,026,454	2,106	—	—	—	Faulted anticline, some lentic. ss.
23	Fruitvale	1928	Surf. geol.	Ceno.	Strat.	3,300	Oil	92,189	12,790	1,794	—	—	—	Monocline
24	Greeley	1936	Geophys.	Ceno.	Struc.	2,100	Oil	99,965	16,248	2,617	—	—	—	Faulted anticline
25	Kern River	1899	Random drlg.	Ceno.	Strat.	8,700	Oil	427,303	187,432	19,625	—	—	—	Monocline

No.	Field	Year	Method of discovery	Age	Class	Oil/Gas	Depth							Type of trap
26	Kern Front	1915	Surf., Subsurf. geol.	Ceno.	Comb.	Oil	4,500	111,117	35,883	2,560				Faulted monocline
27	Kettleman North Dome	1928	Surf. geol.	Ceno.	Struc.	Oil	13,700	444,378	30,639	1,814	2,632.8		39.5	Anticline
28	Lost Hills	1910	Surf. geol., Random drlg.	Ceno.	Comb.	Oil	3,900	99,574	27,293	2,751				Lensing on anticline
29	McKittrick Main Area	1887	Surf. evidence	Ceno.	Comb.	Oil	1,500	124,959	35,575	4,097				Faulted monocline with unconformity
30	Midway-Sunset	1894	Surf. evidence, Random drlg.	Ceno.	Comb.	Oil	24,900	996,210	196,898	25,992				Anticlinal, monoclinal, unconformity
31	Mount Poso	1926	Surf., Subsurf. geol.	Ceno.	Comb.	Oil	2,800	153,565	22,912	1,984				Faulted monocline
32	Rio Bravo	1937	Geophys.	Ceno.	Struc.	Oil	2,000	108,230	27,436	1,563				Faulted anticline
	California, Central (Salinas Valley district)													
33	San Ardo	1947	Surf., Subsurf. geol.	Ceno.	Struc.	Oil	4,200	175,099	101,365	17,529				Anticline
	California, Coastal													
34	Cat Canyon W.	1908	Surf. geol., Random drlg.	Ceno.	Struc.	Oil	2,700	116,237	28,648	4,343				Anticline
35	Elwood	1928	Surf. geology	Ceno.	Struc.	Oil	700	102,258	2,617	188				Anticline
36	Orcutt	1902	Surf. evidence, Random drlg.	Ceno.	Struc.	Oil	3,900	131,539	13,897	2,337				Faulted dome
37	Rincon	1927	Surf. geol.	Ceno.	Struc.	Oil	1,900	94,076	21,882	3,146				Anticline
38	Santa Maria Valley	1934	Subsurf. geol., Geophys.	Ceno.	Strat.	Oil	5,600	142,085	17,896	1,550				Monoclinal
39	South Mountain	1915	Surf. geol.	Ceno.	Struc.	Oil	3,300	113,459	34,507	3,244				Anticline
40	Ventura	1916	Surf. geol.	Ceno.	Struc.	Oil	3,400	720,979	96,949	13,215	1,847.4		29.6	Faulted anticline
	California (Los Angeles basin)													
41	Brea-Olinda	1884	Surf. geol., Random drlg.	Ceno.	Struc.	Oil	2,600	309,180	39,946	4,708				Faulted anticline
42	Coyote East	1911	Surf. geol.	Ceno.	Struc.	Oil	1,500	90,528	9,555	1,328				Faulted anticline
43	Coyote West	1909	Surf. geol.	Ceno.	Struc.	Oil	1,100	212,904	16,932	2,764				Faulted anticline
44	Dominguez	1923	Surf. geol.	Ceno.	Struc.	Oil	1,700	243,140	21,955	3,017				Faulted anticline
45	Huntington Beach	1920	Surf. geol.	Ceno.	Comb.	Oil	6,100	775,076	158,824	22,107	700.2		17.0	Anticlinal and monoclinal traps
46	Inglewood	1924	Surf. geol.	Ceno.	Struc.	Oil	1,200	261,453	40,058	6,591				Faulted anticline
47	Long Beach	1921	Surf., Subsurf. geol.	Ceno.	Struc.	Oil	1,700	851,882	28,019	4,231	1,065.9		4.2	Faulted anticline
48	Montebello	1917	Surf. geol.	Ceno.	Struc.	Oil	1,600	178,291	8,299	1,105				Faulted anticline
49	Richfield	1919	Surf. geol.	Ceno.	Struc.	Oil	1,600	149,334	11,734	2,097				Anticline
50	Sante Fe Springs	1919	Surf. geol.	Ceno.	Struc.	Oil	1,500	591,671	23,256	1,784	824.0		3.1	Faulted anticline
51	Seal Beach	1926	Surf., Subsurf. geol.	Ceno.	Struc.	Oil	900	175,937	18,884	1,995				Faulted anticline
52	Torrance	1922	Surf. geol.	Ceno.	Struc.	Oil	6,700	168,011	21,933	1,619				Faulted anticline
53	Wilmington	1935	Subsurf. geol., Geophys.	Ceno.	Struc.	Oil	7,800	1,098,767	1,502,255	47,116	812.0		7.3	Faulted anticline
	California (North)													
54	Rio Vista	1936	Surf. geol., Geophys.	Ceno.	Struc.	Gas	25,800	—	—	—	2,316	1,200	52.6	Faulted dome
	ILLINOIS (Fig. 8)													
55	Clay City Consolidated	1937	Geophys.	Paleoz.	Struc.	Oil	91,000	253,379	46,621	5,980				Anticline
56	Dale Consolidated	1940	Subsurf. geol., Geophys.	Paleoz.	Struc.	Oil	18,900	83,487	16,513	3,223				Anticline
57	Loudon	1937	Surf. geol., Geophys.	Paleoz.	Struc.	Oil	24,600	320,639	29,361	8,533				Anticline
58	New Harmony Consolidated	1939	Random drlg., Geophys.	Paleoz.	Struc.	Oil	30,000	131,270	43,046	3,939				Anticline
59	Salem Consolidated	1938	Geophys.	Paleoz.	Struc.	Oil	14,900	329,598	20,402	4,326				Anticline
60	Old Illinois	1905	Random drlg.	Paleoz.	Comb.	Oil	130,000	632,994	42,006	10,476				Anticlines & monoclines

(Includes Lawrence Co. & Main Cons. Flds.)

TABLE 1. (Continued)

Field Number Fig. 26	Field	Disc. Year	Discovery Methods	Age of Reserv. Rocks	Type[1] Trap	Approx. Proved Acres	Major Prod. Type	Oil (1,000 bbl) Cum. Prod. to 1/1/67	Oil Est. Res. 1/1/67	Oil Prod. in 1966	Gas (billion cu ft) Cum. Prod. to 1/1/66	Gas Est. Res. 1/1/66	Gas Prod. in 1965	Additional Structure or Trap Data
							INDIANA-OHIO (Fig. 9)							
61	Lima (Ohio)	1884	Random drlg.	Paleoz.	Strat.	550,000 est.	Oil	375,000	(Includes several adjoining fields)			—	—	Anticlines, monoclines, terraces with some por. chgs.
62	Trenton (Indiana)	1886	Random drlg.	Paleoz.	Strat.	2,500,000 est.[2]	Oil, Gas	105,085	—	7.5		—	—	—
							KANSAS (Fig. 10)							
63	Bemis Shutts	1928	Surf., Subsurf. geol.	Paleoz.	Struc.	18,200	Oil	197,802	152,198	3,267	—	—	—	Domal
64	Chase-Silica	1930	Subsurf. geol.; Random drlg.	Paleoz.	Comb.	53,400	Oil	237,642	22,358	2,579	—	—	—	Anticlinal & Monoclinal
65	El Dorado	1915	Surf. geol.	Paleoz.	Struc.	25,000	Oil	266,905	13,095	2,534	—	—	—	Anticline
66	Gorham	1926	Surf. geol.	Paleoz.	Struc.	14,600	Oil	72,647	28,000	1,275	—	—	—	Dome
67	Greenwood	1951	Surf. geol., Geophys.	Paleoz.	Strat.	200,000	Gas	—	—	—	459.5	540.5	37.9	Monoclinal
68	Hall-Gurney	1931	Surf. geol., Random drlg.	Paleoz.	Struc.	25,000	Oil	107,495	29,505	3,290	—	—	—	Anticline
69	Hugoton	1926	Random drlg.	Paleoz.	Strat.	4,000,000[3]	Gas				13,634.4	32,000[4]	860.6	Monoclinal
70	Kraft-Prusa	1937	Surf. geol., Random drlg.	Paleoz.	Comb.	21,300	Oil	101,461	18,539	1,992	—	—	—	Anticline, ss. lensing
71	Trapp	1929	Surf. geol.	Paleoz.	Struc.	32,100	Oil	188,682	31,318	3,055	—	—	—	Anticline
							KENTUCKY (Fig. 11)							
72	Big Sandy	1918	Random drlg.	Paleoz.	Comb.	650,000	Gas				700	—	50	Anticline, domes, monoclines
73	Big Sinking	1917	Random drlg.	Paleoz.	Strat.	14,100	Oil	70,000 est.	30,000	2,500 est.	est.[5]	—	est.	Monoclinal
							NORTH LOUISIANA (Fig. 12)							
74	Black Lake	1964	Subsurf. geol., Geophys.	Meso.	Comb.	16,000	Oil, Gas	1,592	105,000	1,560	—	669.0	—	Reefing across nose
75	Caddo-Pine Island	1905	Random drlg.	Meso.	Struc.	96,100	Oil	293,523	106,477	5,691	—	—	—	Anticline
76	Cotton Valley	1922	Random drlg.	Meso.	Struc.	14,000	Oil	121,415	40,585	3,059	818.8	—	33.7	Anticline, some lentic.
77	Delhi	1944	Subsurf. geol., Geophys.	Meso.	Strat.	17,900	Oil	138,958	86,042	4,634	—	—	—	Monocline, truncation
78	Haynesville	1921	Surf. geol.	Meso.	Struc.	17,400	Oil	145,623	29,377	2,033	—	—	—	Anticline
79	Homer	1919	Surf. geol.	Meso.	Struc.	3,300	Oil	89,777	10,223	546	—	—	—	Domal
80	Lake St. John	1942	Geophys.	Meso.	Struc.	13,000	Oil	85,677	30,323	1,959	—	—	—	Lensing on regional uplift, truncation ss.
81	Monroe	1916	Surf. geol.	Meso.	Comb.	236,300	Gas				6,238.5	—	95.0	Anticline
							SOUTH LOUISIANA (Fig. 13)							
							South Louisiana, Onshore							
82	Rodessa	1930	Random drlg.	Meso.	Struc.	19,600	Oil	131,853	43,147	694	709.6	—	1.6	Anticline
83	Sligo	1922	Subsurf. geol.	Meso.	Struc.	24,800	Gas				1,749.2	—	76.6	Anticline
84	Bastian Bay	1941	Subsurf. geol., Geophys.	Ceno.	Struc.	12,000	Gas				391.5	1,608.5	105.4	Faulted anticline
85	Bay St. Elaine	1929	Geophys.	Ceno.	Struc.	8,400	Oil	86,522	66,478	7,447	530	—	—	Salt dome
86	Bayou Sale	1941	Geophys.	Ceno.	Struc.	10,000	Oil	112,667	87,343	9,325	—	—	—	Anticline
87	Bourg	1952	Geophys.	Ceno.	Struc.	2,500	Gas				340.8	759.2	21.7	Anticline
88	Caillou Island	1930	Geophys.	Ceno.	Struc.	20,700	Oil	289,816	210,184	26,521	—	—	—	Salt dome
89	Cote Blanch Bay W.	1940	Geophys.	Ceno.	Struc.	5,900	Oil	58,166	31,834	6,953	—	—	—	Salt dome

No.	Field	Year	Method	Era		Depth	Type							Trap
90	Deep Lake	1952	Geophys.	Ceno.	Struc.	4,300	Gas	—	—	—	586.0	514.0	50.0	Anticline
91	Delta Farms	1940	Geophys.	Ceno.	Struc.	10,200	Oil	99,158	40,842	2,366	697.5	802.5	—	Anticline
92	Duck Lake	1949	Subsurf. geol., Geophys.	Ceno.	Struc.	6,000	Gas	—	—	—	—	—	63.6	Anticline
93	Erath	1940	Geophys.	Ceno.	Struc.	5,900	Oil, Gas	110,797	39,203	2,890	565.5	934.5	40.4	Anticline
94	Garden Island Bay	1935	Geophys.	Ceno.	Struc.	11,000	Oil	64,116	55,884	8,772	—	—	—	Salt dome
95	Golden Meadow	1938	Geophys.	Ceno.	Struc.	12,200	Oil	92,938	107,062	4,064	—	—	—	Salt dome
96	Grand Bay	1938	Geophys.	Ceno.	Struc.	10,000	Oil	102,919	113,081	6,374	—	—	—	Anticline
97	Hackberry East	1927	Geophys.	Ceno.	Struc.	5,000	Oil	70,682	19,318	4,524	—	—	—	Salt dome
98	Hackberry West	1928	Geophys.	Ceno.	Struc.	5,000	Oil	89,920	44,080	4,085	—	—	—	Salt dome
99	Iowa	1931	Geophys.	Ceno.	Struc.	5,400	Oil	97,354	30,646	1,093	—	—	—	Salt dome
100	Jennings	1901	Surf. evidence	Ceno.	Struc.	5,400	Oil	114,141	17,859	405	—	—	—	Salt dome
101	Lafitte	1935	Geophys.	Ceno.	Struc.	3,600	Oil	133,662	86,337	7,642	—	—	—	Salt dome
102	Lake Arthur	1937	Geophys.	Ceno.	Struc.	8,900	Gas	—	—	—	990.1	—	117.1	Anticline
103	Lake Barre	1929	Geophys.	Ceno.	Struc.	7,600	Oil	103,758	146,242	15,049	—	—	—	Salt dome
104	Lake Pelto	1929	Geophys.	Ceno.	Struc.	5,500	Oil	68,120	17,880	4,286	—	—	—	Salt dome
105	Lake Washington	1931	Geophys.	Ceno.	Struc.	13,600	Oil	125,053	174,947	10,203	—	—	—	Salt dome
106	Leeville	1931	Geophys.	Ceno.	Struc.	9,400	Oil	95,597	54,403	3,907	—	—	—	Salt dome
107	Paradis	1939	Geophys.	Ceno.	Struc.	5,300	Oil	80,130	19,870	3,630	—	—	—	Salt dome
108	Quarantine Bay	1937	Geophys.	Ceno.	Struc.	7,300	Oil	98,062	52,938	6,708	—	—	—	Anticline
109	Timbalier Bay	1938	Geophys.	Ceno.	Struc.	11,700	Oil	159,187	140,813	23,775	—	—	—	Salt dome
110	Venice	1937	Geophys.	Ceno.	Struc.	6,400	Oil	114,644	60,356	5,803	—	—	—	Salt dome
111	Vinton	1910	Surf. evidence	Ceno.	Struc.	3,700	Oil	103,974	28,026	1,789	—	—	—	Salt dome
112	Weeks Island	1945	Surf. evidence, Subsurf. geol.	Ceno.	Struc.	5,900	Oil	144,521	92,479	6,883	—	—	—	Salt dome
113	West Bay	1940	Geophys.	Ceno.	Struc.	11,200	Oil	99,548	110,452	10,692	—	—	—	Salt dome
	South Louisiana, Offshore													
114	Bay Marchand, Blk. 2	1949	Subsurf. geol., Geophys.	Ceno.	Struc.	24,100	Oil	183,534	417,002	27,211	506.6	1,493.4	38.2	Salt dome
115	Eugene Is., Blk. 32	1950	Geophys.	Ceno.	Struc.	3,300	Gas	—	—	—	—	—	—	Salt dome
116	Eugene Is., Blk. 126	1950	Geophys.	Ceno.	Struc.	3,000	Oil	45,800	79,200	5,145	—	—	—	Salt dome
117	Grand Isle, Blk. 16	1948	Geophys.	Ceno.	Struc.	8,100	Oil	65,924	109,076	12,963	—	—	—	Salt dome
118	Grand Isle, Blk. 47	1955	Geophys.	Ceno.	Struc.	3,600	Oil	35,636	61,788	4,069	—	—	—	Anticline
119	Main Pass, Blk. 35	1952	Geophys.	Ceno.	Struc.	4,700	Oil	51,629	48,371	4,393	—	—	—	Faulted nose
120	Main Pass, Blk. 69	1948	Geophys.	Ceno.	Struc.	14,800	Oil	102,962	197,038	11,807	—	—	—	Anticline
121	South Pass, Blk. 24	1950	Geophys.	Ceno.	Struc.	16,400	Oil	256,025	493,975	22,163	—	—	—	Anticline
122	South Pass, Blk. 27	1954	Geophys.	Ceno.	Struc.	20,520	Oil	111,345	199,655	20,179	—	—	—	Salt dome
123	S. Timbalier, Blk. 135	1956	Geophys.	Ceno.	Struc.	5,700	Oil	25,944	—	9,310	—	—	—	Salt dome
124	Vermilion, Blk. 39	1949	Geophys.	Ceno.	Struc.	3,100	Gas	—	—	—	568.3	631.7	50.6	Anticline
125	West Delta, Blk. 30	1949	Geophys.	Ceno.	Struc.	16,000	Oil	121,646	278,354	20,556	—	—	—	Salt dome
126	West Delta, Blk. 73	1962	Geophys.	Ceno.	Struc.	5,800	Oil	18,580	—	10,689	—	—	—	Anticline
	GULF STATES (SOUTHERN) (Fig. 14)													
	Alabama													
127	Citronelle	1955	Subsurf. geol., Geophys.	Meso.	Struc.	16,400	Oil	63,636	56,364	7,281	—	—	—	Anticline
	Mississippi													
128	Baxterville	1944	Geophys.	Meso.	Struc.	16,000	Oil	112,266	87,734	5,399	692.0	708.0	6.2	Dome
129	Cranfield	1943	Geophys.	Ceno., Meso.	Struc.	9,500	Gas	—	—	—	—	—	—	Dome
130	Gwinville	1944	Surf. geol., Geophys.	Meso.	Struc.	22,500	Gas	80,826	29,174	3,830	1,098.0	1,402	35.3	Salt dome
131	Heidelberg	1944	Surf. geol., Geophys.	Meso.	Struc.	9,900	Oil	176,142	43,858	2,325	—	—	—	Salt dome
132	Tinsley	1939	Surf. geol., Geophys.	Meso.	Struc.	12,400	Oil							Anticline

TABLE 1. (*Continued*)

Field Number Fig. 26	Field	Disc. Year	Discovery Methods	Age of Reserv. Rocks	Type[1] Trap	Approx. Proved Acres	Major Prod. Type	Oil (1,000 bbl) Cum. Prod. to 1/1/67	Est. Res. 1/1/67	Prod. in 1966	Gas (billion cu ft) Cum. Prod. to 1/1/66	Est. Res. 1/1/66	Prod. in 1965	Additional Structure or Trap Data
							NEW MEXICO, Eastern (Fig. 15)							
133	Caprock and East	1940	Subsurf. geol.	Paleoz.	Comb.	30,300	Oil	80,619	59,381	4,754	—	—	—	Anticlinal, monoclinal
134	Denton	1949	Subsurf. geol., Geophys.	Paleoz.	Struc.	4,600	Oil	103,561	41,439	3,453	—	—	—	Anticline
135	Empire Abo	1957	Subsurf. geol.	Paleoz.	Comb.	9,800	Oil	37,429	62,571	6,306	—	—	—	Anticline
136	Eunice	1929	Subsurf. geol., Geophys.	Paleoz.	Struc.	20,000	Oil	107,438	34,562	1,427	—	—	—	Anticline
137	Hobbs	1928	Geophys.	Paleoz.	Struc.	est. 14,800	Oil	186,729	29,271	3,679	4,400.0[7]	3,700.0[7]	49.7[7]	Anticline
138	Jalmat	1929	Subsurf. geol.	Paleoz.	Strat.	92,300[6]	Oil, Gas	52,239	48,375	1,161				Monoclinal
139	Maljamar	1926	Subsurf. geol.	Paleoz.	Comb.	23,000	Oil	61,313	38,687	2,941	—	—	—	Anticlinal, monoclinal
140	Monument	1929	Subsurf. geol., Geophys.	Paleoz.	Struc.	25,000	Oil	187,503	62,497	4,482	—	—	—	Anticline
141	Vacuum	1929	Geophys.	Paleoz.	Struc.	est. 21,900	Oil	145,563	139,437	13,812	—	—	—	Anticline
							NEW MEXICO, Western							
142	Blanco Mesaverde	1927	Subsurf. geol.	Meso.	Strat.	630,400	Gas	—	—	—	1,974.6	—	180.9	Monoclinal
							OKLAHOMA (Fig. 16)							
143	Allen	1927	Surf. geol.	Paleoz.	Comb.	10,300	Oil	99,414	20,586	2,636	—	—	—	Anticlinal, monoclinal
144	Avant	1904	Surf. evidence	Paleoz.	Struc.	12,600	Oil	105,178	1,823	240	—	—	—	Anticline
145	Bowlegs	1927	Surf. geol.	Paleoz.	Struc.	3,200	Oil	146,691	13,309	952	—	—	—	Anticline
146	Burbank	1920	Surf. geol.	Paleoz.	Strat.	36,900	Oil	457,881	42,119	10,655	—	—	—	Monoclinal
147	Camrick	1954	Subsurf. geol., Geophys.	Paleoz.	Strat.	17,600	Gas				351.2	2,348.8	133.1	Monoclinal
148	Cement	1917	Surf. geol.	Paleoz.	Struc.	18,500	Oil	120,665	14,335	2,671	—	—	—	Anticline
149	Cushing	1912	Surf. geol., Random drlg.	Paleoz.	Struc.	36,300	Oil	430,955	24,045	3,499	—	—	—	Anticline
150	Earlsboro	1930	Surf. geol.	Paleoz.	Struc.	7,600	Oil	138,530	1,470	623	—	—	—	Anticline
151	Edmund, West	1943	Random drlg.	Paleoz.	Strat.	43,000	Oil	117,494	12,506	1,961	—	—	—	Monoclinal
152	Elk City	1947	Geophys.	Paleoz.	Struc.	12,200	Oil	59,055	40,945	404	—	—	—	Anticline
153	Eola-Robberson	1921	Surf. geol.	Paleoz.	Struc.	5,400	Oil	70,795	54,205	3,632	—	—	—	Anticline
154	Fitts	1932	Surf. geol.	Paleoz.	Struc.	6,000	Oil	122,394	4,606	1,324	—	—	—	Anticline
155	Glenn Pool	1905	Surf. evidence	Paleoz.	Comb.	27,400	Oil	287,464	32,536	4,153	—	—	—	Anticlinal, monoclinal
156	Golden Trend	1946	Subsurf. geol., Geophys.	Paleoz.	Comb.	96,000	Oil	310,300	184,700	13,440	—	—	—	Anticlines, monoclines
	Guymon-Hugoton	1926	Geophys.		—	(see Kansas, Hugoton field)								Monoclinal
157	Healdton	1913	Surf. geol.	Paleoz.	Struc.	10,400	Oil	253,575	16,425	3,036	—	—	—	Anticline
158	Hewitt	1919	Surf. geol.	Paleoz.	Struc.	17,000	Oil	178,554	26,446	3,764	—	—	—	Anticline
159	Keyes	1943	Subsurf. geol., Geophys.	Paleoz.	Strat.	65,000	Gas				170.5	829.5	36.3	Monoclinal
160	Little River	1937	Surf. geol.	Paleoz.	Struc.	est. 3,750	Oil	132,541	2,459	441	—	—	—	Anticline
161	Mocane-Laverne	1952	Subsurf. geol., Geophys.	Paleoz.	Strat.	175,000	Gas				300.0	2,500.0	—	Monoclinal
162	Oklahoma City	1928	Subsurf. geol.	Paleoz.	Struc.	est. 43,000	Oil	729,802	40,198	1,922	—	—	—	Anticline
163	Putnam	1959	Subsurf. geol., Geophys.	Paleoz.	Strat.	52,000	Gas				27.3	1,472.7	—	Monoclinal
164	Red Oak-Norris	1929	Surf. geol.	Paleoz.	Strat.	est.	Gas				73.8	1,926.2	12.0	Monoclinal

TABLE 1. (Continued)

Field Number Fig. 26	Field	Disc. Year	Discovery Methods	Age of Reserv. Rocks	Type[1] Trap	Approx. Proved Acres	Major Prod. Type	Oil (1,000 bbl)			Gas (billion cu ft)			Additional Structure or Trap Data
								Cum. Prod. to 1/1/67	Est. Res. 1/1/67	Prod. in 1966	Cum. Prod. to 1/1/66	Est. Res. 1/1/66	Prod. in 1965	
	Districts 2 and 4 (Lower Texas Gulf Coast) (Fig. 20)													
188	Agua Dulce	1928	Subsurf. geol., Geophys.	Ceno.	Struc.	20,000 est.	Oil, Gas	197,777[10]	63,223[10]	7,326[10]	1,900	—	47.0	Anticline
189	Alazan, North	1958	Geophys.	Ceno.	Struc.	6,000 est.	Oil	18,042	81,958	4,681	—	—	—	Anticline
190	Borregas (all flds.)	1945	Subsurf. geol, Geophys.	Ceno.	Struc.	6,000 est.	Oil	62,799	88,201	7,355	—	—	—	Anticline
191	Government Wells, N.	1928	Subsurf. geol., Random drlg.	Ceno.	Strat.	8,500	Oil	76,288	23,712	914	—	—	—	Monoclinal, ss. lensing
192	Greta	1933	Random drlg.	Ceno.	Struc.	9,300	Oil, Gas	89,475	30,525	1,824	(Inc. in Tom O'Connor field)		50.0	Anticline
193	Kelsey (all flds.)	1938	Geophys.	Ceno.	Struc.	19,800 est.	Oil	67,282	42,718	4,347	—	—	—	Anticline
194	LaGloria and South	1939	Geophys.	Ceno.	Struc.	10,500 est.	Oil, Gas	95,155	54,845	3,415	2,000	900	92.1	Anticline
195	Plymouth	1935	Geophys.	Ceno.	Struc.	4,500 est.	Oil	127,942	72,236	2,187	—	—	—	Anticline
196	Red Fish Bay	1950	Geophys.	Ceno.	Struc.	4,900	Gas	16,718	23,282	1,095	500	500	28.7	Anticline
197	Refugio (all flds.)	1917	Surf. geol.	Ceno.	Struc.	17,000 est.	Oil	90,331	59,669	1,112	—	—	—	Anticline
198	San Salvador	1936	Geophys.	Ceno.	Struc.	16,000 est.	Gas	—	—	—	600	400	17.3	Anticline
199	Saxet	1923	Random drlg.	Ceno.	Struc.	9,900 est.	Oil	78,177	22,450	1,150	—	—	—	Anticline
200	Seeligson (all flds.)	1937	Geophys.	Ceno.	Struc.	19,400	Oil, Gas	288,249	170,329	25,211	—	—	121.0	Anticline
201	Stratton	1937	Subsurf. geol., Geophys.	Ceno.	Struc.	20,500 est.	Oil, Gas	(Inc. in Agua Dulce field)			1,800	700	91.6	Anticline
202	Tijerina-Canales-Blucher	1939	Subsurf. geol, Geophys.	Ceno.	Struc.	10,200	Oil	59,653	40,347	3,212	—	—	—	Anticline
203	Tom O'Connor	1934	Geophysics	Ceno.	Struc.	13,000	Oil, Gas	304,881	145,792	10,295	1,300[11]	1,700[11]	42.2	Anticline
204	Tulsita Wilcox (inc. N. Pettus)	1945	Subsurf. geol.	Ceno.	Struc.	4,400 est.	Gas	—	—	—	500	700	21.7	Anticline
205	West Ranch	1938	Surf. evidence, Geophys.	Ceno.	Struc.	8,200	Oil	186,420	92,391	7,667	—	—	—	Anticline
206	White Point, East	1938	Subsurf. geol., Geophys.	Ceno.	Struc.	4,600 est.	Oil	81,906	28,478	2,308	—	—	—	Anticline
	District 3 (Upper Texas Gulf Coast) (Fig. 21)													
207	Anahuac	1935	Subsurf. geol., Geophys.	Ceno.	Struc.	10,900	Oil	180,484	85,516	4,466	—	—	—	Domal
208	Barbers Hill	1916	Surf. evidence, Random drlg.	Ceno.	Struc.	1,700	Oil	119,253	10,747	1,176	—	—	—	Piercement salt dome●
209	Chocolate Bayou	1939	Subsurf. geol., Geophys.	Ceno.	Struc.	9,500	Gas	36,549	37,661	1,658	1,300	1,200	75.5	Anticline
210	Conroe	1931	Surf. geol, Geophys.	Ceno.	Struc.	17,900	Oil	429,991	174,573	6,632	—	—	—	Anticline

No.	Field	Year	Discovery method	Age	Trap form	Depth	Oil/Gas							Type of trap
165	Seminole	1926	Surf. geol.	Paleoz.	Struc.	8,200	Oil	169,867	7,133	1,115	—	—	—	Anticline
166	Sho-Vel-Tum*	1914	Surf. geol.	Paleoz.	Struc.	75,000	Oil	742,835	158,288	30,712	—	—	—	Anticline
167	Sooner Trend*	1945	Subsurf. geol., Geophys.	Paleoz.	Comb.	346,000	Oil	51,139	49,504	11,496	—	—	—	Anticlinal, monoclinal
168	St. Louis	1925	Surf. geol.	Paleoz.	Comb.	26,100	Oil	199,879	10,121	1,406	—	—	—	Anticline
169	Tonkawa	1921	Subsurf. geol.	Paleoz.	Struc.	4,300	Oil	129,331	4,000	439	—	—	—	Anticline
	ROCKY MOUNTAINS, CENTRAL (COLORADO, WEST NEBRASKA, NEVADA, UTAH) (Fig. 17)													
	Colorado													
170	Rangely[9]	1902	Surf. geol.	Meso., Paleoz.	Struc.	25,000	Oil	329,930	207,070	16,182	—	—	—	Anticline
	Utah													
171	Aneth	1956	Surf. Geol.*	Paleoz.	Comb.	18,000	Oil	193,454	258,546	10,280	—	—	—	Nosing, porosity, perm. chges.
172	Red Wash	1951	Geophys.	Ceno.	Comb.	26,000	Oil	43,104	91,896	6,714	—	—	—	Nosing, porosity, perm. chges.
	ROCKY MOUNTAINS, NORTHERN (IDAHO, MONTANA, NORTH AND SOUTH DAKOTA, WYOMING) (Fig. 18)													
	Montana													
173	Cut Bank	1929	Random drlg.	Paleoz.	Strat.	51,500	Oil	106,050	93,335	4,080	—	—	—	Monoclinal
	North Dakota													
174	Beaver Lodge	1951	Geophys.	Paleoz.	Struc.	16,000	Oil	51,153	51,847	4,718	—	—	—	Anticline
175	Tioga	1952	Geophys.	Paleoz.	Struc.	22,000	Oil	43,394	111,606	2,344	—	—	—	Anticline
	Wyoming													
176	Big Piney	1938	Surf. geol.	Cenoz., Meso.	Strat.	30,000 est.	Gas (Includes Chimney Butte, Hogsback, Tip Top Fields)	—	—	—	376.1	2,423.9	41.1	Monoclinal
177	Elk Basin (also Montana)	1915	Surf. geol.	Paleoz., Meso.	Struc.	5,400	Oil	308,341	91,659	19,889	—	—	—	Anticline
178	Garland	1906	Surf. geol.	Paleoz., Meso.	Struc.	3,300	Oil	86,059	23,941	3,859	—	—	—	Anticline
179	Grass Creek	1914	Surf. geol.	Paleoz., Meso.	Struc.	5,900	Oil	105,854	39,146	5,164	—	—	—	Anticline
180	Hamilton Dome	1918	Surface geol.	Paleoz., Meso.	Struc.	2,400	Oil	123,224	76,776	6,996	—	—	—	Anticline
181	Lance Creek	1918	Surf. geol.	Paleoz., Meso.	Struc.	4,800	Oil	103,622	16,378	521	—	—	—	Anticline
182	Lost Soldier	1916	Surf. geol.	Paleoz., Meso.	Struc.	1,000	Oil	102,573	47,427	4,182	—	—	—	Anticline
183	Oregon Basin (N. and S.)	1912	Surface geol.	Paleoz., Meso.	Struc.	8,500	Oil	139,930	47,070	8,712	—	—	—	Anticline
184	Salt Creek	1906	Surf. geol.	Paleoz., Meso.	Struc.	20,500	Oil	441,585	68,415	12,916	—	—	—	Anticline
185	Wertz	1920	Surf. geol.	Paleoz., Meso.	Struc.	1,000	Oil	54,595	45,405	2,182	—	—	—	Anticline
	*Texas*													
	District 1 (Southwest Texas) (Fig. 19)													
186	Darst Creek	1929	Surf. geol.	Meso.	Struc.	5,600	Oil	124,216	20,784	2,382	—	—	—	Nosing against fault
187	Luling-Branyon	1922	Surf. geol.	Meso.	Struc.	11,300	Oil	130,599	33,227	3,192	—	—	—	Nosing against fault

* This is the last giant field found by surface methods in the United States.

No.	Field	Year	Method	Era	Trap class	Depth	Fluid	Cum.						Trap type
211	Goose Creek	1908	Surf. evidence, Random drlg.	Ceno.	Struc.	3,700	Oil	121,636	38,364	2,090	—	—	—	Domal
212	Hastings	1934	Geophys.	Ceno.	Struc.	7,200	Oil	365,074	134,926	9,918	—	—	—	Anticline (domal)
213	High Island	1922	Surf. evidence, Random drlg.	Ceno.	Struc.	3,700	Oil	109,460	65,540	6,352	—	—	—	Piercement salt dome
214	Hull-Merchant	1918	Surf. evidence, Random drlg.	Ceno.	Struc.	4,800	Oil	173,824	66,388	3,394	—	—	—	Piercement salt dome
215	Humble	1905	Surf. evidence, Random drlg.	Ceno.	Struc.	7,500	Oil	152,125	19,875	1,289	—	—	—	Piercement salt dome
216	Katy and North	1934	Geophysics	Ceno.	Struc.	23,700	Gas	21,507	60,000	9,251	1,400	—	—	Anticline
217	Liberty South	1925	Surf. evidence, Random drlg.	Ceno.	Struc.	5,100	Oil	66,085	76,915	1,983	—	4,600	183.3	Piercement salt dome
218	Magnet-Withers (all flds.)	1936	Subsurf. geol., Geophys.	Ceno.	Struc.	11,800	Oil	52,042	81,547	2,686	—	—	—	Anticline
219	Old Ocean	1934	Subsurf. geol., Geophys.	Ceno.	Struc.	16,600	Oil, Gas	119,681	80,319	2,944	1,500	3,500	110.9	Anticline
220	Pierce Junction	1921	Surf. evidence	Ceno.	Struc.	3,800	Oil	94,163	45,837	1,259	—	—	—	Piercement salt dom
221	Raccoon Bend	1928	Surf. evidence	Ceno.	Struc.	5,100	Oil	75,056	24,944	1,660	—	—	—	Salt dome
222	Sheridan	1940	Geophys.	Ceno.	Struc.	6,100	Gas	49,268	30,732	4,951	700	300	81.6	Anticline
223	Sour Lake	1902	Surf. evidence, Random drlg.	Ceno.	Struc.	5,000	Oil	110,597	34,412	1,169	—	—	—	Piercement salt dome
224	Spindletop	1901	Surf. evidence	Ceno.	Struc.	1,600	Oil	148,091	31,809	1,992	—	—	—	Piercement salt dome
225	Thompson	1931	Subsurf. geol., Geophys.	Ceno.	Struc.	7,400	Oil	274,673	70,327	6,445	—	—	—	Salt dome
226	Tomball	1933	Geophys.	Ceno.	Struc.	19,800	Oil	76,246	27,593	1,920	—	—	—	Domal
227	Webster	1937	Subsurf. geol., Geophys.	Ceno.	Struc.	4,200	Oil	290,223	159,777	6,830	—	—	7.8	Salt dome
228	West Columbia	1915	Surf. evidence, Random drlg.	Ceno.	Struc.	1,900	Oil	147,702	42,298	1,598	(Minor oil prod. in 1904)			Piercement salt dome
	Districts 5 and 6 (East Texas) (Fig. 22)													
229	Bethany-Waskom	1916	Surf. geol.	Meso.	Struc.	62,000	Gas	—	—	—	1,800	700	32.2	Anticline
230	Carthage	1936	Subsurf. geol., Geophys.	Meso.	Comb.	250,000 est.	Gas	—	—	—	5,100	900	179.0	Perm. change on nose, anticlinal
231	East Texas	1930	Random drlg.	Meso.	Strat.	140,000	Oil	3,711,749	1,398,251	45,988	—	—	—	Ss. truncation on west flank Sabine uplift
232	Fairway	1960	Subsurf. geol., Geophys.	Meso.	Comb.	23,500	Oil	25,176	175,967	5,205	—	—	—	Por. & perm. chge. on nose
233	Hawkins	1940	Subsurf. geol., Geophys.	Meso.	Struc.	10,200	Oil	322,708	203,243	10,788	—	—	—	Anticline
234	Joaquin-Logansport[12]	1936	Subsurf. geol.	Meso.	Comb.	30,000	Gas	—	—	—	1,200	100	23.6	Anticlinal & monoclinal
235	Long Lake	1933	Geophys.	Meso.	Struc.	13,900	Gas	32,561 est.	—	450 est.	800	600	1.6	Anticlinal
236	Mexia	1920	Surf. geol.	Meso.	Struc.	3,800	Oil	106,272	3,728	159	—	—	—	Nosing against fault
237	Neches	1953	Subsurf. geol., Geophys.	Meso.	Struc.	9,300	Oil	30,212	154,788	2,078	—	—	—	Anticlinal
238	New Hope	1943	Geophys.	Meso.	Struc.	5,800	Oil	44,649	55,351	2,701	—	—	—	Anticline
239	Powell	1923	Surf., Subsurf. geol.	Meso.	Struc.	2,600	Oil	128,872	6,108	234	—	—	—	Nosing against fault
240	Quitman	1942	Subsurf. geol., Geophys.	Meso.	Struc.	3,200	Oil	61,007	50,475	2,206	—	—	—	Anticline, lenticular ss.
241	Talco	1936	Surf., Subsurf. geol.	Meso.	Struc.	9,000	Oil	202,195	47,805	4,680	—	—	—	Nosing against fault
242	Van (and shallow)	1929	Surf. geol., Geophys.	Meso.	Struc.	4,500	Oil	324,724	81,276	5,748	—	—	—	Anticline

TABLE 1 (Continued)

Field Number Fig. 26	Field	Disc. Year	Discovery Methods	Age of Reserv. Rocks	Type Trap	Approx. Proved Acres	Major Prod. Type	Oil (1,000 bbl) (Fig. 23)			Gas (billion cu ft)			Additional Structure or Trap Data
								Cum. Prod. to 1/1/67	Est. Res. 1/1/67	Prod. in 1966	Cum. Prod. to 1/1/66	Est. Res. 1/1/66	Prod. in 1965	
							Districts 7-C, 8, and 8-A (West Texas)							
243	Andector	1946	Subsurf. geol., Geophys.	Paleoz.	Struc.	5,000	Oil	83,404	56,573	3,381	—	—	—	Anticline
244	Big Lake	1923	Surf. geol.	Paleoz.	Struc.	5,600	Oil	120,370	11,630	459	—	—	—	Anticline
245	Block 31	1945	Geophys.	Paleoz.	Struc.	5,200	Oil	95,527	66,945	6,236	—	—	—	Anticline
246	Brown-Bassett	1958	Surf. geol., Geophys.	Paleoz.	Struc.	16,600	Gas	—	—	—	227	1,673	62	Anticline
247	Cogdell	1949	Geophys.	Paleoz.	Strat.	12,000	Oil	103,284	53,579	6,070	—	—	—	Reef & monoclinal
248	Cowden, North	1930	Subsurf. geol.	Paleoz.	Struc.	38,000	Oil	197,925	62,075	7,271	—	—	—	Anticline
249	Cowden, South	1932	Subsurf. geol.	Paleoz.	Struc.	21,600	Oil	(Included in Foster)			—	—	—	Anticline
250	Coyanosa	1962	Geophys.	Paleoz.	Struc.	5,700	Gas	—	—	—	40.1	3,100	34.3	Anticline
251	Diamond-M	1948	Geophys.	Paleoz.	Strat.	12,500	Oil	136,296	358,704	6,463	—	—	—	Reef & monoclinal
252	Dollarhide	1945	Subsurf. geol., Geophys.	Paleoz.	Struc.	7,700	Oil	89,232	74,768	5,067	—	—	—	Anticline
253	Dora Roberts	1954	Geophys.	Paleoz.	Struc.	3,100	Oil	35,850	65,157	5,443	—	—	—	Anticline
254	Dune	1938	Surf., Subsurf. geol.	Paleoz.	Comb.	34,800	Oil	56,141	95,324	5,786	—	—	—	Anticlinal & monoclinal
255	Emma and Triple N	1937	Subsurf. geol.	Paleoz.	Struc.	7,600	Oil	73,917	46,083	4,402	—	—	—	Anticline
256	Emperor	1935	Subsurf. geol.	Paleoz.	Struc.	10,000	Oil, Gas	34,418	65,582	1,768	475	525	82	Anticline
257	Foster and Johnson	1935	Subsurf. geol.	Paleoz.	Struc.	38,800	Oil	225,291[13]	94,709[13]	8,380[13]	—	—	—	Anticline
258	Fuhrman-Mascho	1929	Subsurf. geol.	Paleoz.	Struc.	35,500	Oil	60,900	39,100	2,600	—	—	—	Anticline
259	Fullerton	1941	Subsurf. geol.	Paleoz.	Struc.	31,600	Oil	186,447	88,553	6,928	—	—	—	Anticline
260	Goldsmith	1935	Subsurf. geol.	Paleoz.	Struc.	66,100	Oil	386,202	84,798	13,608	—	—	—	Anticline
261	Gomez	1963	Geophys.	Paleoz.	Struc.	20,000 est.	Gas	—	—	—	0.6	4,000	0.4	Anticline
262	Grey Ranch	1964	Subsurf. geol., Geophys.	Paleoz.	Struc.	2,600 est.	Gas	—	—	—	—	1,500	—	Anticline
263	Hamon	1965	Subsurf. geol., Geophys.	Paleoz.	Struc.		Gas	—	—	—	—	2,000	—	Anticline
264	Headlee and North	1953	Geophys.	Paleoz.	Struc.	12,800	Oil	66,532	135,120	13,484	—	—	—	Anticline
265	Hendrick	1926	Random drlg.	Paleoz.	Struc.	27,000	Oil	245,228	26,772	1,711	—	—	—	Anticline
266	Howard Glasscock	1925	Surf. geol.	Paleoz.	Struc.	67,900	Oil	240,480	42,850	7,696	—	—	—	Anticline
267	Iatan-East Howard	1925	Surf., Subsurf. geol.	Paleoz.	Struc.	33,400	Oil	57,211	42,789	3,660	—	—	—	Anticline
268	J.M.	1965	Geophys.	Paleoz.	Struc.		Gas	—	—	—	—	—	—	Anticline
269	Jameson	1946	Surf. geol., Geophys.	Paleoz.	Strat.	20,500	Oil	60,325	39,675	2,203	—	—	—	Monoclinal
270	Jordan	1937	Subsurf. geol.	Paleoz.	Struc.	12,700	Oil	77,124	22,876	2,721	—	—	—	Anticline
271	Kelly-Snyder	1948	Subsurf. geol., Geophys.	Paleoz.	Strat.	53,000	Oil	372,349	815,951	22,011	—	—	—	Reef & monoclinal
272	Kermit	1928	Random drlg.	Paleoz.	Struc.	61,600	Oil	95,898	24,102	3,224	—	—	—	Anticline
273	Keystone	1930	Random drlg.	Paleoz.	Struc.	34,300	Oil	215,062	87,938	7,819	—	—	—	Anticline, some perm. trap.
274	Levelland	1945	Geophys.	Paleoz.	Struc.	90,700	Oil	155,084	95,612	7,611	—	—	—	Anticlinal
275	Lockridge	1966	Geophys.	Paleoz.	Struc.	4,500	Gas	—	—	—	—	—	—	Anticline
276	McCamey	1925	Surf. geol.	Paleoz.	Struc.	20,000	Oil	117,259	22,741	1,373	—	—	—	Anticline
277	McElroy	1926	Subsurf. geol.	Paleoz.	Struc.	21,000	Oil	238,639	111,361	8,843	—	—	—	Anticline
278	Means and North	1934	Subsurf. geol., Geophys.	Paleoz.	Struc.	20,100	Oil	87,040	42,960	5,692	—	—	—	Anticline

No.	Field	Year disc.	Method of discovery	Age	Type of trap[1]	Productive area (acres)	Oil or gas							Type of structure
279	Midland Farms (all)	1944	Geophys.	Paleoz.	Struc.	14,900	Oil	125,691	86,128	9,187	—	—	—	Anticline
280	Pegasus	1949	Geophys.	Paleoz.	Struc.	19,200	Oil	84,138	52,862	8,021	—	—	—	Anticline; some perm. trap.
281	Penwell	1926	Subsurf. geol.	Paleoz.	Struc.	17,400	Oil	69,962	32,954	3,146	—	—	—	Anticline
282	Prentice	1950	Subsurf. geol.	Paleoz.	Struc.	9,000	Oil	55,846	44,154	2,595	—	—	—	Anticline
283	Puckett	1952	Surf. geol., Geophys.	Paleoz.	Struc.	37,700	Gas	—	—	—	5,400	—	—	Anticline
284	Rojos Caballos (West)	1960	Geophys.	Paleoz.	Struc.	—	Gas	—	—	—	1,100 / 14.4	165 / 2.6	—	Anticline
285	Russell and North	1943	Subsurf. geol., Geophys.	Paleoz.	Struc.	12,800	Oil	76,870	53,130	5,523	—	—	—	Anticline
286	Salt Creek	1950	Geophys.	Paleoz.	Strat.	4,200	Oil	55,849	44,151	3,918	—	—	—	Reef
287	Sand Hills	1930	Surf., Subsurf. geol.	Paleoz.	Struc.	27,100	Oil	132,555	64,445	8,766	—	—	—	Anticline
288	Seminole and West	1936	Subsurf. geol.	Paleoz.	Struc.	16,400	Oil	137,013	62,987	5,105	—	—	—	Anticline
289	Shafter Lake	1938	Surf., Subsurf. geol.	Paleoz.	Struc.	14,200	Oil	51,641	48,359	2,280	—	—	—	Anticline; some perm. trap.
290	Slaughter	1936	Subsurf. geol.	Paleoz.	Strat.	112,600	Oil	336,965	103,035	13,461	—	—	—	Monoclinal
291	Spraberry trend area	1949	Subsurf. geol., Geophys.	Paleoz.	Strat.	134,100	Oil	224,111	56,033(?)	19,867(?)	—	—	—	Monoclinal, fracturing
292	TXL and North	1944	Subsurf. geol.	Paleoz.	Struc.	19,300	Oil	201,882	83,118	6,466	—	—	—	Anticlinal, some perm. trap.
293	Waddell	1927	Surf., Subsurf. geol.	Paleoz.	Struc.	18,900	Oil	47,427	62,255	3,965	—	—	—	Anticlinal, some perm. trap.
294	Waha, West (Ellenburger)	1965	Geophys.	Paleoz.	Struc.	—	Gas	—	—	—	1,400 est.	—	—	Anticline
295	Ward, South	1929	Subsurf. geol.	Paleoz.	Strat.	41,700	Oil	91,841	33,099	1,818	—	—	—	Monoclinal, some closure
296	Ward-Estes, North	1929	Geophys.	Paleoz.	Strat.	85,800	Oil	224,704	82,296	16,917	—	—	—	Monoclinal
297	Wasson	1936	Subsurf. geol.	Paleoz.	Struc.	73,000	Oil	443,829	206,171	13,170	—	—	—	Anticline
298	Worsham-Bayer	1961	Geophys.	Paleoz.	Struc.	—	Gas	503,414	—	—	1,200 est.	—	—	Anticline
299	Yates	1926	Surf. geol.	Paleoz.	Struc.	25,300	Oil	1,000,000	—	6,273	—	—	—	Anticline
	Districts 9 (North Texas) & 10 (Texas Panhandle) (Fig. 24)													
300	Boonsville	1950	Random drlg.	Paleoz.	Strat.	400,000	Gas	—	—	—	2,100	900	74.2	Monoclinal with lensing
301	Burkburnett	1912	Surf. geol.	Paleoz.	Struc.	11,000 est.	Oil	188,759	—	—	—	—	—	Regional buried ridge
302	Electra	1911	Surf. geol.	Paleoz.	Struc.	12,000 est.	Oil	182,402[14]	—	—	—	—	—	Regional buried ridge
303	Hull-Silk-Sikes	1939	Subsurf. geol.	Paleoz.	Comb.	7,500	Oil	69,785	30,215	4,055	—	—	—	Anticlinal nose, ss. lensing
304	K M A	1931	Surf., Subsurf. geol.	Paleoz.	Struc.	30,400	Oil	148,099	46,913	3,490	—	—	—	Anticlinal, por. variances
305	Panhandle[15]	1910[6]	Surf. geol.	Paleoz.	Struc.	1,725,000 est.	Gas, Oil	1,158,314	488,670	22,976	9,693.2	604.2	(See Hugoton, Kansas)	Highs on regional ridge
306	Walnut Bend	1938	Subsurf. geol.	Paleoz.	Struc.	4,000	Oil	68,914	31,086	3,129	—	—	—	Anticlinal

[1] Types are structural, stratigraphic, or combination. Stratigraphic includes paleogeomorphic traps.
[2] Productive area includes estimated 2,200,000 acres gas, 300,000 acres oil.
[3] Includes approximately 2,500,000 acres in Kansas; 925,000 acres in Oklahoma; and 575,000 acres in Texas Panhandle.
[4] Estimated gas reserves for Hugoton and Texas Panhandle fields combined.
[5] Since 1951.
[6] Field consists of approximately 67,700 gas acres and 24,600 oil acres.
[7] Includes Eunice field gas.
[8] Sooner Trend field consists of several producing areas. Field formed in 1965.
[9] Shallow Mesozoic oil found in 1902; deeper major Paleozoic reserves discovered in 1932.
[10] Includes Stratton field oil.
[11] Includes Greta field gas.
[12] Also North Louisiana.
[13] Includes South Cowden field oil.
[14] Through 1958.
[15] Includes Borger-Pantex and West Pampa oil fields.
[16] Osborne area discovery; Wheeler County major discovery in 1918.

TABLE 2. GIANT OIL AND GAS FIELDS IN UNITED STATES (BY YEAR OF DISCOVERY)

Disc. Year	Field	State and County or Parish	Region	Ref. Chart, Fig. No.
1871	Bradford	Pennsylvania (McKean); New York (Cattaraugus)	Appalachian area	5
1879	Allegany	New York (Allegany)	Appalachian area	5
1884	Brea-Olinda	California (Los Angeles, Orange)	Los Angeles basin	7
	Lima	Ohio (many counties)	Indiana-Ohio	9
1886	Trenton	Indiana (many counties)	Indiana-Ohio	9
1887	McKittrick (Main Area)	California (Kern)	San Joaquin Valley	7
1890	Coalinga Eastside	California (Fresno)	San Joaquin Valley	7
1894	Midway-Sunset	California (Kern)	San Joaquin Valley	7
1899	Kern River	California (Kern)	San Joaquin Valley	7
1900	Coalinga Westside	California (Fresno)	San Joaquin Valley	7
1901	Jennings	Louisiana (Acadia)	South Louisiana, onshore, SW	13
	Spindletop	Texas (Jefferson)	Upper Gulf Coast, Dist. 3	21
1902	Orcutt	California (Santa Barbara)	Coastal area	7
	Rangely	Colorado (Rio Blanco)	Rocky Mountains, central	17
	Sour Lake	Texas (Hardin)	Gulf Coast, Dist. 3	21
1904	Avant	Oklahoma (Osage)	Osage & NE area	16
1905	Caddo-Pine Island	Louisiana (Caddo)	North Louisiana	12
	Glenn Pool	Oklahoma (Creek, Tulsa)	Creek County area	16
	Humble	Texas (Harris)	Upper Gulf Coast, Dist. 3	21
	Old Illinois[1]	Illinois (Crawford, Lawrence, Jasper)	Illinois	8
1906	Garland	Wyoming (Big Horn, Park)	Rocky Mountains, northern	18
	Salt Creek[‡]	Wyoming (Natrona)	Rocky Mountains, northern	18
1908	Cat Canyon West	California (Santa Barbara)	Coastal area	7
	Goose Creek	Texas (Harris)	Upper Gulf Coast, Dist. 3	21
1909	Buena Vista	California (Kern)	San Joaquin Valley	7
	Coyote, West	California (Orange)	Los Angeles basin	7
1910	Lost Hills	California (Kern)	San Joaquin Valley	7
	Panhandle*	Texas (Carson, Collingsworth, Gray, Hartley, Hutchinson, Moore, Potter, Wheeler)	Texas Panhandle, Dist. 10	24
	Vinton	Louisiana (Calcasieu)	South Louisiana, onshore, SW	13
1911	Belridge, South	California (Kern)	San Joaquin Valley	7
	Coyote, East	California (Orange)	Los Angeles basin	7
	Electra	Texas (Wichita, Wilbarger)	North Texas, Dist. 9	24
1912	Burkburnett	Texas (Wichita)	North Texas, Dist. 9	24
	Cushing	Oklahoma (Creek, Payne)	Creek Co. area	16
	Oregon Basin (& South)	Wyoming (Park)	Rocky Mountains, northern	18
1913	Healdton	Oklahoma (Carter, Jefferson)	South-central area	16
1914	Grass Creek	Wyoming (Hot Springs)	Rocky Mountains, northern	18
	Sho-Vel-Tum	Oklahoma (Carter, Garvin, Stephens)	South-central area	16
1915	El Dorado	Kansas (Butler)	Eastern district	10
	Elk Basin	Wyoming (Park) & Montana (Carbon)	Rocky Mountains, northern	18
	Kern Front	California (Kern)	San Joaquin Valley	7
	South Mountain	California (Ventura)	Coastal area	7
	West Columbia[2]	Texas (Brazoria)	Upper Gulf Coast, Dist. 3	21
1916	Barbers Hill	Texas (Chambers)	Upper Gulf Coast, Dist. 3	21
	Bethany-Waskom*	Texas (Harrison, Panola); Louisiana (Caddo)	East Texas (Dist. 6) also North Louisiana	12
	Cymric	California (Kern)	San Joaquin Valley	7
	Lost Soldier	Wyoming (Sweetwater)	Rocky Mountains, northern	18
	Monroe*	Louisiana (Morehouse, Ouachita, Union)	North Louisiana	12
	Ventura	California (Ventura)	Coastal area	7
1917	Big Sinking	Kentucky (Lee, Powell, Estill, Wolfe)	Kentucky, east	11

[1] Includes Main Consolidated and Lawrence fields.
[2] Minor oil production found in 1904.
* Gas fields.
‡ Given erroneously as 1908 by Beck (1929).

TABLE 2. (*Continued*)

Disc. Year	Field	State and County or Parish	Region	Ref. Chart, Fig. No.
	Cement	Oklahoma (Caddo, Grady)	Southwest area	16
	Montebello	California (Los Angeles)	Los Angeles basin	7
	Rufugio	Texas (Refugio)	Lower Gulf Coast, Dist. 2	20
1918	Big Sandy*	Kentucky (Floyd, Knott, Martin, Pike, Johnson, Magoffin)	Kentucky, east	11
	Hamilton Dome	Wyoming (Hot Springs)	Rocky Mountain, northern	18
	Hull-Merchant	Texas (Liberty)	Upper Gulf Coast, Dist. 3	21
	Lance Creek	Wyoming (Niobrara)	Rocky Mountains, northern	18
1919	Elk Hills	California (Kern)	San Joaquin Valley	7
	Hewitt	Oklahoma (Carter)	South-central area	16
	Homer	Louisiana (Claiborne)	North Louisiana	12
	Richfield	California (Orange)	Los Angeles basin	7
	Sante Fe Springs	California (Los Angeles)	Los Angeles basin	7
1920	Burbank	Oklahoma (Osage)	Osage & NE area	16
	El Dorado, South & East	Arkansas (Union)	South Arkansas	6
	Huntington Beach	California (Orange)	Los Angeles basin	7
	Mexia	Texas (Limestone)	East Texas, Dist. 5	22
	Wertz	Wyoming (Carbon, Sweetwater)	Rocky Mountains, northern	18
1921	Eola-Robberson	Oklahoma (Garvin)	South-central area	16
	Haynesville	Louisiana (Claiborne, Webster)	North Louisiana	12
	Long Beach	California (Los Angeles)	Los Angeles basin	7
	Pierce Junction	Texas (Harris)	Upper Gulf Coast, Dist. 3	21
	Tonkawa	Oklahoma (Kay, Noble)	Northern area	16
1922	Cotton Valley	Louisiana (Webster)	North Louisiana	12
	High Island	Texas (Galveston)	Upper Gulf Coast, Dist. 3	21
	Luling-Branyon	Texas (Guadalupe)	S.W. Texas, Dist. 1	19
	Sligo*	Louisiana (Bossier)	North Louisiana	12
	Smackover	Arkansas (Union, Ouachita)	South Arkansas	6
	Torrance	California (Los Angeles)	Los Angeles basin	7
1923	Big Lake	Texas (Reagan)	West Texas, Dist. 7-C	23
	Dominguez	California (Los Angeles)	Los Angeles basin	7
	Powell	Texas (Navarro)	East Texas, Dist. 5	22
	Saxet	Texas (Nueces)	Lower Gulf Coast, Dist. 4	20
1924	Inglewood	California (Los Angeles)	Los Angeles basin	7
1925	Howard-Glasscock	Texas (Mitchell, Sterling, Howard, Glasscock)	West Texas, Dist. 8	23
	Iatan-East Howard	Texas (Howard, Mitchell)	West Texas, Dist. 8	23
	Liberty, South	Texas (Liberty)	Upper Gulf Coast, Dist. 3	21
	McCamey	Texas (Upton, Crane)	West Texas, Dist. 7-C	23
	St. Louis	Oklahoma (Pottawatomie, Seminole)	Seminole area	16
1926	Gorham	Kansas (Russell)	North Central district	10
	Hendrick	Texas (Winkler)	West Texas, Dist. 8	23
	Hugoton*	Kansas (Finney, Haskell, Hamilton, Grant, Kearney, Stanton, Morton, Stevens, Seward)	Western district	10
	Maljamar	New Mexico (Lea-Eddy)	New Mexico, eastern	15
	McElroy	Texas (Crane, Upton)	West Texas, Dist. 8	23
	Mt. Poso	California (Kern)	San Joaquin Valley	7
	Penwell	Texas (Ector)	West Texas, Dist. 8	23
	Seal Beach	California (Los Angeles)	Los Angeles basin	7
	Seminole City	Oklahoma (Seminole)	Seminole area	16
	Yates	Texas (Pecos)	West Texas, Dist. 8	23
1927	Allen	Oklahoma (Pontotoc)	Southeast area	16
	Blanco Mesaverde*	New Mexico (Rio Arriba, San Juan)	New Mexico, western	15
	Bowlegs	Oklahoma (Seminole)	Seminole area	16
	Hackberry, East	Louisiana (Cameron)	South Louisiana, onshore, SW	13
	Rincon	California (Ventura)	Coastal area	7
	Waddell	Texas (Crane)	West Texas, Dist. 8	23

TABLE 2. (*Continued*)

Disc. Year	Field	State and County or Parish	Region	Ref. Chart, Fig. No.
1928	Agua Dulce	Texas (Nueces)	Lower Gulf Coast, Dist. 4	20
	Bemis Shutts	Kansas (Ellis, Rooks)	Western district	10
	Elwood	California (Santa Barbara)	Coastal area	7
	Fruitvale	California (Kern)	San Joaquin Valley	7
	Government Wells, North	Texas (Duval)	Lower Gulf Coast, Dist. 4	20
	Hackberry, West	Louisiana (Cameron)	South Louisiana, onshore, SW	13
	Hobbs	New Mexico (Lea)	New Mexico, eastern	15
	Kermit	Texas (Winkler)	West Texas, Dist. 8	23
	Kettleman, North Dome	California (Kings, Fresno)	San Joaquin Valley	7
	Oklahoma City	Oklahoma (Oklahoma, Cleveland)	Oklahoma City area	16
	Raccoon Bend	Texas (Austin)	Upper Gulf Coast, Dist. 3	21
1929	Bay St. Elaine	Louisiana (Terrebonne)	South Louisiana, onshore, SE	13
	Cut Bank	Montana (Glacier, Toole)	Rocky Mountains, northern	18
	Darst Creek	Texas (Guadalupe)	Southwest Texas, Dist. 1	19
	Eunice	New Mexico (Lea)	New Mexico, eastern	15
	Fuhrman-Masco	Texas (Andrews)	West Texas, Dist. 8	23
	Jalmat	New Mexico (Lea)	New Mexico, eastern	15
	Lake Barre	Louisiana (Terrebonne)	South Louisiana, onshore, SE	13
	Lake Pelto	Louisiana (Terrebonne)	South Louisiana, onshore, SE	13
	Monument	New Mexico (Lea)	New Mexico, eastern	15
	Red Oak-Norris*	Oklahoma (Latimer, LeFlore)	Southeast area	16
	Trapp	Kansas (Russell, Barton)	North Central district	10
	Vacuum	New Mexico (Lea)	New Mexico, eastern	15
	Van	Texas (Van Zandt)	East Texas, Dist. 5	22
	Ward-Estes	Texas (Ward, Winkler)	West Texas, Dist. 8	23
	Ward, South	Texas (Ward)	West Texas, Dist. 8	23
1930	Caillou Island	Louisiana (Terrebonne)	South Louisiana, onshore, SE	13
	Chase-Silica	Kansas (Rice, Barton, Stafford)	North-central district	10
	Cowden, North	Texas (Ector)	West Texas, Dist. 8	23
	Earlsboro	Oklahoma (Seminole, Pottawatomie)	Seminole area	16
	East Texas	Texas (Gregg, Upshur, Cherokee, Smith, Rusk)	East Texas, Dist. 6	22
	Keystone	Texas (Winkler)	West Texas, Dist. 8	23
	Rodessa	Louisiana (Caddo); Texas (Cass, Marion)	North Louisiana; East Texas, Dist. 6	12
	Sand Hills	Texas (Crane)	West Texas, Dist. 8	23
1931	Conroe	Texas (Montgomery)	Upper Gulf Coast, Dist. 3	21
	Hall-Gurney	Kansas (Russell, Barton)	North-central district	10
	Iowa	Louisiana (Calcasieu, Jefferson Davis)	South Louisiana, onshore, SW	13
	K.M.A.	Texas (Wichita, Archer, Wilbarger)	North Texas, Dist. 9	24
	Lake Washington	Louisiana (Plaquemines)	South Louisiana, onshore, SE	13
	Leeville	Louisiana (Lafourche)	South Louisiana, onshore, SE	13
	Thompson	Texas (Ft. Bend)	Upper Gulf Coast, Dist. 3	21
1932	Cowden, South	Texas (Ector)	West Texas, Dist. 8	23
	Fitts	Oklahoma (Pontotoc)	Southeast area	16
1933	Greta	Texas (Refugio)	Lower Gulf Coast, Dist. 2	20
	Long Lake*	Texas (Anderson, Freestone, Leon)	East Texas, Dist. 6	22
	Tomball	Texas (Harris)	Upper Gulf Coast, Dist. 3	21
1934	Edison	California (Kern)	San Joaquin Valley	7
	Hastings	Texas (Brazoria)	Upper Gulf Coast, Dist. 3	21
	Katy & North*	Texas (Waller, Harris)	Upper Gulf Coast, Dist. 3	21
	Means & North	Texas (Andrews, Gaines)	West Texas, Dist. 8	23
	Old Ocean	Texas (Brazoria, Matagorda)	Upper Gulf Coast, Dist. 3	21
	Santa Maria Valley	California (Santa Barbara)	Coastal California	7
	Tom O'Connor	Texas (Refugio)	Lower Gulf Coast, Dist. 2	20
1935	Anahuac	Texas (Chambers)	Upper Gulf Coast, Dist. 3	21
	Emperor	Texas (Winkler)	West Texas, Dist. 8	23
	Foster (and Johnson)	Texas (Ector)	West Texas, Dist. 8	23

TABLE 2. (Continued)

Disc. Year	Field	State and County or Parish	Region	Ref. Chart, Fig. No.
	Garden Island Bay	Louisiana (Plaquemines)	South Louisiana, onshore, SE	13
	Goldsmith	Texas (Ector)	West Texas, Dist. 8	23
	Lafitte	Louisiana (Jefferson)	South Louisiana, onshore, SE	13
	Plymouth	Texas (San Patricio)	Lower Gulf Coast, Dist. 4	20
	Wilmington	California (Los Angeles)	Los Angeles basin	7
1936	Carthage*	Texas (Panola, Harrison)	East Texas, Dist. 6	22
	Greeley	California (Kern)	San Joaquin Valley	7
	Joaquin-Logansport*	Texas (Shelby, Panola); Louisiana (DeSoto)	East Texas, Dist. 6; North Louisiana	22
	Magnet-Withers	Texas (Wharton)	Upper Gulf Coast, Dist. 3	21
	Rio Vista-Isleton*	California (Salano, Contra Costa, Sacramento)	Northern California	7
	San Salvador*	Texas (Hidalgo)	Lower Gulf Coast, Dist. 4	20
	Seminole & West	Texas (Gaines)	West Texas, Dist. 8	23
	Slaughter	Texas (Cochran, Hockley, Terry)	West Texas, Dist. 8	23
	Talco	Texas (Franklin, Titus)	East Texas, Dist. 6	22
	Wasson	Texas (Yoakum)	West Texas, Dist. 8	23
1937	Clay City Consolidated	Illinois (Clay, Richland, Wayne, Jasper)	Illinois	8
	Emma (and Triple N)	Texas (Andrews)	West Texas, Dist. 8	23
	Jordan	Texas (Ector, Crane)	West Texas, Dist. 8	23
	Kraft-Prusa	Kansas (Barton, Ellsworth)	North-central district	10
	Lake Arthur*	Louisiana (Jefferson Davis)	South Louisiana, onshore, SW	13
	Little River	Oklahoma (Seminole)	Seminole district	16
	Loudon	Illinois (Effingham, Fayette)	Illinois	8
	Quarantine Bay	Louisiana (Plaquemines)	South Louisiana, onshore, SE	13
	Rio Bravo	California (Kern)	San Joaquin Valley	7
	Schuler & East	Arkansas (Union)	South Arkansas	6
	Seeligson	Texas (Kleberg, Jim Wells)	Lower Gulf Coast, Dist. 4	20
	Stratton	Texas (Nueces, Kleberg, Jim Wells)	Lower Gulf Coast, Dist. 4	20
	Venice	Louisiana (Plaquemines)	South Louisiana, onshore, SE	13
	Webster	Texas (Harris)	Upper Gulf Coast, Dist. 3	21
1938	Big Piney*	Wyoming (Sublette)	Rocky Mountains, northern	18
	Coalinga Nose (East Ext.)	California (Fresno)	San Joaquin Valley	7
	Coles Levee, North	California (Kern)	San Joaquin Valley	7
	Dune	Texas (Crane)	West Texas, Dist. 8	23
	Golden Meadow	Louisiana (Lafourche)	South Louisiana, onshore, SE	13
	Grand Bay	Louisiana (Plaquemines)	South Louisiana, onshore, SE	13
	Kelsey (all fields)	Texas (Brooks, Jim Hogg, Starr)	Lower Gulf Coast, Dist. 4	20
	Magnolia	Arkansas (Columbia)	South Arkansas	6
	Salem Consolidated	Illinois (Marion, Jefferson)	Illinois	8
	Shafter Lake	Texas (Andrews)	West Texas, Dist. 8	23
	Timbalier Bay	Louisiana (Lafourche)	South Louisiana, onshore, SE	13
	Walnut Bend	Texas (Cooke)	North Texas, Dist. 9	24
	West Ranch	Texas (Jackson)	Lower Gulf Coast, Dist. 2	20
	White Point, East	Texas (Nueces, San Patricio)	Lower Gulf Coast, Dist. 4	20
1939	Chocolate Bayou*	Texas (Brazoria)	Upper Gulf Coast, Dist. 3	21
	Hull-Silk-Sikes	Texas (Archer)	North Texas, Dist. 9	24
	La Gloria & South	Texas (Brooks, Jim Wells)	Lower Gulf Coast, Dist. 4	20
	New Harmony Consolidated	Illinois (White, Wabash, Edwards)	Illinois	8
	Paradis	Louisiana (St. Charles)	South Louisiana, onshore, SE	13
	Tijerina-Canales-Blucher	Texas (Jim Wells, Kleberg)	Lower Gulf Coast, Dist. 4	20
	Tinsley	Mississippi (Yazoo)	Gulf states, southern	14
1940	Caprock & East	New Mexico (Chaves, Lea)	New Mexico, eastern	15
	Cote Blanche Bay, West	Louisiana (St. Mary)	South Louisiana, onshore, SW	13
	Dale Consolidated	Illinois (Hamilton, Saline, Franklin)	Illinois	8
	Delta Farms	Louisiana (Lafourche)	South Louisiana, onshore, SE	13
	Erath	Louisiana (Vermilion)	South Louisiana, onshore, SW	13
	Hawkins	Texas (Wood)	East Texas, Dist. 6	22

TABLE 2. (*Continued*)

Disc. Year	Field	State and County or Parish	Region	Ref. Chart, Fig. No.
	Sheridan*	Texas (Colorado)	Upper Gulf Coast, Dist. 3	21
	West Bay	Louisiana (Plaquemines)	South Louisiana, onshore, SE	13
1941	Bastian Bay*	Louisiana (Plaquemines)	South Louisiana, onshore, SE	13
	Bayou Sale	Louisiana (St. Mary)	South Louisiana, onshore, SW	13
	Fullerton	Texas (Andrews)	West Texas, Dist. 8	23
1942	Lake St. John	Louisiana (Concordia, Tensas)	North Louisiana	12
	Quitman	Texas (Wood)	East Texas, Dist. 6	22
1943	Cranfield*	Mississippi (Adams, Franklin)	Gulf States, southern	14
	Edmund, West	Oklahoma (Oklahoma, Logan, Canadian, Kingfisher)	Oklahoma City area	16
	Keyes*	Oklahoma (Cimarron)	Panhandle-NW area	16
	New Hope	Texas (Franklin)	East Texas, Dist. 6	22
	Russell & North	Texas (Gaines)	West Texas, Dist. 8	23
1944	Baxterville	Mississippi (Lamar, Marion)	Gulf States, southern	14
	Delhi	Louisiana (Franklin, Madison, Richland)	North Louisiana	12
	Gwinville	Mississippi (Jefferson Davis, Simpson)	Gulf States, southern	14
	Heidelberg	Mississippi (Jasper)	Gulf States, southern	14
	Midland Farms	Texas (Andrews)	West Texas, Dist. 8	23
	TXL & North	Texas (Ector)	West Texas, Dist. 8	23
1945	Block 31	Texas (Crane)	West Texas, Dist. 8	23
	Borregas	Texas (Kleberg)	Lower Gulf Coast, Dist. 4	20
	Dollarhide	Texas (Andrews); New Mexico (Lea)	West Texas, Dist. 8	23
	Levelland	Texas (Hockley, Cochran)	West Texas, Dist. 8	23
	Sooner Trend	Oklahoma (Garfield, Logan, Kingfisher)	Northern area	16
	Tulsita Wilcox (inc. No. Pettus)*	Texas (Bee, Goliad, Karnes)	Lower Gulf Coast, Dist. 2	20
	Weeks Island	Louisiana (Iberia)	South Louisiana, onshore, SW	13
1946	Andector	Texas (Ector)	West Texas, Dist. 8	23
	Golden Trend	Oklahoma (Garvin, Grady, McClain)	South-central area	16
	Jameson	Texas (Coke, Sterling)	West Texas, Dist. 7-C	23
1947	Elk City	Oklahoma (Beckham)	Southwest area	16
	San Ardo	California (Monterey)	California, central, Salinas Valley area	7
1948	Diamond-M	Texas (Scurry)	West Texas, Dist. 8	23
	Grand Isle, Blk. 16	Louisiana	South Louisiana, offshore	13
	Kelly-Snyder	Texas (Scurry)	West Texas, Dist. 8	23
	Main Pass, Blk. 69	Louisiana	South Louisiana offshore	13
1949	Bay Marchand, Blk. 2	Louisiana	South Louisiana offshore	13
	Cogdell	Texas (Kent, Scurry)	West Texas, Dist. 8	23
	Cuyama, South	California (Santa Barbara)	San Joaquin Valley	7
	Denton	New Mexico (Lea)	New Mexico, eastern	15
	Duck Lake*	Louisiana (St. Martin)	South Louisiana, onshore, SW	13
	Pegasus	Texas (Upton, Midland)	West Texas, Dist. 7-C	23
	Spraberry trend	Texas (Upton, Reagan)	West Texas, Dist. 7-C	23
	Vermilion, Blk. 39*	Louisiana	South Louisiana, offshore	13
	West Delta, Blk. 30	Louisiana	South Louisiana, offshore	13
1950	Boonsville*	Texas (Wise, Jack, Parker)	North Texas, Dist. 9	24
	Eugene Island, Blk. 32*	Louisiana	South Louisiana, offshore	13
	Eugene Island, Blk. 126	Louisiana	South Louisiana, offshore	13
	Prentice	Texas (Yoakum, Terry)	West Texas, Dist. 8	23
	Red Fish Bay*	Texas (Nueces)	Lower Gulf Coast, Dist. 4	20
	Salt Creek	Texas (Kent)	West Texas, Dist. 8	23
	South Pass, Blk. 24	Louisiana	South Louisiana, offshore	13
1951	Beaver Lodge	North Dakota (Williams)	Rocky Mountains, northern	18
	Greenwood*	Kansas (Morton)	Western district	10
	Red Wash	Utah (Uintah)	Rocky Mountains, central	17
1952	Bourg*	Louisiana (Terrebonne)	South Louisiana, onshore, SE	13
	Deep Lake*	Louisiana (Cameron)	South Louisiana, onshore, SW	13
	Main Pass, Blk. 35	Louisiana	South Louisiana, offshore	13

TABLE 2. (*Continued*)

Disc. Year	Field	State and County or Parish	Region	Ref. Chart, Fig. No.
	Mocane-Laverne*	Oklahoma (Beaver, Harper)	Panhandle-northwestern area	16
	Puckett*	Texas (Pecos)	West Texas, Dist. 8	23
	Tioga	North Dakota (Williams, Mountrail, Burke)	Rocky Mountains, northern	18
1953	Headlee & North	Texas (Ector, Midland)	West Texas, Dist. 8	23
	Neches	Texas (Anderson, Cherokee)	East Texas, Dist. 6	22
1954	Camrick*	Oklahoma (Beaver, Texas)	Panhandle, northwestern area	16
	Dora Roberts	Texas (Midland, Ector)	West Texas, Dist. 8	23
	South Pass, Blk. 27	Louisiana	South Louisiana, offshore	13
1955	Citronelle	Alabama (Mobile)	Gulf States, southern	14
	Grand Isle, Blk. 47	Louisiana	South Louisiana, offshore	13
1956	Aneth	Utah (San Juan)	South Utah, Rocky Mountains, central	17
	South Timbalier, Blk. 135	Louisiana	South Louisiana, offshore	13
1957	Empire Abo	New Mexico (Eddy)	New Mexico, eastern	15
	Swanson River	Alaska	Kenai Peninsula	4
1958	Alazan, North	Texas (Kleberg)	Lower Gulf Coast, Dist. 4	20
	Brown-Bassett*	Texas (Terrell)	West Texas, Dist. 7-C	23
1959	Kenai	Alaska	Kenai Peninsula	4
	Putnam*	Oklahoma (Dewey)	Panhandle-northwestern area	16
1960	Fairway	Texas (Anderson, Henderson)	East Texas, Dist. 6	22
	Rojos Caballos*	Texas (Pecos)	West Texas, Dist. 8	23
1961	Worsham-Bayer*	Texas (Reeves)	West Texas, Dist. 8	23
1962	Coyanosa*	Texas (Pecos)	West Texas, Dist. 8	23
	West Delta, Blk. 73	Louisiana	South Louisiana, offshore	13
1963	Gomez*	Texas (Pecos)	West Texas, Dist. 8	23
	Middle Ground Shoal	Alaska	Cook Inlet	4
1964	Black Lake	Louisiana (Natchitoches)	North Louisiana	12
	Cook Inlet*	Alaska	Cook Inlet	4
	Grey Ranch*	Texas (Pecos)	West Texas, Dist. 8	23
1965	Granite Point	Alaska	Cook Inlet	4
	Hamon*	Texas (Reeves)	West Texas, Dist. 8	23
	J. M.*	Texas (Crockett)	West Texas, Dist. 7-C	23
	McArthur River	Alaska	Kenai Peninsula	4
	Waha, West (Ellenburger)*	Texas (Reeves)	West Texas, Dist. 8	23
1966	Lockridge*	Texas (Ward)	West Texas, Dist. 8	23

EXPLORATION FOR FUTURE GIANTS

As seen on Figure 1, giant field discoveries reached peaks in 1929 and in 1937–1938; also, the right-hand side of the discovery curve for the interval 1938–1966 approximately mirrors the left-hand side which extends from 1929 to 1871. Only new successful exploration methods will prevent the right-hand slope from reaching zero in the near future because obvious traps which can be found by present-day exploratory methods have become very scarce and difficult to find with conventional techniques, *and this situation will grow more acute in the future.*

This *scarcity* of obvious traps in *onshore* regions actually is the impelling force which motivates the present-day obsession with deep and deeper offshore exploration, leasing, and drilling. Companies are spending money at an unprecedented rate in a frenzied effort to find obvious traps in deep offshore waters. This expensive offshore exploration allows them to continue the search for petroleum with the same exploratory methods and thinking which paid off in onshore regions in the past. It is evident that companies intend to find big offshore structures regardless of water depth, cost, and *economics*. To accomplish this, prospective areas are shot and reshot, and the geophysical data are worked and reworked; and, when the mind reaches a limit in "squeezing" information from data and methods available, the operation is turned over to the computers.

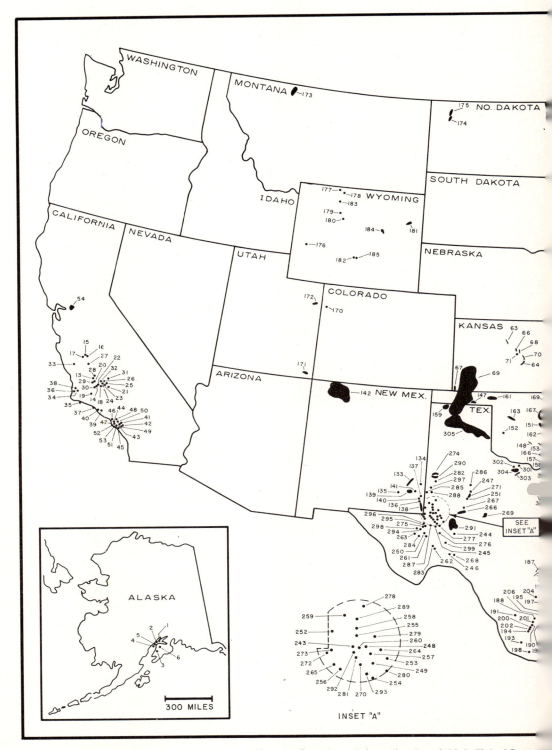

FIG. 26.—Locations of giant oil and gas fields in United States

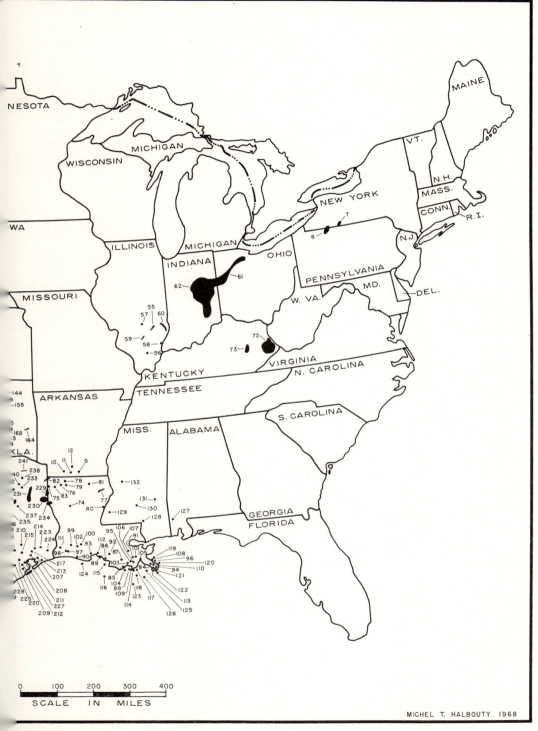

MICHEL T. HALBOUTY 1968

ntity of each field see corresponding number in Column 1, Table 1.

Yet the effort to find new exploratory methods, ideas, and approaches which will lead to the discovery of subtle traps with *reasonable costs* in *onshore regions* is practically ignored. "What was good enough for Dad is good enough for me" (*i.e.,* "don't rock the boat"), seems to be the exploratory motto of most, if not all, major companies.

The writer believes that the new, large, economic petroleum reserves of the future will be found in stratigraphic traps—in features which are not easily found by present-day exploration methods and thinking. *These obscure, subtle, undiscovered traps are in the subsurface.* The known and assumed geology of the country, and the continued, sporadic, usually *accidental* discovery of this type of productive trap through the years are assurances of this fact.

The requirements for finding them are: (1) the focusing of exploratory thinking on the geologic conditions under which such traps and accumulations form; and (2) the purposeful and deliberate drilling of more stratigraphic wildcat tests. The inducement to drill these wildcats must be provided by petroleum explorationists with new ideas which will lead to the mapping of stratigraphic prospects. These explorationists also must have the strength of their convictions which will motivate them or their companies to take a chance by recommending and drilling these prospects.

If geologists are to succeed in finding new reserves for even the near future, both *geologists* and *management* must place a greater emphasis on the search for the stratigraphic trap. Without the cooperation and encouragement of management, the geologist is helpless in his efforts to push new ideas and approaches to exploration. Therefore, management must be educated and reoriented in its exploration thinking and policies so that the geologist will be encouraged to look for the obscure as well as the obvious trap, and the geologist's ideas concerning exploration for the stratigraphic trap will be welcomed instead of discouraged.

The writer is not inferring that finding the obscure trap is easy. It will be difficult to map such prospects and even more difficult to sell the idea of purposely looking for them to many managers. It will take a lot of hard work, imaginative creative thinking, and a sharp eye for small leads to map the prospects, and it will take superb salesmanship to get management to drill them.

The exploratory difference between the structural features for which geologists continue to search and the obscure subtle traps which geologists seem to shun is that the latter are *not obvious* by use of present-day exploratory methods and thinking. Many stratigraphic traps probably can be found by present-day exploratory methods, but only if geologists direct these methods and their thinking *toward* them, not around them. Only by looking *purposely* for such traps will giant fields continue to be found.

Selected References

American Association of Petroleum Geologists, 1929, Structure of typical American oil fields, a symposium, v. 2 : 780 p.

Beck, E., 1929, Salt Creek oil field, Natrona County, Montana, *in* Structure of typical American oil fields, v. 2: Am. Assoc. Petroleum Geologists, p. 589-603.

Braunstein, J., ed., 1965, Oil and gas fields of southeast Louisiana, v. 1: New Orleans Geol. Soc., 195 p.

California Department of Natural Resources, Division of Mines, 1943, Geologic formations and economic development of the oil and gas fields of California: Bull. 118, 773 p.

Clark, J. A., 1963, Chronological history of the petroleum and natural gas industries: Houston, Texas, Clark Book Co., 317 p.

Denham, R. L., ed., 1962, Typical oil and gas fields of southeast Texas: Houston Geol. Soc., 243 p.

Gurley, A. M., II, ed., 1967, Typical oil and gas fields of South Texas: Corpus Christi Geol. Soc., 212 p.

Halbouty, M. T., 1967, Salt domes, Gulf region, United States and Mexico: Houston, Texas, Gulf Publishing Co., 425 p.

Herald, F. A., ed., 1951, Occurrence of oil and gas in northeast Texas: Texas Univ. Pub., no. 5116, 449 p.

———— ed., 1957, Occurrence of oil and gas in West Texas: Texas Univ. Pub., no. 5716, 442 p.

Howell, J. V., ed., 1948, Structure of typical American oil fields, a symposium, v. 3 : Am. Assoc. Petroleum Geologists, 516 p.

Jensen, F. S., H. H. R. Sharkey, and D. S. Stoughton, eds., 1954, The oil and gas fields of Colorado—a symposium: Rocky Mtn. Assoc. Geologists, 302 p.

Kinney, E. E., ed., 1966, The oil and gas fields of southeastern New Mexico, a symposium, 1966 supplement: Roswell Geol. Soc., 185 p.

Levorsen, A. I., ed., 1941, Stratigraphic type oil fields: Am. Assoc. Petroleum Geologists, 902 p.

———— 1966, The obscure and subtle trap: Am. Assoc. Petroleum Geologists Bull., v. 50, no. 10, p. 2058–2067.

McCampbell, J. C., and J. W. Sheller, eds., 1964, Typical oil and gas fields of southwestern Louisiana: Lafayette Geol. Soc., 32 fld. descrip.

McGregor, A. A., and C. A. Biggs, 1968, Bell Creek field, an embryonic giant, Powder River and Carter Counties, Montana (abs.): Am. Assoc. Petroleum Geologists Bull., v. 52, no. 3, p. 541.

McNaughton, D. A., ed., 1953, Guidebook, field trip routes, oil fields, geology, joint annual meeting, AAPG, SEPM, SEG: Houston Geol. Soc., 165 p.

Martin, Rudolf, 1966, Paleogeomorphology and its application to exploration for oil and gas (with examples from western Canada): Am. Assoc. Petroleum Geologists Bull., v. 50, no. 10, p. 2277–2311.

Moore, R. C., ed., 1926, Geology of salt dome oil fields: Am. Assoc. Petroleum Geologists, 797 p.

Nordquist, J. W., and M. C. Johnson, eds., 1958, Montana oil and gas fields: Billings Geol. Soc., 247 p.

Raymond, J. P., Jr., ed., 1963, Salt domes of South Louisiana, v. 1: New Orleans Geol. Soc., 1st revision, 133 p.

Shreveport Geological Society, 1947, Reference report on certain oil and gas fields of North Louisiana, South Arkansas, Mississippi and Alabama: v. 1, 328 p.; 1947, v. 2, p. 317–503; 1951, v. 3, 42 p.; 1958, v. 4, 199 p.; 1963, v. 5, 188 p.

Spillers, J. P., ed., 1962, Salt domes of South Louisiana, v. 2: New Orleans Geol. Soc., 145 p.

Stipp, T. F. et al., eds., 1956, The oil and gas fields of southeastern New Mexico, a symposium: Roswell Geol. Soc., 376 p.

Sweeney, N. N., et al., 1960, The oil and gas fields of southeastern New Mexico, a symposium, 1960 supplement: Roswell Geol. Soc., 229 p.

Ver Wiebe, W. A., 1952, North American petroleum —a complete summary of geological information: Ann Arbor, Michigan, Edwards Brothers, Inc., 459 p.

Wyoming Geological Association Symposium Committee, 1957, Wyoming oil and gas fields symposium: 484 p.

———— 1961, Supplement to Wyoming oil and gas fields symposium: p. 486–579.

Other Sources of Information

AAPG Committee on Statistics of Drilling.

AAPG *Bulletins:* Volumes 1–50.

AIME, TRANSACTIONS, Petroleum Division: Yearbooks, 1934–1946.

International Oil Scouts Association, International Oil and Gas Development: Yearbooks through 1966.

Oil and Gas Journal, Jan. 30, 1967, p. 160–169 (oil-production and estimated oil-reserve figures).

Oil and Gas Journal, correspondence (estimated gas reserves).

Bell Creek Field, Montana: a Rich Stratigraphic Trap[1]

ALEXANDER A. McGREGOR[2] and CHARLES A. BIGGS[3]

Denver, Colorado 80202 and Billings, Montana 59101

Abstract The Bell Creek field, one of the most important oil discoveries in the United States in 1967, is in T8 and 9S, R54, and 55E, in Powder River and Carter Counties, southeastern Montana. Daily production, currently 50,000 bbl of oil, should rise to 65,000 bbl, outstripping that of any other Rocky Mountain oil field. Reserves are estimated at more than 200 million bbl.

The field is a Lower Cretaceous stratigraphic trap uncontrolled by structure, along the gently dipping east flank of the Powder River basin. Located near the intersection of northeast-trending littoral marine bars and a northwest-trending delta system (evident from earlier regional stratigraphic studies), the Muddy Sandstone trap was formed in a complex assemblage of local, shallow-water, nearshore environments during a regressive phase between two major advances of Early Cretaceous seas. Very porous and permeable sandstone bodies pinch out eastward, contain organic matter, and are underlain and overlain by organic-rich marine shale, providing an ideal trap for indigenous oil. The sandstone tongues were tilted upward on the east shortly after deposition as a result of the westward subsidence of the basin in which the overlying Mowry Shale was deposited. Later subsidence of the east limb of the Powder River basin further increased the size and efficiency of the trap. At least four individual reservoirs are delineated in the complex facies assemblage by oil-water and gas-oil contacts.

The Bell Creek field is an excellent example of major oil fields remaining to be found in stratigraphic traps in the sparsely drilled Rocky Mountain region. Drilling, rather than other exploration methods, is the most efficient and conclusive test of evolving economically oriented stratigraphic concepts.

INTRODUCTION

The Bell Creek field, with a daily production rate that will soon exceed that of any other Rocky Mountain oil field, was found on June 6, 1967, at a time when most major oil companies were decreasing both exploration and lease holdings in the region.

[1] Manuscript received, May 28, 1968; accepted, June 8, 1968. Modified from a paper given at the 53rd Annual Meeting of the Association, Oklahoma City, Oklahoma, April 25, 1968. Reprinted from the AAPG *Bulletin,* v. 52, no. 10, p. 1869–1887.

[2] Chief geologist, Samuel Gary, oil producer.

[3] Partner, Sawtooth Oil Company.

The writers are indebted to Steven S. Oriel and Robert R. Berg for constructive suggestions that improved the manuscript.

This paper describes more fully the geologic aspects of Bell Creek field set forth in a preliminary paper (McGregor, 1968), and shows that, despite numerous gaps in available geologic data, the general location of the Bell Creek field could be predicted from earlier regional stratigraphic studies and the application of "imaginative geologic thinking" (Cram, 1966, p. 829). The Bell Creek field may become a classic example of a rich stratigraphic trap.

LOCATION

Bell Creek field is in southeastern Montana on the northeastern flank of the Powder River basin, and northwest of the Black Hills uplift (Fig. 1). The field is mainly in Powder River County but extends into Carter County, as shown in Figure 2, and is in T8 and 9S, R53, 54, and 55E. The town of Broadus, Montana, is 25 mi northwest of the field and Gillette, Wyoming, is 60 mi south of the field. Other oil fields that produce from about the same part of the section along the east flank of the Powder River basin are shown in Figure 2.

REGIONAL STRATIGRAPHY

Producing sandstone beds at Bell Creek are within the Lower Cretaceous sequence, shown in the diagrammatic cross section extending from northwest Montana to central Dakota (Fig. 3). Two widespread marine transgressions and an intervening regressive phase are recorded in the sequence.

Seawater spread eastward very early in Cretaceous time and deposited the sand of the Fall River Sandstone on nonmarine Lakota strata, both of the Inyan Kara Group (Bolyard and McGregor, 1966). Farther seaward, dark clay beds were deposited which are now assigned on the east to the Skull Creek Shale, and on the west to the Thermopolis Shale. Even at the maximum extent of the sea, when the Skull Creek was deposited as far east as eastern South Dakota, sand beds of mixed marine and continental origin,

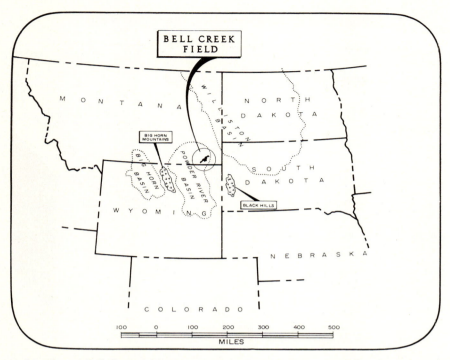

Fig. 1.—Location of Bell Creek field, Montana, Powder River basin, and major structural provinces in western United States.

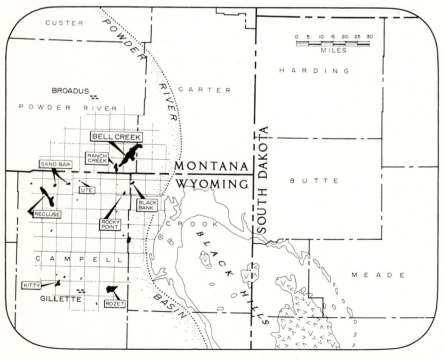

Fig. 2.—Location of Bell Creek field and nearby Muddy and Newcastle oil and gas fields along east side of Powder River basin.

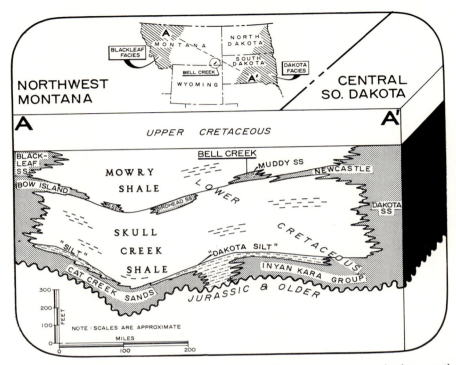

Fig. 3.—Diagrammatic northwest-southeast cross section of Lower Cretaceous rocks from northwest Montana to central South Dakota.

assigned to the Dakota Group, continued to be deposited on the southeast and east margins of the sea.

A regressive phase, critical to the relations at Bell Creek, is recorded in the westward-extending tongue of the Dakota Group (Newcastle Sandstone of the eastern part of the Powder River basin). Discontinuous sandstone bodies in about the same part of the sequence farther west are assigned to the Muddy Sandstone. It is likely that these sandstone bodies are the same age as the Newcastle, but the precise relation has not been established (Cobban and Reeside, 1952). Although the sandstone bodies are discontinuous, and therefore excellent potential traps, they lie within a distinct, literally persistent unit (Eicher, 1960, p. 25; 1962, p. 79). Two other names have been applied to sandstone in about this part of the sequence. The name Birdhead, defined by Thom et al. (1935) as a member of the Thermopolis Shale, has been used by Wulf (1962, Fig. 3, p. 1934) for sandstone beds assigned by others to the Muddy. The name Dynneson was proposed

by Wulf (1962, Fig. 3, p. 1396) for a sandstone member at the base of the Mowry Shale, above a regional disconformity. Although data are inadequate to demonstrate the precise relation of the sandstone beds at Bell Creek to those at the type localities of the Newcastle, Muddy, Birdhead, and Dynneson Sandstones, the name Muddy is used in the rest of this paper.

The second marine transgression is recorded in eastward overlap of the Muddy Sandstone by the Mowry Shale. Farther east in eastern South Dakota the Mowry intertongues with the Dakota.

Various depositional environments have been inferred for the sandstones formed during the regressive phase. Units within the Newcastle Sandstone exposed in the Black Hills have been assigned continental, brackish-water, and marine origins, as summarized by Robinson et al. (1964, p. 45) and Wulf (1962, p. 1392, Fig. 16). Wulf concluded that the Newcastle was deposited in an extensive delta distributary environment. Sandstone beds exposed farther west, as in the Big Horn basin, and assigned to the Muddy, generally are considered to be of shallow-water marine origin (Eicher, 1962, p. 91). The contrast between the environments, therefore, is a clue to the possible presence of shoreline and nearshore shallow-water facies where most organic matter accumulates (Levorsen, 1967, p. 649), and where traps are likely to form during deposition (p. 575).

Even better clues to conditions favorable for oil accumulation are provided by maps prepared in regional stratigraphic studies. The distribution of Muddy sandstone beds and equivalents is shown on a map by Haun and Barlow (1962, p. 21) and is reproduced in Figure 4. Northeast-trending belts of sandstone and shale in northeastern Wyoming and southeastern Montana are conspicuous on this map. The distribution of the Newcastle Sandstone and the Skull Creek Shale is shown on maps prepared by Wulf (1962, Figs. 7, 16) based on data from the Dakotas and eastern

Montana (Fig. 5). Northwest-trending belts of sandstone and shale are dominant in northeasternmost Wyoming and southeastern Montana. If it is assumed that both sets of maps were prepared on the basis of adequate data, the intersection of the two contrasting trends must mark the site of important facies changes produced along the boundary of contrasting environments (Fig. 6). The intersection must mark the site of nearshore environments favorable for the accumulation and entrapment of oil. Moreover, offshore and barrier bars are most common near deltas (Shepard, 1960, p. 220) because the distributaries of the delta system are a source of sand for the bars.

REGIONAL STRUCTURE

The regional structure of the area in which the Bell Creek field lies is relatively simple (Fig. 7). The east flank of the Powder River basin dips rather gently and uniformly west-northwest at

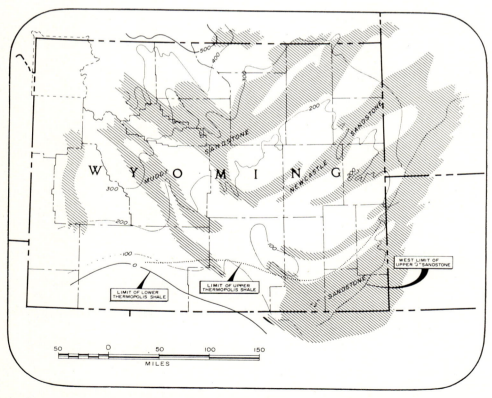

FIG. 4.—Isopach map of Thermopolis Shale showing distribution of Muddy Sandstone and equivalents in Wyoming (from Haun and Barlow, 1962, Fig. 7). CI = 100 ft.

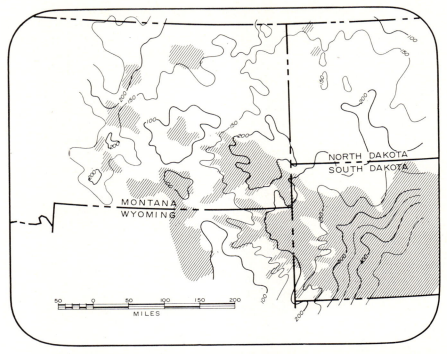

Fig. 5.—Isopach and sandstone-shale ratio map of Skull Creek Shale showing areal distribution of Newcastle Sandstone and equivalents in Dakotas and eastern Montana (from Wulf, 1962, Fig. 7). CI = 50 ft.

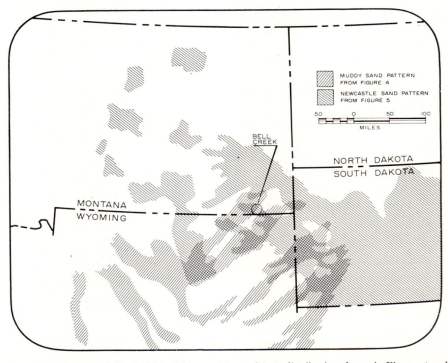

Fig. 6.—Location of Bell Creek field with respect to sandstone distribution shown in Figures 4 and 5.

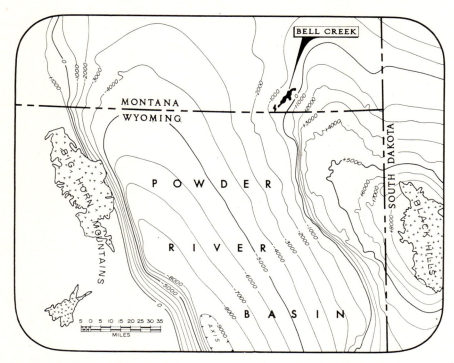

FIG. 7.—Structure contour map of Powder River basin. Datum is top of Fall River Formation. CI = 1,000 ft.

about 100 ft/mi. The structural environment, therefore, is favorable for the entrapment of oil in sandstone bodies that pinch out eastward.

OTHER FAVORABLE FACTORS

Before the discovery of oil at Bell Creek, the presence of commercial quantities of hydrocarbons in the Newcastle and Muddy Sandstones had been demonstrated amply in the small oil fields west of the Black Hills, from Rocky Point to Ranch Creek (Fig. 2).

The exploration program that led to the discovery of Bell Creek uncovered additional favorable factors in southeastern Montana (Gary and McGregor, 1968): (1) locally thick, porous, permeable sandstone beds in the Muddy and Fall River Formations in Powder River County; (2) oil stain and a little free oil observed in thinner and less permeable sandstone beds in some previously drilled holes; and (3) the fact that the Bell Creek area lies along both the structural and stratigraphic strike of the Ranch Creek field. Although four wells completed in the Muddy Sandstone at Ranch Creek averaged about 80 bbl of

oil per day and showed rapid water encroachment and five dry offset tests were drilled in early development, it is very likely that more oil could be found on the northeast.

BELL CREEK OIL FIELD

The Bell Creek field was discovered June 6, 1967 by the Exeter Drilling Co. No. 33-1 Federal-McCarrell test in the NE¼ NE¼ Sec. 33, T8S, R54E (Fig. 8). Samuel Gary and Consolidated Oil and Gas Company supported Exeter in the drilling of the well. A drill-stem test recovered 2,760 ft of 30° API gravity oil from a porous sandstone 27 ft thick in the upper part of the Muddy Sandstone at a depth of about 4,500 ft.

Shallow depths and easy drilling have led to rapid development of the field. By mid-April 1968, 300 productive wells on 40-acre sites had been completed, and the field was producing more than 50,000 bbl of oil per day. If additional facilities are added to the four pipelines serving the field, daily production should be more than 65,000 bbl, which exceeds the production of any other Rocky Mountain oil field. Proved reserves, including sec-

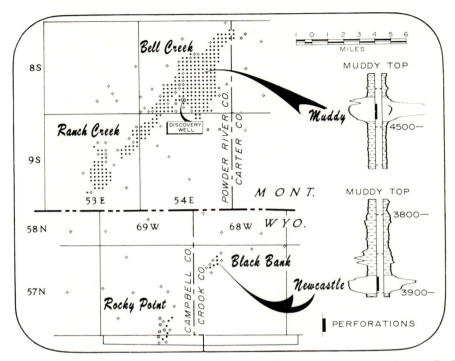

FIG. 8.—Map showing relation of Bell Creek to nearby Ranch Creek, Black Bank, and Rocky Point fields. Electric logs show Bell Creek and Ranch Creek fields produce from upper Muddy Sandstone; Black Bank and Rocky Point fields produce from lower Muddy Sandstone or from Newcastle Sandstone.

ondary recovery, are estimated to be 200 million bbl of oil.

As of April 1968, the Bell Creek field is about 15 mi long and 3.5 mi wide at its widest point (Fig. 8). More than 15,000 acres has been proved productive. Initial oil production rates of wells range from 50 to 1,500 b/d, and the average is 500 b/d of 34° API, sweet, intermediate-base oil. Abnormally low bottom-hole pressures (about 1,190 psi) require artificial lift on all wells.

About 30 wells are being completed each month under current field operations. Active development drilling is concentrated on the east, northeast, and southwest. Only on the northwest, where water is present in the producing sandstone, has the limit of the field been delineated fairly well.

The Bell Creek field has demonstrated amply that it was the most notable discovery made in the United States in 1967. Oil production in Montana has increased by more than 60 percent, contributing greatly to the economy of the state. Moreover, the field is an outstanding example of

giant oil fields remaining to be found in the Rocky Mountain region at reasonable cost by exploration-oriented companies. The minimum return on investment estimated for Bell Creek is 20:1, which makes the field as economically attractive as the larger oil fields of the world.

LOCAL STRATIGRAPHY

Rapid development of the Bell Creek field has resulted in the acquisition of a very large volume of geologic data, only part of which has been sifted and analyzed. Thicknesses of the Muddy Sandstone in the field are shown on the isopachous map in Figure 9. Combined thickness of sandstone units in the Muddy ranges from 20 to 30 ft in most of the field, but locally is more than 40 ft and thickens westward. Thickness trends parallel the structural strike and the sandstone units pinch out southeastward, which suggests that the sandstone beds were parts of northeast-trending barrier bars. Sandstone is also absent along northeast-trending belts within the area, possibly representing lagoonal deposits. Narrow transverse or

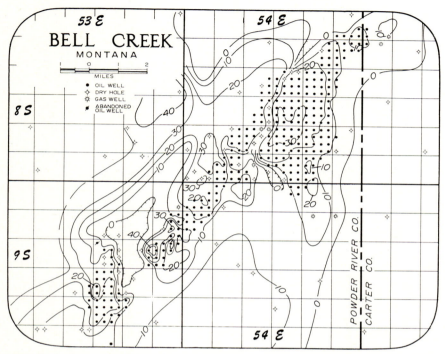

FIG. 9.—Isopach map of gross thickness of Muddy Sandstone in Bell Creek field; based on data available mid-April 1968. CI = 10 ft.

northwest-trending belts of thin sandstone may represent inlets. The southeastward-extending lobes may represent overwash fans into lagoonal areas or they may be part of a delta system.

Properties of the producing sandstone at Bell Creek are well illustrated by the electric log and core analysis shown in Figure 10. The sandstone is 30 ft thick in this well and has porosity greater than 30 percent and a permeability of 10 darcys. Residual oil saturation is about 20 percent, whereas water saturation is markedly varied. However, these oil and water saturation values probably are related more closely to the type and quality of drilling fluid and to the amount of flushing during coring than to properties of the reservoir.

The stratigraphic position of the producing sandstones differs among the nearby fields and within the Bell Creek field itself. The electric logs in Figure 8, for example, show that production in Bell Creek is from sandstone in the upper part of the Muddy Sandstone, whereas at Black Bank it is from the lower Muddy or Newcastle.

Electric logs must be used with caution in stratigraphic and other studies because resistivity is greatly varied among wells within the reservoir or within a single well. Resistivity ranges from 8 ohms to more than 200 ohms within a particular oil reservoir. The resistivity of produced water ranges from 0.25 to 2.2 ohms and is related inversely to the proportion of clay in the sandstone body. Therefore, accurate evaluation of electric logs requires a knowledge of the proportion of clay disseminated through a potential reservoir.

Diagrammatic sections across the Bell Creek field, both from southwest to northeast and from northwest to southeast (Fig. 11), illustrate the lenticular geometry of the main producing sandstone bodies. These bodies may be even more discontinuous than is shown. Drilling indicates at least four different positions for gas caps and two for oil-water contacts. Thus, the field is believed to have at least four separate reservoirs.

Discontinuity of sandstone beds is shown also by the west-to-east electric-log sections in Figures 12 and 13. The sandstone shown in Figure 12 thins eastward and pinches out east of the east-

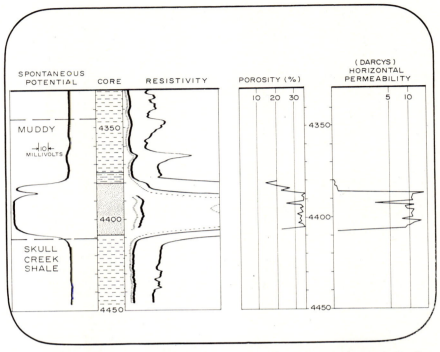

Fig. 10.—Electric-log and core analysis of typical section of productive Muddy Sandstone in Bell Creek field.

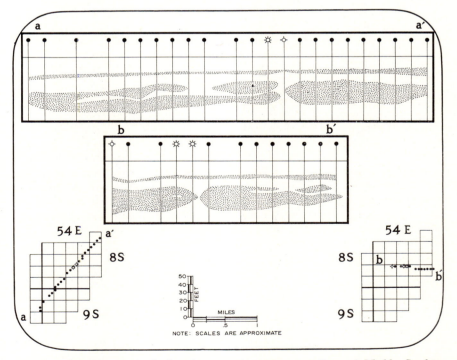

Fig. 11.—Southwest-northeast longitudinal and west-east transverse sections of Muddy Sandstone in Bell Creek field. Illustrates lenticular geometry of main producing sandstone bodies.

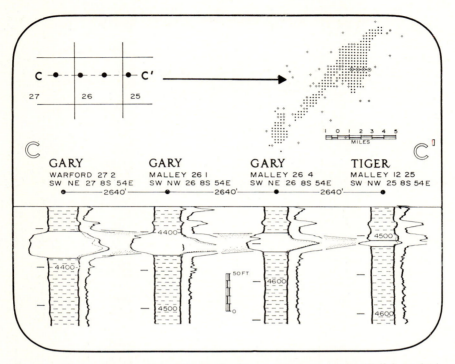

Fig. 12.—West-east cross section of Muddy Sandstone across eastern part of Bell Creek field, showing eastward thinning of reservoir rock.

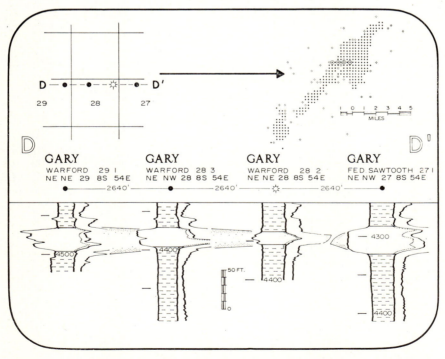

Fig. 13.—West-east cross section of Muddy Sandstone across western part of Bell Creek field showing discontinuity of reservoirs.

Fig. 15.—Southwest-northeast longitudinal cross section of Muddy Sandstone across Ranch Creek and Bell Creek fields.

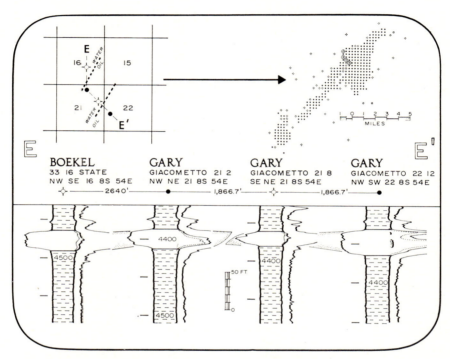

Fig. 14.—North-south cross section of Muddy Sandstone diagonally across two oil-water contacts in western part of Bell Creek field.

ernmost well shown. Figure 13, which is a western continuation of Figure 12, shows two separate reservoirs. The two wells on the west are oil wells with gas-oil ratios of 200:1. The next well east produces gas with a minor amount of condensate, whereas the next well updip on the east is an oil well with a gas-oil ratio of 200:1.

The oil-water contacts along the west side of Bell Creek field are indicated in the electric-log cross section in Figure 14. The northernmost well produced water, but the next well south produced 700 bbl of oil per day. The next well southeast

produced water, whereas the easternmost well also produced 700 bbl of oil per day. These data suggest that continuity of the sandstone bodies is disrupted by pinchouts between the wells in the NW¼ NE¼ and in the SE¼ NE¼ of Sec. 21.

Discontinuity of sandstone bodies is indicated also in the longitudinal section shown in Figure 15, extending from Ranch Creek across Bell Creek.

ENVIRONMENTS OF DEPOSITION

Analysis of data from the Bell Creek field is

Fig. 16.—Diagrammatic map of paleogeography of Bell Creek and Rocky Point areas during deposition of the Muddy Sandstone. Four distinct local environments produced deposits illustrated in Figures 17-20.

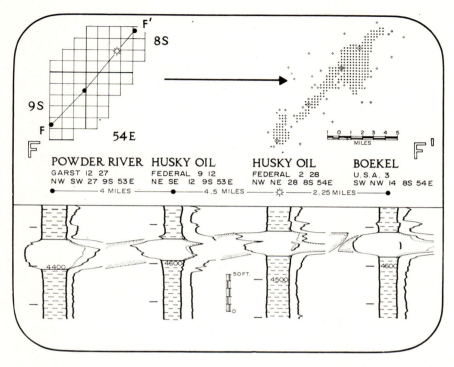

F ── F'

8S

9S

F

54E

POWDER RIVER
GARST 12 27
NW SW 27 9S 53E
├─ 4 MILES ─┤

HUSKY OIL
FEDERAL 9 12
NE SE 12 9S 53E
├─ 4.5 MILES ─┤

HUSKY OIL
FEDERAL 2 28
NW NE 28 8S 54E
├─ 2.25 MILES ─┤

BOEKEL
U.S.A. 3
SW NW 14 8S 54E

4400

4600

50 FT.

4500

4600

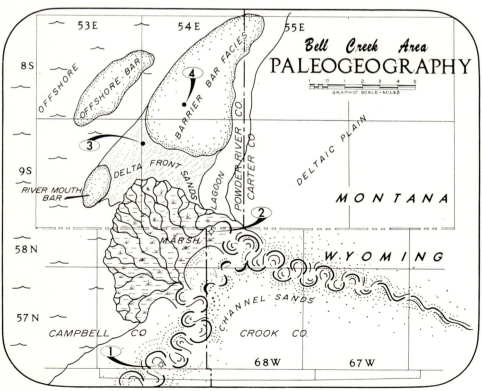

53E   54E   55E

8S

OFFSHORE   OFFSHORE BAR   BARRIER BAR FACIES

Bell Creek Area
PALEOGEOGRAPHY

GRAPHIC SCALE - MILES

④

③

DELTA FRONT SANDS

9S

RIVER MOUTH BAR

LAGOON

POWDER RIVER CO.

CARTER CO.

DELTAIC PLAIN

M O N T A N A

②

58 N

MARSH

W Y O M I N G

57 N

CHANNEL SANDS

①

CAMPBELL   CO

CROOK   CO.

68 W   67 W

FIG. 18.—Electric log and generalized west-east cross section of distributary-channel deposit in delta-marsh facies. Log is well 2, Figure 16.

≫≫→

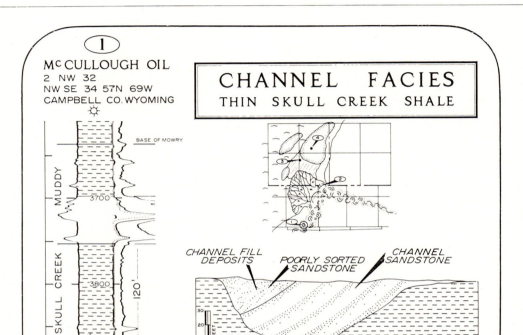

① McCULLOUGH OIL
2 NW 32
NW SE 34 57N 69W
CAMPBELL CO. WYOMING

CHANNEL FACIES
THIN SKULL CREEK SHALE

CHANNEL FILL DEPOSITS   POORLY SORTED SANDSTONE   CHANNEL SANDSTONE

SKULL CREEK SHALE

NOTE: SCALES ARE APPROXIMATE

FIG. 17.—Electric log and generalized west-east cross section of meandering-channel deposit. Log is well 1, Figure 16.

too incomplete to permit firm conclusions regarding the environments of deposition of producing rocks. However, a few tentative ideas may be of interest.

The available data support the general inference drawn from regional stratigraphic studies that the Bell Creek area marks the site of the intersection between northeast-trending offshore bars and a northwest-trending deltaic distributary system. The possible paleogeography for the area during the time of Muddy Sandstone deposition at Bell Creek is represented diagrammat-

ically in Figure 16. At least four distinct local environments are suggested.

Sandstone in the McCullough No. 2 NW 32 gas well in the NW¼ SE¼ Sec. 34, T57N, R69W, Campbell County, Wyoming (Fig. 16, well No. 1), is illustrated in Figure 17 as a possible example of a meandering-channel facies. The sandstone is white with very fine angular to subangular grains, silty and partly clayey, and includes a trace of finely divided plant fragments, some dark shale laminae, scattered coal fragments, and a few tan siderite veins and pellets. Porosity and

≫≫→

FIG. 19.—Electric log and generalized west-east cross section of delta-front deposit. Log is well 3, Figure 16.

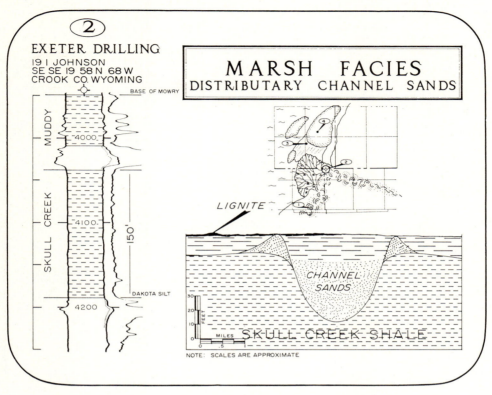

② EXETER DRILLING
19 I JOHNSON
SE SE 19 58 N 68 W
CROOK CO. WYOMING

BASE OF MOWRY

MUDDY

4000

SKULL CREEK

4100

150'

DAKOTA SILT

4200

MARSH FACIES
DISTRIBUTARY CHANNEL SANDS

LIGNITE

CHANNEL SANDS

SKULL CREEK SHALE

FEET
30
20
10
0

MILES
0 .5 1

NOTE: SCALES ARE APPROXIMATE

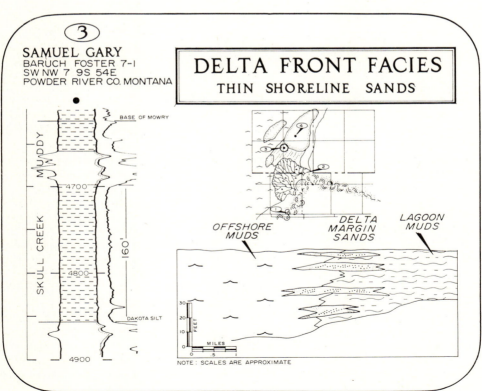

③ SAMUEL GARY
BARUCH FOSTER 7-1
SW NW 7 9S 54E
POWDER RIVER CO. MONTANA

BASE OF MOWRY

MUDDY

4700

SKULL CREEK

160'

4800

DAKOTA SILT

4900

DELTA FRONT FACIES
THIN SHORELINE SANDS

OFFSHORE MUDS

DELTA MARGIN SANDS

LAGOON MUDS

FEET
30
20
10
0

MILES
0 .5 1

NOTE: SCALES ARE APPROXIMATE

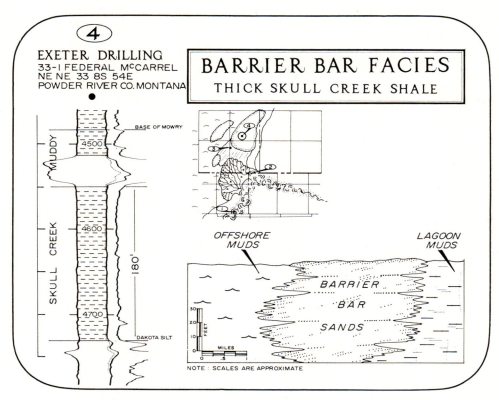

FIG. 20.—Electric log and generalized west-east cross section of barrier-bar deposit.
Log is well 4, Figure 16.

permeability values are poor to fair. The grain size increases downward and the contact with underlying claystone is sharp. The Skull Creek Shale is only 120 ft thick here, 50 ft thinner than in nearby areas. Whether this thinning represents actual removal of shale before filling of a channel by Newcastle Sandstone, as shown in Figure 17, or is associated with facies changes, has not been established. Although the Skull Creek Shale is thinner than in other areas, the total interval between the base of the Mowry and the base of the Skull Creek is remarkably uniform.

A possible marsh facies of the Newcastle Sandstone is illustrated in Figure 18, which shows the sandstone in the Exeter Drilling Company No. 19-1 Johnson dry hole in SE¼ SE¼ Sec. 19, T58N, R68W, Crook County, Wyoming (Fig. 16, well No. 2). The sandstone, possibly deposited along a distributary channel near the delta front, includes interlaminae of siltstone, lignite, and other carbonaceous material.

Sandstone in the Samuel Gary Baruch-Foster No. 7-1 oil well in SW¼ NW¼ Sec. 7, T9S, R54E, Powder River County, Montana (Fig. 16, well No. 3), may be a delta-front deposit (Scruton, 1960) as illustrated on Figure 19. The rock consists of fine, subangular, well-sorted grains, includes variable amounts of clay, and is both indistinctly cross-laminated and ripple marked. The sandstone intertongues abruptly with mudstone and apparently consists of reworked sand detritus from the Newcastle.

Sandstone in the Exeter Drilling Company No. 33-1 Federal-McCarrel discovery well, in NE¼ NE¼ Sec. 33, T8S, R54E, Powder River County, Montana (Fig. 16, well No. 4), is illustrated on Figure 20 as a barrier-island or offshore-bar deposit. The sandstone is 30 ft thick, white, contains fine, very well-sorted, subangular grains of both clear and frosted quartz, and contains about 10 percent kaolin which increases upward. Also present in the indistinctly cross-bedded sandstone

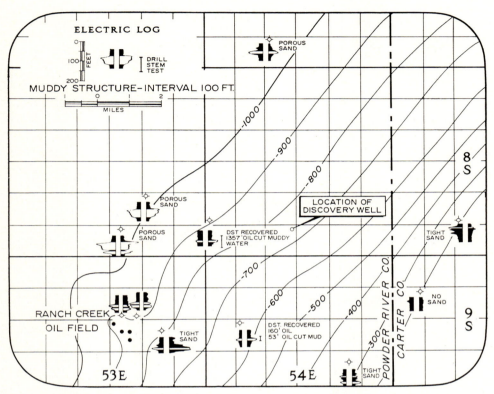

FIG. 21.—Subsurface information available June 1, 1967, before Bell Creek discovery well was drilled. Structural contours drawn with datum at top of Muddy Sandstone. CI = 100 ft. Shows reproductions of Muddy Sandstone electric logs and drillstem test data.

are glauconite, sparse muscovite, wood, and other carbonaceous fragments. The very porous and permeable sandstone grades downward into the underlying shale. The results of a detailed study of cores from Bell Creek field and an interpretation of the origin of the producing sandstone bodies are presented by Berg and Davies (1968).

LOCAL STRUCTURE

Early drilling in the Bell Creek exploration program was influenced by reported local structures along the favorable stratigraphic fairway. Anticlines on the north in the Mizpah and Coalwood areas, shown on surface maps by the U.S. Geological Survey (Parker and Andrews, 1939; Bryson, 1952), had been drilled. Other tests were drilled in Custer, Powder River, and Carter Counties on reported seismic highs (Gary and McGregor, 1968). Structures suggested by photogeologic surveys also were drilled. All the holes

were dry, but useful stratigraphic information was obtained.

Local information available by June 1, 1967, in addition to the published regional stratigraphic studies, is summarized on Figure 21. Four wells in the Ranch Creek area were producing 80 bbl of oil per day with substantial water. One test had been drilled in the SW¼ SW¼ Sec. 30, T8S, R54E, but this dry hole had a show of oil in the Muddy Sandstone, and a drillstem test recovered 1,357 ft of oil-cut water from a 4-ft porous interval. Tests drilled in T8S, R53E, indicated thicker and more porous sandstone. A drillstem test of a hole drilled in the NW¼ NW¼ Sec. 17, T9S, R54E, yielded 213 ft of free oil and oil-cut mud from a 3-ft porous zone.

The elimination of sites where wells could be drilled partly on the basis of structure led to the drilling of a wholly stratigraphic test. The decision was to drill no more than 3 mi from the hole

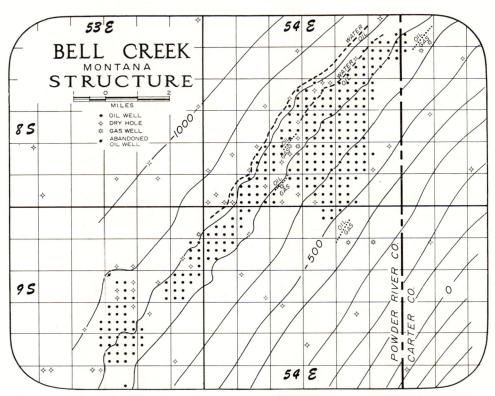

FIG. 22.—Structure-contour map with datum on top of Muddy Sandstone at Bell Creek field. CI = 100 ft. Dip of homocline is remarkably uniform.

that yielded free oil, along structural strike but slightly basinward stratigraphically, to test a delta-front or offshore or barrier-bar environment.

Drilling since then, summarized on the structure contour map (Fig. 22), established the absence of local structural control for the field. The field is on a remarkably uniform homocline sloping northwest 100 ft/mi. The local "nosing" in the Ranch Creek area in T9S, R53E, is accentuated by a change in regional strike from just west of north in northern Wyoming to almost due northeast in this area. Also shown on the map are two relatively sharp oil-water contacts and four distinct gas-oil contacts, which suggest the presence of multiple reservoirs.

A map combining present structure with paleogeography at the time of Muddy Sandstone deposition, shown in Figure 23, perhaps best summarizes the controls for the accumulation of oil in the field. Additional data undoubtedly will lead to modifications of the local environments inferred for the area.

CONCLUSIONS

1. The Bell Creek field now produces 50,000 bbl and is capable of producing more than 65,000 bbl of oil per day. Reserves are estimated at 200 million bbl, and Bell Creek soon will produce more than any other Rocky Mountain field.

2. The field is a rich stratigraphic trap, with no prominent local structural control.

3. The stratigraphic trap formed near the strandline during a regressive phase between two major advances of the Early Cretaceous sea.

4. Local environments of deposition are represented by a complex assemblage of marginal-marine and deltaic facies, comparable with those evident in modern sediments (van Andel and Curray, 1960; Miller, 1965).

5. Updip limits of the overall reservoir are formed by lateral facies change from porous and permeable sandstone to lagoonal claystone, siltstone, and silty carbonaceous claystone.

6. The approximate position of the Bell Creek field could have been predicted by delineating the

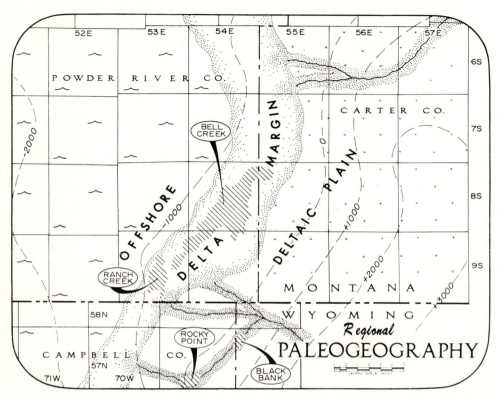

FIG. 23.—Depths of Muddy Sandstone paleoenvironments in Bell Creek and Rocky Point areas. Modified from Figures 16 and 22. Structural datum is top of Muddy Sandstone. CI = 1,000 ft.

intersection of northeast-trending marine-bar sandstones and northwest-trending delta systems that had been established by earlier regional stratigraphic studies.

7. At least four individual reservoirs are suggested by oil-water and gas-oil contacts.

8. The stratigraphic traps formed early. Eastward pinchouts of the sand bodies were enhanced as traps by the depositional basin sinking on the west, as shown by isopachous maps of the overlying Mowry Shale.

9. Oil accumulation in the traps was enhanced further by subsequent development of the Powder River basin during which the sandstone tongues were inclined to still more favorable positions.

10. The sources of hydrocarbons are the local sediments; the subjacent and superjacent marine shales, and the reservoir rock itself are rich in organic material.

11. Regional studies of the Cretaceous System in the western interior indicate that many additional favorable sites comparable with Bell Creek remain to be drilled. Many such sites have been tested by less than one well per township, and it is possible that major oil fields will be found.

12. The estimated return on investment at Bell Creek of 20:1 makes the field as economically attractive as the larger oil fields of the world.

13. Local surface structures in the region are not a clue to the presence of deeper structures, because Tertiary rocks at the surface are separated from underlying Cretaceous strata by an angular unconformity.

14. Apparent seismic highs in the region reflect surface topography rather than subsurface structure.

15. Shallow depths and ease of drilling make drilling the most efficient and unequivocally conclusive exploration method once the favorable geologic environments have been delineated, as at Bell Creek.

REFERENCES CITED

Berg, R. R., and D. A. Davies, 1968, Origin of Lower Cretaceous Muddy Sandstone at Bell Creek field, Montana: Am. Assoc. Petroleum Geologists Bull., v. 52, no. 10, p. 1888–1398.

Bolyard, W. W., and A. A. McGregor, 1966, Inyan Kara Group, Black Hills area: Am. Assoc. Petroleum Geologists Bull., v. 50, no. 10, p. 2221–2245.

Bryson, R. P., 1952, The Coalwood coal field, Powder River County, Montana: U.S. Geol. Survey Bull. 973-B, p. 23–106.

Cobban, W. A., and J. B. Reeside, 1952, Correlation of Cretaceous formations of western interior of United States: Geol. Soc. America Bull., v. 63, no. 10, p. 1011–1044.

Cram, I. H., 1966, The oldest is the newest: Am. Assoc. Petroleum Geologists Bull., v. 50, no. 5, p. 826–829.

Eicher, D. L., 1960, Stratigraphy and micropaleontology of the Thermopolis Shale: Yale Univ. Peabody Museum of Nat. History Bull. 15, 126 p.

——— 1962, Biostratigraphy of the Thermopolis, Muddy and Skull Creek Formations, *in* Symposium on Early Cretaceous rocks of Wyoming and adjacent areas: Wyoming Geol. Assoc. 17th Ann. Field Conf. Guidebook, p. 72–93.

Gary, Samuel, and A. A. McGregor, 1968, Exploration philosophy behind the Bell Creek oil field discovery, Powder River County, Montana: Mountain Geologist, v. 5, no. 1, p. 15–21.

Haun, J. D., and J. A. Barlow, Jr., 1962, Lower Cretaceous stratigraphy of Wyoming, *in* Symposium on Early Cretaceous rocks of Wyoming and adjacent areas: Wyoming Geol. Assoc. 17th Ann. Field Conf. Guidebook, p. 15–22.

Levorsen, A. I., 1967, Geology of petroleum: 2d ed., San Francisco, California, W. H. Freeman and Co., 724 p.

McGregor, A. A., 1968, Bell Creek oil field, Powder River and Carter Counties, Montana: Wyoming Geol. Assoc., Earth Science Bull., v. 1, no. 1, p. 29–36.

Miller, D. N., Jr., 1965, Recognition and classification of marginal marine environments, *in* Sedimentation of Late Cretaceous and Tertiary outcrops, Rock Springs uplift: Wyoming Geol. Assoc. 19th Ann. Field Conf. Guidebook, p. 209–218.

Parker, F. S., and D. A. Andrews, 1939 (1940), The Mizpah coal field, Custer County, Montana: U.S. Geol. Survey Bull. 906-C, p. 85–133.

Robinson, C. S., W. J. Mapel, and M. H. Bergendahl, 1964, Stratigraphy and structure of the northern and western flanks of the Black Hills uplift, Wyoming, Montana and South Dakota: U.S. Geol. Survey Prof. Paper 404, 134 p.

Scruton, P. C., 1960, Delta building and the deltaic sequence, *in* Recent sediments, northwest Gulf of Mexico: Am. Assoc. Petroleum Geologists, p. 82–102.

Shepard, F. P., 1960, Gulf Coast barriers, *in* Recent sediments, northwest Gulf of Mexico: Am. Assoc. Petroleum Geologists, p. 197–200.

Thom, W. T., Jr., *et al.,* 1935, Geology of Big Horn County and the Crow Indian Reservation, Montana: U.S. Geol. Survey Bull. 856, 200 p.

van Andel, T. H., and J. R. Curray, 1960, Regional aspects of modern sedimentation in northern Gulf of Mexico and similar basins, and paleogeographic significance, *in* Recent sediments, northwest Gulf of Mexico: Am. Assoc. Petroleum Geologists, p. 345–364.

Wulf, G. R., 1962, Lower Cretaceous Albian rocks in northern Great Plains: Am. Assoc. Petroleum Geologists Bull., v. 46, no. 8, p. 1371–1415.

Regional Stratigraphy of Frontier Formation and Relation to Salt Creek Field, Wyoming[1]

JAMES A. BARLOW, JR.,[2] and JOHN D. HAUN[3]

Casper, Wyoming 82601, and Golden, Colorado 80402

Abstract   About 420 million bbl of oil has been produced from Salt Creek field, Natrona County, Wyoming. Most production is from the second Frontier sandstone, which is one of many sandstone bodies interbedded with marine shale in the lower part of the Upper Cretaceous in the Rocky Mountain area. The stratigraphic interval between the Mowry Shale and the Niobrara Formation contains the Frontier and equivalent formations. The interval containing this section is more than 1,000 ft thick in central, northeastern, and west-central Wyoming and in southeastern Montana. Another area of thick sediments within this interval is in northwestern Montana and western Alberta. In some areas this section is entirely marine shale; in other areas it contains numerous sandstone bodies. The sand was transported by a series of river systems that formed deltaic complexes or lobate sand concentrations at several places along the margins of the early Late Cretaceous sea. These deltaic deposits are represented by the "D" sandstone of the Denver basin, the Ferron Sandstone Member of the Mancos Shale of Utah, the Cardium and Bad Heart Sandstones of Canada, and the Frontier Formation of Wyoming.

The second Frontier sandstone—producing formation at Salt Creek field—is an offshore bar associated with the eastern terminus of one stage of Frontier deposition. The sandstone body is several miles wide, more than 60 mi long, and as much as 100 ft thick. Salt Creek anticline (formed in late Late Cretaceous or early Tertiary time) is in an area of excellent sandstone development where stratigraphic traps developed in which oil accumulated before secondary migration into present structural positions took place.

There are other sandstone bodies included in the Frontier depositional complex which contain stratigraphically trapped oil, but these are not draped over an obvious anticline. The Wind River and Big Horn basins, and parts of the Green River and Powder River basins, probably contain more "Salt Creeks"!

INTRODUCTION

Salt Creek field, situated on an anticline with 1,500 ft of structural closure, has long been recognized as an outstanding example of anticlinal oil entrapment. More oil has been produced from this than from any other field in the Rocky Mountain region (although several others ultimately will produce more); only 17 fields in the United States have had greater production. The cumulative oil production of 420 million bbl from Salt Creek is 20 percent of the total oil production to date from Wyoming.

Most of the production at Salt Creek is from the Frontier Formation, and the second Frontier sandstone is the most important producing section. The informal terms "first Frontier sandstone" and "second Frontier sandstone" are used in this paper to denote the important producing sections of the formation. Stratigraphic studies have shown that sandstone bodies within the Frontier characteristically are lenticular and have large variations in porosity and permeability. A study of the Cretaceous history of Wyoming indicates that the Salt Creek anticline had very little, if any, structural development prior to the latest Cretaceous and early Tertiary Laramide orogeny. This fact indicates that Salt Creek oil must have accumulated originally in stratigraphic traps, and secondary migration must account for the present structural accumulation. It is the purpose of this paper to outline relation of oil in the Frontier Formation in the southwest Powder River basin to the local and regional stratigraphic framework and depositional environments of the Frontier Formation. Special emphasis is placed on the description of the second Frontier sandstone because it is the major oil-producing reservoir section at Salt Creek field.

PREVIOUS WORK

Reviews of the production history and structural geology of Salt Creek field have been published by Knight and Slossom (1896), Wegemann (1911, 1918), Estabrook and Rader (1925), Beck (1929), Espach and Nichols (1941), Baker (1957), Biggs and Espach (1960), Petroleum Information (1965, and prior years), and others. Recent stratigraphic studies of the Frontier Formation of the Powder River basin, and of Wyoming, include those by Cobban and Reeside (1952), Towse

[1] Read before the Rocky Mountain Section of the Association at Billings, Montana, September 29, 1965. Manuscript received, September 25, 1968. Modified from an article published in the AAPG *Bulletin*, v. 50, no. 10, p. 2185–2196.

[2] Consulting Geologist, Barlow & Haun, Inc., Casper, Wyoming.

[3] Professor, Colorado School of Mines, Golden, Colorado.

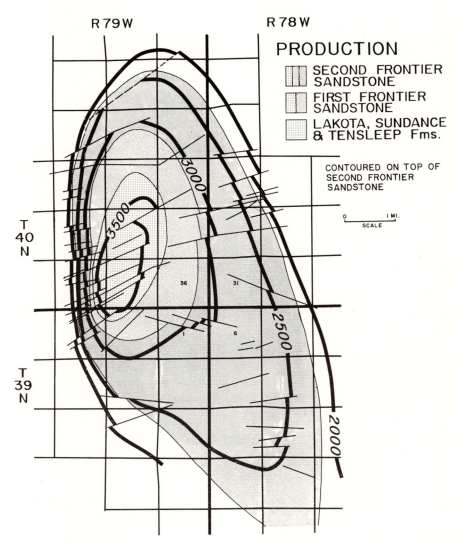

PRODUCTION

SECOND FRONTIER SANDSTONE

FIRST FRONTIER SANDSTONE

LAKOTA, SUNDANCE & TENSLEEP Fms.

CONTOURED ON TOP OF SECOND FRONTIER SANDSTONE

Fig. 1.—Salt Creek field structure and production. Structural datum is top of second Frontier sandstone. Lakota, Sundance, and Tensleep Formations are productive in an area of 1,800 acres; first Frontier sandstone, approximately 2,500 acres; and second Frontier sandstone, more than 22,000 acres. Contours in feet above sea level.

(1952), Masters (1952), Haun (1958), and Goodell (1962). The relation of the Frontier to other Upper Cretaceous rocks of the region has been summarized by Weimer (1960).

PRODUCING INTERVALS AND STRUCTURAL POSITION

Figure 1 illustrates areas of production in and structural configuration of Salt Creek field. There are more than 2,500 wells within the

total productive area, which extends 9 mi north-south and 4 mi east-west. Production from the Lakota, Sundance, and Tensleep Formations is limited to an area of 1,800 acres on the crest of the structure. Production from the first Frontier sandstone is in an area of approximately 2,500 acres in the same crestal area, but extends farther down the flanks of the anticline. Production from the second Frontier sandstone is from an area greater than 22,000 acres and extends southward into Teapot dome, beyond the lowest closing contour on the Salt Creek anticline. There is some production from the shale beds above the Frontier; for simplicity, this shale production is not shown on Figure 1. The Lakota, Sundance, and Tensleep together have produced about 50 million bbl of oil. The first Frontier sandstone also has produced about 50 million bbl of oil. The combined production from these reservoirs of 100 million bbl makes Salt Creek a significant field, but it is the 300 million bbl of oil produced from the second Frontier that puts Salt Creek in the class of giants.

Regional Stratigraphy

Frontier and equivalent sediments were deposited in, or on the margins of, a shallow sea which covered the Western Interior region during the Cenomanian and Turonian stages of the Late Cretaceous. Formations that were deposited during this time generally are 500–2,000 ft thick. The thickness map (Fig. 2) is based on correlations of more than 5,000 electric logs. The mapped interval extends from the top of the underlying Early Cretaceous Mowry Shale (Clay Spur bentonite bed where present) to the base of the Niobrara Formation (or time equivalent where possible). Some of the upper Frontier sandstone beds of western Wyoming and the Bad Heart Sandstone of Alberta (Williams and Burk, 1964) are the time equivalents of the lower Niobrara on the east.

Through most of the area, the mapped interval is dominated by marine shale. Within the shale are significant oil- and gas-producing sandstone beds associated with lobate concentrations of coarse terrigenous clastic rocks along the western, southern, and southeastern margins of the seaway. Bartram (1937) suggested that the Frontier and equivalent formations were parts of a sandstone tongue extending from a delta or series of deltas formed by rivers flowing eastward across the Cordilleran orogenic belt. The more important deltas include the Ferron of central and eastern Utah, the Frontier of western and central Wyoming, the Bad Heart-Cardium of western Alberta (the interval includes the deposits of the Dunvagan Formation), and the "D sandstone of the Denver basin in northeastern Colorado and western Nebraska.

Recognition of the deltaic deposits is based on three factors: (1) an areally restricted, pronounced thickening of the interval in a landward direction, (2) a seaward-projecting lobe

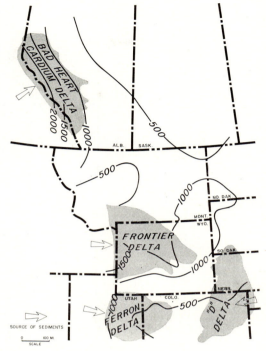

Fig. 2.—Isopach map of total stratigraphic interval between top of Mowry Shale and base of Niobrara Formation (or time equivalent where possible). Formation names in areas of major sandstone concentration are indicated. Contours in feet.

of continental sediments consisting of coal, carbonaceous shale, and channel deposits, and (3) a thick, complex sequence of sandstone deposits. King (1959, p. 108) referred to the sediments of the Frontier and Mesaverde Formations as "clastic wedges." It has been argued by Krumbein and Sloss (1963, p. 532–535) that the sediments of the Frontier Formation should be termed a "clastic wedge" rather than a delta. However, many geologists recently have referred to the sediment complexes of the Frontier as deltas, and this term is used here. A complete discussion of the question is beyond the scope of this paper.

Wyoming Stratigraphy

The west-east stratigraphic diagram (Fig. 3) illustrates the nomenclature and stratigraphic relations between the Frontier and equivalent formations of Wyoming. The Frontier of westernmost Wyoming includes equivalents of lower Cody, lower Niobrara, Carlile, Greenhorn, and Belle Fourche formations farther

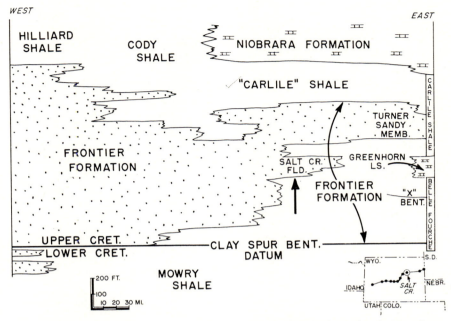

FIG. 3.—West-east stratigraphic diagram of Frontier and correlative formations in Wyoming.

east. The second Frontier sandstone at Salt Creek is equivalent to the upper Belle Fourche and (or) the lower Greenhorn of the Black Hills area (Cobban and Reeside, 1952; Haun, 1958); this correlation is based on the identification of ammonite zones and bentonite beds.

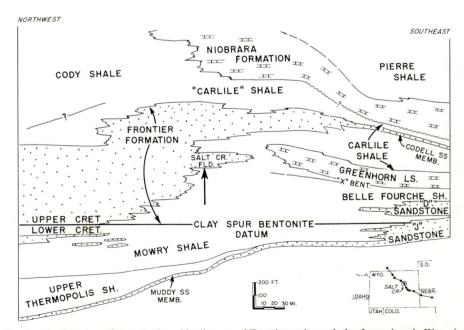

FIG. 4.—Northwest-southeast stratigraphic diagram of Frontier and correlative formations in Wyoming.

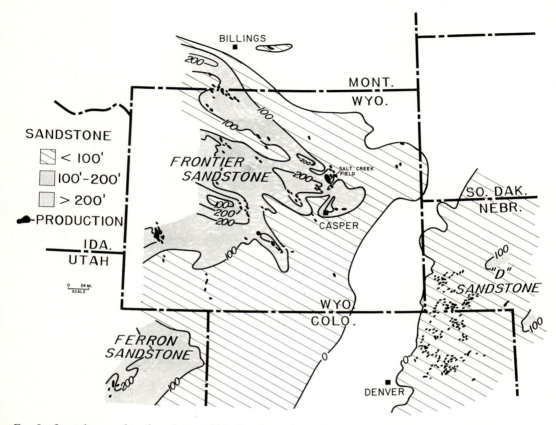

FIG. 5.—Isopach map of total sandstone within Frontier and correlative formations. Codell Sandstone Member of Benton Shale, on southeast, is omitted from computations. Contours in feet.

A northwest-southeast correlation across Wyoming from southern Montana to western Nebraska (Fig. 4) further illustrates the relations between Frontier and equivalent formations. In western Nebraska, the "D" sandstone (within the Dakota Group) was deposited at the same time as the lower part of the Frontier of western and northwestern Wyoming. This age relationship was established by correlating the top of the Mowry Shale and the overlying Clay Spur bentonite bed in outcrop and subsurface sections. Correlations of this part of the section have been discussed by Haun and Barlow (1962). Parts of the Frontier Formation that are indicated by sandstone symbols in Figures 3 and 4 are diagrammatic and contain a varied sequence of marine, littoral, and nonmarine sandstone interbedded with marine shale and siltstone.

Distribution of sandstone within the Frontier and correlative formations is illustrated in Fig-

ure 5. The map is based on lithologic data obtained from more than 2,500 surface and well sections. Comparison with Figure 2 shows that sandstone generally comprises less than 10 percent of the total interval. Southeast of the zero isopachous contour (Fig. 5) of the Frontier sandstone bodies, the Codell Sandstone Member of the Benton Shale has been omitted from thickness computations to permit illustrations of the western limit of the "D" sandstone. The Codell generally is nonporous and ranges in thickness from 10 to 40 ft (Fig. 4).

Salt Creek field, north of Casper, Wyoming (Fig. 5), is near the eastern edge of the thick Frontier sandstone area. Sandstone bodies in central Wyoming were deposited in littoral and offshore environments along the most seaward part of the deltaic deposits. Significant thicknesses of nonmarine sediments are 100–200 mi west of Salt Creek. At times of more rapid basin sinking or diminished supply of coarser

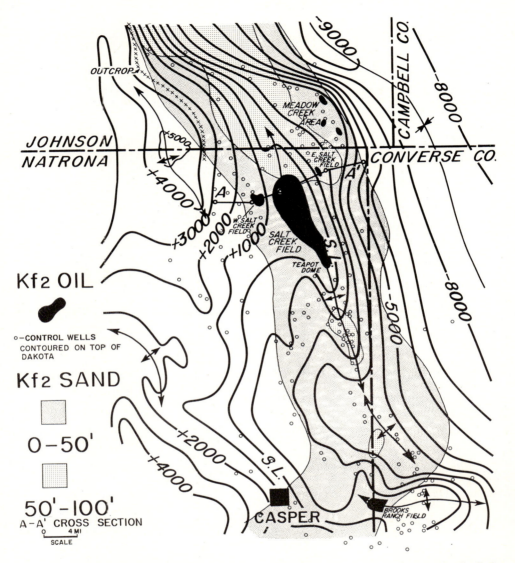

Fig. 6—Structure of southwest Powder River basin (structural datum to top of Dakota), distribuion of second Frontier sandstone (Kf₂), and oil fields in second Frontier sandstone. Contours in feet above and below sea level.

clastics, marine shale and siltstone were deposited across most of Wyoming. Faunal evidence (Cobban and Reeside, 1952) and detailed electric-log correlations indicate that there also were periods of erosion over broad areas. Complex interbedding of shale and sandstone, lenticular development of many sandstones, and disconformities within the sequence make it very difficult to correlate subdivisions of the Frontier from one basin to another.

Local Stratigraphy and Structure

Structural configuration of the southwest flank of the Powder River basin and distribution of the second Frontier sandstone are shown in Figure 6, which was constructed with data from approximately 200 electric logs. The second Frontier in the area including Salt Creek field is an offshore bar, more than 100 ft thick, that thins to zero ft of effective sandstone toward the east and west. The bar is

WEST A – A' EAST

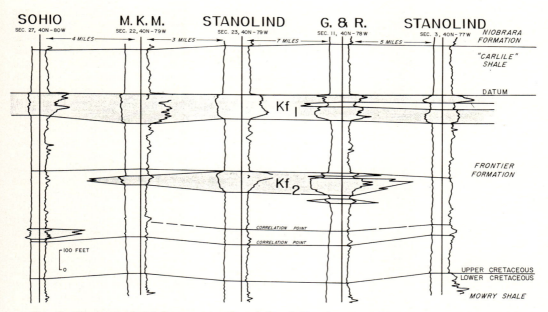

FIG. 7.—West-east electric-log cross section A-A' of Frontier Formation showing correlation of first Frontier (Kf₁) and second Frontier (Kf₂) sandstones, and associated shale and siltstone. See Fig. 6 for location of section.

more than 60 mi long, about 10 mi wide, and is draped across the Salt Creek anticline. An electric-log cross section of the bar is illustrated in Figure 7 (location on Fig. 6).

The relatively uniform thickness of section between correlation lines used in Figure 7 (bentonites?) indicates that the base of the second Frontier sandstone does not occupy a channel—there were no sediments removed prior to deposition. Locally the section is thicker because of sand deposition on the bar, and subsequent differential compaction produced the present configuration. Inlet-channel scouring, indicative of a barrier island, is not evident, and the base of the sandstone appears to grade downward into the underlying shale.

The northwestward extent of the bar is unknown beyond the outcrop along the east flank of the Big Horn Mountains (northwest corner of Fig. 6). In the Kaycee outcrop area (T43 and 44 N, R82 and 83 W; Haun, 1958), there is evidence that the bar was subjected to vigorous wave and current activity: (1) there is marked lateral and vertical change in grain size, but sorting in any particular sample appears to be good, (2) pebbles, as large as 3.5

in. in diameter, are concentrated in short lenses with the exception of a widespread pebble bed at the top of the sandstone, (3) isolated pebbles occur in sandstone which otherwise is well sorted, and (4) bedding in some places is indistinct, whereas, in other places, the sandstone is thin-, thick-, and cross-bedded.

FRONTIER SANDSTONE BODIES

In the western part of the Powder River basin (Fig. 8), Frontier sandstone beds are very fine- to medium-grained; some are silty to shaly, and others are conglomeratic. The average median diameter of grains in 31 samples of sandstone and sandy siltstone from the Kaycee and Buffalo areas (Haun, 1953) is 0.13 mm, though the range in median diameter is 0.007 to 0.26 mm. Sand samples from the southwest part of the basin, analyzed by Towse (1952), have an average median diameter of 0.11–0.12 mm. Size analyses show no apparent lateral correlation or trend in variation in the Kaycee-Buffalo area. The grain size of most sandstone beds increases upward, from a basal part consisting of silty shale and siltstone, grading upward to fine- to medium-grained sandstone at

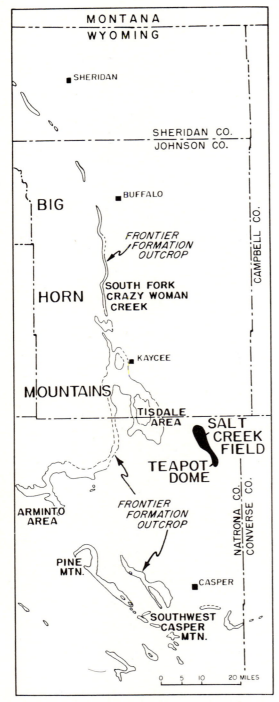

FIG. 8.—West flank of Powder River basin showing Frontier Formation outcrop and localities mentioned in text.

the top. The upper contact of most sandstone beds is sharp, with little or no gradation into the overlying shale.

The average coefficient of sorting of the samples from the Kaycee-Buffalo area is 1.37 with extremes of 1.22 and 1.69. In the southwest part of the basin, Towse (1952) reported sorting coefficients which average 1.30 and range from 1.12 to 1.79. In general, the sandstone beds on the flanks of the Big Horn Mountains have higher sorting coefficients than the sandstone beds of other parts of the basin.

Frontier sandstone beds ("salt and pepper") are composed predominantly of white quartz. Other constituents are black and gray chert, clear and pink quartz, feldspar, biotite, glauconite (common in the subsurface), and traces of carbonaceous fragments and heavy minerals. Many of the sand grains show the addition of secondary quartz. Fragments of carbonaceous matter including coal are present in sandstone units of the southwest part of the basin and are especially concentrated in sandstone units correlative with the Belle Fourche Shale. The chief matrix and cementing materials are clay, calcite, and silica, with minor amounts of iron oxide. Insoluble residue of the matrix of the sandstone samples analyzed ranges from 65.63 to 97.90 percent and averages 84.20 percent.

Sand grains in the Frontier range in shape from angular to well rounded; most are subrounded to subangular. The majority of the sandstone beds contain a small percentage of well-rounded frosted grains. The frosted grains are slightly larger than the surrounding grains, occur in the sandstone bodies of larger grain size, and are not concentrated in particular beds. (*See* Goodell, 1962, for an extensive analysis of Frontier petrology.)

Pebbles form a characteristic, but very small part of the Frontier deposits. Their source areas, method of transport, and environment of deposition are important to an understanding of Frontier sedimentation. They may occur in any part of the formation, but are most common and reach their maximum size in both the second Frontier sandstone and its approximate equivalents. From a maximum diameter of 3.5 in. in the north Kaycee area, the pebbles decrease in size to 3 in. southwestward in the Tisdale and Arminto areas, and 2 in. at Pine Mountain and southwest Casper Mountain (localities are shown on Fig. 8). North of the area of maximum size near Kaycee, the maximum diameter decreases to 1 in. at the South Fork of Crazy Woman Creek, 0.5 in. in the Buffalo area, and to 0.25 in. west of Sheridan. The most prominent pebble rock-type is black-coated chert; quartzite and quartz are common. In the north Kaycee area there is a high percentage of andesite porphyry pebbles and cobbles.

The constituent percentage of pebbles and small cobbles at the top of the second Frontier sandstone in the Kaycee area is shown in Table 1. Most of the chert pebbles are chatter-marked, but most of the other pebbles are smooth. Many of the chert pebbles contain corals, fusulines, crinoid columnals, and bryozoans of the types to which Hunter (1952, p. 64) assigned a Pennsylvanian or Mississippian(?) age.

The majority of the pebble-bearing beds contain fresh (unabraded); shark teeth. The freshness of these teeth indicates that they were transported only a short distance. Their presence in the pebble-bearing beds is believed to indicate periods of slow deposition during which currents were not strong enough to transport the teeth. They were, therefore, added to the pebble-bearing beds after the pebbles had been deposited. It is difficult to ascertain whether or not the teeth are concentrated at the top of the beds, because many of the teeth have slipped down between the pebbles. The teeth of the skate *Ptychodus,* a bottom feeder, commonly are present with the pebbles, but the vast majority of teeth are the sharp-pointed type of the more active predatory sharks.

The chief source of sediments for the Frontier was a land area along the western border of Wyoming—in Utah and Idaho—where Paleozoic sedimentary rocks were being eroded. Andesite porphyry cobbles, as large as 10 in. in diameter, were deposited in the Big Horn basin southeast of Lovell (Hunter, 1950). A fauna reported by Haas (1949) indicates that the Big Horn basin cobbles are the age equivalent of those in the Kaycee area. The most logical

source of the porphyry cobbles is the Beartooth-Yellowstone Park region of southwestern Montana and northwestern Wyoming, or an area farther west; this area was one of orogenic and volcanic activity during the Cretaceous. Large andesite porphyry cobbles have not been reported from Frontier conglomerate beds on the west side of the Big Horn basin (Van Houten, 1962), but their route of transport to the area of study may have been narrow, and restricted to the northeast side of the area of thickest sandstone development (Fig. 5). Through a distance of 95 mi between the two areas on opposite sides of the Big Horn Mountains, the maximum size increases from 3.5 on the east to 10 in. on the west. If the size increases at the same rate across the remaining 80–95 mi into the Beartooth-Yellowstone Park area, the size at the source would be 16–17 in. —a size which is well within the ability of waves to transport. In summary, the probable source, direction of movement, and size of the cobbles indicate that there was strong wave and current activity in a shoal area (or series of offshore bars) extending northwest-southeast along the northeast side of the Frontier lobe of sand deposition.

FORMATION FLUIDS

Second Frontier Sandstone

Production data are summarized from Biggs and Espach (1960, p. 225–235). Oil was discovered in the second Frontier sandstone at Salt Creek in 1917.[4] The average initial daily production from 339 wells drilled between 1917 and 1921 was 669 bbl; subsequent wells had smaller initial oil production. The oil contained an appreciable quantity of solution gas. Producing depths are 1,330–2,900 ft and the original shut-in wellhead pressure was about 850 psi. Average porosity is about 20 percent. The average gravity of oil was originally 37–38° API except in a zone 50–100 ft above the oil-water contact in which there was little solution gas and the gravity was 33–35° API. Waters of the second Frontier contain 8,000 to 15,000 ppm total solids (Crawford and Davis, 1962). In contrast, the waters of the first Frontier have 2,000 to 13,000 ppm total solids; this reflects the greater regional extent of the first

Table 1. Constituent Pebbles in Top of Second Frontier Sandstone

Mineral or rock type	Percent
Black-coated chert	40
Uncoated chert	27
Andesite porphyry	21
White quartz	5
Pink quartz	3
Quartzite	3
Clear quartz	1
	100

[4] According to Beck, oil first was exploited commercially in 1889 from very shallow wells. The first truly commercial well was in the first Frontier sandstone, and was discovered in 1906. Major production, however, did not begin until the second Frontier sandstone discovery in 1917.

Frontier and its greater dilution is meteoric water.

In a cross section published by Beck (1929, p. 600) there is an indicated tilt of the oil-water contact in the second Frontier of approximately 130 ft from south to north across the field. Other producing intervals also are depicted as having south-to-north tilted oil-water contacts. The exact position of the cross section in the field is not designated, but the indicated tilt in the second Frontier is approximately 15 ft/mi. More recent detailed maps (*i.e.*, the 1954 map by Wyoming Geological Association, 1957) do not show a clearly defined tilt of the oil-water contact. In fact, the lowest producing wells are at the south end of the field in the structural saddle that separates Salt Creek from Teapot dome. If the second Frontier sandstone bar is a reasonably restricted aquifer extending from the outcrop south of Casper (Fig. 8) to a lower outcrop northwest of Salt Creek, then there should be a northward slope of the piezometric surface of 4–6 ft/mi through the Salt Creek area. The many transverse faults in the Salt Creek–Teapot area may modify greatly the hydrodynamic pattern. The lack of accurate pressure information during the early development of the field limits the complete hydrodynamic analysis of the area.

Gussow (1954, p. 833–834) used the Salt Creek–Teapot accumulations as an example of differential entrapment of oil and gas. Teapot contains a gas cap (which Salt Creek does not), yet the structurally highest part of Teapot dome is 1,300 ft lower than the structural culmination of Salt Creek. The implication from Gussow's hypothesis is that secondary oil migration was from south to north along the anticlinal trend.

Oil-producing fields in the second Frontier are shown on Figure 6. The production at Brooks Ranch and West Salt Creek is from small stratigraphic traps. Production at Meadow Creek, East Salt Creek, and Salt Creek is from structural traps.

Source beds for oil in the Frontier appear to be Frontier shale beds. Hunt and Jamieson (1958) have estimated that within an 800-sq-mi area, including Salt Creek and the other Frontier fields in the southwest Powder River basin, there still are 3 billion bbl of hydrocarbons within the Frontier shale beds.

SUMMARY

Salt Creek field commonly has been selected as an outstanding example of an anticlinal accumulation. It has been cited as an example of differential entrapment by Gussow (1954) and as an example of fields with tilted oil-water contacts by Levorsen (1954, p. 295). We have attempted to show that the major accumulation, in the second Frontier sandstone, is within a sand bar that was deposited at the seaward margin of a lobate concentration of coarse, terrigenous clastic rocks (a delta?), which were derived from a land area on the west and northwest. The bar was subjected to strong wave and current action which produced relatively high porosity through sorting processes. The bar became a stratigraphic trap during the early stages of sediment compaction and accumulated petroleum derived from the surrounding source beds in the Frontier. During Laramide folding, the oil migrated to approximately its present structural position. Subsequent hydrodynamic gradients may have modified slightly the structural position of the oil.

If the second Frontier bar had been 15 mi east of its present position, in a synclinal area, would the field have been discovered? Are there other such bars remaining to be discovered in the Rocky Mountain area?

REFERENCES CITED

Baker, F. E., 1957, History of the Salt Creek oil field, *in* Wyoming oil and gas fields symposium: Wyoming Geol. Assoc., p. 387–388.

Bartram, J. G., 1937, Upper Cretaceous of Rocky Mountain area: Am. Assoc. Petroleum Geologists Bull., v. 21, no. 7, p. 899–913.

Beck, Elfred, 1929, Salt Creek oil field, Natrona County, Wyoming, *in* Structure of typical American oil fields, v. 2: Am. Assoc. Petroleum Geologists, p. 589–603.

Biggs, Paul, and R. H. Espach, 1960, Petroleum and natural gas fields in Wyoming: U.S. Bur. Mines Bull. 582, p. 224–235.

Cobban, W. A., and J. B. Reeside, Jr., 1952, Frontier Formation, Wyoming and adjacent areas: Am. Assoc. Petroleum Geologists Bull., v. 36, no. 10, p. 1913–1961.

Crawford, J. G., and C. E. Davis, 1962, Some Cretaceous waters of Wyoming: Wyoming Geol. Assoc. 17th Ann. Field Conf. Guidebook, p. 257–267.

Espach, R. H., and H. D. Nichols, 1941, Petroleum and natural-gas fields in Wyoming: U.S. Bur. Mines Bull. 418, p. 86–96.

Estabrook, E. L., and C. M. Rader, 1925, History of production of Salt Creek oil field, Wyoming, *in* Petroleum development and technology in 1925: Am. Inst. Mining Metall. Engineers, p. 199–254.

Goodell, H. G., 1962, The stratigraphy and petrology of the Frontier Formation of Wyoming: Wyoming Geol. Assoc. 17th Ann. Field Conf. Guidebook, p. 173–210.

Gussow, W. C., 1954, Differential entrapment of oil and gas, a fundamental principle: Am. Assoc. Petroleum Geologists Bull., v. 38, no. 5, p. 816–853.

Haas, Otto, 1949, Acanthoceratid Ammonidea from near Greybull, Wyoming: Am. Mus. Nat. History Bull., v. 93, art. 1, 39 p.

Haun, J. D., 1953, Stratigraphy of Frontier Formation, Powder River basin, Wyoming: Unpub. Ph.D. thesis, Wyoming Univ.

———— 1958, Early Upper Cretaceous stratigraphy, Powder River basin, Wyoming: Wyoming Geol. Assoc. 13th Ann. Field Conf. Guidebook, p. 84–89.

———— and J. A. Barlow, Jr., 1962, Lower Cretaceous stratigraphy of Wyoming: Wyoming Geol. Assoc. 17th Ann. Field Conf. Guidebook, p. 15–22.

———— and ———— 1966, Regional stratigraphy of Frontier Formation and relation to Salt Creek field: Am. Assoc. Petroleum Geologists Bull., v. 50, no. 10, p. 2185–2196.

Hunt, J. M., and G. W. Jamieson, 1958, Oil and organic matter in source rocks of petroleum, *in* Weeks, L. G., ed., Habitat of oil: Am. Assoc. Petroleum Geologists, p. 735–746; reprinted from Am. Assoc. Petroleum Geologists Bull., v. 40, no. 3, March, 1956, p. 477–488.

Hunter, L. D., 1950, Evidence of uplift in the Bighorn Mountains during Upper Cretaceous time (abs.): Geol. Soc. America Bull., v. 61, no. 12, p. 1554.

———— 1952, Frontier Formation along the eastern margin of the Big Horn basin, Wyoming: Wyoming Geol. Assoc. 7th Ann. Field Conf. Guidebook, p. 63–66.

King, P. B., 1959, The evolution of North America: Princeton Univ. Press, 190 p.

Knight, W. C., and E. E. Slossom, 1896, The petroleum of Salt Creek, Wyoming: Univ. Wyoming School of Mines Bull. 1, 47 p.

Krumbein, W. C., and L. L. Sloss, 1963, Stratigraphy and sedimentation: 2d ed., San Francisco, W. H. Freeman and Company, 660 p.

Levorsen, A. I., 1954, Geology of petroleum: San Francisco, W. H. Freeman and Company, 703 p. [Copyright 1956.]

Masters, J. A., 1952, The Frontier Formation of Wyoming: Wyoming Geol. Assoc. 7th Ann. Field Conf. Guidebook, p. 58–62.

Petroleum Information, 1965, 1964 resume, oil and gas operations in the Rocky Mountain region: p. C-252.

Towse, Donald, 1952, Frontier Formation, southwest Powder River basin, Wyoming: Am. Assoc. Petroleum Geologists Bull., v. 36, no. 10, p. 1962–2010.

Van Houten, F. B., 1962, Frontier Formation, Big Horn basin, Wyoming: Wyoming Geol. Assoc. 17th Ann. Field Conf. Guidebook, p. 221–231.

Wegemann, C. H., 1911, The Salt Creek oil field, Natrona County: U.S. Geol. Survey Bull. 452, p. 37–83.

———— 1918, The Salt Creek field, Wyoming: U.S. Geol. Survey Bull. 670, 52 p.

Weimer, R. J., 1960, Upper Cretaceous stratigraphy, Rocky Mountain area: Am. Assoc. Petroleum Geologists Bull., v. 44, no. 1, p. 1–20.

Williams, G. D., and C. F. Burk, Jr., 1964, Chapter 12 Upper Cretaceous, *in* Geological history ot western Canada: Alberta Soc. Petroleum Geologists, p. 169–189.

Wyoming Geological Association, 1957, Wyoming oil and gas fields symposium: 579 p.

Geology and Development of California's Giant—Wilmington Oil Field[1]

M. N. MAYUGA[2]

Long Beach, California 90801

Abstract The Wilmington oil field is near the southwestern margin of the Los Angeles basin of southern California, one of the most prolific oil-producing basins of the world and considered to be an example of optimum conditions in the habitat of oil. The Wilmington structure, discovered in 1936, is a broad, asymmetric anticline broken by a series of transverse normal faults which divided the producing reservoirs into many separate pools. The seven major producing zones range in age from late Miocene (Puente) to early Pliocene (Repetto). Production is primarily from sandstone beds of varied thickness and character, but some production is obtained from the basement schist and overlying conglomerate beds. Approximately 1,800-2,000 ft of nearly horizontal beds on top of the unconformity between the lower Pliocene Repetto Formation and the upper Pliocene middle Pico Formation conceals the Wilmington anticline from the surface. The effectiveness of the faults as barriers to communication between fault blocks is shown by significant variations in edgewater conditions, subsurface pressure, gas-oil ratio, and oil gravity from one fault block to another. Generally the development program in the field has been based primarily on segregation of the pools by fault blocks and zones.

The problem of land subsidence in the Wilmington oil field has been attributed by many investigators to the reduction of pressures in the reservoirs as a result of the withdrawal of oil and gas. Total subsidence to date in the center of the bowl of subsidence is 29 ft. A massive water-injection program has increased oil recovery and reduced subsidence at the center of the bowl from an annual rate of 2.37 ft in 1951 to 0.1 ft in 1967. The area of subsidence has been reduced from 20 to less than 4 mi^2.

By April 1, 1968, the Wilmington oil field had produced more than 1.156 billion bbl of oil, primarily from the old developed area. With water flooding, it is estimated that 500–700 million bbl will be recovered from the old area of the field. An estimated 1.0–1.2 billion bbl will be produced from the new area on the east (known as the Long Beach Unit or East Wilmington) within the next 35–40 years under a pressure-maintenance program. Recent developments in the eastern area revealed horizontal lithologic changes in the strata. To date, six of the seven known productive zones in the old area are also productive in the new area, but are somewhat limited in extent. Recent discovery of production from sandstones and fractured shale above the basement schist could add another commercial zone in the eastern area.

INTRODUCTION

The Wilmington oil field is in southern California, on the southwest coast of continental United States. It is in the Los Angeles basin, one of the most prolific oil-producing basins in the world. Approximately 20 mi south of the Civic Center of Los Angeles, the Wilmington oil field extends southeast from the industrial and residential areas of the Wilmington District of Los Angeles, through the Long Beach Harbor, and beyond the offshore area of the City of Long Beach. The entire structure is approximately 11 mi long and 3 mi wide, covering a productive area of approximately 13,500 acres; about 6,400 acres is in tidelands or submerged areas being operated by the City of Long Beach and its field contractors. From the discovery in 1936 to the end of 1967, the Wilmington oil field had produced more than 1.156 billion bbl of oil and 840 billion ft^3 of gas.

The Wilmington oil field has occupied a unique position in the oil industry of the nation. Since 1938 the field has been the largest oil producer in California, and since the middle of 1966 it has maintained the highest daily oil-production rate among the oil fields in North America. Wilmington's unique location in the hub of the highly industrialized and densely populated section of southern California and its proximity to all major oil refineries in the Los Angeles area have made it of great economic significance to the State of California. Since the early 1940s, the Wilmington oil field has attracted wide attention because of the unusual and costly problem of land subsidence believed to be related to the withdrawal of oil and gas. The threat to the multimillion-dollar industrial, port, and naval facilities within the area of subsidence spurred the City of Long Beach and the various industries and agencies in the area to seek a solution to the problem. On the basis of many years of study, it was concluded that

[1] Read before the 53rd Annual Meeting of the Association, Oklahoma City, Oklahoma, April 25, 1968. Manuscript received, May 7, 1968; revised and accepted, September 26, 1968.

[2] Assistant Director, Department of Oil Properties, City of Long Beach.

Appreciation is extended to the Director, Department of Oil Properties, City of Long Beach, for permission to publish this paper. The valuable assistance of members of the department in the preparation of this paper is gratefully acknowledged. The very helpful suggestions of Martin Van Couvering and U. S. Grant, IV, who reviewed this paper, are acknowledged with deep gratitude.

restoration and maintenance of subsurface pressures by water injection would prevent further subsidence. Today, the Wilmington oil field has the distinction of being the site of the largest water-injection operation in the country. The project has been a great economic success, as well as an effective subsidence-abatement program.

DEVELOPMENT HISTORY

The original discovery of the Wilmington oil field may be credited to the Ranger Petroleum Corporation's Watson No. 2, drilled in January 1932, in the northwestern part of the field. Initial production from a zone later known as the Ranger was 150 bbl per day of 14.1° API clean oil. At that time, the discovery was interpreted as merely an extension of the Torrance oil field and therefore did not stimulate a drilling boom.

In 1936, on the basis of a seismic survey, the General Petroleum Corporation (now Mobil Oil Corporation) drilled Terminal No. 1 to 6,814 ft, penetrating 27 ft of the basement schist complex. The well, on the north side of Cerritos Channel near Ford Avenue, was completed in what later was designated as the Upper Terminal zone, flowing 1,350 bbl per day of 20.5° API oil. The discovery started a drilling boom which extended southeastward to the Long Beach Harbor District.

By January 1942, approximately 1,000 wells were producing from five zones: Tar, Ranger, Upper Terminal, Lower Terminal, and Ford. On January 6, 1942, the Union Pacific Railroad Company well No. 215 was completed in the Upper Terminal zone and in a new productive zone below, which subsequently was named Union Pacific zone. In January 1946, the Long Beach Harbor Oil Company drilled "Cedarholm" No. 2 to the basement schist and placed the sandstones between the Ford and basement schist on production. The well flowed 1,370 bbl of 29.5° API oil with a cut of 3 percent and gas production of 737 Mcf on the first day. This zone, discovered 2 years before by Union Pacific well No. 237 but not produced, was named the 237 zone. The completion of "Cedarholm" No. 2 in the 237 zone was followed by an intensive and very competitive drilling activity which lasted for 3 years. Because of the small size of individual holdings in the townlot area, close spacing (2.3 acres average per well) became the general practice; as a result, peak production from the zone was reached within a relatively short period of time

and individual well production declined sharply.

Beginning in the early 1940s, land subsidence was observed in the Wilmington oil field in the shape of a bowl with the center or area of deepest subsidence on the east end of Terminal Island. Because of the increasing danger created by land subsidence, the City of Long Beach delayed further development of the field on the east until a solution to the problem was found. In June 1953, the city and its contractor, the Long Beach Oil Development Company, started a pilot water-injection operation in the upper HX sand of Fault Block V, Upper Terminal pool, to determine the feasibility of water injection for repressurization to control subsidence and increase oil recovery. Other pilot injection operations were started by the Union Pacific Railroad Company and Phillips Petroleum Company in 1954 and 1956, respectively. Experience gained in these pilot repressurization operations proved the feasibility of water injection both for subsidence control and secondary oil recovery. To expand the water-injection program for fieldwide application, it was necessary for the individual holdings in the field to be pooled under unit or cooperative agreements. After intensive work by participants, four fault-block unit agreements and a cooperative agreement were completed and put into effect between January 1959 and April 1964 in the previously developed area of the field.

Early in 1954, the City of Long Beach planned and supervised an offshore seismic survey to determine the southeasterly extension of the Wilmington anticline. The survey showed that the Wilmington structure extends at least 4 mi southeast and beyond the city's eastern boundary. The city, however, delayed new development on the east until the subsidence problem was brought under control. Late in 1959 and early in 1960, it became evident that the field repressurization program was successful. At the request of the Long Beach City Council, the petroleum staff of the Long Beach Harbor Department prepared and submitted, in November 1961, a comprehensive plan for the development of the eastern part of the Wilmington oil field by pressure-maintenance operation. On the basis of the plan, the voters of the City of Long Beach, by referendum on February 27, 1962, lifted the ban on offshore drilling east of the Harbor District. In accordance with the restrictions imposed under the plan, the number of drill-site islands was limited to four, each island not to exceed 10 acres nor to be built closer than 1,500 ft from the city's shoreline. It was required further that the development be under a single pressure-maintenance operation. Protection of the natural beauty of the shoreline and offshore areas of the city also was required.

Eight core holes drilled in 1962 verified the presence of producing zones similar to those found in the older area of the field. In general, core-hole information confirmed the accuracy of previous geologic and seismic interpretations.

Between late 1962 and early 1964, the City of Long Beach, the private landowners, and townlot operators in the upland area concluded a unit agreement under which the entire eastern area could be developed. The City of Long Beach also prepared a field contractor's agreement, which it intended to put out for bid, to select the field contractor who would perform the drilling and production operations for the city and the new unit. The proposed unit agreement was completed and presented for approval to the California State Lands Commission by the City of Long Beach, together with a draft of the field contractor's agreement. Both documents were approved during the latter part of 1964 by the commission and bids were called by the city in early 1965 to select the field contractor for the City of Long Beach, as well as for the new unit. THUMS Long Beach Company, a joint venture of Texaco, Inc., Humble Oil & Refining Company, Union Oil Company, Mobil Oil Corporation, and Shell Oil Company, was the successful bidder. The first well of the new unit (officially named the Long Beach Unit) was completed by THUMS in August 1965 from the Pier J drill site in the Harbor District. Four 10-acre drill-site islands were built by THUMS. Each island is capable of accommodating as many as 200 wells. As of April 1, 1968, the Long Beach Unit has completed a total of 279 oil-production and 60 water-injection wells.

REGIONAL GEOLOGY

Oil first was discovered in the Los Angeles basin in 1880 and, as of January 1, 1968, the Conservation Committee of California Oil Producers estimated that total production from the Los Angeles basin had reached approximately 5,706,644,300 bbl of oil from 35 fields covering approximately 47,670 productive acres. The Wilmington field produced about 20 percent of this volume. It is helpful to discuss the regional aspects and geologic features of the Los Angeles basin and their relation to the Wilmington oil field before describing the geology of the oil field itself.

Coastal southern California includes parts of three geomorphic provinces: the Coast Ranges, Transverse Ranges, and the Peninsular Ranges (Fig. 1). The western parts of all three provinces are submerged under the Pacific Ocean. The present Los Angeles basin is at the north end of the Peninsular Ranges province and is bounded on the north by the Santa Monica and San Gabriel Mountains; on the east, in part, by the Santa Ana Mountains; and on the south and west by the Pacific Ocean and the Palos Verdes Hills. The basin comprises about 1,250 mi^2. It is roughly rectangular with the long side extending about 50 mi northwest-southeast and the shorter side extending about 25 mi (Fig. 2). Edwards (1951) has estimated that the Los Angeles basin contains 2,250 mi^3 of sedimentary strata, but Barbat (1958) suggested an estimate of 1,600 mi^3 if strata unrelated to oil production were eliminated and only the beds within the drainage area of the basin oil fields were included.

The dominant structures in the Peninsular Ranges province are the northwest-southeast-trending fault zones which divide the province into roughly parallel block faults with varied histories of deformation and sedimentation. In

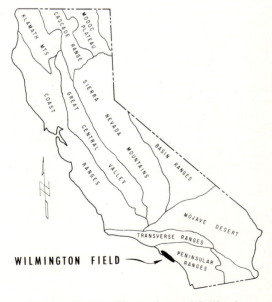

FIG. 1.—Geomorphic provinces of California.

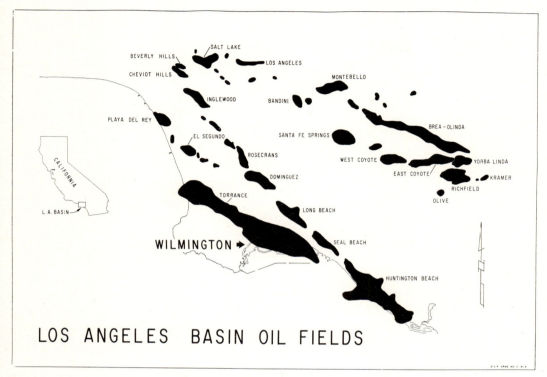

LOS ANGELES BASIN OIL FIELDS

Fig. 2.—Los Angeles basin oil fields.

the Los Angeles basin, this structural trend is reflected by topographic prominences parallel with or *en échelon* to these major faults. The topographic highs range from mountains along the eastern border to hills and knolls within the plain area. Most of the knolls within the plain reflect anticlinal structures which now are developed as oil fields.

Yerkes *et al.* (1965) interpreted the evolution of the basin in five major phases, each of which is represented by a distinctive rock assemblage (Fig. 3): (1) predepositional phase (basement rocks), (2) the prebasin phase of deposition (Upper Cretaceous to lower Miocene rocks), (3) the basin-inception phase (middle Miocene rocks), (4) the principal phase of subsidence and deposition (upper Miocene to lower Pleistocene rocks), and (5) the basin-disruption phase (middle Pleistocene to Holocene deposits).

The prebasinal strata of the Los Angeles basin, ranging from Upper Cretaceous to lower Miocene, appear to be marginal to some older basin unrelated geographically to the present site of the Los Angeles basin. Neither their distribution pattern nor lithologic character suggests that they are basinal. The beds were folded, perhaps faulted, and deeply eroded in most areas. Commercial oil production has not been developed from these strata. During the middle Miocene, deposition took place in some parts of the basin but a significant and widespread episode of emergence and erosion characterized the latter part of the epoch. Most of the middle Miocene rocks in some sections of the basin, including the Wilmington oil field area, probably were removed during this period. During late Miocene time a phase of accelerated subsidence and deposition began and ultimately gave the Los Angeles basin its present form. At the lowest part of the basin, subsidence and deposition continued without interruption into the late Pliocene.

In the marginal areas, tectonic activity continued, and unconformities within the lower and upper Pliocene sequence indicate geologic unrest in the southern and southwestern parts of the basin, including the Wilmington oil field area. It is evident that crustal movements have not ceased in parts of the basin since middle

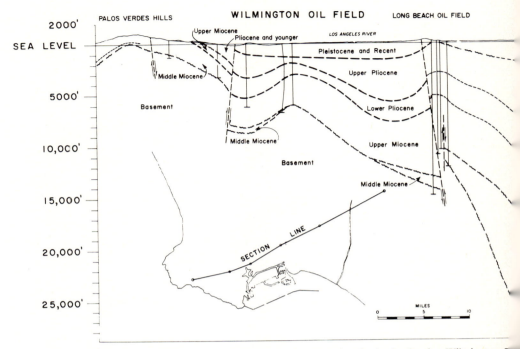

Fig. 3.—Generalized WSW-ENE structure section across Los Angeles basin showing Wilmington fi

Miocene time; these movements culminated during middle Pleistocene time and continue today. The most conspicuous tectonic movements of the middle Pliocene were the downfolding into a major syncline of the central basin area and the upward movements of the margins, followed by a certain degree of erosion of the exposed materials. The middle Pliocene events were important to the Wilmington-Long Beach area, because the long, gradual development of the Wilmington structure culminated during that time. Generally, most of the middle Pleistocene and post-Pleistocene movements in the Los Angeles basin took place along lines of weakness that had formed during middle Miocene time. Although the Pleistocene crustal movements affected the basinal strata in most of the Los Angeles basin, they apparently did not affect the Wilmington structure, which had been quiet since late Pliocene time. This is an interesting phase of the geologic history of the area, because the Pleistocene and Holocene strata in the Wilmington oil field are almost flat, but in nearby Signal Hill and Seal Beach oil fields they are steeply tilted and folded by recent movements along the Inglewood-Newport uplift.

In the Los Angeles basin, oil is produced mainly from lower Pliocene and upper Miocene rocks, and relatively minor production comes from the middle Miocene and upper Pliocene beds. Figure 4 shows a generalized correlation of the coastal fields on the southwest margin of the Los Angeles basin, where production is mainly from Miocene and Pliocene beds. Oil is produced also from the fractured, brecciated Jurassic(?) schist in the Wilmington, Playa del Rey, and El Segundo oil fields. With minor exceptions, oil has come from sandstone and conglomerate reservoirs. Ample source rocks probably were provided by the many thousands of feet of fine-grained strata interbedded with the reservoir rocks.

A nearly unique combination of factors and timing of events account for the productivity of the basin. The petroliferous sediments accumulated rapidly in stagnant, cool water more than 1,600 ft deep (Yerkes et al., 1965). The initially high organic content of the sediment was preserved because of poor circulation in the constricted basin and rapid filling. The interlayering provided a favorable conduit between finer grained source rocks and coarser grained reservoir sands. Among the many conditions enumerated by Barbat (1958) which contributed to the high productivity of the Los An-

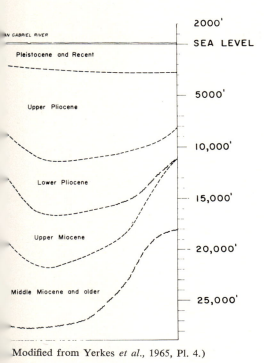

2000'

SEA LEVEL

5000'

10,000'

15,000'

20,000'

25,000'

Modified from Yerkes *et al.*, 1965, Pl. 4.)

geles basin area are the following.

1. Ample deposition of organic materials.
2. Adequate protection from chemical and biologic destruction.
3. Ample load compression, particularly in the cen-

ter of the basin, to squeeze entrapped fluids from the fine-grained rocks.

4. Ample interfingering of carrier and reservoir sands of lateral persistence with the fine-grained rocks.

5. Available traps of considerable size near the margins of the basin to remove hydrocarbons from the fluids before they were expelled.

6. Relative geologic youth of the area, absence of a longstanding load of superjacent rock, and a single moderate orogeny, resulting in little postdepositional alteration of sediments.

7. Absence of large-scale uplift and major erosion, with consequent preservation of reservoir fluids and pressures.

The broad Wilmington structure, near the southwestern margin of this great basin, with its many sandstone reservoirs, provided one of the ideal traps for hydrocarbons, which probably migrated from the area adjacent to the structure and other parts of the basin.

STRATIGRAPHY AND PETROLOGY

Overlying the Jurassic basement complex in the Wilmington oil field is 6,000–10,000 ft of Miocene, Pliocene, Pleistocene, and Holocene strata. The entire sedimentary section consists of varied sandstone-shale thicknesses, with a few thin beds of extremely well-cemented sandstone locally referred to as "shells" (Fig. 5). The degree of compaction of sediments generally is related closely to the depth of burial. The hard, dense shale at deeper levels grades to siltstone, soft claystone, and mudstone in the shallow beds. Sandstones that are partly indurated in the deeper zones are friable, less indur-

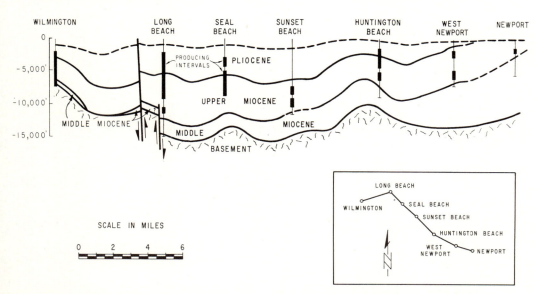

FIG. 4.—Generalized correlation of coastal oil fields, southern California. Vertical scale in feet.

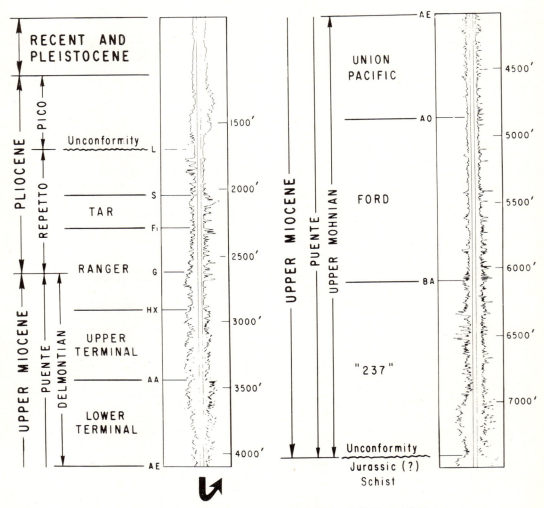

FIG. 5.—Composite log and stratigraphic units, Wilmington field.

ated, and unconsolidated at shallow depths. There are several local unconformities in the stratigraphic column of the field, but the most significant are those between the Jurassic(?) schist and the middle Miocene beds and between the lower Pliocene Repetto Formation and the upper Pliocene middle Pico Formation.

Basement Rocks

The oldest rock in the Wilmington field is the Jurassic(?) Catalina Schist at a depth of approximately 5,700 ft at the crest of the structure. Examination of several core specimens showed generally fine-grained, gray-green, chlorite-bearing schist, composed chiefly of chlorite,

quartz, muscovite, albite, glaucophane, lawsonite, and epidote in varied combinations and proportions. The schist appears to be a low-grade metamorphic facies. Some specimens show intense weathering. Veinlets of calcite and dolomite are common. Several core samples examined appear to be slightly metamorphosed basalt and shale. Many samples show well-developed schistose texture, and some contain feldspar porphyroblasts surrounded by a fine groundmass of siliceous materials and calcite. Weathering seems to have been intense in places and probably produced the pore spaces and fractures favorable for oil accumulation. Other samples are massive, granular, crystalloblastic rocks with lawsonite as the most prominent mineral, and glaucophane, muscovite, and chlorite as accessory minerals.

Similar schist is found underlying other oil fields in the southwestern part of the Los Angeles basin. Schoellhamer and Woodford (1951) described outcrops of schist in the Palos Verdes Hills and on parts of Catalina Island, which indicate that an extensive area is underlain by these rocks. However, the Catalina Schist is not known east of the Inglewood-Newport fault zone.

The schist is overlain by basal conglomerate, ranging in thickness from a few inches to several tens of feet, that consists of subangular to rounded fragments of schist and volcanic rocks.

Miocene Rocks

The Miocene rocks in the Los Angeles basin are of middle and late Miocene age. The Wilmington oil field has a thin middle Miocene section and a thick upper Miocene section. Approximately 3,700 ft of Miocene strata overlies the basement complex at the crestal part of the structure in the older area of the field. The Miocene section increases in thickness on the southeast as the basement slopes to greater depths. In the vicinity of Pier A of the Long Beach Harbor, the section increases to approximately 4,600 ft, and it is estimated that it may thicken to 7,000 ft along the crest of the structure in the eastern part of the field. Thicker Miocene sections are present on both flanks of the structure.

Most of the beds belong to the upper Miocene Puente Formation, except for approximately 350 ft of middle Miocene Topanga Formation overlying the basement schist in the General Petroleum Corporation Terminal No. 1. Previous workers in the area recognized the presence of Foraminifera belonging to the Luisian, Mohnian, and Delmontian stages of the late Miocene.

The entire Miocene section of the Wilmington structure is of great economic importance because it includes the following oil zones: the lower part of the Ranger, Upper Terminal, Lower Terminal, Union Pacific, Ford, and 237 (Fig. 5).

Middle Miocene (Topanga Formation).— The middle Miocene at the crest of the structure consists of alternating layers of coarse-grained, poorly sorted sandstone and dark-brown or gray, distinctly stratified shale. The lower sandstones are coarse grained, arkosic in

places, and range from brown, gray, to greenish, the last color caused by the presence of chlorite and greenish quartz, probably from the basement schist. The shale members, ranging from a few inches to several feet in thickness, are generally brown or dark gray, massive, and well stratified, with pyrite in places. Nodular shale near the base of upper Miocene beds contains abundant concretions or nodules which are generally greenish and brown, amorphous, phosphatic materials. Similar phosphatic concretions are described in other oil fields of the Los Angeles basin. In the Wilmington field, the appearance of these concretions during the coring of a deep well indicates proximity of the drill to basement rock.

Wissler (1943) classified the beds with the nodular shale in his Division E of the upper Miocene, and placed only 100–130 ft of beds just above the basement on the crest of the structure in his Division F, middle Miocene. He observed that the contact between the foraminiferal shale of his Division F and the overlying phosphatic shale of Division E is unconformable, on the basis of the distinct lithologic change and abrupt faunal break.

Upper Miocene (Puente Formation).—The upper Miocene Puente beds in the Wilmington oil field consist of poorly sorted, fine- to coarse-grained sandstones interbedded with claystone, siltstone, shale, and a few hard sandstone members locally referred to as "shells." Many sandstones show high feldspar content. The lower part of the section consists predominantly or medium- to coarse-grained sandstones with a few pebbles and of hard, well-stratified shale. The upper sandstone beds are less consolidated and generally finer grained than those of the lower section. Shale beds become more silty toward the top of the section and generally the upper beds are less indurated and softer than the lower members of the formation. The shale and siltstone are rich in Foraminifera and other organic and plant remains.

The upper Miocene sandstone, shale, and siltstone members range in thickness from a few inches to several tens of feet. A very few homogenous shale beds exceed 50 ft. The upper part of the Puente Formation has approximately 50–70 percent sand, the middle part has 25–35 percent, and the lower section has approximately 40 percent. The top 100–150 ft of the formation is characterized by a series of thin layers of shale locally referred to as "poker-chip shale" because of their resemblance, in core samples, to stacks of poker chips. These shale beds, which have a high diatom content and are referred to locally as "diatomaceous shale," are recognized easily on electric logs by their low resistivity and high self-potential values, similar in character to those of permeable sandstones saturated with salt water. The upper Miocene–lower Pliocene contact is placed approximately 40–60 ft above these beds near electric-log marker G of the Ranger zone.

Pliocene Rocks

Between 2,000 and 7,000 ft of Pliocene strata overlies the Miocene beds of the Los Angeles basin. The Pliocene rocks of the basin consist of a series of alternating beds of sand, silty shale, claystone, shale, and siltstone ranging in thickness from a fraction of an inch to several hundred feet. The shale and siltstone grade from light grayish green in the upper part of the section to dark brown in the lower part. The color gradation is accompanied by a gradual corresponding increase in the quantity of organic matter. Foraminiferal remains are abundant in most of the shale members of the section. In 1930 the Pacific Section of the Society of Economic Paleontologists and Mineralogists (Wissler, 1943) proposed the subdivision of the Pliocene beds into two distinct formational units: the Repetto Formation for the lower Pliocene and Pico Formation for the upper Pliocene. Wissler (1943) further subdivided the two formations into lower, middle, and upper subdivisions based on diagnostic foraminiferal assemblages. The Pliocene rocks in the Los Angeles basin are important economically because they contain some of the very productive oil and gas reservoirs in most of the oil fields.

In the Wilmington oil field there is 1,600–1,900 ft of Pliocene strata similar in character to those described in the Los Angeles basin. Between 800 and 1,100 ft of the Pliocene section belongs to the Repetto Formation and approximately 700–1,100 ft belongs to the Pico Formation.

The Repetto shale beds are soft and poorly indurated, ranging from brown to greenish gray and grading to very micaceous siltstone toward the upper part of the formation. The sands are generally unconsolidated, friable, and fine to medium grained with varied amounts of silt. The shale and siltstone are generally massive with abundant Foraminifera, mica, and some carbonaceous material. The Repetto section of the Pliocene includes one of the most extensive

oil zones in the field—the upper part of the Ranger zone—and therefore has great economic significance (Fig. 5). The less extensive Tar zone is in the upper section of the Repetto.

The Pico Formation unconformably overlies the Repetto. The lower Pico and perhaps the uppermost part of the Repetto, as well as part of the middle Pico, probably were eroded from the top of the structure, but there may have been no deposition at the top of the Wilmington oil field during early Pico deposition at the flanks of the structure and elsewhere in the basin. The Pico Formation consists of a series of sand and siltstone strata, with some claystone and hard shale beds. Both the sand and siltstone beds range from a fraction of an inch to several hundred feet in thickness. The sands are generally unconsolidated, poorly sorted, and range from fine to coarse grained. The Pico Formation is nonproductive in the Wilmington oil field and most of the sands are freshwater bearing.

Pleistocene and Holocene Rocks

Pleistocene and Holocene beds consist of unconsolidated sand, gravel, clay, and marl, with abundant shell fragments in places. Most of the sand and gravel beds were freshwater bearing, but because of rapid freshwater withdrawal inland, seawater has intruded the formation in the Wilmington-Long Beach Harbor area. The upper part of these beds, known as the "Gaspur zone," is the principal source of injection water in the Wilmington oil field. Water produced from the beds is seawater with very low oxygen content and devoid of marine life because of the filtration effect of the Gaspur sands. No oil or hydrocarbon gas has been found in these young sediments.

Structural Geology

The Wilmington oil field structure is a broad, gently folded, asymmetric anticline with a northwest-southeast axis (Fig. 6). During a period of emergence between the deposition of the Repetto and the upper Pico, the upper part of the Wilmington structure was eroded to a flat surface. Subsequent submergence of the area resulted in almost horizontal deposition of about 1,800–2,000 ft of upper Pliocene and Pleistocene beds on top of the anticline; these beds buried the structure from surface observation. The structure was broken by a series of normal faults transverse to the axis of the anticline. Bed dips range from a maximum of 20° on the north flank to approximately 60° on the

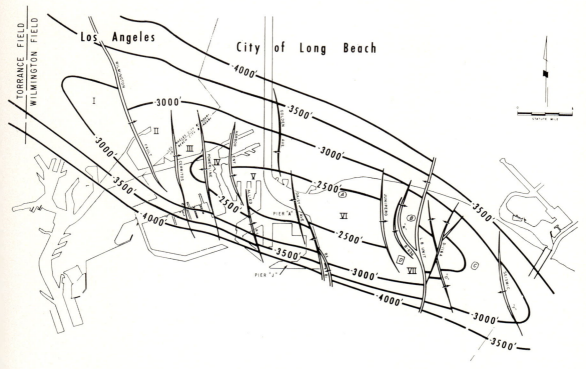

FIG. 6.—Structural contours; datum is top of Ranger zone, Wilmington field. CI = 500 ft.

south flank. The anticline plunges slightly to the northwest and a low saddle separates the Wilmington structure from the Torrance anticline. The structure also plunges gradually southeast, and a low nonproductive structural saddle separates the Wilmington structure from the Huntington Beach offshore oil field (Fig. 2). The upper Miocene and lower Pliocene beds appear to be drape folded over a basement high, with thinner beds at the crest of the structure and thicker beds on the flanks, probably as a result of both differential compaction and deposition on a rising structure. An exception, however, is the several hundred feet of strata directly overlying the basement at the highest part of the structure. These beds appear to be thinner on the southwest flank than on the crest (Fig. 9). This relation suggests that the top of the structure probably was tilted southwestward after deposition of middle Miocene and lower beds of the upper Miocene.

The Wilmington anticline is broken by a system of transverse normal faults which apparently were formed primarily by vertical components of tensional forces acting along the axis

of the anticline. Many of these faults terminate toward the flanks of the structure, although several seem to extend across the entire structure beyond the limits of development.

Some new faults appear *en échelon* with other faults toward the flanks of the structure and form a system of faults. An example of an *en échelon* fault system is the Harbor Entrance–Allied–Allied A-1 fault system (Fig. 6) which is discussed hereafter. Most of the major faults in the western part of the field dip east; however, some dip west and intersect the east-dipping faults, forming several horst and graben-type structures (Fig. 7). The strike of the faults ranges from north-south to N25°W, with dips between 45° and 65°. Several faults show evidence of lateral movements. None shows displacement above the Pico-Repetto unconformity; hence movements along the faults took place before late Pliocene deposition. Lower Pliocene Repetto sections are relatively thicker on the downthrown side of most faults, suggesting gradual displacements along the fault planes as the Repetto sediments were being deposited. Many faults divide the field

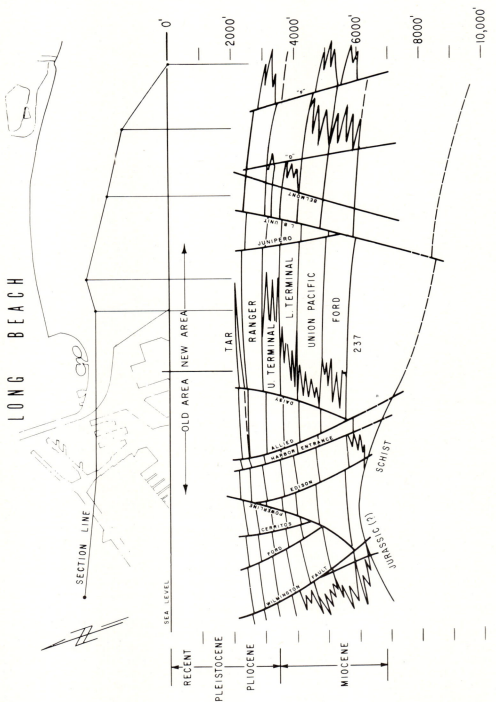

FIG. 7.—West-east geologic section along axis of Wilmington structure, showing approximate oil-water limits in producing zones. Vertical scale in feet. Length of section about 8 mi.

into separate fault blocks and form barriers to fluid migration and pressure communication. The zones within each block are classified as separate pools. Seven major fault blocks are recognized in the field, and the block boundary faults or fault systems from west to east are: the Wilmington fault, the Cerritos fault, the Powerline fault, the Harbor Entrance–Allied– Allied A-1 fault system, the Daisy Avenue– Golden Avenue fault system, the Junipero fault, and the Long Beach Unit fault (Fig. 6).

The Wilmington fault is an east-dipping fracture with a known maximum throw or vertical displacement of about 350 ft and average dip of 55°. It separates Fault Block I on the west from Fault Block II on the east. The Wilmington fault extends to an unknown distance on the north flank of the structure but terminates a short distance south of the axis.

The Cerritos fault dips east, with a maximum throw of 230 ft and an average dip of 45°. In contrast to the Wilmington fault, the Cerritos fault extends to an unknown distance south but terminates a short distance north of the crest of the structure. It is the recognized boundary between Fault Block II on the west and Fault Block III on the east.

The Powerline fault dips west, with a maximum throw of 270 ft and an average dip of about 55°. It diminishes in magnitude northward and terminates a short distance beyond the crest of the structure. On the south the amount of vertical displacement diminishes but the fracture extends beyond the present limit of development. The Powerline fault is the recognized boundary between Fault Block III on the west and Fault Block IV on the east.

The Harbor Entrance–Allied–Allied A-1 fault system consists of three east-dipping faults in *en échelon* pattern. The maximum vertical displacement of the Harbor Entrance fault is 325 ft and average dip is 50° east. This fault shows lateral movement. The apparent vertical displacement of the Harbor Entrance fault gradually disappears toward both flanks of the structure. On the south, the Allied and the Allied A-1 faults appear to have replaced the Harbor Entrance fault as stress-relief planes across the top of the structure. The Allied fault has a maximum displacement of 175 ft and an average dip of 45° east. It dies out at the crest of the structure but continues for a short distance on the south flank, where it appears to merge with the Allied A-1 fault.

The Allied A-1 fault has a maximum displacement of 200 ft and dips 47° east. It dies out at the crest of the structure but continues to an unknown distance on the south flank and has been considered to be the southern extension of the Harbor Entrance–Allied–Allied A-1 fault system, which separates Fault Block IV from Fault Block V on the east.

The Daisy Avenue–Golden Avenue fault system consists of three west-dipping *en échelon* faults. The Daisy Avenue fracture has a maximum throw of 110 ft and average dip of 60° west. It apparently terminates at the top of the structure and is present only on the south flank. The Golden Avenue fault, which has an approximate displacement of 50 ft and 65° dip westward, is present only on the north flank of the structure and appears to die out toward the top. The relatively minor Maine Avenue fault, with limited extent at the crest of the structure, is the third fault in the system. Together, as a system, these three faults separate Fault Block V from Fault Block VI.

The new development in the eastern area (Long Beach Unit) shows the presence of several north-south-trending faults of varied vertical and horizontal extent. Several of them had been identified in the seismic geophysical survey of 1954. The most significant faults discovered to date are the Junipero fault (formerly designated as Seismic B fault) and the Long Beach Unit fault. The Junipero fault, recently designated as the eastern boundary of Fault Block VI, is an east-dipping normal fault with an average dip of 60° and 200 ft of maximum vertical displacement. It is of limited extent across the top of the structure.

The Long Beach Unit fault (formerly referred to as the Seismic J fault) is a west-dipping normal and lateral fault with an average dip of 70° and a range of 200–400 ft of vertical displacement. It is the largest discovered to date in the Wilmington field and apparently traverses the entire width of the structure. The Long Beach Unit fault and the Junipero fault recently have been designated as the eastern and western boundaries, respectively, of Fault Block VII.

The major structural blocks are complicated further by secondary or minor faults, which subdivide them into sub-blocks. Fault Block II is divided into two segments by the east-dipping Ford fault. Between the Cerritos and the Powerline faults are numerous minor faults of relatively small displacement and limited horizontal extent. Within Fault Block IV, between the

Powerline fault and the Harbor–Allied–A-1 fault system, are the west-dipping Powerline B-1 and B-2 faults and the east-dipping Edison fault system which consists of three separate faults that subdivide Fault Block IV into several sub-blocks. Other faults also have been recognized east of the Long Beach Unit fault, but development in the area is insufficient to show their extent.

GEOLOGIC HISTORY

The Wilmington oil structure developed in a relatively unstable area in the southwestern margin of the Los Angeles basin. The Wilmington area was probably above sea level during Cretaceous and early Tertiary time but, like certain parts of the Los Angeles basin, it was submerged below sea level and received some sedimentary deposits during middle Miocene time. During the middle Miocene, compressive stresses forming a north-south couple folded the strata and established the present northwest-southeast trend of the Wilmington structure. A period of emergence followed and the area was the site of erosion or nondeposition during the rest of the middle Miocene. During the late Miocene, the Wilmington area, as well as most of the region occupied by the present Los Angeles basin, subsided and received basinal sediments. The early Wilmington structure on the southwestern margin of the Los Angeles basin apparently remained in prominent structural relief, and sediments were drape folded over the high.

The upper Miocene Puente mud and sand were deposited in relatively quiet water, perhaps near the shore. During the end of late Miocene time, however, the Wilmington area was apparently in deeper waters and a large percentage of mud and silt was deposited. Deposition continued uninterrupted through early Pliocene (Repetto) time. The southern and southwestern parts of the Los Angeles basin, including Wilmington, underwent geologic unrest during the early and late Pliocene. There probably were several periods of raising and lowering of the Wilmington structure during Repetto deposition, but the area apparently was below sea level and continued to receive sediments. The net effect of the oscillatory vertical movements of the Wilmington structure was a gradual uplifting of the area, increasing arching of the strata, and progressive movements along the fault planes at the crest of the structure. As movements along the fault planes were taking place gradually, Repetto sediments apparently were being deposited contemporaneously in greater volume on the downthrown side of the faults. The culmination of the emergence of the structure probably occurred during the early part of middle Pico (upper Pliocene) deposition. During this time, probably part of the middle Pico, all of the lower Pico, and perhaps part of the Repetto sediments were eroded off the top of the Wilmington structure.

The emergence of the area during lower Pico deposition culminated in the formation of the present anticlinal structure, and probably most of the movements along planes of weakness transverse to the anticlinal axis took place during the emergence. The area remained above sea level for some time and erosion cut an almost flat surface across the Wilmington structure. Lower Pliocene Repetto beds on top of the structure are truncated by the erosion surface. During upper Pico deposition, the Wilmington area submerged below sea level; upper Pliocene, Pleistocene, and Holocene sediments covered the flat erosion surface and buried the anticline under approximately 1,800–2,000 ft of horizontal younger beds (Fig. 9). Except for a slight seaward tilting of the area during the Pleistocene epoch, the Wilmington structure has been relatively quiet and unaffected by the tectonic forces that raised and faulted Pleistocene and Holocene strata along the nearby Inglewood-Signal Hill-Newport uplift.

OIL-PRODUCING ZONES

The eight major producing zones in the Wilmington oil field are lithologic units with generally similar reservoir characteristics and water tables. The zones are designated Tar, Ranger, Upper Terminal, Lower Terminal, Union Pacific, Ford, 237, and Basement. Figure 5 shows the electric-log markers identifying each zone and their positions in the stratigraphic column.

The reservoir rocks of all zones except the Basement are sand or sandstone in different degrees of consolidation, arkosic in places, and with varied silt content. The sandstones range from a few inches to several tens of feet in thickness. The percentage of sand in the zones ranges from 23 percent in the Union Pacific zone to 70 percent in the Upper and Lower Terminal zones.

Mechanism of primary production has been essentially solution-gas drive by pressure depletion. Several of the shallower reservoirs have a partial water drive, evidenced by some water encroachment and a more moderate pressure decline. The more permeable zones containing all but the lowest gravity crudes also have shown some gravity segregation, with a consequent formation of secondary gas caps.

Most of the faults have provided effective barriers to communication between fault blocks. However, the competency of the fault systems apparently dies out on the flanks, where there is pressure communication between most blocks through the aquifer. This interblock communication has created pressure gradients in previously unproduced reservoirs ranging from hydrostatic to two-thirds hydrostatic.

The geologic cross sections (Figs. 8–11) show the decreasing areal extent of the productive beds from west to east. Many productive units in the old area on the west are either nonproductive or of limited extent in the eastern area. The Ranger zone, however, has consistently shown a large productive areal extent on both flanks of the structure, although the oil-water contacts of the several units in the zone also have decreased in extent in the easternmost part of the field.

Table 1 summarizes the general reservoir characteristics and other data for the oil-producing zones in the field.

RELATION OF OIL ACCUMULATION
TO STRUCTURE

The major faults traversing the Wilmington structure have formed separate fault blocks and subpools. Oil gravities, edgewater limits, subsurface pressures and gas-oil ratios in similar oil zones differ from one fault block to another. Although the primary structural trap is the anticline, the accumulation of hydrocarbons apparently occurred independently in each of the major fault blocks. In many places, clean oil sandstones are directly across the fault from water sandstone. Oil gravity depends on structural position, with the higher gravities at higher structural positions and lower gravities at the lower positions. The segregation apparently took place independently within each fault block, as no relation in gravity is evident in adjacent fault blocks.

Productive capacities of wells from correlative zones across major faults show marked differences. Well productivity also is related to structural position. Subsurface pressures, which differed across faults by as much as 400 psi, in-

dicate the effectiveness of certain faults as seals or barriers. Fault Blocks I-IV appear to have pressure communication through the aquifer where fault displacements die out at the flank of the structure.

The subdivision of the Wilmington oil reservoirs into many independent subpools and individual reservoirs led to the development of the field by fault blocks. Wells usually are completed to produce from one fault block to avoid pressure intercommunication.

CHARACTER OF RESERVOIR FLUIDS

Generally, the character of the reservoir fluids and the pressure and temperature vary with depth. The oil and gas gravity, solubility of gas in crude oil, formation volume factors, viscosity of gas, and reservoir pressure and temperature increase with depth. In contrast, oil viscosity decreases with increasing depth. Figure 12 shows the temperature gradient in the Wilmington oil field to average 3.06°F per 100 ft of depth. The high temperature gradient has significantly affected the viscosity of the oil and contributed to a better mobility ratio. As was mentioned, oil gravity ranges from 12° to 34° API. Winterburn (1943) reported that oil below 17° gravity contains 1.7–2.5 percent sulfur. The sulfur content decreases with increase in gravity, to range from 0.6 to 0.9 percent for 29°-gravity oil. Gas gravity ranges from 0.58 to 0.90 (air = 1.0) and gas solubility from 90 to 530 ft³ per barrel. Reservoir characteristics are summarized in a previous section and are shown on Table 1.

PRODUCTION HISTORY

At the end of 1967, cumulative production of the Wilmington oil field reached 1.156 billion bbl of oil and 840 billion ft³ of gas. Figure 16 shows an interesting history of production since the discovery of the field in 1936. Initial flush production was followed by a slight decline as a result of market conditions. Production increased moderately during World War II and increased significantly after the war as deeper reservoirs were developed. Production declined steadily from 1951 to 1959, after which the water-injection program began to increase oil production. It is estimated that approximately 150 million additional barrels of oil had been recovered by waterflood stimulation by the end of 1967. Since 1965, oil production has increased sharply with the new development of the Long Beach Unit. It is antici-

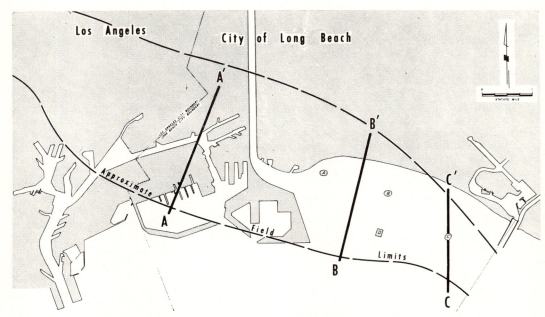

FIG. 8.—Cross-section index. **A-A′** is Figures 9 and 13; **B-B′** is Figure 10; **C-C′** is Figure 11.

pated that the Long Beach Unit will reach a maximum rate of about 150,000 bbl of oil per day sometime in 1970 and the entire Wilmington oil field probably will reach a peak production of 245,000 bbl per day in 1970 or 1971.

RESERVE ESTIMATE

If it is assumed that the entire western area of the Wilmington field will be subjected to waterflood operations in the future, it is estimated that 500–700 million bbl of additional

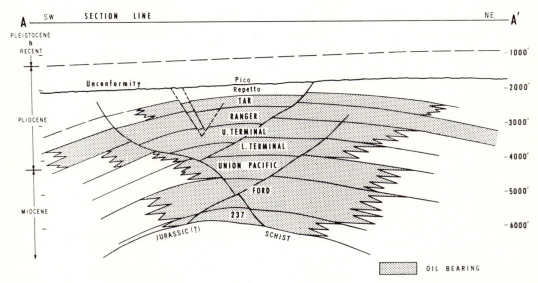

FIG. 9.—SW-NE transverse cross-section **A-A′** showing approximately edgewater limits of producing zones in old area of Wilmington field. Location on Figure 8. Vertical scale in feet. Length of section nearly 3 mi.

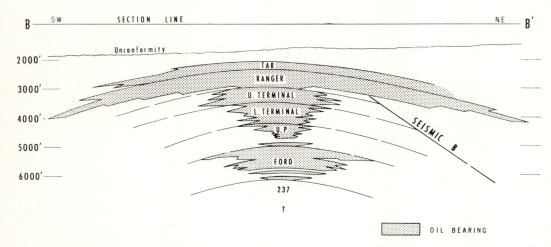

FIG. 10.—SW-NE transverse cross-section **B-B'**. Location on Figure 8. Vertical scale in feet. Length of section about 2.5 mi.

reserves will be recovered from the old area before the economic limit is reached. In the new Long Beach Unit area, it is estimated that 1–1.2 billion bbl of oil will be recovered under a pressure-maintenance operation during the next 35–40 years. The entire Wilmington oil field therefore could reach a total cumulative production of about 3 billion bbl of oil during its entire production history.

LAND SUBSIDENCE AND WATER-INJECTION PROGRAM

The land-subsidence problem in the Wilmington oil field has attracted considerable interest because of its effect on the multimillion-dollar industrial, port, and naval shipyard facilities within the subsidence area. Small-scale regional subsidence related to groundwater withdrawal or other causes was noted in the area as

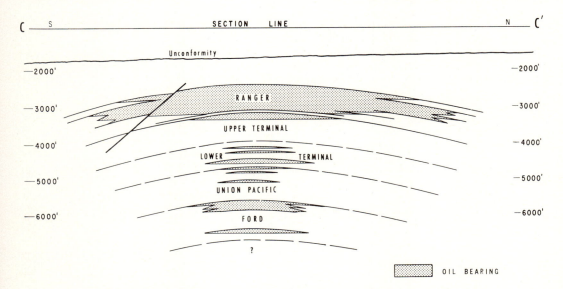

FIG. 11.—South-north transverse cross-section **C-C'**. Location on Figure 8. Vertical scale in feet. Length of section 2 mi.

Table 1. Summary of Reservoir Data, Oil-Producing Zones, Wilmington Oil Field

Zone	Electric-Log Markers	Approx. Zone Thickness (Ft)	% Sand In Zone	Gravity Range (°API)	Av. Porosity (%)	Weighted Av. Permeability (Md)	% Interstitial Water	Formation Volume Factor
Tar	S to F (old area)	250–400	40	12–15	30–40	1,000+	20–22	1.07
	T to F₁ (new area)	650–750						
Ranger	F to HX (old area)	400–600	30+	12–25	35	700 to 1,500	23	1.10
	F₁ to HX₁) (New area)	650–750						
Upper Terminal	HX-AA	400–850	50–70	14–25	35	700	25	1.11–1.18
Lower Terminal	AA-AE	500–800	50–70	20–31	30	450	25–29	1.11–1.18
Union Pacific	AE-AM (western area) AO-BA (central and eastern area)	400–900	20–25	27–32	20–25	150	35–45	1.33
Ford	AO (or AM)-BA	750+–1,200	25–35	28–32	25	100	40–45	1.28
237	BA-basement	200–1,200	40	28–32	25	275	43	1.30

early as 1928. However, significant subsidence did not occur until after oil-field development began in 1938 and 1939. The subsidence rate and the size of the affected area increased steadily during the following years, and the threat of inundation of the land area was serious because of the low harbor elevations.

The subsidence area is in the shape of an elliptical bowl superimposed on the Wilmington oil field structure. The deepest part of the bowl is over the crest of the structure, where the largest gross production per unit surface area has been obtained to date (Fig. 13). Cumulative subsidence between 1926 and 1967 was 29 ft at the center of the bowl on the eastern end of Terminal Island in the Long Beach Harbor District (Fig. 14). Areas along the edge of the bowl consequently were stretched; as much as 10 ft of horizontal surface movement toward the center of subsidence was recorded. The horizontal movements have caused extensive damage to wharves, pipelines, buildings, streets, bridges, and oil wells, and have necessitated costly repairs, replacements, razing of structures, and redrilling of oil wells. Total remedial costs may reach $150 million. Maximum rate of subsidence at the center of the bowl (2.37 ft; Fig. 15) was recorded in 1951, approximately 9 months after the then-developed area of the field had attained its maximum rate of oil and gas production (Fig. 16).

Early predictions by investigators of ultimate subsidence at the center of the bowl ranged from 7 to 12 ft, but were exceeded within a few years. Later projections were between 30 and 45 ft, although one investigator predicted

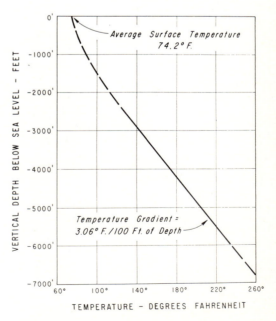

FIG. 12.—Temperature gradient, Wilmington Field.

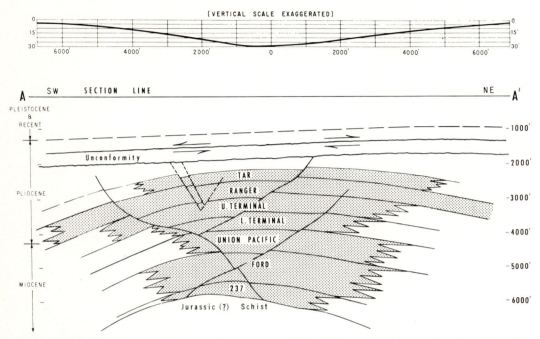

FIG. 13.—SW-NE transverse cross section **A-A'** showing profile of subsidence bowl on top of Wilmington anticline. Location on Figure 8. Vertical scale in feet.

70 ft. These predictions were made before it was found that repressurization of the reservoirs by water injection would ameliorate subsidence.

The following possible causes of subsidence were investigated: (1) lowering of hydraulic head by groundwater withdrawals; (2) oil-reservoir compaction from fluid and gas withdrawals; (3) compaction of clay and silt interbedded with the oil reservoirs; (4) surface loading by structures; (5) vibrations due to land usage; (6) regional tectonic movements; (7) lack of structural rigidity of the Wilmington structure; (8) movements along the known faults in the field; and (9) lack of preconsolidation in the sediments.

Most investigators agreed that withdrawals of fluids from the oil zones and the consequent lowering of pressure resulted in compaction of the oil-bearing sandstones and the interfingered siltstone and shale. The relative amounts contributed to be subsidence by compaction are still in question, but the land subsidence has placed strata at the center of the bowl under horizontal compression and those on the flanks under tension, with the amount of tension decreasing toward the margin of the subsidence.

These stresses were relieved several times by sudden horizontal movements along claystone and soft shale beds between 1,500 and 2,000 ft below the surface. The small shallow earthquakes generated by these sudden movements were recognizable because of their pattern on the seismograph. As a result of the movements, casings of several hundred oil wells were sheared or severely damaged along the planes of movement. Five such earthquakes were recorded between November 1949 and April 1961. A movement of 9 in. was observed along one of the subsurface horizons at about 1,550 ft after one of the earthquakes. A slow continuing horizontal creeping was also evident between earthquakes, as many oil wells were continually being damaged along suspected planes of movement. Cores recovered between 1,500 and 2,000 ft showed slickensides along several bedding planes at the horizons where most of the damage to the wells occurred (Fig. 13).

To determine the location and magnitude of compaction of the subsurface strata, a measuring method, using a magnetic "collar-counting" device, was developed by a consultant of the City of Long Beach in conjunction with Lane

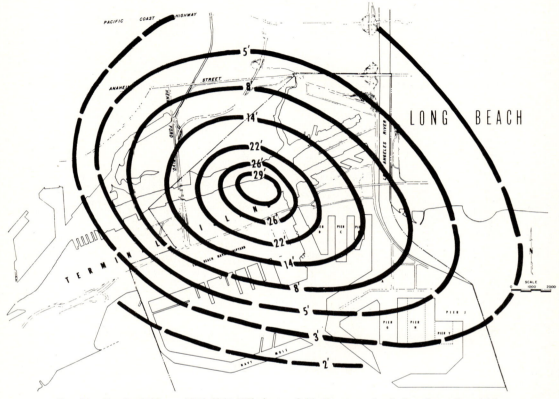

Fig. 14.—Total subsidence 1926–1967, Wilmington field. Contour value in feet, interval variable.

Wells Electric Logging Company (Allen, 1968; Brock, 1957). The method involves precision location of collar joints in a well and comparisons of the joint lengths with those measured before placing the casing in the well. Any compaction occurring between the surface and the bottom of a well must be reflected in a shortening of casing lengths opposite the compacting interval. To reflect zone compaction, the casing must be firmly gripped by the stratum, and the shortening per joint must exceed the accuracy of the measuring device. Many wells surveyed by this method showed shortening of 1 ft per 40 ft in many joints penetrating the oil zones, particularly opposite the four upper producing zones of the field. Figure 17 illustrates the results of "collar-count" measurements made on an oil well during three surveys covering a period of 20 years. Shortening or compression of the casing opposite the oil zones (Ranger and Upper Terminal) and the lengthening caused by tension in the upper part of the well are evident in the illustration. Each bar represents the

difference in hundredths of a foot between measured length of a joint of casing after compaction and the original casing tally. Accuracy of the joint-length measurement is estimated by Allen (1968) to be about 0.05 ft. Shooting radioactive bullets into the underground strata is another technique being used for direct measurement of formation compaction. No satisfactory results are available at this time.

If oil and gas production is, in fact, the primary cause of subsidence, the question of why the Wilmington oil field has subsided far more than surrounding oil-producing fields must be answered. Although no single answer may be satisfactory, a combination of the following probably explains this unusual phenomenon.

1. Higher percentage of unconsolidated or poorly consolidated sediments in the producing zones, where most compaction has taken place at Wilmington.
2. Lack of structural rigidity of the broad Wilmington structure, which was formed primarily by deposition and drape folding of sediments over a high, followed by relatively gradual uplift of the area, in contrast with intense folding and faulting that formed the

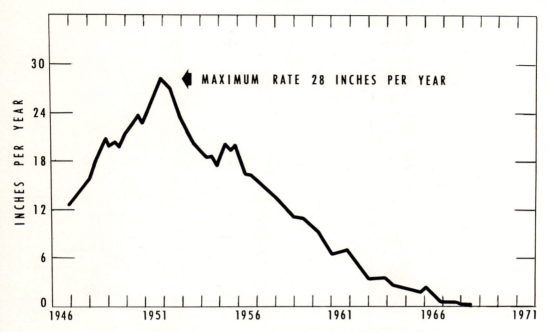

FIG. 15.—Subsidence rate, center of subsidence bowl, Wilmington field.

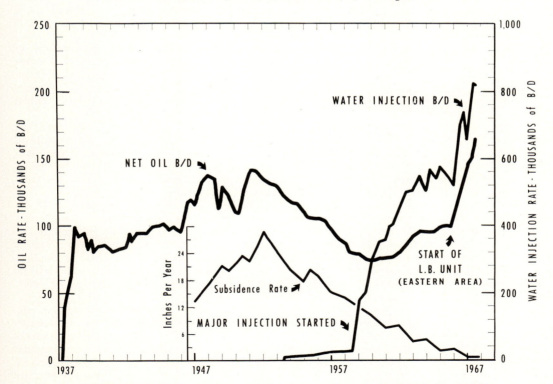

FIG. 16.—Oil-production rate showing subsidence and water-injection rates, Wilmington field.

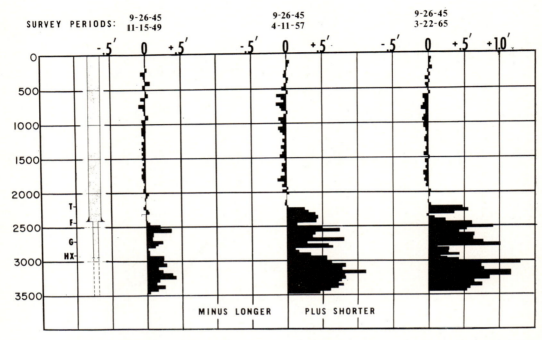

FIG. 17.—Joint-length difference between casing tally and measurement. From Allen (1968). Vertical scale in feet. For explanation of **T, F, G,** and **HX** see Figure 5.

neighboring oil structure along the Inglewood-Newport uplift.

3. Burial of the producing zones by incompetent and loosely consolidated strata incapable of supporting the weight of the overburden when subsurface pressures in the oil reservoirs were substantially lowered and compaction of the sediments occurred.

On the basis of the results of studies, it was concluded that solution of the subsidence problem depended on restoration and maintenance of pressure in the oil reservoirs. The first major water-injection operation was started in 1958, when the City of Long Beach instituted a 360,000-bbl-per-day water-injection program on the south flank of the structure. The city, as owner and operator of the tidelands part of the Wilmington field, and the upland oil operators extended water injection to other parts of the field. Four units and one cooperative operation were formed in the previously developed area and by April 1, 1968, approximately 750,000 bbl of salt water per day was being injected. The cumulative amount of water injected to that date was approximately 1.8 billion bbl.

Subsidence now has been stopped in much of the field and the area affected has been reduced from 20 mi² to approximately 3 mi². The present rate of subsidence at the center of

the bowl is 0.1 ft per year. A small surface rebound has occurred in the areas of greatest water injection (Fig. 18), by as much as 10 in. Figure 19 shows a surface rebound of more than 6 in. at Bench Mark No. 1350, established in 1952 on Pier A.

In addition to ameliorating subsidence, the water-injection program in the Wilmington oil field has been a great economic success. Approximately 70 percent of the present daily production in the old area of the field is credited to waterflood stimulation.

In the Long Beach Unit area on the east, development is based on a pressure-maintenance program designed to prevent significant reduction in pressure in the reservoirs. Water is being injected concurrently with production. Under present plans, 1.25 bbl of water will be injected for each barrel of voidage in the reservoirs.

As a subsidence-surveillance method, the City of Long Beach maintains a constant survey of the elevations of about 900 bench marks within and adjacent to the oil-development area. First-order level surveys are conducted on a quarterly basis. Reservoir pressures also are being monitored closely by periodic surveys of

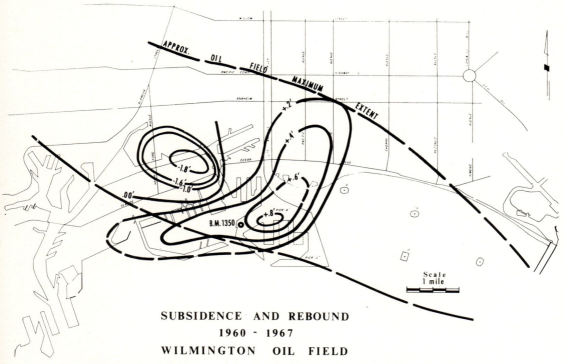

SUBSIDENCE AND REBOUND
1960 - 1967
WILMINGTON OIL FIELD

Fig. 18.—Subsidence and rebound 1960–1967, Wilmington field. From Allen (1968). CI = 0.2 ft
(except between —1.0 and —1.6).

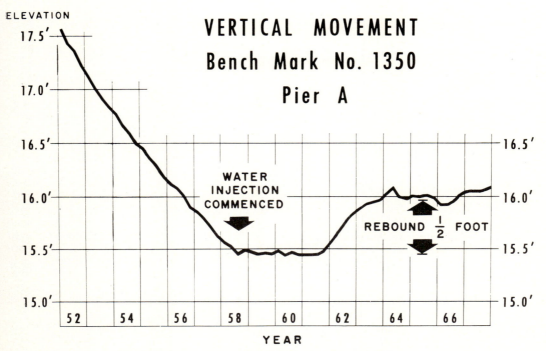

Fig. 19.—Vertical movement, Bench Mark No. 1350, Pier A.

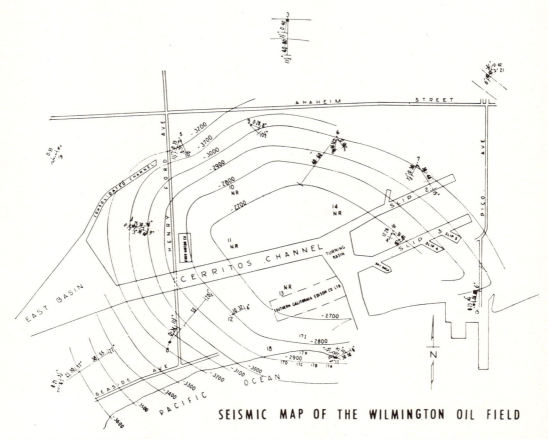

SEISMIC MAP OF THE WILMINGTON OIL FIELD

FIG. 20.—Seismic map of Wilmington oil field before discovery in 1936. CI = 100 ft. (After Salvatori, 1945.)

selected wells. Tidal gauges have been installed on the drilling islands off Long Beach as a means of detecting subsidence. Several strategically located wells also are logged on a yearly basis by the "collar-counting" technique to detect changes in casing-joint lengths which should reflect compaction in the oil reservoirs.

GEOPHYSICAL SURVEYS

Early in 1936, a seismic survey was conducted by the Western Geophysical Company of America for the General Petroleum Corporation, to outline the structure underlying the Terminal Island area within the harbors of Long Beach and Los Angeles (Salvatori, 1945). Generally, reflections from 10 shot points were obtained from several horizons ranging from 900 ft to more than 6,000 ft deep. The best reflections were from horizons ranging in depth from 2,000 to 3,000 ft. The

seismic-contour map in Figure 20 shows a large domal structure with a northwest-southeast trend. Shortly after the completion of the survey, the General Petroleum Company drilled the discovery well G. P. Terminal No. 1 and completed it on December 6, 1936. Subsequent development of the oil field showed a remarkable correlation between the seismic information and the geologic information from well data. No faults were shown on the seismic map of 1936 because the few shot points employed did not permit a detailed correlation of the horizons. The 1936 seismic survey extended only to Long Beach Harbor west of the Los Angeles River channel, but it was evident that the Wilmington structure extended an unknown distance southeast.

In anticipation of possible development of the eastern part of the Wilmington oil field, the City of Long Beach planned and supervised an offshore seismic geophysical survey, which was conducted for the city by the Western Geophysical Company of America in January 1954. It was necessary to establish generally the location of the anticlinal axis and the possible extent of the Wilmington structure east of the developed area. To obtain maximum coverage, a grid pattern of 28 seismic traverse lines was laid out with a total length of 65 mi, covering approximately 4,000 acres of submerged lands in the bay inside the 3-mi limit.

The survey operations used floating detector cables to form an L spread with 1,000-ft active segments on each leg. The recording boat towed the array from a yoke at the intersection of the legs. Tailboats at the far ends of the "L" maintained the configuration and orientation of the spread within preset tolerances. The shot was fired, as nearly as possible, halfway between the centers of both legs of the "L". A State Department of Fish and Game Inspector was present during the survey operations. As required by the California regulations of that period, the only explosive used was seismo-

graph-grade black powder, and the maximum permissible charge was 90 ib. Detectors used on the floating cables were gimbal-mounted, velocity-sensitive seismometers in streamlined mahogany fairings. Configuration and orientation of the spreads and shot point were determined by graphic means from the known dimensions of the cables, arrival times of shock wave through water at the detectors, and bearings of the yoke, cable ends, and shot point taken from a gyrocompass slave station on the deck of the recording ship. Locations of the recording ship were determined by shoran. All seismic data obtained were interpreted by means of plotted cross sections and dip-strike symbols on maps. The final results were presented in the form of maps of two horizons—the **a** horizon, which was correlated near the top of the Tar zone (middle Repetto), and the **b** horizon, which was estimated to be near the top of the Ranger zone of the lower Repetto (Fig. 21). The seismic cross section in Figure 22 shows the Repetto-Pico unconformity previously recognized in the old area on the west.

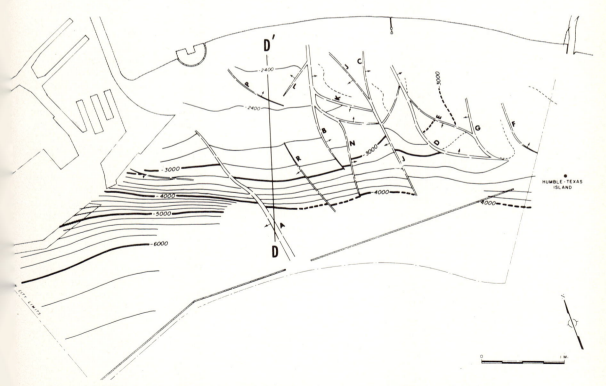

FIG. 21.—Seismic contours on **b** horizon near top of Ranger zone (1954). CI = 200 ft. **D-D′** is Figure 22.

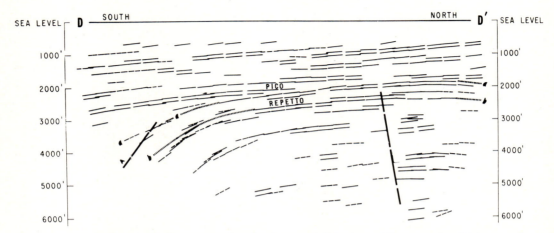

Fig. 22.—South-north seismic cross section showing **a** and **b** horizons. Note unconformity between lower Pilocene Repetto and upper Pilocene Pico beds. Location is **D-D'**, Figure 21. Vertical scale in feet. Length of section about 2¼ mi.

On the basis of the seismic data and information from the developed area of the field, structural-contour maps of the major horizons were prepared. Early estimates of oil reserves and preliminary planning for the development of the eastern area of the field were based on these data. Subsequent development showed that the geologic interpretation based on the seismic information was fairly accurate.

RECENT DEVELOPMENTS

Since July 1965, the Long Beach Unit, on the eastern part of the Wilmington oil field, has been the site of the most active oil-development operation in California. As of April 1, 1968, the City of Long Beach (unit operator) and its field contractor, THUMS Long Beach Company, had completed 279 production wells and 60 water-injection wells; daily production was approximately 100,000 bbl of oil. Four drill-site islands offshore from the City of Long Beach were completed, as well as the laying of 141,000 ft of submarine pipelines and approximately 55,000 ft of submarine electric cables connecting the islands to mainland facilities. Each island, which is approximately 10 acres, is capable of accommodating up to 200 wells and associated facilities.

Earlier drilling in the new area had confirmed the presence of at least six oil-productive zones similar to those in the old area on the west. Commercial production has been established from the Tar, Ranger, Upper Terminal, Lower Terminal, Union Pacific, and Ford zones. Recent completion of a deep test (D-

118) by THUMS established the presence of productive beds below the Ford zone in 1,290 ft of section above the basement schist east of the Long Beach Unit fault. Except for the Ranger zone, the areal extent of the productive oil zones in the eastern area of the field is narrower and more limited than in the western area.

Initial development of the Ranger zone in the Long Beach Unit indicated an average reservoir pressure of approximately 85 percent of hydrostatic (1,120 psi at −3,000 ft). The reservoir pressure in the Upper Terminal zone was only 60 percent of hydrostatic (870 psi at −3,300 ft). The low pressures were due to drainage caused by production in the older area of the field on the west.

The overall development of the new Long Beach Unit involves a water-injection, pressure-maintenance program under a single unit plan, with water-injection volumes and rates equal to at least 1.25 times the volumetric withdrawal from each producing zone. The program is designed to prevent land subsidence caused by reduction of pressures in the oil reservoirs and to obtain maximum economic recovery. The Ranger zone, which has the widest areal extent and contains 80 percent of the total reserves, is being developed initially on a modified staggered-line pattern, with the lines of injectors separated by three equidistant lines of producers, a producer-to-injector ratio of 3:1 (Fig. 23). The average well spacing is at a density of 13.5 acres per producing well or 10 acres per total well. The other zones are being

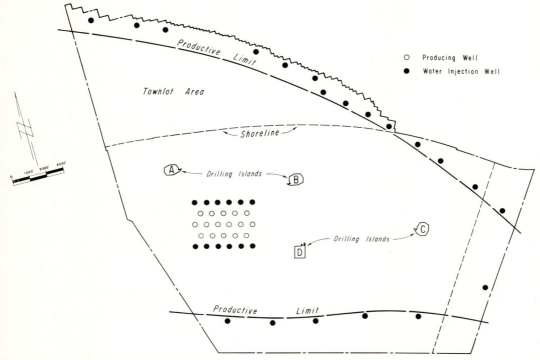

○   Producing Well

●   Water Injection Well

Fig. 23.—Long Beach Unit of Wilmington field, showing location of islands and a part of the injection pattern.

developed in a peripheral pattern of water injection, with full-zone producers at the crest of the structure and intermediate partial-zone producers along the flanks, surrounded by aquifer injectors. Under the present plan, it is estimated that 600–700 wells will be drilled by the Long Beach Unit, of which approximately one third will be water injectors.

In the older part of the field, all the upper four zones within the unitized areas of Fault Blocks II, III, IV, and V and in Fault Block VI are under various stages of waterflooding. Parts of the deep zones also are under a pilot waterflood, but poor water injectivity into these tight formations is presenting problems. Operators of nonunitized areas on the west, in Fault Block I and north of the present units, are planning to unitize or enter into cooperative agreements for water-injection operations. On April 1, 1968, 370 injection wells were operating in the old area of the Wilmington oil field, injecting approximately 750,000 bbl per day. Water-injection plants owned by the City of Long Beach have a total capacity of 1.6 million bbl per day and are capable of delivering water with surface pressures ranging from 800 to 3,000 psi.

Water now being used for injection is filtered seawater produced from shallow beds (100–400 ft deep) that are connected directly with the ocean. Plans also are under way to use fresh water, "renovated" sewage water, and filtered produced water for injection.

SELECTED REFERENCES

Allen, D. R., 1968, Physical changes of reservoir properties caused by subsidence and repressuring operations: Jour. Petroleum Technology, v. 20, no. 1, p. 23–29.
Barbat, W. F., 1958, The Los Angeles basin area, California: in L. G. Weeks, ed., Habitat of oil: Am. Assoc. Petroleum Geologists, p. 62–78.
Brock, L. W., 1957, Summary of down hole casing joint measurements since 1950: Petroleum Div., Long Beach Harbor Dept., unpub. rept., December 1957.
Conrey, B. L., 1967, Early Pliocene sedimentary history of the Los Angeles basin, California: California Div. Mines and Geology Spec. Rept. 93.
Driver, H. L., 1948, Genesis and evolution of Los Angeles basin, California: Am. Assoc. Petroleum Geologists Bull., v. 32, p. 109–125.
Edwards, E. C., 1951, Los Angeles region: Am. Assoc. Petroleum Geologists Bull., v. 35, p. 241–248.
Gilluly, J., and U. S. Grant, 4th, 1959, Subsidence in the Long Beach harbor area, California: Geol. Soc. America, Bull., v. 60, p. 461–529.
Grant, U. S., 4th, 1954, Subsidence of the Wilmington

oil field, California: California Div. Mines and Geology Bull. 170, p. 19–24.

Hoots, H. W., 1943, Origin, migration and accumulation of oil in California: California Div. Mines Bull. 118, p. 253–274.

Mayuga, M. N., 1963, Geologic highlights—easterly extension of the Wilmington oil field (abs.): Am. Assoc. Petroleum Geologists Bull., v. 47, p. 1774.

———— 1965, How subsidence affects the City of Long Beach: Landslides and Subsidence, Geologic Hazards Conference, The Resources Agency, State of California, May 26–27, 1965, p. 122–129 and 163–164.

———— and D. R. Allen, 1966, Long Beach subsidence, in Engineering geology in Southern California: Los Angeles Sec., Assoc. Eng. Geologists, Spec. Pub., October 1966, p. 281–285.

Reed, R. D., 1933, Geology of California: Am. Assoc. Petroleum Geologists, xxiv + 355 p. Reprinted and bound with Reed and Hollister (1936), in 1951, 1958.

———— and J. S. Hollister, 1936, Structural evolution of Southern California: Am. Assoc. Petroleum Geologists Bull., v. 20, no. 12, p. 1529–1704. Reprinted and bound as special volume, 1936, xix + 157 p.;

reprinted and bound with Reed (1933) in 1951, 1958.

Salvatori, Henry, 1945, Early reflection seismograph exploration in California: Geophysics, v. 10, no. 1, p. 17–33.

Schoellhamer, J. E., and A. O. Woodford, 1951, The floor of the Los Angeles basin, Los Angeles, Orange and San Bernardino Counties, California: U.S. Geol. Survey Oil and Gas Inv. Map OM-117.

Winterburn, Read, 1940, Effect of faulting on accumulation and drainage of oil and gas in the Wilmington oil field: Am. Inst. of Mining Metall. Engineers Tech. Publication 1154, Petroleum Technology, February 1940.

———— 1943, Wilmington oil field, in Geological formations and economic development of the oil and gas fields of California: California Div. Mines Bull. 118, p. 301–304.

Wissler, S. G., 1943, Stratigraphic formations of the producing zones of the Los Angeles basin oil fields: California Div. Mines Bull. 118, p. 209–233.

Yerkes, R. F., T. H. McCulloh, J. E. Schoellhamer, and J. G. Vedder, 1965, Geology of the Los Angeles basin, California—an introduction: U.S. Geol. Survey Prof. Paper 420-A, p. A1–A57.

Oil Fields of Pennsylvanian-Permian Horseshoe Atoll, West Texas[1]

E. L. VEST, JR.[2]

Houston, Texas 77001

Abstract The Horseshoe atoll is an arcuate chain of reef mounds, composed of mixed types of bioclastic debris, that accumulated in the interior part of the developing intracratonic Midland basin during late Paleozoic time. The atoll is 175 mi (282 km) long and locally is almost 3,000 ft (914 m) thick. The reef environment was established early in basin history and retained because of the limited amount of terrigenous clastic material transported to the basin interior. About 1,800 ft (549 m) of limestone accumulated during the Pennsylvanian, and primary dips commonly as great as 8° developed along the margins of the atoll. During earliest Permian time the reef was restricted to the southwest side of the complex, where more than 1,100 ft (335 m) of additional limestone accumulated before death of the reef. Reef mounds were buried by prograding Early Permian terrigenous clastic material which progressively covered the atoll from northeast to southwest. Westward tilting of the reef complex after burial elevated Pennsylvanian mounds along the east side of the atoll 1,400 ft (428 m) higher than Permian mounds along the southwest side. The updip migration of hydrocarbons was uninhibited in the lower part of the reef, and most mounds along the eastern half of the atoll are full to the spill point. Some mounds along the trend are not productive because of Wolfcamp sandstone contacts with the upper surface of the mounds. Fifteen individual fields, containing 2.54 billion bbl of recoverable oil, are present along the crest of the atoll. The Scurry field is the giant of the trend. It includes approximately 73,000 productive acres (295 km²), has a maximum oil column of 765 ft (233 m), and ultimately will yield 1.72 billion bbl. This field has no active water drive, so pressure maintenance was initiated early to achieve maximum efficient recovery. Scurry field has produced 521 million of the 857 million bbl of oil produced from reef rocks along the crest of the atoll. Scurry was discovered in 1948 with reflection-seismic methods, but only a small part of the field was mapped before the drilling of the discovery well.

INTRODUCTION

The late Paleozoic Horseshoe atoll (Adams *et al.*, 1951, Fig. 1) is a sinuous chain of limestone mounds that extends for a distance of

[1] Read before the 53rd Annual Meeting of the Association, Oklahoma City, Oklahoma, April 24, 1968. Published by permission of Chevron Oil Company. Manuscript received May 7, 1968; accepted, May 1, 1969.

[2] Staff Exploration Geologist, Chevron Oil Company, Corpus Christi District, Southern Exploration Division.

The writer gratefully acknowledges suggestions and criticisms of J. E. Adams, M. L. Rhodes, G. A. Risley, and R. H. Shepard. G. C. Hoelscher and O. G. Ward drafted the illustrations; D. M. Goin typed the manuscript.

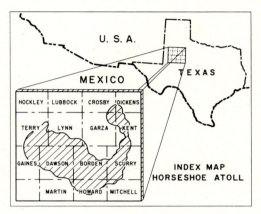

FIG. 1.—Index map showing location of Horseshoe atoll, West Texas.

175 mi (282 km) in the subsurface of western Texas (Fig. 1). The atoll and underlying limestone platform, termed jointly the Horseshoe reef complex (Myers *et al.*, 1956, p. 30), include an area of approximately 6,000 mi² (15,540 km²), occupy most of the northern end of the Midland basin, and contain the thickest Middle Pennsylvanian to Early Permian isolated limestone reef sequence known in North America.

The terms *atoll, reef, reef complex, bioherm,* and *mound* have been used by different writers to describe the horseshoe-shaped limestone complex, but none describes the limestone buildup adequately. Most petroleum geologists in West Texas, including the writer, use the term *reef,* but some object because of a lack of the remains of frame-building organisms and prefer either *bioherm* or *mound.* I do not consider these terms applicable to a limestone buildup as large as the Horseshoe, but use them interchangeably to describe small buildups along the trend. Objection is made to the application of the term *atoll* to the Horseshoe on the basis of the absence of a backreef or lagoonal facies during the time of major upbuilding after deposition of the lower Strawn platform. The term is not used commonly by West Texas geologists, but it has been used extensively in the literature and is used herein for convenience. In this paper a paleogeographic division of the limestone complex is made, yet there is no significant facies difference between the margin of the nearly circular ancestral platform and the younger members which have a more restricted, horseshoe-shaped pattern of distribution. Each part contains facies which have been called "reef" by many geologists. In order to define the Horseshoe in terms of its external geometry, the term *atoll* is used herein with reference to the almost horizontally bedded post-lower Strawn members of the complex, the term *platform* is applied to the tabular-bedded, more widespread lower Strawn limestones, and the term *reef complex* is used to describe the total paleotopographic entity including all forereef detritus, detritals, and fringing mounds.

Emphasis in this paper is on the regional conditions and physical factors which permitted the development of a significant limestone buildup that ultimately would serve as a major oil reservoir. The geologic history of any one field along the trend is intimately related to that of the other fields and inseparable from the history of the reef complex as a whole. The geometry of parts of Scurry and Cogdell fields is presented as typical of reef mounds along the trend. Reserve data and exploitation techniques are discussed for the Scurry field only.

Previous Work

Many reports have been published on parts of the Horseshoe atoll since 1951. The most comprehensive published studies of the reef complex were made by members of the U.S. Geological Survey during the period 1952–1959. Petrographic work by Bergenback and Terriere (1953) and stratigraphic studies by Heck *et al.* (1952), Rothrock *et al.* (1953), Myers *et al.* (1956), Stafford (1955, 1959), and Burnside (1959) probably have been the most widely distributed and are excellent sources for detailed information on the complex near the Scurry field.

ECONOMIC SIGNIFICANCE

Petroleum Reserves[3]

Fifteen significant individual reservoirs ranging in depth from 6,100 ft (1,859 m) to 9,900

ft (3,018 m) are present along the crest of the Horseshoe atoll, and jointly the fields contain approximately 2.54 billion bbl of recoverable oil. Only three of the fields have an estimated ultimate recovery in excess of 100 million bbl, and each is along the structurally high eastern side of the atoll at crestal depths ranging from 6,100 ft (1,859 m) to 6,600 ft (2,012 m). Ten of the 15 fields represent productive reef mounds that comprise parts of one major stratigraphic trap and contain 90 percent of the recoverable oil. The giant of the trend is the Scurry field, with an estimated ultimate recovery of 1.72 billion bbl.

Additional reservoirs have been discovered from limestone beds representing the forereef talus, from isolated mounds fringing the complex, from the limestone platform underlying the atoll, and from nonreef rocks overlying the reef complex. However, the ultimate recovery from all of these fields combined is probably little more than 100 million bbl.

Exploration History

The first well drilled into the Horseshoe atoll was the Gulf Oil Corporation No. 1-B Swenson Land and Cattle Company, in northeasternmost Garza County, 9 mi (15 km) northwest of Salt Creek field. This well was abandoned March 29, 1939, and, though unknown at the time, Gulf had drilled near the top of the only reef mound void of commercial production on the eastern side of the atoll. Nine years elapsed before production was established from the reef complex at the Seaboard Oil Company of Delaware No. 1-B J. C. Caldwell (OWDD completed January 5, 1948) which opened the Vealmoor field of northern Howard County at the southern edge of the atoll. Later that year the "reef boom" began with the completion of the Standard Oil Company of Texas No. 1 J. W. Brown 2 at the north end of the Scurry field in Scurry County, Texas.

The pace of exploration and development was frenzied from 1949 to 1953, and by the end of 1953 all individual reservoirs that produce from reef rocks along the crest of the Horseshoe atoll with reserves in excess of 5 million bbl had been discovered. Exploratory drilling continued at a high level through 1958 as efforts were made to establish production from limestone along the flanks of major

[3] The estimated ultimate recovery figures quoted herein are based largely on public testimony presented in hearings before the Railroad Comission of Texas. An estimate of ultimate recovery generally is available for fields for which secondary-recovery hearings have been held. Most of the smaller fields are produced by primary methods only, and estimated ultimate recovery for many of them is based on calculations made from the reservoir parameters presented at MER (Maximum Efficient Recovery) hearings for those fields.

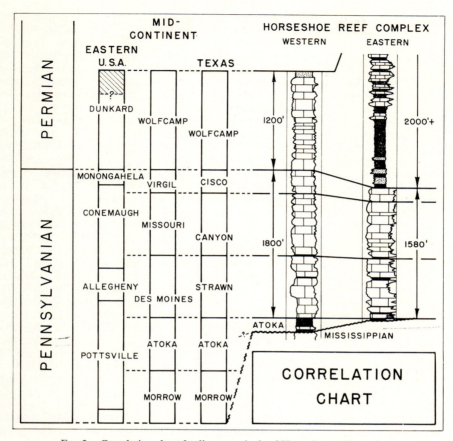

FIG. 2.—Correlation chart for limestone beds of Horseshoe reef complex.

mounds and from other rocks associated with the reef complex. A general decline in activity occurred from 1958 to 1968, but additional successful completions continue to be made along the edges of many of the older fields that were thought to be thoroughly exploited years ago. Approximately 5,500 wells have been drilled into the top of the Horseshoe reef complex, and almost 1,500 wells penetrate the entire reef mass. More data probably are available on the Horseshoe reef complex, though it is an entirely subsurface feature, than on any isolated limestone reef sequence of comparable size in the world.

GEOLOGY

The Horseshoe reef complex is composed of massively bedded bioclastic limestone and thin intercalated shale beds representing the Strawn, Canyon, and Cisco Groups of the Middle and Upper Pennsylvanian and the Wolfcamp Series

of the Lower Permian (Fig. 2). The maximum known thickness of the reef sequence is 2,920 ft (890 m). Pre-Strawn Pennsylvanian rocks in the eastern Midland basin consist generally of time-transgressive, onlapping, thin-bedded limestone, shale, and a few sandstone beds of Atokan age. These rocks unconformably overlie Mississippian limestone and shale and record the initial phase of a major marine transgression. They do not constitute a part of the Horseshoe reef complex.

Most of the limestone in the reef complex along the eastern side of the atoll in the areas of major oil accumulation is of Pennsylvanian age. The thickest reef section in a single well there is 1,535 ft (468 m) in the Standard Oil Company of Texas No. 5 J. W. Brown 2 at the north end of Scurry field. The maximum Pennsylvanian reef section penetrated in any well on the Horseshoe atoll is 1,800 ft (549 m) in the Standard Oil Company of Texas No. 1 G. F.

Pool in western Dawson County 75 mi (121 km) west. This gradual thickening attests to the uniform subsidence in the central part of the Midland basin during the Pennsylvanian. Pennsylvanian offreef shale beds are very thin by comparison with the time-equivalent reef facies. The top of the Pennsylvanian is difficult to define in the shale sequence, but the total thickness of the Pennsylvanian certainly is less than 500 ft (152 m) in most of the lagoonal area and along the southern margin of the atoll.

Permian limestone on the Horseshoe atoll is of Wolfcampian age only and generally is confined to the southern and southwestern edges of the complex. The limestone is best developed at the southwestern edge, and the maximum thickness in a single well is 1,120 ft (341 m) in the Standard Oil Company of Texas No. 1 G. F. Pool. Wolfcampian offreef shale and reef detritus are only slightly thinner than the reef facies, but most of this material is believed to be younger than the reef facies.

Limestone is the dominant lithologic type in the reef complex; shale represents less than 3 percent of the total section (Myers *et al.*, 1956, p. 21). Many types of limestone are present. Most is light colored and consists of skeletal remains of organisms which flourished and were buried without appreciable transportation from the areas in which they had lived. Limestone containing skeletal remains preserved in various degrees of sorting in matrices of carbonate mud and crystalline calcite is interlensed complexly with detrital limestone along the high parts of the atoll and with both micrite and poorly sorted limestone detritus in the inter-mound areas and along the basinward margin of the atoll. There is a conspicuous absence of the remains of organisms, other than algae, that are capable of erecting a rigid structural framework, and no core facies of this type is present. Correlation on the basis of lithologic types over appreciable distances is questionable. Thin shale members, commonly 1–10 ft (0.3–3 m) thick and in very few places exceeding 20 ft (6 m), are present throughout the complex and are important stratigraphic markers. The shale members are more persistent in the older part of the atoll.

Depositional Environment

The Horseshoe reef complex occupied a large part of the northern end of the Midland basin, and through most of its history the complex was in an open-marine area frontal to the undaform edges of shelf areas bordering the basin (Fig. 3). The position of the shelf edges remained generally stable during most of Middle and Late Pennsylvanian time. Channel areas between the reef complex and bordering shelves, and all of the Midland basin south of the atoll, were relatively starved for terrigenous sediment and contain thin time-equivalent rock units.

Limestone accumulation during upper Strawn deposition generally was restricted to the eastern, southern, and western edges of the ancestral platform. This accumulation probably represents the response of organisms living on the platform to the increasing nutrient supply along its margins, but perhaps partly to a temporary introduction of minor amounts of clay to the environment from a northerly source. In geometric configuration, the upper Strawn limestone bank resembles a horseshoe-shaped atoll, and the depositional topography established by the bank controlled the areas of deposition of younger reef limestone. Throughout the rest of the history of the Horseshoe reef complex, the upper surface of the reef was maintained near sea level as the Midland basin continued to subside. Minor amounts of clay introduced into the deeper parts of the basin accumulated at levels well below that of flourishing organic communities along the crest of the atoll.

Many terrigenous clastic wedges surrounding the atoll can be mapped readily. All thicken independently of the atoll northward and eastward toward the Matador archipelago and the Eastern shelf, where the upper surfaces of the wedges form recognizable shelf margins. The stratigraphic and present structural strike of the upper surfaces of these wedges differs from that of the reef facies in the Horseshoe atoll which they abut.

Early Tectonic Influences

The base of the Horseshoe reef complex has a regional westerly to southwesterly dip of 35 ft/mi (0°23'). The dip rate approximates 25 ft/mi (0°16') in the central and eastern parts of the complex and increases to 55 ft/mi (0°36') along the southwestern margin. Restoration of the top of the lower Strawn platform, isopachous maps of pre-Strawn rocks, and the distribution of small Permian(?)-age, tectonically controlled structures outline an area of broad, low-relief, intermittent structural arching under the ancestral platform in the atoll interior area. Maximum possible relief on this arch (Fig. 3) at any time during the formation

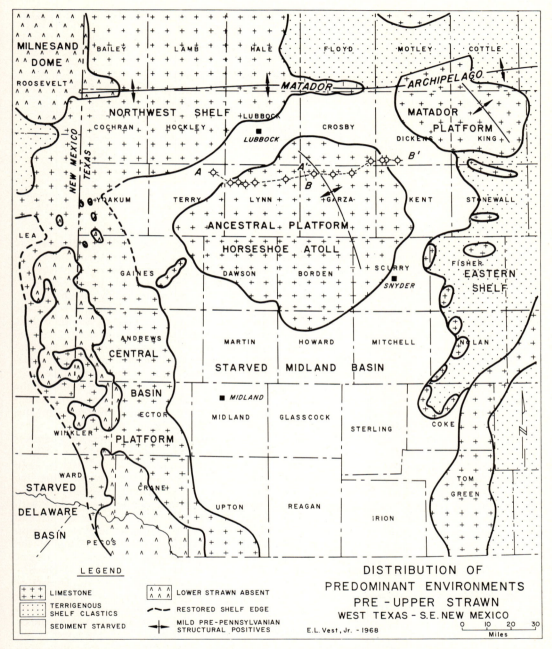

FIG. 3.—Predominant environments in shelf and platform areas bordering sediment-starved part of Midland basin, West Texas, near end of lower Strawn deposition, just before deposition of atoll limestone. Section **A-A'** is Figure 4; Section **B-B'** is Figure 5.

of the basal platform was about 150 ft (46 m) (Figs. 4, 5). At least five members can be traced across a large part of the basal platform (Fig. 6), and none shows evidence of time transgression. It is probable, therefore, that topographic relief on the arch did not exceed

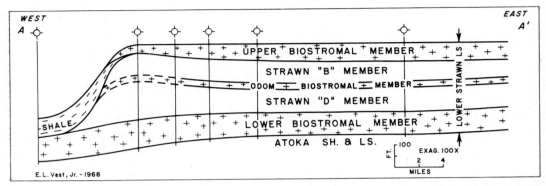

FIG. 4.—West-east restored cross-section **A-A′** near northwest side of pre-atoll platform, showing topographic relief along platform edge and tectonically controlled thinning toward platform interior. Location of section is shown on Figure 3.

50 ft (15 m) at any specific time, but the arch was important in providing ideal water depths for limestone accumulation over an area of 6,000 mi² (15,540 km²) near the center of the Midland basin. Limestone accumulated during periods when little clay was introduced into the environment, and 50–300 ft (15–91 m) of bank-to-basin relief was present along the margin of the platform when lower Strawn deposition ended.

Stratigraphy

Strawn Group.—Strawn limestone beds in the reef complex range in thickness from 340 to 760 ft (104–232 m) and are present in two distinct areas of distribution. Lower Strawn limestone makes up most of a basal platform and ranges in thickness from 340 ft (104 m) in the central part of the platform to an average of 500 ft (152 m) along the edge. Significant thicknesses of upper Strawn limestone are confined to the eastern, southern, and southwestern margins of the platform and represent the initial cycle of atoll formation. In areas of maximum development these limestone beds are about 250 ft (76 m) thick and commonly are referred to as the Strawn "A" zone. In many localities adjacent to the basinward edge of the Strawn banks less than 100 ft (30 m) of limestone is present. The total thickness of all Strawn offreef rocks ranges from as little as 100 ft (30 m) near the western and southern edge of the atoll to about 350 ft (108 m) near the northeastern edge.

Canyon Group.—Canyon reef limestone reaches a maximum thickness of 680 ft (207 m). Lower Canyon limestone, averaging about 400 ft (122 m) in thickness, is present along

the atoll wherever upper Strawn reef rocks are present. A notable exception is in the low area between the north end of Scurry field and the south end of Cogdell field, where only thin lowermost Canyon limestone is preserved. Upper Canyon limestone beds are of erratic distribution along the axis of the atoll and, where present, probably represent erosional remnants resulting from a prolonged low sea-level stand in post-Canyon time. Canyon limestone talus intertongues with thin shale beds low on the forereef and backreef flanks of the atoll, and in some areas talus is spread sufficiently far from the reef to merge with talus from bordering carbonate provinces.

Cisco Group.—Cisco limestone deposits have a maximum thickness of about 500 ft (152 m) and unconformably overlie both lower and upper Canyon limestone along the crest of the atoll. A significant volume of Cisco limestone is in stratigraphic juxtaposition with upper Canyon limestone along the western end of the atoll at structural levels that are commonly 250 ft (76 m) below the top of the Canyon. Thin Cisco limestone strata are present similarly along the crest of much of the eastern side of the atoll where the upper part of the Canyon Group is absent, as well as along the tops of high mounds where upper Canyon limestone has been preserved. Precise intra-group differentiation of Cisco limestones is difficult, but a large part of the limestone must have accumulated during a period of low sea-level stand. Much of the limestone is reworked indurated upper Canyon limestone, and mixed Canyon and Cisco faunas are common both along the crest of the atoll and along the fore-reef flanks. Only in a very few places, however,

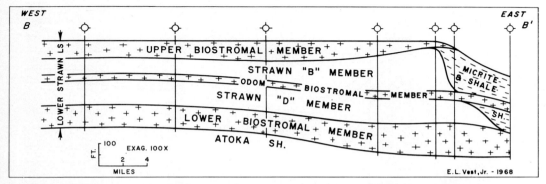

FIG. 5.—West-east restored cross-section **B-B'** near northeast side of pre-atoll platform showing topographic relief along platform edge and tectonically controlled thinning toward platform interior. Location of section is shown on Figure 3.

does Cisco detritus extend beyond the basinward limits of Canyon detritus.

Wolfcamp Series.—Atoll limestone representing the Wolfcamp Series is at maximum development (Fig. 7) along the southwestern edge of the complex, where it is 1,120 ft (341

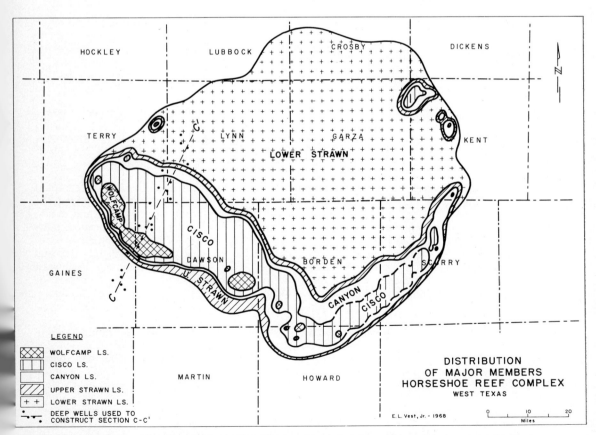

FIG. 6.—Distribution of major members, Horseshoe reef complex, West Texas, based on undaform edges or stratigraphic limits of major members. Offreef facies and minor fringing mounds are not shown. **C-C'** is Figure 7.

---

192   E. L. Vest, Jr.

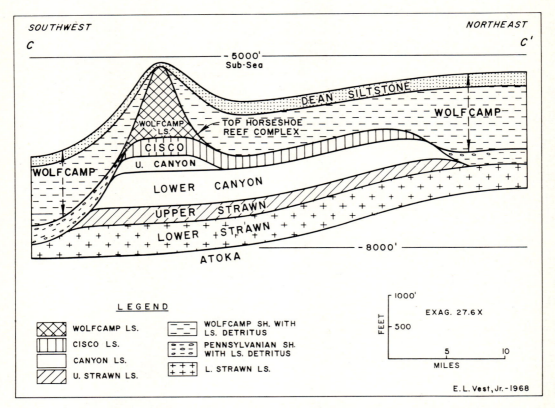

FIG. 7.—Southwest-northeast schematic cross-section **C-C'** through thickest known part of Horseshoe atoll. Offreef talus is included with basinal facies. Location of section and locations of wells that penetrated base of reef complex within 2 mi of line of section are shown on Figure 6.

m) thick. Unlike Pennsylvanian limestones, Wolfcamp reef rocks thin markedly eastward from mound to mound (Figs. 6, 8) and are absent at the top of mounds northeast of the East Vealmoor field. Wolfcamp limestone is present between mounds in this area and is sporadically present along the flanks of the atoll, but these lenses are believed to represent detritus that was moved eastward by longshore currents.

Burial

Burial of the reef mounds began in Early Permian time along the northeasternmost corner of the atoll as terrigenous clastic sediment in significant quantities started filling the previously sediment-starved parts of the Midland basin (Fig. 3). The clastic material prograded southwestward toward the deeper part of the basin and progressively covered younger parts of the atoll, until the highest reef pinnacles along the southwest side were buried by the

Dean siltstone near the end of Wolfcampian time. Terrigenous clastic sediment entered the basin cyclically from the northeast and north. Wolfcamp barrier reefs and associated talus around the margin of the Midland basin recorded the advance of the prograding deposits (Figs. 10, 11) by forestepping basinward across the clastic wedges during periods of clear water. Facies maps of many of the terrigenous sedimentary wedges show patterns typical of alluvial fans (Fig. 12).

Active building of limestone mounds along the atoll ceased long before burial, and it can be demonstrated that most mounds had foundered to depths of 1,000 ft (305 m) or more below sea level at the time crestal peaks were covered by Wolfcamp clinoform and fondaform sand and clay (Fig. 10). Reef mounds along the atoll generally are not productive where Wolfcamp sandstones are in contact with the upper surfaces of the mounds.

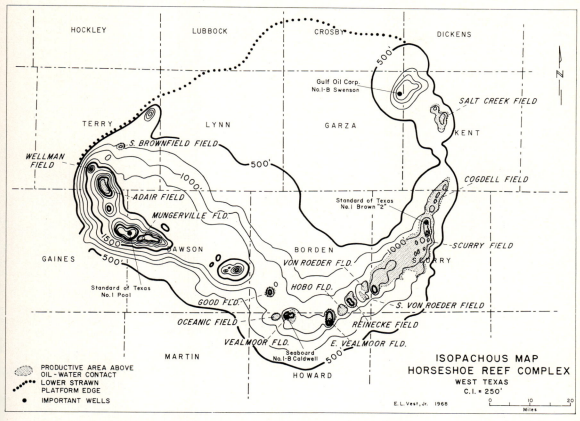

FIG. 8.—Isopachous map of Horseshoe reef complex, West Texas, showing location of significant production from reef limestone along crest of atoll. On this map basinal limestone detritus of Wolfcampian age is not considered part of Horseshoe complex.

Stratigraphic Discontinuities

Stratigraphic discontinuities throughout the Horseshoe reef complex are numerous and widespread. The discontinuities are more strikingly apparent as progressively younger rocks are involved, but most resulted from normal sedimentary processes in a continuously subsiding basin, where variations in bottom topography were substantial. Nowhere do these discontinuities appear to be related to local tectonic events. A significant unconformity with about 250 ft (76 m) of relief is present along the crest of the atoll at the top of Canyon limestone (Fig. 7). It is recognizable in provinces bordering the Horseshoe atoll and is attributed to a period of eustatically lowered sea level. The degree to which many other minor discontinuities within the reef complex are related to possible periods of lowered sea level is uncer-

tain. A second obvious discontinuity is present between the reef rocks of the atoll and the sand and shale which overlie the atoll. In all areas, with a possible exception of Salt Creek field, the nonreef rocks overlying the higher parts of the reef complex are of Wolfcampian age, and locally these rocks overlie limestone as old as lower Canyon. Nonreef rocks overlying the atoll limestone are successively younger southwestward.

Petroleum Entrapment

Burial of the Horseshoe reef complex was completed by the end of Wolfcampian time, but the Midland basin continued to founder until Late Permian. Maximum subsidence occurred west of the Horseshoe reef and resulted in a significant westward tilting of the entire reef complex (Fig. 8). Pennsylvanian reef

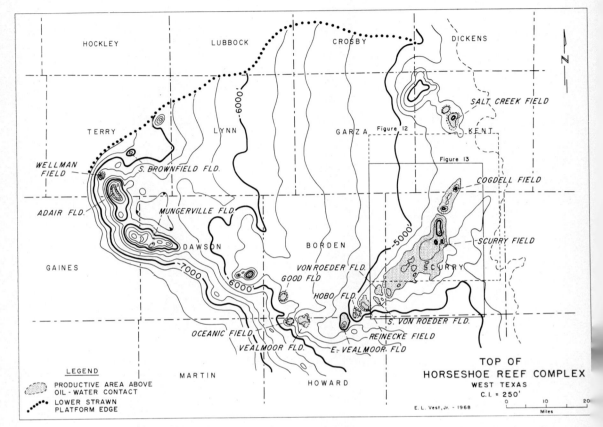

FIG. 9.—Structural contours of top of Horseshoe reef complex, West Texas, showing location of significant production from reef limestone along crest of atoll. Basinal limestone detritus of Wolfcampian age is not considered part of Horseshoe complex. Locations of Figs. 12 and 13 are shown.

mounds along the eastern side of the atoll are now 1,400 ft (427 m) higher structurally than Permian reef mounds along the western edge of the atoll (Fig. 9).

Good reservoir communication in the laterally more continuous lower parts of the reef is indicated by uninhibited updip migration of hydrocarbons. Pennsylvanian reef mounds along the eastern side of the atoll generally are full to the spill point, except for one buildup at the northeast end of the atoll, where the caprock is imperfect. Despite the fact that porous and permeable limestone members are separated by impermeable beds throughout the reef complex, the vertical migration of hydrocarbons also seems to have been uninhibited, though more tortuous. At the southwestern edge of the atoll several large Wolfcamp reef mounds exhibit more than 1,000 ft (305 m) of closure at the limestone-shale interface after the tilting, but

are either water saturated or contain reserves that are minor in relation to the volume of potential reservoir rock. Fondaform sandstone or siltstone (Fig. 7) overlies each of these mounds and evidently has not served as an adequate caprock. One small Wolfcampian mound (Wellman field) in this area, where the upper surface of the mound is covered by shale, contains 817 ft (249 m) of oil column (Anderson, 1953, p. 511). This indicates that adequate source rocks are present locally and that the absence of hydrocarbons in some mounds must be the result of leakage at the top of the mounds.

SCURRY FIELD—CANYON RESERVOIR

Definition

The term *Scurry field* is used in this paper to include the "Canyon" reef production of both Kelly-Snyder and Diamond M fields, as now

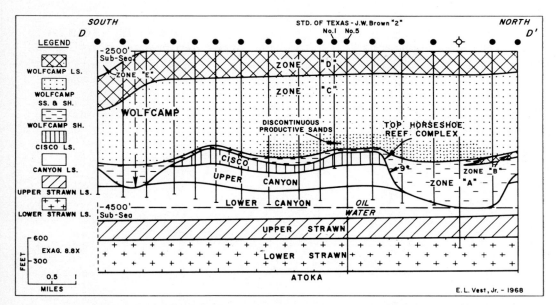

FIG. 10.—South-north cross-section **D-D'** along crest of Horseshoe atoll at north end of Scurry field, West Texas. Datum is sea level. Location of section shown on Figures 12 and 13.

recognized by the Railroad Commission of Texas. The "Canyon" reservoir, as recognized by the Commission, includes the limestone reef reservoir of Canyon, Cisco, and in places ear-liest Wolfcampian age. The Kelly-Snyder field includes most of the central and northern parts of the reservoir and contains approximately 85 percent of the reservoir volume. The Diamond

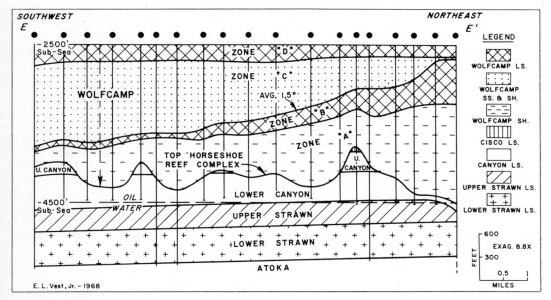

FIG. 11.—Southwest-northeast cross-section **E-E'** along crest of Horseshoe atoll through Cogdell field, West Texas, showing gradual westerly regional dip and relation of atoll limestone beds to overlying prograded Wolfcamp terrigenous clastic rocks. Datum is sea level. Location of section shown on Figures 12 and 13.

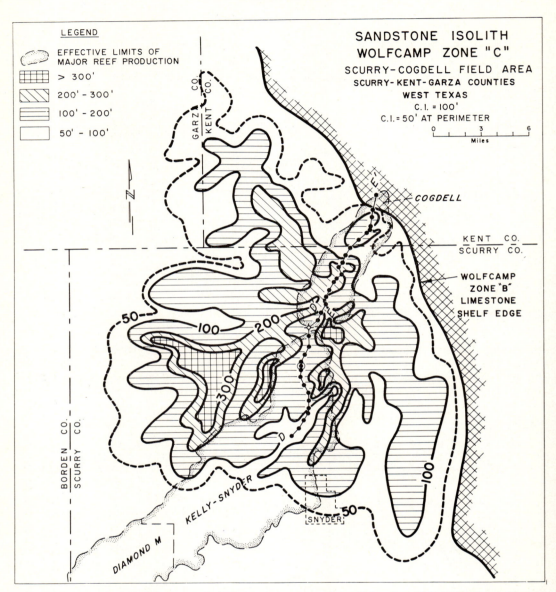

FIG. 12.—Sandstone isolith map. Map includes sandstone members thicker than 10 ft in Wolfcamp Zone **C** terrigenous clastic wedge, illustrated in Figures 10 and 11, overlying Horseshoe reef complex. Map is modified from unpublished map by R. E. Danielson. Locations of Figures 10 and 11 are shown.

M field includes the southernmost part of the reservoir and contains approximately 15 percent of the reservoir volume. More than 99 percent of Kelly-Snyder has been unitized for pressure maintenance and is operated as the SACROC (Scurry Area Canyon Reef Operators Committee) Unit. Most of the Diamond M field is operated by two pressure-mainte-

nance units, the Lion Diamond M Unit and the Sharon Ridge Canyon Unit.

The Scurry field (Fig. 13) is a productive part of the Horseshoe atoll that is approximately 25 mi (40 km) long and 2–9 mi (3–15 km) wide. It contains 88,000 acres (356 km²) above the approximate oil-water contact, which is placed at a subsea depth of 4,500 ft (1,372

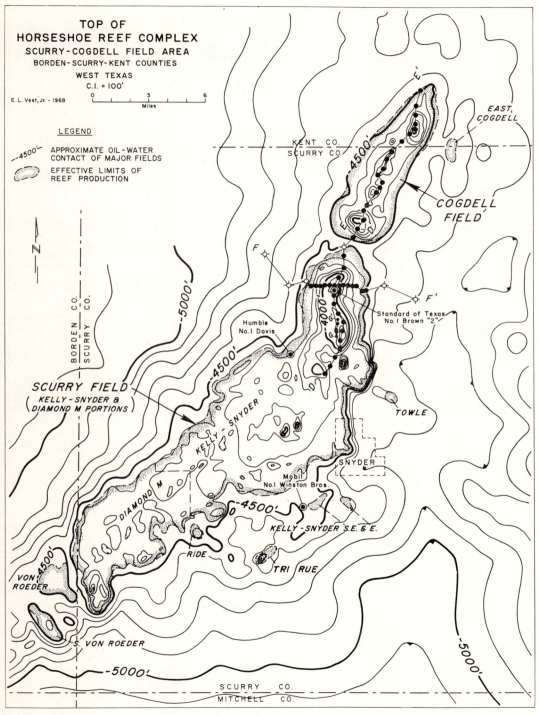

FIG. 13.—Structural contours of top of Horseshoe reef complex, Scurry and Cogdell field areas, Scurry and Kent counties, Texas. Map is modified after Stafford (1957, Map 1, p. 301). **D-D′** is Figure 10, **E-E′** is Figure 11, and **F-F′** is Figure 14.

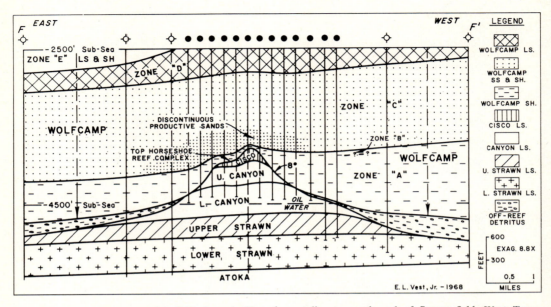

FIG. 14.—West-east cross-section **F-F'** through Horseshoe atoll near north end of Scurry field, West Texas. Pennsylvanian shale is interbedded with talus low on flanks of atoll. Location of section is on Figure 13.

m) (Fig. 14). Approximately 73,000 acres (295 km²) of the 88,000 acres (356 km²) is productive. Most of the *nonproductive* impermeable limestone *above* the oil-water contact lies along the northwest and southeast flanks of the reef in the central and southern parts of the field. The field contains 1,825 productive wells and was developed with an approximate well spacing of 40 acres (16 ha.). At the present time only about half these wells are used to produce the field allowable.

No data are presented for the many other individual pools associated with the primary "Canyon" reef pay in the Scurry field area. Though many of the pools are economically significant, the total estimated ultimate recovery from all these individual reservoirs is minor compared with the ultimate recovery from the "Canyon" reef reservoir.

Discovery Method

The first well drilled through the "Canyon" reef pay in the Scurry field was the Humble Oil and Refining Company No. 1 T. C. Davis, near the northwest edge of the field (Fig. 13). It penetrated approximately 200 ft (61 m) of gross pay section, but the interval was not tested before abandonment on January 28, 1947, at a total depth of 8,027 ft (2,447 m) in Ordovician Ellenburger dolomite. This well was reentered by the Standard Oil Company of

Texas in April 1950 and completed as a field extension; it flowed 1,509 b/d through a ¾-in. choke. On November 20, 1948, almost 2 years after abandonment of the No. 1 T. C. Davis by Humble, and 10 months after oil first was discovered from reef rocks in Howard County at the southern edge of the atoll, Standard Oil Company of Texas opened prolific Canyon reef production in the Scurry field with the completion of their No. 1 J. W. Brown 2 well near the center of the high pinnacle at the north end of the field (Fig. 13). The well was drilled to a total depth of 6,414 ft (1,955 m) and penetrated 155 ft (47 m) of reef section. It was completed from the reef for 532 b/d through a ¼-in. choke. The No. 1 J. W. Brown 2 completion followed by 2 weeks the completion of the official discovery well for the field, the Magnolia Petroleum Company (now Mobil Oil Corporation) No. 1 Winston Brothers, which was drilled low on the southeast flank of the reef complex but was completed from limestones underlying the atoll.

The No. 1 J. W. Brown 2 well was drilled after a reflection-seismic survey, and the area of the seismic anomaly closely approximates the north half of the large reef mound at the north end of the field. The continuous event mapped was believed to represent the approximate top of the Ordovician Ellenburger Group 150 ft below the reef complex; however, discontinuous events also were mapped at a depth which approximates the top of the reef limestone. Subsequent drilling suggests that most of the west, north, and east dips mapped from the deeper, more continuous seismic event resulted from the high interval velocity of the upper Canyon and Cisco reef section, and it can be demonstrated that the south dip which effectively closed the seismic anomaly was spurious. It was not evident that the field ultimately would extend more than 20 mi (32 km) southwest of the seismic anomaly.

External Geometry

The trend of the Horseshoe atoll through the Scurry field area (Fig. 13) is approximately northeast-southwest. The northwest flank of the reef is of generally uniform strike and dip, but the southeast flank is very erratic, and the effects of destructional processes are evident. The top of the reef complex along the crest of the atoll is unpredictable from well to well. Most of the flank surface is concave upward (Fig. 14), and flank dip generally is less than 8° but locally may exceed 20°. Dip rates are consistently steeper along the southeast flank, and most of the local reef promontories are along the east side of the atoll.

Three irregular depressions trending approximately 90° from the long axis of the reef divide the upper surface of the reef mass into four erratically shaped areas of similar size. Three low cols more definitely separate upper limestone members in the Cogdell field into four smoothly shaped mounds. The most pronounced transverse depression is between the Scurry and Cogdell fields.

Early workers (Van Siclen, 1950, p. 72) compared the reef topography in the Scurry field to that of an eroded divide and noted that many of the valleys cut more deeply into the reef flank than the crest. In many places topographic noses extend basinward from the saddle or valley areas. The similarity of these noses to alluvial fans is striking. A very broad basinward protuberance of the reef mass is present along the southeastern side of the Scurry field (note −4,700-ft contour southwest of the town of Snyder, Fig. 13) adjacent to the central part of the field where much of the upper Canyon limestone is absent. This protuberance contrasts sharply with the steep flank present farther north, where upper Canyon limestone has been preserved along the crest of the atoll.

Maximum relief is along the northern end of

the field, where the top of the complex is 765 ft (233 m) above the oil-water contact which is the spill point for the field. Average relief above water in the broad central section is 250 ft (76 m). An additional 100 ft (30 m) of reef section below the oil-water contact is closed, but important production has not been found in this interval. The upper 50 ft (15 m) of this interval produces locally where reservoir quality is adequate, especially at the northern end of Cogdell field. The lower 50 ft (15 m) apparently is nonproductive because of downward structural readjustments after entrapment. This is also evident at the northern end of Cogdell field, which represents the updip limit of a chain of 10 separate reservoirs that are full to the spill point everywhere south of Cogdell field. Areas of local closure along the flanks of the atoll between −4,500 and −4,600 ft subsea (−1,372 and −1,402 m) in the Scurry-Cogdell fields area are productive where adequate porosity and permeability are present.

Internal Geometry

Internal correlations within the reef complex are difficult only for upper Canyon and Cisco limestones. Thin shale beds are present throughout the complex and are invaluable for subdivision of the reef mass, but they are discontinuous in the upper part of the reef. Group boundaries generally coincide with shale members. Near the reef margin the thin shale beds are discontinuous at all horizons, and paleontologic data are necessary if group boundaries are to be traced. Accurate intragroup correlations can be made for short distances by the use of microlog surveys, and this technique was used extensively by Stafford (1955, 1959), Myers et al. (1956), and Burnside (1959). Nonporous members generally are either shale, carbonate mud, or detrital limestone. Porous members are mostly light-colored limestone and consist largely of skeletal remains preserved as calcarenite. The correlation of generalized limestone types to porous and nonporous beds (Myers et al., 1956, p. 24) and the apparent continuity of porous and nonporous beds over moderate distances on the basis of microlog correlations attest to the rhythmic nature of limestone accumulation (Fig. 15), even though specific cycles are not recognized. Core data are available from less than 5 percent of the wells drilled, but are sufficient to document the flat-bedded nature of the buildup and the absence of massive, nonbedded "reef core."

Most of the porosity in the Scurry field is

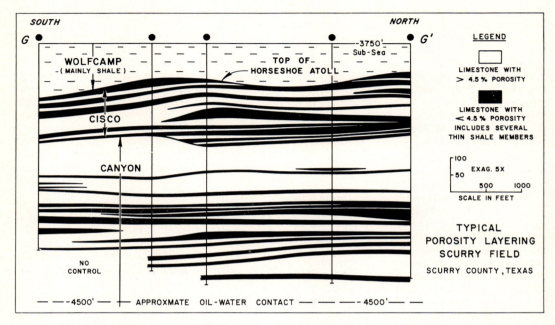

FIG. 15.—South-north cross section **G-G'** showing typical porosity layering in Scurry field (after Myers *et al.*, 1956, Pl. 2). Thin members with low porosity and low permeability are not effective pay and inhibit vertical communication of reservoir fluids. Location of section shown on Figure 13.

secondary and consists of small interconnected vugs (Bergenback and Terriere, 1953, p. 1023), and in many places leaching apparently occurred without regard to grain size. The alternation of leached and nonleached members may be the result of periodic exposure to meteoric water, but most of the leaching could have occurred during a period of protracted erosion in post-Canyon time. The repetitive nature of porous and nonporous beds throughout the reef complex, and the reasonably flat bedding of these members relative to the upper surface of the reef, led to the institution of a unique center-to-edge plan of waterflooding when pressure-maintenance operations were initiated.

Production and Reserves

Through January 1, 1968, the Scurry field had produced approximately 521 million bbl of oil. Water has been approximately 9 percent of the total liquids produced, and casinghead-gas production is estimated to be 0.5 Tcf (14.2 billion m³). Average monthly production during 1967 was approximately 2.9 million bbl but a much greater rate is possible under emergency conditions.

The estimated ultimate recovery for the field, with current methods of pressure maintenance, is 1.72 billion bbl based on an estimated recovery of 50 percent of the original oil in place. A recovery factor of less than 20 percent is estimated if only primary reservoir energy were utilized.

The common original completion practice for field wells was to bottom the well 50–100 ft (15–30 m) above the approximate oil-water contact and complete naturally from open hole. Many operators set pipe through the pay section and selectively perforated porous intervals. A few operators acidized the pay section, but the type of acid treatment was variable.

KELLY-SNYDER PART OF SCURRY FIELD

Reservoir Parameters

Determination of reservoir parameters that are applicable to the entire Scurry field is not possible from the incomplete data collected from many individual operating units within the field. The Kelly-Snyder part of the field (Fig. 13) includes the northeasterly 73 percent of the productive area and contains 85 percent of the effective reservoir volume. The reservoir parameters presented in Table 1 apply to the Kelly-Snyder area only.

Pressure Maintenance

Early in the history of the Scurry field it was recognized that the producing mechanism was a solution-gas drive. The formulation and early execution of plans to prevent waste and perhaps triple the estimated ultimate recovery by pressure maintenance illustrate the cooperation achieved among the members of the oil fraternity. Three significant pressure-maintenance projects are operative within the Scurry field area. The largest is the SACROC Unit which operates the Kelly-Snyder segment. In the Kelly-Snyder area alone there are more than 300 working-interest owners and more than 2,500 royalty owners, yet 99.5 percent of the reservoir in the area is unitized. Most of the remaining 0.5 percent represents parts of the reservoir that were developed after unitization agreements had been executed.

The average reservoir pressure in the Kelly-Snyder area is plotted against the production rate per calendar day in Figure 16, to illustrate reservoir performance. A rapid pressure decline occurred during the early life of the field, and by the end of 1951, shortly after most of the development drilling had been completed, average reservoir pressure fell below the bubble-point pressure of 1,800 psi (127 kg/cm²). At the time pressure-maintenance operations were begun in September 1954, average reservoir

Table 1. Reservoir Parameters for Kelly-Snyder Part of Scurry Field[1]

Rock Properties	
Porosity (%)	
Net (>3%)	10.03
Gross	7.11
Permeability (md)	
To air horizontally	30.6
Fluid Properties	
Gravity of oil (°API)	42
Paraffin base	
Sulfur (%)	0.183
Formation volume factor (bbl/bbl @ IRP)	1.500
Formation volume factor (bbl/bbl @ SP)	1.525
Viscosity of oil (cp @ SP)	0.375
Gas cap	None
Gravity of gas	1.13
Solution gas-oil ratio (CFPB @ SP)	1,000
Water salinity (ppm)	159,000
Water saturation (%)	28.2
Pressures	
Initial reservoir pressure (psig @ −4,300 ft)	3,122
Bubble-point pressure (psig)	1,800
Critical gas-saturation pressure (psig)	±1,600
Drive Mechanism	
Original	Solution gas drive
Present	Artificial water drive
Pay Section	
Maximum oil column (ft)	765
Average pay thickness (ft)	229
Est. oil in place (billion bbl)	2.83
Est. recovery (%)	51.7

[1] These data were presented in public testimony in an MER hearing before the Railroad Commission of Texas on July 28, 1967, by Standard Oil Company of Texas acting as unit operator for the SACROC Unit.

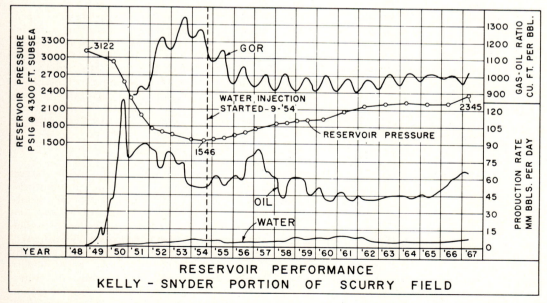

FIG. 16.—Reservoir performance for Kelly-Snyder part of Scurry field, West Texas. Variation in production since 1954 is function of monthly state allowable which generally reflects market demand.

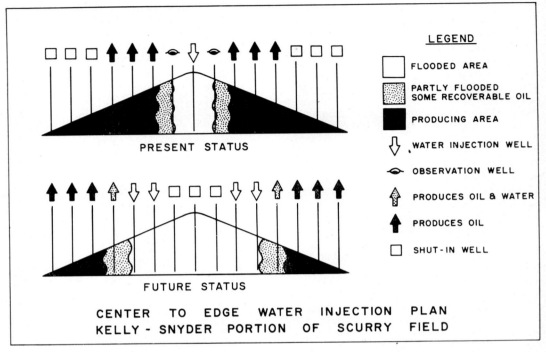

FIG. 17.—Center-to-edge water-injection plan for Kelly-Snyder part of Scurry field, West Texas.

pressure was 1,546 psi (109 kg/cm^2), and only 1 percent of the reservoir volume remained above the bubble point.

A center-to-edge water-injection plan (Fig. 17), similar to plans used in the southern part of the Scurry field, was selected for use in the Kelly-Snyder area. Conventional methods of waterflooding, either from the base of the pay section upward or from the edges of the field inward, were rejected because of the presence of relatively impermeable limestone members, bedded almost horizontally, throughout the pay section. The center-to-edge water-injection plan resulted in an immediate pressure response throughout most of the field (Fig. 16), and reservoir conditions have continued to improve during the 13 years in which the plan has been in operation. Pressure data are available through April 1967, and at that time average reservoir pressure was 2,355 psi (166 kg/cm^2). Approximately 85 percent of the reservoir is above the bubble-point pressure of 1,800 psi (127 kg/cm^2), and 97 percent is above the critical gas-saturation pressure of approximately 1,600 psi (112 kg/cm^2); almost no withdrawals are made where reservoir pressure is below 1,800 psi (127 kg/cm^2). Water injected since September 1954 has exceeded the amount of oil withdrawn during the life of the field by approximately 34 percent.

With the start of pressure-maintenance operations, axial wells were converted to water-injection wells and edge wells were shut in. Intermediate wells are produced, and applicable per-well allowables have been transferred to the producing wells. As flooding proceeds, additional wells near the flood front will be converted to water-injection wells, and concurrently additional edge wells now shut in will be placed on production.

CONCLUSION

The Horseshoe atoll is one of the largest subsurface limestone reef mounds in the world. It formed in an environment that was established early in Midland basin history, and continued to build upward because the basin interior remained starved of terrigenous clastic sediment during a period of slow and generally continuous subsidence. Unlike most modern oceanic reefs with sessile organisms, the Horseshoe atoll consists essentially of a series of imbricated biostromes with varied amounts of detritus along the flanks of the complex. There is no core facies, and few remains of organisms capable of erecting a rigid vertical framework, other than algae, are observed. Flank dip, commonly less than 8°, is gradual compared with the steep dip associated with modern reefs. Most of the effective porosity and permeability in the reef limestone are secondary and resulted from one or more periods of exposure to meteoric water.

Most of the relief along the crest of the atoll in Scurry field, although modified slightly with thin Cisco limestone beds, was formed by destructional processes. The relief along the crest farther west, where Wolfcamp limestone is present in significant thickness, is believed to be the result of rapid foundering, with consequent restriction of the sites of reef building, even though the areas of restriction are at least partly controlled by topography created by Late Pennsylvanian erosion.

The reef complex was buried in Early Permian time by terrigenous clastic sediment which prograded cyclically toward the west and southwest. Reef mounds south of Salt Creek field foundered to water depths in excess of 1,000 ft before complete burial by Wolfcamp clinoform and fondaform sand and clay.

Lateral communication of reservoir fluids takes place throughout the reef complex, and vertical communication apparently occurs in the long term in reef rocks younger than Strawn. Vertical communication of fluids during normal field life is not effective.

The entrapment of major oil reserves is due to the lateral migration of oil within the reef complex as a result of regional westerly tilting after burial, irregularities along the crest of the atoll, and critical spill points adjacent to mounds with inadequate caprock.

The Horseshoe atoll is a unique product of a starved basin. If the concept of the starved basin had been formulated and widely recognized before the discovery of the Horseshoe atoll, perhaps the possible presence of such a complex could have been anticipated, and the 9-year gap between discovery of the atoll and discovery of commercial production could have been shortened.

REFERENCES CITED

Adams, J. E., H. N. Frenzel, M. L. Rhodes, and D. P. Johnson, 1951, Starved Pennsylvanian Midland basin: Am. Assoc. Petroleum Geologists Bull., v. 35, no. 12, p. 2600–2607.

Anderson, K. C., 1953, Wellman field, Terry County, Texas: Am. Assoc. Petroleum Geologists Bull., v. 37, no. 3, p. 509–521.

Bergenback, R. E., and R. T. Terriere, 1953, Petrography and petrology of Scurry reef, Scurry County, Texas: Am. Assoc. Petroleum Geologists Bull., v. 37, no. 5, p. 1014–1029.

Burnside, R. J., 1959, Geology of part of the Horseshoe atoll in Borden and Howard Counties, Texas: U.S. Geol. Survey Prof. Paper 315-B, p. 21–35.

Heck, W. A., K. A Yenne, and L. G. Henbest, 1952, Pennsylvanian and Permian(?) contact in subsurface Scurry reef, Scurry County, Texas: Am. Assoc. Petroleum Geologists Bull., v. 36, no. 7, p. 1465–1466.

Myers, D. A., P. T. Stafford, and R. J. Burnside, 1956, Geology of the late Paleozoic Horseshoe atoll in West Texas: Texas Univ. Bur. Econ. Geology Pub. 5607, 113 p.

Rothrock, H. E., R. E. Bergenback, D. A. Myers, P. T. Stafford, and R. T. Terriere, 1953, Preliminary report on the geology of the Scurry reef in Scurry County, Texas: U.S. Geol. Survey Oil and Gas Inv. Map OM 143.

Stafford, P. T., 1955, Zonation of the late Paleozoic Horseshoe atoll in Scurry and southern Kent Counties, Texas: U.S. Geol. Survey Oil and Gas Inv. Chart OC 53.

—— 1957, Scurry field, in Occurrence of oil and gas in West Texas: Texas Univ. Bur. Econ. Geology Pub. 5716, p. 295–302.

—— 1959, Geology of part of the Horseshoe atoll in Scurry and Kent Counties, Texas: U.S. Geol. Survey Prof. Paper 315-A, p. 1–20.

Van Siclen, D. C., 1950, Reef-type oil fields, Scurry County, Texas, in Geological contributions: Abilene (Texas) Geol. Soc., p. 70–79.

Panhandle-Hugoton Field, Texas-Oklahoma-Kansas—the First Fifty Years[1]

LLOYD PIPPIN[2]
Amarillo, Texas 79105

Abstract A detailed study of the geometry and an understanding of the mechanics of entrapment are essential in order to unravel the complexities of the Panhandle-Hugoton field. The reservoir rocks in the Panhandle-Hugoton field generally are considered to be Wolfcampian. Gas and oil appear to have migrated from Pennsylvanian marine shale in the Anadarko basin, through "granite wash" into the Panhandle field.

In the Panhandle field, the trap is mainly structural, but in the Hugoton field, it is stratigraphic. A hydrodynamic component is present in both.

Red Cave (Leonardian) reservoirs above the Wolfcamp and Pennsylvanian reservoirs below it usually are not considered part of the Panhandle-Hugoton field pay but, in the writer's opinion, could be so considered, because they appear to have had the same source, same initial pressure, and similar hydrocarbon-water contacts.

LOCATION, DISCOVERY, AND DEVELOPMENT

The Panhandle-Hugoton field is the largest gas field in the United States. Figure 1 shows its location and size on a map of the United States. It covers parts of 19 counties in 3 states, is 275 mi long, and ranges in width from 8 to 57 mi. There are approximately 5 million productive acres within the field, and it is still being extended.

The Panhandle field discovery well in Texas, designated by an arrow on Figure 2, was drilled in northern Potter County, Texas, 21 mi north of Amarillo, in December 1918. Charles N. Gould mapped a surface feature called the "John Ray dome" on which the well was drilled. A group of Amarillo businessmen formed the Amarillo Oil Company, to test the structure. Amarillo Oil Company's No. 1 Mas-

[1] Read before the Association at Oklahoma City, Oklahoma, April 23, 1968. Manuscript received, May 7, 1968; accepted August 13, 1968.
[2] Phillips Petroleum Co.

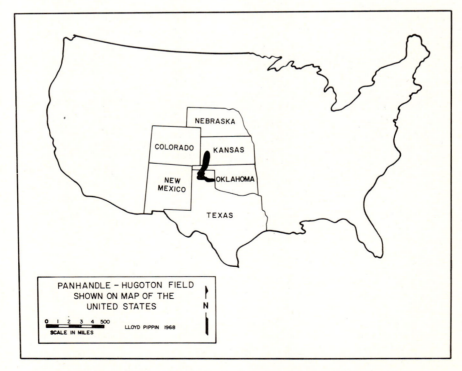

FIG. 1.—Map of United States showing location of Panhandle-Hugoton field.

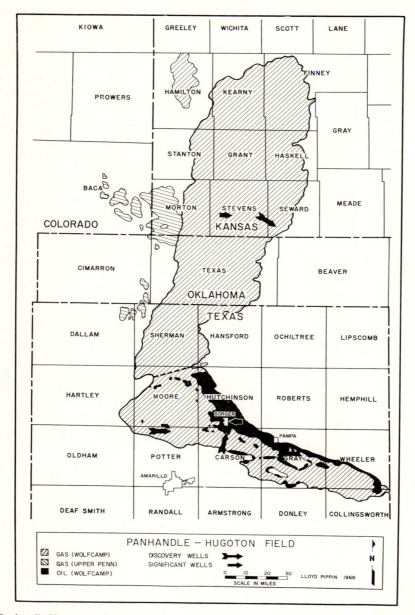

Fig. 2.—Panhandle-Hugoton field, showing discovery wells, significant wells, and oil- and gas-bearing areas.

terson C in Sec. 65, Block 0-18, D&P Survey, was completed as a gas well at a depth of 2,395 ft, in rhyolite (Flawn, 1956). The initial potential was 5 million cu ft of gas per day, and the shut-in pressure was 420 psi. It was worked over in 1950 and recompleted for 15 million cu ft of gas per day with a shut-in pressure of 350 psi. After 48 years, it has a poten-

tial of 7.74 million cu ft of gas per day, and a shut-in pressure of 190 psi. Cumulative production as of January 1, 1967, was 9,675,524 Mcf, and the well is still producing more than 1 million cu ft of gas per day.

Gulf Producing Company drilled the No. 1 Burnett in Sec. 118, Blk 5, I&GN Survey, Carson County, Texas, in September 1920. It was

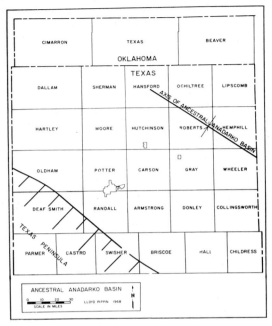

FIG. 3.—Map of Texas and Oklahoma panhandles showing part of ancestral Anadarko basin.

completed as a gas well with an initial potential of 75 million cu ft of gas per day (Rogatz, 1961).

The Hugoton discovery well in Kansas was Defender's No. 1 Boles, Sec. 3, T35S, R34W, 3 mi west of Liberal in Seward County. It was completed in December 1922. Few people realized the significance of this discovery and, as a result, activity was not great. Several wells were drilled in Texas County, Oklahoma, in 1925 and 1926, but it was not until May 1927 that the Hugoton field became important (Hemsel, 1939).

Realization that Hugoton was a single big field came when Independent Oil Company drilled the No. 1 Crawford in Sec. 31, T33S, R37W, Stevens County, Kansas. This well, 3 mi southwest of Hugoton, Kansas, was the first well in Stevens County (Hemsel, 1939). Seventy-five wells were drilled in southwestern Stevens County by 1930. The first concentrated field development was near the town of Hugoton, whence the field acquired its name.

Gulf Producing Company's No. 2 Burnett was the discovery oil well for the Panhandle field of Texas. It was drilled in northern Carson County, Texas, in Sec. 106, Block 5, I&GN Survey, in May 1921 as a confirmation well 2 mi southeast of the No. 1 Burnett. The initial

potential was 200 b/d of oil from the "granite wash." Drilling for oil continued, but it was not until April 1925 that the oil "boom" really began (Rogatz, 1961)—when Dixon Creek Oil Company completed the No. 1 Smith in Sec. 14, Block Y, H&GN Survey, Hutchinson County, Texas, for an initial potential of 1,000 b/d from the "arkosic dolomite" (Rogatz, 1961).

Since discovery, more than 20,000 oil wells have been drilled in the Panhandle field. As of January 1, 1967 there were 11,827 still producing. Activity has diminished during the last few years, but there were still approximately 20 wells drilled in the Panhandle field each month during 1966.

REGIONAL GEOLOGY

The ancestral Anadarko basin, as it existed until post-Mississippian time, was bounded on the south by the Texas peninsula (Fig. 3) and on the north by a broad flat cratonal shelf (Huffman, 1959). Regional uplift in northeast New Mexico and southeast Colorado at the end of Devonian time (and of Hunton deposition) resulted in truncation of Hunton and older rocks in the Texas and Oklahoma Panhandles.

Conditions remained stable through Mississippian time. Post-Mississippian diastrophism formed the Dalhart basin, Cimarron-Stratford arch, Kathryn arch, Amarillo mountains, and two major faults just north of the Amarillo mountains (Fig. 4). The south edge of the ancestral Anadarko basin was shifted north-

FIG. 4.—Map of Texas and Oklahoma Panhandles showing post-Mississippian tectonic features.

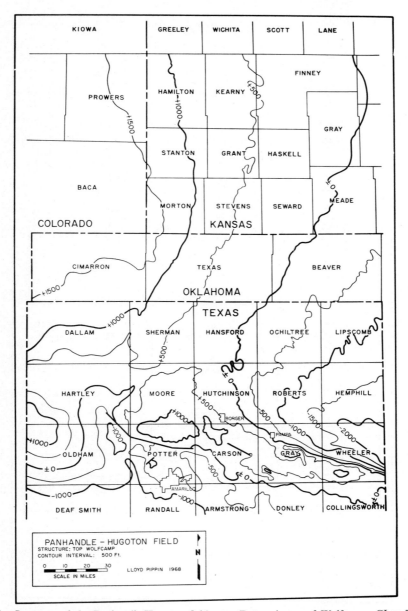

FIG. 5.—Structure of the Panhandle-Hugoton field area. Datum is top of Wolfcamp; CI = 500 ft.

ward from the Texas peninsula to the Amarillo mountains, thus forming the Anadarko basin as it is today. Pre-Pennsylvanian sediments were eroded from the Cimarron-Stratford arch and Amarillo mountains.

The effects of post-Mississippian and post-Morrowan diastrophism are shown on section *A-A'* (Fig. 7), which extends from the south edge of the Amarillo mountains to the axis of

the Anadarko basin (Fig. 6). Mississippian sediments were partly eroded on the most basinward horst block and sediments as old as Ordovician were exposed on the other two blocks. The most basinward block was covered in late Morrowan time. Maximum stresses occurred during Atokan time, resulting in major uplift of the Amarillo mountains. All sediments were removed from the mountains, and the granite

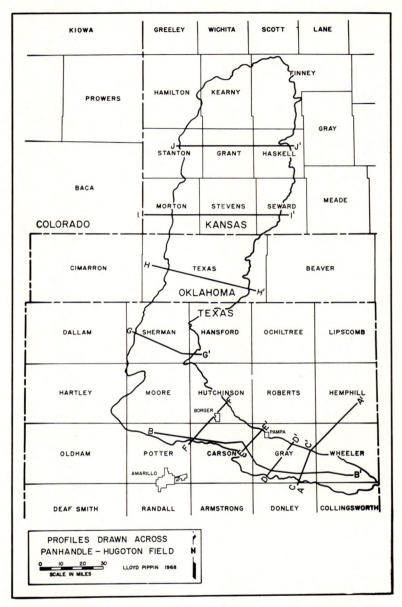

FIG. 6.—Index to sections across Panhandle-Hugoton field (Figures 7, 10, 11, 12, 15, and 16).

core was exposed. Erosion of granite resulted in basinward deposition of granite wash over the Atoka unconformity. Granite wash was interbedded with marine mud and carbonate, the basin filled, and the Amarillo mountains were covered by Wolfcampian time.

Regional uplift in the northwest part of the Anadarko basin in Atokan time caused reversal of the dip direction from west to southeast in southwest Kansas. A structure map, with the top of the Wolfcamp as datum (Fig. 5), shows the regional southeast dip component. Southeast tilting of the basin caused updip wedgeout of Permo-Pennsylvanian carbonates in a westerly direction, forming the trap along the west edge of the Hugoton field. The Wichita Forma-

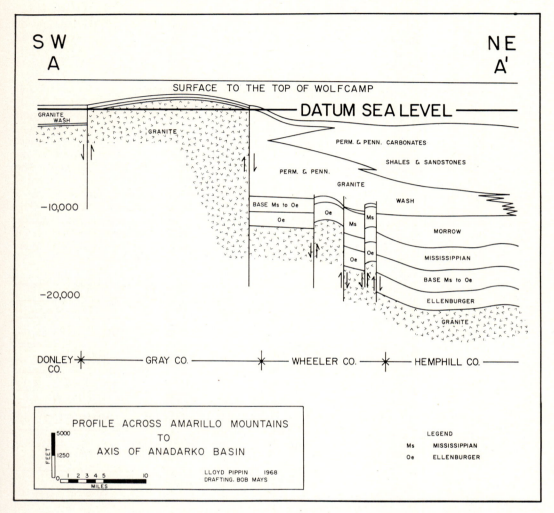

Fig. 7.—Section across Amarillo mountains to axis of Anadarko basin. Trace is line *A-A'*, Figure 6.

tion was then deposited, forming a seal over the Wolfcampian reservoir beds.

Oil and gas probably were generated in marine shale, then migrated into the granite wash, which served as a conduit into the Panhandle field. Some of the oil and gas were trapped in transit, but approximately 7 billion bbl of oil and 70 trillion cu ft of gas accumulated in the Panhandle-Hugoton field.

Panhandle Field, Texas

Figure 8 is a correlation chart for the Panhandle-Hugoton field. Most Panhandle-Hugoton pays are in the Wolfcampian Stage of the Permian System. Local nomenclature for the

Panhandle field pay zones is, in descending stratigraphic order: "Brown dolomite," "White dolomite," "Moore County limestone," "arkosic dolomite," "arkosic lime," and "granite wash."

Figure 9 is a map with top of the Cambrian-Precambrian granite as datum; 5,000-ft contours are shown. The axis of the Anadarko basin and the axis of the Amarillo mountains are prominent. There is very little throw on the fault zones along the south side of the mountains except in western Potter County. The throw on the north side, however, increases from 5,000 ft in eastern Carson County to approximately 18,000 ft in eastern Wheeler County. Two grabens present in Gray County within the mountain complex are referred to commonly as the Lefors and Deep Lake basins.

Figure 5 is a structure contour map, using the top of the Wolfcamp as datum; contour interval is 500 ft. The Anadarko basin and Amarillo mountains are prominent, even though the mountains were covered in Wolfcampian time. The steep dip on the north side of the mountains in Gray and Wheeler Counties is above the fault zone shown on Figure 9.

The angle of dip in southern Hutchinson and northwestern Gray Counties is low compared with the steep dip in eastern Gray and Wheeler Counties. This is important, for the productive area of the field is much broader where the dip rate is low because the oil column is about the same thickness and is relatively flat. Normally, gas, oil, and water are consistent in relation to a specified reservoir, This is not true in the Panhandle field, where

gas, oil, and water cut across formational boundaries. Hence, in this field, the relations of the reservoirs to the gas, oil, and water must be considered. The width of the field is determined by the angle of dip and the aggregate thickness of reservoir beds.

Figure 10 (section B-B') is a structural and stratigraphic section along the axis of the Amarillo mountains (Fig. 6). The highest unit shown is the Leonardian Red Cave, which consists of red shale and fine-grained sandstone. The Wichita Formation, below, consists of anhydrite and dense anhydritic dolomite. The "Brown dolomite" is a buff, cherty, saccharoidal dolomite. The "White dolomite" is a white, vuggy, coarsely crystalline dolomite. The lowermost sedimentary rock is the "granite wash," which ranges from loose, unconsolidated gravel to fine-grained arkosic red shale; it overlies Precambrian granitic rocks.

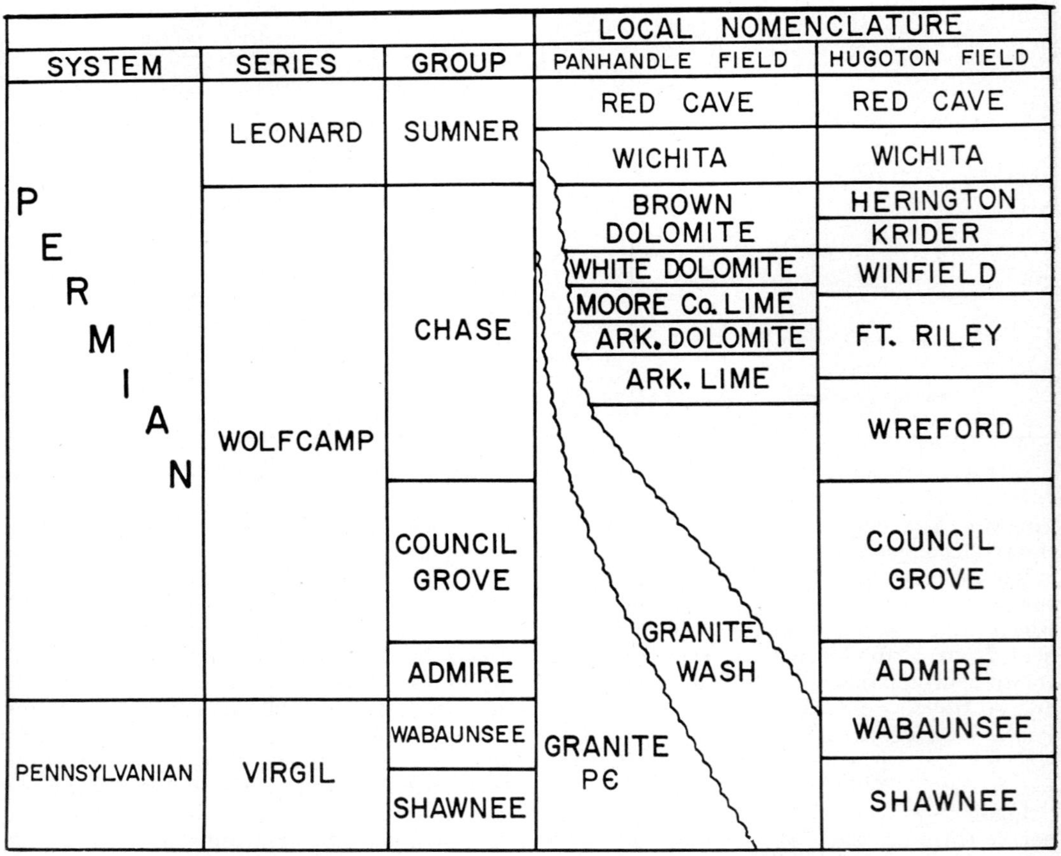

SYSTEM	SERIES	GROUP	LOCAL NOMENCLATURE	
			PANHANDLE FIELD	HUGOTON FIELD
PERMIAN	LEONARD	SUMNER	RED CAVE	RED CAVE
			WICHITA	WICHITA
	WOLFCAMP	CHASE	BROWN DOLOMITE	HERINGTON
				KRIDER
			WHITE DOLOMITE	WINFIELD
			MOORE Co. LIME	FT. RILEY
			ARK. DOLOMITE	
			ARK. LIME	
				WREFORD
		COUNCIL GROVE		COUNCIL GROVE
		ADMIRE	GRANITE WASH	ADMIRE
PENNSYLVANIAN	VIRGIL	WABAUNSEE	GRANITE PЄ	WABAUNSEE
		SHAWNEE		SHAWNEE

FIG. 8.—Stratigraphic column of Panhandle-Hugoton field.

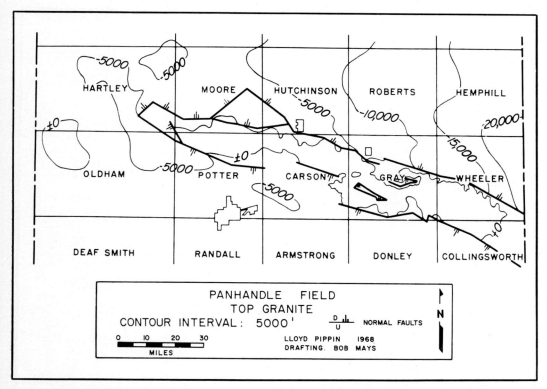

Fɪɢ. 9.—Structure map of Panhandle field area, Texas. Datum is top of granite; CI = 5,000 ft.

The highest structural point on the cross section (Fig. 10) is at the west end in Potter County, near the Panhandle field discovery well. The Wichita, "Brown dolomite," and "granite wash" are thin in this area. As the strata thicken eastward across Carson County the "White dolomite" and "Moore County limestone" are developed. The "Brown dolomite" thins eastward and pinches out against a granite peak in western Wheeler County, but is present again east of the peak. The peak was buried near the end of the time of Wichita deposition.

Figure 11 (sections C-C', D-D') shows two southwest-northeast sections across the Panhandle field. These sections demonstrate the relative thickness of the reservoir beds present and the relation of the oil and gas columns to them at various places across the Panhandle field. Section C-C' passes through eastern Gray County, Texas (Fig. 6). The reservoir beds are "Brown dolomite," "granite wash," and granite. The oil column at the north end of the field is approximately 50 ft below to 120

ft above sea level; the column intersects the "Brown dolomite" and "granite wash," but is limited by the presence of granite, except in local areas where intense fracturing has made the granite a reservoir, also. Gas overlies water on the south side of the field at approximately +160 ft.

Section D-D', approximately 9 mi west of section C-C' (Fig. 6), crosses the Lefors and Deep Lake grabens. Both grabens were filled with granite wash, then covered during Wolfcampian time. Additional later movement and compaction produced steep dips along the fringes. These two depressions are commonly called the Lefors and Deep Lake basins. The oil column is in the interval from +50 ft to +150 ft on the north edge of the field; it rises abruptly so that it occupies the interval from +90 ft to +200 ft on the north edge of the Lefors basin. It narrows gradually southward to the south side of the Deep Lake basin, where it terminates against granite. Gas overlies water on the south edge of the field at approximately +220 ft.

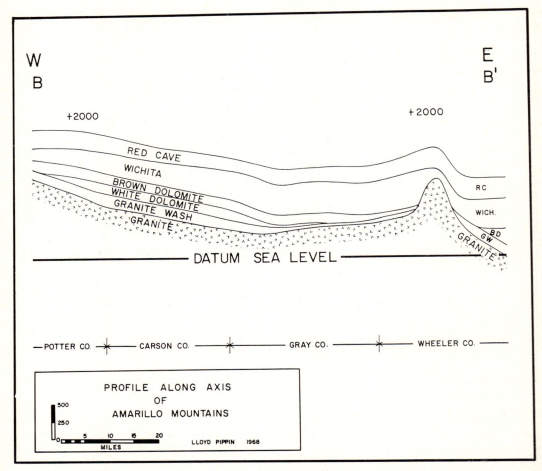

FIG. 10.—West-east section along axis of Amarillo mountains, showing structure and stratigraphy. Trace is line B-B', Figure 6.

Section *E-E'* (Fig. 12) extends from southeast of White Deer to northwest of Pampa (Fig. 6). The dip rate is less than that on previous cross sections and the aggregate thickness of reservoir beds is greater. As a result, the width of the band of oil pay is greater. The oil column is similar to that shown in previous sections, with gas overlying water on the south edge of the field.

Section *F-F'* (Fig. 12) extends from northeastern Potter County into northwestern Carson County, passing near the discovery oil well in the Panhandle field, and terminates in central Hutchinson County (Fig. 6). There are more reservoir beds present in this area than in any other part of the Panhandle field. They are, from top to bottom, "Brown dolomite,"

"White dolomite," "Arkosic dolomite," "Arkosic limestone," and "granite wash." The angle of dip is low in this area, so that the intersection of the oil column with all these reservoir beds produces a wide band of oil pay. Migration of oil was limited southward by intersection of the oil column with granite.

If a profile were drawn across northwestern Hutchinson County, it would show the oil column intersecting "Brown dolomite," "White dolomite," and the "Moore County limestone" from east to west. Migration of oil would have been halted by the dense "Moore County limestone." A band of oil production is present in the "Moore County limestone" along the edge of the field as a result of porosity development in the top of that unit. It is productive in very few

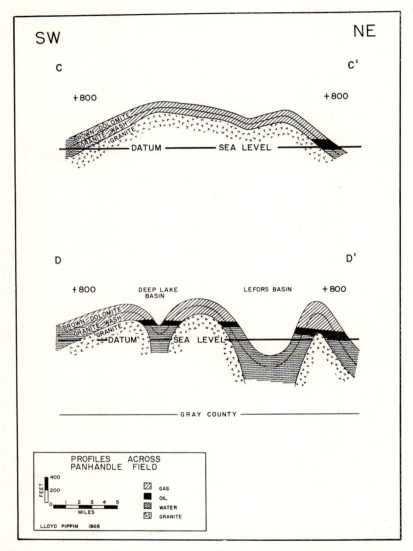

Fɪɢ. 11.—Sections *C-C'* and *D-D'* show relations of oil, gas, and water to reservoirs in Panhandle field. Oil column rises and thins from north to south across the mountains; gas overlies water on south side of field. Traces are lines *C-C'* and *D-D'*, Figure 6.

places beyond this contact, except where intense fracturing has created a local reservoir. The dip rate is low in this area; hence the band of oil production is relatively wide in northwestern Hutchinson County.

Figure 13 shows the oil-productive areas for the reservoirs of the Panhandle oil field. Each symbol represents the intersection of a reservoir bed with the oil column. Progressively older reservoirs are found in an updip direc-

tion. The width of the productive band and the number of reservoirs are greatest where the angle of dip is least, as in southwestern Hutchinson County. Conversely, the productive band is narrowest where the reservoirs are the fewest and the dip is steep, as in Wheeler County.

The most productive reservoir to date has been the oölitic zone in the "Brown dolomite." Production in this zone has been approximately 10,000 bbl of oil per acre in Hutchinson

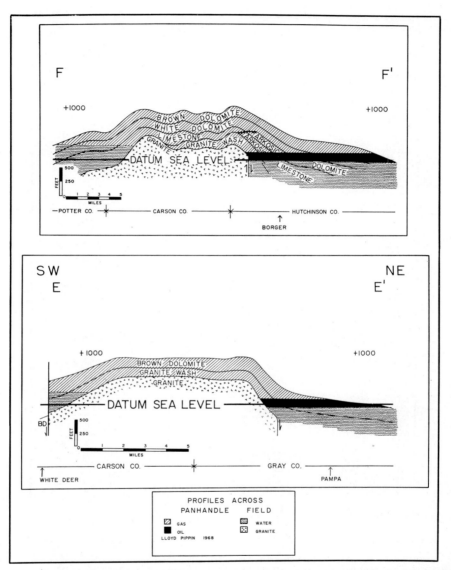

Fig. 12.—Sections *E-E'* and *F-F'* show relations of oil, gas, and water to reservoirs in Panhandle field. More reservoir beds are present, and reservoirs are thicker than in sections *C-C'* and *D-D'* (Figure 11); therefore, width of oil field is much greater. Traces are lines *E-E'* and *F-F'*, Figure 6.

County. The average Panhandle-field well is expected to yield from 1,500 to 2,000 bbl of primary oil per acre.

Cumulative oil production for the Panhandle field as of January 1, 1967, was 1,138,284,070 bbl of oil (Dwight's Production Reports, 1967). Of this, primary production is estimated at 1,081,934,070 bbl, and secondary at 56,350,000 bbl.

Secondary-Recovery Projects

Secondary-recovery projects in the Panhandle field (Fig. 14) include gas injection, waterflood, and thermal projects. Gas repressuring began in 1946 in the West Pampa area (Rogatz, 1961), where oil production had been declining for several years. Gas injection brought about a substantial increase in production rates and in ultimate recoveries.

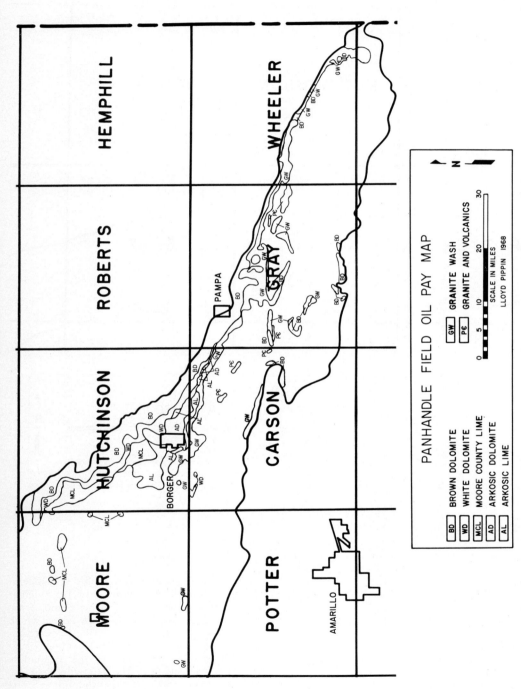

PANHANDLE FIELD OIL PAY MAP

BD	BROWN DOLOMITE
WD	WHITE DOLOMITE
MCL	MOORE COUNTY LIME
AD	ARKOSIC DOLOMITE
AL	ARKOSIC LIME
GW	GRANITE WASH
P€	GRANITE AND VOLCANICS

SCALE IN MILES

LLOYD PIPPIN 1968

FIG. 13.—Oil pay map of Panhandle field showing intersection of oil column with various reservoir beds.

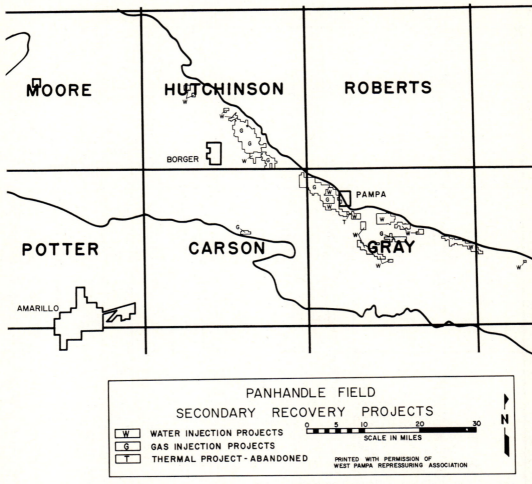

FIG. 14.—Secondary-recovery projects in Panhandle field. Modified from Watkins' Operators Report, 1966.

Waterflooding began in 1946 in Hutchinson County (Rogatz, 1961); early results were disappointing. There are, however, several successful waterfloods in Gray County. The most successful are the Kewanee "Webb" and "Morse" floods on the Gray-Wheeler county line and the "Little Seminole" flood south of Pampa. Most of the Gray County waterfloods are responding favorably.

A thermal flood was attempted south of Pampa, in Gray County, Texas, but proved to be unsuccessful.

Secondary recovery from the Panhandle field as of January 1, 1967, was estimated at 5.6 percent of the cumulative production. Of

this, gas repressuring accounted for 4.3 percent, and waterflood 1.3.

Hugoton Field

Figure 8 shows the stratigraphic sequence in the Panhandle and Hugoton fields. The Wichita Formation, consisting of anhydrite and dense dolomite, forms the seal for both fields; which are productive from the underlying Wolfcampian carbonates. The "Brown dolomite" is at the top of the Wolfcamp in the Panhandle field; its equivalents in the Hugoton field are the Herington and Krider dolomites. The Herington is a saccharoidal, buff, slightly anhydritic, and cherty dolomite. The Krider is a sandy, buff to black, crinoidal dolomite with vuggy porosity. The Krider is the primary pay in the Hugoton field.

All units from the Herington through the Topeka are gas productive in areas of steep dip, such as western Sherman and Morton Counties, Texas and Kansas, respectively. Reservoirs below the Council Grove, however, generally are not considered to be part of the Hugoton pay, even though they have the same source, same original pressure, and same water contact.

The northwestern terminus of the Panhandle field is in northern Moore County, Texas (Fig. 2). The Panhandle and Hugoton fields originally had the same 435 psi bottom-hole pressure, and the source of hydrocarbons is the same for both fields. They differ in that the Panhandle field has an associated oil column, but the Hugoton field does not. The trap is mainly structural in the Panhandle field but stratigraphic in Hugoton. A hydrodynamic component forms part of the trapping mechanism in both fields. Pressure and elevation data afford conclusive evidence that water moves eastward, and that the inclination of the gas-water contact is the result of hydrodynamic tilting (Hubbert, 1967).

Study of Figures 2 and 5 shows that the Panhandle field overlies the Amarillo mountains, whereas the Hugoton field overlies a monocline.

Figures 15 and 16 contain a series of four west-east cross sections through the Hugoton field. The datum of each cross section is sea level, and the gradient of the water table can be seen in relation to the reservoir rocks. The red beds and marine carbonates are shown because their interfingering relation is critical in the creation of the trap.

Section *G-G'* (Fig. 15), which extends from western Sherman County to western Hansford County, Texas (Fig. 6), illustrates several points related to the geologic history of the area. Throughout Late Pennsylvanian time and extending into Early Permian (Admire) time, marine carbonates were being deposited in the Hugoton area. After deposition of the Admire, a slight uplift on the west was followed by erosion and an influx of red clay and sand. Redbed deposition dominated in the western part of the Hugoton field from the close of Admire to the beginning of Krider deposition (Figs. 8 and 15). Marine carbonate deposition was resumed and continued until the

close of Herington time. The Wichita Formation thins markedly westward, but extended far enough toward the west to form an effective seal over the Wolfcamp in the Hugoton field. Locally, the Stratford arch forms the trap on the western edge of the field. There is vertical fluid communication between the Herington, Krider, and Winfield.

The water table stabilized at +170 ft on the east edge of the field. In central Sherman County it is +200 ft. The gradient there rises sharply until the water table is at +800 ft on the Stratford arch, presumably by dynamic flow of formation water from west to east.

The water table stabilized at about the same level in the Lower Permian and Upper Pennsylvanian reservoirs under the Stratford arch. Redbeds formed the seal above the Upper Pennsylvanian and basal Permian reservoirs.

Cross section *H-H'* (Fig. 15), from eastern Cimarron County, Oklahoma, to the Beaver County line (Fig. 6), shows that marine carbonates were formed through the time of Admire deposition. Subsequent uplift on the west brought an influx of red sand and clay into the west side of the Hugoton area. The seas gradually retreated westward until the end of Chase deposition, then regressed, became landlocked during Wichita deposition, and finally withdrew. Only red clay and sand were deposited above the Wichita.

The Herington dolomite generally is thin in the area of section *H-H'*. As a result, the principal reservoirs are the Krider and Winfield of the Chase Group. There is a general gas-water contact at approximately +170 ft on the eastern edge of the field; this contact rises to approximately +850 ft at the western edge. Updip wedgeout of the carbonate reservoirs, plus downdip flow of water caused entrapment. The critical updip point beyond which the reservoir bears only water appears to be determined by an abrupt local change of porosity and permeability. This trapping mechanism exists from central Sherman County, Texas, to northwest Texas County, Oklahoma (Fig. 2). From Texas County northward, a similar, though slightly different, situation exists. This can be seen on Figure 16.

Section *I-I'* (Fig. 16) extends from Baca County, Colorado, to Seward County, Kansas (Fig. 6). There appears to be vertical communication between the Herington, Krider, and Winfield formations, as there was on section *G-G'* (Fig. 15). As a result of this com-

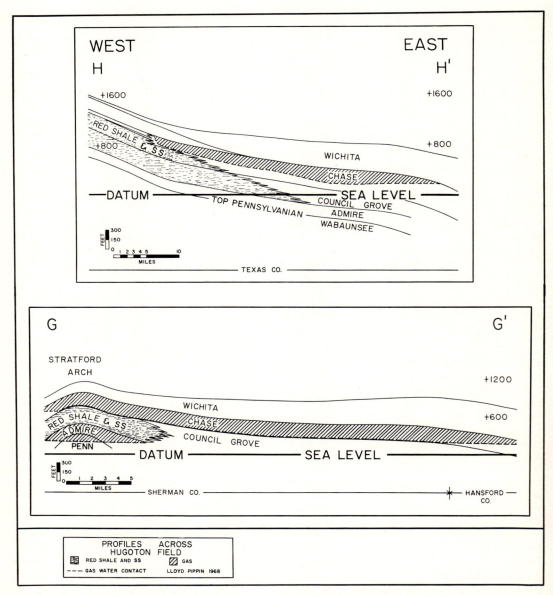

FIG. 15.—Cross sections G-G' and H-H', west-east across southern part of Hugoton field. Note tilted gas-water interface, and facies change from nonmarine on west to marine on east. Traces are lines G-G' and H-H' on Figure 6.

munication, there is a common gas-water contact for all three reservoirs. The lowest is at +160 ft on the eastern edge of the field; it rises gradually to central Stevens County, Kansas, then abruptly to central Morton County on the western edge of the field, where the water table is +1,300 ft.

The Pennsylvanian pool at the western end of section I-I' is called the Greenwood. Steep dip in western Morton County placed the porous Pennsylvanian reservoir beds in the proper structural position to be the site of hydrocarbon accumulation.

At Greenwood, at the time of migration,

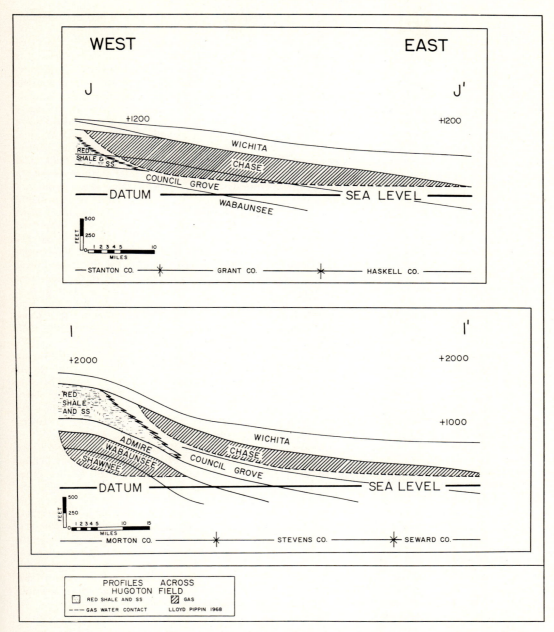

FIG. 16.—Cross sections *I-I'* and *J-J'*, west-east across northern part of Hugoton field. Note thickening of pay northward between the two sections. Traces are lines *I-I'* and *J-J'*, Figure 6.

communication between Wolfcamp reservoirs, above, and the Pennsylvanian reservoirs, below, provided the gas to Greenwood, and abrupt changes in porosity and permeability westward produced the trap. This pool is not considered

part of the Hugoton pay, but it has approximately the same gas-water contact (+180 ft), and approximately the same formation pressure (435 psi), as the Wolfcamp pay above.

Section *J-J'* (Fig. 16), from Stanton County

on the west to Gray County, Kansas, on the east (Fig. 6), is similar to *G-G'* (Fig. 15) in that the upper Wolfcamp reservoir beds have vertical communication and, as a result, a single gas-water contact. The area of section *J-J'* is different in that the Hugoton pay is thickest along this section. All reservoirs from the top of the Chase through the Council Grove are productive. The gas-water contact is +140 ft at the eastern end of the section; it rises gradually across most of the Hugoton field, but abruptly on the western edge, owing to the marked facies change there.

The Hugoton field is terminated on the north, as on the west, by thinning of upper Wolfcamp carbonates and an abrupt rise in the gas-water contact.

The field has been developed within the confines of the broad outline shown on Figures 2 and 6. Extension drilling is continuing north and west of the limits shown, where conditions are similar to those which produced the major trap.

Cumulative gas production for the Panhandle-Hugoton field, including casinghead gas, was 41,540,415,334,000 cu ft as of January 1, 1967 (Dwight's Production Reports, 1967). It is estimated that the ultimate recovery will be 70 trillion cu ft.

Helium

Natural gas in the Panhandle-Hugoton field has an unusually high helium content. It ranges from 0.2 percent, and averages 0.5 percent. Helium content in the Bush dome, 9 mi south of the Panhandle field, is 1.86 percent. In 1929 a plant began extracting helium from the Bush dome which supplied most of the Free World's helium during World War II. The Panhandle-Hugoton field is now the primary United States source of helium.

Shallower and Deeper Pays

The search for shallower and deeper pays is continuing in the Panhandle-Hugoton area. Figure 17 shows the location of production from shallower and deeper pays, as of January 1, 1967.

One of the more recent shallow developments is the Leonardian Red Cave gas pay in Hutchinson, Moore, Potter, and Carson Counties, Texas. The first well in the Red Cave was completed in northern Potter County in 1919. Shows were common, but until 1960 the zone generally was bypassed for the more lucrative Wolfcamp pay zone. Most of the development of the Red Cave took place from 1960 to 1965.

Figure 8 shows the position of the Red Cave in the rock column. In the producing area, it is underlain by the Wichita Formation and overlain by the Clearfork. The Red Cave consists of red shale, siltstone, and fine-grained sandstone. Limits of the field are determined largely by development of porosity. The sandstones are porous and permeable in the productive area, but pore spaces are filled with salt and anhydrite around the perimeter of the field. The lack of porosity in contemporaneous strata around the field indicates that the gas did not migrate laterally into the field, but probably migrated vertically from the Wolfcamp below, through fractures. This hypothesis is supported by the fact that formation pressure originally was the same in both Red Cave and Wolfcamp. The only adverse evidence is that a chemical analysis shows some difference in the composition of Red Cave and Wolfcamp gases. If it is assumed that gas did migrate into the Red Cave from the Wolfcamp, then the Red Cave could also be considered as a Panhandle field pay.

The average depth of Red Cave producing wells is 1,600 ft. Reserves were estimated at 500 billion cu ft of gas for the pool (Rogers, 1961). More recent estimates made in May 1966 on the basis of pressure decline, place reserves at 420 billion cu ft of gas (industry figures—source unavailable).

All other pools shown on Figure 17 are from reservoirs below the Hugoton pay. These include Council Grove, Toronto, Collier limestone, Lansing, Des Moines, Morrow, Chester, and St. Louis. Of these, the Morrow is the most important; the St. Louis is second in importance. The focal point of activity in recent years has been in Texas County, Oklahoma, with significant, but lesser amounts in southwestern Kansas. Exploration and development are continuing, and it is expected that much of the northern two thirds of the Hugoton area will be productive in these older units.

Summary

The Hugoton pay generally is considered to consist of late Wolfcampian carbonate rocks. Late Pennsylvanian pools such as Greenwood in western Morton County, Kansas, are not considered to be a part of the Hugoton field; yet the source, trapping mechanism, pressure, and general gas-water contact are the same for

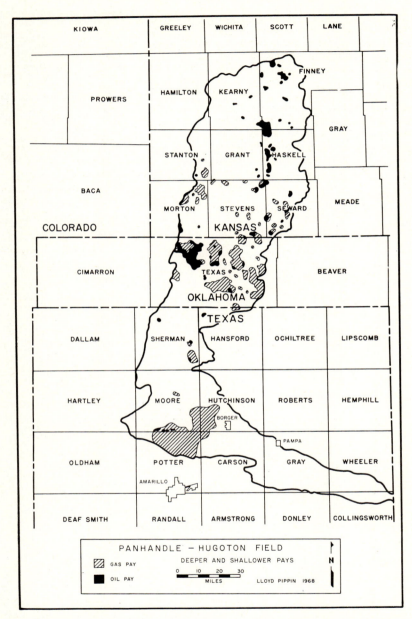

FIG. 17.—Areas producing from units above and below main Panhandle-Hugoton pay. Area shown in Hutchinson, Moore, Potter, and Carson Counties, Texas, produces from Red Cave Formation (Leonardian), above Wolfcamp. All others are from reservoirs below Hugoton pay and range in age from Lower Permian (Council Grove) to Mississippian (St. Louis).

both late Pennsylvanian and Wolfcampian reservoirs. Late Pennsylvanian pays thus could be considered to be Hugoton field pays.

Rocks of Wolfcampian age are the pay in the Panhandle field. Although the Red Cave is separated from the Wolfcamp by the Wichita Formation, gas apparently migrated from the Wolfcamp into the Red Cave through vertical fractures in the Wichita. This hypothesis is supported by the absence of source beds in the Red Cave field and a common formation pressure for both the Wolfcamp and Red Cave. The Red Cave thus could be considered to be a Panhandle field pay.

The Hugoton field long has been considered to be a classic example of a stratigraphic trap. It is true that carbonate reservoirs thin updip and wedge out westward, but the critical trapping mechanism appears to be the southeastward, downdip, dynamic flow of formation water.

References Cited

Dwight's Production Reports, 1967: Box 7041, Amarillo, Tex., Jan. 1.

Flawn, P. T., 1956, Basement rocks of Texas and southeast New Mexico: Texas Univ. Pub., no. 5605, 261 p.

Hemsell, C. C., 1939, Geology of Hugoton gas field of southwestern Kansas: Am. Assoc. Petroleum Geologists Bull., v. 23, no. 7, p. 1054–1067.

Hubbert, M. King, 1967, Application of hydrodynamics to oil exploration: 7th World Petroleum Cong. Proc., Mexico City, v. 1B, p. 59–75.

Huffman, G. G., 1959, Pre-Desmoinesian isopachous and paleogeologic studies in central Mid-Continent region: Am. Assoc. Petroleum Geologists Bull., v. 43, no. 11, p. 2541–2574.

Rogatz, H., 1961, Shallow oil and gas fields of the Texas Panhandle and Hugoton, *in* Oil and gas fields of the Texas and Oklahoma Panhandles: Panhandle Geol. Soc., p. 8–37.

Rogers, R. G., 1961, The Red Cave formation of the Texas Panhandle, *in* Oil and gas fields of the Texas and Oklahoma Panhandles: Panhandle Geol. Soc., p. 38–44.

Watkins' Operators Report, 1966: Watkins Operators Committee, September, p. 9, 11.

Oklahoma City Field—Anatomy of a Giant[1]

LLOYD E. GATEWOOD[2]

Oklahoma City, Oklahoma 73102

Abstract Oklahoma City field, Oklahoma County, Oklahoma, is truly a giant oil field. It is a billion-barrel field, having already produced more than that amount of oil and oil-equivalent gas. The Wilcox sandstone alone has produced 50 percent of the estimated 1.07 billion bbl of oil in place. Today the field ranks among the 10 largest oil fields in the United States.

From the very dramatic beginning of the Oklahoma City field 40 years ago, myths have been created from half-remembered tales, but very little of its recent history is known. The purposes of this paper are to update and recount what has happened to Oklahoma City field in the last 36 years and to answer the questions most commonly asked about its origin, growth, size, influence, and destiny.

The field was discovered in 1928 by the drilling of a wildcat well on a 100-ft surface closure. Today the field is 12 mi long and 4½ mi wide. Its 1,000 ft of producing closure is confined in a 32-mi² area and production has come from at least 30 different producing zones.

The discovery well produced from the Ordovician Arbuckle dolomite, the oldest pre-Pennsylvanian rocks on the crest of the structure beneath the unconformity. The most prolific production has been from the oil-wet youngest Ordovician Simpson Wilcox sandstone on the lowest part of the west flank of the structure. Production from the Wilcox has been 537.5 million bbl, of which 187.5 bbl of oil was produced by natural gravity drainage.

The field is near the south end of a buried mobile basement feature—the Nemaha ridge—at its intersection with the northeast rim of the Anadarko basin. The structural intersection coincides with an environmentally favorable sedimentary section of thick porous Arbuckle dolomite and alternate sandstone and shale of Simpson Group in a series of shelf-edge hinge zones. In number, variety, and production history, the reservoir beds have not been equaled or surpassed since Oklahoma City's discovery.

The field's structural growth was allied closely with the stages of evolution of the Anadarko basin. Growth probably began in Cambrian time, but surely was in progress from Ordovician through Pennsylvanian time, as a result of subsidence in the Anadarko basin. This subsidence caused compression folding, but the culminating influence on Oklahoma City was a differential vertical displacement in the stronger folds and faulting near the northeast rim of the basin.

The structure was folded, faulted, and truncated contemporaneously. Approximately 2,000 ft of Ordovician-Pennsylvanian strata was removed from the top. A 2,000-ft down-to-the-east fault prevented lateral migration of oil from the fold. The unconformity and the overlying Pennsylvanian shale allowed only limited upward migration. Relief was so prominent, even after truncation and burial, that the fold provided an ideal environment for development and accumulation of oil and gas in the numerous shallow Pennsylvanian zones on its irregular surface. Accumulations within the Pennsylvanian are in pinchouts, fault traps, and channel deposits.

The field has been a model and proving ground for exploration techniques and production technology, modern proration rules and laws, drilling and testing techniques in deep rotary wells, and establishment of standards for formation evaluation and reserve estimates. Developments within a major city furnished excitement caused by many "wild" wells like "wild Mary Sudik," but the field was also an economic benefit during the worst days of the depression. Geologists for the past 40 years have found and developed great quantities of oil and gas in many other areas by using it as a case history for the dating of structural growth as the basis of exploration.

Introduction

From its very dramatic beginning in December 1928, the Oklahoma City field has been surrounded by an aura of greatness, mystery, a sense of history, and excitement. In the 40 years since its discovery, myths have been created from half-remembered tales, but very little of its recent history is known.

The purposes of this paper are to update and reveal what has happened to Oklahoma City field since McGee and Clawson (1932) described its discovery, early history, and growth, and Hill *et al.* (1937) reported the development and engineering investigations of the field. The writer will deal in detail with the pre-Pennsylvania stratigraphic section because

[1] Read before the 53d Annual Meeting of the Association, Oklahoma City, Oklahoma, April 25, 1968. Manuscript received, March 26, 1969; accepted, April 15, 1969.

[2] Consultant and independent, 718 Cravens Building.

Appreciation is extended to Pat McDonald, engineer and historian of the Oklahoma City field, for his valuable assistance in providing data, authenticating the events, and rolling back time with his stimulating conversation and helpful suggestions.

I also thank Mrs. Pat McDonald, who acted as arbitrator and typed the manuscript, Jack Armbrust for his help and the generous use of his facilities, Ed Barrett and Ralph Espach for their helpful suggestions in editing and correcting the manuscript, and many geologists for their contributions and assistance. Other sources of assistance and information, both written and by personal communication, are acknowledged in the text and references.

that is the source of most of the production (76 percent) and of the remaining reserves which made it a giant.[3]

Questions

The following questions are asked most frequently about the Oklahoma City field.

1. Why is Oklahoma City field a giant?
2. What facts, which were recognized as characteristic of Oklahoma City field during its early history, have proved to be of practical value in application to subsequent exploration, production, and political activities in adjacent areas and throughout the world?
3. How reliable are isopachous changes as a tool for dating structural growth?
4. Are the shallow and deep structural axes coincident?
5. Have we adequately assessed the significance and potential of this giant in solving subsequent exploration problems?
6. Have we profited from the use of the information Oklahoma City field has provided?

The following discussion consists of a personal interpretation of what the facts indicate. If there are fallacies in interpretation, the sole responsibility for them is mine.

Unique Aspects

Oklahoma City field is unique in that it has served for 40 years as a model and proving ground for exploration techniques and production technology, modern proration rules and laws, drilling and testing techniques in deep rotary wells, and establishment of standards for formation evaluation and reserves estimates.

Specifically, the following aspects of Oklahoma City field are unique.

1. It is a major field where the development was carried on within the confines of a large city.
2. The field has produced from at least 30 different zones.
3. The reservoirs had a great variety of drive mechanisms for one field—water drive, gas in solution, and gravity drainage.
4. Gravity drainage for the Wilcox sandstone was a unique drive mechanism which has resulted in the production of large volumes of oil not obtainable by secondary-recovery methods in this field.
5. The Wilcox sandstone is so loosely consolidated that much sand was produced with the oil.
6. The Wilcox sandstone was an oil-wet sand.
7. The field had a common water level.
8. The high part of the structure flooded with water first.
9. Proration was applied at the very beginning of production in the Oklahoma City field.
10. It was one of the first fields in the Mid-Conti-

nent with production at depths of more than 6,200 ft where volumes of produced oil were so large and gas pressures so high that drilling with cable tools was impractical. At those depths rotary equipment was used almost exclusively, and large volumes and high pressures dictated the use of drilling muds at those depths.

11. Weight indicators were generally a part of the regular drilling equipment and consequently the percentage of crooked holes drilled in the field was small.
12. The development of the field entailed the extensive catching of samples on rotary wells, and such innovations as the shale shaker and other sample-settling devices were used for the first time.
13. The Geolograph, a drilling-time indicator, was developed.
14. New formation-temperature devices were developed.
15. The Oklahoma City Commission required a licensed stationary engineer to be in charge of the steam boilers when pressure was maintained, and steam fittings had to be installed by a steam fitter licensed by the city.
16. City inspectors had power to shut down any production and drilling wells when conditions warranted such action.
17. Development of the field was rapid and the large volume of production came on the market at the same time as East Texas oil. The oversupply forced prices down to 16¢ and 20¢/bbl. The price was so low that the Governor of Oklahoma forcibly shut in the field with state troops to prevent waste, as a conservation measure, and to stabilize the price at $1.00/bbl.
18. The structure is so big, the structural growth periods so distinct, and the periods of rejuvenation in the folding and faulting so pronounced that geologists have used the field as a case history for dating of structural growth as a basis of exploration in many other areas.

Regional Setting

The Oklahoma City structure is on the southernmost end of the old buried mobile basement feature—the Nemaha ridge. The Nemaha ridge is a major family of vertical-uplift-type local structures arranged in a north-south belt trending across eastern Kansas and central Oklahoma (Figs. 1, 2).

The Oklahoma City field is in the southwest-central part of Oklahoma County adjacent to the north line of Cleveland County, in Ts 11, 12N, Rs 2, 3W, in the central part of the state.

Origin and Growth of Structure

The structural adjustments that were the principal factors in the formation of the Oklahoma City structure were in close harmony with the stages of evolution of the Anadarko basin. Growth probably began in Cambrian time, but surely occurred from Ordovician through early Pennsylvanian time as a result of subsidence in the Anadarko basin. Subsidence caused compressional folding, but the culminating influence on Oklahoma City was differential vertical displacement in the stronger folds, and

---

[3] A giant oil field is one of a select group which has 100 million bbl or more of ultimately recoverable reserves and which has a production rate of at least 3,000 bbl/day (1,095,000 bbl/year).

Fig. 1.—Location map showing Oklahoma City field area (from R. R. Wheeler, 1951). Star indicates Oklahoma City area.

faulting near the northeast rim of the basin. There its juncture with the southern end of the Nemaha ridge, influenced not only its position, but also its size, shape, and structural complexity. The oil in the Oklahoma City field is found in a pronounced, faulted anticlinal fold. The history of the structure includes at least five stages of structural adjustments, with intervening periods of complete or partial submergence and fairly continuous deposition.

The diagrammatic regional structure and pool map (Fig. 3) shows that the Oklahoma City uplift dominates a four-county area. It shows also how the prominence of the uplift provided the environment for the development of very large flanking stratigraphic traps in the Siluro-Devonian Hunton Limestone, which is absent from the structure at Oklahoma City. The composite production map (Fig. 4) illustrates and contrasts three Ordovician oil- and

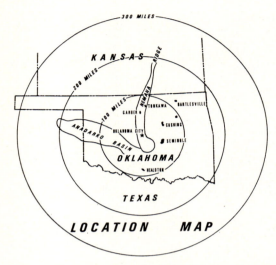

FIG. 2.—Location map showing Oklahoma City on south end of Nemaha ridge in central part of Oklahoma, and relation to other significant fields in 1928.

gas-producing zones, and leaves no doubt as to the dominance of the Wilcox sandstone as the major producing zone that made the Oklahoma City field a true giant (Fig. 5). Recoveries of 600 bbl/acre-ft already have been obtained from the 902 wells in the Wilcox sandstone, with 50 percent recovery of the original oil in place of 1,072 million bbl.

There were at least three important stages in the structural development of the anticline— folding, faulting, and truncation. All three processes occurred more or less contemporaneously. The fault which terminates the anticline on the east obviously had its inception before the structure had been elevated long enough for erosion to remove an appreciable amount of material from the crest or upthrown side, because Mississippian limestone, Hunton limestone, Sylvan Shale, Viola Limestone, and the entire Simpson Group are preserved in full thickness on the downthrown side of the fault.

In the early drilling stages, structural growth at the Pennsylvanian level was indicated by a noticeable shortening of the intervals from the Pawhuska downward. This thinning was apparent over the crest of the fold. It was also evident that the "high" or axis of the structure shifted about ¾ mi eastward with depth—from the E½, Sec. 24, T11N, R3W, the axis of the Permian surface closure, to the W½, Sec. 19, T11N, R2W, the axis at the Arbuckle level.

The east-bounding fault is normal and vertical, with a probable maximum displacement at

the Arbuckle level of 2,400 ft, but an average of about 2,000 ft near the upper Simpson level. The fault was an important factor in the accumulation of the oil on the high side of the structure. It is believed that the oil and gas migrated into the growing structure while the fault was developing along the east side of the field.

Considerable weight is given, in this hypothesis, to the action of water in concentrating the oil in the high part of the structural trap. There are at least eight major Ordovician producing formations (Arbuckle and Simpson) in the Oklahoma City field, and they all had a somewhat common water level at 5,400 ft subsea (Fig. 6). That the uplift of the anticline was gradual is shown by the fact that on the downthrown side of the fault, in certain areas which are now 300 ft below the top of the fault scarp, the Pennsylvanian beds overlie the Hunton limestone. Evidently erosion had removed the Mississippian rocks and part of the Hunton Group before this area was protected by the rising fault scarp. There are indications that shearing or lateral movements accompanied the faulting. Several small "highs" along the axis of the structure are aligned and connected with the main fault at *en echelon* or transverse angles and in sinuous trends, especially along the crest where maximum displacement of the main fault occurred. These smaller shear or transverse faults affect the behavior of fluid movements from one sector of the field to another. It may be that the shearing stresses contributed to the high intensity of fracturing of the Arbuckle carbonate rocks. The fractures provided avenues for water movement and facilitated development of honeycombed solution cavities from which prolific quantities of oil were produced; also large quantities of water were produced from the established channels through the process of coning.

There must have been some folding in Permian or post-Permian time, as it is unlikely that the formation of the very pronounced surface structure was due entirely to differential compaction of the Permian-Pennsylvanian sediments over the pre-Pennsylvanian topography.

Repeated rejuvenation is indicated in strata up to the surface beds (Permian) at Oklahoma City. The Desmoinesian Cherokee, at the base of the Pennsylvanian, lies in contact with the Ordovician Arbuckle limestone on the crest of the structure, and is in contact with progressively younger Simpson formations westward on the flank. On the crest of the structure,

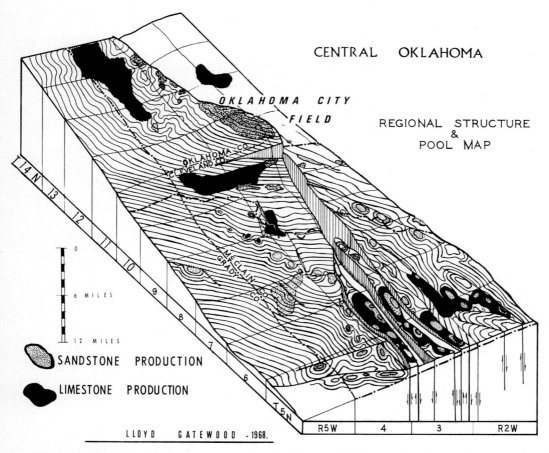

CENTRAL OKLAHOMA

OKLAHOMA CITY FIELD

REGIONAL STRUCTURE
&
POOL MAP

SANDSTONE PRODUCTION

LIMESTONE PRODUCTION

LLOYD GATEWOOD - 1968.

FIG. 3.—Regional structure and pool map showing prominence of Oklahoma City uplift, which also provided environment for development of large flanking Siluro-Devonian Hunton limestone stratigraphic traps indicated by darker areas.

post-Cherokee erosion or truncation removed several hundred feet of the Desmoinesian strata, which by contrast are present on the east, on the downthrown side of the fault, as well as on the west flank of the structure. The pre-Pennsylvanian subcrop map shows the enormity of the uplift and truncation—8 mi across from the Hunton surface on the west to the eastern limits of erosion (Fig. 7).

The truncated form of the Simpson subcrop over much of the area and the unconformity at the base of the Pennsylvanian strata suggest that appreciable quantities of pre-Pennsylvanian oil may have escaped, possibly through seepage, during erosion of the Simpson and Arbuckle Groups, and through faults, fissures, crevices, and fractures during later Pennsylvanian periods of structural rejuvenation. The ac-

cumulation of asphalt in the Simpson sandstone and Arbuckle limestone near the crest at the Pennsylvanian-Arbuckle-Simpson contact, and redeposited asphaltites cored in the basal few feet of the Pennsylvanian, indicate a significant accumulation of oil before Pennsylvanian time.

WHY IS OKLAHOMA CITY FIELD A GIANT?

Oklahoma City field is a giant because of its location near the south end of a buried mobile basement feature—the Nemaha ridge—and its intersection with the northeast rim of the Anadarko basin (Fig. 1). The structural intersection coincides with an environmentally favorable, thick sedimentary section beginning with the porous, massive Arbuckle dolomite, topped by a 700-ft oil column on one of many shelf-edge hinge zones (Fig. 8). Other advantageous

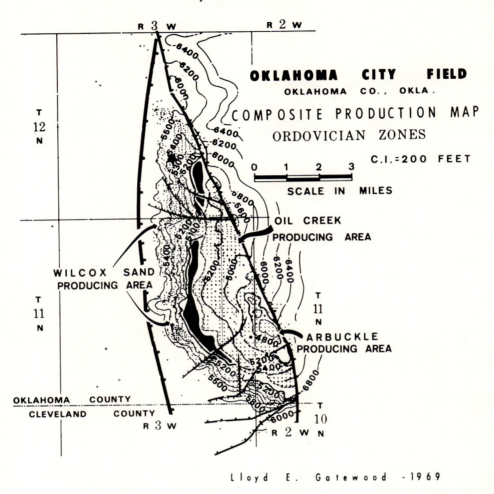

FIG. 4.—Composite production map contrasting size of three Ordovician producing zones—Arbuckle carbonate, smallest area and oldest rocks on crest; Oil Creek sandstone, second largest in size; and younger Wilcox sandstone, lower on west flank of structure. Black represents a stream channel cut into Simpson strata below the Wilcox sandstone, on the unconformity surface (*see* Fig. 26).

factors resulted from recurrent growth and movement during deposition of alternate sandstone and shale beds of the Simpson. These beds either contained indigenous hydrocarbon source material or were in communication with sources in the nearby Anadarko basin. In number, variety, and production history the reservoirs have not been surpassed or equaled in any field found since Oklahoma City's discovery 40 years ago. This locality was affected by well-timed structural-growth intervals followed by rejuvenating movements. Early trapping in the fold was effected by a major fault with a throw of more than 2,000 ft, which prevented updip or lateral migration out of the fold. As

growth accelerated and the upthrown west side rose, erosion removed from its crest approximately 2,000 ft of strata, including some Arbuckle dolomite, the Simpson Group, Viola Limestone, Sylvan Shale, Hunton Group, Woodford Shale, and Mississippian limestone (Fig. 9). These strata are preserved in their normal thickness only on the downthrown side of the fault. The overlapped unconformity preserved the form of the remaining Ordovician Arbuckle and Simpson strata, and prevented upward migration from the reservoirs of much of the oil and gas already accumulated. The resultant trap was big. The combination of pronounced relief and continuous growth provided

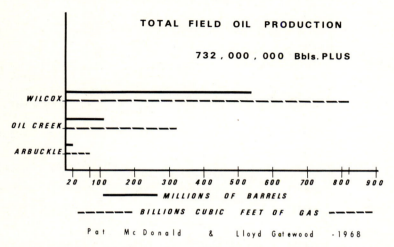

FIG. 5.—Summary of cumulative production to Jan. 1, 1968 from three principal producing zones that made Oklahoma City field a giant.

an ideal environment for development and trapping of oil and gas in the numerous shallow Pennsylvanian zones. The accumulations are on the irregular surface of the fold above the unconformity in pinchouts, rejuvenated fault traps, and channel deposits which followed the drainage pattern of a graben zone along the longitudinal axis of the structure.

The field is 12 mi long and 4½ mi wide. Its 32 mi² contains a total productive area of 13,770 acres. There is a producing closure of 1,000 ft upon which 1,810 wells have been drilled. A total in excess of 1 billion bbl of oil and oil-equivalent gas has been produced from at least 30 different zones, 20 of which are distinctly significant and are noted in Figure 10.

DISCOVERY

The discovery well for Oklahoma City field —the Indian Territory Illuminating Oil Company and Foster Petroleum Company No. 1 Oklahoma City—was drilled on a surface structure in C SE ¼ SE ¼, Sec. 24, T11N, R3W. This location was in the center of the outcrop of the Garber Sandstone, which protrudes prominently through the surrounding red Hennessey Shale (Fig. 11). The surface structure indicated about 100 ft of closure, but it extended about 10 mi from the Capitol on the north into Cleveland County on the south. The well was about 2 mi outside the Oklahoma City limits at that time and 5 mi southeast of the central part of the city. The only other production in the county in 1928—a 4,390-ft Penn-

sylvanian sandstone well—was north of the Capitol in Sec. 10, T12N, R3W. John Bunn (personal commun., 1967) had induced Cromwell and Franklin to drill it in 1926 and had proved that production could be found beneath the redbeds. The ITIO No. 1 Oklahoma City well was drilled on a 10,000-acre block which had been assembled by an ITIO landman, Homer Keegan, in the fall of 1927 and 1928 at an average cost of $5.00 per acre. It came in as a gusher from the Arbuckle dolomite on December 4, 1928, when the tools were blown from the hole after drilling out cement at 6,402 ft (Fig. 12).

ARBUCKLE DOLOMITE
Reservoir Characteristics

The maximum rate of flow from the Arbuckle dolomite in the No. 1 Oklahoma City well was 6,564 bbl/day of oil. The final total depth was 6,624 ft. The high-gravity, amber-green oil came from the upper zone of banded gray and brown, finely crystalline, slightly porous dolomite containing solution cavities and numerous fractures (Fig. 13). The solution cavities are of varied size and commonly the drill bit would drop several feet in the crevices; lost circulation was a problem during drilling. Below this are irregular zones of medium- to coarsely crystalline dolomite honeycombed with solution cavities, many of which are large enough to permit the formation of drusy surface by secondary crystallization of dolomite. The more porous zones alternate irregularly

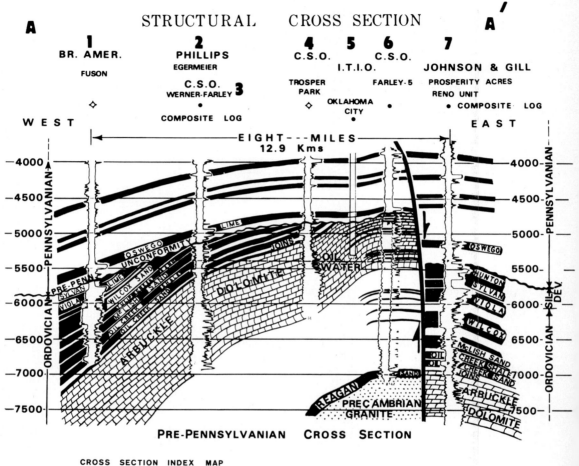

STRUCTURAL   CROSS   SECTION

Pre-Pennsylvanian   Cross   Section

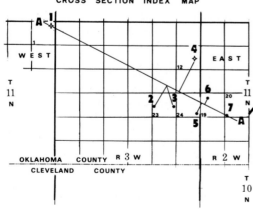

CROSS SECTION INDEX MAP

FIG. 6.—Pre-Pennsylvanian structural cross section showing maximum displacement of east-bounding fault. Strata absent from upthrown west side are preserved in normal thickness on downthrown side. Note position of common oil-water contact at −5,400 ft. Location of section shown in Figure 7.

with zones of coarsely crystalline dolomite and a few streaks of fine Simpson-type sandstone. The oil and gas production from the Arbuckle

dolomite was confined largely to porous zones below the upper 200–250 ft of the more finely crystalline upper dolomite. Production rates in-

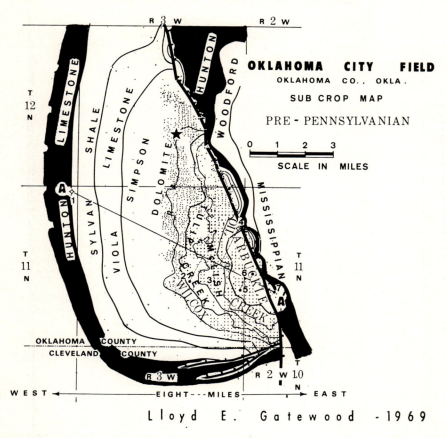

FIG. 7.—Pre-Pennsylvanian subcrop map illustrating large areal extent of erosion and truncated shape of Ordovician Simpson and Arbuckle preserved at unconformity surface. A-A' is line of sections in Figures 6 and 9.

creased appreciably in many high-volume wells upon penetration from 250 ft to more than 500 ft of the dolomite.

Identification Significance

The age and identity of the producing zone were in considerable doubt for some time. Geologists and paleontologists were reluctant to accept the correlation indicated by the characteristic lithology of the producing formation because of the magnitude of the structure which would be indicated if it were Arbuckle. As additional wells were drilled it became apparent that the production was in fact from the Ordovician Arbuckle Group.

Lithology

The Arbuckle Group contains a wide variety of lithologic types in a depositional sequence that is predominantly dolomite, but contains thin streaks 8–10 ft thick of sandstone with fine to coarse, rounded to subangular grains, some pale-green shale, sandy dolomite, and many varieties of dark-brown, tan, gray, white, blue, and oölitic chert. Porosity types are as varied as the lithologic types—intergranular, saccharoidal vuggy, cavernous, pinpoint, and fracture. Fossil assemblages appear throughout the section but are most commonly in fragments.

The Arbuckle dolomite is found at the base of the Pennsylvanian-Ordovician unconformity at 6,000-ft depth on the crest of the structure. Approximately 300 ft of the upper Arbuckle was removed by erosion. The upper 200–250 ft of in-place Arbuckle, where it is not truncated, consists of gray to tan, finely crystalline, saccharoidal dolomite with a few nodules of gray, translucent chert, very thin partings of laminated, waxy, grayish-green shale, and thin

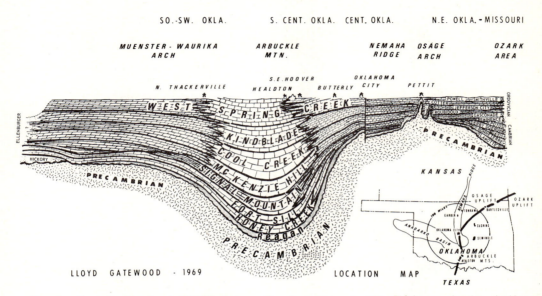

FIG. 8.—Generalized regional SW-NE cross section from Texas to NE Oklahoma, showing significant Arbuckle-producing fields including Oklahoma City, and their relation to carbonate facies of the Arbuckle Group in line of cross section.

beds of white, oölitic dolomite. Cores (Fig. 13) show a distinct but discontinuous banded or laminated effect of light to darker dolomite with mottled coloration similar to the stromatolite or algal mats on West Spring Creek Formation beds in the outcrops of the Arbuckle Mountains, and in cores of Healdton field Arbuckle in Carter County, Oklahoma.

Below the finely crystalline dolomite are irregular zones of medium- to coarsely crystalline dolomite, honeycombed with solution cavities. Cavities as large as 7 ft in diameter were indicated by drilling action and by the loss of circulation during drilling in this lower zone, about 400–500 ft below the top. Production is confined largely to the porous parts of this lower zone of West Spring Creek-Kindblade strata. A spicular chert zone is present just below the oil-water contact and probably marks the lower Kindblade boundary, which occupies an analogous position in the outcrop section in the Arbuckle Mountains.

Only one well penetrated the entire Arbuckle section to the Precambrian in the Oklahoma City field—Cities Service No. 5 Farley, drilled in 1947, in SW ¼ NE ¼ NW ¼, Sec. 19, T11N, R2W. It is on the very crest of the structure, where 2,300 ft of Cambrian-Ordovician carbonate section, with about 20 ft of Reagan Sandstone with subrounded to coarse, calcare-

ous, glauconitic to subrounded, large, frosted to clear grains, was found overlying the Precambrian pink granite. The well has been producing since completion from 150 ft below the top of the Arbuckle. Total depth was 8,344 ft. Approximately 300 ft of the upper Arbuckle is absent by erosion in this well. The No. 5 Farley has produced more than 135,000 bbl of oil and now is averaging 25 bbl/day.

Development History and Reservoir Performance

It was found that the Arbuckle high is ¾ mi east of the surface high and that there was a 700-ft oil column with production confined largely to the porous parts of the lower zone. Development and production were terminated abruptly on the east by a 2,000-ft, down-to-the-east, nearly vertical fault (Fig. 14). Although the Arbuckle oil-producing zone was on the highest part of the structure, it was the least productive and the first zone to become flooded with water. The productive area consisting of 2,460 acres ultimately produced 7,400 bbl/acre—162,500 bbl/well from an average penetration of 450 ft. The first wells were started by ITIO on 40-acre spacing, but because of the complex ownership of mineral rights in the townsite locations, much smaller tracts soon were being drilled. The Arbuckle

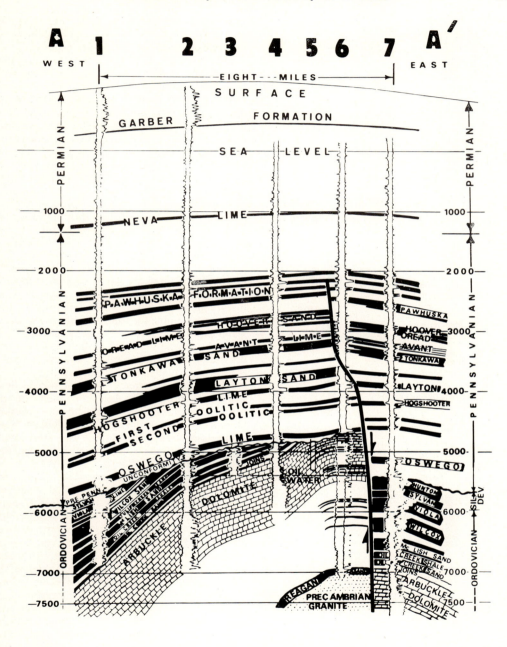

FIG. 9.—Structural cross section of Oklahoma City field emphasizing prominence of fold from surface down, plane of fault, and magnitude of pre-Pennsylvanian unconformity. Trace of section shown on Figures 6 and 7.

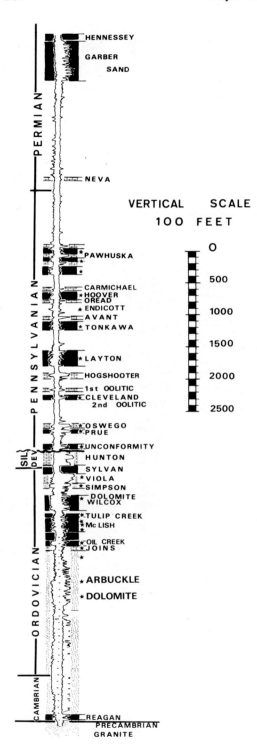

eventually averaged 22 acres per well. Cumulative production to 1968 was 18.2 million bbl of oil. Two Arbuckle wells still are producing.

One of the largest wells tested more than 43,000 bbl of oil in 24 hours. The largest cumulative production—1,223,868 bbl—was from an ITIO well in Sec. 19, T11N, R2W. Several wells, including the discovery well, produced more than 1 million bbl. Peak monthly production of 1,643,224 bbl was obtained from 35 wells in December 1929, and by December 1930 the entire area had been tested with 111 wells, proving an area of approximately 2,460 acres. By 1933 cumulative production was 17.5 million bbl, but the practice of producing erratically at high rates and then shutting in for periods of time caused water to cone in and drown the Arbuckle oil zones in the heterogeneous dolomite reservoir.

Table 1 shows the reservoir statistics for the Arbuckle dolomite at Oklahoma City field.

Unique water encroachment.—The water history in the Arbuckle zone presents an outstanding example of the rapidity with which water can move into a honeycombed and fractured limestone reservoir following the withdrawal of oil. By December 1930 the Arbuckle zone was virtually flooded, and in March 1932 only one well completed in the Arbuckle limestone was producing clean oil. Wells on the highest structural part of a field are generally the last to show water, but the Arbuckle limestone formation, constituting the structurally highest part of the reservoir, was flooded first in the Oklahoma City field.

In general, under the restricted production program, the wells were flowed at relatively high rates for short periods of time. Conclusions drawn from the study of the water conditions were that considerable volumes advanced rapidly and irregularly updip and spread both vertically and horizontally. The water probably had a strong tendency to cone and bypass the oil, and shut-in periods were too short to permit flattening of the cone. The intermittent operation of wells, theoretically at least, was conducive to the formation of irregular drainage channels in the formation around each borehole and ultimately caused interconnection be-

Fig. 10.—Composite log section of Oklahoma City field, from Permian surface to Precambrian with most significant producing zones marked by star.

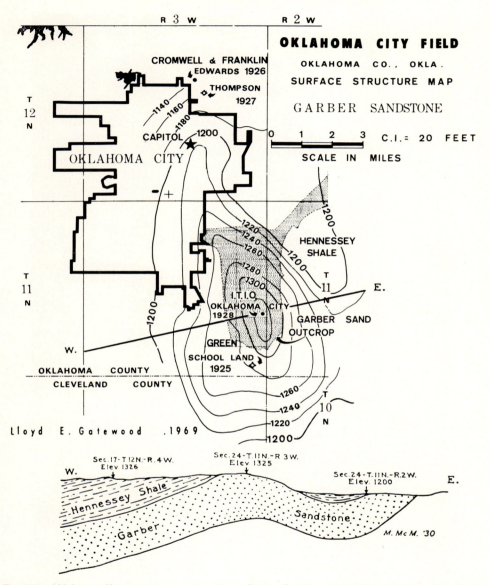

Fig. 11.—Oklahoma City structure as mapped on Garber Sandstone by Indian Territory Illuminating Oil Company in 1928. Generalized cross section shows arching of Garber Sandstone from west to east across structure. Line of section is **W.-E.** on map.

tween adjacent wells. Rapid withdrawal of oil from these channels established an extensive drainage system that facilitated rapid and irregular water encroachment. As soon as the oil was drained from these more or less open channels, which may have been fissures, crevices, or even cavities in the limestone, the water found easy access into the borehole.

Another deduction is that the water moved generally updip from the 5,400-ft subsea level into the oil reservoir. The movement was relatively rapid toward the low-pressure areas where one or more wells were producing. After the allowed production of oil was obtained and the wells shut in, the direction of movement changed toward another low-pressure area. The

Fig. 12—Oklahoma City field discovery well. ITIO and Foster Petroleum Company No. 1 Oklahoma City came in as "gusher" on December 4, 1928. Maximum rate of flow from Arbuckle dolomite was 6,564 bbl/day at 6,405-ft depth. Location shown on Figure 14.

continual changing and shifting of direction of water movement did not permit density separation of the oil and water; therefore, little, if any, leveling effect took place in the reservoir. When a well was shut in, the position of the water with respect to that well remained unchanged until the well again was allowed to produce. Each time the well was opened and large production obtained even for a short pe-

riod of time, the water advanced rapidly toward the well along channels of low resistance until it finally reached the borehole. Once the water channel was established, little or nothing could be done to keep the water from entering the borehole.

Theoretically, oil production from the more permeable parts of the formation and the wide-open flow of the wells for short periods favored heavy drainage in areas where formation conditions offered little resistance to the movement of fluid into the wells.

Lesson in conservation and higher ultimate recovery.—The relatively low average oil recovery obtained from the Arbuckle zone leads to the conclusion that operation of the wells at a constant low rate of flow might have given higher yields by retarding water movement and allowing more efficient drainage of the tighter parts within the effective drainage areas contiguous to the boreholes.

Operators at Healdton, in the Arbuckle sector, probably have learned a lesson from Oklahoma City. They have elected to perforate and complete the Brown zone of the Arbuckle near the oil-water contact, even though there may be as much as 300–400 ft of the zone above water. As the well begins to yield water, the perforations are plugged and a higher interval is perforated. This procedure will be repeated until the entire zone is watered out, the idea being that, although the zone has excellent vertical communication through fractures, less permeable streaks indicated on porosity logs might act as temporary barriers and retard vertical migration of oil. During reservoir depletion the encroachment of water could override the local oil accumulations, making them very difficult if not impossible to recover. Healdton Arbuckle development in the 1960s, however, is aided by excellent porosity logs from which the more porous and fractured zones can be picked. Operators at Oklahoma City did not have such logs in the early 1930s.

EARLY DEVELOPMENT ACTIVITY

Early development was rapid at Oklahoma City. The field soon became the first major field in the Mid-Continent in which the producing wells were more than a mile deep and where development was with rotary tools. Before the structural features of the reservoir had been disclosed, every outpost well was considered a wildcat because of the apparent geologic complexity of the subsurface. The wells were not only deep and the reservoir pressures high, but the production from the early wells consisted of high-API-gravity oil accompanied by large volumes of gas. Several blowouts and fires occurred, and the proximity of these now unusual events to a large city gave the field much publicity, both local and national. Later, when its production threatened to demoralize the market for oil and its products, the Oklahoma City field again attracted wide publicity. The inade-

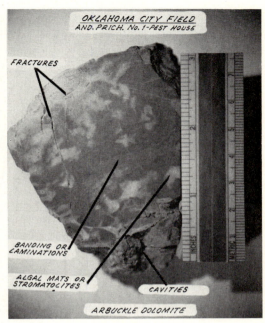

FIG. 13.—Core of Arbuckle dolomite from Oklahoma City field showing fractures connected with honeycombed cavities. Distinctive banding or laminations and algal mats are characteristic of upper West Spring Creek Formation of Arbuckle Group.

quate understanding of the geology of the underground reservoir and the diversity of ownership of the land overlying the producing structure made orderly development and proration difficult to enforce. The rate of production from the field continued to increase.

LOWER SIMPSON ZONES

Oil Creek Sandstone

The discovery well for the Oil Creek sandstone was the Coline Oil Corporation No. 1 Olds, C SE¼ NW¼ Sec. 24, T11N, R3W. It was completed on June 27, 1929, with initial production of 4,173 bbl of oil in 24 hours, natural flow. The average porosity of the Oil Creek sandstone ranges from 18 to 20 percent; that of the upper sandstone bodies in the lower Simpson (McLish and Tulip Creek) zone is varied, but the average is approximately 15 percent.

Reservoir characteristics.—Some of the outstanding features of the Oil Creek sandstone were (1) a free gas cap, (2) high gas-oil ratios, (3) good water drive—upward migration of water from the west edge along or near the Oil Creek-Arbuckle contact, and (4) a water-drive

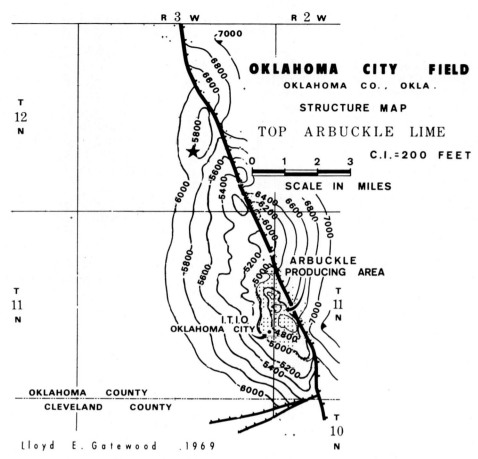

FIG. 14.—Arbuckle structure map and producing area.

mechanism which still causes wells to flow by downdip migration of water through fissures and faults connecting the Arbuckle dolomite with the Oil Creek sandstone.

Peak monthly Oil Creek production was 2,238,000 bbl in June 1930. Significantly, there also has been one good Oil Creek producer on the downthrown side of the major fault bordering the east side of the field. Evidently the sandstone members of the lower Simpson, including the Oil Creek and the McLish sandstones, contain free gas in the tip or high part of the pay strata where they were continuous across the structure, because they all produced large quantities of gas with the oil. Oil production from each Oil Creek sandstone well has averaged approximately 277,778 bbl.

Table 2 gives reservoir statistics for the Oil Creek sandstone at Oklahoma City field.

Lithology.—The Oil Creek sandstone ranges in thickness from 20 to 80 ft, the average being about 40 ft. The sand is poorly sorted at the top where large rounded frosted grains are included in the matrix of medium subangular to subrounded smooth grains. The lower part of the sandstone is relatively free of large grains and is more uniform. Thin dolomitic, tightly cemented zones are irregularly present from the top to the bottom. The basal part of the sandstone grades into the sandy dolomite below without a marked break.

Development history.—Development in the Oklahoma City field started in the Arbuckle limestone in the structurally high part of the field; the lower Simpson zone flanking the Arbuckle formation was the next to be developed. As stepout wells were drilled westward from the discovery well, they penetrated the basal or lower Simpson Oil Creek sandstone wedgeout on the structure's west flank (Fig. 15). Three hundred sixty wells covering a much larger area (about 4,500 acres) than the Arbuckle have had a cumulative production to 1968 from the Oil Creek sandstone in excess of 100 million bbl of oil. An estimated 320 billion ft^3 of gas has been produced from the Oil Creek sandstone (Fig. 5), which, next to the Wilcox sandstone, is the most prolific oil- and gas-producing unit in the field.

WILCOX SANDSTONE

Discovery and Early Surprises

Development drilling was extended southward into Sec. 31, T11N, R2W and on March 26, 1930, as the crew was pulling drill pipe out of the hole of the ITIO No. 1 Mary Sudik, it blew in with a roar, carrying 20 joints of heavy pipe into the derrick. The crew had failed to keep the hole full of mud as they withdrew the drill pipe (Fig. 16). The Mary Sudik was brought under control after 11 days, but in the meantime, carried by a strong north wind, it had sprayed 200 million ft^3 of gas and 20,000 bbl of oil per day as far south as the university town of Norman, 12 mi away. Later the wind shifted and parts of Oklahoma City received a shower of oil. Twice a day Floyd Gibbons, a news commentator, broadcast Wild Mary's antics to a radio-listening world. Later, other wells blew out[4] and there were some spectacular well fires.

More important, however, was the realization of the tremendous flow of oil and gas which came from an entirely new, untapped pay zone—the Ordovician Wilcox sandstone. Perhaps the greatest surprise in the development of the Oklahoma City field was the prolific recovery from the Wilcox zone—the lowest part of the field, structurally nearest the common water table, and the last area to be developed. The greatest drilling campaign of all was begun. The No. 1 Oklahoma City discovery well produced from the Arbuckle dolomite, the oldest rocks on the crest of the structure. The *youngest* Ordovician Simpson Wilcox sandstone, the lowest structurally, on the west

[4] The loosely bonded Simpson sandstones were readily disaggregated, and tons of sand were brought to the surface by the great volumes of gas under extremely high pressures (2,300–2,700 psi), creating a sand blast that "ate up" surface fittings as "though they were made of cheese," resulting in "wild wells," and creating a serious fire hazard in the metropolitan area. This was one of the unique operating problems at Oklahoma City.

Table 1. Reservoir Statistics, Arbuckle Dolomite, Oklahoma City Field

Productive acreage (above 4,950 ft subsea)	2,460
Av. thickness, productive zone (above 4,950 ft subsea)	109
Pay zone (acre-ft)	267,820
Producing mechanism	water
Formation volume factor	1.4
Recovery (% oil in place)	24.1
Bbl/acre-ft in place (STB)	282
Bbl/acre-ft recovered (STB), Jan. 1, 1968	68
Total oil in place (STB)	75,500,000
Total oil recoverable (STB)	18,250,000
Total oil recovered, Jan. 1, 1968	18,200,000
Av. prod. (bbl/acre)	7,400
Av. prod. (bbl/well)	162,500
No. productive wells	112
Av. acres per well	22
Percent of field total producing acreage	17.86

Cumulative prod., Jan. 1, 1968		
	Oil (bbl)	18,200,000
	Gas (ft^3)	68,250,382,680

Est. initial oil in place	75,500,000
Est. ultimate prod.	18,250,000

flank, was destined to be by far the most prolific reservoir and made a true giant of the Oklahoma City field.

This Wilcox sandstone body was found to cover 7,880 acres of the 13,770 producing acres in the field; thus, 57 percent of the acreage in the field produced from the Wilcox. Of the total 1,810 wells drilled in the field, 902 or 50 percent produced from this unit (Fig. 17).

Lithology

The Wilcox sandstone has an average thickness of 225 ft where not truncated. The upper

Table 2. Reservoir Statistics, Oil Creek Sandstone, Oklahoma City Field

Productive acreage	4,500
Av. thickness, productive zone (ft)	40
Pay zone (acre-ft)	180,000
Av. porosity (%)	18.0
Connate water (%)	10.0
Producing mechanism	water (partial)
Formation volume factor	1.4
Recovery—(% oil in place)	70
Bbl/acre-ft in place (STB)	898
Bbl/acre-ft recoverable (STB)	630
Total oil in place (STB)	161,640,000
Total oil recoverable (STB)	113,400,000
Total oil recovered, Jan. 1, 1941	100,000,000+
Av. prod. (bbl/acre)	22,222.0
Av. prod. (bbl/well)	277,778.0
No. productive wells	360.0
Av. acres per well	12.5
Recovery (bbl/acre-ft), Jan. 1, 1941	556
Percent of field total producing acreage	33

Cumulative prod., Jan. 1, 1941		
	Oil (bbl)	100,000,000.0
	Gas (ft^3)	320,000,000,000.0

Est. initial oil in place (STB)	161,640,000
Est. ultimate prod. (STB)	113,400,000

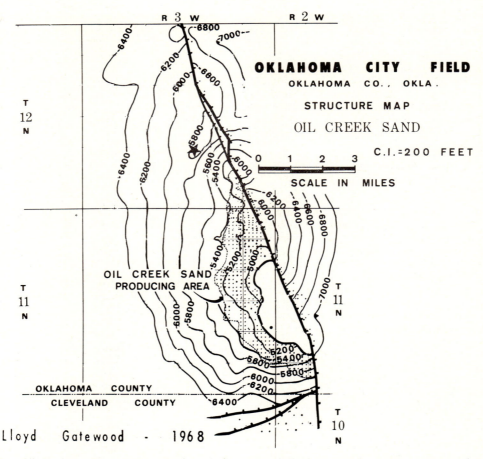

FIG. 15.—Oil Creek sandstone structure and producing area, Oklahoma City field. It is second to Wilcox as most prolific oil- and gas-producing zone in field.

45 ft is composed of large rounded and frosted sand grains embedded in a matrix of subrounded grains of medium size. The large "golf-ball" sand grains are present in largest percentage at the very top of the sandstone and in the limestone and dolomite just above. Where the top of the Wilcox is eroded, green shale and asphaltic inclusions are commonly present in the upper part of the remaining sandstone. The upper section is generally less porous than the middle section, which is made up of medium subrounded to subangular sand grains with very little cementing material.

The middle 100 ft is the most porous of the total 225 ft of Wilcox sandstone. Sand grains in the lowest 25 ft of the middle section commonly show recrystallization. The lower 30 ft contains thin beds of light-green shale; the sand

ranges from loosely cemented to white, tight sandstone.

The core of the Mary Sudik indicated porosity to be between 20 and 25 percent, with an average of 22.5 percent. Much of the Wilcox sandstone is so loosely consolidated that it was nearly impossible to obtain satisfactory recoveries in coring. The loosely bonded parts probably have greater porosity than that shown in cores recovered. Much of the loose sand flowed from wells during the productive life of the Wilcox sandstone, and porosity ranged as high as 30 percent in numerous areas, especially at the north end of the field where the average porosity is highest. The low dip on the unconformity surface was indicated by the fact that the upper Wilcox sandstone was found at approximately the same depths (6,200–6,500

FIG. 16.—ITIO No. 1 Mary Sudik blowing wild. This was discovery well for Ordovician Wilcox zone. Location shown on Figure 17.

ft) as the stratigraphically lower producing zones nearer the top of the anticline. The productive zone in the Wilcox ranges in width from 1½ mi where the dips are relatively slight to ½ mi where the dips are steep (Fig. 17).

Reservoir Characteristics

Early performance.—Generally, the natural flow for the early drilled part ended in 1934. Gas lift followed for about 1 year and then pumping was begun. In 1935 and 1936 wells in

the newly drilled part of the north end of the field flowed for some time, with reservoir pressures of 200 psi or less at 6,500 ft. The first known reservoir-pressure measurement was made by Pat McDonald on December 6, 1931, and regular surveys began in June 1932. Proration has been in effect throughout the life of the pool; allowables ranged from less than 1 percent, in the flush period, to maximum at the present time.

Water entered the oil-bearing sandstone in a

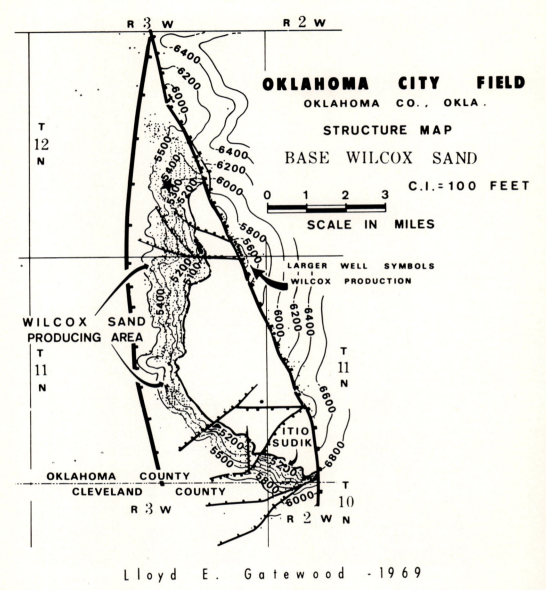

R 3 W   R 2 W

OKLAHOMA CITY FIELD
OKLAHOMA CO., OKLA.
STRUCTURE MAP
BASE WILCOX SAND

C.I.= 100 FEET

0   1   2   3

SCALE IN MILES

LARGER WELL SYMBOLS
WILCOX PRODUCTION

WILCOX SAND
PRODUCING AREA

OKLAHOMA COUNTY
CLEVELAND COUNTY

Lloyd E. Gatewood - 1969

Fig. 17.—Wilcox (Bromide) structure map. This zone, although on lowest part of structure, was the most extensive, most prolific, and last to be developed. Note Wilcox production on downthrown side of fault.

rather peculiar manner, having detrimental effects on wells completed below −5,300 ft. (The original oil-water contact was estimated to be at −5,424 ft. The intersection of the top of the Wilcox with this plane limits the west edge of production.) Water entered the most closely drilled areas in the west-edge wells first, with a gradual rise to about −5,325 ft in the part first drilled. The encroaching water was very noticeable to 1938, but since 1939 no apparent rise of water level has occurred.

During 1940 there were few changes in the oil-water contact. However, the gas-oil contacts were lowered as much as 35 ft in the more productive areas with little change elsewhere. From January 1 to July 1, 1941, an average lowering of 12 ft was observed for the entire Wilcox in the field.

Table 3 gives reservoir statistics for the Wilcox sandstone at Oklahoma City field.

Early concepts.—The presence of an initial gas cap was demonstrated in the Mary Sudik discovery well. Core data on some of the newly drilled Wilcox wells showed a gas-oil contact over a range of a few feet. The small gas cap was found to be present on the south end of the field, where the gas-oil contact was at −5,116 ft.

At that time it was thought that gas depletion with the removal of oil from sandstone indicated a liquid saturation of 40–50 percent in the upper part of the reservoir, with increasing saturation in the lower part. Subsequent examination of the facts has proved otherwise. Oil saturation of 1–25.9 percent in the gas zone contrasted with oil saturations of 52.7–92.8 percent of the pore space in the sandstone below the oil-gas contact. The probability of a gas drive was diminished because of the low oil-saturation values of 20 percent in the gas zone and 85 percent in the oil zone. These values indicated that slightly less than 50 percent of the pore space below the top of the water intrusion contained crude oil.

Early pressure decline.—Original reservoir pressure was 2,686 psia at −5,260 ft. As reservoir pressure declined with the withdrawal of oil and gas, water from the aquifer invaded the reservoir. As the water moved updip, it displaced some oil ahead of it. By 1938 water encroachment had stopped, having reached equilibrium with the aquifer at the subsea 5,300-ft datum. At this level it had moved through 60 percent of the original reservoir. The water has

been practically dormant since. At about the same time the reservoir pressure at the top of the Wilcox sandstone became equal to atmospheric pressure, and the only energy left to move oil into the wells was the force of gravity plus what energy could be supplied by gas injection. The oil, which had been displaced ahead of the water moving up structure, was at first virgin reservoir oil with all its gas in solution. As it kept moving up structure, however, the pressure constantly decreased, gas was lost, and by the time the water movement stopped at the −5,300-ft datum, it had accumulated enough oil ahead of it to saturate completely most of the part of the reservoir above that datum. It is from this bank of inert oil that the reservoir has been producing since about 1939; the producing mechanism is gravity drainage (Fig. 18).

Wilcox Reservoir Producing Mechanism

Estimation of the future oil production has been made by use of decline curves since proration is no longer applicable to the dwindling production. The conclusion (based on an average porosity of 21.5 percent for the field) is that an estimate of 1,072 million bbl is in accord with the sandstone volume and porosity information, and is the most probable quantity of initial crude in the Wilcox reservoir. The natural water drive was very irregular and detrimental to natural production, hence presented a difficult problem.

Gravity-Drainage Experiment

With many factors adverse to water flooding, tests were made to investigate the wettability of the Wilcox sandstone by oil and water, and the recovery of oil by gravity drainage.

During 1938 interest in the recovery of oil in the Wilcox sandstone by gravity drainage and the feasibility of water flooding induced the management of the Indian Territory Illuminating Oil Company to authorize a large-scale experiment to be conducted by the company's engineering department. A site was selected on the Wisel lease where the terrane sloped about 5°—the average dip of the Wilcox sandstone. The experiment consisted of cleaning (Fig. 19) and packing Wilcox sandstone in 300 ft of 6-in. OD pipe (Fig. 20). The final layout is shown in Figure 21.

Because of the difficulty of cleaning the sand produced from Wilcox wells, several thousand pounds of commercially produced glass sand was obtained from a plant operating on an out-

Table 3. Reservoir Statistics, Wilcox Sandstone, Oklahoma City Field

Productive acreage	7,880
Av. sand thickness (ft)	114
Pay zone (acre-ft)	900,000
Av. porosity (%)	21.5
Connate water (%)	1–0
Producing mechanism (since 1940)	Gravity Drainage
Formation volume factor	1.4
Recovery (% oil in place)	50.1
Bbl/acre-ft in place (STB)	1,192
Bbl/acre-ft recovered (STB), Jan. 1, 1968	600
Total oil in place (STB)	1,072,000,000
Total oil recovered, Jan. 1, 1968	537,465,000
Av. prod. (Bbl/acre, 1-1-68)	68,206.2
Av. prod. (Bbl/well, 1-1-68)	595,859.0
No. productive wells	902.0
Av. acres per well	8.74
Percent of field total producing acreage	57.2

Cumulative prod., Jan. 1, 1968

Oil (bbl)	537,465,000
Gas (ft³)	820,000,000,000
Est. initial oil in place (STB)	1,072,000,000

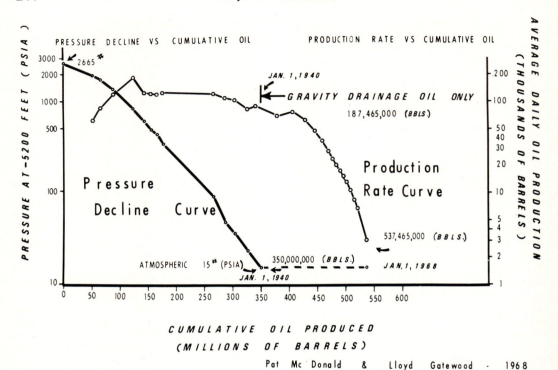

FIG. 18.—Wilcox sandstone pressure-decline and production-rate curves, Oklahoma City field.

FIG. 19.—Oklahoma City field Wilcox sandstone gravity-drainage experiment. Cleaning and drying Oklahoma glass sand.

crop of Simpson sandstone in the Arbuckle Mountains. The sand was cleaned and dried (Fig. 19) and packed into 20-ft joints of 6-in. OD pipe (Fig. 20). The sand used was weighed so that the porosity of the system could be calculated. In each 20-ft joint a 1-in. pipe connection was made to simulate a producing well. The "wells' could be opened and a sample of the sand could be obtained by means of a cork bore; the oil saturations were determined by colorimetric methods.

Live crude oil was introduced through the 1-in. pipe at the lower end of the system until the sand was saturated. A miniature tank battery was installed at the lower end. The system was opened and the first production was obtained by expansion of the gas from the 50-psi saturation pressure. After this brief burst of oil, the lower outlet was left open and the system produced oil by gravity alone. The test covered a period of several months. Periodic sampling of the 15 "wells" indicated that the oil saturation in the upper part of the system had been reduced to less than 10 percent by gravity drainage. The notes on this experiment are not

FIG. 21.—Oklahoma City field Wilcox sandstone gravity-drainage experiment. Final layout and operation during 1938. (Photo by Pat McDonald.)

available, but the results were comparable with those obtained in laboratory experiments made later by the Oklahoma City Wilcox Secondary Recovery Association, as reported in a paper by Katz (1942, p. 38–39). Before the start of the large-scale experiment, a test of the wettability of the Wilcox sandstone was made in the ITIO laboratory and the results were similar to those of the later experiment reported by Katz.

It was the opinion of the engineers that secondary recovery measures would not increase the ultimate oil recovery that could be expected by gravity drainage alone. It is possible that present *in situ* combustion techniques could speed the gravity drainage, and recover additional oil from the Wilcox oil zone.

Wettability Experiments

In no case could Wilcox water be made to displace Wilcox crude oil from clean or reservoir sand. Experiment showed that washing dry, clean sand in Oklahoma City Wilcox crude oil eliminated the preferential wetting of tap-

FIG. 20.—Oklahoma City field Wilcox sandstone gravity-drainage experiment. Packing sand in 6-in. OD pipe. (Photo by Pat McDonald.)

water over kerosene wetting. The Oklahoma City Wilcox crude oil altered the surface of dry, clean, outcrop sandstone to the extent that extraction with benzine for 30 hours or washing with natural gasoline did not clean the sandstone sufficiently to permit tap water to displace kerosene from the sandstone surface. It was concluded that oil production by water drive would not be a feasible means of secondary recovery from this sandstone.

The sandstone is apparently *oil wet,* the reservoir has a definite gas-oil contact or *fluid level,* there is less than 1 percent connate water, and gravitational drainage has reduced the saturation in the gas cap greatly. The logical conclusion from these observations is that natural gravity drainage will deplete the reservoir to a point beyond which depletion by gas drive will be impossible. In addition to this oil, the oil in the water-permeated zone has possibilities of recovery because both water and oil will drain from the sandstone in a manner similar to drainage of oil alone.

LATER DEVELOPMENT ACTIVITY

In May 1930, development reached the southeast edge of Oklahoma City, and in October production was expanded into Oklahoma City proper. City encroachment created a thriving business for lease hounds on city lots and presented the City Commission with many knotty problems. Organized chaos (Fig. 22) was the rule during 1931 in the Oklahoma City field. It was normal practice to let the wells flow over the crown and also over the countryside when they were completed, as evidenced by the Oil Creek sandstone producing well, Oil Incorporated No. 1 Thompson, Sec. 2, T11N, R3W, completed in January 1931 (Fig. 22).

The glut on the market, plus the huge oil production from East Texas in the summer of 1931, was decisive in causing posted prices to drop from 89¢/bbl at the beginning of 1931 to 16¢ and 20¢ in Oklahoma City during the summer of 1931. Governor "Alfalfa Bill" Murray of Oklahoma had been watching development and acted promptly. On August 4, 1931, he proclaimed a state of emergency, declared martial law within a 50-ft zone around each well, and ordered the militia to take control of 29 Oklahoma oil fields and to close all prorated wells until the price of oil hit $1.00/bbl. In his executive order, Governor Murray stated that it was unlawful to take crude oil from its stratum when there was no market. The market was strengthened on August 24 when one

FIG. 22.—Organized chaos—development activity in center of Oklahoma City during January 1931. Order ! ? , Confusion ! ? , Boom — Glut (From Shale Shaker, December 1966, Phil C. Withrow).

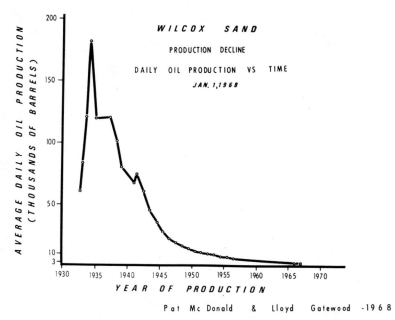

FIG. 23.—Decline curve, Wilcox production, Oklahoma City field.

major purchaser announced a willingness to pay $1.00/bbl for crude while others kept their quotations at 70¢/bbl. Field control was lifted in October and the troops were withdrawn. By 1935 Oklahoma City had the second largest single American oil field, embracing 11,000 acres; 1,700 wells had been drilled and from these had come almost 300 million bbl of oil.

By January 1, 1940, production from the Wilcox sandstone had amounted to 350 million bbl of oil and 820 billion ft³ of gas. By that date the Wilcox zone pressure was reduced to atmospheric. Since then natural water drive has

not been effective, but natural gravity drainage has resulted in the production of an additional 187,465,000 bbl of oil (Fig. 18). Total yield of the Wilcox has been 537,465,000 bbl. Highest average daily production was in August 1933— 212,794 bbl/day (Fig. 23). At the same time maximum monthly production was 6,383,837 bbl from 451 wells. Production has declined in 1968 to 3,000 bbl/day, and there is no way to estimate how long it will continue at that rate.

About 70 wells, completed since January 1, 1940, in areas abandoned because of the fingering of water, produced 400–500 bbl/day. Observations made during the years following bear out the general validity of the foregoing descriptions of reservoir conditions. The influence of the 70 wells is shown by the fact that they account for the greatest percentage of current production.

Table 4 gives Oklahoma City field statistics.

PENNSYLVANIAN ZONES

Development History

Because of the eagerness to get to the main objectives—the Arbuckle, Oil Creek, McLish, Tulip Creek, and Wilcox—many distinct Pennsylvanian zones were drilled through hastily with no attempt to develop production from

Table 4. Oklahoma City Field Statistics

Field Cumulative, Jan. 1, 1969	
Oil (bbl)	733,706,000[1]
Gas (trillion ft³)	1.7
Total wells producing	384
Total daily oil prod. (bbl), Dec. 1968	5,445
Total prod. 1968 (bbl)	1,963,817.0
Total daily Wilcox prod. (bbl), Dec. 1968	2,813
Total Wilcox prod. 1967 (bbl)	1,095,000
Total Wilcox producing wells (1-1-69)	187
Total productive acres	13,770
Total no. wells drilled in field	1,810
Length of field (mi))	12
Width of field (mi)	4.5
Sq mi within field	32

[1] Source: Oil and Gas Journal.

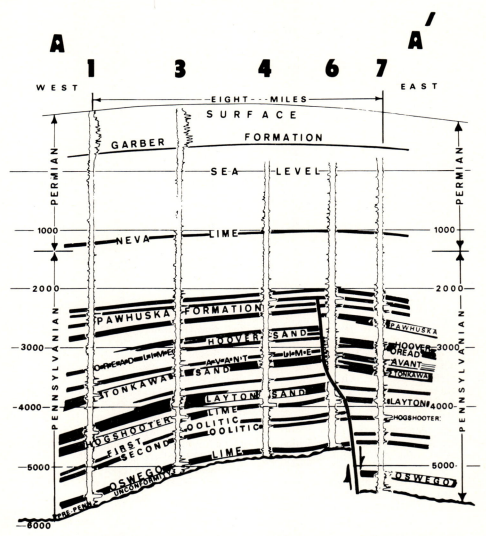

FIG. 24.—Pennsylvanian structural cross section. Location shown on Figures 27 and 6.

them. Many of the zones were limited to gas production, but substantial quantities of oil were found in several. Generally, the Pennsylvanian zones were tested (1) by drilling or coring into the sandstone or limestone and drill-stem testing or (2) by plugging back with cement from pre-Pennsylvanian zones after their depletion. Production potential was indicated in many of the Pennsylvanian zones by the presence of gas or by oil shows during drilling. A study of the cuttings from Upper Pennsylvanian and Permian strata showed that the first reliable and easily recognized key beds for

structural mapping are below 3,000 ft. The upper zones were drilled so rapidly, however, that it was difficult for the drilling crews to catch samples at short intervals. Limestone beds were more easily recognized and more commonly used as structural markers during the process of drilling. Some of these are the Neva, Pawhuska, Oread, Checkerboard, and Oswego limestones. The Neva Limestone marks the base of the Permian (Figs. 9, 24).

The Pennsylvanian producing zones are in both limestone and sandstone strata. Large volumes of gas have been produced, and lesser

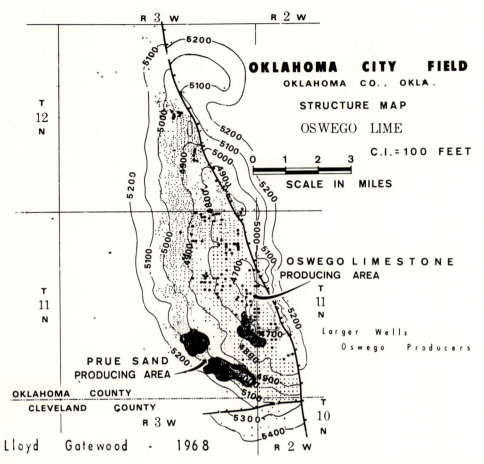

FIG. 25.—Structure-contour map and producing wells. Structural datum is top of Oswego limestone. Prue sandstone production is indicated in darker areas on southwest flank of structure.

quantities of oil, but the production is in no way comparable to that from the underlying pre-Pennsylvanian zones.

Lithology

One of the most productive zones was the Oswego-Prue in the Cherokee Shale just above the unconformity. It was the most prolific and the principal "free-gas" producer of the Pennsylvanian.

The Oswego limestone consists of two members—an upper thin member that ranges from almost zero to 15 ft thick, and a lower massive member with an average thickness of about 60 ft. The upper and lower members are separated, where both are present, by about 20 ft of gray shale. The Oswego is gray and white, finely to coarsely crystalline limestone, generally very fossiliferous near the base. The basal part of the lower member commonly is honeycombed with small solution cavities and contains much gas. The Oswego is greatly varied in thickness on the anticline, and the consequent porosity-thickness variation probably accounts partly for the erratic pattern of producing wells on the structure (Fig. 25). This limestone was productive of gas in probably 130 wells, almost all of which are now nearly depleted, having produced 185 billion ft³ of gas. The Oswego limestone is the first really good marker bed above the pre-Pennsylvanian unconformity. The distribution pattern shows that the Oswego limestone production is mostly on the axis of the anticline.

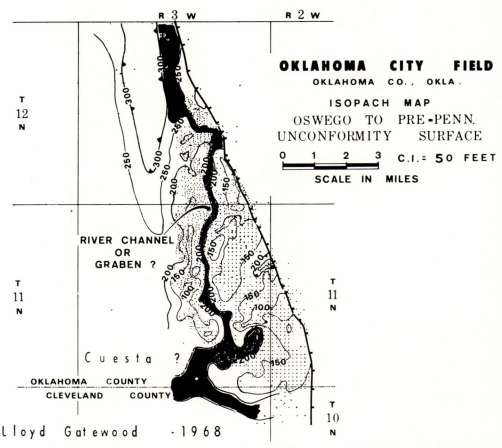

OKLAHOMA CITY FIELD
OKLAHOMA CO., OKLA.
ISOPACH MAP
OSWEGO TO PRE-PENN.
UNCONFORMITY SURFACE

0   1   2   3   C.I.= 50 FEET

SCALE IN MILES

Lloyd Gatewood   -1968

FIG. 26.—Isopach map, Oswego limestone to pre-Pennsylvanian unconformity surface. Varied thickness of Cherokee strata is shown. Prominent north-south stream channel (black) runs length of structure. Prue sandstone is productive in channel on south and west part of structure. Possibly channel is a fault-graben zone.

The Prue sandstone directly under (and in places in communication with) the Oswego limestone is mainly productive of oil. The interval between the Oswego–Prue and the pre-Pennsylvanian unconformity surface is very irregular (Fig. 26). The variable thickness of the Cherokee strata is due to deposition on a prominent and irregular topography (Fig. 27). The great stratigraphic variation is the result of continued vertical growth of the structure, redistribution of sediments from the old pre-Pennsylvanian surfaces by differential uplift, and lateral structural movement which rejuvenated old zones of weakness and reoriented them in transverse patterns.

Stream Channel or Graben

Most Oklahoma geologists are familiar with the old east-west pre-Woodford stream channel which cut completely through the Bois d'Arc in the middle of the West Edmond pool—14 mi northwest of the State Capitol, in the south part of T14N, R4W—dividing it into two parts. A younger superimposed stream, the result of Early Pennsylvanian movement, followed the same drainage pattern. Similar processes were undoubtedly at work on the Oklahoma City uplift itself, for there are indications of an old north-south stream channel running the length of the structure. It lies between the contact line of the lowermost Simpson sandstone and shale and the easternmost producing limits of the Wilcox sandstone (Figs. 4, 26). An isopach of the interval from the Oswego limestone to the pre-Pennsylvanian unconformity surface (Fig. 26) suggests that a channel may have been cut at least 50 ft below the Wilcox topography on the west and 80 ft below the

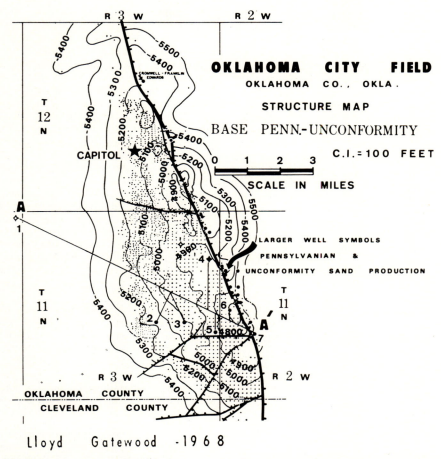

OKLAHOMA CITY FIELD

OKLAHOMA CO., OKLA.

STRUCTURE MAP

BASE PENN.-UNCONFORMITY

C.I.=100 FEET

0   1   2   3

SCALE IN MILES

LARGER WELL SYMBOLS

PENNSYLVANIAN &

UNCONFORMITY SAND PRODUCTION

Lloyd   Gatewood -1968

Fig. 27.—Structure contour map with datum at base-Pennsylvanian unconformity surface showing prominent and irregular topography, fault displacement, and Pennsylvanian production on downthrown side. **A-A'** is line of section in Figure 24.

older Simpson topography on the east. A younger superimposed stream, as a result of Early Pennsylvanian movement, could have followed the same drainage pattern and deposited the Prue sandstone in the channel. Much of the Prue sandstone, developed and productive on the south end of the structure, lies within the channel; it appears to be reworked Simpson sandstone. The distribution pattern of the Prue sandstone on the southwesternmost part of the structure suggests cuesta-type topography as its environment of deposition. Inflection profiles based on available data indicate that this feature could be a fault graben zone. Also, it is almost in alignment with the prominent north-south graben zone which extends through Cleveland and McClain Counties (Fig. 3).

FAULTS

Production is not confined to the upthrown west side of the major NW-SE bounding fault. The fault bears N 25° W and has a maximum throw of 2,400 ft at its midpoint near the crest of the structure. It is very straight, generally normal and vertical, but in some sectors is appears to become a high-angle reverse fault.

On the downthrown side of the fault, strata of Pennsylvanian age or younger generally dip less than 10° eastward, whereas the pre-Pennsylvanian strata below the unconformity generally dip more than 50° east.

Almost 9.5 million bbl of oil has been produced from at least six different zones on the downthrown side of the fault. Most of this— 7.73 million bbl—has come from the Wilcox sandstone, which had a strong water drive on this side of the fault. Production from the Pennsylvanian Skinner sandstone and the unconformity sandstone accounts for most of the rest. Other zones of limited production are the Cleveland (1st oölitic) sandstone, Simpson dolomite, and Oil Creek sandstone.

A few significant faults were instrumental in controlling production and migration of fluids. The fault that limits Wilcox production on the west (Fig. 17) is inferred from inflection profiles across the dip of the structure, which indicate a decided steepening of dip near the west flank. The original oil-water contact was estimated to be 5,425 ft subsea. The intersection of the top of the Wilcox sandstone with this water level limits the west edge of production. There is a steepening of dip at this point, certainly more than the normal 5° at the base of the sand. This steeper dip does not match in profiles; the projection of dip from west to east accounts for up-to-the-west fault displacement and, consequently, the down-to-the-east segment, which includes the west flank of the productive Wilcox sandstone. This fault may have been the factor which kept the generally good water drive from being effective, as it is in most of the other Wilcox producing fields in the province.

DATING STRUCTURE BY ISOPACHOUS RELATIONS

Probably the first structural growth recognizable in central Oklahoma by isopachous thinning was noted by the many geologists who worked in the Oklahoma City field. McGee and Clawson (1932) noted, "Structural growth and thinning, indicated by a noticeable shortening of the intervals from the Pawhuska down, occurs over the crest of the fold." They also noted ". . . a shifting of the 'high' axis eastward with depth."

The most prolific production in this part of Oklahoma is from low-relief structural traps in Ordovician Simpson sandstone beds. It has been found mostly by subsurface geology, and later substantiation by seismic work. To date consistent and reliable seismic data have been unobtainable below the Viola, and it is known that the Viola does not necessarily define Simpson structure. Some of the more significant structural-growth increments of the better know fields are measured in only tens of feet, and commonly are too small to be resolved by seismic methods alone. Subsurface work has been integrated with the seismic data so that

structural growth can be dated on the basis of isopachous thinning. The results are used as criteria for defining structural traps and orientation in the Simpson environment.

Some of the most significant producing structures in the area, where isopachous thinning over the structure reveals the critical age of growth for entrapment, are shown in Figure 28. These periods of growth were: during lower Simpson (Oil Creek to Bromide), Bromide to Viola, post-Hunton, and post-Mississippian deposition.

A relatively simple method for dating productive structures and geologic features involves the use of isopachs and cross sections. The arithmetic differences between the off-structure and on-structure intervals are noted and plotted as a "growth bar graph." By this method the highest peaks would represent periods of greatest growth and would show direct anomalous structural relations. The same method may be used for dating critical growth of faults.

The direct differential plotted on a bar graph is suitable for the Oklahoma City field area, which has undergone periods of uplift, erosion, and truncation known to affect Simpson production. "Growth curves" which utilize the percentage of change of like intervals between an up-structure or upthrown-side well and an off-structure (regional) or downthrown-side well

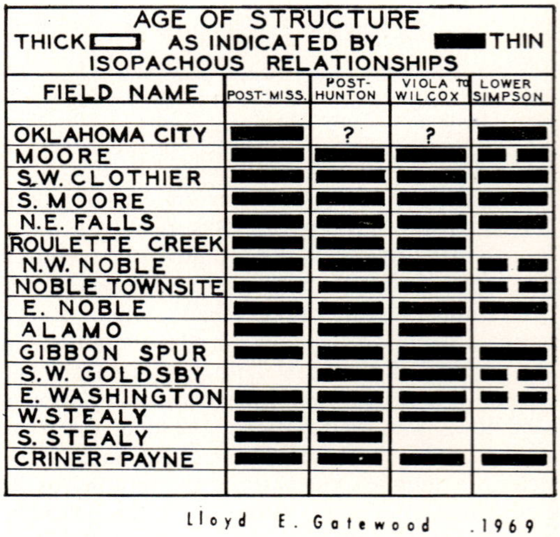

AGE OF STRUCTURE AS INDICATED BY ISOPACHOUS RELATIONSHIPS THICK ▭ ◼ THIN				
FIELD NAME	POST-MISS.	POST-HUNTON	VIOLA TO WILCOX	LOWER SIMPSON
OKLAHOMA CITY	◼	?	?	
MOORE	◼	◼	◼	◼ ▪
S.W. CLOTHIER	◼	◼	◼	
S. MOORE	◼	◼	◼	
N.E. FALLS	◼	◼	―	
ROULETTE CREEK	◼	◼	◼	
N.W. NOBLE	◼	◼	◼	▪ ▪
NOBLE TOWNSITE	◼	◼	◼	
E. NOBLE	◼	◼	◼	
ALAMO	◼	◼	◼	
GIBBON SPUR	◼	◼	◼	
S.W. GOLDSBY	◼	◼	◼	
E. WASHINGTON	◼	◼	◼	▪
W. STEALY	◼	◼	◼	
S. STEALY	◼	◼	▪	
CRINER-PAYNE	◼	▪	◼	

Lloyd E. Gatewood .1969

FIG. 28.—Oklahoma Simpson fields where "isopachous thinning" gives recognition of amount and age of structural growth periods as key indicators of time of oil trapping, and amount of oil fill-up in structure.

can be used in areas not affected by truncation. Both the growth bar graph and the growth curve add the dimension of time to structural and stratigraphic maps. The most interesting intervals are generally those in which deposition occurred during periods of structural growth and faulting. It has been found in many areas that the age of growth is a key factor controlling the amount of oil fill-up in the structure. It is also a factor in whether or not the structure will produce, as well as a key to which units will produce.

Conclusion

There may be nothing new under the sun, for *new* may not mean the first time, but may simply mean recognition and acceptance. The events that were new and unique at Oklahoma City 40 years ago are commonplace now. Oklahoma City field is a giant oil field, but was almost hidden under the redbeds. The Garber Sandstone was the window through which it was revealed, but the recognition of what it was and why it was there could come only bit by bit over a long period of time. It is an obvious understatement to say that it was a giant oil field because it was in the right place at the right time and in the right environment. Environment can be a state of mind, as well as a part of the stratigraphic parlance, because the acceptance of new concepts, methods, and techniques, which can validate the ideas so necessary in exploration, is almost as much of the battle as recognition and evaluation of the anomaly after the art is mastered.

Probably three fourths of the Arbuckle oil and one half of the Wilcox oil is still in the reservoirs. Ultimately there is an end for all oil fields. Improving technology of reservoir recoveries and need for the oil can be factors dictating when the end will come.

Only 24 percent of the Arbuckle oil has been recovered. Whether 36 years is enough time to allow the density separation of oil and water to permit a leveling effect within the inhomogenous fractured dolomite reservoir is uncertain. It will take new Arbuckle wells to determine its ultimate fate. It is almost certain that operation of the original wells at constant low rates of flow might have resulted in higher yields by retarding the water movement and allowing more efficient drainage of the tighter parts within the heterogeneous drainage area of the Arbuckle dolomite. Operators in the Arbuckle sector of the Healdton field have probably learned a lesson from Oklahoma City.

Wilcox oil recovery estimates have been revised continually since the beginning, and always upward. Mechanical failures of corroded and leaky pipe and cement, and inundation by water from other zones may curtail production before the gravity drainage mechanism is complete. One ultimate end to the Wilcox production could be a gigantic "fire" flood.

The geologists who are advocates of shear mechanics will still wonder and look for the other part of the Oklahoma City field. Many trend geologists have continued to look for another Oklahoma City structure. They haven't found another giant structure, but they have found nearby giant accumulations in stratigraphic traps and many smaller structural traps.

An oil accumulation is possible in the postulated upthrown block off the west flank of the Wilcox producing area, but its location in the very center of Oklahoma City poses more and bigger problems.

All the interesting history of Oklahoma City field was not necessarily geologic. There was the history of huge volumes of "hot oil," hidden flow lines, wrong-way valves, "blind" Christmas trees, and crooked holes. Chicanery was practiced in reporting producing zones for the obvious higher oil allowables. Also, the economic infusion to Oklahoma City during the depression and since has left an eternal impact upon not only the city but the state, as well.

For one who lived through three major oil booms—Seminole, Oklahoma City, and East Texas—the memories will always be as sharp and clear as the differences in the giants. One feels that the old giants never die, for in our hearts and in our minds they are always very much alive.

Appendix

Evaluation of Effective Displacement Pressures

The relative wettability of reservoir rocks by crude petroleum oils and by water indicates that certain constituents of the oils, even though present in minute amounts, may greatly influence the degree of wetting of the solid by the oil. An evaluation of relative wetting of reservoir rock by reservoir fluids has been attempted by the measurement of contact angles formed by petroleum oils and water on solid-silica plates. Displacement pressures calculated from the contact-angle data were found to be in reasonable agreement with displacement pressures actually observed with the petroleum oils and waters in packed silica powders. A study was made of both advancing and receding contact angles for a series of petroleum oils and water on silica. Hysteresis effect of advancing interfacial contact angles and receding interfacial contact angles was very pronounced in nearly all of the systems studied. For most of the systems, the silica appears to have dual wetting characteristics, inasmuch as the receding contact angle is less than 90° and the advancing contact angle is greater than 90°. The results from this research indicate that spontaneous displacement of oil from the underlying solid by water should occur only if both the advancing and the receding angles are less than 90° and that spontaneous displacement of water by oil should occur if both angles are greater than 90°. Also, no spontaneous movement should occur if the two angles are on opposite sides of 90°.

It appears that in most cases measurement of the contact angles formed on a silica plate between a petroleum oil and water will serve as a fairly reliable guide for the determination of the displacement tendencies of oil by water or of water by oil. The results obtained give fairly definite proof that crude petroleum contains constitutents which become adsorbed on siliceous surfaces, causing those surfaces to assume hydrophobic properties. The contact angle of Oklahoma City crude oil increased to 180° measured through the water phase. This abnormally high angle of contact indicated that the Wilcox sand is preferentially wet by oil.

Effect of Polar Impurities on Capillary and Surface Phenomena in Petroleum Production

It is presumed that the crude oil contains certain polar compounds that are responsible for this very striking behavior. Of great interest is the observation that although the crude oils do not displace water from a surface initially wet by water, the oils do alter the solid surface on contact, so that after a short time the water is incapable of displacing the oil from the solid surface. It seems reasonable to assume that this alteration of the solid surface is due to polar constituents tenaciously adsorbed from the crude oils. Various polar compounds, containing oxygen, sulfur, and nitrogen, some acidic and others basic, have been isolated from crude oils. The compounds are generally surface active, as they consist of a polar or hydrophilic part and a nonpolar or hydrophobic part, and may alter many of the tensions by adsorption at interfaces in the reservoir. The effects of polar impurities upon capillary and surface phenomena, in altering the contact angle of the Oklahoma City Wilcox oil with the sand (and thus making it preferentially wet by oil instead of by water), may be operative at Oklahoma City field. What caused these impurities to be present, making an oil-wet

sand of the Oklahoma City Wilcox, when the productive Wilcox sandstones found elsewhere are almost all considered preferentially water-wet sands? There are three possible explanations: (1) perhaps the Oklahoma City Wilcox was exposed to erosion, (2) the vast amount of pure, clean, unconsolidated sand with unusually high porosity and permeability may have undergone flushing and refilling numerous times under such conditions of structural growth at the unconformity surface, or (3) contamination may have occurred by distant migration into the Oklahoma City Wilcox sandstone body.

SELECTED REFERENCES

Abernathy, J. H., 1943 Engineering report on Oklahoma City zone of the Oklahoma City field: Oklahoma City Wilcox Pool Engineering Assoc., 110 p.

Benner, F. C., and F. E. Bartell, 1944, The effect of polar impurities upon capillary and surface phenomena in petroleum productions: Am. Petroleum Inst. Ann. Rept. Progress 1943, p. 79–84.

Dahlgren, E. G., 1931, Oklahoma City oil field in pictures: Oklahoma City, Jock Lance, Inc., 64 p.

Gatewood, L. E., 1966, Some aspects of the geologic history of Cleveland, McClain, and Oklahoma Counties, Oklahoma: Shale Shaker, v. 16, no. 10, p. 227–244.

Hill, H. E., E. L. Rawlins, and C. R. Bopp, 1937, Engineering report on Oklahoma City oil field, Oklahoma: U.S. Bur. Mines Rept. Inv. 3330, 242 p.

Katz, D. L., 1942, Possibilities of secondary recovery for the Oklahoma City Wilcox sand: Am. Inst. Mining Metall. Engineers Trans., v. 146, p. 28–53.

Latham, J. W., 1968, Petroleum geology of Arbuckle Group (Ordovician), Healdton field, Carter County, Oklahoma: Am. Assoc. Petroleum Geologists Bull., v. 52, no. 1, p. 3–20; reprinted, 1970: this volume.

McDonald, P., 1941, Total recovery and statistical report, Oklahoma City district: Indian Territory Illuminating Oil Co. unpub. rept.

McGee, D. A., and W. W. Clawson, Jr., 1932, Geology and development of Oklahoma City field, Oklahoma County, Oklahoma: Am. Assoc. Petroleum Geologists Bull., v. 16, no. 10, p. 957–1020.

Rister, C. C., 1949, Oil! Titan of the southwest: Norman, Oklahoma, Oklahoma Univ. Press, 467 p.

Small, A. D., 1941, Engineering report on Wilcox zone of the Oklahoma City field: Oklahoma City Wilcox Secondary Recovery Assoc., 100 p.

———— 1942, Engineering report on Wilcox zone of the Oklahoma City field: Oklahoma City Wilcox Pool Engineering Assoc., 110 p.

Travis, A., 1930, Oil and gas in Oklahoma County, Oklahoma: Oklahoma Geol. Survey Bull. 40-SS, 32 p.

Turk, L. B., 1937, Resume of the Oklahoma City field: a study of minor folds, ultimate production and production problems: Tulsa Geol. Soc. Digest, p. 11–14.

Wheeler, R. R., 1951, Tectonic features of the Southwest: Shale Shaker, v. 2, no. 2, p. 20.

Petroleum Geology of Healdton Field, Carter County, Oklahoma[1]

JACK W. LATHAM[2]

Calgary 2, Alberta

Abstract The Healdton field, in western Carter County, Oklahoma, is largely confined to the northeast half of T4S, R3W, but extends into adjacent townships. The townsite of Healdton lies within the field's limits. Oil production is principally from the Hoxbar Group (Missourian) of Pennsylvanian age and the Arbuckle Group (Canadian) of Ordovician age.

Production was first established in 1913 with subsequent field development resulting in oil production from four shallow Pennsylvanian sandstone bodies—the Healdton sandstones. All of them can be recognized over most of the field with some local discontinuities. Approximately 2,600 wells had been drilled by 1955, covering a production area of more than 7,100 acres.

Several of the earlier development wells were drilled into the pre-Pennsylvanian section where minor amounts of Ordovician oil was found.

In 1960, the discovery of a commercial reservoir within the Arbuckle brought renewed importance to this already prolific field. The new production is from three dolomite zones—Wade, Bray, and Brown. These zones are restricted to the upper 1,600 ft of a 5,000-ft carbonate section. The Brown zone, the lowermost, has proved to be the only zone of significance. It is crystalline dolomite, approximately 600 ft thick, with good intercrystalline porosity and excellent permeability caused by a highly developed fracture system. The Arbuckle produces from 43 wells within an area of 1,800 acres.

Entrapment of hydrocarbons is attributed to a northwest-southeast structural trend which originated in Early Pennsylvanian time and was again activated during Late Pennsylvanian. The Healdton area was subjected to intense uplift and faulting in Morrowan time by the Wichita orogeny. Associated high-angle faulting with displacement of 10,000 ft placed Pennsylvanian shales and sandstones in juxtaposition with Ordovician carbonates. These younger sediments are believed to be the source and means of migration for most, if not all, Arbuckle oil in the Healdton structure. Following an extensive period of erosion, Hoxbar sandstones and shales were laid down over truncated pre-Pennsylvanian rocks, and later folded by the Arbuckle orogeny.

Owing to the magnitude of forces affecting pre-Pennsylvanian strata, the Arbuckle producing structure has closure in excess of 1,500 ft whereas the overlying Pennsylvanian closure is approximately 600 ft.

Hoxbar sandstones, from an average depth of 1,000 ft, have yielded about 250 million bbl of oil, and secondary recovery methods are now being employed. The Arbuckle produces from an average depth of 3,900 ft and had a cumulative in excess of 3 million bbl to January 1968.

INTRODUCTION

The Healdton field is one of the early discoveries in southern Oklahoma and has since proved to be one of the state's most prolific producers. Although the field was not the first to be discovered in the area, it did mark the beginning of extensive use of geology in the search for oil in southern Oklahoma. With the realization that this was truly oil country, several geologic field parties were organized from 1913 to 1915 for the purpose of detailed mapping of surface anticlines and associated oil seeps, leading to methodical and expeditious oil development in this part of Oklahoma.

Oil production in the Healdton field is derived from two types of reservoirs discovered and developed during different periods of the field's history. In 1913 production was established from shallow Pennsylvanian—Hoxbar (Missourian)—sandstone beds which until recently have been the field's prime contributors of oil. In 1960 renewed importance was brought to this already prolific field by discovery of oil in commercial quantities from underlying carbonate rocks of Ordovician age. Pennsylvanian and Ordovician oil entrapment in the Healdton field was effected by anticlinal closures that were created during different orogenic movements, separated in time by much of the Pennsylvanian Period (pre-Atokan [Wichita orogeny] to post-Missourian [Arbuckle orogeny]). Wichita orogeny produced a sharp fold that was deeply truncated and subsequently was accentuated by Arbuckle orogeny. Hydrocarbon accumulations in the older and younger reservoirs appear to have differences in origin and migration.

The two reservoirs at Healdton are seemingly associated only by their proximity of location; however, the very presence of preexisting struc-

[1] Read before the 53d Annual Meeting of the Association, Oklahoma City, Oklahoma, April 24, 1968. Reprinted with revisions from an article, "Petroleum Geology of Arbuckle Group (Ordovician), Healdton Field, Carter County, Oklahoma": Am. Assoc. Petroleum Geologists Bull., v. 52, no. 1, p. 3–20, 1968. Manuscript received, May 7, 1968; accepted, October 20, 1968.

[2] Geologist, Atlantic Richfield Canada Ltd.

The writer expresses his appreciation to the drafting department of Atlantic Richfield Company (formerly Sinclair Oil & Gas Co.) for assistance in preparing the illustrations. Thanks also are extended to James H. Kempf for his criticism, and to Miss Bernice Gentry for typing the manuscript.

FIG. 1.—Location map of Healdton field, south-central Oklahoma, showing Wichita-Criner structural trend and related major structural features.

tures in southern Oklahoma commonly predestined the folding of overlying strata during subsequent orogenic movement. Healdton is only one example.

LOCATION AND REGIONAL GEOLOGIC SETTING

The Healdton field is confined mostly to the northeast half of T. 4 S., R. 3 W., western Carter County, Oklahoma, but it extends into adjacent townships on the north, west, and east (Fig. 2). The townsite of Healdton is within its limits.

The field was developed on what is known as the "Healdton uplift." This anticlinal feature, aligned northwest-southeast, is on trend with and is a part of the Wichita Mountains-Criner Hills anticlinorium, formed by the Wichita orogeny in Morrowan and Atokan time. Other fields associated with this extensive structural belt are the Loco and Asphaltum fields northwest of Healdton and the Hewitt, Bayou, and Lone Grove fields on the southeast (Fig. 2).

Major structural trends paralleling the Wichita-Criner uplift are the Muenster arch, Marietta syncline, Ardmore basin, and Arbuckle uplift (Fig. 1). The Muenster arch began its emergence in Late Mississippian time while the Ardmore basin on the north began its subsidence. The Wichita-Criner uplift, occurring in Early Pennsylvanian (post-Morrowan), formed an abrupt south shoreline for the deepening Ardmore basin and caused development of the Marietta syncline on the south. Several thousands of feet of sediments were removed by erosion on the positive areas, throughout Morrowan and Atokan time.

With advancement of Atokan seas the Marietta syncline took the form of a depositional basin and continued transgression caused deposition of Pennsylvanian sediments over the entire region.

Uplift by the Arbuckle orogeny in Late Pennsylvanian time created compressional forces that effected considerable lateral shortening and rejuvenation of ancestral structures. Thus, sediments overlying these old positive trends were commonly uplifted and folded.

Pennsylvanian oil production was developed on one of these gently dipping anticlines superimposed over the Wichita-Criner anticlinorium. Ordovician oil accumulation is associated with a more complex anticlinal fold that has approximately 1,500 ft of known closure. Cut by numerous small normal faults, this feature is bounded on the north by a down-to-the-north major normal fault and on the south by up-to-the-south reverse faulting.

Topography of the Healdton area is characterized by gently rolling hills with elevations ranging from 885 to 1,070 ft above sea level.

The Healdton structure is revealed at the surface as a topographic high with flat to low-dipping redbeds which are a part of the Pontotoc Group of Late Pennsylvanian and Early Permian age. The Garber Sandstone is the youngest rock unit encircling this breached anticline.

DISCOVERY AND DEVELOPMENT

PENNSYLVANIAN

Discovery of the Healdton field is credited to a prospector named Palmer who, early in the 1890s, drilled a well in Sec. 5, T. 4 S., R. 3 W., near an oil seep. Oil flowed to the surface from a depth of 425 ft. However, an existing Federal treaty with the Chickasaw and Choctaw Nations prohibited development of Indian lands for oil. This restriction was removed, and, prompted by the Palmer well and significant oil shows found in water wells drilled subsequently, prospecting was resumed in the early 1900s.

The Healdton field was discovered in 1913 by drilling of the No. 1 Wirt Franklin, NE¼, Sec. 8, T. 4 S., R. 3 W. Although initial development of the field was completed by 1919, drilling continued periodically to the early 1950s. Oil production is from four Healdton sandstone bodies which are part of the Hoxbar Group and are of Missourian age. Approximately 2,600 wells have

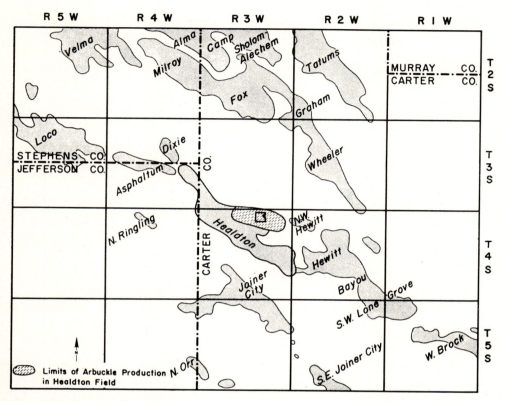

FIG. 2.—Area map showing Healdton field, Carter and Jefferson Counties, Oklahoma, recent Arbuckle development in Healdton, and location of the field in relation to other fields.

been drilled with an average total depth of 1,000 ft. The field is approximately 8 mi long and from 1 to 3 mi wide, and covers a productive area of more than 7,100 acres (Fig. 2). These Healdton sandstone bodies have yielded more than 230 million bbl of oil and secondary-recovery methods are being employed.

ORDOVICIAN

Birdseye Limestone.—The Birdseye Limestone of the Simpson Group has contributed minor quantities of oil from a few wells within the field.

Arbuckle Group.—Although its potential has been recognized only recently, Arbuckle production at Healdton is almost as old as the field itself. During development of the field, several wells were drilled into the pre-Pennsylvanian section in the hope of finding deeper pays. A few Birdseye Limestone completions were effected from this drilling but, because of unpredictable reservoir conditions, Simpson exploitation was not pursued.

At least seven of the deeper tests penetrated a part of the Arbuckle and with varied results (Fig. 3 and 3A). In 1924, the first Arbuckle oil was found when Pure Oil Co. drilled the No. 31 Lowery, SW¼ NW¼ SE¼, Sec. 4, T. 4 S., R. 3 W., to a total depth of 4,092 ft. Because of water at 4,005 ft, the well was plugged back to 3,637 ft and completed flowing 35 bbl of 40°-gravity oil a day from open hole. Although the well was reported as a Simpson completion, examination of records shows that production actually is from the upper and middle members of the West Spring Creek Formation, the Arbuckle top being at 3,076 ft. After several thousand barrels of oil had been produced by January 1962, the daily production declined to 3 bbl of oil and 85 bbl of water. A diesel cement squeeze was performed in open hole, and the well was recompleted for 18 bbl of oil a day.

In 1937, the Sun No. 17-A Mullen, NW¼

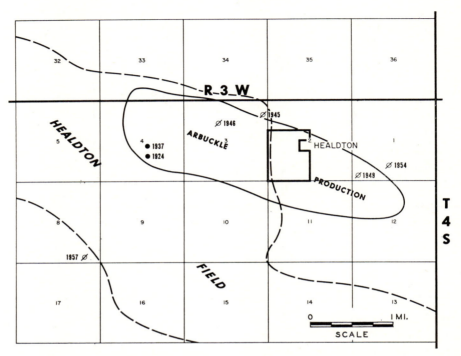

FIG. 3.—Healdton field showing Arbuckle and part of Hoxbar producing limits. Also shown are Arbuckle penetrations and dates prior to 1960.

NW¼ SE¼, Sec. 4, drilled 200 ft of Arbuckle and shows of oil were recorded. This well was plugged back to 2,345 ft and was completed in the Birdseye Limestone for 24 bbl of oil a day. In 1945, the Tomlinson No. 1 Schermerhorn, SE¼ NE¼ NE¼, Sec. 3, tested water from the Arbuckle. In 1946, the Tomlinson No. 5 Bell, SE¼ NW¼, Sec. 3, blew out in the Arbuckle and some gas was produced after the drill pipe was cemented and perforated. Three other tests were drilled in 1949, 1954, and 1957. The first two, in Sec. 1, T. 4 S., R. 3 W., penetrated less than 100 ft of Arbuckle and no shows were reported.

The third venture, the Seaboard No. 1 Apple, SW¼ SE¼ SE¼, Sec. 8, found the Arbuckle at 1,750 ft and was drilled to a total depth of 7,013 ft in Arbuckle. As a result of post-Springeran erosion, the upper 2,500 ft of Arbuckle is absent. This well was drilled on a severely breached pre-Pennsylvanian structure that is an overthrust block adjacent to the Arbuckle productive anticlinal feature which is truncated to a much lesser degree. No shows of hydrocarbons were reported in this test.

In the late 1950s several companies began to view more seriously the Arbuckle carbonates of southern Oklahoma as potential reservoirs. Be-

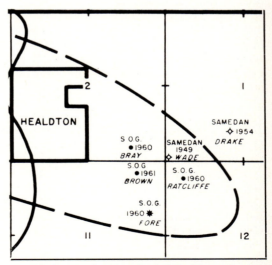

FIG. 3A.—Southeast end of Arbuckle production, showing recent wells which led to discovery of the Arbuckle reservoir (Brown zone).

cause of the shallow depths involved, Healdton was considered to be an ideal feature on which to test the Arbuckle section. After acquiring available acreage, Sinclair Oil and Gas Co. began drilling the No. 1 Ratcliffe in the spring of 1960. Located in the NW¼ NW¼, Sec. 12, T. 4 S., R. 3 W., this test was drilled to a total depth of 11,251 ft after reaching the Arbuckle (West Spring Creek Formation) at 2,985 ft. Although the well originally was intended as a Reagan Sand test, drilling was terminated after reverse faulting was found that repeats part of the Arbuckle. Numerous shows of oil and gas were recorded throughout the penetrated section and, after cementing of pipe to a plugged-back total depth of 6,800 ft, several intervals were perforated and tested. In October 1960, the well was completed for 47 bbl of oil and 49 bbl of salt water from the West Spring Creek, Kindblade, and Cool Creek Formations. On the basis of this completion it would have appeared that further exploration for Arbuckle in this area was not warranted. Drilling did continue but, because of poor porosity in deeper zones and a probable oil-water contact present in the Ratcliffe well, subsequent tests were not drilled below 4,000 ft.

Two offsets to the No. 1 Ratcliffe were drilled simultaneously. The Sinclair No. 1 Modell Fore, SW¼ NE¼ SE¼ NE¼, Sec. 11, was completed dually on November 8, 1960 for 3,250 Mcf of gas a day from the upper West Spring Creek Formation and 8 bbl of oil a day from the middle West Spring Creek. The pays are the Gas or Wade zone and Bray zone, respectively (Fig. 4). The well was drilled to a total depth of 4,000 ft after reaching the top of the Arbuckle at 3,027 ft. The Sinclair No. 1 Geneva Bray, SW¼ SE¼ SE¼, Sec. 4, was completed on December 20, 1960 in the Bray zone flowing 189 bbl of oil a day on a 14/64-in choke. The West Spring Creek Formation, found at 2,970 ft, was cored from 2,986 ft to a total depth of 3,996 ft.

After three wells had been drilled, all resulting in completions in the Arbuckle, the commercial quality of the reservoir still was not certain because the productive Wade and Bray zones discovered in the first wells consist of thin-bedded, very fine saccharoidal dolomite characterized by low permeability. All doubt was removed, however, by the fourth test. The Sinclair No. 1 Myrtle Brown, NW¼ NE¼ NE¼, Sec. 11, with a total

depth of 3,999 ft, was completed flowing (natural) 336 bbl of oil a day and 150 Mcf of gas a day on a 12/64 in. choke from a coarsely crystalline porous dolomite at 3,853 ft, 953 ft below the top of the Arbuckle. Of 140 ft penetrated, 65 ft is porous, but an oil-water contact at 2,985 ft (subsea) limits the net effective pay thickness to 34 ft. This new productive member is known as the "Brown zone" and is considered by the writer to be the top of the Kindblade Formation (Fig. 4). The same dolomite section was observed in its entirety in the No. 1 Ratcliffe, where it has a thickness of 560 ft, of which 200 ft has indicated microlog porosity; however, being structurally low, the No. 1 Ratcliffe is water-bearing.

The discovery of Brown-zone production established the Arbuckle as a worthwhile objective and set the stage for its orderly development. This was effected on the basis of two fundamental assumptions: (1) the Brown zone has a somewhat broad lateral extent and (2) there is a common oil-water contact; therefore, drilling locations were contingent only on structural position.

The present area of Arbuckle production is 3.5 mi long and less than 1 mi across, a productive area of 1,800 acres. This small areal extent is more than compensated by the presence of the porous Brown-zone dolomite that has a field average gross thickness of 580 ft, of which an average of 250 ft has effective porosity. Some wells drilled in Secs. 3 and 4, T. 4 S., R. 3 W., have found the entire zone above the oil-water contact. Using 40-acre spacing, 55 Arbuckle tests have been drilled at Healdton since 1960, resulting in 45 completions and 10 dry holes. Thirty-five wells produce from the Brown zone and eight are Wade and/or Bray completions. Cumulative production for all three zones at the end of 1965 was 1,080,883 bbl of oil. The Birdseye Limestone is also productive in four of the wells.

STRATIGRAPHY

The stratigraphic section is described in the order in which it is penetrated by drilling. No attempt is made to calculate true thickness from the apparent or drilled thickness. Therefore, intervals are of an average drilled section such as is shown by the type log (Fig. 4).

PERMIAN SYSTEM

Pontotoc Group.—The Healdton field is over-

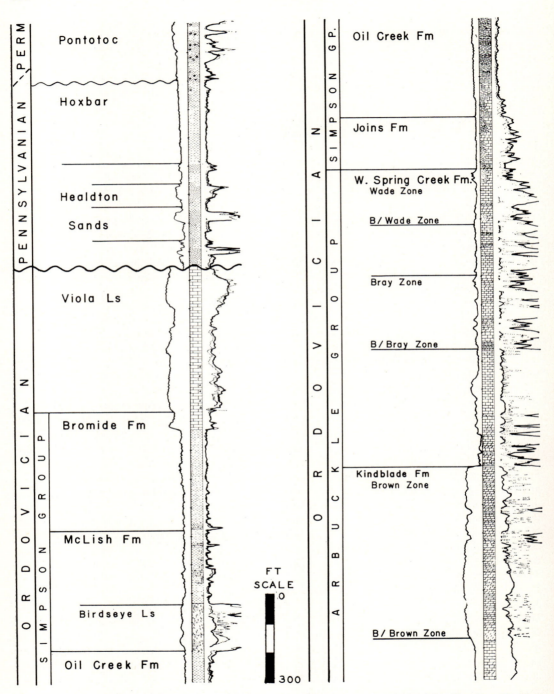

FIG. 4.—Composite electric and lithologic log of stratigraphic section at Healdton field. The four Healdton sandstones (Pennsylvanian) are numbered in descending order—*see* Figure 10.

lain by a redbed sequence belonging to the Pontotoc Group of Early Permian and Late Pennsylvanian ages. However, no differentiation is attempted.

The Pontotoc section consists predominantly of red to varicolored shale and sandy shale with thick conglomeratic sandstone containing chert and limestone fragments. Thickness ranges from 500 ft over the feature to more than 2,000 ft on the flanks. The Pontotoc lies unconformably on Hoxbar strata.

PENNSYLVANIAN SYSTEM
Missourian Series

Hoxbar Group.—This group consists of a sandstone and shale sequence which ranges in thickness from 600 to 1,200 ft. The average thickness is approximately 800 ft in much of the Healdton field. The upper 400 ft is a light- to medium-gray, finely micaceous, soft, locally silty, smooth-textured shale. The rest of the section is predominantly sandstone interbedded with gray shale. Four main sandstone units have been recognized for their oil production and are the Healdton sandstones which are characterized by marked lateral changes in thickness and porosity. Locally, a lenticular basal sandstone conglomerate lies unconformably on pre-Pennsylvanian rocks.

The Healdton sandstones are fine grained and poorly cemented, with fossiliferous and oölitic limestone beds present locally in the three lowermost sandstone bodies. Because of the similarity and lenticularity of these sandstones, correlations on the basis of samples are commonly difficult; therefore, much dependence is placed on electric-log correlations. Positions of the four Healdton sandstone bodies are shown on Figure 4.

The first or uppermost Healdton sandstone commonly is a poorly developed, fine-grained, micaceous, carbonaceous sandstone with interbedded gray silty shale.

The second Healdton sandstone, approximately 40 ft thick, is light gray, fine grained, and friable. It grades downward to very thinly laminated sandstone and shale and ultimately to a gray shale.

The third Healdton sandstone, with an average thickness of 30 ft, is buff to gray, fine grained, friable to loose, and maintains a more persistent development. It is underlain by 40 ft of very thinly laminated sandstone and shale.

The fourth Healdton sandstone is characterized by a much more varied lithology, grading from a gray, pale-green silty shale to interbedded fine-grained sandstone and gray to pale-green shale. There is also a prevalence of limestone conglomerates associated with coarse-grained, fossiliferous sandstones that are intermixed with pale-green shale. Average total thickness is 60 ft.

An abrupt facies change on the north and east flanks of the Healdton structure, coinciding with the development of a much thicker Hoxbar section, makes lithologic and electric-log correlations extremely difficult to nearly impossible. As a result, there are many different opinions about the regional equivalent of the Healdton sandstones. The flank strata have an increased Hoxbar thickness of approximately 2,000 ft and a sandstone sequence more similar to that found in fields on the east and south.

Desmoinesian Series

Deese Group.—Although the Deese is on the flanks and areas adjacent to the Healdton structure, its deposition over the uplifted part is questionable. Some geologists have proposed that part of the Healdton sandstone sequence is of Desmoinesian age. The writer suggests that this area was likely to have maintained a somewhat positive attitude throughout Desmoinesian time. Therefore, for the purpose of this paper, all Healdton sandstone units are included within the Hoxbar Group discussed above.

ORDOVICIAN SYSTEM
Champlainian Series

Viola Group.—The Viola Limestone of the Viola Group, because of truncation, is the youngest pre-Pennsylvanian rock unit present on the Healdton feature. Although the Viola is partly eroded, approximately 600 ft of Viola is present in Sec. 4 and this thickness probably is within 200 ft of being a complete section.

The upper 250 ft preserved in the Healdton field is a buff to white, chalky to finely crystalline limestone with traces of tan chert. The lower 350 ft is a buff, tan, light-gray, finely crystalline limestone with abundant white chert. Shows of oil and fracture-type porosity are common in the uppermost part.

Bromide Formation (Simpson Group).—The Bromide Formation has an apparent thickness of

400 ft in Sec. 4, but is absent elsewhere over the Arbuckle producing part of the Healdton field. The upper 50 ft is a pale-gray to green, firm, dense, lithographic limestone that in places is slightly argillaceous. This upper unit is the "Bromide dense" of subsurface terminology. The other 350 ft is an apple-green, waxy, brittle shale with traces of white to gray, fossiliferous, crystalline limestone. The Bromide sandstone beds, known for their productivity in southern Oklahoma, are represented here only by gray, white, very fine-grained, tight, calcareous sandy zones.

A lack of sandstone and limestone, which are typical of other areas, does not permit a clear identification of the Tulip Creek Formation.

McLish Formation (Simpson Group).—The McLish has an average thickness of 450 ft. The upper 250 ft is principally medium- to dark-green, waxy, splintery, soft shale, interbedded with white, buff to gray, coarsely crystalline, fossiliferous limestone that commonly is associated with thin, white, phosphatic-speckled, fine-grained, tight, calcareous sandstone.

The basal unit of the McLish Formation is the Birdseye Limestone Member. It is a buff to white, semidense, homogeneous limestone with characteristic clear calcite inclusions for which it is named. Its thickness ranges from 150 to 250 ft. Some geologists have placed the Birdseye in the Oil Creek Formation, but the writer considers it to be a part of the McLish (Decker and Merritt, 1931).

Oil Creek Formation (Simpson Group).—Underlying the Birdseye Member is 400–500 ft of dark-green, waxy, soft shale that is interbedded and intermixed with off-white, dark-gray-mottled, very fossiliferous, nodular limestone containing abundant ostracods and bryozoans. The index fossil *Aparchites perforata* is found in the upper 100 ft.

Near the base of the Oil Creek Formation there is a slight increase in the amount of limestone with traces of white, rounded, "floating" sand grains. This is the equivalent of the "Basal Oil Creek sandstone," found productive in other areas of Oklahoma.

Joins Formation (Simpson Group).—The Joins is 150–200 ft of buff, dark-gray-mottled, ostracod-bearing, crystalline to buff dense, homogeneous limestone with some thin interbeds of green waxy shale. The upper half is predominantly

gray-mottled, coarsely crystalline, and fossiliferous, and the lower half is dense limestone and less fossiliferous. At the base, a 20-ft green shale lies on the Arbuckle without apparent disconformity.

Canadian Series

Arbuckle Group.—As originally defined, the Arbuckle section has included strata of both Late Cambrian and Early Ordovician ages. However, it was proposed later that the Arbuckle Group be restricted to the Ordovician System, and that the Upper Cambrian units be included in the Timbered Hills Group (Harlton, 1964). The latter usage of the Arbuckle is adopted in this paper.

The Arbuckle Group is a massive carbonate section approximately 5,000 ft thick, and is subdivided into four formations. The presence or absence of a wide variety of minor lithologic types is invaluable in the recognition of these formations by visual study alone. The importance of these accessory lithologic types is not always the result of their presence or absence, but of their preponderance. Because expediency commonly necessitates determination in the field through sample examination and electric-log correlations, a thorough knowledge of the criteria set forth here is absolutely necessary in formation identification and unit correlations.

West Spring Creek Formation (Arbuckle Group).—This formation is the uppermost unit of the Arbuckle Group and is found at an average depth of 2,500 ft within the Arbuckle productive limits of the Healdton field. The formation contains a wide variety of lithologic types in a depositional sequence that is not present in other Arbuckle formations. It is predominantly limestone, but contains zones of thin-bedded saccharoidal dolomite associated with pale-green shale, anhydrite, and sandy dolomite. Thin pelletoid zones and chert are randomly present throughout. The pellets do not have the typical concentric structure, but otherwise resemble oölites. Megafossil assemblages appear in thin restricted units. The formation has an average thickness of 1,000 ft in the Healdton area, but thickens to 1,200 ft on the Fox-Graham structure in the north part of T. 3 S., R. 3 W.

The West Spring Creek has been subdivided further by the recognition of two dolomite zones, separated by more massive homogeneous limestone bodies. These zones have a broad areal extent and, through detailed study of their cyclic patterns, are recognizable for several miles along depositional strike. Therefore, regional correlations along strike are in many places more obvious than local ones. The dolomite zones are considered to be of primary origin, deposited in a shallow-water evaporitic environment.

Wade zone: This zone composes the upper 200 ft of the West Spring Creek Formation and is a buff, tan to resinous-brown, finely to very finely saccharoidal dolomite. This lithology is found thinly bedded with green dolomitic shale to pale-green, subdense, argillaceous dolomite. The dolomites have varied quantities of fine- to medium-grained, well-rounded quartz sand which locally is predominant in the rock. A considerable quantity of anhydrite is interbedded to intermixed throughout the zone. These saccharoidal dolomites commonly are oil saturated and have pinpoint porosity but rarely significant permeability.

The Wade zone is correlative with what is termed the "Joins dolomite" of other areas. However, the lithology is so characteristic of the West Spring Creek that it seems very impractical to exsert the Wade from a suite of rocks whose environment of deposition was approximately the same. There is not only an abrupt depositional change between the Wade and the overlying Joins Formation, but also a marked absence of ostracods which are prevalent in the Joins.

The Wade zone is underlain by approximately 200 ft of buff, gray to brown, finely crystalline to dense, massive limestone. A few streaks of shale, anhydrite, or dolomite are observed.

Bray zone: The Bray zone, 250 ft thick, has a sandy dolomite sequence much like the Wade zone. It differs only by being interbedded with finely crystalline to semidense limestone ranging in color from buff to creamy, greenish-gray, tan, gray, or brown, and commonly chalky. The dolomites are oil saturated and locally oil productive.

The lower part of the West Spring Formation is a buff, tan, creamy, and gray, finely crystalline to subdense, massive limestone. A few shale and saccharoidal dolomite laminations are present, but there seems to be a complete absence of anhydrite and only traces of sand. Cores show a distinct banded effect similar to the stromatolite structures seen on rocks cropping out in the Ar-

buckle Mountains on the northeast.

The basal 50–75 ft is a thin-bedded, buff, gray, gray-green, very finely crystalline to saccharoidal, sandy dolomite, interbedded with green shale and finely crystalline limestone. Pelletoid structures are common.

Kindblade Formation (Arbuckle Group).—The Kindblade Formation has an apparent thickness of 1,600 ft and is recognized by homogeneity and the relative absence of noncarbonates. Although this is predominantly a subdense limestone with some interbedded crystalline dolomite, the unit grades laterally into predominantly dolomite west and south of the Healdton field. In T. 3 S., R. 7 W., and T. 4 S., R. 6 W., the Kindblade is almost wholly finely to coarsely crystalline dolomite.

Brown zone: At Healdton the upper 500–600 ft of Kindblade is the Brown zone, which is the primary Arbuckle objective. It is a tan, gray to brown, finely to coarsely crystalline dolomite in which well-formed rhombs are common. This zone was cored in the Sinclair Oil and Gas No. 1 Freeman, SW¼ SW¼ NW¼, Sec. 2, T. 4 S., R. 3 W. Studies of the cores resulted in a much greater appreciation of carbonates as reservoirs. Although drill cuttings show some intergranular and vuggy porosity, cores show pore spaces to be of such a large scale that small drill cuttings are inadequate in determining the type of porosity. Porosity types include intergranular, vuggy, cavernous, pinpoint, and fractured. Intense fracturing which caused the core barrel to jam made it impossible to cut more than 10 ft of section per run and the rock was gathered "by the bucket full." So complete is the fracturing that the rock bears a striking resemblance to an unconsolidated intraformational breccia. A rock crusher could not have done a better job of fracturing. However, carbonate rocks above and below this zone are fractured to a much lesser degree and commonly appear to be healed by calcite. Although cores from the No. 1 Freeman have excellent reservoir properties, they are not the best examples of the Brown zone as a reservoir. This is supported by the fact that porosity values from electric-log calculations of this well are lower than the values from many of the other wells in the field.

The writer originally considered this zone to be a part of the lower West Spring Creek Formation. However, a more detailed study of the nature of Brown-zone development at Healdton and other areas of southern Oklahoma indicates that this dolomite facies is more closely related to the Kindblade Formation.

The remaining 1,000 feet of Kindblade is typically buff, tan to gray, finely crystalline to hard, dense, homogeneous limestone. This monotonous section is interrupted in places by thin units of finely crystalline dolomite. Abundant pseudo-oölitic or pelletoid limestone is present throughout. The limestone locally appears to be slightly siliceous, and minor quantities of oölitic and spicular white chert occur at the base of the formation.

Cool Creek Formation (Arbuckle Group).— Only three wells in the Healdton field have penetrated the Cool Creek Formation. In the Sinclair No. 1 Ratcliffe, NW¼ NW¼, Sec. 12, T. 4 S., R. 3 W., the top was reached at 5,630 ft. The unit has a drilled thickness of 1,100 ft. Characterized by a conspicuous abundance of white oölitic chert and quartz sandstone, this is the most easily recognized unit of the Arbuckle Group and it has an average thickness of approximately 1,200 ft in much of southwestern Oklahoma.

The Cool Creek Formation consists of interbedded buff, tan, gray, finely to medium-crystalline limestone and dolomite. White oölitic chert, oölitic to sandy limestone, and conglomeratic sandstone are typical of the subordinate rock types. "Floating" quartz sand grains and a minor quantity of pale-green shale also are common.

The No. 1 Ratcliffe well produced some oil from this formation, but depth and marginal porosity do not justify its exploitation.

McKenzie Hill Formation (Arbuckle Group). —The McKenzie Hill Formation, penetrated only by the No. 1 Ratcliffe, has an apparent thickness of 1,050 ft and is dominated by brown to gray, finely crystalline limestone interbedded with thin, finely to medium-crystalline, granular-textured dolomite. Thin ostracod concentrations can be observed and intraformational limestone conglomerate occurs throughout the section, associated with mottling.

Chert and sandstone are present in the upper 500 ft but in smaller quantities than in the overlying Cool Creek Formation. The only glauconite in the Arbuckle Group is in the lower few hundred feet of the McKenzie Hill Formation, where it is present as very fine fragments disseminated in the finely crystalline limestone.

CAMBRIAN SYSTEM
Croixan Series

Signal Mountain Formation (Timbered Hills Group).—This formation has a drilled thickness of 1,900 ft in the No. 1 Ratcliffe. The excessive thickness here probably is the result of faulting and a steepening of dip. In contrast to the McKenzie Hill Formation, limestone consistently is more finely crystalline and darker; chert is absent and only a very few oölites occur. There is an increase in intraformational conglomerate and glauconite, but the glauconite is not nearly as abundant as it has been observed to be in equivalent rocks which crop out in the Arbuckle Mountains and elsewhere in the subsurface of southern Oklahoma. Thin, dark greenish-gray shale layers are interbedded locally.

Fort Sill Formation (Timbered Hills Group).—This is the oldest rock unit penetrated at Healdton and its thickness also is affected by faulting and steep dip.

The Fort Sill Formation is distinctively more dense and lighter colored than the Signal Mountain Formation, but otherwise is very similar. Locally it contains pebbly limestone conglomerate, thin streaks of pale-green shale, and a fine sprinkling of glauconite.

STRUCTURE

The Healdton field includes two anticlinal structures formed during different orogenic movements. Entrapment of hydrocarbons in the shallow Hoxbar sandstone beds is attributed to a northwest-southeast-trending anticline which evolved during Late Pennsylvanian time (Fig. 5). High-angle faulting terminates the field on the north and has approximately 2,000 ft of throw (Fig. 6). Although four main sandstone bodies are present throughout much of the field, their lenticular nature (described under Stratigraphy) and a common lack of adequate well data make correlations highly questionable in some parts of the field. Therefore, continuity of mapping horizons are difficult to maintain. With the top of the first Healdton sandstone zone as a datum, Figure 5 represents the structure of all four Healdton sandstones. Control provides a known productive closure of 600 ft, the highest part being in the northwest and central parts of the structure.

Underlying and paralleling the gently dipping

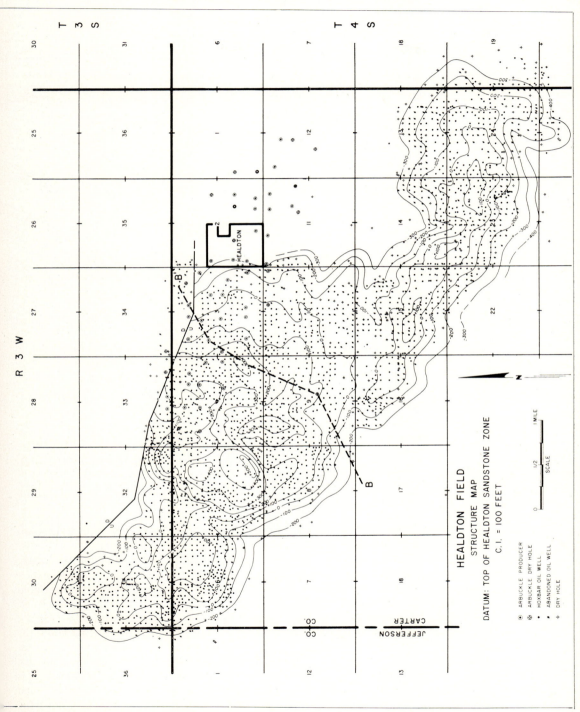

FIG. 5.—Structure-contour map of Healdton field. Datum is top of Healdton sandstone zone. Depths subsea, C.I. = 100 ft. Arbuckle wells drilled since 1960 are circled. **B-B′** is line of section in Figure 6.

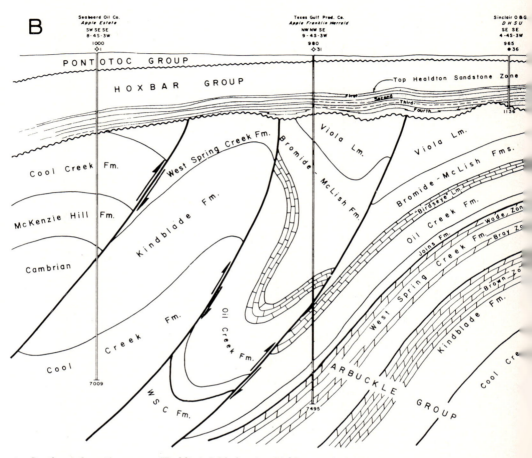

FIG. 6.—South-north section across Healdton field showing highly complex structure of pre-Pennsylvanian rocks, and broad anticlinal attitude of overlying Pennsylvanian rocks. All four productive Healdton sandstones are present over much of the field although lenticularity is common locally. Difficulty in correlating these sandstone units on north edge of field indicates a facies change or faulting, but a combination of both is likely. Line of section is B-B' (Fig. 5).

anomaly in the Hoxbar, the pre-Pennsylvanian consists of two more tightly folded and deeply breached anticlines that are separated by thrust faults. The southern fold is the higher block and underwent severe erosion after being folded and faulted by the Wichita-Criner movements of Early Pennsylvanian time. Truncation has resulted in the removal of more than 2,000 ft of Arbuckle, exposing the Cool Creek Formation at the old erosional surface.

The adjacent fold on the northern block, being downthrown, was protected from truncation of Arbuckle rocks. The deepest erosion was to the Oil Creek Formation. It is this part of the pre-

Pennsylvanian complex that now produces from the Arbuckle dolomite at Healdton.

The configuration of the Healdton productive structure is that of an elongate anticlinal fold extending northwest-southeast through the Healdton townsite. Structural closure exceeds 1,500 ft, and there is 700 ft of productive closure (Fig. 7). Subsequent to folding and partial truncation, the anticline appears to have been subjected to considerable tilting. This is best illustrated by the east-west cross section (Fig. 8). Although it is a southeast-plunging anticline, the eastern end was eroded more deeply to expose rocks of the Oil Creek Formation. Erosion at the west end

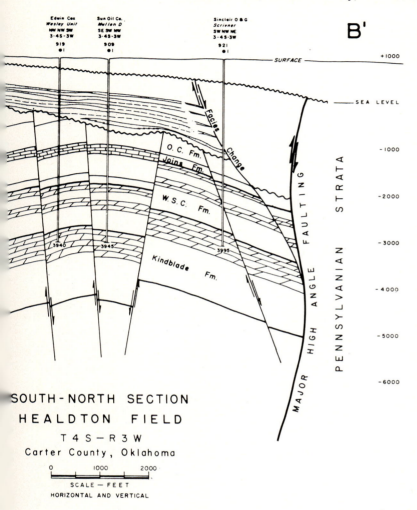

B'

SURFACE   +1000

SEA LEVEL

O. C. Fm.   -1000
Joins Fm.

W. S. C. Fm.   -2000

-3000

Kindblade Fm.   -4000

-5000

-6000

SOUTH-NORTH SECTION
HEALDTON FIELD
T 4 S — R 3 W
Carter County, Oklahoma

0   1000   2000
SCALE — FEET
HORIZONTAL AND VERTICAL

did not extend below the Viola Limestone. The pre-Pennsylvanian erosional surface also slopes from west to east, which indicates that the highest element of the anticlinal trend was east of its present location.

The anticline is terminated on the north by a high-angle fault, down to the north, that originated in Early Pennsylvanian time and continued to be active through much of Missourian time (Fig. 6). This fault has been considered to be a normal fault, but drilling has proved that locally the fault has a reverse attitude with a throw of approximately 2,000 ft. Associated faults on the north have a probable total displacement of more than 10,000 ft, placing Springeran strata of the

Ardmore basin in juxtaposition with the Arbuckle.

The Springer section is composed primarily of shale, but several well developed sandstone bodies are present which are productive in parts of the Ardmore basin. The direct contact of these excellent source beds with the Arbuckle reservoir has given rise to much speculation on the true place of origin for Arbuckle oil at Healdton. Many geologists have questioned indigeneity of the Arbuckle oil and suggest a Springeran origin. Although studies of oils from Arbuckle reservoirs in Oklahoma and equivalents in Texas indicate hydrocarbons to be indigenous at some localities, conditions in the Healdton field are excellent for a

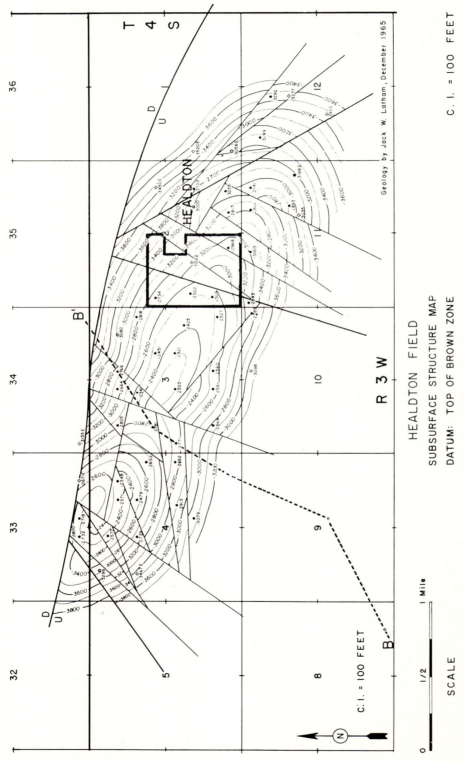

HEALDTON FIELD

SUBSURFACE STRUCTURE MAP

DATUM: TOP OF BROWN ZONE

C. I. = 100 FEET

Geology by Jack W. Latham, December 1965

FIG. 7.—Structural map of Healdton field. Structural datum, top of Brown zone. Wells shown either produce from Arbuckle or are wells drilled since 1960. **B-B'** is line of section in Figure 6.

probable Springeran origin, with subsequent migration into the Arbuckle reservoir.

Several pre-Pennsylvanian horizons have been contoured and, except for minor irregularities, all appear to have a similar configuration. Because the Brown zone is the principal objective, contours at the top of this zone were chosen to depict the pre-Pennsylvanian structure (Fig. 7). Core and dipmeter-survey data indicate that this is a simple anticline plunging 15°–30° southeast.

However, drilling has found numerous normal faults which have displacements of 10–500 ft. Of the 55 tests drilled, 36 are known to have found faults and some wells crossed more than one. A total of 58 occurrences are known where wells crossed faults that have more than 50 ft of displacement. Faults having less than 50 ft of displacement were ignored for clarity.

Because of their abundance, it seems reasonable to assume that the structure possibly is shattered by faults. However, it also appears that many of the faults are compensatory in nature because structural contour interpretations which utilize the known faults would be very awkward unless the presence of additional faults were assumed. If *no* faults are assumed to be present—with the exception of a few which have 400 ft or more displacement—stratigraphic horizons can be contoured very well using only well tops and dipmeter data. The structural contour map presented with this report (Fig. 7) is one of many interpretations that can be made by a realignment of fault systems.

Ordinarily, faulting would affect continuity of reservoir beds, but in the case of the Brown zone there appears to be complete lateral communication of fluids from fault block to fault block. This is illustrated best by Figure 8, a northwest-southeast cross section through the Arbuckle productive structure showing major cross faults that cut the field, some of which have vertical throws of more than 500 ft. However, because the Brown zone is approximately 600 ft thick and has excellent vertical permeability resulting from fracturing, it is everywhere in contact with itself across faults. This condition has made possible the existence in the Brown zone of a common oil-water contact of 2,985 ft subsea (shown on the cross sections in Figure 8). The Wade and Bray zones, however, are characterized by very thin reservoir beds and even small faults could restrict lateral communication of hydrocarbons.

GEOLOGIC HISTORY

The formation of a deepening geosyncline in early Paleozoic time provided the depositional site for several thousand feet of carbonates of the Arbuckle and Timbered Hills Groups of Cambro-Ordovician age. Subsidence was contemporaneous with deposition, and these sediments in the Healdton area accumulated to a thickness of approximately 7,000 ft, comparable with that exposed at the surface in the Arbuckle Mountains on the northeast (Fig. 9).

The Simpson Group and Viola Limestone of this area appear to be similar to the same units in other areas of southern Oklahoma, except that Simpson sandstone is lacking at Healdton. Although the Upper Ordovician, Devonian, Silurian, and Mississippian units are absent over the producing structure, their deposition and subsequent removal by erosion are assumed because they are developed in the area just northwest.

At the end of the time of Caney deposition, uplift was begun along a line from Healdton northwest to the Wichita Mountains area and southeast through the Criner Hills. This probably was the beginning of the pulsations of the Wichita orogeny, and the uplift formed a somewhat abrupt southern shoreline for the subsiding Ardmore basin. Although Goddard and Springer sediments were deposited close by on the north, they are not believed to have been laid down over the then-positive Healdton feature.

Pronounced uplift accompanied by folding and thrust faulting was effected by the Wichita orogeny, and culminated after lower Dornick Hills (Morrowan) deposition and before upper Dornick Hills (Atokan) deposition. This was followed by extensive erosion, in conjunction with adjustment faulting; erosion deeply breached the anticlinal feature.

The area was positive throughout Atokan and Desmoinesian times and deposition was confined to the flanks. It was not until late Desmoinesian or early Missourian time that transgressing seas covered the area completely. Beginning with the deposition and reworking of shale and lenticular conglomeratic sandstone around the base of ridges formed by resistant Ordovician beds, Hoxbar sandstone and shale were deposited over the Healdton uplift. High-angle faulting along the north side, originating in Springeran time, probably was active during Deese and Hoxbar deposition. This is reflected by a marked difference in

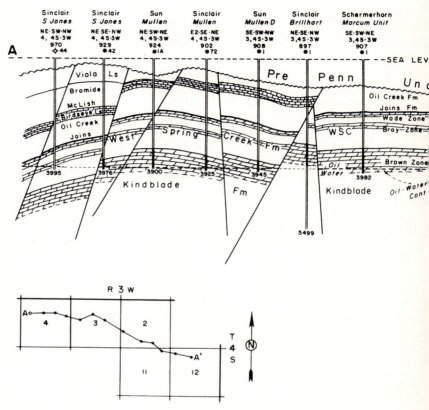

FIG. 8.—Northwest-southeast cross section traversing Arbuckle structure of Healdton field, showing effects of faulting on reservoir beds.

thickness on either side of the fault zone.

After Hoxbar deposition, there was renewed uplift in the Healdton area. Elevation of the Arbuckle Mountains by the Arbuckle orogeny (Late Pennsylvanian) created compressional stresses sufficiently great that the old pre-Pennsylvanian structure at Healdton was rejuvenated by broad anticlinal uplift.

Subsequent regional and local erosion provided the source of a Permian redbed sequence which was laid down unconformably on a Hoxbar surface. This was followed by regional emergence, and the area probably has been positive since Permian time.

RESERVOIR CHARACTERISTICS

PENNSYLVANIAN

The Healdton sandstones are commonly poorly consolidated and possess a wide range of porosity and permeability values (Fig. 10). A core analysis of a well in Sec. 5, T. 4 S., R. 3 W., has a porosity range of 10 to 33 percent with permeabilities from 0 to 3,800 md. Effective thickness of individual sandstone bodies also varies greatly, attaining more than 100 ft locally.

Reservoir energies primarily responsible for production are dissolved gas and gravity drainage. Increased water production in localized spots along the edges of the field indicates some water encroachment and the possibility of a slightly active water drive affecting some of the sandstones. However, the field's longevity has been due principally to permeability and thick sandstone, which are ideal reservoir conditions for effecting gravity drainage.

Large quantities of gas have been produced also from these Pennsylvanian sandstones, but records of volumes produced were not kept initially. Ap-

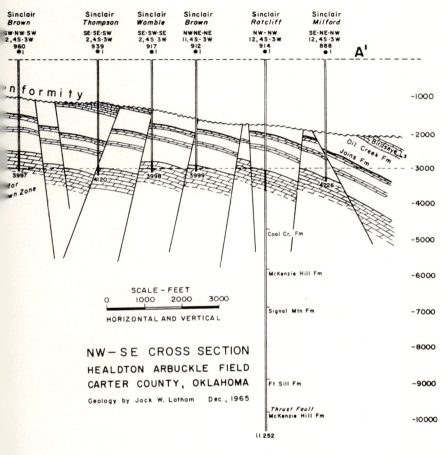

NW—SE CROSS SECTION
HEALDTON ARBUCKLE FIELD
CARTER COUNTY, OKLAHOMA
Geology by Jack W. Latham   Dec., 1965

proximately 50 gas wells were completed during the field's initial development. Most of these wells began producing oil after several years and many others were abandoned early in the life of the field. The first significant gas well was drilled in 1913 on the Sinclair Oil & Gas Co.'s. Million and Thomas lease, in Sec. 4, T. 4 S., R 3 W., and completed for a potential of 30,000 Mcf/day of gas.

The Healdton sandstone production is presently being influenced by secondary recovery (water injection). Although some increases in oil production have been realized locally, it is too early to predict the effectivenes of this method.

ARBUCKLE

Production from the Arbuckle, from three dolomite zones, is found in the upper 1,600 ft of section. The Wade and Bray zones are identical in physical character and have poor permeability

and limited lateral continuity. However, the Brown zone seems to lack nothing as a reservoir and is capable of yielding large quantities of oil. The produced oil is a sour crude containing high concentrations of hydrogen sulfide gas.

Wade zone.—Drilling in the Healdton field has resulted in four Wade zone completions; production is gas with some condensate under assigned 160-acre spacing.

Because of the thin-bedded, heterogeneous nature and low permeability, the quantity of recoverable hydrocarbons is questionable. This is indicated not only by low completion potentials, but also by porosity-log calculations. Figure 11 represents typical Wade and Bray zone electric-log surveys. The dolomitic zones are recognized readily on the induction log by a reduction of resistivity and an increase on the self-potential curve. Although the microlog rarely indicates significant

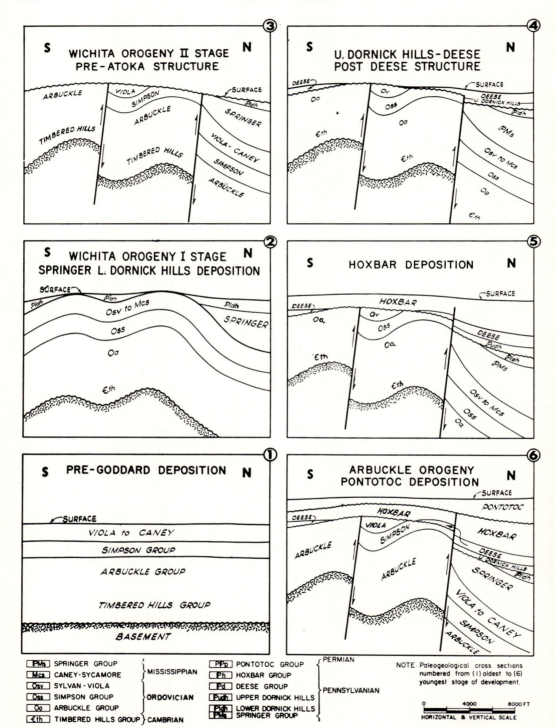

FIG. 9.—Paleogeologic north-south cross sections of Healdton area, T. 4 S., R 3 W., Carter County, Oklahoma No. 1 is oldest stage of development; No. 6 is youngest.

permeability, the sonic log shows porosity values to range from 10 to 18 percent. However, because of associated shale and anhydrite in the Wade and Bray zones, difficulty in accurately isolating lithologic types causes effective porosity calculations from the sonic log to be questionable to completely unreliable. The thin-bedded nature of these zones also prevents correlations of lithologic types (determined from well cuttings) with electric-log character. Therefore, because lithologic identification is necessary before effective porosity can be recognized, proper evaluation of these dolomite zones requires the running of several different porosity-log surveys and then the cross-plotting of these surveys to determine lithology.

Bray zone.—Eight wells are Bray-zone completions, producing oil on 40-acre spacing. As in the case of the Wade zone, information from evaluation tools does not coincide with completion results. The Sinclair Oil and Gas No. 1 Bray, SW¼ SE¼ SE¼, Sec. 2, T. 4 S., R. 3 W., was

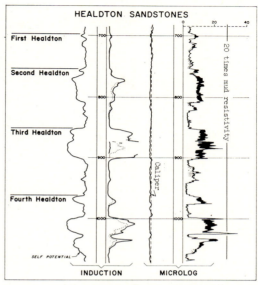

FIG. 10.—Typical induction-log and microlog curves of productive Healdton sandstones. Indicated microlog porosity shown in black. Depths from surface.

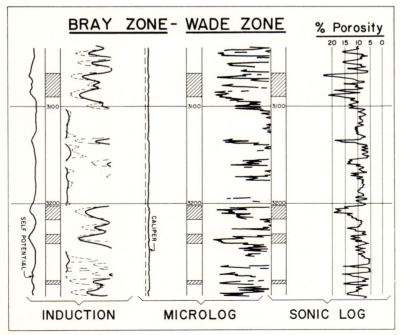

FIG. 11.—Typical induction-log, microlog, and sonic-log curves of Wade and Bray zones in Healdton field. Saccharoidal dolomite units which potentially are productive are shown by diagonal lines. Depths from surface.

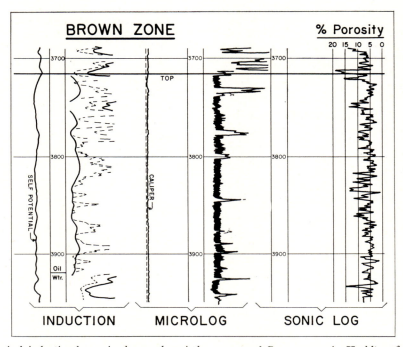

FIG. 12.—Typical induction-log, microlog, and sonic-log curves of Brown zone in Healdton field. Indicated microlog porosity is shown in black. Depths from surface.

completed from the Bray zone flowing 189 bbl of oil a day on a 16/64-in. choke. The Sinclair No. 2 Clark, NW¼ NE¼ NW¼, Sec. 11, was completed from this zone for only 1 bbl of oil a day. Both wells showed oil saturation and have comparable porosity analyses taken from the sonic and micrologs. These completions serve as a prime example of the difficulty in determining productive potentials of the Wade and Bray zones with a limited electric-logging program.

Brown zone.—The Brown zone is the lowest productive Arbuckle unit and has proved to be the most prolific. Since its discovery by the Sinclair No. 1 Brown, the zone has been found by drilling to have as much as 700 ft of oil column in Sec. 3, T. 4 S., R. 3 W. Spacing is on 40-acre units with daily allowables originally set at 30 bbl of oil.

This zone was first drill-stem-tested in the Sinclair Oil and Gas No. 1 Brillhart, NE¼ SE¼ NW¼, Sec. 3, and flowed at a daily rate of 2,541 bbl of oil and 1,519 Mcf of gas on a 5/8-in. choke. This test is indicative of Brown-zone productivity throughout the field which is made possible by the

excellent reservoir characteristics described in the section on stratigraphy. Permeability is illustrated (Fig. 12) by a set of electric-log surveys showing a part of typical or average Brown-zone development.

The reservoir energies now in force are fluid expansion and solution gas. It is believed that there also might be an active water drive. However, because of the small pressure decline, proper evaluation of water influx cannot be ascertained at this time. No estimate of reserves is presented here, but the Arbuckle is expected to approach major-field status.

Operators have elected to perforate and complete the Brown zone near the oil-water contact, even though there might be as much as 300 or 400 ft of the zone above water. As the well begins to yield water, these perforations are plugged and a higher interval perforated. This procedure will be repeated until the entire zone is "watered out," the idea being that, although the zone has excellent vertical communication through fractures, less permeable streaks indicated on porosity logs might act as temporary barriers and retard vertical migration of oil. Thus, as the reservoir is being depleted, the encroachment of water could override the local oil accumulations, making them very difficult, if not impossible, to recover.

RESERVOIR DATA

The reservoir data for the Hoxbar and Arbuckle Groups are summarized in Tables 1 and 2, respectively.

DRILLING AND COMPLETION

Early development drilling of the shallow Pennsylvanian was accomplished by means of cable tools. Rotary tools have been used for wells drilled to productive pre-Pennsylvanian reservoirs as well as more recent infill drilling of the Healdton sandstones.

With Healdton sandstone completions, use of three strings of casing was common practice—10-in. at approximately 300 ft, 8¼-in. at 400 to 600 ft, and 6⅝-in. at 800 ft. A 5¾₆-in. slotted liner was set on the bottom and extended approximately 20 ft up into the 6⅝-in. casing.

Development of Arbuckle production at Healdton commonly has necessitated drilling within the townsite limits. Some drill sites were in alleys of residential areas or on 75-ft lots with dwellings on either side.

For the protection of fresh-water sandstones, operators set approximately 500 ft of surface casing. However, where drilling is in the proximity of old shallow production, a protective string set through productive sandstones is preferred.

Native mud is used above the Arbuckle, but for drilling into and through potential pay zones,

a chemical mud is employed with a maintained weight of 9–10 lb/gal. The purpose of this weight of mud is to contain gas found in the Wade zone. The presence of anhydrite in the Wade and Bray zones necessitates special treatment of drilling fluids to prevent fluffing of the mud which would result in a mud-weight loss and possibly well blowout.

Completion practices involve setting a production string through the lowest potential zone, generally 5.5-in. or 7-in. casing. Productive units are perforated and are treated if necessary. The Wade and Bray zones require a stimulant in the form of acid fracturing, but Brown zone completions, with few exceptions, have been natural and all are capable of flowing.

CONCLUSIONS

The shallow Pennsylvanian production at Healdton was discovered on the basis of oil seeps and water wells contaminated with oil in association with a surface anticline. Accumulations of this type were exploited during the oil boom of southern Oklahoma and are now virtually nonexistent. It might therefore appear that this field has little to contribute toward present-day exploration.

However, the Healdton field serves as a prime example of possible reservoir potentials in highly complex structures that have undergone subsequent burial. The geologic relation of the pre-Morrowan at Healdton with younger strata is typical of many oil fields in southern Oklahoma and recently these less revealing features beneath fields of shallow production have been the stimulant for a majority of petroleum exploration effort in this region.

TABLE 1. SUMMARY OF RESERVOIR DATA, HOXBAR GROUP, HEALDTON FIELD, OKLAHOMA

Healdton Sandstones	First	Second	Third	Fourth
Proved acreage		7,160—field		
Depth range (ft)	600–1100	700–1200	800–1400	900–1600
Av. pay thickness (ft)		72—all four sandstones		
Av. porosity (%)	23	23	23	20
Av. permeability (md)	515	900	800	82
Av. water saturation (%)	19	18	23	27
Gravity of oil (°API)	32	28	31	30
Original bottom hole pressure (est.) (PSIA)	270	270	270	270
Bottom hole temp. (est.) (°F)	80	80	80	80
Type of drive	Solution gas and gravity with water encroachment locally in 2d–3d and 4th sandstones.			
Original oil reserves in place (bbl)		674,195,000		
Original mobile oil reserves (bbl)		422,000,000		
Cumulative production to date (bbl)		250,000,000		

TABLE 2. SUMMARY OF RESERVOIR DATA, ARBUCKLE GROUP, HEALDTON FIELD, OKLAHOMA

	Wade Zone	Bray Zone	Brown Zone
Proved acreage	640	320	1,400
Vertical productive limits Lowest depth and subsea (ft) Highest depth and subsea (ft)	— —	3,793(−2,905) 2,680(−1,773)	3,897(−2,985) 3,207(−2,303)
Total average pay thickness Gross (ft) Net (ft)	200 25	275 30	580 250
Average porosity (%)	13	14	10
Average permeability (md)	0.1	0.1	1,500 max.; 150 av.
Average water saturation (%)	34	38	20
Type of drive	Gas expansion	Solution gas	Solution gas
Oil-water contact (ft)	—	—	2,985 subsea
Gravity of oil (API)	Gas	38°	37.6°
Bottom-hole pressure (psia)	1,641	1,625	1,758
Bottom-hole temperature (°F)	92	96	100

REFERENCES CITED

Decker, C. E., and C. A. Merritt, 1931, The stratigraphy and physical characteristics of the Simpson Group: Oklahoma Geol. Survey Bull. 55, 112 p.

Harlton, B. H., 1964, Symposium on the Arbuckle: Tulsa Geol. Soc. Digest, v. 32, p. 38.

Riggs, C. H., et al., 1953, Petroleum engineering study of Healdton oil field, Carter County, Oklahoma: U.S. Bur. Mines Rept. Inv. 4917, 76 p.

Tomlinson, C. W., and W. McBee, Jr., 1959, Pennsylvanian sediments and orogenies of Ardmore district, Oklahoma, in Petroleum geology of southern Oklahoma, v. 2: Am. Assoc. Petroleum Geologists, p. 3–52.

Bay Marchand–Timbalier Bay–Caillou Island Salt Complex, Louisiana[1]

M. G. FREY[2] and W. H. GRIMES[2]
New Orleans, Louisiana 70112

Abstract The Bay Marchand–Timbalier Bay–Caillou Island salt complex, Louisiana, more than 28 mi long and up to 12 mi wide, may be part of a much longer salt feature that extends both east and west. The mother salt bed, of probable Late Triassic–Early Jurassic age, is now buried to depths of 40,000–50,000 ft, whereas the tops of the individual domes along the trend are within 2,000–3,000 ft of the surface.

Production to date on this three-field complex has been in excess of 0.7 billion bbl of oil. Oil reserves are estimated to range from another 0.75 billion to 1 billion bbl. In addition, significant gas reserves are present.

Hydrocarbon accumulations are in sands of Pleistocene through late Miocene age and range in depth from 1,000 to more than 20,000 ft. A wide variety of traps is found, including supradomal arching, shale and salt truncations, stratigraphic traps, and traps resulting from faults.

Production was established on the complex in 1933. The total hydrocarbon production for 1967 was approximately 97 million bbl.

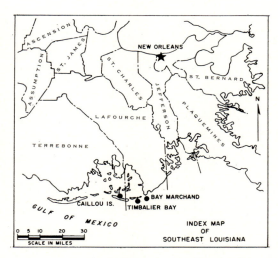

FIG. 1.—Map showing location of Bay Marchand–Timbalier Bay–Caillou Island trend, Louisiana.

GEOGRAPHIC SETTING

The Gulf Coast region of the United States has been a very active oil province for several decades. Because of the general lack of structural expression at the surface, the real impetus to exploration was the introduction of geophysical mapping in the late 1920s and early 1930s, first by means of the torsion balance and refraction seismograph, then by the use of the gravity meter and reflection seismograph. Exploration activity has progressed from the interior marshes and swamps seaward, and now the concentrated effort is along the coastline and into the open gulf to water depths of about 300–400 ft.

Bay Marchand–Timbalier Bay–Caillou Island salt complex generally is regarded as the largest salt feature in the South Louisiana–offshore Louisiana region. It extends east-west along the South Louisiana coastline of Lafourche and Terrebonne Parishes. The fields are approximately 65 mi south-southwest of New Orleans, as shown on Figure 1. The origin of such a long salt feature, punctuated by a series of individual domes, is related closely to

[1] Read before the 53d Annual Meeting of the Association, Oklahoma City, Oklahoma, April 25, 1968. Manuscript received, May 7, 1968; accepted, July 25, 1968. Published with permission of Chevron Oil Co.

[2] Chevron Oil Company, The California Company Division.

the rapid deposition of a flood of silicate clastic sediments under a generally regressive depositional regime in which the Tertiary shelf prograded seaward.

DEVELOPMENT HISTORY

Caillou Island (Fig. 1).—Early geophysical exploration along the trend resulted in the discovery of Caillou Island dome in 1928, by means of torsion balance and seismograph surveys. However, 5 years elapsed before Texaco established enough production to warrant exploitation of the dome. After a series of dry holes and marginal wells, the No. 8 State was drilled to a total depth of 5,981 ft and was completed as an oil well in July 1933. Drilling and completion of the well took 79 days, an operation which would require less than a week with today's techniques. This initial well in the field produced a total of 3.25 million bbl of oil before abandonment in 1953.

Timbalier Bay.—The Gulf Oil Corporation, another very active company in the early exploration of the Gulf Coast, used the gravity meter to good advantage. Many salt domes were located in the 1930s on the basis of their characteristic minimum anomalies. The Timbalier Bay dome was discovered in 1936 by Gulf because of the pronounced gravity minimum above it. In 1937, its identification as a salt dome was confirmed by a reflection-seismograph survey. On the north and east flanks, seismic data were of sufficient quality to establish the presence of the dome beneath East Timbalier Island and to define the northeastern part of the structure. However, an abandoned buried channel of the Mississippi River crosses the southwestern part of the dome, and the seismograph was not able to yield usable data where the channel is present. The channel, about 4 mi wide, is filled with soft, unconsolidated sediments which attenuate all seismic energy abnormally and effectively obscure deeper reflections. It formed across the dome during a Pleistocene glacial epoch when sea level was 300–400 ft lower than it is now, and can be traced for many miles in southern Louisiana. Many such buried channels, up to several thousand feet deep and 1 mi or more wide, are known in the coastal area and in the shallow offshore waters of the Gulf of Mexico.

Gulf's discovery well on Timbalier Bay dome, the Louisiana State Lease 192 No. 3, was located on April 6, 1937. It was completed January 25, 1938 from the interval 6,955–6,959 ft for initial production of 304 bbl/day of oil. The well was drilled to a total depth of 8,728 ft; the top of salt was penetrated at 7,741 ft.

Bay Marchand.—The Bay Marchand dome first was detected in 1927 by the Gulf Oil Corporation and was mapped as a high-velocity area by a refraction-seismic survey. Exploratory drilling on the dome was conducted initially by Gulf, which drilled two dry holes. The first unsuccessful test was drilled in 1930 on the north flank of the dome, and the second was drilled near the crest in 1941. The Gulf shallow test drilled in 1941 was a small gas-producing well until a hurricane destroyed the rig and platform during the summer of 1941. Gulf subsequently abandoned its interest in the area. Nine additional unsuccessful wildcat wells were drilled on the north flank by Union Oil Company of California, Placid Oil Company, and William Helis (now Estate of William G. Helis) from 1937 to 1946.

In 1947 and 1948, The California Company (now Chevron Oil Company), after conducting a reflection-seismograph survey of the area, acquired 27,484 acres on the dome. After the abandonment of the Calco A-1 well, which reached salt at 2,109 ft on top of the dome, the

B-1 and C-1 wells, drilled simultaneously high on the east and west flanks, respectively, proved successful and were completed in early 1949. The B-1 well, State Lease 1366 No. 1, credited as the discovery well, was completed as an oil producer from the interval 2,873–2,892 ft on March 3, 1949.

Ridge between Caillou Island and Timbalier Bay.—For several years, geologists had been aware of a ridge and possible subordinate dome between the Caillou Island and Timbalier Bay domes. Stanolind (Pan American) drilled two dry holes along this part of the ridge, one in 1947 and the other in 1948. Gulf, exploring west from Timbalier Bay dome, drilled two dry holes in 1954. Union Oil Company of California made the initial discovery on the "King prospect" by completing their State Lease 2826 No. 1 on December 9, 1956, as a dual oil well. Subsequently, Union drilled 42 additional producers and 5 dry holes.

Subsequent development.—After the discovery of the individual domes, the development of the oil fields provided additional data on the configurations of the salt cores. This information and the continuous improvement in the resolution of seismic data by the introduction of magnetic tape, stacking, digital recording, and the more sophisticated digital processing techniques enabled the industry to map the trend in more detail. It became apparent that the linear nature of the gravity anomaly as shown in Figure 7 truly expressed the continuity of the salt ridge from Bay Marchand to Caillou Island.

SOUTH LOUISIANA STRATIGRAPHY

The Gulf Coast area of South Louisiana is a downwarped sedimentary basin, in which great thicknesses of sediments were deposited. The Tertiary along the Louisiana coast and the inner part of the continental shelf is one of the world's thickest sections of terrigenous clastic sediments. There, sediments of Miocene age and younger, which are estimated to be more than 40,000 ft thick, constitute one of the world's greatest deltaic accumulations (Limes and Stipe, 1959, p. 80). The shoreline fluctuated across the area many times in response to the rates of sedimentation and subsidence. The dominant movement has been seaward, resulting in the outbuilding of the continental shelf. Thus, the overall sedimentation was regressive.

Figure 2 is a regional diagrammatic cross section, oriented normal to the axis of the Gulf

Coast geosyncline. The arrow denoting the present coastline is the approximate position of the Bay Marchand–Timbalier Bay–Caillou Island complex. Gulfward progradation of depocenters and gross lithologic facies of each successively younger stratigraphic unit are illustrated. The interpretive southward thinning, as illustrated in Figure 2, is believed to have been caused by sediment starvation as distance from the source area increased. Three gross lithologic facies generally are recognized in each biostratigraphic unit.

1. The massive sandstone facies ranges from 50 to 75 percent sandstone with interbedded thin shale units. This facies was deposited in a nearshore, inner-neritic, and continental environment; however, the presence of marine fossils within the shale section indicates that minor marine transgressions took place.
2. The interbedded facies consists of alternating sandstone and marine shale. The sandstone content ranges from 10 to 50 percent. The sediments were deposited across a shelf of varying width and at predominantly neritic depth. Regional correlations based on electric logs and paleontologic markers are considered to be very reliable in this zone.
3. The massive shale facies contains less than 10 percent sandstone and was deposited in outer-neritic and bathyal depths. The facies is mainly dark marine shale, commonly interbedded with thin "erratic" sandstone beds in which hydrocarbons, where found, generally are under higher pressure and of small volume.

Within an individual time-rock unit, the lithofacies grade from essentially nonmarine sandstone with some claystone southward into an interbedded sandstone and shale sequence and then into a marine shale sequence. The continental and nearshore deposits of the massive sandstone facies extended across large geographic areas and consequently represent a sizable total sedimentary volume. The largest volume of sediments is in the interbedded and massive shale facies. This section is of primary importance because it contains the source and

reservoir beds for most hydrocarbon accumulations in the area. The hydrocarbon reserves in the Bay Marchand–Timbalier Bay–Caillou Island fields are in the upper Miocene, Pliocene, and Pleistocene strata. It is estimated that 90 percent of the known reserve is in the interbedded facies of the upper Miocene, 9 percent is in the Pliocene, and 1 percent is in the Pleistocene. Figure 3, an electric log of the Placid Oil Company 2857 No. 1 well (on the west flank of Caillou Island), is used as a type log for this trend. The scale on the right side of the log shows the estimated depth of water in which sediments were deposited. The Pleistocene and Pliocene section to a depth of approximately 8,500 ft is primarily massive sandstone interbedded with shallow-water shale. The Miocene section in this well is characterized by a favorable ratio of alternating sandstone and deep-water marine shale. This interbedded facies can be projected to depths below 20,000 ft. Figures 4 and 5 are photographs of conventional cores of upper Miocene oil-bearing sandstone in the Bay Marchand field. The shallower core (8,000 ft) shows fine- to very fine-grained sandstone interlaminated with very clayey siltstone, whereas the deeper core (11,000 ft) shows a massive, fine-grained, slightly clayey sandstone.

STRUCTURE

Regional.—The regional structure of South Louisiana is essentially monoclinal with south dip that gradually increases at depth. The south dip ranges from 30 ft/mi in shallow beds to approximately 300 ft/mi in the interbedded sandstone and shale facies at intermediate depths and to more than 800 ft/mi in the deeper shale facies. These dip estimates are generalized; dips differ from one structural province to another and are modified and interrupted by salt domes, salt ridges, and major normal faults crossing the area.

The lower Gulf Coast area where the Bay Marchand, Timbalier Bay, and Caillou Island fields are located is characterized by numerous diapiric salt structures and other domal or anticlinal structures which, on the basis of geophysical and geologic information, are considered to be salt controlled or salt related. The terms "diapiric shale" and "diapiric salt" were discussed and defined by Atwater and Forman (1959, p. 2594) as follows.

Diapiric salt is the lithologic term applied to the intrusive salt of the domal core, in contrast to the bedded salt encountered in its normal stratigraphic posi-

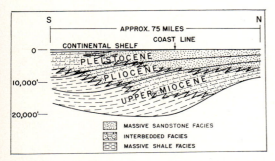

FIG. 2.—Diagrammatic cross section of South Louisiana, showing gross lithologic facies of Gulf Coast geosyncline.

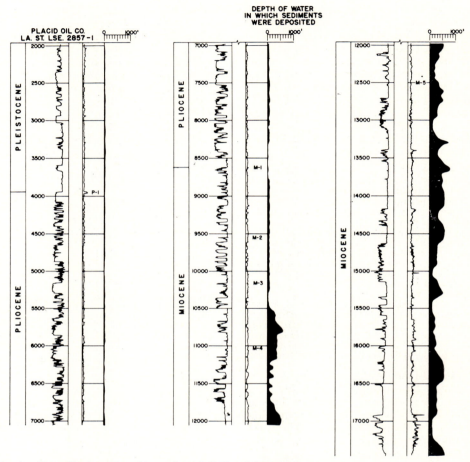

Fig. 3.—Generalized stratigraphic section of Caillou Island–Timbalier Bay–Bay Marchand fields, Terrebone and Lafourche Parishes, Louisiana. Shows S.P. curve on left, and depth of water in which sediments were deposited on right.

tion. . . . The diapiric salt has intruded and moved upward through the overlying sediments, driven by the buoyant force derived from the density contrast between the lighter salt and the heavier sedimentary (and in some places, igneous) overburden.

Diapiric shale is the lithologic term applied to the intrusive shale of the domal core, in contrast to the bedded shale encountered in its normal stratigraphic position.

Diapiric shale is identified on the basis of microfauna and physical characteristics (low resistivity, low density, abnormally high pressure, low velocity, and a tendency to flow into hole where penetrated). Dipmeters have been very useful for identifying these anomalous shale masses. Dips in diapiric shale generally are steeper than those in the normally bedded sediments above, and are also consistent in

magnitude and direction. Dips observed at the top of a diapiric shale mass appear to prevail through greater depths. In contrast, dips recorded in the gouge or brecciated zone adjacent to salt are random in both magnitude and direction (Gilreath, 1968, p. 141). The diapiric shale appears to be associated with, and to accompany, the upward, piercing movement of the salt core, and may act as a truncating or warping element as does the salt. The amount of displacement or movement is varied in the domes in the Gulf Coast area. On the basis of paleontologic correlations, the diapiric shale mass on the southeast flank of the Caillou Island field intruded or lifted upward as much as 5,000–6,000 ft (Fig. 10, cross-section B-B').

The salt domes of South Louisiana are var-

Fig. 4.—Photograph of conventional Bay Marchand Miocene core showing laminated sandstone and siltstone.

Core analysis of Miocene sandstones at approximately 8,000 ft in Bay Marchand field:

Sandstone—tannish-gray, oil-stained, fine-grained, moderately well-sorted and winnowed, finely interlaminated with medium-gray, very clayey, very fine-grained sandstone to siltstone.
Sandstone porosity range—21–29 percent.
Sand permeability range—48–1,000 md (air permeability).

ied in size and shape; however, a typical dome is generally circular to slightly elongate, steep sided, and complexly faulted. Other salt masses appear to have originated as linear ridges extending for many miles, with spines or upward piercement structures forming individual salt domes along the ridge. Geologic knowledge de-

rived from deep drilling along the Bay Marchand–Timbalier Bay–Caillou Island trend indicates that the individual domes are on a common salt ridge. Figure 6 is a structure map of the top of the salt which demonstrates the linear pattern of the salt complex. Cross-section A-A′ (Fig. 6) graphically displays the subterra-

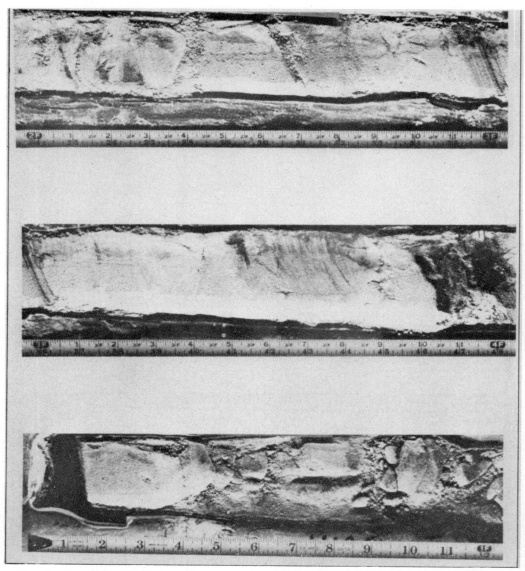

Fig. 5.—Photograph of conventional Bay Marchand Miocene core showing more massive nature of some sandstone units.

Core analysis of Miocene sandstones at approximately 11,000 ft in Bay Marchand field:

Sandstone—medium-gray, oil-stained, fine-grained, slightly clayey, well-sorted, faintly laminated.
Sandstone porosity range—29–32 percent.
Sand permeability range—600–1,600 md (air permeability).

nean mountains of salt rising on both ends of the ridge more than 18,000 ft and interconnected by smaller salt uplifts. The salt ridge is oriented east-west and is approximately 28 mi long and more than 11 mi wide in the east at Bay Marchand. The Bay Marchand dome is considered to be the largest salt dome in the Louisiana Gulf Coast area. At a depth of 20,000 ft, the salt complex encompasses an area of more than 140 mi², and to that depth contains approximately 200 mi³ of salt. If the salt is projected vertically downward from

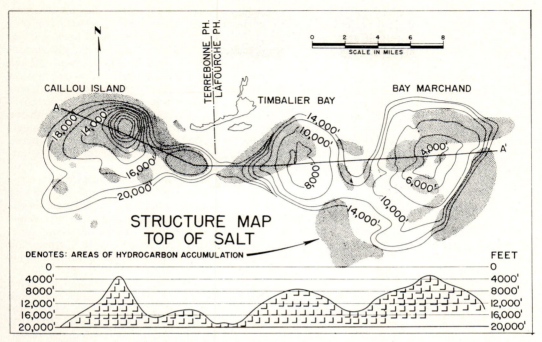

FIG. 6.—Structure map and longitudinal profile section showing top of salt, which is datum for structure contours. CI = 2,000 ft. Stippling indicates hydrocarbon production as related to the salt massif.

20,000 ft to its probable base at 40,000 ft, the total complex is computed to contain more than 700 mi³ of salt. The Bay Marchand and Caillou Island domes are shallow piercements, on which salt was reached at 2,100 and 2,700 ft, respectively, whereas the Timbalier Bay dome is classed as an intermediate piercement with shallowest salt reached at 6,400 ft. The deepest salt reached by drilling (16,300 ft) is on the southwest flank of Caillou Island. Deep drilling along what appeared to be reentrants or synclines on the interconnecting salt ridge has demonstrated the productive and the economic importance of this vast structural closure. The domes along the salt ridge are varied in growth history; the Bay Marchand dome probably is the most complicated both structurally and stratigraphically. Thickness studies of stratigraphic units readily demonstrate periods of pronounced domal growth as well as periods of dormancy.

The origin of salt ridges such as Bay Marchand-Timbalier Bay–Caillou Island is not fully understood. It is probable that salt structures start to form in some areas with as little as 1,000 ft of overburden. This conclusion is based on studies of isopachous thinning of

beds above salt uplifts. It is also evident that only the salt deposits (exclusive of those in orogenic belts) that later are covered by a flood of silicate clastic sediments develop the various types of salt swells, domes, and ridges. In evaporite basins such as the Michigan and Williston basins, thick salt layers are buried by thousands of feet of sediments. There, however, the evaporites are covered by a series of widespread deposits of cratonic or platform-type sediments. In those basins the salt behaves in a manner similar to that of overlying and underlying beds, and shows no appreciable thickening or thinning in the crests or limbs of the folds. Solution structures around the margins of the basins provide the only structural evidence of the presence of salt within the basins. In contrast, the salts of the Gulf Coast, like those of the Paradox basin, were buried by rapidly accumulating silicate clastic sediments in a series of deltas. The key to the origin of the salt flowage seems to be the differential loading of the salt by a deltaic or similar local mass of younger sediments.

The rapid outpouring of the Late Jurassic Cotton Valley, Early Cretaceous Hosston, and younger Mesozoic sediments into the North

Louisiana-Mississippi salt basin, and the dumping of the Cenozoic deltaic deposits into South Louisiana and the offshore gulf area, gave rise to considerable differential loading of the Late Triassic–Early Jurassic salt. The more deeply buried salt moved out from beneath the prograding mass of deltaic and related sediments toward a less deeply buried position in front of those sediments. There, depending upon the straightness of the deltaic front, the salt formed linear or arcuate swells and ridges, with as little as 1,000 ft of sediment cover, or perhaps even less. As the growing deltas shifted and merged with neighboring deltas, the salt ridges and swells were overridden, or bypassed, and a new series of ridges developed at the new delta front. They, in turn, underwent similar burial, growth, and development. Ewing and Antoine (1966, p. 498) suggested "that the Sigsbee Scarp is the front of a chain of diapirs with salt cores and that this chain probably forms ridges parallel with the scarp."

When first formed, the crest of a salt ridge probably is irregular, some parts of the crest rising higher than others. As sediments continue to be deposited above this ridge, the higher axial regions grow at the expense of the lower levels because sediment loading is greatest in the lower parts of the ridge, and salt migrates into the zone with the least overburden. In this "cannibalistic" fashion, a linear series of piercement domes with interspersed saddles form from the ridge.

The Bay Marchand ridge was developed in this manner. The cross sections (Figs. 8, 10, 11) show that the domes were growing throughout the period of time represented by the strata penetrated in the wells. For details of earlier growth, one must rely largely on seismic information.

Figure 7 is a regional Bouguer gravity map showing the linear alignment of the massive minimum ridge and local closed minimum anomalies. The closed gravity-minimum anomalies directly overlie the salt piercements as demonstrated by well control. Gravity information and subsurface data strongly suggest that the ridge extends eastward through the Grand Isle 16 and 18 domes, and westward to the Lake Pelto dome; thus the salt ridge possibly extends a total of 60 mi.

Many salt domes in South Louisiana have grown periodically or in stages, commonly causing a shift in the structural apex, erosional unconformities, local stratigraphic variations, and complex structural conditions. A detailed structural analysis, involving the construction of isopachous maps, may demonstrate a geographic shift of domal structural apex and the presence of what are termed "relic structures." At times the domal structural closures have been modified or completely eliminated by subsequent salt movement centering at a different geographic position on the dome. The ancestral highs historically have been associated with the principal hydrocarbon reserves on salt domes. The larger salt domes, such as Bay Marchand, generally have the most structural and stratigraphic irregularities.

The Gulf Coast area is transected by large regional fault systems (normal or gravity type), which commonly were contemporaneous with deposition (growth faults). These faults have had a pronounced effect on stratigraphy, structure, and production. Most of them are downthrown to the south, with displacements as great as 5,000 ft and lengths up to 50 mi. Regional down-to-the-north faults are much less common in the Gulf Coast area, but where they are present, they have the same pronounced effects on the sedimentation, structure, and production. An example of major down-to-the-north faulting of semiregional magnitude is the large fault extending more than 12 mi from the west flank of the Timbalier Bay field to the southeast flank of the Caillou Island field (Fig. 9).

Traps.—Five recognized types of hydrocarbon traps are associated with the Bay Marchand–Timbalier Bay–Caillou Island complex. They can be identified on the Bay Marchand northwest-southeast structural cross-section D-D′ (Fig. 8).

1. Supradomal-closure entrapment generally is just above the crestal position of the salt dome, where the strata dip gently in all directions (well 1366 No. 16).
2. Salt-truncation entrapment results where the upward-intrusive salt mass truncates and displaces strata (wells 1367 C-1, 1366 G-1).
3. Shale-truncation entrapment results where deep-water marine shale breaches overlying strata in a manner similar to that of the intrusive salt. The shale is believed to accompany and derive its piercing impetus from the upward-moving salt mass (well 1488 R-1).
4. Stratigraphic pinchout of sandstone is present on the flanks of the domes, generally in a downdip or off-structure position. The sandstone grades into an updip shale equivalent, thus providing excellent entrapment potential. Control gained from intensive drilling throughout South Louisiana has indicated that vast reserves result from this type of entrapment (well 1488 R-1).
5. Faults provide the entrapment mechanism for a large percentage of hydrocarbon reserves in various positions on domal structures. Faults which are contemporaneous with deposition commonly have an im-

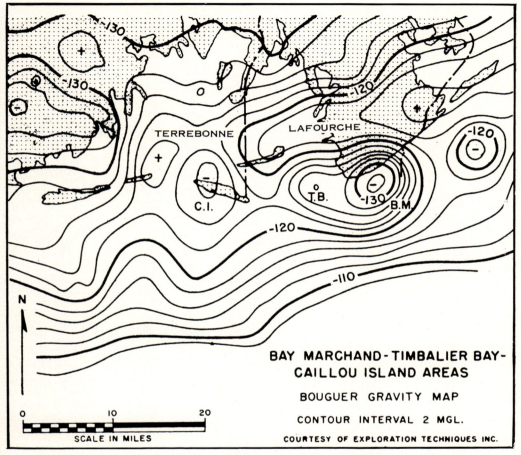

FIG. 7.—Gravity map of Bay Marchand–Timbalier Bay–Caillou Island trend. Note strong minima caused by salt trend.

portant and pronounced effect on the stratigraphy. Generally, sandstone on the downthrown fault blocks of expansion or depositional faults is much better developed (wells 1366 No. 16, 1367 C-1).

An additional but uncommon type of entrapment, which is more difficult to recognize and controversial in definition, is caused by onlapping younger strata deposited on a beveled or eroded surface of older strata. Because of the cyclic growth of most salt domes, local unconformities are common. It also should be noted that major hydrocarbon reserves are present in the area between the Caillou Island and Timbalier Bay domes "King area," and between the Timbalier Bay and Bay Marchand domes, over a deeply buried part of the interconnecting salt ridge which appears to have been synclinal in the early development of the trend.

Structure map (Fig. 9).—The structure of the strata adjacent to the salt differs considerably in the three major piercements composing the salt-ridge complex. The shallower or supradomal section above the salt generally is arched gently over the salt and is further complicated by a network of graben-type fault systems. Radial fault patterns are present above the salt masses and on the flanks of the domes. Most of the faults reaching the shallow section are not present at depth, because of compensatory displacement into other faults or because they extend into the salt mass. Many of the radial faults decrease in throw outward from the domes and generally die out at the outer limits of domal influence. Generally, structure maps of intermediate or deeper datums show less faulting attributable to domal growth.

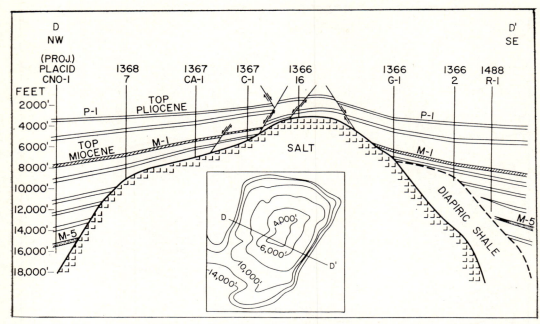

Fig. 8.—NW-SE cross section of Bay Marchand field. Note diapiric shale mass and many types of traps present. No vertical exaggeration.

The datum for the structure map (Fig. 9) is an upper Miocene sandstone bed which is part of the primary producing section of the trend. Structural contours of sedimentary strata are essentially concordant with the salt face except at Timbalier Bay, where strikes on the north and east flanks of the dome are almost tangential to the salt face. The discordance in structural strike of sedimentary strata and the salt becomes more pronounced at depth and extends to the west flank of the Timbalier Bay dome. This discordance and the essentially east-west strike of the deep section indicate a strong and continuing influence of the ancestral salt ridge on the structure of the Timbalier Bay field. Subsequent upward movement from the ridge of the Timbalier Bay salt dome pierced and displaced parts of the relic structure, but failed to warp or incline the strata to conform with the younger salt outline. Other examples of discordance in structural strike of sedimentary strata and salt are present throughout the trend, differing in depth and areal extent.

Bay Marchand.—A large down-to-the-south growth-type fault strikes northeast across Bay Marchand approximately parallel with the ancestral salt ridge. The fault continues southwestward across the south flank of Timbalier Bay, where it splits into two large down-to-the-

southeast faults. The major hydrocarbon reserves on this flank of Timbalier Bay are on the downthrown sides of both segments of this important trap-forming fault system. Major hydrocarbon reserves are also present on the downthrown side of this fault in Bay Marchand.

By means of detailed thickness studies and the construction of isopachous maps of various stratigraphic wedges, a comprehensive history of salt movement at Bay Marchand can be developed. A prominent ancestral ridge was present throughout early and late Miocene time, and extended across the south and southeast flanks westward through South Timbalier Block 21 field (southeast flank of Timbalier Bay dome) and into Timbalier Bay. The structural effect of the ridge was eliminated or tilted out by later cyclic Miocene–Pliocene salt movement, which occurred as much as 3 mi north and west. However, structure maps of deeper Miocene sandstone beds demonstrate the distinct influence which the ridge maintained despite the younger piercing salt movement toward the northwest. The shallow sedimentary strata adjacent to salt in domes of this trend generally are upwarped extensively, the dips being very steep to vertical.

A large diapiric shale mass of low velocity is

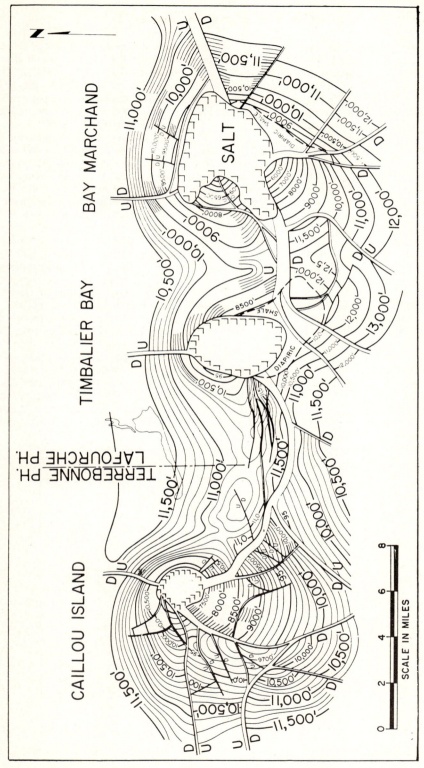

Fig. 9.—Structure map showing complex fault pattern associated with salt trend. Datum is top of upper Miocene sandstone. CI = 100 ft. King area is along crest between Caillou Island and Timbalier Bay domes.

identified on the southeast flank of Bay Marchand (Fig. 8, cross-section D-D'). It was not recognized when an extensive refraction survey was conducted in 1950. To account for the long travel times at intermediate depths, a salt overhang was interpreted to be present on this flank of the dome. Subsequent drilling proved the presence of the low-velocity shale mass and the absence of the interpreted salt overhang.

Because of the numerous types of entrapment, the large areal extent of the dome, and the thick section of interfingering sandstone and shale, the Bay Marchand field has several hundred separate producing reservoirs. It is also significant that more than 125 separate and individual sandstone bodies ranging in depth from 1,200 to 16,000 ft are known to contain hydrocarbon accumulations capable of production. Deeper drilling in the future could add significantly to this number.

Timbalier Bay.—Timbalier Bay is approximately 8 mi west of Bay Marchand along the same ancestral ridge. Fault patterns and incidence are not as complex and extensive as those in Bay Marchand. A large north-south-striking, down-to-the-west fault enters the north flank of the dome and extends southward the full length of the dome, where it changes to an essentially east-west-striking, down-to-the-north fault paralleling the salt ridge. Near its termi-

nus it enters the southeast flank of the Caillou Island dome. In the Caillou Island dome, the fault appears to split into several smaller segments and to extend south and west. The fault causes a readily identifiable escarpment on the Timbalier Bay salt mass. Cross-section B-B' (Fig. 10) on the eastern flank of Caillou Island shows a thin upthrown Miocene section and an expanded downthrown section with well-developed sandstone bodies. From the top of the Pliocene to the top of the Miocene, the fault exhibits little or no evidence of growth; however, from the top of the Miocene to the top of the M-5 sand interval, major growth or expansion of the downthrown section is obvious. This interval (M-1 to M-5) is approximately 2,000 ft thick in the upthrown block and approximately 6,000 ft thick in the downthrown block —a three-fold expansion from the upthrown section. The principal reserves of both Caillou Island and Timbalier Bay are on the downthrown side of this fault system. South Timbalier Block 21 is an exception; however, the accumulation in this area appears to be related to a much-faulted, structurally complex area adjacent to a diapiric shale mass and partly within the reentrant between Bay Marchand and Timbalier Bay. The hydrocarbon accumulation in the Block 21 area is believed to be trapped by transverse faulting, updip pinchouts, and sand-

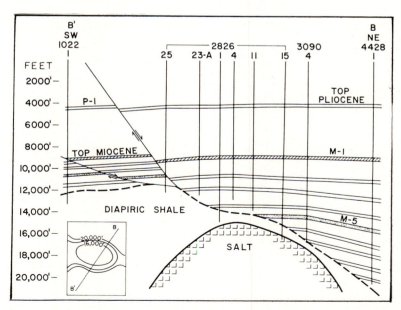

Fig. 10.—NE-SW structural cross section of Caillou Island field. Note expanded M-1 to M-5 section on downthrown side, and presence of diapiric shale.

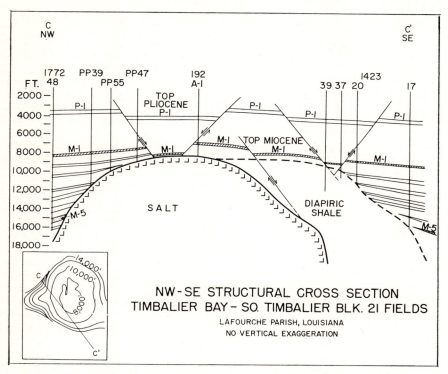

NW-SE STRUCTURAL CROSS SECTION
TIMBALIER BAY – SO. TIMBALIER BLK. 21 FIELDS
LAFOURCHE PARISH, LOUISIANA
NO VERTICAL EXAGGERATION

FIG. 11.—NW-SE structural cross section of Timbalier Bay–So. Timbalier Block 21 fields. Hydrocarbon production is from Miocene sandstones beyond diapiric shale mass. *See* Figure 6.

stone truncation by diapiric shale movement (Fig. 11, cross-section C-C'). Cross-sections C-C' and D-D' show that sands on the northwest flank in both Timbalier Bay and Bay Marchand are truncated by salt, whereas on the southeast flank sand pinchouts and shale truncations are common. An expansive diapiric shale mass is believed to be present on the east flank of the Caillou Island dome (Fig. 10).

Caillou Island.—The Caillou Island salt dome is approximately 10 mi west of Timbalier Bay on the western end of the salt complex. An interdomal, salt-related anticlinal closure previously referred to as the "King area" is approximately midway between Caillou Island and Timbalier Bay. The shallowest salt penetration in the interdomal area is 16,000 ft. Steep and persistent north dips extend from this closure. East-west structural relief is very gentle, but becomes critical at depth. Hydrocarbon entrapment is practically continuous along the ridge between Caillou Island and Timbalier Bay.

The Caillou Island salt mass is essentially circular at shallow depths. However, there is a pronounced southwestward elongation at depth. Structural contours of sedimentary strata generally conform with the salt-dome outline, with the exception of an anomalous structural irregularity on the west flank. The irregularity may be the result of deep salt movement and later tilting and warping by salt movement toward the northeast. There is a small synclinal separation between this area and the main crestal area of the dome. A complex fault network is present primarily on the south and west flanks, some of which provide the traps for the scattered hydrocarbon accumulations. Some accumulations are as far as 6 mi from the crest of the dome, and thus demonstrate the importance of faults to hydrocarbon accumulation. The larger upthrown segment on the southwest flank, which was formed by large down-to-the-northeast and down-to-the-northwest growth-type faulting, is sandstone-poor in comparison with all other flanks of the dome. In cross-section B-B' (Fig. 10), the 1022 No. 1 well on the southeast end of the section penetrated a thin upthrown section and massive diapiric shale, whereas wells on the northeast have a greatly

Table 1. Production and Reserve Data for Fields[1]

Field	Cumulative Production to 1/1/68 (Bbl)	Crude Oil and Condensate Reserves (1/1/68 Est.)	1967 Production (Bbl)
Caillou Island (Incl. offshore)	322,856,000	500,000,000	33,040,000
Timbalier Bay (Incl. onshore)	192,220,000	300,000,000	33,033,000
Bay Marchand (Incl. onshore)	214,442,000	600,000,000	30,908,000
Total	729,518,000	1,400,000,000	96,981,000

[1] Source: Oil and Gas Jour. (1968).

expanded section with favorable objective sandstone beds to a depth of at least 20,000 ft. Accumulation in the upthrown segment is relatively minor and limited primarily to fault traps. A deep Texaco well (State 1021 No. 1) drilled in the south part of this segment reached shale at 13,400 ft and drilled approximately 4,000 ft farther in a high-pressure, deep-water shale which contains no sandstone. Significantly, less than 5 mi north, Texaco has wells producing oil from well-developed sandstone beds below 20,000 ft. The lowest depth limit of favorable sandstone objectives has not been established along the north flank of the complex. As drilling technology improves and it becomes economically attractive to drill and produce wells from depths greater than 4 mi, it is certainly conceivable that deeper drilling will add significantly to the known overall reserves of the complex.

PRODUCTION AND RESERVES

Figure 6 is a composite salt map of the Bay Marchand–Timbalier Bay–Caillou Island complex. The stippled areas represent the positions of known hydrocarbon accumulations. The almost uninterrupted hydrocarbon accumulation extending along the entire north flank from Caillou Island to Bay Marchand should be noted. Hydrocarbons also are present adjacent to reentrants along the salt face. There are more than 1,900 producing wells along this structural trend.

The production chart (Table 1) shows estimated ultimate recoverable reserves for the individual fields of the salt-ridge complex; it also records cumulative production (as of January 1, 1968) and total production for the year 1967. It is believed that present and future deeper exploration drilling along the salt-ridge

complex will increase significantly the ultimate recoverable reserves.

Most of the wells in the Caillou Island and Timbalier Bay fields were drilled with inland drill barges. In Bay Marchand, which is just outside or gulfward of the Louisiana shoreline in water depths ranging from 20 to 50 ft, most of the wells are drilled as directional holes from central platform locations. The well clusters range from 4-well satellites to 25-well platforms. Wells drilled directionally from platforms are extended outward an average of approximately 2,500 ft, but some have been extended outward more than 1 mi.

Hydrocarbon accumulations range in depth from 1,200 to more than 20,000 ft. The typical reservoir rock is generally fine-grained sandstone, well sorted, loosely consolidated, and in many places interlaminated with medium-gray to black marine shale. The porosity ranges from 15 to 33 percent, air permeability ranges from 0.5 to 2,430 md, and the sand-grain density ranges from 2.62 to 2.98 g/cc. It is estimated that hydrocarbon recoveries from producing sandstone beds range from 150 to 1,200 bbl/acre-ft.

SUMMARY

The Bay Marchand-Timbalier Bay–Caillou Island salt domes are diapiric intrusions or spines on an east-west-oriented ancestral salt ridge. The ridge extends approximately 28 mi and encompasses more than 140 mi^2; gravity and subsurface data suggest that the ridge continues eastward to the Grand Isle 16 and 18 domes, and westward to the Lake Pelto dome. Thus the salt ridge may be more than 60 mi long. Detailed structural and interval-thickness studies indicate that domes of the trend grew periodically or in stages. The cyclic growth patterns commonly caused a geographic shift in the structural apex, erosional unconformities, local stratigraphic variations, and complex structural conditions.

Faults in the salt-ridge area are believed to be exclusively of normal or gravity type. The large growth faults which were active contemporaneously with deposition have a pronounced effect on the stratigraphy, structure, and production.

Production along the trend is almost continuous on the north flank from Caillou Island to Bay Marchand. Deeper drilling has demonstrated the economic and productive significance of the ancestral salt ridge connecting these domes.

REFERENCES CITED

Atwater, G. I., and M. Forman, Jr., 1959, Nature of growth of southern Louisiana salt domes and its effect on petroleum accumulation: Am. Assoc. Petroleum Geologists Bull., v. 43, p. 2593–2622.

Ewing, M., and J. Antoine, 1966, New seismic data concerning sediments and diapiric structures in Sigsbee Deep and upper continental slope, Gulf of Mexico: Am. Assoc. Petroleum Geologists Bull., v. 50, p. 479–504.

Gilreath, J. A., 1968, Electric-log characteristics of diapiric shale, *in* J. Braunstein and G. D. O'Brien, eds., Diapirism and Diapirs: Am. Assoc. Petroleum Geologists Mem. 8, p. 137–144.

Limes, L. L., and J. C. Stipe, 1959, Occurrence of Miocene oil in South Louisiana: Gulf Coast Assoc. Geol. Socs. Trans., v. 9, p. 77–90.

Oil and Gas Journal, 1968, Where are the reserves?: v. 66, Feb. 5, p. 162–163.

Geology of Tom O'Connor Field, Refugio County, Texas[1]

HERBERT G. MILLS[2]
New York, N.Y. 10020

Abstract The Tom O'Connor field, discovered in 1934, is in Refugio County, Texas, on the Gulf of Mexico coastal plain. Large accumulations of oil are present in the Oligocene Frio Formation, with increasing percentages of associated gas in the progressively younger sandstone beds of this regressive sequence; primarily gas reservoirs with minor oil columns are present in the transgressive Oligocene Anahuac Formation; and dry gas is found in the regressive Pliocene-Miocene Fleming sandstone. Ultimate recovery is expected to be more than 500 million bbl of oil and 1 trillion ft³ of gas.

Entrapment is the result of anticlinal folding on the downthrown side of the Vicksburg fault zone. Faulting originated in large-scale gravity slumping of incompetent clay at the continental shelf edge or on the continental slope. The faults' arcuate strike, concave toward the basin, is important to the location of the structure. Deep-seated plastic shale, slumping away from the faults in paths normal to them, converged to form a supportive shale core which makes Tom O'Connor the dominant structure in the area.

Because the massive, permeable, inner- to middle-shelf Frio Formation sandstone beds are structurally uninterrupted for 10–15 mi into the basinward source areas, a large hydrocarbon drainage system was available to yield oil and gas to the trap.

Thus, Tom O'Connor is the dominant structure in the area, the first and largest trap in the path of hydrocarbons migrating updip from deeper basin source areas.

INTRODUCTION

Tom O'Connor field, on the Texas Gulf Coast 150 mi southwest of Houston and 30 mi inland from the Gulf of Mexico, is in the Gulf of Mexico Tertiary basin (Fig. 1). Production is from Pliocene, Miocene, and Oligocene strata which have been folded by gravity slumping of sediments along the Vicksburg fault zone. Figure 1 shows the field and other giant Gulf Coast oil and gas fields containing more than 100 million bbl of oil and 1 trillion ft³ of gas.

[1] Read before the 53d Annual Meeting of the Association, Oklahoma City, April 23, 1968. Published by permission of Humble Oil & Refining Company and Quintana Petroleum Corporation. Manuscript received May 7, 1968; accepted July 12, 1968.

[2] Standard Oil Company (N.J.).

Appreciation is expressed to the Humble Oil & Refining Company's South Texas Division staff for their assistance in gathering data and their helpful suggestions. Appreciation is also due to the Quintana Petroleum Corporation for supplying maps, reports, and samples, and to Allan Bramlette for drafting the illustrations.

The field was discovered in October 1934 with the completion of the Quintana No. 1-A Thomas O'Connor on the southeast margin of the field. The well was drilled on acreage selected by the late H. R. Cullen as a result of a gravity survey which he made through much of the Texas Gulf Coast. Cullen also located his well within a newly discovered oil- and gas-field trend, locally defined by the Refugio field 10 mi west-southwest, the Greta field 4 mi northwest, and the McFaddin field about 14 mi northeast.

By early 1936 the field had produced more than 1 million bbl and was recognized as a giant oil field. Today it is the "Frio" trend's largest oil field. The 3-mi-wide by 10-mi-long anticline contains approximately 15,000 productive acres and 902 wells have been drilled within its limits. Ultimate recovery is estimated to be more than 500 million bbl of oil, 1 trillion ft³ of gas, and 15 million bbl of natural gas liquids. Accumulated production to January 1, 1968, was 301 million bbl of oil, 775 billion ft³ of gas, and 11 million bbl of condensate.

PHYSIOGRAPHY

The surface is a featureless coastal plain which slopes seaward at the rate of a few feet per mile. There is no topographic evidence of the local structure; however, the outcrop pattern of the contact between the Beaumont Clay and underlying "Lissie sand" generally traces the outlines of a broad structural nose enclosing the Refugio, Greta, and Tom O'Connor fields. Moreover, subsurface structural folding is very evident in the shallowest mappable beds.

STRATIGRAPHY

The stratigraphic section consists of strata of the Oligocene, Miocene, Pliocene, and Pleistocene Series. Figure 2 is a typical electric log of the productive strata.

Oligocene

Vicksburg Formation.—The Vicksburg, not shown in Figure 2, is the oldest unit reached by wells in the Tom O'Connor area. In this local-

···SCALE···

0   40 Miles

N

Tom O'Connor Field

VICKSBURG FAULT ZONE

GULF of MEXICO

TEXAS

MEXICO

LA. TEX.

Location Map

MIDLAND •   •DALLAS
AUSTIN
SAN ANTONIO•   •HOUSTON

Fig. 1.—Texas Gulf Coast giant fields. Map shows Tertiary fields with more than
100 million bbl of oil or 1 trillion ft³ of gas.

ity only a few hundred feet has been penetrated because the formation contains very little reservoir-quality rock, being composed of deep-water marine shale with thin impermeable sandstone near the top. The upper contact of the Vicksburg with the Frio Formation is marked by the first appearance of *Textularia warreni.*

Frio Formation.—Conformably overlying the Vicksburg is the Frio Formation, about 3,500 ft thick in this area. Its upper contact with the overlying Anahuac Formation is at 4,500 ft near the middle of the massive "Greta sands." The contact is placed where the gray "salt-and-pepper" sandstone of the Anahuac gives way to sandstone containing much pink and milky quartz.

In the productive section between 4,500 and 6,000 ft there are 15 oil reservoirs and 4 non-associated gas reservoirs containing 96 percent of the field's oil and 50 percent of its gas. Nearly all of the formation's oil is in the series of sandstone beds from 5,500 to 6,000 ft and most of its gas is in the sandstone bodies grouped near 5,200 ft. Oil produced from res-

ervoirs below 5,500 ft is a sweet, light-green, basic fuel-products, 36°-API-gravity crude.

The Frio was deposited in a regressive period of sedimentation during which a large sediment supply caused outbuilding of the land. Evidence for regression is apparent in the succession of sedimentary environments represented in wells at Tom O'Connor. The basal few hundred feet of middle-neritic, interbedded thin sandstone and shale bodies, containing middle-shelf fauna, transitionally grades upward to the massive sandstones typical of those shown on the bottom 700 ft of the type log. The thin shale interbeds of this massive sandstone section are barren of microfauna but commonly contain abundant oysters, characteristic of bay or lagoonal deposits. Although not common, lignite is present as thin beds in both the sandstone and shale. The massive quartzose sandstone bodies are very clean, free of interstitial clay, very permeable, well sorted, and friable though compacted—a typical product of the high-energy environment of a beach or shore face. They culminate at 5,900 ft with the

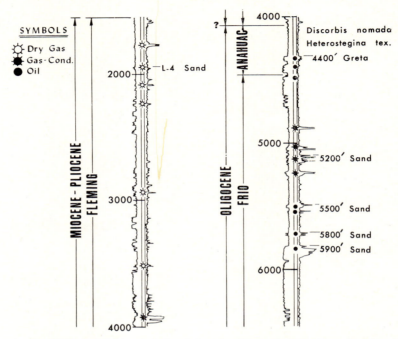

FIG. 2.—Tom O'Connor field typical electric log, showing stratigraphy of productive section.

field's largest oil reservoir (Fig. 2).

Above the 5,900-ft sand the formation is predominantly shale which was deposited in a lagoonal environment. The shale is barren of fauna except for numerous beds of oysters; it is mostly green, in places gray, or red to brown where oxidized. It is called colloquially "poker-chip" shale because of its brittle nature and tendency to fracture parallel with bedding upon drying after removal from a core barrel. Lignite is scarce in the shale and sandstone of the interval.

The sandstone beds from 5,000 to 6,000 ft are the principal reservoirs of the field. Thickness ranges from 10 to 100 ft; porosity averages 31 percent; permeability is as high as 6,500 md but normally ranges from 500 to 2,300 md. Generally, the sandstones are free of shale or interstitial cement except at the top or base, where there is commonly a transition to the enclosing shale. The reservoir sandstone beds extend in thickening sheets downdip to the southeast to form a massive sandstone section similar to that below 5,900 ft. This large aquifer was important to oil migration and today supplies the reservoir energy to a predominantly water-drive mechanism which is producing recoveries anticipated to range from 45 to

70 percent of the original oil in place.

The sandstones of the upper Frio represent periods of more rapid subsidence and minor encroachments of the beach and shore-face environment on the lagoon. Where the sandstones are discontinuous and patchy, they probably are tidal-delta or washover-fan deposits laid in the lagoon behind a barrier island.

Anahuac Formation.—Overlying the Frio is the Anahuac Formation, a 400-ft-thick transgressive marine unit composed of middle- to outer-neritic shale, with shallow-water sandstone at its top and base. The formation's top is determined by the presence of *Discorbis nomada*. The middle of the formation's three faunal zones, *Heterostegina texana,* is present at Tom O'Connor but the basal *Marginulina* zone was not deposited that far updip.

The principal pay of the Anahuac is the 4,400-ft "Greta sand," a very soft, unconsolidated, silty, very fine-grained sandstone commonly containing thin beds of sandy fossiliferous limestone. The semi-consolidated nature of the sand causes severe production problems.

The oil produced from the "Greta sand" is a dark-brown, sweet, basic fuel-products crude with an API gravity of 24°, considerably heavier than the Frio crude.

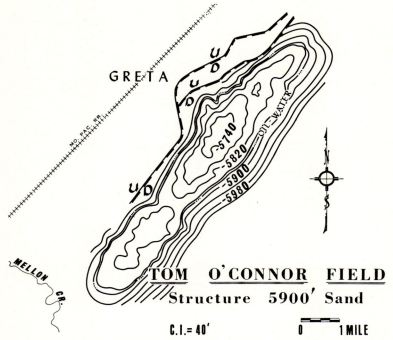

FIG. 3.—Structure map of Frio 5,900-ft sand. Normal faults separating Tom O'Connor from Greta field show increased throw with depth.

Miocene-Pliocene

The next shallower unit is the Fleming group, the formations of which are locally undifferentiated. The sandstones of the group are an important source of dry gas in the Tom O'Connor field. They are unconsolidated, very fine grained, and commonly contain thin beds of impermeable calcareous sandstone. The shales are green, in places mottled brown to red, brittle to waxy, barren of microfauna, but contain oysters and other miscellaneous pelecypods. The Fleming is a regressive sequence, progressing from inner neritic to littoral at the base to mixed lagoonal and fluvial near the top.

Pleistocene

The Lissie Formation overlies the Fleming. It is a section of massive sandstones about 1,400 ft thick, a profuse source of fresh water suitable for human consumption and irrigation.

LOCAL STRUCTURE AND RESERVOIR DESCRIPTION

5,900-ft sand (Fig. 3).—The Tom O'Connor structure is an elongate, asymmetric anticline paralleling the regional strike. The basinward limb dips 2.5° and the northwest limb dips 3°.

Total vertical closure is 250 ft; the oil-column height of the 5,900-ft reservoir is 150 ft. Two normal faults with 200–250 ft of combined throw strike along the northwest flank and separate the Tom O'Connor from the Greta field. They are depositional or growth faults showing increasing throw with depth.

The 5,900-ft sand, the field's largest oil reservoir, will yield more than 50 percent of the field's ultimate recovery. It is a massive, uniform sand throughout the productive area, but is replaced by lagoonal shale on the northwest in the Greta field. Basinward, the sandstone thickens and is laterally and vertically in communication with a large aquifer which maintains reservoir pressure.

5,800-ft sand (Fig. 4).—The second principal reservoir of the field illustrates the erratic areal distribution of some of the Frio reservoirs. The 5,800-ft sand zone is composed of five discontinuous members, each 5–15 ft thick, which successively lap onto the structure. The basal contact of each member with underlying shale is sharp, whereas the upper contact is transitional to shale. Grain size diminishes upward in the well-sorted sandstone; lignite commonly is seen in the base of the sandstone, and clay

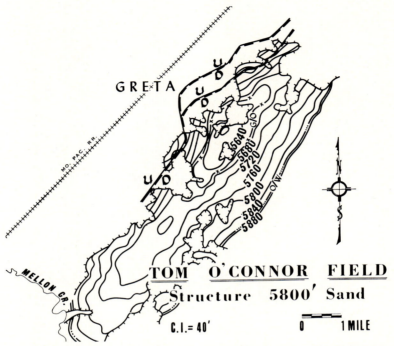

Fig. 4.—Structure map of Frio 5,800-ft sand. Hachured line is demarcation of sand and illustrates erratic areal distribution of some Frio sandstone reservoirs.

clasts are present near the top. Like the 5,900-ft sand, the reservoir is clean, very porous, and permeable. Its areal distribution and association with lagoonal clay indicate deposition by tidal deltas, succeeded downdip by barrier-island and shore-face environments. Basinward the sandstone beds are known to thicken and vertically merge near the field's southeast margin.

5,200-ft sand (Fig. 5).—This reservoir is characteristic of a group of four sandstone beds which are major producers of gas-condensate, yielding 6 bbl of condensate per million cubic feet. The sandstones are commonly somewhat shaly or calcareous and have lower permeability than the deeper reservoirs. The widespread presence of oysters and lignite, and the lack of open-marine fauna indicate a continuation of the lagoonal to littoral environment.

At the 5,200-ft horizon the flank dips are diminished to 1° basinward and 1.8° on the northwest flank. The domal Greta field structure is shown on the upthrown side of the separating faults, the combined throw of which has decreased to between 100 to 150 ft.

4,400-ft Greta sand (Fig. 6).—By the time of Anahuac deposition the faults between Tom

O'Connor and Greta were no longer active and only the prominent saddle separated the two structures. Decrease of the flank dips to less than 1° southeastward and 1.3° northwestward reflects the slowing of subsidence in the general area of the Vicksburg fault zone.

The 4,400-ft reservoir at Tom O'Connor holds an 80-ft gas column and a 10-ft oil column, whereas at Greta field the oil column is 60 ft high. The two fields have a common oil-water contact, 10 ft below the spill elevation of the saddle area; the Tom O'Connor field gas-oil contact is exactly at the spill elevation between the fields. This is a classic illustration of the distribution of oil and gas resulting from successive filling of structures by updip migration of gas- and liquid-phase hydrocarbons. Refugio field, the next higher structure on the west, is water bearing in this sand.

L-4 sand (Fig. 7).—At this shallow depth in the upper Fleming, the saddle area dividing the Tom O'Connor and Greta fields is essentially filled and the structure has become a large dome with a southwestward-plunging nose marking the strike of the Tom O'Connor anticline. The dry-gas reservoir is confined to one domal structure, the crest of which is centered over the

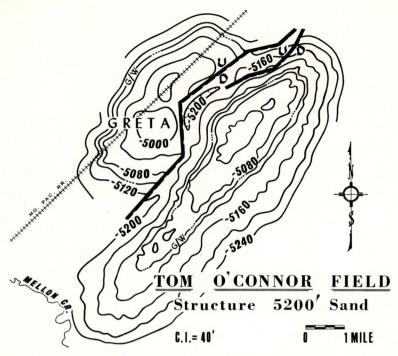

TOM O'CONNOR FIELD
Structure 5200' Sand

C.I.= 40'   0 ___ 1 MILE

FIG. 5.—Structure map of Frio 5,200-ft sand. Map shows relation to Greta field and illustrates diminishing flank dips with shallower depths.

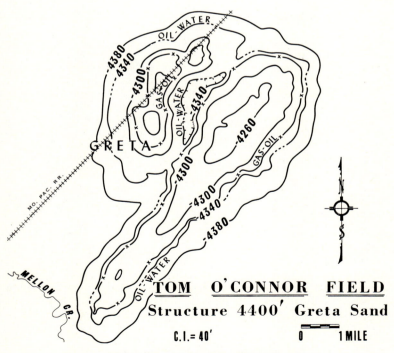

TOM O'CONNOR FIELD
Structure 4400' Greta Sand

C.I.= 40'   0 ___ 1 MILE

FIG. 6.—Structure map of Anahuac 4,400-ft Greta sand. Depositional faults were no longer active at this time.

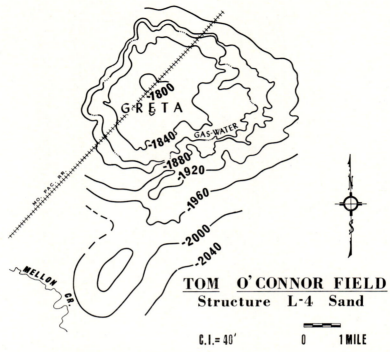

FIG. 7.—Structure map of upper Fleming L-4 sand showing saddle area filled and greatly diminished flank dips. Anticlinal crest is shifted to Greta field.

Greta field. Flank dips have diminished to 0.5° southeastward and 1° northwestward.

OIL MIGRATION AND ACCUMULATION

The present regional structure of the upper Frio in Refugio County is shown in Figure 8. The map covers approximately 580 mi².

The Vicksburg fault zone, a band of faulting about 6 mi wide, consists of 2–4 faults. The major northwesternmost fault is essentially continuous and has a throw ranging from 500 to more than 1,000 ft at the map horizon. Smaller basinward faults are discontinuous and have successively decreasing throws toward the basinward margin of the fault zone. All faults are contemporaneous with deposition and show increasing throw with depth. Their areal trace pattern is arcuate, concave toward the basin, as is typical of depositional gravity faults.

A second zone of faulting, the Frio fault zone, is shown partly on the southeast margin of the map. It is wider, much more complex, and composed of more and smaller faults than the older Vicksburg fault zone. Although the overall zone is the site of pronounced thickening of the Frio Formation, the faults shown on the map have small throws of less than 400 ft

and do not show downbending in the upper Frio. They probably were not a barrier to oil migration in this interval.

Shelf-break gravity slumping seems to account adequately for the origin of the fault zones. Faulting had its inception in slumping of incompetent clay at the continental shelf edge and on the continental slope. It continued upward and was accentuated by sediment loading as younger sediments were built outward across the older shelf edge. Faulting ceased when shelf-edge progradation proceeded sufficiently to buttress against further basinward creep.

Between the two fault zones the Tom O'Connor anticline dominates the structure. On the northwest, the beds dip 3–4° toward the Vicksburg faults; basinward they dip away from Tom O'Connor at a uniform 2° without structural interruption for 10 mi to the first small Frio fault zone displacements. In both directions along strike there is a basic plunge away from Tom O'Connor, each successive closure being somewhat lower. The dominance of the field results from a supportive shale core derived from the convergence of clay masses slipping along paths normal to the faults' strike. If lines were drawn perpendicular to the arcu-

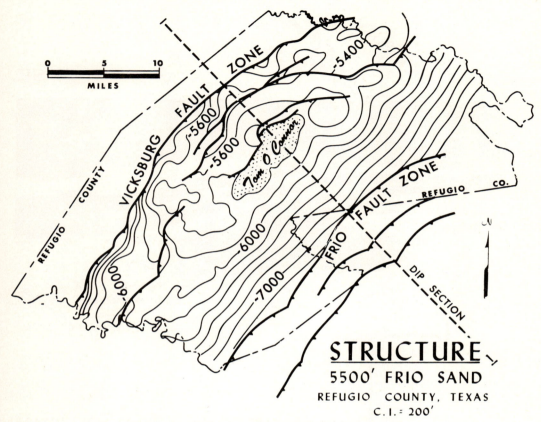

Fig. 8.—Structure map of 5,500-ft Frio horizon in Refugio County. Tom O'Connor field dominates areawide structure.

ate fault trace, they would focus just in front of Tom O'Connor. These are the paths of slippage of the underlying plastic shale which converged to form the positive core supporting the structure. Thus Tom O'Connor was the focal feature in the path of hydrocarbons migrating updip from deeper water source areas through the thick, permeable sandstones of the middle to upper Frio.

A simplified stratigraphic-dip section is shown in Figure 9. Faults are omitted but the fault zones are indicated by marked sedimentary thickening. The position of Tom O'Connor is shown by the column of hydrocarbon symbols.

The Frio Formation is a regressive sequence deposited on a continental shelf 20–25 mi wide. At Tom O'Connor the productive upper third of the formation is predominantly lagoonal shale and shoreline sand. Basinward the sandstone beds thicken and merge in the inner-neritic environment to form an almost mas-

sive sandstone body. At the middle-shelf area the sandstone thins and more shale is present. Finally, shale becomes predominant in the outer-neritic and bathyal environments. Hydrocarbons, which probably originated in fine-grained sediments rapidly deposited in the deeper water areas, could move through the permeable inner-shelf sandstone bodies to the Tom O'Connor trap.

Basically the same situation, an easy path of migration through massive sandstone from a basinward hydrocarbon source, was present in the overlapping transgressive Anahuac and regressive basal Fleming. The dry gas of the middle to upper Fleming is probably the product of local marsh or swamp deposits.

In conclusion, Tom O'Connor is the dominant structure in the area, the first and largest trap in the path of a hydrocarbon drainage system extending perhaps 15 mi downdip and reaching a like distance or farther along strike in both directions from the field.

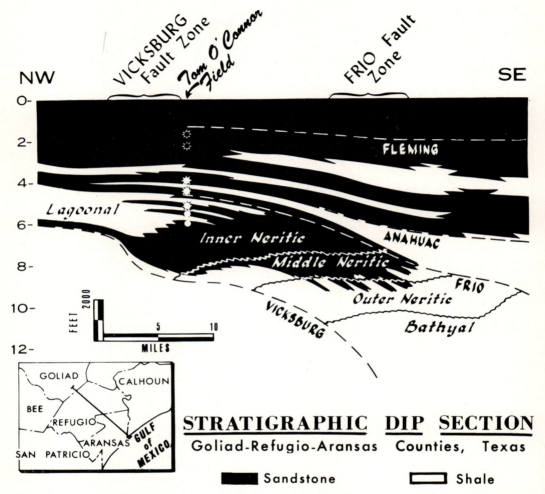

STRATIGRAPHIC DIP SECTION
Goliad-Refugio-Aransas Counties, Texas

■ Sandstone   □ Shale

Fig. 9.—Simplified stratigraphic-dip section. Position of section is shown on Figure 8. Hydrocarbons originating in deeper water sediments moved through massive inner-shelf sandstone to Tom O'Connor trap.

SELECTED REFERENCES

Bornhauser, Max, 1958, Gulf Coast tectonics: Am. Assoc. Petroleum Geologists Bull., v. 42, no. 2, p. 339–370.

Boyd, D. R., and B. F. Dyer, 1964, Frio barrier bar system of south Texas: Gulf Coast Assoc. Geol. Socs. Trans., v. 14, p. 309–322.

Burke, R. A., 1958, Summary of oil occurrence in Anahuac and Frio Formations of Texas and Louisiana: Am. Assoc. Petroleum Geologists Bull., v. 42, no. 12, p. 2935–2950.

Cloos, Ernst, 1968, Experimental analysis of Gulf Coast fracture patterns: Am. Assoc. Petroleum Geologists Bull., v. 52, no. 3, p. 420–444.

Hardin, F. R., and G. C. Hardin, Jr., 1961, Contemporaneous normal faults of Gulf Coast and their relation to flexures: Am. Assoc. Petroleum Geologists Bull., v. 45, no. 2, p. 238–248.

Holcomb, C. W., 1964, Frio Formation of southern Texas: Gulf Coast Assoc. Geol. Socs. Trans., v. 14, p. 23–33.

Johnson, R. B., and H. E. Mathy, 1957, The South Texas Frio trend: Gulf Coast Assoc. Geol. Socs. Trans., v. 7, p. 207–218.

Murray, G. E., 1961, Geology of the Atlantic and Gulf Coastal province of North America: New York, Harper and Bros., 692 p.

Quarles, Miller, Jr., 1953, Salt-ridge hypothesis on origin of Texas Gulf Coast type of faulting: Am. Assoc. Petroleum Geologists Bull., v. 37, no. 3, p. 489–508.

Steinhoff, R. O., 1967, Continental slope origin of Gulf Coast faulting: Oil and Gas Jour., May 22, p. 178–182.

Weaver, Paul, 1955, Gulf of Mexico: Geol. Soc. America Spec. Paper 62, p. 269–278.

Vicksburg Fault Zone, Texas[1]

T. B. STANLEY, JR.[2]

Houston, Texas 77001

Abstract One of the most petroliferous structural features of the Gulf of Mexico basin is the Vicksburg fault zone of southern Texas and northeastern Mexico. For almost 300 mi the downthrown side of these down-to-basin faults is characterized by an alignment of oil and gas fields, representing a total ultimate reserve of more than 3 billion bbl of oil and 20 trillion ft³ of gas.

Although the Vicksburg fault-zone trend has been explored over a period of more than 30 years, it still is not clearly understood. The principal reasons for the lack of understanding are that the displacements are buried deeply and have been interpreted almost entirely from seismic records and well data.

The accumulation of the hydrocarbons associated with the fault zone is largely a result of two notable characteristics of the zone: (1) downbending of the downthrown block provided dip reversal necessary for the formation of anticlinal closures; and (2) faulting contemporaneous with deposition resulted in the formation of structural closure during deposition and, thus, in early entrapment of migrating oil and gas.

Throughout the zone the Vicksburg faults show a history of post-Eocene movement. However, data collected at widely separated places indicate the possibility of earlier movements. Maximum activity occurred during Oligocene and early Miocene time, when the terrigenous clastic sediments of the Vicksburg and lower Frio were being deposited. Faulting decreased and ceased during or shortly after deposition of the upper Frio.

In places, where the fault zone consists of multiple faults, movement shifted progressively from one fault to another.

INTRODUCTION

If economic value can be used to evaluate the importance of a tectonic feature, then one of North America's most important tectonic feature is the Vicksburg fault zone of the Gulf of Mexico basin (Fig. 1) on the coastal plan of southern Texas and northeastern Mexico. This zone of down-to-basin (downthrown toward the Gulf) faults has localized and preserved more than 3 billion bbl of oil and 20 trillion ft³ of gas in Oligocene and Miocene strata (Fig. 2). During the late 1930s and the postwar 1940s, the Vicksburg fault zone was the site of one of the Texas Gulf Coast's most ambitious exploratory programs. Despite this intense activity and the acquisition of great quantities of geologic and geophysical data—through a period of more than 30 years—the faulting itself still is not clearly understood. The faults and associated structures are difficult to map because they are deeply buried and have little or no surface expression. Subsurface displacements of several thousand feet decrease and die out upward and are completely covered by younger deposits. Analyses of the fault zone, therefore, have been based on the study of well logs and seismic records. Although these data have led to an orderly development of the associated oil and gas fields, they are ambiguous in that they are subject to more than one interpretation.

Early drilling operations discovered oil along the fault zone long before the trapping mechanism was interpreted correctly. At first the traps were thought to be simple anticlines, and the presence of faulting was not suspected until deeper drilling revealed evidence of major fault throw in the subsurface. The principal evidence was the discovery that specific stratigraphic units thicken abruptly within very short distances. Thus the presence of downthrown blocks with greatly thickened sections was established (Fig. 3). Many geologists attributed the thickening to flexing (*e.g.,* Colle *et al.,* 1952, p. 1196), and the term "Vicksburg flexure" was introduced by workers who named it after the principal formation involved (for discussion, *see* Colle *et al.,* 1952; Hardin and Hardin, 1961). Although it was not long until faulting, and not flexing, was found to be the cause of the thickening, the name "Vicksburg flexure" has remained popular in South Texas, where the correct name, Sam Fordyce–Vanderbilt fault zone, is seldom used. However, some workers, such as the writer, prefer the term "Vicksburg fault zone."

As mapped today, the greatest displacement of the zone is near the Mexican border, where more than 5,000 ft of abrupt stratigraphic thickening has been measured. From the border the zone extends northeastward 240 mi, where it becomes poorly defined and difficult to identify in an area of widely separated and probably unrelated faults. It has been traced

[1] Read before the 53d Annual Meeting of the Association, Oklahoma City, Oklahoma, April 23, 1968. Manuscript received, May 31, 1968; accepted, September 25, 1968. Published by permission of Humble Oil & Refining Company.

[2] Senior Geological Scientist, Humble Oil & Refining Company.

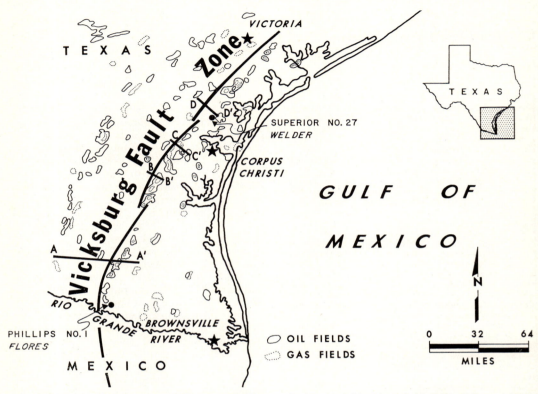

Fig. 1.—Vicksburg fault zone, south Texas, and adjacent oil and gas fields. Lines **A-A'**, **B-B'**, **C-C'**, and **D-D'** show locations of Figures 4-7.

southward into Mexico for more than 50 mi. In places the zone may consist of only one fault, but elsewhere it is known to consist of two or more faults.

The downthrown side of the zone is characterized by an almost continuous alignment of multipay oil and gas fields. Included are such giants as Kelsey deep, Seeligson, Agua Dulce, Refugio, and Tom O'Connor (Fig. 3). Ultimate reserves of more than 100 million bbl per field are common, and one field, Tom O'Connor, is expected to yield more than 500 million bbl.

The accumulation of the large quantities of hydrocarbons associated with the fault zone is largely a result of two notable characteristics of the zone: (1) downbending provided the dip reversal necessary to form anticlinal closures, and (2) faulting contemporaneous with deposition resulted in the formation of structural closure during deposition and, thus, in early entrapment of migrating oil and gas. Other factors may have contributed to the large accumulations, but were less important.

CHARACTERISTICS

Thirty miles north of the Mexican border, the Vicksburg fault zone appears to consist of three step faults, two of which are opposed by smaller graben-forming antithetic faults (Fig. 4). The sediments are terrigenous clastic, and the fact that faulting occurred during deposition of these beds is indicated by the thickened section in the downthrown block. The Oligocene Vicksburg alone thickens from about 600 ft to more than 4,000 ft, but thickening is less pronounced in the overlying Frio, and is absent in younger strata. If similar thickening is present in the underlying late Eocene Jackson Formation, it has not been detected.

Well data show that the westernmost down-to-basin fault of Figure 4 is the youngest of the three shown. It is the only one which displaced the top of the Miocene Frio. Frio strata, which downbend into the fault, are unaffected by the deeper displacements of the two eastern down-to-basin faults. Also, the faultward dip or downbending increases with depth and is associated with small antithetic faults. The portrayal of the antithetic faults as single fractures probably is an oversimplification, because abundant slickensides in cores recovered from the area indicate the presence of multiple planes along which antithetic movement could have taken place. Wells shown in Figure 4 and the other cross sections produce oil and gas from Frio sandstone where the beds are folded anticlinally on the downthrown side of the Vicksburg fault zone.

The master fault plane is shown in Figure 4 to be curved and to decrease in dip downward. Actually, the fault attitude has not been measured here or elsewhere because of the practice of deliberately avoiding the faults in making well locations. Wells which do cross the faults generally are too widely spaced to provide useful information, and lack of well control makes it imposible to determine the attitude of the beds in the deeper fault blocks.

Farther north, deeper well control shows the attitude of the older beds in greater detail. Well data in the vicinity of Figure 5 permit determination of the structure in each of the three downthrown blocks shown. The beds in each block bend into the adjacent fault on the west, but the greatest bending is in the easternmost fault block in beds of Jacksonian age. The highest Vicksburg beds are not displaced by this fault, but thicken and downbend toward the central fault. The displacement on the central fault affects younger beds, and the fault dies out in the lower Frio. The upper Frio is displaced by the westernmost fault. This westward or updip progression of displacement produced an overall asymmetric anticlinal structure, the crest of which shifts upward in the same direction.

The upward decrease in downbending is related to the contemporaneous nature of the faulting and reflects a decrease in fault activity with time. Faulting in this area is believed to be post-Jackson. If so, the top of the Jackson in the easternmost block is downbent in response to the total throw along all three faults of the zone. In contrast, the gentle bending of the shallow Frio strata is related only to movements along the westernmost fault that occurred after deposition of the upper Frio.

Throw along the fault zone decreases northward. The sevenfold thickening of the Vicksburg Formation shown in Figure 4 is only a twofold thickening in Figure 6. Moreover, the Vicksburg fault zone in Figure 6 seems to consist of only one fault, along which movements were less complex and are more easily inter-

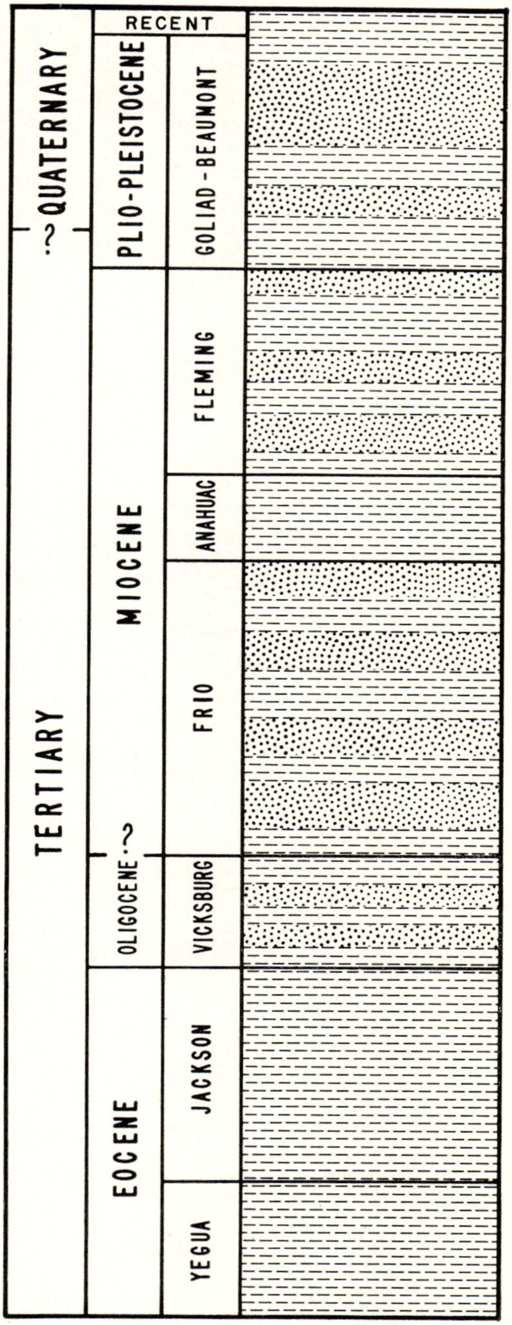

Fig. 2.—Columnar section for south Texas.

preted. The pattern of increased downbending with greater depth is apparent. The antithetic fault shown probably is a zone of multiple faults. There is no evidence in the thicknesses

T. B. Stanley, Jr.

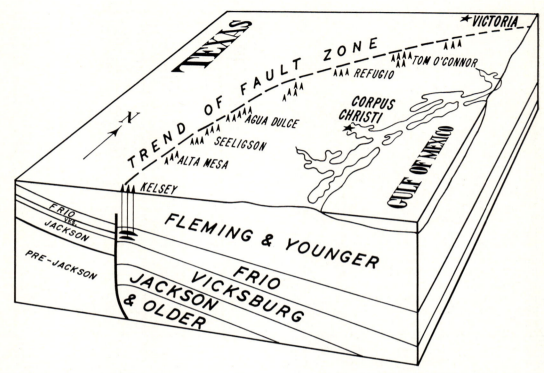

Fig. 3.—Generalized block diagram of Corpus Christi area, south Texas, showing part of Vicksburg fault zone and related oil and gas fields. Note abrupt thickening of section across fault zone.

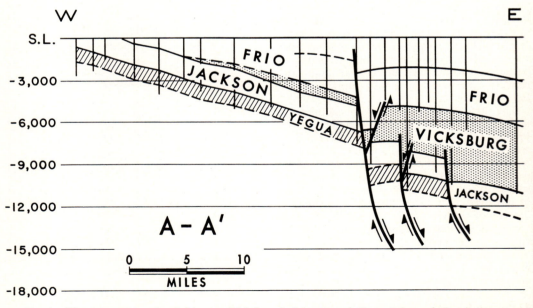

Fig. 4.—West-east cross-section A-A' across Vicksburg fault zone, south Texas. Primary fault activity occurred during Vicksburg deposition. Location of section shown in Figure 1. Vertical lines show well control. Vertical scale in ft.

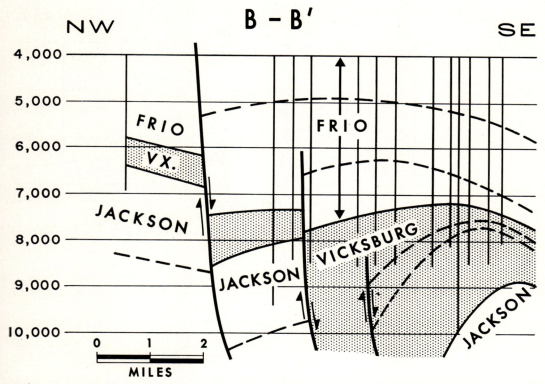

FIG. 5.—Northwest-southeast cross-section **B-B′** across Vicksburg fault zone south Texas. Note structural asymmetry in downthrown block. Location of section shown in Figure 1. Vertical lines show well control. Vertical scale in ft.

of the strata of upward movement in the up-thrown block or in the folded downthrown block. Instead, all major movement seems to have been downward.

Many deep wells on the downthrown side of the fault zone reached Jackson strata, and some penetrated this formation completely, but only one or two penetrated what may be a normal downthrown thickness. One of them is shown in Figure 6 (with an arrow at the base of the well trace). In this well, the downthrown Jackson shows no thickening and apparently was deposited before faulting occurred. Other sections of the Jackson are abnormally thin, but these also can be interpreted as faulted parts of unthickened sections which predate the faults. Such an interpretation is shown in Figure 4. Other Jackson sections, however, are abnormally thick. For example, in San Patricio County the Superior No. 27 Welder (Figs. 1, 7) penetrated almost 7,000 ft of Jackson strata in an area where no more than 1,800 ft was expected. In Figure 7, the thickening of the Jackson is shown to have been caused by fault

movement during deposition. This interpretation is strengthened by a similar occurrence in the Rio Grande Valley, where the Phillips No. 1 Flores (Fig. 1) penetrated more than 5,200 ft of the same formation.

Thus, faulting may have begun earlier in some places than in others, but this conclusion is tentative. Additional penetrations are needed before final conclusions can be reached, and consideration must be given to the fact that Jackson thicknesses can be misleading, because this shale is known to have flowed "plastically" in the Gulf Coast region. In places it seems to have formed shale domes and to have moved diapirically through younger, overlying beds.

Activity During Deposition

Lithologic studies of the strata along the Vicksburg fault zone shown no appreciable differences in equivalent strata on the upthrown and downthrown sides. Sedimentation apparently kept pace with fault movement because there is no indication in the facies that a topographic or submarine fault scarp ever was pres-

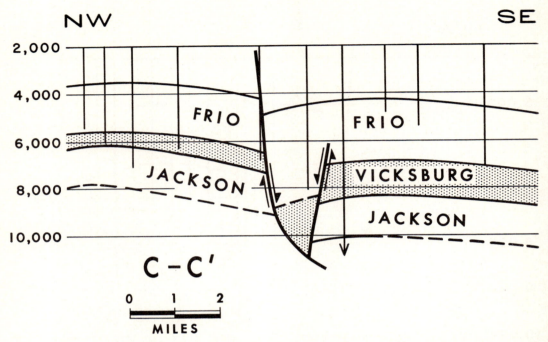

FIG. 6.—Northwest-southeast cross-section **C-C′** across Vicksburg fault zone, south Texas. Location of section shown in Figure 1. Vertical lines show well control. Vertical scale in ft.

ent. In fact, the depositional environments seem to have been relatively unaffected by the faulting. Consequently, a conclusion to be drawn from a comparison of thicknesses and depositional environments across the fault zone is that both sides subsided during deposition, but that the downthrown side subsided more rapidly. However, fault movement was restricted to a rather short geologic time interval, and it was therefore contemporaneous only with sedimentation during that interval.

Structural Closure in Downthrown Block

The productive domes and anticlines which parallel the downthrown side of the Vicksburg fault zone are found only where downbending reversed the regional dip and provided structural closure. This phenomenon of downbending ("rollover" and "reverse drag") has been described by many investigators, but its cause is not known. Field data collected in one area are not necessarily consistent with those collected elsewhere. As a result, various hypotheses have been advanced to explain the downbending.

Some of the best examples of downbending are in the Colorado Plateau, where the folds are associated with postdepositional faults and can be studied in outcrops (Fig. 8). They first were reported by Powell (1875, p. 184) after his historic trip down the Colorado River. He wrote that beds downthrown by faults in the Grand Canyon district are commonly "turned down." In 1882, Dutton (1882a, p. 127, 132; 1882b, p. 113–115, 162, 186) confirmed this observation, and wrote that the beds "flex downward." Later, Noble (1914, p. 76) wrote that the same beds "bend downward." After studies in the same area, Gardner (1941) and Strahler (1948) referred to the feature as "downbending."

Although some Colorado Plateau workers believed the fold to be the remnant of a monocline, Gardner (1941) interpreted it to be a result of gravity sagging of strata, a conclusion also reached by Hamblin (1965). Hamblin related it to normal movements along a fault plane that decreases in dip with depth, and he documented his conclusions with field observations.

Elsewhere the phenomenon was thought to be a result of double movements (Mears, 1951; Sears, 1929) involving complete reversal of throw, or a result of a modified form of elastic rebound (Brucks, 1929), or a shallow reflec-

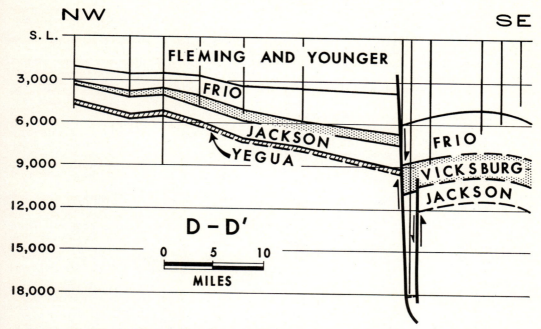

FIG. 7.—Northwest-southeast cross-section **D-D'** across Vicksburg fault zone, south Texas. Location of section shown in Figure 1. Deep well in downthrown block is Superior 7 Welder. Vertical lines show well control. Vertical scale in ft.

tion of an underlying salt anticline (Quarles, 1953).

Only the gravity-sagging hypothesis explains movements along the Vicksburg fault zone, because the variable sedimentary thicknesses there record only downward movements which were greatest in the downthrown block adjacent to the fault. Furthermore, this zone of maximum subsidence is characterized by antithetic faults and grabens, and all evidence sug-

gests that it is a zone of incipient gapping—gapping which could have been caused by fault movements along a curved fault plane similar to those observed by Hamblin (1965) in Arizona.

CONCLUSIONS

The economic importance of the Vicksburg fault zone and the possibility that it represents a basic structure that can be expected else-

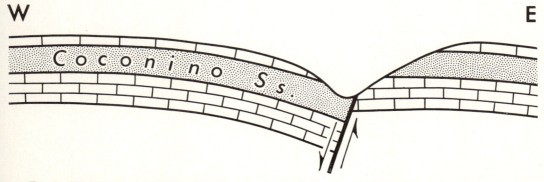

FIG. 8.—Phenomenon of downbending ("reverse drag") in Pennsylvanian and Permian strata of Colorado Plateau area. West-east cross section across West Kaibab fault north of Grand Canyon. (Sketched from photograph by writer.)

where have encouraged numerous attempts to relate the fault zone to basin development. An understanding of such a relation, it was believed, might allow extrapolation of predictions to other undeveloped basins of the world. However, because available data have been inconclusive, workers have supplemented them with observations made elsewhere and, consequently, hypotheses to explain the Vicksburg and similar fault zones have differed considerably. In South Texas, fault data became interwoven with assumptions which included salt-dome tectonics, basinward sliding of sediments, subsidence under sedimentary load, flexures, and depositional belts. As deeper wells are drilled and improved seismic techniques provide information on deeper structure, the origin of fault zones such as the Vicksburg will be greatly clarified. If so, fundamental principles may be learned that will be of great value in petroleum exploration.

REFERENCES CITED

Brucks, E. W., 1929, Luling oil field, Caldwell and Guadalupe Counties, Texas, *in* Structure of typical American oil fields, v. 1: Am. Assoc. Petroleum Geologists, p. 256–281.

Colle, J., *et al.*, 1952, Sedimentary volumes in Gulf coastal plain of United States and Mexico. Pt. IV. Volume of Mesozoic and Cenozoic sediments in western Gulf coastal plain of United States: Geol. Soc. America Bull., v. 63, p. 1193–1200.

Dutton, C. E., 1882a, The physical geology of the Grand Canyon district: U.S. Geol. Survey 2d Ann. Rept., 1880–1881, p. 47–166.

———— 1882b, Tertiary history of the Grand Canyon district: U.S. Geol. Survey Mono. 2, 264, p.

Gardner, L. S., 1941, The Hurricane fault in southwestern Utah and northwestern Arizona: Am. Jour. Sci., 5th ser., v. 239, p. 241–260.

Hamblin, W. K., 1965, Origin of "reverse drag" on the downthrown side of normal faults: Geol. Soc. America Bull., v. 76, p. 1145–1164.

Hardin, F. R., and G. C. Hardin, Jr., 1961, Contemporaneous normal faults of Gulf Coast and their relation to flexures: Am. Assoc. Petroleum Geologists Bull., v. 45, p. 238–248.

Honea, J. W., 1956, Sam Fordyce-Vanderbilt fault system of southwest Texas: Gulf Coast Assoc. Geol. Socs. Trans., v. 6, p. 51–55.

Mears, Brainerd, Jr., 1951, "Reversed drag" on high angle faults in Oak Creek Canyon, Arizona: Geol. Soc. America Bull., v. 62, p. 1539–1540.

Noble, L. F., 1914, The Shinumo quadrangle, Grand Canyon district, Arizona: U.S. Geol. Survey Bull. 549.

Powell, J. W., 1875, Exploration of the Colorado River of the West: Washington, D.C., U.S. Govt. Printing Office, 291 p.

Quarles, Miller, Jr., 1953, Salt-ridge hypothesis on origin of Texas Gulf Coast type of faulting: Am. Assoc. Petroleum Geologists Bull., v. 37, p. 489–508.

Sears, J. D., 1925, Geology and coal resources of the Gallup-Zuni basin, New Mexico: U.S. Geol. Survey Bull. 767, 52 p.

Strahler, A. N., 1948, Geomorphology and structure of the West Kaibab fault zone and Kaibab Plateau: Geol. Soc. America Bull., v. 59, p. 513–540.

Golden Lane Fields, Veracruz, Mexico[1]

FRANCISCO VINIEGRA O.[2] and CARLOS CASTILLO-TEJERO[3]

Mexico, D.F., Mexico

Abstract Even though exploration, exploitation, and the discovery of new fields continue along the perimeter of the Golden Lane, the origin of this Cretaceous atoll is still a subject of controversy. The most widely accepted hypothesis is that it is a biohermal reef that began to grow during late Neocomian time on a late Kimeridgian (Late Jurassic) positive element called the "Isla de Tuxpan." At some localities within the atoll rim, evaporite, calcarenite, and dolomite are present. Elsewhere, however, the sedimentary rocks are rudistid-bearing limestone or miliolid limestone.

The atoll rim consists of a prominent belt of structural culminations underlain by rudistid and/or miliolid limestone. The Jardín No. 35, drilled in 1930, and Mesita No. 1 and Cañas No. 101, drilled in 1968, are the only wells which have penetrated the reef core. Data from these wells, however, can be interpreted in more than one way. Notwithstanding the depositional environment in which rudistids are believed to have lived (rudistids are the principal component of this great reef), its morphology (i.e., is it a bioherm, a biostrome, or a combination of the two?) will continue to be a subject of conjecture.

Through December 1967 the cumulative production from the Golden Lane fields was approximately 1.25 billion bbl of oil.

History

The history of petroleum development in Mexico began in 1869 when a well, 28 m deep, was drilled near the oil seeps of Cerro de Furbero, Veracruz State, and found minor production. Years later, between 1881 and 1883, the presence of oil seeps on the Cerro Viejo and Chapopote Núñez haciendas, north of the Río Tuxpan, revived interest in petroleum. Two shallow wells were drilled near these seeps, and one was a small oil producer. However, it was soon abandoned because the operation was uneconomical.

The first significant events in the development of a Mexican petroleum industry began at the start of this century when two investment groups, Doheny and Pearson, began to explore the country. The Doheny interests began exploration in the Ébano-Pánuco petroleum dis-

[1] Adapted from a paper given before the 53d Annual Meeting of the Association, Oklahoma City, Oklahoma, April 24, 1968. Manuscript received, January 1, 1969. Translation from Spanish text by A. A. Meyerhoff.

[2] Sub-Gerente de Geología, Gerencia de Exploración, Pemex.

[3] Geologist, Pemex.

trict west of Tampico, and the Pearson interests worked in the Isthmus of Tehuantepec area of southern Mexico. The Doheny group discovered oil in 1904 at Ébano-Pánuco on completion of the La Pez No. 1 well. The initial production was 1,500 bbl/day. This well generally is considered to be the first true discovery well of Mexico.

In the Isthmus of Tehuantepec region, the Pearson group had little commercial success, and moved its operations northward to Veracruz State—to the area which later was called the Golden Lane; Pearson's organization was attracted to the northern area because of Doheny's success. In the Golden Lane area, the Pearson group discovered oil in May 1908 with the San Diego de la Mar No. 1 well. Initial production was estimated at 2,500 bbl/day. This well proved conclusively that the Golden Lane was an important district which ultimately would bring to Mexico much fame as a petroleum-producing nation.

The Golden Lane in the State of Veracruz underlies Mexico's eastern coastal plain along the shore of the Gulf of Mexico (Fig. 1). The Golden Lane is a series of aligned fields which forms an arc concave toward the east. The length of this arc is approximately 180 km, and the average width of each field is approxi-

Table 1. Identification of Wells Shown in Figure 1

Cross-Section No. 1 (Fig. 6)	Cross-Section No. 3 (Fig. 7)
A Palo Blanco-110	O Zapatalillo-3
B Agua Nacida-101	P Plan de Ayala-1
C Jardín-105	Q Reparo-1
D Jardín-35	R Reparo-2
E Sta. Marta-1	S Buenaventura-1
F Tuxpan-3	T Mesita-1
G Esturión-1	U Cazones-2
Cross-Section No. 2 (Fig. 6)	**Cross-Section No. 4 (Longitudinal; Fig. 8)**
H Miquelta-125	V Dos Bocas
I Tincontlán-1	W Naranjos-24
J Horcón-102	X Canas-101
K Zapotal-10	
L Muro-2	
M Tumilco-2	
N Bagre-1A	

Note: locations on Fig. 8 are identified principally by field names. Wells V-X are shown on Fig. 1.

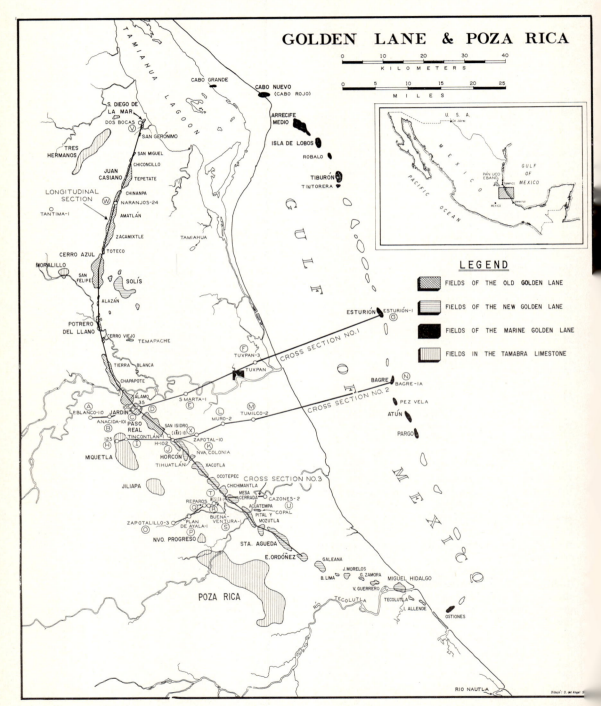

Fig. 1.—Index map, Golden Lane fields, eastern Mexico, showing lines of cross sections in Figures 6–8. Individual wells used in cross sections (A-V) are listed in Table 1.

mately 2,000 m. The range in the field widths is 300 to 2,700 m.

The region had attracted the attention of petroleum interests from the very beginning because of the presence of numerous oil seeps; this interest increased when production was discovered in the Ébano-Pánuco district (Fig. 1, inset map). In fact, oil seeps in the Ébano-Pánuco area were the basis for locating Doheny's La Pez No. 1 well. Because of the presence of the many seeps, open competition developed between the groups which operated in the area. Each group tried to acquire the largest possible concessions on lands that were associated with seeps.

The San Diego de la Mar No. 1 well (Fig. 1), considered to be the Golden Lane discovery well, also was drilled on a site selected because of oil seeps. During the early exploratory days, it was customary to make maps showing the locations of all oil seeps, together with various strike and dip measurements from the exposed strata; no attention was given to the differences among rock formations exposed at the surface.

On July 4, 1908, after completion of the San Diego de la Mar No. 3 well (better known as the Dos Bocas well—**V** on Fig. 1), the well went "wild" and caught fire. Estimates made while the well was blowing wild suggest that more than 10 million bbl of oil escaped during the 56 days that the well burned. The fire died when the oil pool was invaded by salt water. Tales about this well spread widely and contributed to the legend of the giant oil potential of Mexico.

Activity continued within the same area, and in July 1910 the Juan Casiano No. 6 well (Fig. 1), also drilled near oil seeps, was completed with an initial daily production of 15,000 bbl of oil. Two months later, the Juan Casiano No. 7 was brought in for 35,000 bbl/day. The latter well alone has been credited with a total production of 70 million bbl.

In December 1910 the Potrero del Llano No. 4 well (Fig. 1) was brought in, but went out of control, blowing more than 100,000 bbl of oil a day. It took 3 months to bring the well under control, and earth dams were built in the vicinity to recover several million bbl of oil which flowed from it while it was uncontrolled. The well is credited with having produced more than 95 million bbl—believed to be the world's record cumulative production from a single well.

After the early epoch of exploration, a pe-

riod of geologic exploration began which differed from company to company according to individual ideas and interest. Nevertheless, and despite the fact that geologic field work showed in many areas the possible presence of surface structures, the basis for all drilling in the district that led to new-field discoveries was the numerous oil seeps which extended in a narrow belt overlying the present fields. The fields were named after the haciendas in the regions, whether or not the haciendas coincided with the local highs on the Cretaceous reef which forms the Golden Lane.

In March 1916, Cerro Azul No. 4 (Figs. 1, 5), located by Ezequiel Ordóñez, was completed with an estimated initial daily production of 260,000 bbl. The well "ran wild" for a period of time, and the daily production figure is probably the largest recorded for any well in the world. The total production of this well, 87 million bbl, is exceeded only by the 95 million bbl credited to Potrero del Llano No. 4. The enormous production from Cerro Azul No. 4 and the results from Juan Casiano No. 7 and San Diego de la Mar No. 3 (Dos Bocas) focused the attention of the entire world on the Mexican petroleum industry.

Within this productive belt, now known as the "Old Golden Lane" (Fig. 1), more than 1,000 wells have been drilled of which more than half were producers. Depths ranged from 500 m in Cerro Azul to 800 m in San Isidro. The oil fields of the district, the discovery wells of which are shown on Table 2, had produced through December 1967 approximately 1.25 billion bbl of oil. Large reservoir permeability values permitted rapid exploitation of the fields. As a result, production increased steadily to a peak in 1921. However, the same factor—the great reservoir permeability—led to an extraordinarily rapid production decline. In 1919, salt water began to invade the principal fields.

In 1920 exploration by geophysical methods began. Gravimetric studies of the Golden Lane were made first with the torsion balance. None of these studies led to the discovery of a new field, but the continued use of gravimetric methods did lead to the discovery, south of the Golden Lane, of Poza Rica field, the largest petroleum field in Mexico to date.

On the basis of geologic reasoning, the extension of the Cretaceous reef toward the south already had been inferred. Moreover, regional geologic studies had become customary by 1923, and fossil Foraminifera were used for the first time during this period to make sys-

Table 2. Golden Lane Field Discoveries, 1949–1968

Field Name and No. of Discovery Well[1]	Year Disc.	Top of El Abra (Producing Fm.) (m below S. L.)	°API	Initial Prod.[1a] (m³/day)
Acuatempa No. 3	1955	1,171.0	23.0	872.0
Alazán No. 101	1949	582.0	23.0	100.0
Allende No. 1-A	1955	2,365.0	28.0	320.0
Arrecife Medio No. 1[2]	1963	2,045.0	24.0	115.5
Atún No. 3[2]	1967	2,833.0	38.0	238.0
Bagre No. 1-A[2]	1966	2,693.0	38.5	110.0
Boca de Lima No. 1	1956	1,681.0	14.5	68.0
Cabo Nuevo No. 1101	1966	1,655.0	20.0	35.0
Cabo Rojo No. 1010	1953	1,198.2	13.0	50.0
Copal No. 1	1957	1,313.0	17.5	50.0
Chichimantla No. 102	1955	1,071.0	21.5	77.0
Esturión No. 1[2]	1965	2,597.0	34.0	99.0
Ezequiel Ordóñez No. 1	1952	1,399.8	—	362,000.0 (gas)
Gutiérrez Zamora No. 2	1956	1,927.0	20.0	62.0
Horcón No. 103	1952	849.4	22.5	70.0
Isla de Lobos No. 1-B[2]	1963	2,087.0	40.0	179.0
Las Cañas No. 2	1956	828.0	21.0	42.0
Mesa Cerrada No. 101	1956	1,084.0	19.5	52.0
Miguel Hidalgo No. 1	1954	2,333.0	30.5	477.0
Mozutla No. 1	1953	1,189.0	21.5	755.0
Nueva Colonia No. 107	1956	859.4	22.0	13.0
Ocotepec No. 101	1953	973.0	22.5	800.0
Pez Vela No. 1[2]	1968	2,745.0	—	not measured (gas)
Robalo No. 1-A[2]	1964	2,183.0	34.0	29.0
Santa Agueda No. 1	1953	1,361.0	16.0	1,100.0
Tamiahua No. 102	1951	1,045.3	19.5	35.0
Tecolutla No. 2	1956	2,303.0	22.5	72.0
Temapache No. 119	1955	751.5	—	7,416.0 (gas)
Tiburón No. 2-A[2]	1965	2,228.0	29.5	68.0
Tihuatlán No. 102	1957	848.0	22.0	15.0
Tintorera No. 1[2]	1968	2,227.0	19.5	9.0
Vicente Guerrero No. 1	1955	2,007.0	24.5	260.0
Xacotla No. 101	1953	901.0	20.5	750.0

[1] Total fields =33.
[1a] × 6.3 = bbl oil; × 35.3 = cu ft gas.
[2] Discoveries in the Marine Golden Lane.

tematic stratigraphic studies. The southern extension of the Golden Lane was investigated by refraction-seismograph methods beginning in 1930, but the structural axis located by seismic methods was north of the actual position of the reef trend. The seismic studies demonstrated that the reef top gradually deepened toward the southeast. Despite this knowledge and continued exploration there were no new-field discoveries because of local geologic conditions, lack of sufficiently deep drilling in places, and the absence of structural closure in other locations. As a result, it was believed for a long time that the southern reef extension, because of the deepening toward the south, probably was invaded by salt water.

Beginning in 1948, seismograph exploration was renewed, but only the reflection method was utilized. However, techniques had improved—including instrumentation and shothole methods. Practically the whole area south

of the "Old Golden Lane" to the Nautla River was explored seismically. The results of this exploration led to the discovery of numerous fields. In 1952, Ezequiel Ordóñez No. 1 (Fig. 1) was drilled on a structure located by the reflection seismograph. Production was found in the middle Cretaceous El Abra Limestone—the same producing formation as that of the "Old Golden Lane." With this discovery, the southern extension at last had been found.

The discovery of Ezequiel Ordóñez field was followed by a period of exploratory drilling and development on other seismologically delineated structures which comprise the southern extension of the Golden Lane reef trend. The fields are referred to as the "New Golden Lane," or the "southern extension of the Golden Lane." The discovery wells of the fields which were found along this trend, mainly from 1949 through 1968, are shown on Table 2 and Figure 1.

Marine seismic studies in the offshore region of the Golden Lane, in the Gulf of Mexico, began in 1957. They showed the presence of a platform (Continental shelf) on the east (Fig. 2); beneath this platform is a giant Cretaceous atoll. Thus it was believed that the line of fields comprising the Old and the New Golden Lanes might extend in a loop offshore. The magnetometer was utilized during the offshore exploration and, though the techniques of processing magnetometric data are not perfected, it was possible to interpret, on the basis of previous experience, that there is an alignment of structural highs in the offshore area. The magnetic data showed further that these highs actually form a prolongation of the Golden Lane trend, from the New Golden Lane area on the south in an arc convex toward the east that joins with the Old Golden Lane on the north. The first drilling of the offshore reef trend began in 1963 (Isla de Lobos No. 1-B, Table 2, Fig. 1) and soon confirmed that middle Cretaceous reefs are present in the northern part of what now is called the "Marine Golden Lane."

The first productive structures in the Marine Golden Lane correspond to reef highs. Isla de Lobos No. 1-B, the first well, was drilled from a barge in 1963. It was completed as a potential oil-producing well in the El Abra Limestone. Initial production was estimated at 1,100 bbl/day. The top of the El Abra was reached at a depth of 2,087 m below sea level. The next discovery, Arrecife Medio No. 1 (Fig. 1), also was completed in 1963 as a potential oil producer; initial production was esti-

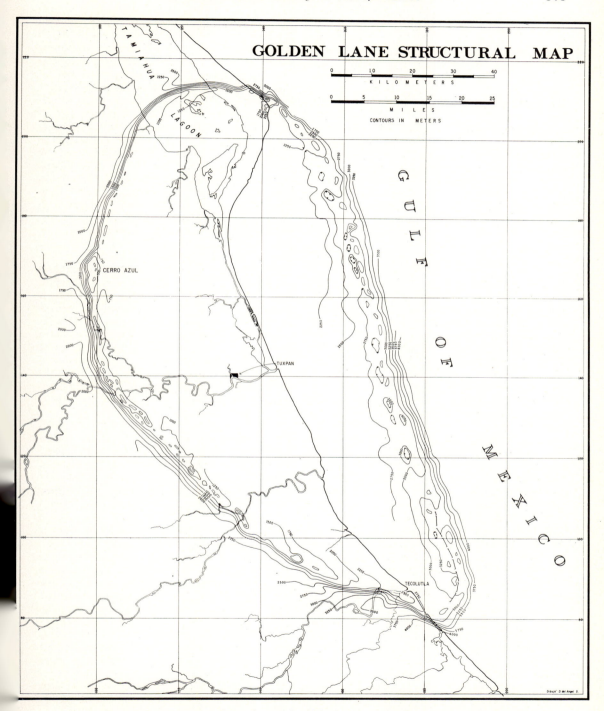

FIG. 2.—Structural contour map; datum is top of middle Cretaceous El Abra Limestone. CI = 250 m.

mated at 750 bbl/day. In this well the top of the El Abra was reached at 2,045 m below sea level. At the Isla de Lobos structure, several wells have been drilled from a drilling platform, with the result that the first Marine Golden Lane field has been developed. The estimated reserves of the field are about 10 million bbl.

Other highs have been drilled which confirm the continuation of the Marine Golden Lane reef toward the south. Among these wells which have been completed as potential producers are the Esturión No. 1, Tiburón No. 2-A, Bagre No. 1-A, and, the most recent of all, Atún No. 3, with initial production estimated at 1,500 bbl/day (*see* Fig. 5).

GEOLOGY

Background

During the 1930s, the fields discovered south of Tampico were known by the name "southern fields." It was not until later that they became known as the "Golden Lane," "Dos Bocas-Álamo structure," or simply, "The Ridge." Most geologists believed that the producing formation consisted of strata of reef origin, folded like those of the Sierra del Abra (in the Sierra Madre Oriental orogenic system west of the Golden Lane; west of the map area of Fig. 1). It was believed widely that the Golden Lane was one of a series of asymmetric anticlines with western flanks dipping more steeply than the eastern flanks (Muir, 1936). The age assigned to the uppermost of the El Abra facies (reef limestone) in parts of "The Ridge" was early Cenomanian (Buda). This age assignment was based on the presence of *Pecten roemeri* Hill and it was suspected that Late Cretaceous (Coniacian) strata in other parts of The Ridge directly overlay reef facies of Albian age. This relation is observed in places in the Sierra del Abra, west of the Golden Lane.

Stratigraphy
(Fig. 3)

Reef facies.—The reef facies which forms the principal mass of the Golden Lane has been correlated with the biogenic limestone of the Sierra del Abra, and the assignment of the Golden Lane reef facies to the El Abra Limestone has persisted to the present. The term includes all of the reef-carbonate rocks of Albian and Cenomanian ages which are present, not just in the Sierra del Abra, but also on the south in the Sierra Madre Oriental and on the

east in the Golden Lane. Muir (1936) divided the El Abra into two facies (Fig. 4)—the El Abra facies proper (miliolid-bearing limestone) of Albian and Cenomanian ages, equivalent to the Georgetown through lower Edwards of Texas; and Taninul facies (rudistid-bearing limestone) which is partly at least of Albian age (equivalent to the Buda Limestone [Cenomanian] and probably to part of the Georgetown Formation [Albian] of Texas). The Taninul facies is limestone characterized by massive bedding, cream-white to gray colors, subporcelaneous texture, and abundant fossils. The facies is composed partly of shell fragments, cemented in a carbonate matrix with numerous fossils such as *"Toucasia* cf. *T. texana* (Roemer), *Pecten* sp., *Caprinula* sp., gastropods, pelecypods, corals, and many other forms.

The El Abra facies is more massive and does not show stratification. The limestone is cryptocrystalline, cream to light gray, and contains lenses or "pockets" of rudists which are cemented in a compact carbonate matrix. This facies contains numerous miliolids. In the Sierra del Abra, as in the Golden Lane, bentonite laminae commonly are found near the top of the formation; also, near the top of the El Abra are fractures and vugs filled with bentonite. The bentonite ranges in color from gray blue to emerald green.

The stratigraphic range of the El Abra facies is not settled. In some wells on the Golden Lane, the San Felipe Formation (Coniacian) has been penetrated and overlies the El Abra directly; Turonian strata are absent. The absence of Turonian could be explained in three ways: (1) a hiatus is present; (2) the uppermost part of the reef is actually of Turonian age; and (3) the Turonian strata are absent because of erosion. Paleontologically, there is no evidence to prove or disprove any of the three possible explanations. However, the presence of bentonite—which characterizes the base of the Turonian Agua Nueva Formation in the entire Tampico-Tuxpan basin and which is present in the central part of the Golden Lane atoll—leads one to suspect that Turonian time may be represented in the reef facies, in the upper part of the El Abra, in the areas where the upper part has not been eroded.

The Jardín No. 35 well (Figs. 1, 7) was drilled in 1930 by Penn Mex[4] south of Álamo

---

[4] Penn Mex: a company, now defunct, which operated formerly in Mexico.

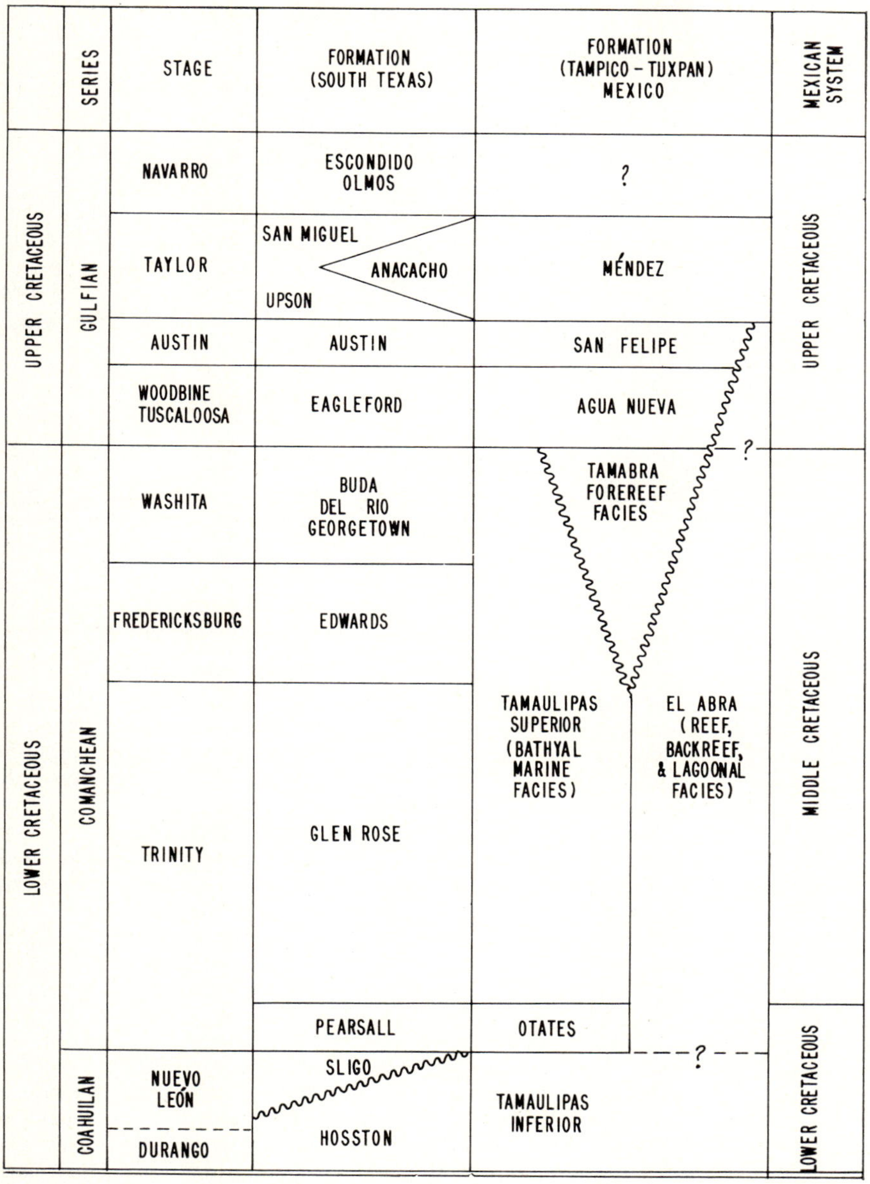

SERIES	STAGE	FORMATION (SOUTH TEXAS)	FORMATION (TAMPICO – TUXPAN) MEXICO	MEXICAN SYSTEM
UPPER CRETACEOUS / GULFIAN	NAVARRO	ESCONDIDO OLMOS	?	UPPER CRETACEOUS
	TAYLOR	SAN MIGUEL / ANACACHO / UPSON	MÉNDEZ	
	AUSTIN	AUSTIN	SAN FELIPE	
	WOODBINE TUSCALOOSA	EAGLEFORD	AGUA NUEVA	
LOWER CRETACEOUS / COMANCHEAN	WASHITA	BUDA DEL RIO GEORGETOWN	TAMABRA FOREREEF FACIES ?	MIDDLE CRETACEOUS
	FREDERICKSBURG	EDWARDS		
	TRINITY	GLEN ROSE	TAMAULIPAS SUPERIOR (BATHYAL MARINE FACIES) EL ABRA (REEF, BACKREEF, & LAGOONAL FACIES)	
		PEARSALL	OTATES ?	LOWER CRETACEOUS
COAHUILAN	NUEVO LEÓN	SLIGO	TAMAULIPAS INFERIOR	
	DURANGO	HOSSTON		

FIG. 3.—Stratigraphic correlation chart, eastern Mexico and South Texas (from Boyd, 1963, Fig. 3).

and west of Tuxpan, and apparently is the only onshore Golden Lane well which may have penetrated the reef core. The well was drilled to a total depth of 10,585 ft (3,493 m). The lithologic log, which is in the Petróleos Mexicanos files, indicates that the lowest reef-type limestone is at 7,377 ft (2,263 m). Below 7,377 ft to total depth, there is no additional information except that the well penetrated some thin-bedded dark-gray limestone. If one assumes that the depth of 7,377 ft (2,263 m) is the base of the El Abra reef facies, then the total thickness of the reef is 5,108 ft (1,467 m). This interpretation is supported by data from wells drilled in the backreef or lagoonal area, where the base of the lagoonal facies

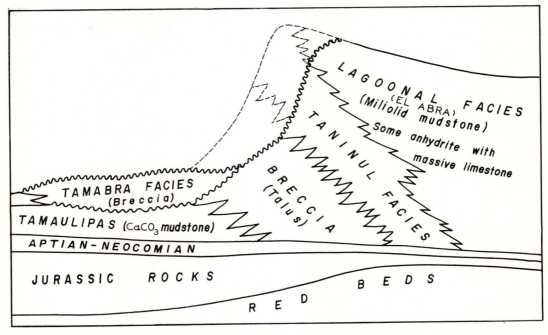

Fig. 4.—Diagrammatic west-east cross section of Golden Lane reef, showing interpreted facies relations.

corresponds to the estimated thickness of the reef facies in the Jardín No. 35 well (Figs. 1, 6).

The Old Golden Lane, its southern continuation to the Tecolutla River, and its marine extension discovered during the 1950s and 1960s form a massive elliptical reef platform dipping markedly southeast. The entire platform is circumscribed by a belt of structural highs of varied elevations (Fig. 2) which contain rocks of diverse lithologic character, according to the descriptions of investigators. The strata have been described by some as compact limestone containing abundant miliolids, bentonite, calcarenite, and related rock types; in other places conglomerate and breccia have been observed at the crest of the highs; elsewhere the rocks have been described as very fossiliferous, permeable limestone containing pelecypod, rudistid, and benthonic microfaunal remains.

The phenomena of erosion and solution are found almost everywhere in the upper part of the reef. For example, in the Mozutal No. 1 well, a cavern found in the limestone was filled with marine Oligocene sediments. This karst topography which apparently was developed in the upper part of the reef confirms the statement of Ezequiel Ordóñez, discoverer of the

Cerro Azul No. 4 gusher, that as the well came in and started to blow, fragments of limestone and great pieces of stalactites were thrown into the air.

The erosion which affected this enormous atoll was most intense in the Old Golden Lane which forms the northern and northwestern part of the atoll. There, upper Oligocene beds directly overlie the reef, and the entire section from the top of the Eocene to the base of the Late Cretaceous is absent. Early Tertiary strata probably were not deposited in this region. However, Late Cretaceous beds probably were present but were removed by erosion, or possibly they are unrecognized because of facies change. In the central or lagoonal part of the atoll, the Agua Nueva (Turonian), San Felipe (Coniacian), and Méndez (Maestrichtian)—all three of deep-water or pelagic facies—are found in places.

Nevertheless, there is evidence from wells drilled on top of the highs or along the flanks of their outer periphery that the structural highs are only erosion remnants, or a topographic form produced by erosion. In most places where wells have penetrated the limestone highs, the lithologic character of the limestone corresponds to the typical description of the El Abra facies of Muir (1936); in modern reef terminology, this is a backreef facies with abundant miliolids and a compact porcelaneous matrix, and is interbedded with calcarenite containing the remains of megafossils. In the backreef area, strata of the Taninul facies are found in very few places; it is the facies which contains a biocoenosis of rudists, caprinids, and corals. The absence or near absence of the Taninul facies in the structural highs leads one to believe that possibly subaerial erosion during intermittent periods of time between late Albian and Cenomanian, and possibly marine and subaerial erosion in the Tertiary, may have destroyed and partly eliminated the original forereef section which fringes the outer periphery of this great atoll (Figs. 3–5). Some workers have proposed normal faults to explain the presence of limestone bodies with rudist, pelecypod, caprinid, and echinoid remains in the Tamabra facies which is more than 1,000 m down the forereef flank—at the foot of the reef and around it (Figs. 6–7).

Tamabra facies.—After Poza Rica and Moralillo fields were discovered in 1932 and 1948, respectively (Fig. 1), on the west flank of the Golden Lane (the two fields are 80 km apart), a new exploration trend was revealed—the carbonate-rock trend which is called the Tamabra facies (Figs. 6–7).

New discoveries of oil fields in the Tamabra[5] facies demonstrated the presence of a broad belt of permeable (primary permeability) carbonate rocks at the foot of the forereef side of the atoll. This facies is characterized by caprinid, rudistid, pelecypod, and coral fragments in a microcrystalline matrix (micrite) with a few intercalations of limestone containing globotruncanids of Cenomanian age (determination by F. Longoria). Thus this facies contrasts sharply with the rocks of the atoll which are biosparites containing very few fragments of megafossils and a few beds with abundant miliolids.

The Tamabra and El Abra have been analyzed in core samples and thin sections. Petrographic analyses of this facies in the Tantima No. 1 and Naranjos No. 24 (Fig. 1, W) wells are shown in Table 3.

The origin of the Tamabra facies has been the subject of controversy. Some geologists consider the Tamabra to be an *in situ* reef extension of the Golden Lane (along its west

flank) which, because of deepening water, could not continue its growth together with the reefs of the atoll on the east. Other geologists, including the writer, consider the Tamabra to be the debris from the atoll on the east—debris deposited at the base of the reef as a result of erosion of the reefs (Fig. 4). As heretofore indicated, other geologists have suggested that normal faults are present along the west flank of the Golden Lane. These faults are postulated to explain, among other things, the steep escarpment of the west flank and the difference in elevation (more than 1,400 m) between the reef summit on the east and the Tamabra facies top on the west.

Little or nothing is known about the genesis of the Golden Lane reef, partly because information is sparse or incomplete, but mainly because no detailed analytic studies have been made to summarize the information from all of the wells, most of which have penetrated only a few meters into the top of the reef structures.

Bonet (1952) described the Sierra del Abra reef (west of the Golden Lane and cropping out at the surface) as a tabular reef. He noted certain characteristics, especially the rudist occurrences. The rudists appear to form lenses or pockets which are distributed erratically within the biosparite limestone beds.

No one had any idea of the magnitude of the Golden Lane reef body until the recent drilling of two wells (similar to Jardín No. 35 —Fig. 1, **D**; Fig. 6) which penetrated the entire thickness of the reef. The two wells are the Mesita No. 1 (Fig. 1, **T**; Fig. 7) and Cañas No. 101 (Fig. 1, **X**). The cores recovered and studied during the drilling of both wells showed that the upper part of the section is equivalent to the El Abra facies—the upper 318 m in the Mesita No. 1 well and the upper 185 m in the Cañas No. 101 well. The Taninul facies in the Mesita No. 1 well is 444 m thick and in the Cañas No. 101 well is 989 m thick. These thickness figures should be regarded with caution because the samples have not been analyzed carefully in the laboratory.

If the Taninul facies is present within the belt of the Golden Lane, then Muir's (1936) observations in the Sierra del Abra fit well with those found in the wells along the western side of the Golden Lane atoll. Determination of the age of the two facies must await detailed laboratory analysis.

The wells which have penetrated the reef limestone show that the reef directly overlies a body of dolomitic limestone (Figs. 6, 7), be-

[5] The word "Tamabra" is combined from the two words "Tamaulipas" and "El Abra," and is designed to show the mixed nature of the Tamabra facies.

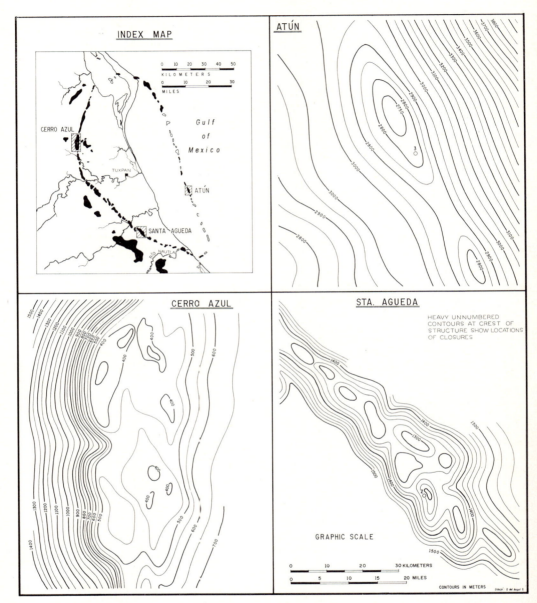

Fig. 5.—Structural contour maps and index map of Cerro Azul (Old Golden Lane), Santa Agueda (New Golden Lane), and Atún (Marine Golden Lane) fields, showing structural types. Structural datum is top of El Abra Limestone. Contours in meters. Heavy, unnumbered contours at crest of structures show locations of closures (Cerro Azul and Santa Agueda). Outlines (not filled in black) on Index Map are undrilled seismic structures.

low which are some cryptocrystalline limestone beds, with intercalations of bentonite and, at the base, calcarenite. The dolomitic limestone body is believed to correlate with the lower part of the Lower Cretaceous. The Upper Ju-rassic is represented by a thin unit of shaly limestone of Tithonian age and by limestone and shale of the Tamán Group (Kimeridgian). Beneath the Tamán, which is marine, are red-beds of Jurassic or Triassic age—the exact age

Table 3. Petrographic Descriptions of Tamabra Facies in Tantima No. 1 and Naranjos No. 24 Wells[1]
(Locations on Fig. 1.)

Depth (m)	Chevron Research Co.[2] Sample No.	Description
TANTIMA NO. 1 (elev. 195 m)		
2,095		**Top Tamabra**
2,132–2,270	8421	Pinkish-gray porous ls., mottled by irregular patches of hydrocarbons; rexld., poorly sorted shell debris in microcrystalline ls. matrix; caprinids, pelecypods, echinoderms, gastropods
2,270–2,272	8422	Very light-brownish-gray dense ls.; carbonate grains, few oölites, and thin shell fragments in microcrystalline ls. matrix; a few textularids, miliolids, echinoid spines, ostracods, pelecypods, *Dictyoconus* sp., *Globotruncana*(?) sp.
2,272–2,273	8423	Very pale-brown dense ls.; few carbonate grains in microcrystalline ls. matrix
NARANJOS NO. 24 (elev. 86m)		
598		**Top El Abra**
1,037	8340	Very pale-yellowish-brown dense ls.; few foraminifers, and some shell debris in microcrystalline ls. matrix; irregular rexld. areas may be *Girvanella* sp. and miliolids
1,038	8352	Very pale-yellowish-brown dense ls.; few foraminifers and shell fragments in microcrystalline ls. matrix; abundant disseminated dolo. rhombs; miliolids, gastropods
1,045	8341	Pale-yellowish-brown dense vuggy ls.; poorly sorted shell debris in microcrystalline ls. matrix; gastropods, miliolids, textularids, pelecypods, *Dictyoconus* sp., *Acicularia* sp., ostracods, echinoderms
1,047	8342	Very pale-yellowish-brown dense ls.; few gastropods and miliolids in microcrystalline ls. matrix
1,048	8343	Pale-yellowish-brown dense ls.; poorly sorted shell debris in microcrystalline ls. matrix; *Dictyoconus* sp., *Solenopora* sp., pelecypod; corals?, encrusting algae?
1,056	8344	Pale-yellowish-brown dense fossilif. ls.; poorly sorted shell debris and carbonate rock fragments in rexld. microcrystalline ls. matrix; *Dictyoconus* sp., *Solenopora?* sp., miliolids, gastropods, pelecypods, echinoderms
1,073	8345	Very pale-orange, dense, slightly vuggy ls.; poorly sorted carbonate rock and shell fragments in partly rexld, microcrystalline ls. matrix; *Dictyoconus* sp., *Solenopora?* sp., *Osagia?* sp., *Girvanella?* sp., miliolids, pelecypods, textularids, echinoderms
1,084	8346	Pale-yellowish-brown dense fossiliferous ls.; poorly sorted shell debris in partly rexld. microcrystalline ls. matrix; caprinids, pelecypods, corals?, miliolids, encrusting algae?
1,196	8347	Medium-gray to white-mottled, porous ls. and brownish-gray porous ls., several large vugs; poorly sorted shell and carbonate rocks debris in partly rexld. microcrystalline ls. matrix; approximately ¼ of thin section consists of colonial corals, echinodems, pelecypods
1,245	8348	Yellowish-gray, dense, vuggy ls. (vugs contain hydrocarbons); moderately sorted, rounded, carbonate grains in partly rexld. microcrystalline ls. matrix; colonial corals, encrusting algae?, pelecypods
1,259	8349	Pinkish-gray, porous, granular dolo.; pores contain hydrocarbons
1,270	8350	Pinkish-gray, dense, slightly vuggy ls. and light-olive-gray ls.; poorly sorted shell and carbonate rock debris in partly rexld. microcrystalline ls. matrix; pelecypods, corals?
1,313	8351	Pinkish-gray porous, granular dolo.; interstitial hydrocarbons

[1] Descriptions are given through the courtesy of Chevron Research Corp., La Habra, California.
[2] Formerly California Research Co.

has not been determined. Where the redbeds are absent, the marine Tithonian overlies plutonic igneous and metamorphic rocks which may be an island remnant or part of a large land area which emerged from the Gulf of Mexico and on which the Golden Lane reef platform grew during Early Cretaceous (Neocomian) time; the reef atoll developed subsequently on this platform. This land area is termed the "Isla de Tuxpan."

To date there has been no complete geologic analysis of the genesis, tectonic evolution, and geomorphology of the Golden Lane atoll. Studies are incomplete, and most are unpublished.

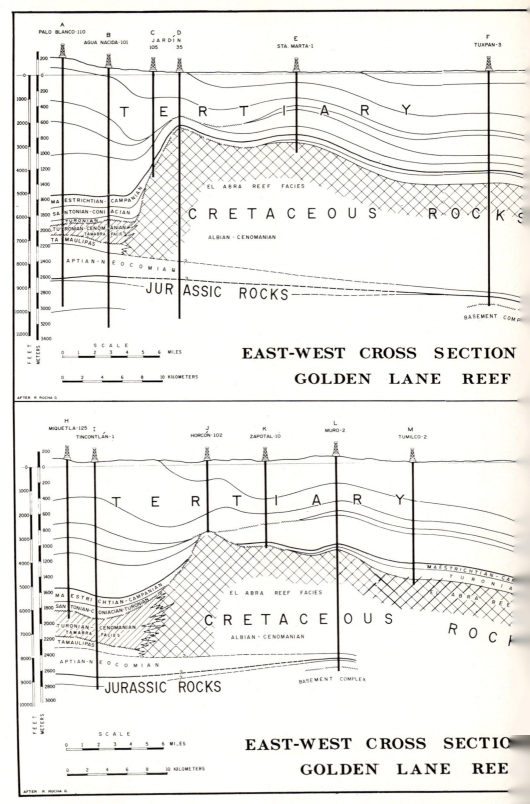

Fig. 6.—West-east cross sections. No. 1, Palo Blanco well 110 to Esturión well 1; No. 2, Miquetla well to Bargre well 1-A. Trace of sections shown on Figure 1.

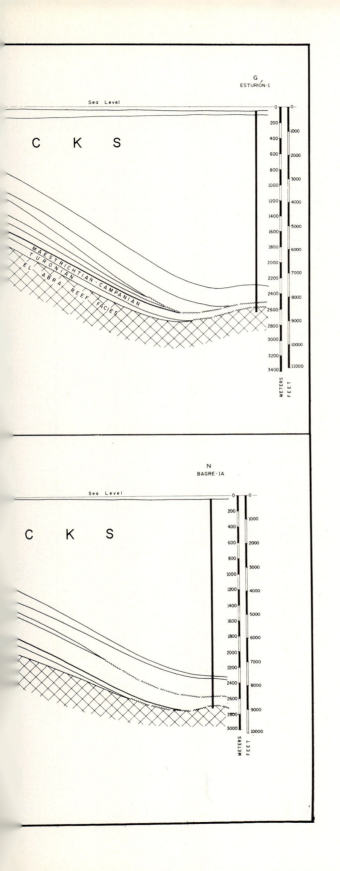

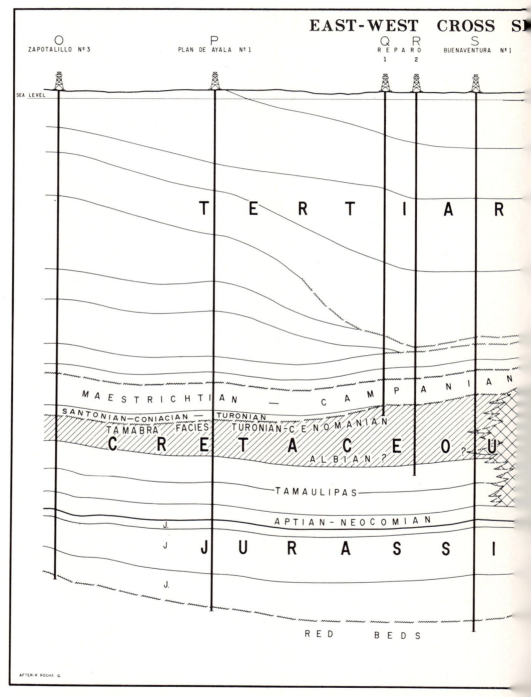

Fig. 7.—West-east cross-section No. 3, Zapatalillo well 1 to Cazones well 2. Trace of section shown on Figu

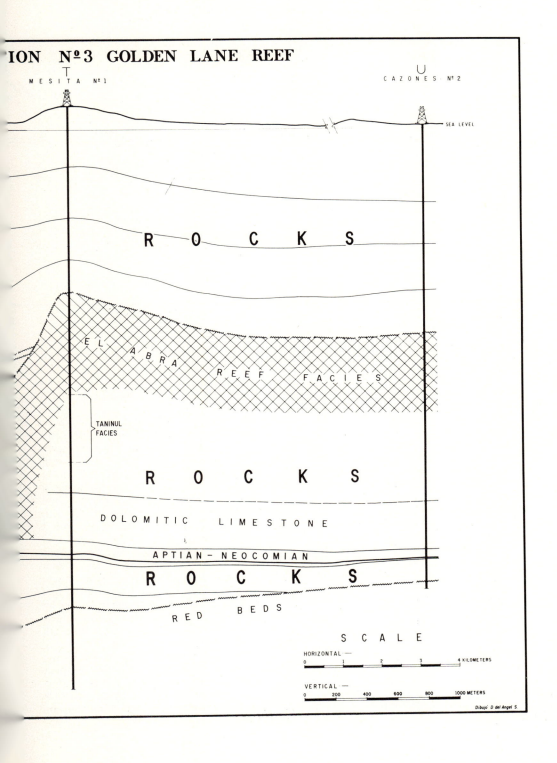

ION Nº3 GOLDEN LANE REEF

MESITA Nº1

CAZONES Nº2

SEA LEVEL

R O C K S

EL ABRA REEF FACIES

TANINUL FACIES

R O C K S

D O L O M I T I C   L I M E S T O N E

A P T I A N − N E O C O M I A N

R O C K S

R E D   B E D S

S C A L E

HORIZONTAL —

0   1   2   3   4 KILOMETERS

VERTICAL —

0   200   400   600   800   1000 METERS

Dibujó D del Angel S

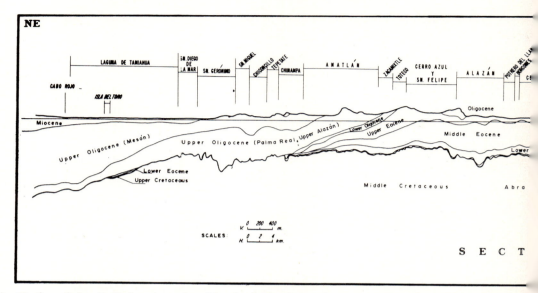

FIG. 8.—North-south longitudinal cross section (No. 4), Golden Lane of Mexico. Note different stratigraphic units which overlie El Abra Limestone. Trace of section shown on Figure 1.

Though relatively incomplete, the information available leads one to conclude that the Golden Lane atoll, after Cenomanian or perhaps Turonian time, underwent vertical movements which involved temporary emergence from the sea and, at other times, partial to total subsidence below sea level. This type of vertical movement seems to have continued from the end of Coniacian time until the end of late Eocene time. These movements of the reef and reef platform probably are related to the epeirogenic movements which affected Tertiary sedimentation along the Gulf Coast.

The intense fracturing which characterizes the atoll limestone bodies may be related to normal faulting, of large or small scale. The fractures may have been caused by tilting and other adjustments by the limestone during the repeated vertical movements that affected the atoll from its inception until the end of Tertiary time. By the end of Tertiary time, the atoll had been buried completely, with a dip toward the southeast.

The southeastward tilt is in the order of 1°3′ from the northwest part of the atoll at San Diego de la Mar to Miguel Hidalgo near the coast on the southeast. The principal tilting occurred during middle Tertiary time and was completed by the beginning of the Miocene.

Accompanying the tilting is a gradual thickening of the section overlying the reef. The post-reef thickness increases from 500 m on the north (Cerro Azul field) to 2,250–2,500 m on the south (Figs. 1, 8).

The migration of the oil in the reef appears to be related closely to the tilting of the atoll. The oil is heavier in the north (specific gravity 0.920 at Cerro Azul) and lighter in the southeast (0.816 at Atún field). There is also an increase in the quantity of gas from northwest to southeast. In the San Diego de la Mar field, Oligocene sedimentary rocks overlie the El Abra directly. Southward the Tertiary section is more complete and parts of the Late Cretaceous section are preserved as erosion remants (Coniacian and Maestrichtian). These remnants of Late Cretaceous strata overlie directly the topographic or structural highs on the reef.

Many geologists have believed that the hydrocarbons of the reef migrated from southeast to northwest during the deposition of the Oligocene and much of the Miocene.

In order to understand better the true economic value of the Golden Lane, Pemex plans a general study of the atoll this year (1969). During the course of this study all work by previous investigators will be summarized and synthesized.

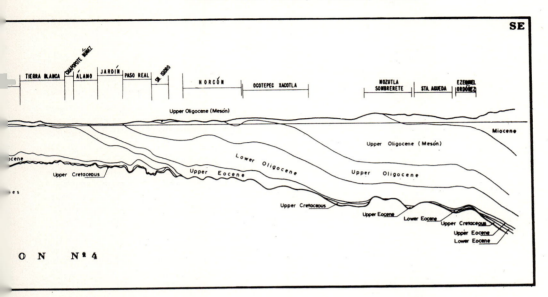

SE

TIERRA BLANCA CHAPOPOTE NÚÑEZ ÁLAMO JARDÍN PASO REAL SN ISIDRO HORCÓN OCOTEPEC XACOTLA MOZUTLA SOMBRERETE STA. AGUEDA EZEQUIEL ORDÓÑEZ

Upper Oligocene (Mesón)

Miocene

Upper Oligocene (Mesón)

Lower Oligocene

Upper Oligocene

Upper Eocene

ocene

Upper Cretaceous

Upper Eocene

Upper Cretaceous

Upper Eocene

Lower Eocene

Upper Cretaceous

Upper Eocene

Lower Eocene

O N N° 4

SELECTED REFERENCES

Acuña, G. A., 1957, El distrito petrolero de Poza Rica: Asoc. Mexicana Geólogos Petroleros Bol., v. 9, p. 505–553.

Barnetche, Alfonso, 1963, Lobos Island—Mexico's most challenging discovery: Petroleum Management, v. 35, no. 13, p. 67–73.

Basurto G., Jesús, 1955, Miguel Hidalgo field discovery on the southeastern extension of the Golden Lane, México: Pemex, unpub. rept.

Bonet, Federico, 1952, La facies urgoniana del Cretácico medio de la region de Tampico: Asoc. Mexicana Geólogos Petroleros Bol., v. 4, p. 152–162.

Boyd, D. R., 1963, Geology of the Golden Lane trend and related fields of the Tampico embayment; in Peregrina Canyon and Sierra de El Abra: Corpus Christi Geol. Soc. Ann. Field Trip Guidebook, p. 49–56.

Figueroa H., Santos, 1953, El progreso de las exploraciones geofísicas petroleras en la Republica Mexicana y el descubrimiento del Campo "Ezequiel Ordóñez": Asoc. Mexicana Geólogos Petroleros Bol., v. 5, p. 85–93.

Flores R., José V., 1955, Los arrecifes de la cuenca de Tampico-Tuxpan: Unpub. prof. thesis, ENI-Univ. Nac. Aut. México.

Flores V., Aurelio, 1955, Estudio estratigráfico y estructural del campo petrolero de Santa Agueda, Veracruz, México: Unpub. prof. thesis, ESIA-Inst. Nac. Petroleo, México.

Guzmán, E. J., 1955, Reef type stratigraphic traps in México, in Geología petrolera de México: Asoc. Mexicana Geólogos Petroleros Bol., v. 7, p. 137–172.

——— E. López Ramos, and R. Suárez, 1953, Petroleum geology of Mexico: 19th Internat. Geol. Cong., Algiers, sec. 14, fasc. 16, p. 35–63.

López Ramos, E., 1950, Geología de subsuelo de tres pozos de exploración al S-E de Poza Rica, Veracruz, México: Asoc. Mexicana Geólogos Petroleros Bol., v. 7, p. 381–395.

Mena R., Enrique, 1955, Estudio geológico-económico del Cretácico superior y medio al este de la Faja de Oro: Asoc. Mexicana Geólogos Petroleros Bol., v. 7, p. 327–366.

Muir, J. M., 1936, Geology of the Tampico region, Mexico: Am. Assoc. Petroleum Geologists, 280 p.

Sotomayor, C. A., 1954, Distribución y causas de la porosidad en las calizas del Cretácico medio en la región de Tampico-Poza Rica: Asoc. Mexicana Geólogos Petroleros Bol., v. 6, p. 157–206.

Giant Fields of Venezuela[1]

ANIBAL R. MARTÍNEZ[2]

Caracas, Venezuela

Abstract Only 44 of the nearly 300 oil fields discovered in Venezuela to the end of 1965 can be classified as giant, that is, having resources[3] of at least 100 million bbl of oil. The average time lag for the recognition of an oil field as a giant is 8 years and 2 months. Giant fields have been discovered in three of the sedimentary basins of the country. The estimated resources of all giant fields in Venezuela are $27,730 \times 10^9$ bbl. The average for the country is 630 million bbl per giant. Even if the resources of the Bolívar Coastal field were only 500×16^9 bbl, the average resources would be 267 million bbl of oil per giant field. If the data on the giant Venezuelan fields are combined with estimates of the oil resources of the country, estimated ultimate number of giants is 78. Ten giant fields had been discovered but not recognized by the end of 1965, leaving 24 giants undiscovered. It is estimated that 40 percent of the oil resources of Venezuela are in the giant fields.

INTRODUCTION

The first organized and successful effort to exploit the oil resources of Venezuela was by Don Manuel Antonio Pulido and his Petrolia company. Operations were carried on beginning in 1878 in La Alquitrana near San Cristóbal, near the Colombian border on the fringes of the Maracaibo sedimentary basin, in a region which today is far from the established petroleum areas of the country.

A wildcat drilled during 1913 near the Guanoco asphalt lake in eastern Venezuela produced a genetically related subsurface accumulation of very heavy, asphaltic-base crude. Significantly, the next discovery, on April 15, 1914, was Mene Grande, also in an area of seepages and other favorable surface indications of petroleum. The Mene Grande oil field was the first large petroleum deposit discovered in Venezuela, but more than 10 years elapsed before the fact was recognized.

After five decades of development, nearly 300 oil fields had been discovered in Venezu-

ela by the end of 1966. But only a few had resources large enough to classify them as "giant."

A "giant," as defined in the United States, is an oil field with ultimate production estimated to be 100 million bbl or more. There is no particular significance to the selected volume, which was introduced by *The Oil and Gas Journal*.[4] Hubbert (1962) found such a classification useful in an analysis of the giant oil fields in the U.S. and their relation to crude oil resources, and the same definition is used herein for the sake of uniformity.

All other oil fields with resources of less than 100 million bbl thus would be classified as small. Lahee (1958) and others have used a distribution of oil and gas fields according to size, in an analysis of the results of exploratory drilling in the United States during the period 1943–1957. Their Class A, the largest oil fields, includes those with ultimate recoverable reserves of more than 50 million bbl. Hence, all giants, as well as many small fields with resources of between 50 and 100 million bbl, would be included in Class A.

The giant oil fields of a country or sedimentary basin constitute a small fraction of the total number of oil accumulations. Furthermore, several years generally pass before a newly discovered oil field is recognized as a giant. This is the normal development process, because one successful and very promising completion alone cannot be taken as indicative of the size and capabilities of the field. However, although very scant data may be available, certain discoveries now commonly are considered to be of major importance even before a single step-out appraisal well has been drilled after the discovery.

The classification of an oil field as a giant, then, is a time process within the oil industry. Only detailed geologic, geophysical, and reservoir engineering studies should, and could, be the basis of the definition. It is not out of place to insist on a sound scientific analysis of the

[1] Read before the 53d Annual Meeting of the Association, Oklahoma City, Oklahoma, April 25, 1968. Manuscript received July 25, 1967; accepted November 28, 1967.

[2] Apartado 50514.

M. King Hubbert and D. C. Ion read the manuscript; I am grateful for their very valuable and competent criticisms.

[3] "Resources" means estimated volume of the cumulative production.

[4] *The Oil and Gas Journal* has published for the past 30 years a listing of the U. S. giant fields in the *Review-Forecast*, which appears each January.

available data as the all-important deciding factor.

For the purpose of this report, I have made a detailed study of the yearly field data as published in the "Memoirs" of the Ministry of Mines and Hydrocarbons, and of similar information reported by the Venezuelan Oil Scouting Agency in its "Yearly Drilling and Production Reports for All Venezuela by Operators and Fields" and the "Catalogue of Venezuelan Oilfields and Wells" (1878–1961, 1962, 1963, 1964, and 1965 supplements). The MMH "Anuario" (1952) and the "World Wide Oil" and "Review-Forecast" issues of *The Oil and Gas Journal* also were consulted. Nevertheless, some readers may disagree with a few of my estimates, or find discrepancies between their figures and mine. The differences would be due not to lack of reliability of my sources and analysis, but rather to the unending variation, for certain areas in Venezuela, in the definition of fields and producing areas, or to

differences of interpretation of nongeologic factors.

I cannot claim precision in this paper beyond that which is inherent in the assumptions on which it is based. I have not attempted to stretch the figures, or to obtain from facts more than they can render. The use of statistics and statistical methods should not hide the fact that the only guiding and limiting factors are the geologic realities and the natural physical framework.

OIL GEOLOGY OF VENEZUELA

There are six sedimentary basins in Venezuela—Maracaibo, Gulf of Venezuela, Falcón, Barinas, Cariaco, and Maturín (Fig. 1). Only a very general discussion of the oil geology of Venezuela is necessary for the purpose of this paper.

By far the most important petroliferous areas are in the Maracaibo basin, which has been described in detail by Miller *et al.* (1958). It at-

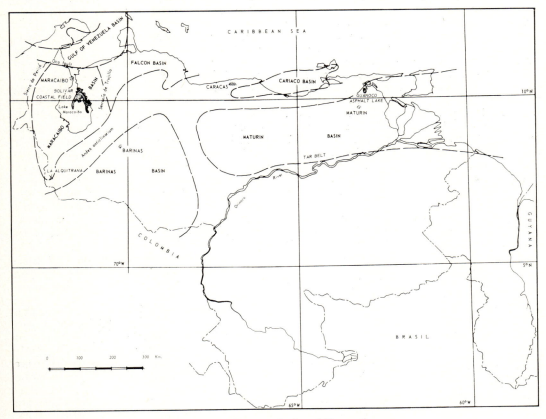

FIG. 1.—Sedimentary basins of Venezuela.

tained its present configuration at the end of Miocene time, after a short period of orogenesis probably related to the Cascadian revolution in North America (Rodanian orogeny of Europe). Generally, the structural axes within the basin are parallel with the nearest major tectonic element—the Sierra de Perijá, the Serranía de Trujillo, the Oca fault, or the Andes anticlinorium. The basin is markedly asymmetric in cross section, with the deepest zones on the southwest nearer the Andes. The Cretaceous sediments were deposited in a typically progressive transgression period, similar to the Urgonian cycle of the same age in Europe. They included the most likely source beds for the petroleum found in the basin. During Eocene time, conditions of sedimentation were not uniform. Later, the marine inundation from the east that deposited Miocene sediments on the peneplaned surface of broken and folded Eocene rocks was a very important event with respect to oil.

A shallow body of water, the Maracaibo lake, is contained within the basin. Along its eastern shore, and partially under water, is the Bolívar Coastal field (BCF), one of the largest petroleum deposits in the world. During early development, almost all production was from Miocene sandstone beds; Eocene traps were tapped in 1940, the most important oil reservoir being a well-indurated, medium-grained sandstone locally designated "B-6-X." In recent years reservoirs in deeper Eocene sandstone beds, the "C" sands, have been discovered. Some Paleocene and Cretaceous limestone tests were completed successfully in 1965 and 1966. The BCF has been, and will continue for many decades to be, the focus of development of the Venezuelan oil industry.

The exploration of the last concessions granted in Venezuela (during 1956 and 1957) led in a short time to the discovery of at least four giant fields in the geographic center of Maracaibo lake.

Another group of giant fields is west of the city of Maracaibo, including the La Paz field, third largest in the country. Production is from fractured Cretaceous limestone and basement igneous and metamorphic rocks.

Immense volumes of very heavy oil are found southwest of Maracaibo, in Boscán, Los Claros, and Urdaneta fields. Oil is in stratigraphic traps just above and below the Eocene unconformity, in the updip wedges formed by major converging faults. The Mene Grande oil field is in a large faulted anticline, south and

east of the BCF. Most production is from Miocene sandstone beds, but Eocene strata also are productive. On the south end of the basin, completely isolated from the other giants, is the Las Cruces field.

The Maturín basin is asymmetric in cross section, with a gentle southern rim against the Guayana shield and a steeply rising northern limb. As a result of isostatic adjustments, there was a pronounced parallel shift of the depositional axis southward (Hedberg, 1950). The strata are less marine on the west and southwest, as communication with the open sea was only through the eastern end. Within every transgression period there were many facies changes, and intervals of erosion caused sharp formational contacts. The abrupt lateral facies changes and structural developments make it difficult to determine the actual stratigraphic relations. More than 150 names for groups, formations, members, and facies have been proposed in the literature. The oil appears to have originated in the center of the subbasins and to have migrated later to the nearshore beds, where it is exploited today. A geologic description of the basin has been given by Renz et al. (1958).

There is a prominent cluster of giant as well as small fields in the geographic center of the basin. Only two giant fields—Las Mercedes on the west and Pedernales on the east—are at appreciable distances from the central cluster. It is a coincidence that the largest single accumulation in the basin—the Quiriquire oil field—is within the limits of the leases of one concessionaire. Entire fields generally are shared by several operators or companies.

Superimposed on the complex geology is a very confusing nomenclature developed by the many operators. The same name commonly is used to designate the whole and a part of an area, or an altogether different producing unit; for instance, the Jusepín giant oil field comprises the Jusepín, Muri, Travieso, Mulata, Santa Bárbara, Pirital, and Mata Grande operating areas.

The Barinas basin was in communication with the Maracaibo basin during Cretaceous time. The Andean orogeny at the beginning of the Tertiary severed the communication. Upper Oligocene and Miocene rocks in the basin are mostly of continental origin. Oil accumulation has been influenced greatly by the Cretaceous unconformity; pools have been found only in areas of favorable structural conditions, where there is also a thick sedimentary cover above

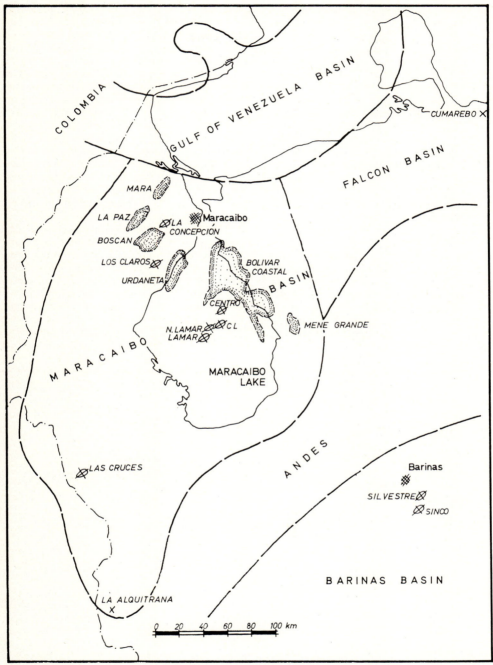

FIG. 2.—Giant fields, Western Venezuela.

the main producing sandstone reservoirs.

The stratigraphic relations and structural features of the Barinas basin have been investigated in detail only during the last two decades. The basin does not appear to be an important petroliferous region. Two giant fields, Silvestre and Sinco, have been discovered, but overall results of exploration are disappointing. Three new producing areas have been found near Silvestre by the Venezuelan state oil agency, the CVP.

The Falcón basin, though relatively large and extensively explored, has yielded only minor volumes of crude oil. Petroleum indications are not favorable in the Cariaco basin.

The Gulf of Venezuela basin is practically unexplored. About 10 wells have been drilled onshore along its margins. The state oil company (CVP) has published the first geologic paper on the basin and, near the end of 1967, invited any interested company to participate in a seismic survey of the offshore area, with a view to concluding service contracts for exploration and eventual development. It is a very promising new oil territory.

Giant Field Data

Until the end of 1965, a total of 44 giant oil fields had been discovered and recognized in Venezuela. Figures 2 and 3 are index maps showing the locations of the giant oil fields in western and eastern Venezuela, respectively.

Of the 44 fields, 13 are in the Maracaibo basin, 2 are in the Barinas basin, and 29 are in the Maturín basin. Half of the giants were

discovered in the 1914–1945 period (31 years), and the rest were discovered in the last 21 years. Four giant fields were discovered in both 1941 and in 1954, and three were found in both 1951 and 1958 (Table 1).

The cumulative curve of discoveries of giant fields in Venezuela is shown in Figure 4. A rather abrupt change in the slope of the curve occurred about 1937. Before that year, there had been an average discovery of a giant every 35 months—that is, one nearly every 3 years. After 1947 the tempo of discovery increased, and the average since then has been one discovery every 8 months. The marked decrease in exploration activities in Venezuela since 1958 has had no apparent effect on the curve of the discovery of giants.

The cumulative curve of recognition of the Venezuelan giant fields is shown in Figure 5, with the curve of discoveries. As could be expected, the two curves are roughly parallel. The cumulative curve of recognition steepens after 1944. During the initial stage, there was an average recognition of one giant every 43 months (compared with 35 months for a discovery during the corresponding period), whereas there has been one recognition every 7

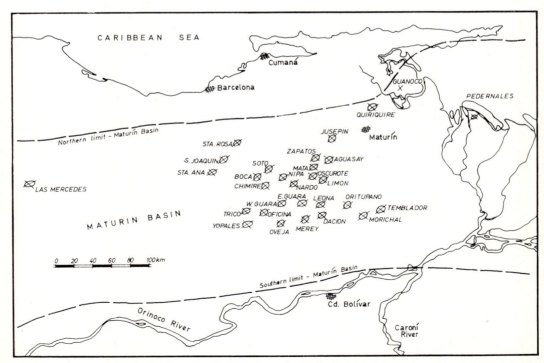

Fig. 3.—Giant fields, Maturín basin, Eastern Venezuela.

months during the later stage. There seems to be no particular event in the evolution of the Venezuelan petroleum industry that accounts for, or is related to, the turning point in the rates of discovery and recognition.

Table 2 shows the giant oil fields arranged alphabetically by basins, the dates of discovery and recognition, and the time lag in years. Coincidentally, the two giants which took the longest to be recognized are in the West Maracaibo group of fields and are similar in exploitation history. The La Concepción and La Paz fields were discovered in 1924 and 1922 and were recognized 26 and 23 years later, respectively, when their prolific Cretaceous deeper pools were tapped.[5]

[5] One might be tempted, on the basis of the immensity of the Cretaceous resources and the smallness of those of the Tertiary, to consider the time of discovery of the field to be that of the deeper and much larger reservoirs. To do so, however, would be wholly inconsistent with the clear and useful definitions of Lahee, widely accepted in the petroleum industry.

Table 1. Discoveries

Year	Giant Fields
1914	Mene Grande
1916	Las Cruces
1917	Bolívar Coastal
1922	La Paz
1924	La Concepción
1928	Quiriquire
1933	Oficina, Pedernales
1937	Temblador
1938	Jusepín
1939	San Joaquín
1940	East Guara, Leona
1941	Las Mercedes, Santa Ana, Santa Rosa, Trico
1942	West Guara, Oveja
1945	Mara, Nipa
1946	Boscán
1948	Silvestre, Chimire
1950	Dación
1951	Soto, Limón, Boca
1952	Oscurote
1953	Sinco
1954	Mata, Merey, Oritupano, Nardo
1955	Aguasay-Zapatos
1956	Urdaneta
1957	CL, Centro
1958	Lamar, North Lamar, Morichal
1960	Los Claros

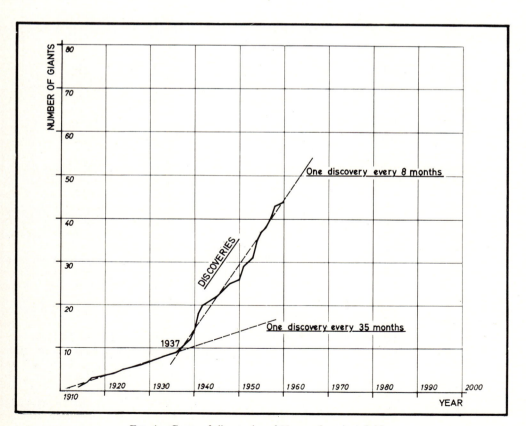

Fig. 4.—Curve of discoveries of Venezuelan giant fields.

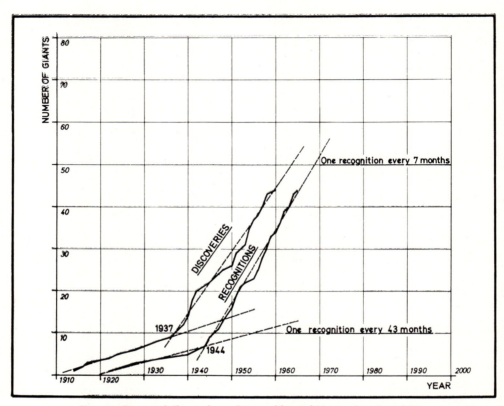

Fig. 5.—Curves of discoveries and recognitions of Venezuelan giant fields.

The longest time lag for the giants in the Maturín basin was 8 years (Pedernales). It took 6 years to recognize Quiriquire as a giant (5 years for the BCF). The shortest delay, actually less than one full year, was for the Lamar field. At the end of 1965, the average delay of recognition was 8 years and 2 months.[6] Figure 5 shows that the time lag from discovery to recognition has remained constant for about the last 15 years.

At the end of 1965, 20 of the giant oil fields had produced the prescribed volume of 100×10^6 bbl of oil, as shown chronologically in Table 3. The discovery and recognition years are also shown. Seven of the 13 giants of the Maracaibo basin, 12 in the Maturín basin, and the Sinco field in the Barinas basin have produced more than 100 million bbl. The first giant oil field to pass the mark, the BCF, had

produced more than 13×10^9 bbl at the end of 1967; La Paz was second with cumulative production of nearly 710×10^6 bbl. Lamar had produced more than 300 million bbl 3 years after passing the 100-million mark.

I have estimated the ultimate volume of the cumulative production of the individual giant fields on the basis of all available literature (Table 4). The "x" indicates fields which have produced more than 100 million bbl.

The BCF is one of the largest petroleum accumulations in the world with total resources of 16,500 million bbl. Six other giant Venezuelan oil fields have estimated resources in excess of 500 million bbl, four in the Maracaibo basin (Lamar, La Paz, Boscán, and Mene Grande) and two in the Maturín basin (Quiriquire and Oficina). Four other giants have estimated resources of between 300 and 500 million bbl (Mara in the Maracaibo basin; Jusepín, Nipa, and Chimire in the Maturín basin). More than half of the giants—27— have estimated resources of between 100 and 200 million bbl. The largest field in the Falcón

6 Hubbert (1962) found that half the giants discovered during a certain year in the USA have been recognized after 15 years. This mention of Hubbert's finding does not imply any comparison with mine.

basin—the Cumarebo oil field—has resources estimated at 62 million bbl.

The estimated resources of the giant oil fields in the Maracaibo basin are $21,405 \times 10^6$ bbl, or an average of 1,647 million bbl per giant. Even if the resources of the BCF were only 500×10^6 bbl, the average volume per giant field would still be 416 million bbl.[7] That figure is one fourth the average, including the BCF.

The total estimated resources of the two giant fields of the Barinas basin are 355×10^6

[7] Excluding the BCF, the average resources per field of the other 12 giant fields in the Maracaibo basin are 409×10^6 bbl.

Table 2. Recognitions

Giant Field	Year Discovered	Year Recognized	Time Lag for Recognition (years)
Maracaibo basin			
Bolívar Coastal	1917	1922	5
Boscán	1946	1952	6
Centro	1957	1959	2
CL	1957	1960	3
La Concepción	1924	1950	26
La Paz	1922	1945	23
Lamar	1958	1959	1
Las Cruces	1916	1928	12
Los Claros	1960	1964	4
Mara	1945	1948	3
Mene Grande	1914	1925	11
Urdaneta	1956	1959	3
North Lamar	1958	1961	3
Barinas basin			
Silvestre	1948	1953	5
Sinco	1953	1956	3
Maturín basin			
Aguasay	1955	1961	6
Boca	1951	1964	13
Chimire	1948	1952	4
Dación	1950	1956	6
East Guara	1940	1945	5
Jusepín	1938	1942	4
Las Mercedes	1941	1951	10
Leona	1940	1955	15
Limón	1951	1962	11
Mata	1954	1957	3
Merey	1954	1963	9
Morichal	1958	1965	7
Nardo	1954	1964	10
Nipa	1945	1950	5
Oficina	1933	1940	7
Oritupano	1954	1962	8
Oscurote	1952	1958	6
Oveja	1942	1957	15
Pedernales	1933	1951	18
Quiriquire	1928	1934	6
San Joaquín	1939	1944	5
Santa Ana	1941	1947	6
Santa Rosa	1941	1948	7
Soto	1951	1958	7
Temblador	1936	1950	14
Trico	1941	1957	16
West Guara	1942	1947	5
Yopales	1937	1951	14
Zapatos	1955	1962	7

Table 3. Giant Fields that Have Produced More than 100×10^6 Bbl

	Giant Field	Year Discovered	Year Recognized	Year Produced 100 Millionth bbl
1	Bolívar Coastal	1917	1922	1928
2	Mene Grande	1914	1925	1932
3	Quiriquire	1928	1934	1938
4	Jusepín	1938	1942	1945
5	Oficina	1933	1940	1945
6	Las Cruces	1916	1928	1946
7	La Paz	1922	1945	1948
8	East Guara	1940	1945	1948
9	West Guara	1942	1947	1951
10	Mara	1945	1948	1951
11	Nipa	1945	1950	1955
12	Chimire	1948	1952	1955
13	Boscán	1946	1952	1956
14	Santa Rosa	1941	1948	1959
15	Las Mercedes	1941	1951	1961
16	Dación	1950	1956	1961
17	Soto	1951	1958	1961
18	Mata	1954	1957	1961
19	Lamar	1958	1959	1963
20	Sinco	1953	1953	1966

bbl, or an average of 178×10^6 bbl. The estimated resources of the giants in the Maturín basin are $5,850 \times 10^6$ bbl, or an average of 202 million bbl per field.

Table 4. Size of Giant Fields

Giant Fields	Resources[1] (10^6 bbl)	
Bolívar Coastal	16,500	x
Lamar	1,000	x
La Paz	905	x
Quiriquire	810	x
Boscán	675	x
Mene Grande	615	x
Oficina	610	x
Mara	425	x
Urdaneta	350	
Nipa	340	x
Chimire	315	x
West Guara	260	x
Santa Rosa	240	x
Mata	230	x
Los Claros	220	
Sinco	215	x
Jusepín	210	x
Centro, Dación (x)	200	
East Guara	195	x
Merey	190	
San Joaquín	185	
Las Mercedes (x), Soto (x)	180	
Morichal	170	
CL, Oscurote	150	
Leona	145	
Silvestre	140	
Nardo	135	
La Concepción	130	
Las Cruces	125	x
Oveja, Zapatos, Aguasay	120	
Temblador, Trico, Pedernales	115	
Limón, Santa Ana, North Lamar	110	
Boca, Oritupano. Yopales	100	

x = Production in excess of 100×10^6 bbl.
[1] "Resources" means estimated ultimate volume of the cumulative production.

The estimated resources of all giant fields in Venezuela are 27,730 × 10⁶ bbl. The average for the country is 630 million bbl. If only a token volume of 500 million bbl is assigned to the BCF, rather than the actual amount, the average resources are 267 million bbl of oil per giant field. The average size of giant fields in the United States, a figure given here for purely informative purposes, is 247 million bbl (Hubbert, 1962).

ANALYSIS OF DATA

The data on the giant oil fields of Venezuela can be used to project the petroleum resources of the country.

Martínez (1959, 1966) has used the technique developed by Weeks to estimate the magnitude of the Venezuelan oil resources on the basis of the stratigraphic and structural development of the sedimentary basins of the country, as related to petroleum generation, migration, and accumulation. He applied the integral technique of prediction developed by Hubbert, to estimate the duration of the volume of resources based on its relation to the past evolution of the industry. Martínez analyzed the theoretical results and concluded that the method cannot be applied because of the artificial controls on normal development. Since 1958, for example, exploration activity and the discovery rate have dropped sharply because of governmental policy limiting the increase in the rate of growth of production to no more than 4 percent per year.

The magnitude of the Venezuelan oil resources is estimated by Martínez (1959) to be 69,200 million bbl of oil. That volume refers only to liquid hydrocarbons; natural gas and solid or semisolid asphaltic minerals are excluded. The Gulf of Venezuela basin was not considered. Figure 6 shows the prediction of the conditioned growth of cumulative discoveries, cumulative production, and proved reserves of oil in Venezuela. The lines of points show, for comparative purposes, the unrestricted theoretical growth.

The data on the giant Venezuelan fields are combined with the study of the oil resources for a double purpose. First, it would be possible to extend the analysis of the giants on the basis of (1) the data to the end of 1965 on the discovery and recognition of the fields and (2) the theoretical and conditioned prediction of their evolution to the respective hypothetical final year of production (2016 and 2055). Second, the combined analysis would serve as an appraisal of the validity and as a further check on the reliability of both studies.

Figure 7 shows the theoretical and conditioned prediction curves for the discovery and recognition of the giant oil fields of Venezuela; Table 5 is a breakdown of the data by basins. The ultimate number of giant fields of Venezuela is estimated to be 78. Of the 34 giants not

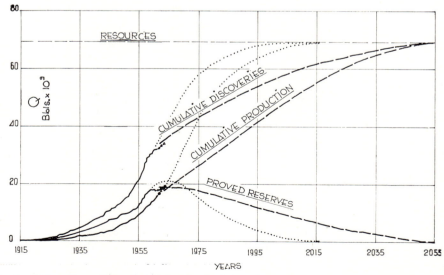

FIG. 6.—Prediction of conditioned growth of cumulative discoveries, cumulative production, and proved reserves of oil in Venezuela (Martínez, 1966).

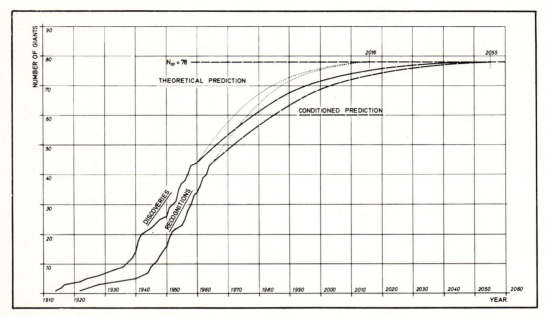

Fig. 7.—Prediction of discoveries and recognitions of giant oil fields in Venezuela.

recognized as such, 10 were presumably discovered by the end of 1965, as shown by the intersection of the curve of the theoretical prediction of discoveries with the abscissa corresponding to year 1965. There are still 24 giants undiscovered in the Venezuelan sedimentary basins.

During the last 10 years, the average size of giant oil fields in Venezuela has remained constant at about 300 million bbl, and during the last two decades, the average size has varied only between 300 and 350 million bbl. Thus there is little possibility that the average size of the giants would change appreciably in the future. It is possible, therefore, to estimate the volume of resources in the reservoirs of the giant oil fields alone. The result is shown in

Table 6, in which the total volume of the resources by basins (from Martínez, 1966) is compared with the resources of the giant oil fields alone.

As can be seen, 42 percent of the oil resources of the Maracaibo basin are estimated to be in the giant oil fields of the basin. The corresponding figure for the Maturín basin is 39 percent. In contrast, the percentage of resources of the Falcón and Barinas basins in giant oil fields are only 30 and 15 percent, respectively. In the case of the Barinas basin, this might indicate that the resources of the basin have been estimated too optimistically. In the case of the Falcón basin, it is impossible to analyze rationally the resource figure, which is very small in comparison with the country's

Table 5. Number of Giant Fields by Basins

Basin	Discovered 1965	Undiscovered	Total
Maracaibo	13	12	25
Barinas	2	1	3
Falcón	—	1	1
Cariaco	—	—	—
Maturín	29	20	49
Total	44	34	78

Table 6. Resources of Basins and of Giant Fields

Basin	Resources of Basin (10⁶ Bbl)	Resources of Giants Alone (10⁶ Bbl)	Percentage
Maracaibo	51,200	21,405	42
Barinas	2,400	355	15
Falcón	400	120	30
Cariaco	100	—	—
Maturín	15,100	5,850	39
Total	69,200	27,730	40

total. It is estimated that 40 percent of the Venezuelan resources are in the giant oil fields.[8]

If a statistical analysis could be made also of the small oil fields of Venezuela—those with resources of less than 100 million bbl—complete verification would be possible. An estimate of resources on the basis of data from giant and small fields is not as exact and reliable as the integral technique of prediction, but it would give an order of magnitude that might be useful, together with further evidence, in supporting the estimates already obtained.

REFERENCES CITED

Hedberg, H. D., 1950, Geology of the Eastern Venezuela basin: Geol. Soc. America Bull., v. 61, no. 11, p. 1173–1216.

Hubbert, M. K., 1962, Energy resources: Natl. Acad. Sci.—Natl. Research Council Pub. 1000-D, 141 p.

Lahee, F. H., 1958, Statistics on natural gas discoveries: Am. Assoc. Petroleum Geologists Bull., v. 42,

[8] Hubbert (1962) estimates that all U.S. giants account for 65 percent of the resources of that country.

no. 9, p. 2037.

Martínez, A. R., 1959, Técnicas de predicción aplicables a la industria petrolera de Venezuela: 3d Cong. Geol. Venezolano Mem., v. 4, p. 1531–1554; also, 1961, Caracas, Div. Geol., Ministerio Minas e Hidrocarburos Pub. Esp. 3

—— 1966, Estimation of petroleum resources—discussion: Am. Assoc. Petroleum Geologists Bull., v. 50, no. 9, p. 2001–2008.

Miller, J. B., et al., 1958, Habitat of oil in the Maracaibo basin, Venezuela, in Habitat of oil, a symposium: Am. Assoc. Petroleum Geologists, p. 601–640.

Ministerio de Minas e Hidrocarburos, 1957, Anuario petrolero y minero de Venezuela—datos de 1952: p. 374.

Renz, H. H., et al., 1958, The Eastern Venezuelan basin, in Habitat of oil, a symposium: Am. Assoc. Petroleum Geologists, p. 551–600.

ADDITIONAL SOURCES CITED

Ministerio de Minas e Hidrocarburos, Memoria, and Apéndice estadístico de la memoria: Caracas, published various years.

Oil and Gas Journal, World wide oil, and Review-forecast: Tulsa, Oklahoma, published various years.

Venezuelan Oil Scouting Agency (VOSA), 1963, Catalogue of Venezuelan oilfields and wells, 1878–1961. Also supplements for 1962, 1963, 1964, 1965, 1966.

General Geology and Major Oil Fields of Recôncavo Basin, Brazil[1]

JOÃO ÍTALO GHIGNONE[2] and **GERALDO DE ANDRADE**[2]

Salvador, Bahia, Brazil

Abstract   The Recôncavo basin, on the Atlantic Coast near the city of Salvador, has an area of about 10,000 km² and is the principal petroleum province of Brazil. Since 1939, approximately 255 wildcats have been drilled and have discovered 43 accumulations which total 942 million bbl of producible oil and 992 billion ft³ of gas. API gravity of most oil ranges from 35 to 40°.

The Bahia Supergroup, the main objective for petroleum exploration, has a maximum thickness of 6,500 m. This nonmarine unit ranges in age from Late Jurassic(?) to Early Cretaceous. The Upper Jurassic(?) typically consists of a redbed sequence (Aliança Formation) overlain by a blanket sandstone (Sergi Formation). The Sergi is the best reservoir rock of the basin. The Lower Cretaceous (Wealden) strata are mainly dark-gray and grayish-green shale of the Itaparica, Candeias, and Ilhas Formations and are considered to be the oil and gas source rocks. The **A** Sandstone, the lenticular sandstone bodies of the Candeias Formation, and the São Paulo and Santiago Sandstones of the Ilhas Formation are the best reservoir rocks of the Lower Cretaceous section.

The Recôncavo basin is an intracratonic half graben. Intensive faulting occurred during the deposition of Candeias and lower Ilhas when the basin became a rapidly sinking trough. Accelerated growth of the Salvador and Mata-Catu uplifts, the most prominent structural features of the basin, produced the two principal (NE- and NW-trending) sets of normal faults. A late post-São Sebastião tectonic phase reactivated ancient faults and caused new ones to form. Consequently, the basin is characterized by a complex system of faulted blocks.

The six major fields, containing 96 percent of the total producible oil, are related to the structural evolution of the basin. It is believed that the early period of faulting, contemporaneous with the deposition of the Candeias and lower Ilhas, was a decisive factor in the control of petroleum migration to, and accumulation in, the Sergi and **A** sandstones. The horst blocks of Água Grande, Buracica, and Dom João fields (the first two having been partly uplifted during this tectonic phase) trapped about 623 million bbl of recoverable oil in these two sandstone bodies. Accumulation in Ilhas reservoirs was controlled mainly by the later, post-São Sebastião phase of faulting. Folds developed in the downthrown blocks of normal faults, but the folding was not caused by compressional stress. These folds form traps for accumulations in the São Paulo, Santiago, and other Ilhas sandstones. Examples of such traps, in which about 187 million bbl of producible oil accumulated, are the Miranga and Taquipe fields. The genesis of the lenticular sandstone reservoirs in the Candeias field—a stratigraphic trap—is related to syntectonic deposition of the Candeias Formation; fractured shale and limestone also are reservoirs in this field. Candeias trapped about 93 million bbl of producible oil.

[1] Read before the 53d Annual Meeting of the Association, Oklahoma City, Oklahoma, April 25, 1968. Manuscript received, November 8, 1968; accepted February 28, 1969.

[2] Petróleo Brasileiro (Petrobrás), S.A.

INTRODUCTION

The Recôncavo basin on the Atlantic Coast near the city of Salvador, State of Bahia, is one of the eight Cretaceous coastal basins of Brazil and is its principal oil-producing province (Fig. 1). It is a southern extension of the Tucano basin. The geographic limits are placed arbitrarily at the Precambrian Aporá salient on the north and at the southern extremity of Itaparica island on the south. The Recôncavo basin is limited on the west by the Maragogipe fault system and on the east by the Salvador horst block (Salvador high). The basin has an area of about 10,000 km².

Petroleum exploration began in 1939 when a shallow well was drilled near Salvador on the basis of oil seeps. This well discovered the small Lobato field. The first geophysical work—a torsion balance survey—was carried out in 1937 by the Departamento de Fomento da Produção Mineral, a government agency. Subsequently, geophysical and geologic work was done intermittently by the Conselho Nacional de Petróleo. Exploration activity increased and the quality of geologic and geophysical work improved when Petrobrás began to operate in 1954.

About 255 wildcats have been drilled since 1939, resulting in the discovery of 43 oil and gas fields which contain 942 million bbl and 992 billion ft³, respectively, of producible oil and gas (Fig. 2). About a third of the recoverable oil already has been produced.

This paper reviews the regional geology of the basin and its six major fields. The Candeias field was discovered in 1941, Dom João in 1947, Água Grande in 1951, and Taquipe in 1958. Each of these discoveries resulted from surface geologic mapping. Geophysical surveys led to the discovery of Buracica field in 1959 and Miranga in 1965. The six fields contain 96 percent of the oil found in the basin. In 1967, combined geophysical and subsurface studies discovered the promising Aracás field which probably will become the seventh major oil field of the basin. Present production from the basin is approximately 140,000 bbl/day of oil.

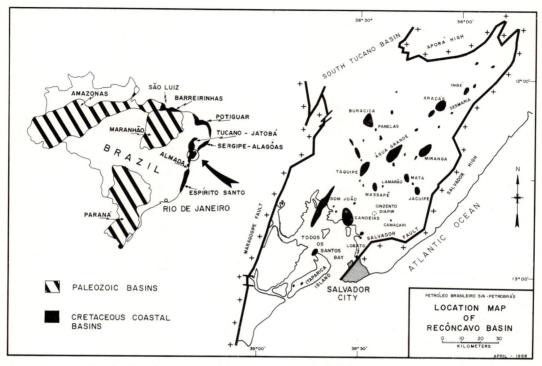

Fig. 1.—Location map, Recôncavo basin, Brazil. Oil and gas fields are shown.

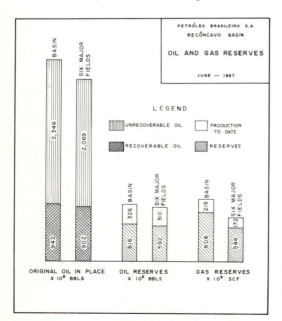

Fig. 2.—Oil and gas reserves, Recôncavo basin.

STRATIGRAPHY

General.—The first references to the sedimentary beds of the Recôncavo basin, in 1828, are by von Spix and von Martius (*in* Oliveira and Leonardos, 1943, p. 570). Hartt (1870, p. 264) proposed the name Bahia Group for the strata within the limits of the Todos os Santos Bay. The first formal stratigraphic nomenclature, proposed by Pack and Almeida (1946), was revised by Taylor (1947), Barnes (1950), and Luz (1955). The present stratigraphic division is based on the "Basin Study" prepared by the geological staff of Depex Petrobrás (Depex, 1958).

Figure 3 shows the stratigraphic section of the basin. The Bahia Supergroup, principal objective for petroleum exploration, has a maximum thickness of 6,500 m and consists of six formations—Aliança, Sergi, Itaparica, Candeias, Ilhas, and São Sebastião. These beds are nonmarine, and range in age from Late Jurassic(?) to Early Cretaceous. The Late Jurassic(?) (Brotas Group) consists of a typical redbed sequence—the Aliança Formation,

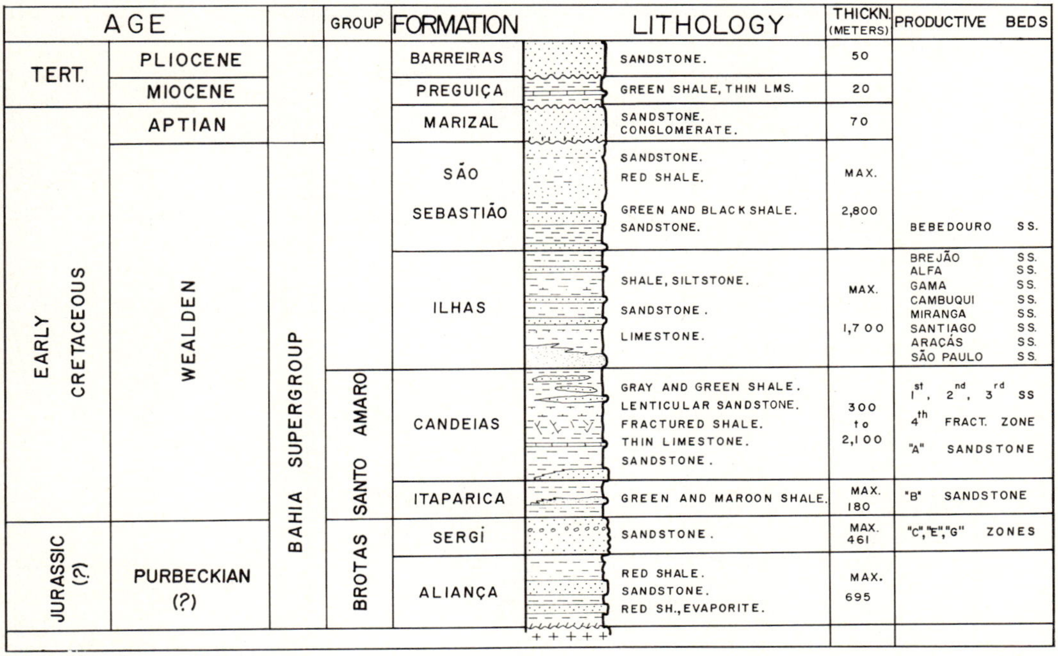

AGE		GROUP	FORMATION	LITHOLOGY	THICKN. (METERS)	PRODUCTIVE BEDS
TERT.	PLIOCENE		BARREIRAS	SANDSTONE.	50	
	MIOCENE		PREGUIÇA	GREEN SHALE, THIN LMS.	20	
	APTIAN		MARIZAL	SANDSTONE. CONGLOMERATE.	70	
EARLY CRETACEOUS	WEALDEN	BAHIA SUPERGROUP	SÃO SEBASTIÃO	SANDSTONE. RED SHALE. GREEN AND BLACK SHALE. SANDSTONE.	MAX. 2,800	BEBEDOURO SS.
			ILHAS	SHALE, SILTSTONE. SANDSTONE. LIMESTONE.	MAX. 1,700	BREJÃO SS. ALFA SS. GAMA SS. CAMBUQUI SS. MIRANGA SS. SANTIAGO SS. ARAÇÁS SS. SÃO PAULO SS.
			CANDEIAS	GRAY AND GREEN SHALE. LENTICULAR SANDSTONE. FRACTURED SHALE. THIN LIMESTONE. SANDSTONE.	300 to 2,100	1st, 2nd, 3rd SS 4th FRACT. ZONE "A" SANDSTONE
		SANTO AMARO	ITAPARICA	GREEN AND MAROON SHALE.	MAX. 180	"B" SANDSTONE
JURASSIC (?)	PURBECKIAN (?)	BROTAS	SERGI	SANDSTONE.	MAX. 461	"C","E","G" ZONES
			ALIANÇA	RED SHALE. SANDSTONE. RED SH., EVAPORITE.	MAX. 695	

FIG. 3.—Stratigraphic column of Recôncavo basin.

overlain by a blanket sandstone, the Sergi Formation. The Lower Cretaceous section (Wealden) is composed largely of dark-gray and grayish-green shale of Itaparica, Candeias, and Ilhas Formations and the predominantly sandy section of the São Sebastião Formation. Near the Salvador high the last three formations cannot be differentiated lithologically because of interfingering with a thick section of syngenetic conglomerate (Fig. 4).

Fossils of the Bahia Supergroup are chiefly ostracods which have been used successfully for stratigraphic correlations, because the biozones correspond fairly well to rock-stratigraphic units.

Striking lithologic and paleontologic similarities between the Recôncavo basin and the Gabon basin of West Africa have been confirmed by Fonseca (1965) and Viana (1965). In fact, a study of the Recôncavo basin tempts one to believe that it is only half of the basin, the other half being the Gabon basin. The two would have been separated by continental drift since Early Cretaceous time. Thus the Salvador horst block, the present eastern limit of the

basin, might not have been the eastern limit of Bahia Supergroup deposition.

A thin post-tectonic sequence overlies the Bahia Supergroup and is represented by the Marizal Formation of Aptian age and the Pliocene Barreiras Formation; both consist of coarse material of fluviatile deposition. The shale and thin limestone beds of the Preguiça Formation, which are present locally along the eastern border of the basin, represent a brief Miocene marine invasion.

The basement rocks on both sides of the Rocôncavo basin are early Precambrian migmatite and granulite of sedimentary and igneous origins. Subsurface evidence indicates that at least part of the Recôncavo basin was occupied by a Cambrian epicontinental sea, in which limestone and sandstone of the Estância Formation (known from surface exposures in the Tucano basin) were formed.

Aliança Formation.—The basal formation of the Brotas Group is divided into three members. The lower consists of red, purple to maroon, silty micaceous shale with scattered green spots. Thin white sandy layers are repeated cyclically throughout the section. Near the base,

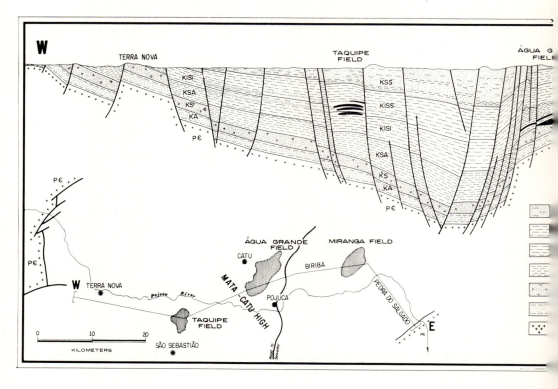

lenses of light-gray arkosic sandstone, shale, and laminae of gypsum and anhydrite are present. Thin-bedded dense limestone and chert also are found. The middle member is composed mainly of arkosic, fine- to medium-grained sandstone with well-developed cross-beds. The color varies according to the nature of the argillaceous matrix—buff, white, yellow, red, *etc.* Some sandstone bodies are conglomeratic and contain pebbles of reworked maroon shale. The upper member is very similar to the lower one. Sandstone is present at the top, and grades compositionally and in grain size into the Sergi Formation.

A maximum thickness of 695 m of Aliança Formation was found in a dry hole at Itaparica island (Fig. 1). Generally, the formation is thicker in the southern part of the basin. The three members thin out toward the northeastern border of the basin near the Sesmaria area. Good surface exposures are present along the western border.

The Late Jurassic (Purbeckian) age assigned to the formation is questionable. The fossil content of the formation is very poor in the Recôncavo basin but on the north, in the Tucano basin, ostracods are abundant in the red shale. The only recognized species are nondiagnostic *Darwinula* sp. and *Metacypris* sp. 2, 3, 4, and 5. Fish and reptile remains were reported by Ghignone (1963) in the northern part of the Tucano basin.

Sergi Formation.—The Sergi Formation at the top of the Brotas Group is the best reservoir in the basin. It is productive at Dom João, Água Grande, and Buracica, as well as in some smaller fields.

The formation has three members. The upper one consists of fine-grained to conglomeratic sandstone, clean to dirty, locally cemented by silica and carbonate. At the top, pink garnet, rutile, and chert are common. At the base the sandstone characteristically is conglomeratic and feldspathic and contains quartz pebbles up to 70 mm in diameter. The middle member is characterized by cut-and-fill structures, flow rolls, cross-beds, and distinctive laminations. The sandstone is fine to coarse grained, and locally is conglomeratic; grains generally are well rounded. Weathered feldspar, mica, pink garnet, and opaque minerals are common. The lower member consists of red to

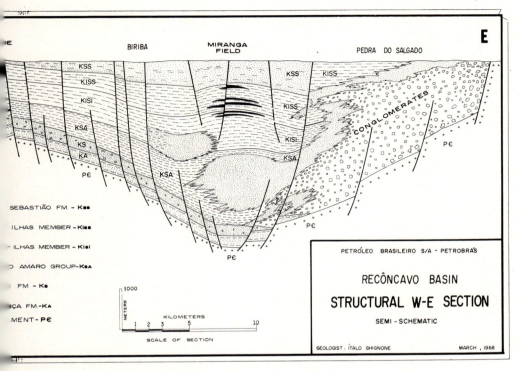

FIG. 4.—Semi-schematic structural cross section of Recôncavo basin.

tan sandstone with fine to very fine subangular and subrounded grains in a chloritic and silty matrix.

The Sergi Formation crops out along the western and northeastern borders of the basin; in the subsurface it is present across the entire basin, except near the eastern border where there is evidence that it was eroded before deposition of the Itaparica Formation. The Sergi Formation is somewhat thicker in the southern part of the basin. The maximum thickness penetrated is 461 m in Dom João field.

Only the nondiagnostic ostracod, *Darwinula* sp., was found in the Sergi Formation; the Late Jurasic(?) age assigned to it is based on the gradational contact with the underlying Aliança Formation.

Itaparica Formation.—The Itaparica Formation is a transitional phase between the Brotas Group and the overlying formations.

Three distinctive and fairly uniform sequences compose the Itaparica Formation. At the base is a sequence of greenish-gray, calcareous, thin-bedded shale and siltstone with one or more beds of maroon shale. Near the top of this sequence is a thin limestone bed which is a

widespread marker in the basin. In the middle of the formation is the **B** Sandstone—a gray, very fine- to fine-grained, argillaceous sandstone. A section similar to the basal sequence overlies the **B** Sandstone. Maroon shale near the top of the Itaparica is an excellent marker bed.

The thickness ranges from 18 m near the Aporá salient to 180 m in the southern part of the basin. Along the northeastern border, the Itaparica Formation is absent but it is found in the Almada basin, about 300 km south of Salvador (Fig. 1).

The guide ostracods of the Itaparica are *Cypridea kegeli* Wicher, *C. bisculturata* Moos, *C. brevicornis* Peck, *C. armata, C. acuta,* and *C. depressa.*

Candeias Formation.—Dark-gray to green calcareous shale containing thin beds of hard, tan to brown limestone constitutes most of the monotonous Candeias section. The dark shale is considered by many geologists to be the source of much of the oil in the basin.

The **A** Sandstone, which conformably overlies the Itaparica, generally is placed at the base of the Candeias Formation. However, the lithologic composition and the fossils found recently in Araçás field suggest that the **A** Sandstone belongs more logically in the Itaparica Formation. The **A** Sandstone is productive in the Água Grande and Buracica fields, as well as in several small fields. Other sandstones of the Candeias Formation generally are fine to coarse grained and tight; their distribution is erratic and they are everywhere lenticular. Sandstone lenses in the middle part of this formation are the oil reservoirs in the Candeias field.

Tectonic activity contemporaneous with deposition caused marked thickness variations in the Candeias Formation within relatively short distances. In the past, an unconformity was assumed by several geologists to be at the top of the Candeias section. However, there is no conclusive evidence for such an unconformity. In this paper, the upper part of the Candeias and the lower member of the Ilhas Formation are considered as a format (Forgotson, 1957) with lateral lithologic variations related to differential subsidence of faulted blocks during deposition. Because of the difficulty in determining the top of the Candeias, the isopach map in Figure 5 also includes the lower member of the Ilhas Formation.

Ostracod guides characterize two zones—the **R-8** and **R-9** (**R** for Recôncavo). The **R-8** zone is indicated by the *Cypridea (Morininoides) candeienses, C. (M) hadronodosa, C. sellata,* and *Reconcavona imitatrix.* The **R-9** zone includes the *Metacypris* sp. 1 and *Cypridea ambigua.*

Ilhas Formation.—The siltstone and gray-green shale with thin interbeds of limestone of the Ilhas Formation are similar to those of the Candeias. However, the amount of sandstone present in the Ilhas section generally makes it possible to differentiate the two units.

The formation is divided into two members. The upper is extremely important for oil exploration. Four intervals, each containing productive sandstone, are, from top to bottom, I, II, III, and IV (Passos, 1968). The Brejão, Alfa, and Gama Sandstones of interval I are productive in Miranga, Imbé, and Panelas fields, respectively. Interval II includes the Miranga and Cambuqui Sandstones which are productive, respectively, in Miranga and Taquipe fields. A more attractive objective is the Santiago Sandstone of interval III which is productive in Taquipe and Miranga fields. The productive sandstone of interval IV—the Araçás—is named for the recently discovered field.

The areal distribution of each interval is not the same throughout the basin. It is possible to correlate them only in the northern part of the basin, and all fields that produce from sandstone beds of the Ilhas Formation are in this northern area. In the southern part of the Recôncavo basin, the upper Ilhas is more argillaceous and has fewer sandstone bodies.

The lower member is similar in lithologic composition to the upper, but the lateral variations are more unpredictable. Only the top of the lower member has a fairly uniform lithologic character. The São Paulo Sandstone, near the top, is a very good reservoir and is very productive in Taquipe and Miranga fields. Massive sandstones, the distribution of which is not well understood in the Ilhas, yielded hydrocarbons in the Massapê area (Fig. 1).

Three main zones identified by distinctive species of ostracods are recognized in the Ilhas Formation. The uppermost part of the upper member, indicated by *Cypridea (Morininoides) bibullata,* is the **R-4** zone. The *Paracypridea obovata obovata* characterizes the **R-5** zone (lower part of upper member) and is subdivided into two subzones, the **R-5.1** with *Reconcavona gastracantha* and **R-5.2** with *Paracypridea elegans.* The index ostracod of the lower member (**R-6** zone) is the *P. brasiliensis.*

São Sebastião Formation.—The youngest formation of the Bahia Supergroup is predominantly arenaceous. Generally, the sandstone is coarse grained, argillaceous, and reddish. Green and black shale is in the lower member; red silty shale is common in the middle member. The upper member is composed almost entirely of thick-bedded to massive sandstone. The three members are distinguished easily in the subsurface because of fairly constant electric-log characteristics throughout the basin.

Interest in petroleum production from the São Sebastião Formation is restricted to the deeper part of the basin. In the Lamarão area (Fig. 1) the basal Bebedouro Sandstone yielded very good shows of gas.

The paleontologic zonation of the São Sebastião includes the *Coriacina coriacea* or **R-3** zone (lower member) and the *Cypridea (Sebastianites)* **KR** or **R-2** zone (middle member), which is subdivided into the **R-2.1** subzone of *C. (S.) fida* and the **R-2.2** subzone of *C. (S.) sostensis.*

BASIN EVOLUTION

For a clear understanding of the depositional and structural history of the Recôncavo basin,

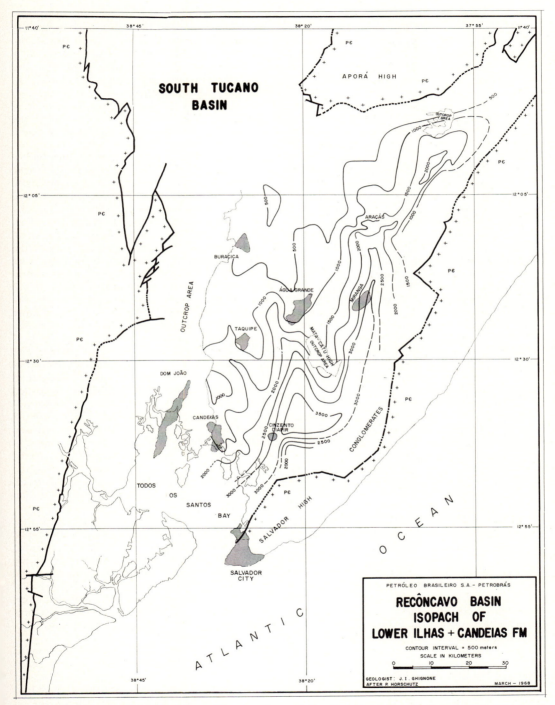

Fig. 5.—Isopach map of combined lower Ilhas member and Candeias Formation, Recôncavo basin. Thickness variation was controlled mainly by tectonic activity contemporaneous with deposition.

much geologic work remains to be done. The tentative reconstruction that follows is based on the features shown by isopach maps. These features are considered to be essentially depositional, except for the Brotas Group which was eroded near the eastern border of the basin.

The initial sedimentary sequence—Brotas Group—has two phases. The first, the Aliança Formation, records a shallow-water environment in a semiarid and warm climate; sediments were laid down as floodplain deposits on a fairly even surface of basement rocks. The second phase—the Sergi Formation—represents a deposit of coarse terrestrial clastic material on an alluvial plain.

Although the Brotas Group is considered to represent a pretectonic phase because of its remarkably uniform lithologic character across the Recôncavo and Tucano basins, there is evidence that incipient movements of the Salvador high took place at the end of Jurassic(?) time. Such movements caused the erosion of the Sergi and Aliança Formations in a narrow band near the eastern border of the basin (Fig. 6). Hydrodynamic studies in this area show that the salinity and chemical composition of the Sergi waters are anomalous in relation to the depth of occurrence; this suggests that an unconformity is present.

During deposition of the Itaparica Formation, southeastward tilting of the basin caused a general but gradual increase in sediment thickness in that direction. The absence of Itaparica beds in the Imbé area (Fig. 1) may be the result of nondeposition rather than erosion. The Itaparica strata record a predominantly lacustrine environment.

The Mata-Catu high (Fig. 5; this is the major structural feature inside the basin) and the Salvador high began to develop during the time of Candeias and lower Ilhas deposition. The isopach contours of Figure 5 show two main trends of abrupt thickening, one parallel with and west of the Mata-Catu region, and the other adjacent to the Salvador high. These two depositional trends coincide with the development of two normal-fault sets, oriented northeast and northwest, which subdivide the basin into a complex system of faulted blocks (Fig. 4). Thus, deposition of the Candeias Formation and the lower member of the Ilhas took place during tectonism. Thickening and thinning of this interval are related to uplift and subsidence during deposition.

The major source area for sediments during Candeias deposition was on the north. Obstruction of the basin on the south resulted in a lacustrine environment. During Ilhas deposition, the lacustrine environment was replaced by a predominantly fluvial regime. Candeias and Ilhas depositional cycles apparently were continuous. The simultaneous action of the tectonic and environmental changes had a marked effect on thickness and lithology, and is the principal reason why the Candeias-Ilhas boundary cannot be identified precisely.

During deposition of the upper member of Ilhas Formation, the rate of subsidence of the basin was greater in the south than in the north. The Mata-Catu high divided the basin into two sedimentary provinces. On the north a relatively stable area extended to the western border, including the Taquipe field; on the south correlations are very difficult and many of the productive sandstone beds of upper Ilhas were not deposited. The thick argillaceous Ilhas-Candeias section in the southern province was the source of diapiric material that caused the Cinzento structure. This diapir penetrated more than 1,000 m through the overlying São Sebastião Formation.

Differential movements ceased during São Sebastião deposition, but general subsidence of the basin as a whole continued. Consequently, a thick alluvial-lacustrine section was deposited in the Camaçari area (see Fig. 6 for location).

The accelerated growth of the Salvador horst block, which separates the basin from the sea, caused the accumulation of large volumes of conglomerate along the eastern border of the basin; the conglomerate bodies interfingered with the sediments of the Bahia Supergroup during the time of Candeias to São Sebastião deposition.

Final tectonic movements took place after São Sebastião deposition. Some ancient faults were reactivated and new ones were formed until complete stability was established at the end of the Wealden Stage. Figure 6 shows the present structural configuration of the basin. The basin then was eroded until Aptian time, but the topographic expression of the structure was preserved by the thin veneer of post-tectonic fluviatile Marizal beds.

The depositional history of the Recôncavo basin ended in Cenozoic time. The terrigenous clastic beds of the Pliocene Barreiras Formation represent coastal-plain deposits. The Preguiça Formation, the only marine section in the basin, is not important for hydrocarbon exploration on the continent because of its

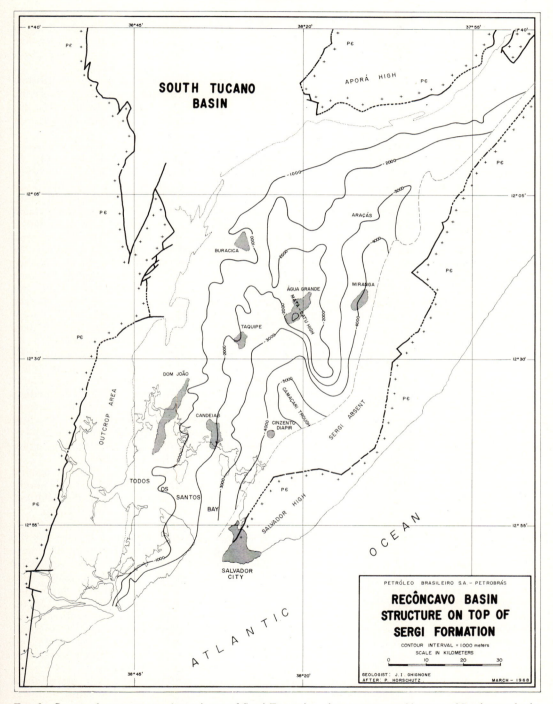

FIG. 6.—Structural contour map, datum is top of Sergi Formation, shows present architecture of Recôncavo basin. Faults not shown.

limited extension and thickness. However, the brief marine interlude during Miocene time may become important when Miocene prospects are extrapolated onto the continental shelf.

FIELDS

Água Grande Field

The Água Grande field, in the central part of the basin and on the Mata-Catu structural high (Fig. 6), was discovered in January 1951 by the Conselho Nacional de Petróleo, the government agency then in charge of oil exploration. The discovery well, located on a supposed surface anticlinal feature where Ilhas shale was exposed, was completed successfully with an initial production rate of 700 bbl/day of oil. The field trends northeast, and is about 8 km long and 3 km wide. Its limits have been established by a total of 231 wells, of which 27 were abandoned as dry holes. Up to 1967 Água Grande contributed 47 percent of the total oil produced in the basin and was the best oil field discovered.

Productive beds.—The main producing reservoirs are sandstone of the Sergi Formation and the A Sandstone of the Candeias Formation; oil is found within a depth range of 1,050–1,350 m. Less important accumulation is found in the shallow São Paulo Sandstone of the Ilhas Formation.

The Sergi Formation, containing about 63 percent of the recoverable oil of the field, is a massive sandstone sequence 240 m thick with a few thin layers of siltstone and shale. The sandstone is fine grained to conglomeratic, poorly to well sorted, quartzose to locally arkosic, and has a variable content of argillaceous matrix which directly affects the porosity and permeability. Porous sandstone, slightly argillaceous to clean, composes about 75 percent of the total thickness of the formation.

The A Sandstone, having an average thickness of 15 m, is the second-best reservoir. It is composed of two lithologically distinct sandstone beds separated by a very thin layer of greenish silty shale. The upper bed is predominantly fine to medium grained, clean, well sorted, and quartzose; the lower bed is argillaceous, fine to very fine grained, and less porous than the upper sandstone bed because of a high clay content. The A Sandstone has a maximum thickness of 35 m, and pinches out locally in the western part of the field.

The São Paulo Sandstone—the uppermost producing reservoir—consists of two sandstone bodies, 10 and 40 m thick, separated by a shale unit. The sandstone is fine to very fine grained, quartzose, and well sorted; the argillaceous content averages 6 percent.

The average porosity of the pay sandstone is 17.5 percent for the A and Sergi reservoirs and 25 percent for the São Paulo Sandstone. The maximum net oil-sandstone thickness is 112 m.

Structure.—The structural map with contour datum at the base of the A Sandstone (Fig. 7) shows that the Água Grande field is a SW-NE-trending horst block. Major normal faults limit the horst on the southwest (Mata-Catu fault, striking N35°W with 1,000 m of throw) and on the east (a fault striking N60°E to N5°E with 300–400 m of throw). Several NE- and NW-trending faults bound the horst on the northwest and divide it into secondary blocks. In the vicinity of the Mata-Catu fault, some faulted blocks have a southwestward dip, but most blocks are tilted west-northwest. In the central part of the field, a slight domal closure in one block—resulting apparently from drag on the limiting faults—can be observed.

Structural closure in the field is provided by the boundary faults and/or associated faults and also by the tilting of blocks. The structural configuration at the top of the Sergi Formation conforms fairly well with that at the base of the A Sandstone; at the surface, the structure does not everywhere reflect that at depth because the early pre–upper Ilhas faulting phase did not affect the overlying strata.

Accumulations in the São Paulo Sandstone are found only in small blocks in the southwestern and northeastern parts of the field and are related to post-São Sebastião faulting.

The oil-water contact for both the A Sandstone and the Sergi Formation is level for the entire field, at −1,200 m. This indicates intercommunication between the two reservoirs through some of the faults. Four gas-oil contacts have been identified in the A Sandstone, ranging in depth from −1,070 to −1,150 m, the higher one extending to the Sergi reservoir.

The hydrocarbon columns in the A Sandstone and Sergi Formation are 300 and 182 m, respectively, but structural closure apparently is greater for both reservoirs.

Structural evolution.—Isopach maps show that, until the end of the deposition of lower Candeias, there was no structural movement in the area. The early faulting phase was contemporaneous with the deposition of middle-

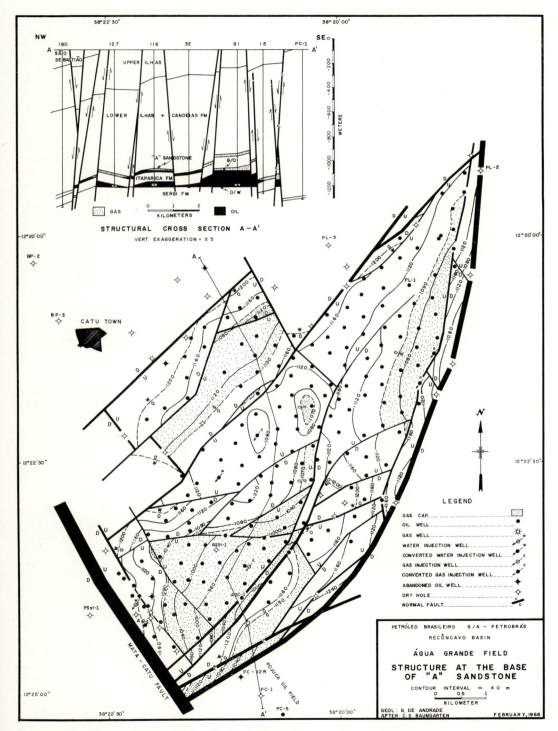

FIG. 7.—Structural map of Água Grande field, Recôncavo basin. Datum is base of **A** Sandstone.

upper Candeias and lower Ilhas, as indicated by abrupt changes in the thickness of that interval. The eastern and southwestern boundary faults formed during this phase. Thickening of upper Ilhas and São Sebastião beds toward the west indicates that during the deposition of these units the area generally was tilting in that direction. The second faulting phase, after deposition of the São Sebastião, produced the present structure. The northwestern boundary faults, as well as most of the faults that divide the horst into secondary blocks, formed during this faulting phase.

Production.—Cumulative production in the Água Grande field through May 31, 1967, was 152.6 million bbl, which is about 27 percent of the oil in place. Ultimate recovery is expected to be 260 million bbl. About 99 percent of the oil produced is from the **A** and Sergi reservoirs.

After reaching a peak of 53,000 bbl/day in 1960, oil production today is 36,000 bbl/day, the second-ranking field in the basin. Large-scale secondary recovery is to be used in 1969, but small amounts of gas and water have been injected in the Sergi reservoir.

Average properties of the oil, which is of paraffinic base, are: API gravity, 40–41°; viscosity at 54.4°C, 45 SSU; pour point, 32°C; paraffin content, 24 percent; sulfur content, 0.04 percent. Average analysis of the crude is 22 percent gasoline, 16 percent gas oil, and 8 percent fuel oil.

Dom João Field

The Dom João field, discovered in March 1947 by the Conselho Nacional de Petróleo, is in the northwestern part of the Todos os Santos Bay, about 40 km from the city of Salvador, near the western border of the basin (Fig. 6). The wildcat was located on the basis of surface geologic mapping, and was staked on the axis of a supposed SW-plunging anticlinal feature in an area of Candeias shale exposure. The field now is completely developed. It extends for 23.5 km from southwest to northeast, and is 1–3.5 km wide. Approximately two thirds of the field is in the Todos os Santos Bay, and it is the only offshore field in the basin. A total of 671 wells has been drilled in this field, of which 29 were abandoned as dry holes.

Productive beds.—The producing reservoirs are sandstone in the Sergi Formation. Oil is found at a depth of 160–374 m. Some small noncommercial accumulations have been found in the **A** and **B** Sandstones.

The Sergi Formation consists almost entirely of reddish- to whitish-gray, clean to argillaceous, very fine-grained to conglomeratic sandstone with scarce thin layers of reddish to greenish shale and brownish to whitish sandy limestone. The thickness ranges from 280 m at the north end of the field to 461 m at the south end. The upper 192 m of the formation, which is gas- and oil-bearing, is divided into 8 production units named the **C** through **J** zones. Each zone is recognized easily on the basis of its lithologic and electric-log characteristics.

Laboratory studies confirmed the great vertical lithologic variation in the sandstone that already had been observed from core descriptions. The major variations are in clay and heavy-mineral contents, degree of sorting, types of clay minerals, roundness and sphericity of the grains, *etc.* The argillaceous-matrix percentage, which is very important to the storage capacity of the reservoir, ranges from 1.2 percent in the clean porous sandstone to 37.5 percent in the tight argillaceous sandstone where montmorillonite clay predominates.

Most of the Sergi sandstones in this field are tight argillaceous types. The porous, less argillaceous layers are concentrated in the **C, E, G,** and **J** zones, where more than 90 percent of the oil volume is found.

The total net oil-sandstone thickness has a maximum of 75 m and the average porosity ranges from 18 percent in the north to 22.4 percent in the south.

The **B** Sandstone of the Itaparica Formation is poorly developed, and is very fine grained to silty, argillaceous, and tight. In the southeastern part of the field, this sandstone has some porosity and is oil bearing, but production tests have not been encouraging.

The **A** Sandstone crops out beneath the water on the structurally highest part of the field. Thus most of the oil trapped in this reservoir was lost, except in the DJx-1 area where a small accumulation remains; preliminary production tests have been unsuccessful. The sandstone is fine grained to conglomeratic, generally porous, clean to somewhat argillaceous, and calcitic at the base. The thickness ranges from 0 to 45 m.

Structure.—The structural contour datum at the top of the Sergi Formation shows a SW-NE-trending horst block (Fig. 8). Steep major normal faults striking N10°E to N30°E, with 150–500 m of throw, bound the structure on the west and east. Two other important faults cross the horst. One (170 m of throw) forms the southern limit of the field; the other (150

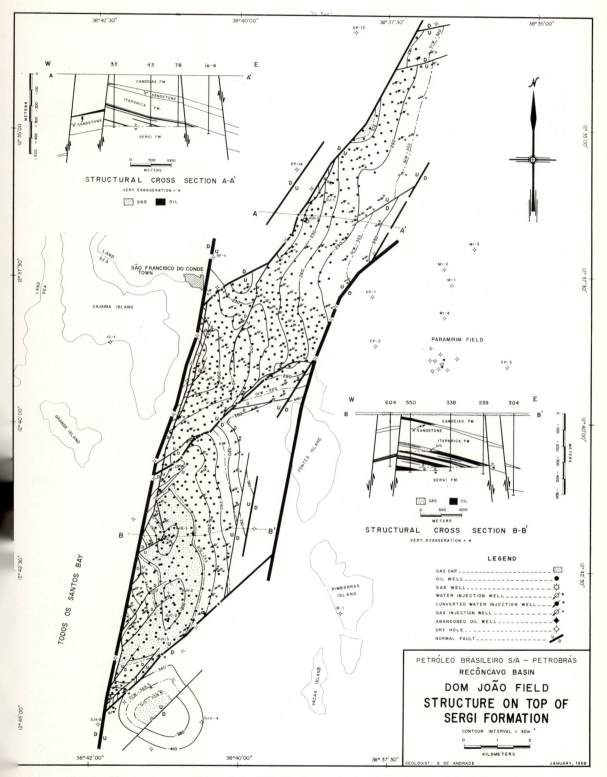

FIG. 8.—Structural map of Dom João field, Recôncavo basin. Datum is top of Sergi Formation.

m of maximum throw) divides the field into two major blocks with the downthrown block on the south. Subsidiary step faults are very common at the field borders.

A well-shaped anticline developed on the horst, probably as the result of drag along the major boundary faults. Local domal closures are present in the southern part of the field. On the north the Sergi beds dip southeast along the eastern limb of the aforementioned anticline. Generally, the structure of the Sergi datum shows gentle dips ranging from 3 to 5°. Steeply dipping strata are found only close to the step-fault blocks near the major faults.

Structural closure, which is at least 200 m at the level of the Sergi Formation, is the result of both the presence of the boundary faults and the dips of the anticlinal feature. Because the latter is related genetically to the faults, the Dom João field is basically a fault trap.

The position of the oil-water contact in the Sergi Formation is controlled mainly by structure and somewhat by lateral variation in sandstone properties. In the northern block, the contact ranges in depth from −320 to −332 m, whereas in the southern block, it generally is below −335 m, reaching a maximum of −369 m. Three gas-oil contacts have been identified in field, the highest being at −233 m in the southern block.

Production.—The field has produced 30.5 million bbl of oil through May 31, 1967. This is only 3.6 percent of the oil in place. Ultimate oil recovery is expected to be 247 million bbl. Present production is 15,000 bbl/day, and Dom João is the fourth-largest producing field of the basin.

Dom João is essentially a low-energy field. Original pressure dropped sharply soon after production started, and only the **J** zone showed less pressure decline because of a large underlying aquifer. In all zones except **J**, the following geologic factors are adverse to production— lenticularity of permeable sandstones, presence of argillaceous matrix even in the best pay zones, and fault sealing in all directions (this prevents lateral water influx).

Secondary recovery by saltwater injection is being carried out in the northern block and, in the near future, this operation will be extended to the entire field to at least double the present production. Recently, hydraulic fracturing was used successfully to improve both injection and production rates.

Average properties of the paraffinic oil are: API gravity, 36–38°; viscosity at 54.4°C,

50.8–57 SSU; pour point, 29°C; paraffin content, 15 percent in weight; and sulfur content, 0.05 percent. An average analysis of the crude is 17 percent gasoline, 14 percent gas oil, and 7 percent fuel oil.

Miranga Field

Miranga field, in the deep east-central part of the basin (Fig. 6), was discovered in July 1965 by Petrobrás. The discovery well, completed in Santiago Sandstone, flowed 600 bbl/day of oil. The location was selected on the basis of a seismic-reflection interpretation which delineated an elongate domal feature on a shallow mapping horizon. The field has a SW-NE trend, and is 6.5 km long and 5 km wide. Of the 120 wells drilled, seven were abandoned as dry holes. About 10 infill locations remained to be drilled in December 1967.

Productive beds.—Miranga is a multipay field, producing from several sandstone beds of the Ilhas Formation, mainly the Santiago and São Paulo Sandstones which are found between 1,000 and 1,500 m and contain about 90 percent of the oil reserve. The São Paulo consists of several sandstone bodies, 5–70 m thick, separated by shaly intervals; the Santiago, with an average gross thickness of 40 m, also has thin shale layers, mainly at the top. Less important accumulations are found in the Miranga, upper Miranga, and Brejão Sandstones.

In lithologic character all the productive beds are similar. They are fine- to very fine-grained, quartzose, well-sorted, and generally clean and porous sandstones. Laboratory analysis by Fonseca *et al.* (1966) indicated a very low clay and carbonate content, the former averaging 3 percent. Permeability barriers caused by "shaleout" are present in the upper Miranga and Miranga Sandstone and in some bodies of the São Paulo Sandstone.

The porosity of the reservoirs generally ranges from 19 to 25 percent and the total net oil-sandstone thickness is as much as 70 m.

Structure.—A structure map with the top of the Santiago Sandstone as datum (Fig. 9) shows a well-defined, elliptical dome which trends northeast and is cut by several strike and diagonal faults. Two distinct fault sets are evident in the field; approximately N45°E and N35°W. The accumulation in the several producing beds is controlled mainly by the domal closure and partly by the NE-trending faults. The domal closure on top of the Santiago Sandstone is 180 m.

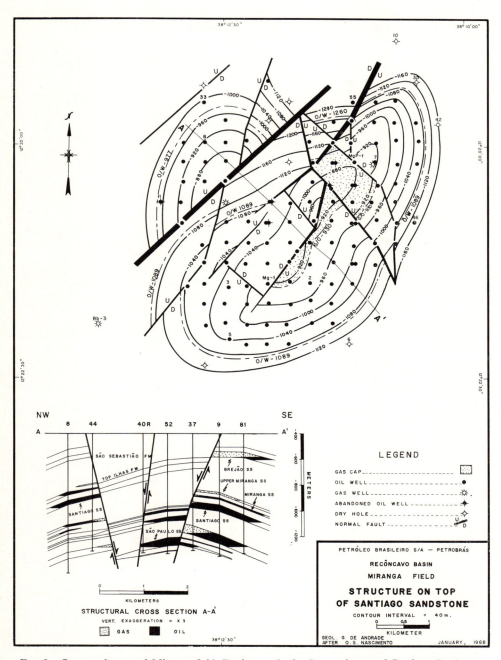

FIG. 9.—Structural map of Miranga field, Recôncavo basin. Datum is top of Santiago Sandstone.

The field is divided into two major blocks by a N45°E-striking normal fault which dips 43° southeast and has 100–300 m of throw. This fault provides the southeastern closure for the accumulation in the northwestern upthrown block.

Almost nothing is known about the projection of this shallow structural feature on the deep Sergi horizon because only one well reached that formation. At the surface, structure is concealed almost entirely by deep weathering of the outcropping São Sebastião strata.

Several gas caps have been identified in the field. The position of both gas-oil and oil-water contacts for each productive sandstone is controlled mainly by structure. In the northwestern upthrown block the contacts average 90 m higher than those in the southeastern downthrown block.

Structural evolution.—From the information supplied by the shallow Miranga wells, the development of the present field structure began before the deposition of the Miranga Sandstone when contemporaneous faulting formed the transverse horst block in the northeastern part of the field. An isopach map of the upper member of the Ilhas Formation shows that, at the end of the deposition of this sequence, the domal feature at the level of the São Paulo Sandstone was fairly well delineated on the east, south, and west, but open on the north. The post–São Sebastião faulting phase completed the present structure, the northern domal closure probably resulting from tilting of the faulted blocks in that direction.

Production.—Miranga field produced 14.4 million bbl through May 31, 1967; this is only 2.7 percent of the oil in place. About 99 percent of the oil production has been from the São Paulo and Santiago Sandstones. Primary oil recovery is expected to be 104 million bbl.

Production of 40,000 bbl/day makes Miranga the largest producing field of the basin. Double and triple completions are the normal procedure in the field, because most wells were drilled through several productive sandstone bodies. At present the reservoir pressure is great enough to assure flowing production, but secondary-recovery studies are being carried out for future use.

Average properties of the oil, which is of paraffinic base, are: API gravity, 38–41°; viscosity at 48.9°C, 52 SSU; pour point, 32°C; and sulfur content, less than 0.1 percent. An average analysis of the crude is 21 percent gasoline, 14 percent gas oil, and 7 percent fuel oil.

Candeias Field

Candeias field, in the south-central part of the basin, is on the western flank of the deep Camaçari trough and about 35 km northwest of Salvador (Fig. 6). It was discovered in December 1941 by the Conselho Nacional de Petróleo. The wildcat was completed with an initial production rate of 75 bbl/day of oil. The discovery well was located on the basis of surface mapping, with support from seismic data, on the axis of a north-south-trending anticline; however, later development of the field revealed the presence of a typical stratigraphic-trap accumulation. The field strikes NW-SE, and is 8 km long and 1.5–4 km wide. It is completely developed; of the total 165 wells drilled, 29 were abandoned as dry holes or noncommercial.

Productive beds.—The productive beds are lenticular sandstone and fractured shale and limestone of the Candeias Formation ranging in depth from 690 m to 2,400 m. Minor accumulation also has been found in stray sandstones of the Ilhas Formation and in the basal **A** Sandstone of the Candeias Formation.

The producing beds of the Candeias Formation are grouped into four production units named 1st, 2d, 3d, and 4th zones (Fig. 10). The shallow 1st and 2d zones are developed in the northwestern part of the field and are very similar lithologically. They consist of sandstone lenses interbedded in shale and limestone. The sandstone is fine to medium grained, quartzose, calcitic, and locally contains shale and limestone fragments. At the base of both zones there is an intraformational conglomerate of limestone and shale pebbles in a sandstone matrix. The primary porosity and permeability values are variable, and depend on the amount of carbonate cement. Laboratory studies by Leroy *et al.* (1964) showed a wide variation in the cement content of the sandstone, with an average of 10–12 percent but a maximum of 35 percent. Argillaceous matrix is very scarce. Fractures, in places oil stained, have been observed in cores from several wells. The gross thickness of each zone is about 50 m, whereas the net oil-sandstone thickness averages 10 m.

The 3d zone is the best reservoir and is developed in the central part of the field. It also consists of sandstone lenses and interbedded shale, limestone, and siltstone. The sandstone is fine to coarse grained, calcitic, hard, commonly

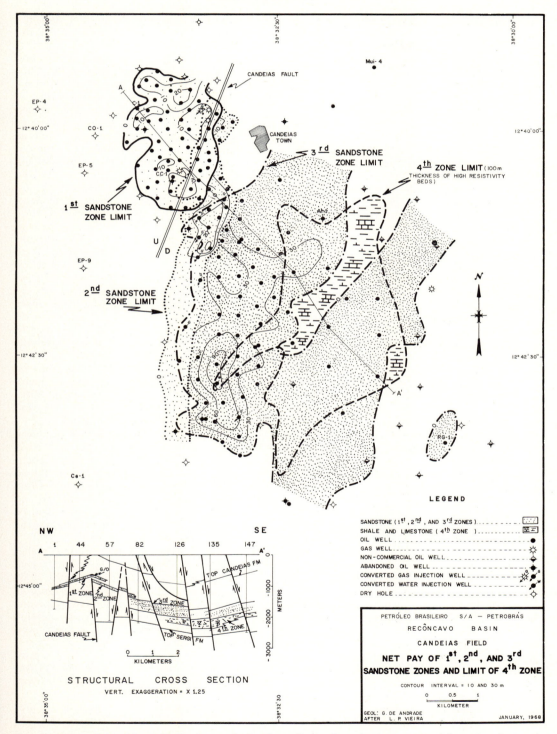

Fig. 10.—Map of Candeias field, Recôncavo basin, shows limits of productive zones and isopachs of net pay sandstones in Candeias sandstone bodies.

fractured, and in places has limestone and shale inclusions. Primary porosity and permeability values depend on the amount of calcareous cement. The gross thickness of the zone is 350 m and the net oil-sandstone thickness is as great as 100 m.

The 4th zone comprises fractured shale and limestone. No sandstone has been found in continuous cores taken from two wells. The shale is dark gray to grayish black, very hard, and grades to grayish-brown, cryptocrystalline limestone. These hard rocks show high resistivity on electric logs. Vieira (1964) studied the fractures and obtained the following data from cores: average thickness (where filled with calcite), 1 mm; predominant dips, 70–90°; preferred orientation, N65°E (principal) and N25°W (secondary). The 4th zone, 250–300 m thick, is overlain by the 3d zone and is developed in the southeastern part of the field.

The average porosity values of the sandstone are 16 percent for the 1st and 2d zones, and 13 percent for the 3d zone.

Structure.—The four productive zones in the field are arranged in overlapping steps, beginning with the shallower 1st zone in the northwest, and extending to the deepest (4th) zone in the southeast (Fig. 10). The structure shown at the top of the 1st zone is a faulted, closed anticline trending N13°E. This structural pattern is not reflected at the top of Sergi Formation. The 2d zone is a faulted homocline dipping southeast, whereas the 3d and 4th zones are broad faulted synclines whose axes trend southwest-northeast.

The most important structural feature is the Candeias fault, striking N20°E, in the northwestern part of the field. At the top of the Sergi Formation this fault has about 650 m of throw. The throw dies out upward before reaching the 1st zone level. Faulting was active during deposition of the Candeias sediments and was important in the distribution of the sands, because most sandstone lenses of the field are in the downthrown block.

Later structural movement, after deposition of the São Sebastião Formation, affected the area and is represented by several faults, some of them dying out downward. No oil-water contact has been found in the four zones, but a small gas cap is present in the highest part of the 1st zone anticline.

Origin of sandstone lenses.—Although oil accumulation in the Candeias field is not related to the present structure, because the trap is stratigraphic, the deposition of the reservoir lenses resulted from special events that occurred during the structural evolution of the area.

Isopach maps show that the first faulting phase took place during the deposition of Candeias Formation. The activity of the Candeias fault during that time created two depositional environments; on the southeastern downthrown block a paleotrough was the depositional site of the thick sand lenses of the 3d zone, while at the same time on the northwestern upthrown block no sand was deposited. Sedimentologic studies by Leroy *et al.* (1964) indicate that the sand was transported from the northwest and probably originated from outcrops of the Sergi Formation.

The sandstone lenses of the 2d zone, also in the downthrown block of the Candeias fault, formed during a new sand-deposition cycle when lower energy currents caused deposition closer to the source area. At the time of deposition of the 1st zone, a new base level was established because the Candeias fault no longer was active and the southeastern tectonic trough already was filled with sediments. According to Leroy *et al.* (1964), the 1st zone represents shallow-water fluvial deltaic deposits.

Later structural adjustments of the blocks, probably after compaction of the Candeias sediments, fractured the more competent rocks such as the calcitic sandstone, limestone, and hard calcareous shale of the four zones.

Production.—The Candeias field produced 47.3 million bbl of oil through May 31, 1967, which is 13 percent of the oil in place. About 80 percent of the oil produced was from the 3d zone. The remaining oil reserve (46 million bbl) is distributed equally between the 3d and 4th zones; consequently, the latter zone will be more significant in future production. Both the 1st and 2d zones are almost depleted.

Present production of 8,000 bbl/day makes Candeais the sixth-largest field of the basin. Peak production was 18,000 bbl/day in 1959. The production of the 4th zone is erratic, as may be expected from a fractured reservoir. Three wells are good producers, averaging 600–1,000 bbl/day, but several others have been tested without success.

Gas is being injected in the 3d zone for pressure maintenance and a pilot water project was carried out, but large-scale secondary recovery was not started before 1969.

Average properties of the paraffinic oil are: API gravity, 29–32°; viscosity at 54.4°C, 116–208 SSU; pour point, 30°C; and sulfur

content, 0.09 percent. An average analysis of
the crude is 13 percent gasoline, 8 percent gas
oil, and 10 percent fuel oil.

Buracica Field

Buracica field, discovered in April 1959 by
Petrobrás on the basis of a seismic survey, is in
the northwestern part of the basin, on the Mata-
Catu structural trend (Fig. 6). Because this
field is similar to Água Grande, it is described
only briefly.

Buracica is a typical horst divided into three
major blocks by east-west-striking faults (Fig.
11). The producing reservoirs are the **A** Sandstone (southern block) and the Sergi Formation (central and northern blocks); these reservoirs range in depth from 520 to 740 m. In the
Sergi reservoir, which contains about 88 percent of the recoverable oil of the field, the
structural closure is provided by the major
western and eastern limiting faults (200–300
m of throw) and by the tilting of the blocks toward the south-southeast. The oil column is
about 210 m in the central block. The accumulation in the **A** reservoir is controlled partly by
a pinchout of the sandstone. The maximum net
oil-sandstone thickness per well is 150 m, and
no gas cap has been found. The Sergi oil-water
contact in the central and northern blocks is
level, at −523 m; in the southern block the
contact is at −575 m for both **A** and Sergi reservoirs.

The evolution of the Buracica structure took
place in two distinct phases. The early tectonic
movement, contemporaneous with the deposition of the Candeias Formation, is represented
by the western limiting fault and the northern
east-west-striking fault. The eastern boundary
faults and the southern east-west-striking fault
belong to the post–São Sebastião faulting phase
during which the ancient faults were reactivated.

Cumulative production through May 31,
1967, was 32.6 million bbl of oil, which is only
7.3 percent of the oil in place. The remaining
recoverable oil is about 83 million bbl. Present
production is 20,000 bbl/day, third greatest in
the basin. Large-scale secondary recovery from
the Sergi reservoir is to begin in 1969, but the
A reservoir has been waterflooded since 1963.

Average properties of the paraffinic oil are:
API gravity, 33–35°; viscosity at 54.4°C, 70
SSU; and pour point, 36°C.

Taquipe Field

Taquipe field, discovered in September 1958

by Petrobrás on the basis of surface mapping
with seismic-data support, is in the central part
of the basin about 60 km north of Salvador
(Fig. 6).

As in Miranga field, the producing reservoirs
are several sandstone beds of the Ilhas Formation, mainly the Santiago Sandstone which contains about 72 percent of the recoverable oil.
Less important accumulations are found in the
São Paulo, Cambuqui, and Intermediário Sandstones.

The structural contour map with the top of
the Santiago Sandstone as datum (Fig. 12)
shows a doubly plunging anticline trending
north-south and faulted along the flanks; this
structure was developed in the downthrown
block of a major fault (Riacho da Boa Espe-
rança fault). The São Paulo accumulation, with
an oil column of 45 m, is in the central part
of the field only and is controlled by a domal
closure. Additional closure in the other reservoirs is provided by pinchout and normal
faults. The maximum net oil-sandstone thickness per well is about 55 m. A gas cap is present in the Santiago Sandstone.

The Taquipe structure is largely the result of
the post-São Sebastião faulting phase. Isopachs
of the upper Ilhas, indicating pre-São Sebastião
paleostructure of the São Paulo Sandstone,
delineate a small, open-to-the-east, domal feature in the central part of the field.

Cumulative production through May 31,
1967, was 32.9 million bbl, 14.7 percent of the
oil in place. The remaining recoverable oil is
estimated to be 50 million bbl. Present production is 10,500 bbl, making this field the fifth
largest in the basin. Some gas and water have
been injected into the Santiago Sandstone, but
large-scale secondary-recovery operations were
not begun before 1969. API gravity of the paraffinic Taquipe oil ranges from 35 to 39°.

Conclusions

1. The Recôncavo basin can be considered to
be a fairly prolific basin, because of the volume
of the recoverable oil discovered in relation to
the number of wildcats drilled and the extent
of the basin.

2. Six major shallow accumulations contain
most of the oil reserves and provide most of
the production. These major accumulations are
related to the structural evolution of the basin
and can be classified into three basic types.

The *first type* is represented by the Água
Grande, Dom João, and Buracica fields. They
are well-defined horsts developed along re-

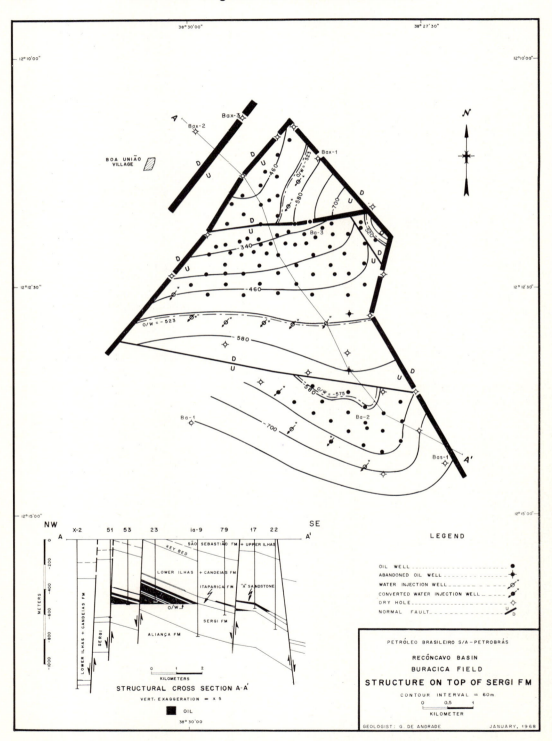

FIG. 11.—Structural map of Buracica field, Recôncavo basin. Datum is top of Sergi Formation.

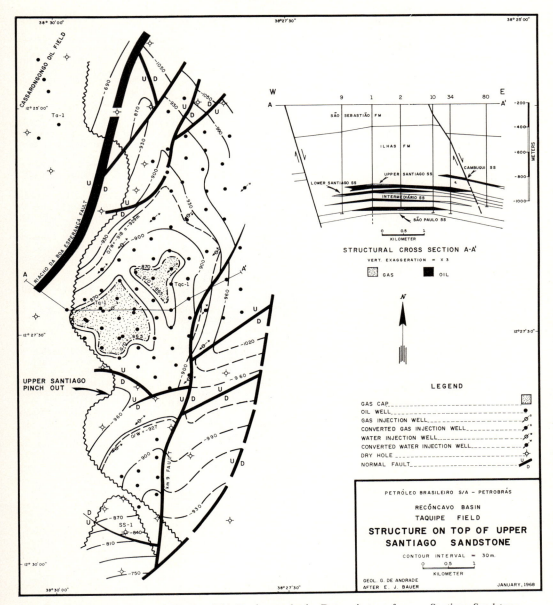

FIG. 12.—Structural map of Taquipe field, Recôncavo basin. Datum is top of upper Santiago Sandstone.

gional basement highs. The oil entrapment in these fields is related to the early faulting phase in the basin. The producing reservoirs are the Jurassic(?) Sergi Formation and the Lower Cretaceous A Sandstone.

The *second type* includes Miranga and Taquipe fields; both are in deep basin troughs and were produced by structural conditions developed during the late faulting phase in the basin. Folds formed on the downthrown blocks of major faults are the main traps in these fields; faults and local pinchouts of reservoir sandstones also were important in oil accumulation. The main producing beds are the Lower Cretaceous Santiago and São Paulo Sandstones.

The *third type* is stratigraphic and is represented by the Candeias field on the west flank of a deep regional trough; this field has no

major counterpart in the basin. The reservoir
rocks are sandstone lenses and fractured shale
and limestone of the Lower Cretaceous Candeias Formation. In this field the distribution of
the sandstone lenses is related to faulting contemporaneous with deposition of the Candeias
sediments.

SELECTED REFERENCES

Barnes, B. E., 1950, Progress report of surface geo-
ogical work in Recôncavo during 1949: Salvador,
unpub. rept., Conselho Nacional de Petróleo.
Bauer, E. J., 1962, A structural-stratigraphic analysis
of the Taquipe field, Recôncavo basin: Salvador,
unpub. Petrobrás rept., Divigeo.
——— 1967, Genesis of Lower Cretaceous "A" Sandstone, Recôncavo basin, Brazil: Am. Assoc. Petroleum Geologists Bull., v. 51, no. 1, p. 28–54.
Depex, 1958 (revised, 1961), Recôncavo baiano: Rio
de Janeiro, unpub. basin study rept., Petrobrá
Explor. Dept.
Figueredo, N., 1967, Relatório semi-anual das reservas
de óleo e gás da RPBA: Salvador, unpub. Petrobrás
rept., Dirpro.
Fonseca, J. I., 1965, Viagem à África (Congresso e
Contatos): Rio de Janeiro, unpub. Petrobrás rept.,
Divex.
Fonseca, J. R., L. D. Jobim, and L. E. Neves, 1966,
Estudo sedimentológico dos arenitos da Formação
Ilhas na Bacia do Recôncavo: Salvador, unpub.
Petrobrás rept., Direx.
——— and ——— 1967, Estudo sedimentológico do
Arenito Sergi no campo de Dom João: Salvador,
unpub. Petrobrás rept., Direx.
Forgotson, J. M., Jr., 1957, Nature, usage, and defi-
nition of marker-defined vertically segregated rock
units: Am. Assoc. Petroleum Geologists Bull., v.
41, no. 9, p. 2108–2113.
Ghignone, J. I., 1963, Geologia do flanco oriental do

Tucano Norte (do Vaza Barris ao São Francisco):
Salvador, unpub. Petrobrás rept., Setex.
Hartt, C. F., 1870, Geology and physical geography
of Brazil: Boston, Fields Osgood & Co., 620 p.
Humphrey, F. L., D. Mathias, and F. Shiguemi, 1961,
Petrographic study of the "A" Sandstone from well
AG-156-BA, Recôncavo basin, Brazil: Salvador,
unpub. Petrobrás rept., Cenap.
Jenkins, C. B., 1960, Progress report on Água Grande:
Salvador, unpub. Petrobrás rept., Divigeo.
Leroy, M., J. R. Fonseca, C. V. Delboux, 1964,
Estudo sedimentológico das areias da Formação
Candeias médio, Campo de Candeias: Salvador,
unpub. Petrobrás rept., Setex.
Luz, A. A., 1955, Geologia de área norte de S. Sebas-
tião do Passé—Estrutura do Rio Taquique: Salvador,
unpub. rept., Conselho Nacional de Petróleo.
Nascimento, O. S., and L. J. Passos, 1966, Análise
estrutural, estratigráfica e econômica do Campo de
Miranga: Salvador, unpub. Petrobrás rept., Divigeo.
Oliveira, A. I., and O. H. Leonardos, 1943, Geologia
do Brasil, 2d ed.: Rio de Janeiro, Serviço de In-
formação Agrícola do Ministério da Agricultura,
813 p.
Pack, O., and L. A. Almeida, 1946, Geological developments during the year 1945: Salvador, unpub.
rept., Conselho Nacional de Petróleo.
Passos, L. J., 1968, Tentativa de uniformização da
nomenclatura dos arenitos da Formação Ilhas
(membro superior): Salvador, unpub. Petrobrás
rept., Direx.
Taylor, E. F., 1947, Proposed geological nomenclature for the Recôncavo, Bahia: Salvador, unpub.
rept., Conselho Nacional de Petróleo.
Viana, C. F., 1965, Estudo preliminar de Ostracodes
da Série Cocobeach, Gabon: Salvador, unpub.
Petrobrás rept., Setex.
Vieira, L. P., 1964, Relatório geológico do Campo
de Candeias: Salvador, unpub. Petrobrás rept.,
Divigeo.

Geology of Groningen Gas Field, Netherlands[1]

A. J. STÄUBLE[2] and G. MILIUS[2]

The Hague, Netherlands

Abstract The Slochteren No. 1 well discovered in 1959 what now is known as the Groningen gas field in the northern Netherlands. The field is on a culmination of the large, regional northern Netherlands high which was formed during the late Kimmerian tectonic phase (Late Jurassic–Early Cretaceous). However, there is some evidence that the structure was a positive element during earlier periods, *i.e.,* during Triassic and possibly even late Carboniferous times.

The reservoir overlies unconformably the truncated and strongly faulted coal-bearing Pennsylvanian strata which are considered to be the main source of the Groningen gas. The reservoir, 300–700 ft thick, consists of fluviatile and eolian sandstone and conglomerate of the Rotliegendes Formation (Lower Permian). These coarse clastic beds are overlain by a few thousand feet of Permian Zechstein evaporites, notably rock salt and to a lesser extent anhydrite and dolomite, which constitute the very effective reservoir seal. Because of intensive salt movements, the thickness of the overlying Mesozoic and Cenozoic strata ranges from 3,000 ft to more than 6,500 ft.

The field covers 180,000 acres, and the reserves now are estimated at 58 trillion ft^3. Present production potential is 2 billion ft^3 of gas per day from nine "clusters" of about six closely spaced wells each. The favorable reservoir properties of the sandstone allow, at least for the present, drainage of the field from the structurally highest southern part.

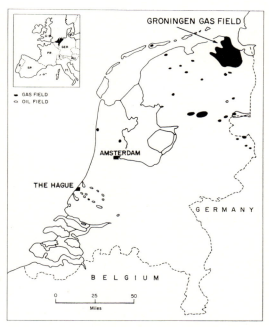

FIG. 1.—Location map of Groningen gas field, The Netherlands.

INTRODUCTION

The Groningen gas field is in the northeastern part of The Netherlands near the most densely populated and highly industrialized areas of northwestern Europe (Fig. 1). Its discovery in 1959 by the N. V. Nederlandse Aardolie Maatschappij (NAM) and the realization of the size of the field triggered the current intensive search for gas in The Netherlands, in the adjoining countries, and on the continental shelf bordering the United Kingdom, The Netherlands, Germany, Denmark, and Norway.

In The Netherlands this sudden keen competition revealed the inadequacy of the mining legislation, the Napoleonic law of 1810, and

threatened to create a chaotic situation. A temporary drilling halt was imposed by the government in 1965 and new legislation is being worked out. Interim laws were passed in 1967 to allow for an orderly continuation of exploration activity on land, and a mining law was passed by the Parliament to regulate exploration and production on the Dutch part of the North Sea shelf.

The purpose of this paper is to outline the geology of the Groningen gas field. It is based on the work of numerous present and past staff members of NAM whose valuable efforts are acknowledged by the writers.

HISTORY

Exploration for hydrocarbons in The Netherlands was governed for many years by an antiquated mining law.

In 1933, the Bataafsche Petroleum Maatschappij, one of the Royal Dutch/Shell compa-

[1] Read before the 53d Annual Meeting of the Association, Oklahoma City, Oklahoma, April 23, 1968. Manuscript received, July 10, 1968; accepted, October 7, 1968.

[2] Bataafse Internationale Petroleum Mij.

We thank the management of the N. V. Nederlandse Aardolie Maatschappij and the parent companies, Shell Petroleum N. V. and Standard Oil Company of New Jersey, for permission to publish this information.

nies, obtained temporary exclusive exploration rights in the eastern Netherlands. Gravity surveying was started in 1935, and was followed soon by seismic-refraction shooting and strathole drilling. Oil first was struck at Schoonebeek, near the end of the war.

In 1947, a joint venture was founded by Shell and Standard Oil Company (New Jersey) under the name of the "N.V. Nederlandse Aardolie Maatschappij" (NAM). Initially the search mainly concentrated on the Lower Cretaceous oil prospects of the eastern Netherlands, but later included the gas prospects in the Permian carbonate rocks, also in the eastern Netherlands, and the Lower Cretaceous oil prospects of the western Netherlands.

The first exploration well in the province of Groningen was drilled in 1952. The target of this well, Haren 1, was the Permian Zechstein carbonate rocks, which were found in an anhydritic, tight, basinal facies. With deeper drilling the well penetrated for the first time 600 ft of water-bearing Lower Permian Rotliegendes sandstone and thus proved the presence of a potentially important reservoir underlying the Permian evaporites. A follow-up well, Ten Boer 1, was stopped for technical reasons in the predominantly shaly upper part of the Lower Permian. It was completed as a small gas producer from Zechstein carbonate rocks (Fig. 6).

Drilling of two additional wells at Delfzijl and Slochteren was deferred when the Suez crisis shifted attention to development drilling in proved oil fields. It was not until 1959 that the Slochteren 1 well found the Lower Permian Rotliegendes Formation to be gas bearing. The next well, Delfzijl 1, revealed only the upper part of the sequence to be gas bearing on what then was thought to be a separate structure. Seismic detailing, additional appraisal wells, and the use of improved velocity data led to drastically altered maps showing the separate structures to be part of one giant gas field.

A careful review of all pertinent data in 1966 showed that the 100-percent water contour encloses about 180,000 acres with an estimated proved reserve of 58×10^{12} ft^3.

STRATIGRAPHY

From late Carboniferous time, the stratigraphy of The Netherlands has been governed by three orogenic phases, i.e., the Variscan tectonic phase in Carboniferous time, the Kimmerian tectonic phase in Mesozoic time, and the Laramide tectonic phase in Late Cretaceous–

early Tertiary time. These phases created the paleogeographic framework which controlled the depositional history of The Netherlands, including the Groningen area (Fig. 2).

Carboniferous

In the northern Netherlands the oldest strata penetrated are Pennsylvanian. The facies consists of alternating deltaic shale, sandstone, and coal; the marine intercalations become less important upward. These clastic deposits were derived from a mountain range in the south of The Netherlands which was formed during the Variscan orogeny in Carboniferous time. In late Pennsylvanian time the deltaic facies changed gradually into a redbed facies without coal, a sequence which in the Groningen area either was not deposited or was removed during uplift and truncation associated with the late Variscan (Asturian and/or Saalian) tectonic phases (Bartenstein, 1968; Thiadens, 1963).

Permian

Unconformably overlying the differentially truncated Pennsylvanian strata are redbeds of the Lower Permian Rotliegendes Formation. Because of the lack of faunal evidence, an exact age determination has not been possible, but on the basis of regional considerations the strata are believed to belong to the upper part of the Lower Permian. The important reservoirs of the Groningen gas field and of several other fields in The Netherlands belong to the Rotliegendes Formation, which extends across large areas of northwestern Europe.

In the Groningen field and neighboring areas, the overall development of the Rotliegendes Formation is rather uniform and allows a distinct subdivision into two rock-stratigraphic subunits, the Slochteren Member and the Ten Boer Member.

The Slochteren 4 well has been selected as the type section for both members. This development well was drilled in 1963, 7.5 mi eastsoutheast of the city of Groningen. The well reached a total depth of 9,507 ft. The Rotliegendes Formation was cored completely, and the cores are kept by NAM.

Slochteren Member.—The gamma-ray log of the type-section interval 8,850–9,370 ft in the Slochteren 4 well is shown in Figure 3.

The base of the Slochteren Member is defined by the unconformity separating the coarse clastic beds of the Rotliegendes from the fine

TIME-STRATIGRAPHY			THICK-NESS	LITHOLOGY	ENVIRONMENT OF DEPOSITION	GAS OCC.	ROCK STRATIGRAPHY		MAIN TECTONIC EVENTS
CENOZOIC	Quaternary		900'		Continental				
CENOZOIC	Tertiary		1000'-3000'		Deltaic to Marine				Laramide / Early Alpine
MESOZOIC	Cretaceous	U	2200'-3500'		Marine				Late Early Kimmerian
MESOZOIC	Cretaceous	L			Marine				
MESOZOIC	Triassic		0-1200'		Arid Continental to Restricted Marine		Muschelkalk / Buntsandstein	M Triassic L	
PALEOZOIC	Permian	U	2000'-4800'		Restricted Marine		IV Leine / III Aller / II Stassfurth / I Werra	Zechstein	
PALEOZOIC	Permian	L	400'-900'		Arid Continental to Coastal	☼	Ten Boer Member / Slochteren Member	Rotliegendes	Saalian / Asturian
PALEOZOIC	Pennsylvanian		>3000'		Humid Deltaic to Marine				Variscan

Legend:
- ▭ Shale/Clay
- ⬚ Sand
- ⦾ Conglomerate
- ⟋⟋ Anhydrite
- ⊠ Salt
- ▭ Coal
- ⊞ Limestone
- ⊟ Dolomite
- III Chalk
- ⟋⟋ Removed by diff. truncation

Fig. 2.—Generalized stratigraphy of Groningen gas field, The Netherlands.

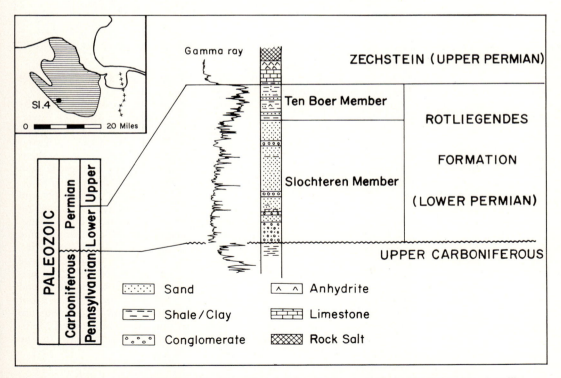

Fig. 3.—Type section of Rotliegendes Formation for Groningen gas field (in Slochteren No. 4 well).

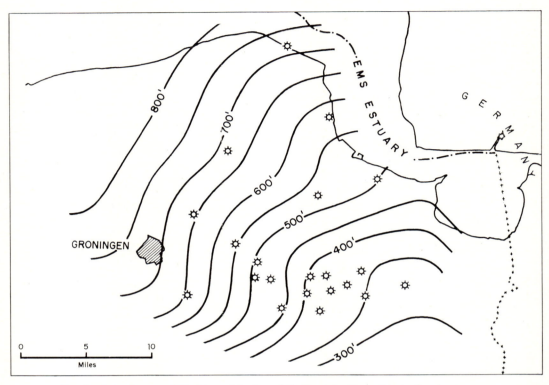

FIG. 4.—Isopach map of Slochteren Member of Rotliegendes Formation in Groningen area.

clastic facies of the Pennsylvanian. The top of the Slochteren Member is defined by the top of the uppermost massive sandstone underlying the Ten Boer Member. The thickness of the sequence ranges from 270 ft in the south of the field to more than 660 ft on its northern edge (Fig. 4). The member extends in all directions beyond the limits of the field.

The beds of the Slochteren Member range from coarse reddish conglomerate to poorly to very well-sorted, pebbly to fine, brown-gray sandstone (commonly loosely consolidated). The number of conglomerate bands decreases upward. Several shale intercalations, mostly insignificant, are present. The *in situ* porosity of the sandstones ranges from 15 to 20 percent, and their permeability ranges from 0.1 to 1 darcy, with a few streaks of up to 3 darcys.

The sandstone strata described are typical of Slochteren 4 and of the whole southern part of the Groningen field.

To the north, the number of conglomeratic beds decreases and the content of fine sand in the lower part, associated with shale, increases (Fig. 5). This indicates the main source of the

Rotliegendes clastic material to be to the south.

Genetically, the Slochteren Member may be subdivided into two major groups, (1) fluviatile conglomerate, sandstone, and clay; and (2) eolian sandstone. Both were deposited under severe arid conditions, the fluviatile clastic beds representing wadi deposits which alternate with dunes. A distinction between the types is possible on the basis of sedimentary structures and the type and quantitative distribution of detrital fragments. The rock fragments in fluviatile deposits consist primarily of epimetamorphic quartzite, schist, granite, and volcanic material, whereas the dunes contain a relatively large amount of katametamorphic material. The origin of these highly metamorphic fragments and thus of part of the eolian clastic material is not known. In the Groningen field, dune sands are more common in the Upper Slochteren Member, and fluviatile sands predominate in the lower part.

Ten Boer Member.—The gamma-ray log of the type-section interval, 8,743–8,850 ft, is given in Figure 3. The top of the member is defined by the first occurrence of red shale underlying the dark-gray to black bituminous Copper Shale. The base is defined by the first occurrence of coarse sandstone of the Slochteren Member. The beds consist predominantly of reddish-brown, hard, silty to fine sandy claystone, with white to green-gray anhydrite nodules. They are interspersed with numerous inch-thick siltstone or sandstone beds and a few sandstone layers 1–2 ft thick.

Porosity and permeability of these commonly clayey or silty sandstone stringers are very low, whereas the values for the thicker beds equal those of the Slochteren Member.

The thickness of the Ten Boer Member ranges from 100 ft in the south of the field to 250 ft in the north. On all sides the unit extends beyond the limits of the field. The change from the high-energy coarse clastic beds of the Slochteren Member to the clayey Ten Boer Member is the first indication of a change from the arid continental environment of deposition and erosion, which had prevailed from latest Carboniferous onward, to the restricted marine environment of the Upper Permian evaporite series. The Ten Boer Member is believed to be

a sabkha deposit laid down along the margins of a subsiding continental to coastal basin.

Upper Permian

The Upper Permian evaporite series began with a marine transgression marked by the deposition of the ubiquitous 4–5-ft-thick black bituminous Copper Shale, which overlies, apparently conformably, the redbeds of the Ten Boer Member. The transgression initiated the four evaporite cycles of the Zechstein which are well known in Germany—from oldest to youngest: I, the Werra-cycle; II, the Stassfurt-cycle; III, the Leine-cycle; and IV, the Aller-cycle, all deposited in a differentially subsiding restricted marine basin. In the Groningen field each of these cycles is characterized by a more or less pronounced upper and lower anhydritic to dolomitic sequence and (except for cycle I) a middle unit consisting predominantly of rock salt. In cycles III and IV, also potassium and magnesium salts were precipitated (Visser, 1963). Because of intensive salt movements, salt thickness ranges from 2,000 to 4,800 ft. The estimated depositional thickness in the Groningen area is believed to be of the order of 3,000

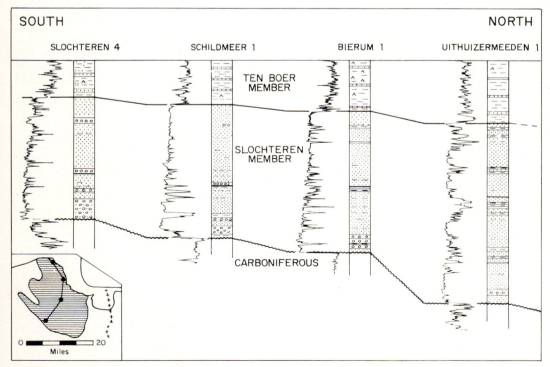

FIG. 5.—South-north stratigraphic cross section through Rotliegendes Formation in Groningen gas field showing thickening toward north.

ft, covering the Rotliegendes reservoir with an effective seal which, in the Groningen gas field, never was broken during the subsequent geologic history.

Triassic

In The Netherlands, the Triassic is developed in the so-called Germanic facies, with the characteristic threefold lithostratigraphic subdivision into Buntsandstein, Muschelkalk, and Keuper. The lower part of the Buntsandstein is a monotonous sequence of possibly shallow-marine anhydritic shale up to 1,200 ft thick. It is environmentally comparable with the sabkha deposit of the Ten Boer Member. Mild tectonic movements, regionally known to have occurred in the Middle Buntsandstein, resulted in a depositional thinning and intraformational truncation in the Groningen area. This tectonic episode was followed by subsidence and by the deposition of up to 300 ft of rock salt (Rötsalt) in the Upper Buntsandstein. Dolomite and green shale characterize the Muschelkalk. Both Upper Buntsandstein and Muschelkalk are preserved only in isolated marginal deeps around salt domes in the southern part of the Groningen province.

The youngest rock-stratigraphic unit of the Triassic, the continental Keuper, has not been found. The absence of Keuper and Muschelkalk, and in parts of the field even of the Buntsandstein is due to late Kimmerian truncation of Triassic strata previously folded by salt movements.

Jurassic

From regional considerations, it appears that the Groningen area, and The Netherlands as a whole as shown by Harsveldt (1963), belonged to the northwestern European sedimentary area. Within this area, about 2,000 ft of Liassic beds and an equal amount of Dogger and Malm beds were deposited. However, no Jurassic strata are left in the Groningen area. They were removed completely during the late Kimmerian tectonic phase in Late Jurassic-Early Cretaceous time.

Cretaceous

Overlying the differentially truncated Triassic strata are transgressive gray to brown, silty shale beds of Early Cretaceous age, ranging in thickness from 200 to 450 ft. The onlap onto the late Kimmerian erosion surface was differential. The low areas were flooded at earlier stages in the Early Cretaceous than the relatively high areas.

At the beginning of Late Cretaceous time, uniform sedimentary conditions were restored over wide areas of northwestern Europe, and an important sequence of marl grading upward into chalk with chert nodules was deposited. On the Groningen structure, halokinetic movements resumed in the Late Cretaceous and account for the variation in thickness from 2,000 to 3,300 ft.

Tertiary and Quaternary

After a period of regional mild uplift and truncation associated with the Laramide tectonic phase in Late Cretaceous–early Tertiary time, deposition resumed in the Eocene with glauconitic sand followed by clay. Calcareous sandstone and shale became characteristic of the middle and late Eocene. Tectonic movements related to the late Alpine tectonic phases took place in the Oligocene. In the Groningen area the lower Oligocene is everywhere absent and, consequently, sandy to shaly middle Oligocene beds overlie an eroded Eocene section. The sandy to shaly Miocene section is incomplete or absent. The Quaternary is developed in a sandy to conglomeratic facies with few clay intercalations.

The thickness of the combined Tertiary and Quaternary ranges from 1,100 to 2,900 ft. The thickness variations are due less to the regional tectonic events than to strong local halokinetic movements.

TECTONICS

The Groningen structure forms part of the large regional North Netherlands high which originated during the late Kimmerian tectonic phase in Late Jurassic–Early Cretaceous time. In the process of uplift, tilt, and truncation, this regional feature was strongly faulted and broken into individual elements. The Groningen structure is such an element, which itself is a gently folded wedge-shaped feature with strongly faulted east, south, and west flanks and a gently dipping north flank.

Figure 6, depicting the outline of the field at the level of the Slochteren Member, shows that most faults trend northwest-southeast, as is characteristic for the late Kimmerian tectonic phase. The throws of the boundary faults range from a few hundred to more than 1,000 ft. In the Groningen field, the faults may be grouped into three categories based on their main periods of activity: (1) faults related to the Variscan

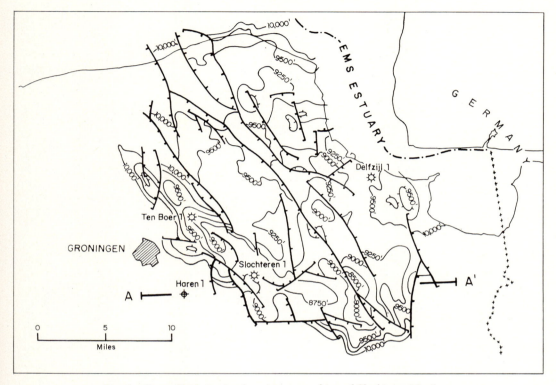

Fig. 6.—Generalized structural contour map of top of Slochteren Member.

orogeny, (2) faults related to younger tectonic phases, and (3) faults related to halokinesis.

The first category comprises faults which were active during Asturian and Saalian deformation in late Carboniferous to earliest Permian time. The evidence for the presence of these faults is scanty and circumstantial and consists of the variation in the amount of truncation of Pennsylvanian strata underlying the Lower Permian Rotliegendes reservoir. Differential truncation of Pennsylvanian strata in the Groningen gas field indicates that late Carboniferous faulting was considerable and that the cumulative throw of some of these faults is in the order of 2,000–3,000 ft and perhaps even more.

The second category contains most of the NW-SE-trending faults which dissect the Groningen structure (Fig. 6). They may be partly rejuvenated Paleozoic faults and possibly are associated with the late Kimmerian tectonic phase during which the Groningen structure attained its present form. An exact determination of the age of faulting is not possible, because all the faults terminate in the incompetent salt

layer of the Upper Permian. The throw of most faults within the field does not exceed 200 ft and is therefore smaller than the thickness of the Slochteren reservoir. Additional displacement is likely to have taken place in the course of movements related to the Laramide and late Alpine orogenies. Such movements probably caused mainly the rejuvenation of older subsalt faults.

The third category, post-Permian faults, is associated primarily with halokinesis in Mesozoic, Tertiary, and Quaternary times (Fig. 7). Compensatory movements around salt diapirs and pillows are common and, although the throw of some is considerable, they have not affected the subsalt geometry of the Groningen field.

GEOLOGIC HISTORY

Only a general reconstruction of the geologic history of the Groningen gas field is possible. Because of erosion, the more subtle, gradual stratigraphic changes and the effects of minor tectonic events during the long depositional periods remain unknown, whereas the effects of

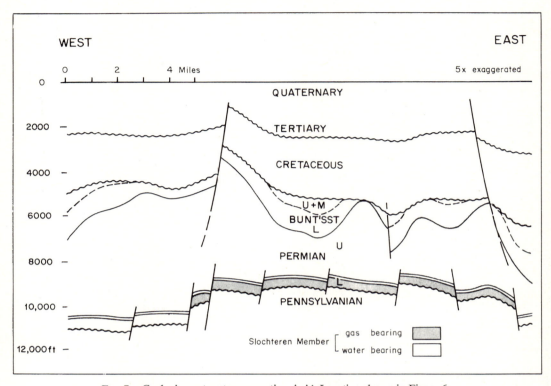

WEST   EAST

0   2   4 Miles   5x exaggerated

QUATERNARY

TERTIARY

CRETACEOUS

U + M
BUNT'SST
L
U

PERMIAN

L

PENNSYLVANIAN

Slochteren Member ⌈ gas bearing
                  ⌊ water bearing

FIG. 7.—Geologic west-east cross-section A-A'. Location shown in Figure 6.

major tectonic events are disproportionally emphasized.

The first evidence of a structurally high position of the Groningen area is found in the upper Carboniferous–Lower Permian record. During this time, the last tectonic phases of the Variscan orogeny strongly affected the area, causing uplifts, faulting, and important differential truncations.

On the Groningen structure an unconformity separates the coal-bearing Pennsylvanian from the Rotliegendes Formation. East, south, and southwest of the field, however, two unconformity surfaces are present, one separating the coal-bearing Pennsylvanian from an important sequence of Late Pennsylvanian redbeds (Asturian unconformity), and the other separating these redbeds from the Rotliegendes Formation (Saalian unconformity). Two explanations are possible. One is that the Groningen structure, or a large tectonic element of unknown outline, attained a structurally high position during the Asturian tectonic phase and maintained it until the Early Permian Saalian phase. In the surrounding depressions, the Pennsylvanian redbeds were deposited and preserved during the subsequent Saalian truncation. Another explanation would be that the Groningen area also was part of the post-Asturian sedimentary basin, and was strongly uplifted and truncated during the Saalian tectonic phase. The well evidence from the area is too scanty to support either of the two hypotheses. However, both suggest that a Paleozoic high existed in the Groningen area in late Carboniferous time.

During the rest of the Permian and until Late Jurassic time, the Groningen area was part of the northwest European sedimentary basin which was formed during the Saalian tectonic phase. Because of the removal of all the Jurassic and a large part of the Triassic section during the late Kimmerian tectonic phase, the structural history of the field during those times is unknown. In the Lower Triassic section preserved in the Groningen field, some evidence for mild epeirogenic movements has been found. According to Wolburg (1962) and

Trusheim (1961), these movements caused the formation of regional shallow north-south-striking swells and depressions. The most complete sequence is found in the depressions, whereas reductions in thickness and possibly even intraformational truncation are evident on the swells. In the southern part of the Groningen structure, where the Middle Buntsandstein is preserved, no sandstones are present and the thickness of the sequence is reduced greatly by intraformational truncation; thus, the Groningen structure may have formed part of a Lower Triassic swell or may have been an independent high.

No evidence of the early Kimmerian tectonic phase which caused pronounced uplift and truncation in the eastern Netherlands (Harsveldt, 1963), has been found on the Groningen structure. Regional evidence indicates a quiet depositional environment during Jurassic time. The late Kimmerian uplift resulted in the formation of the North Netherlands high and thus also of the Groningen structure in its present configuration.

In the basins surrounding the North Netherlands high, sedimentation resumed in the late Jurassic after a relatively short period of nondeposition and erosion. The progressive onlap on the Groningen structure was gradual and started only in the Early Cretaceous. Deposition then continued uninterruptedly until the Laramide tectonic phase in the Late Cretaceous–early Tertiary. Locally, however, salt movements complicated the depositional pattern.

In contrast to the other parts of The Netherlands, where uplift, faulting, and truncation associated with the Laramide tectonic phase were severe, the northern Netherlands remained a stable area with relatively mild, uniform regional uplift and truncation not deeper than the uppermost Cretaceous. Thus the structural outline of the Groningen gas field established in Late Jurassic–Early Cretaceous time was not basically altered, although rejuvenation of the subsalt faults and a slight northward tilt of the structure may have taken place. Movements related to the late Alpine orogeny in Oligocene time had a similarly small influence.

In summary, it may be said that the presence of the Groningen structure is proved from Late Jurassic time onward. Information from earlier periods is less conclusive, but nevertheless indicates the presence of a structurally positive element, a possible Groningen high, from as early as Late Pennsylvanian time.

GENERATION AND MIGRATION OF GRONINGEN GAS

The origin of the dry Groningen gas, with its composition of 81 percent methane, 4 percent ethane and heavier components, 14 percent nitrogen, and 1 percent CO_2, is still a matter of discussion.

The main organically rich rocks from which the hydrocarbons may have originated primarily comprise the Carboniferous marine and deltaic coal-bearing beds directly underlying the Rotliegendes Formation plus, possibly, the bituminous Copper Shale overlying the Ten Boer Member.

It is now generally agreed that the Carboniferous coal and shale are a likely source for the Groningen gas. Differences in opinion regard mainly the mechanisms and time of generation. An attractive, though not undisputed, hypothesis was presented by Patijn (1963). On the assumption that the thermal gradient in the Carboniferous was about a half of the present gradient, he postulated that in order to start recarbonization after Carboniferous time, the coal layers would have to be buried about twice as deep as their deepest position before the Saalian truncation.

Table 1. Paleodepths of Burial, Top Carboniferous, Groningen Gas Field

Time	Tectonic Event	Burial Depth (ft)	Approx. Truncation (ft)
Holocene		9,000–10,000	
Beginning of Tertiary		±6,600	
	Laramide phase		±2,300
End of Cretaceous		±8,900	
Beginning of Cretaceous		±4,900	
	Late Kimmerian phase		±6,100
Late Jurassic		±11,000	
End of Permian		±4,000	
	Saalian / Asturian phases		0
End of Carboniferous		3,000–6,000	

On Table 1 the depth of burial of the top of the Carboniferous section is given for various times. Because of the uncertainties concerning the original thickness of the now eroded beds and the depth of the differential truncation, these figures can be only approximations. It is nevertheless interesting to note that sometime in the Jurassic, and possibly in the Tertiary, the present-day top Carboniferous strata may have been buried twice as deep as they were before the onset of the Asturian tectonic phase. Dur-

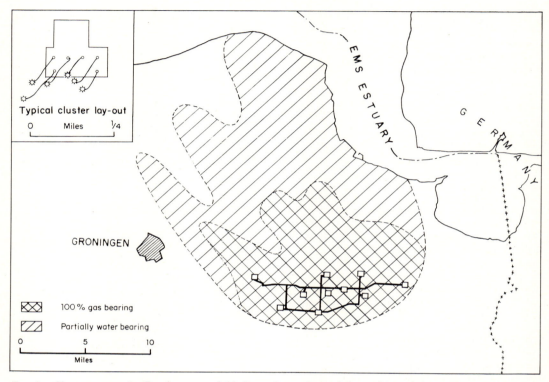

Fig. 8.—Cluster pattern in Groningen gas field. Inset shows cluster layout with surface positions of wells and subsurface positions at top of Slochteren Member.

ing these periods, according to Patijn, recarbonization may have taken place.

The hypotheses involving the Carboniferous as a likely source of the gas have two very attractive points in their favor. First, the Carboniferous source rock is thick and ample coal is available for recarbonization; second, the generated gas would have had simple migration paths in the Carboniferous sands, the numerous faults, and the differentially truncated Carboniferous surface directly overlain by the reservoir.

Summary Geological Considerations

On the basis of the stratigraphy and the geologic history of the Groningen area, four main factors are apparent: (1) the presence of a thick reservoir rock of high quality, (2) the presence of a perfect, unbroken seal from Late Permian time onward, (3) the presence of a structural high from late Paleozoic time onward, and (4) the proximity of the source rock and availability of carrier beds. They form the unique combination which allowed the accumu-

lation of the vast amounts of gas now present in the Groningen field.

Development

The most economic method of development was pursued from the start. The first step toward optimal economic development of the field was the concentration of wells in so-called "clusters" (Fig. 8), instead of the drilling of regularly spaced single wells. The second step was the concentration of all the clusters over the structurally high southern part of the field, where the complete sandstone section is gas bearing and coning effects of water could be ruled out (Fig. 8).

The clusters are about 1.5 mi apart. The number of wells drilled per cluster depends on the amount of gas that can be produced safely. On the basis of the capacity of the gas-drying plants (210×10^6 ft³/day) and an assumed sand-free production of 26×10^6 ft³/day per well, 8 wells were needed initially per cluster. Further testing at high rates and a careful selection of the intervals to be perforated led to an increased production of 42 million ft^3 per day per well. At present and for at least the next 5 years, only 5 wells are needed per cluster, until the pressure in the reservoir drops to a point at which the high production rates cannot be maintained.

The present producible reserves of the Groningen gas field are in the order of 58 × 10^{12} ft^3. The cumulative production so far is only 530 × 10^9 ft^3. The present producible potential of the field is in the order of 3 × 10^9 ft^3/day and will continue to be increased in the future as the field is further developed. Because of the many variables involved, a computer study has been initiated to determine further the optimum development program (de Ruiter *et al.,* 1967).

References Cited

Bartenstein, H., 1968, Paläogeographische Probleme beim Aufsuchen von Kohlen-wasserstoff-Lagerstätten im Paläozoikum und in der Untertrias von Mittel und Nordwest-Europa einschliesslich des Nordsee-Raumes: Erdöl u. Kohle, v. 21, no. 1, p. 2–7.

Groen, D. M. W. te, and W. F. Steenken, 1968, Exploration and delineation of the Groningen gas field: Koninkl. Nederlandse Geol. Mijnb. Gen. Verh., Geol. Ser., v. 25, p. 9–20.

Haanstra, U., 1963, A review of Mesozoic geological history in the Netherlands: Koninkl. Nederlandse Geol. Mijnb. Gen. Verh., Geol. Ser., v. 21, pt. 1, p. 35–55.

Harsveldt, H. M., 1963, Older conceptions and present view regarding the Mesozoic of the Achterhoek, with special mention of the Triassic limestones: Koninkl. Nederlandse Geol. Mijnb. Gen. Verh., Geol. Ser., v. 21, pt. 2, p. 109–130.

Laan, G. v.d., 1968, Physical properties of the reservoir and volume of gas initially in place: Koninkl. Nederlandse Geol. Mijnb. Gen. Verh., Geol. Ser., v. 25, p. 25–33.

Patijn, R. J. H., 1963, De Vorming van gas ten gevolge van nainkoling in het noordoosten van Nederland: Geologie en Mijnbouw, v. 42, p. 349–358.

Ruiter, H. J. de, G. v.d. Laan, and H. G. Udink, 1967, Development of the north Netherlands gas discovery in Groningen: Geologie en Mijnbouw, v. 46, no. 7, p. 255–264.

Thiadens, A. A., 1963, The Palaeozoic of the Netherlands: Koninkl. Nederlandse Geol. Mijnb. Gen. Verh., Geol. Ser., v. 21, pt. 1, p. 9–28.

Trusheim, F., 1961, Uber Diskordanzen im Mittleren Buntsandstein Nordwest-deutschlands: Erdöl-Zeitschr., v. 77, Heft 9, p. 361–367.

Udink, H. G., 1968, Reservoir behaviour and field development: Koninkl. Nederlandse Geol. Mijnb. Gen. Verh., Geol. Ser., v. 25, p. 35–41.

Visser, W. A., 1963, Upper Paleozoic evaporites: Koninkl. Nederlandse Geol. Mijnb. Gen. Verh., Geol. Ser., v. 21, pt. 2, p. 61–71.

Wolberg, J., 1962, Uber Schwellenbildungen im Mittleren Buntsandstein des Weser-Ems-Gebietes: Erdöl Zeitschr., v. 78, Heft 4, p. 183–190.

Lacq Gas Field, France[1]

E. WINNOCK[2] and Y. PONTALIER[3]

Pau, France

Abstract Lacq, near the southern edge of the Aquitaine basin, is France's most important gas field. Hydrocarbons are trapped beneath two unconformities—shallow Lacq (pre-Tertiary) produces oil at 2,100–2,300 ft (640–700 m), and deep Lacq (pre-Aptian) produces gas at 11,650–13,300 ft (3,530–4,060 m). The field is 10 mi (16 km) long, 6 mi (10 km) wide, and covers 25,000 acres (10,000 ha.); gas production of 700 million ft³/day comes from 31 wells.

Trapping is primarily anticlinal, although porosity and permeability have considerable influence; the highest values are on the crest of the anticline. The anticline was formed progressively in the course of deposition of the Cretaceous and Tertiary, as were many other structures in the Aquitaine. The present structure at Lacq shows 4,500 ft (1,400 m) of closure at the top of the Neocomian; the south flank is considerably steeper than the north.

The gas reservoir is divided into an upper calcareous part (650–1,000 ft or 200–300 m thick) comprising Neocomian and Valanginian (Early Cretaceous), and a lower dolomitic part (latest Jurassic; 500–650 ft or 150–200 m thick). Porosity range for the upper part is 0.1 to 6 percent, with permeability less than 0.1 md; porosity in the lower part is 5–6 percent and permeability is 0.1 to a few millidarcys. Fracturing is very important to both parts of the reservoir. Initial reservoir pressure was 9,670 psi at 13,120 ft (4,000 m) and temperature was 260°F (130°C). Abnormally high pressure indicates a closed reservoir of constant volume, without a water drive. Cumulative production to the end of 1967 was 1.76 trillion ft³, with a pressure drop of 3,370 psi. Gas reserves are estimated at 8.8 trillion ft³.

Lacq gas is sour; volumetric percentage of H₂S is 1.25 lb of sulfur per 100 ft³ gas. It constitutes important sulfur reserves and production, and France now ranks third among western sulfur producers.

GEOGRAPHIC AND GEOLOGIC SITUATION

The Lacq gas field, France's most important, is in the Lacq basin in the southern part of the Aquitaine basin, 100 mi (160 km) south of Bordeaux and 20 mi (30 km) northwest of Pau (Fig. 1). The Aquitaine basin is now a large, low-lying area of about 35,000 mi² (90,000 km²). It opens toward the Atlantic Ocean on the west and is bounded by the Armorican massif on the north, the Central mas-

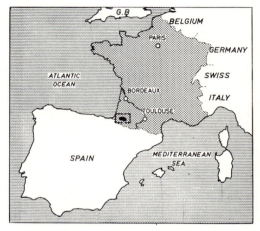

FIG. 1—Location of Lacq field, France, in Western Europe.

sif on the east, and the Tertiary Pyrenees Mountains on the south (Fig. 2).

The sedimentary section of the Aquitaine basin ranges in age from Triassic to Quaternary. The Paleozoic substratum crops out only along the borders, particularly in the east adjacent to the Central massif and in the south in the axial zone of the Pyrenees. Mesozoic sedimentary rocks crop out along regular concentric belts in the north and northeast and within a strip of complex folds along the Pyrenees Mountains. Between these areas, the basin is underlain by Tertiary and Quaternary strata.

Geologic History

Knowledge of the geologic history of the area during the Paleozoic is limited because little information is available. It appears, however, that marine conditions prevailed from Cambrian to middle Carboniferous time. The major phase of the Hercynian (late Paleozoic) orogeny then folded the whole Aquitaine area, resulting in the uplift and emergence of the basin to form part of the central European continent.

Continental sediments were deposited during late Paleozoic time. Middle and upper Carboniferous strata contain coal, whereas Permian rocks are predominantly sandstone and con-

[1] Read before the 53d Annual Meeting of the Association, Oklahoma City, Oklahoma, April 23, 1968. Manuscript received, May 7, 1968; accepted, July 19, 1968. Published by permission of Société Nationale des Pétroles d'Aquitaine, Pau, France.

[2] Senior geologist, SNPA.

[3] Chief geologist, South-West Metropolitan Division, SNPA.

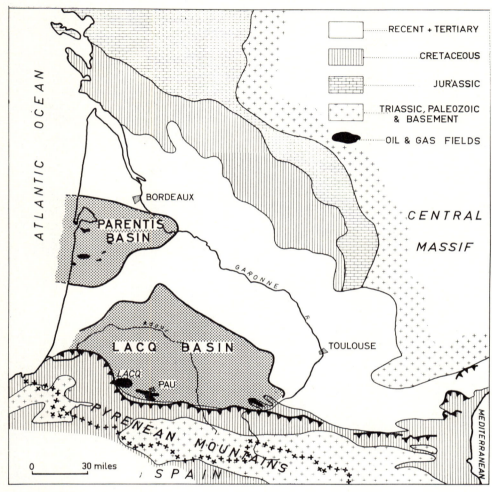

FIG. 2.—Lower Cretaceous basins within Aquitaine basin, France. Oil and gas fields are in Lower Cretaceous basins.

glomerate. These beds are present in several small, isolated basins whose configurations are related to Hercynian structural trends. Metamorphic rocks, either Precambrian formations or associated with Hercynian granitic intrusions, are widespread within the substratum of the Aquitaine basin.

Triassic deposits in the Aquitaine basin are of a continental-lagoonal facies. In the south-central part the Triassic is greatly downwarped, and a complete section is present. It includes a lower sandstone and argillaceous sandstone unit, a middle dolomitic unit, and an upper unit consisting of thick shale and evaporite, with volcanic rocks (Fig. 3). On the north and along margins of the Aquitaine basins, more

stable conditions prevailed, and only Upper Triassic—variegated shale and sandstone—is present. Upper Triassic salt produced halokinetic tectonics (salt diapirism) in the southern part of the basin. The same paleogeographic setting persisted through Rhaetian (latest Triassic) and Hettangian (earliest Jurassic or Liassic) times.

After Triassic time, the configuration of the Aquitaine basin was more clearly defined and during the Sinemurian widespread normal marine conditions existed for the first time. The Sinemurian and Domerian limestone sequence, dolomitic at the base and argillaceous in the upper part, and the late Liassic (late Early Jurassic) argillaceous limestone have a constant

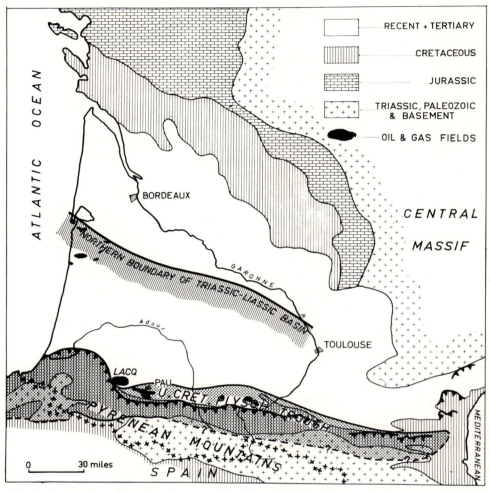

FIG. 3.—Triassic and Upper Cretaceous features of Aquitaine basin. Limit between platform (north) and folded basin (south) coincides with northern boundary of Triassic and Liassic thick evaporite-bearing deposits.

thickness and facies throughout the basin. Beds deposited from the Bajocian (early Middle Jurassic) to the end of the Kimeridgian (middle Late Jurassic) do not change significantly in thickness, but different lithofacies are present on either side of a roughly north-south line. West of the line the sequence is predominantly limestone, whereas east of it the rocks are entirely dolomitic and provide the main reservoir of the gas fields near Meillon. Dolomitization was general throughout the basin during the Portlandian (latest Jurassic).

At the end of Jurassic and the beginning of Cretaceous time, orogenic movements caused the emergence of the basin borders and folding; continental deposits (lacustrine limestone and argillaceous sandstone) of the Purbeckian (latest Jurassic) and Wealdian (earliest Cretaceous) were formed. The sea returned after a period of erosion but transgression was limited to the southwestern part of Aquitaine; Lower Cretaceous deposits are in isolated smaller basins (Fig. 2). Sedimentation began in a normal marine environment as indicated by organic limestone of the Valanginian, but was restricted in late Neocomian time, which is represented by deposits of sandstone, freshwater limestone, dolomite, and anhydrite.

Shale and limestone were deposited during an early Aptian transgression. During the rest of Early Cretaceous time subsidence was active, especially in the central syncline, called Arzacq basin. Lacq gas field is on its southern border. There are abrupt facies changes in upper Aptian and Albian strata, which are mostly calcareous claystone with minor sandstone and reef limestone; carbonate rocks are more abundant along the edges of the basin.

Renewed orogeny affected the Aquitaine basin between Albian and Cenomanian times, particularly in the southern part, along the belt which later became the Pyrenees. New or old folds emerged, forming sharp uplifts that subsequently were deeply eroded. This embryonic phase of Pyrenees orogeny was followed by a general sinking of the basin and a transgression of the Cenomanian sea over most of the formerly emergent areas. The area underwent important paleogeographic changes; on the north was a stable epicontinental platform, with the Parentis and Arzacq basins still evident as zones of relatively greater subsidence. Essentially carbonate rocks were deposited—chalky, organic, or detrital limestone, and dolomite. The same platform facies is present on the northern edge and on the western plunge of the present axial zone of the Pyrenees.

Between the platform and the southern uplift was a long, narrow, rapidly subsiding trough extending east-west, in which were many local uplifts. Important masses of flysch-facies sediments accumulated. Subsidence shifted with time, affecting the southern part of the trough during the Cenomanian and the northern part during the Senonian. Thus the flysch trough continued to accumulate deposits through Senonian time.

Lacq field is on the southern limit of the epicontinental platform, in the vicinity of the hinge line bordering the flysch trough, where thickness of strata is much reduced.

The sea receded during early and middle Eocene time, but the main paleogeographic outline of the Late Cretaceous remained—with a platform facies on the north, and flysch-facies deposition continuing in the flysch trough on top of the late Senonian flysch. Conditions changed near the end of middle or the beginning of late Lutetian time, as one of the major phases of the Pyrenean orogeny took place. The rise of the Pyrenean chain was accompanied by regression of the sea, which then covered only the westernmost Aquitaine basin. Regression continued through late Eocene, Oli-

gocene, and Miocene times. In the east, molassic formations derived from the erosion of the Pyrenees are equivalent to the late Eocene-Miocene marine strata in the west.

Structural Outline

Aquitaine may be divided into three large structural areas, from north to south: the Aquitaine platform, folded area, and northern Pyrenean foreland.

Aquitaine platform.—This area, remaining constantly high and stable, has been affected little by folding. It is characterized only by gentle folds which probably reflect buried Hercynian structure.

Folded area.—This part of Aquitaine presents definite structural features under a molassic cover. It corresponds to the areas of Triassic and Lower Cretaceous subsidence and contains all the petroleum fields. Anticlines generally trend northwest-southeast and are complicated at depth by unconformities, relief inversions, and diapiric intrusions of Triassic salt. The latter is particularly important along the Atlantic Coast, where numerous diapirs are present.

Northern Pyrenean foreland.—The entire stratigraphic sequence from the base of the Triassic to the Tertiary crops out in east-west-trending belts in the area just north of the Paleozoic axial zone of the Pyrenees. In its northern part, the Upper Cretaceous flysch, almost everywhere vertical, shows little relief, whereas in the south, the Lower Cretaceous and the Jurassic rocks are tightly folded and form the foothills of the Pyrenees.

The Lacq anticline is in the southern part of the folded area, at the foot of the zone of overthrusts which separates it from the north Pyrenees foothills (Fig. 4).

Petroleum Production

The Aquitaine basin is by far the most important petroleum province in France. Its annual production is 2.5 million tons (17.5 million bbl) of oil and 195 billion ft^3 of gas, or, respectively, 80 percent and 100 percent of France's national hydrocarbon production. A new gas field—Meillon-Rousse—is being developed, with reserves nearly half those of Lacq. The main pay zones are in the Upper Jurassic and Lower Cretaceous. Gas also is produced from the "Cenomanian breccia" at St. Marcet, and some oil from lower Senonian limestone and dolomite at Lacq and Meillon.

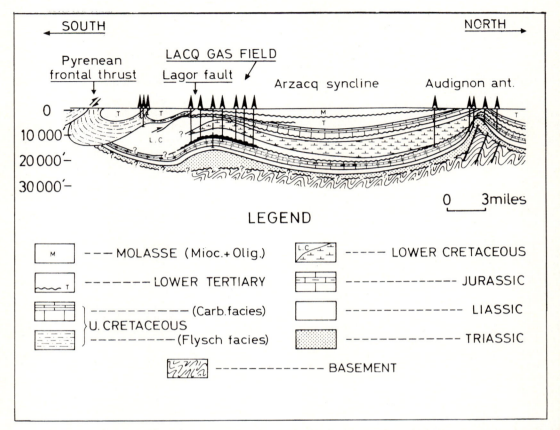

FIG. 4.—South-north cross section of Lacq gas field and Arzacq syncline. Note different facies of Upper Cretaceous on either side of Lagor fault. Lacq gas field is just north of this hinge line. Audignon anticline never produced commercially.

DISCOVERY HISTORY

The Lacq anticline is not evident at the surface because it is unconformably overlain and obscured by Cenozoic strata; it was found by a geophysical investigation. Electric methods (two electric lines in 1942 and a telluric survey in 1943) first were used. Despite widespread interference from industrial installations, these methods detected an anomaly. Although not very pronounced, this anomaly significantly was located on the southern flank of the Arzacq syncline, just east of an important fault that cuts the surface anticline of Ste. Suzanne, where asphaltic oil shows were well known. Later, in 1944–1945, a reconnaissance gravimetric survey showed a weak-amplitude anomaly of small extent in the Lacq area.

Finding of the electric and gravimetric anomalies led to a reflection seismic survey in 1947–1948. Despite the simple technique used,

this seismic work soon resulted in the correct interpretation of the structure on which the first well was staked in October 1949 (Fig. 5). The main objective was an Upper Cretaceous breccia, overlain by the flysch cover, which was the gas reservoir of St. Marcet—the first discovery in Aquitaine, made in 1939. Other objectives were in the Cretaceous and Jurassic, which had yielded many oil shows in wells already drilled on other structures of the western Aquitaine.

Planned for drilling to 10,000 ft (3,000 m), Lacq No. 1 was spudded in December 1949 and found the lower Senonian to be oil impregnated between 2,100 and 2,300 ft (640 and 700 m). The well was put into production with a potential flow of 1,000 bbl/day. This zone —the upper Lacq oil field—was delineated by 1950.

Lacq No. 3 resumed deep exploration

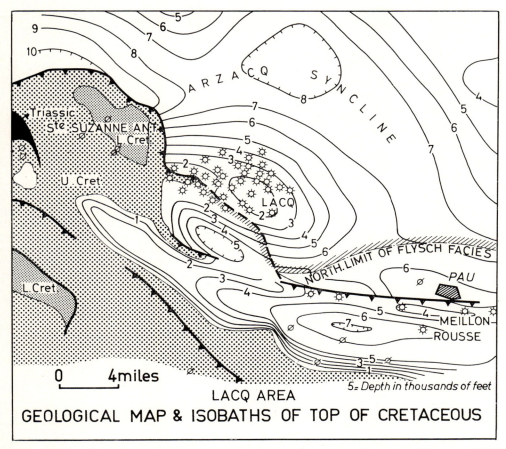

FIG. 5.—Lacq area, structural contour map; datum is top of Cretaceous. Locations of Lacq and Meillon-Rousse gas fields, on either side of a sedimentary and tectonic hinge line, are clearly shown. Depths in thousands of feet. CI = 1,000 ft.

which had been temporarily neglected. In December 1951 the well blew out from a depth of 11,650 ft (3,530 m) after drilling 330 ft (100 m) of basal Cretaceous. The high pressure of the gas and its toxic and corrosive nature (due to the pressure of H_2S) caused great anxiety among the local population, and the well was killed by the American specialist Myron Kinley. By blowing out 350 million ft³ of gas during the killing operations, Lacq No. 3 proved the presence of a major gas reserve. However, numerous and complex problems arising from the very corrosive action of the H_2S on drill pipe, casing, tubing, and valves had to be solved before commercial exploitation could be planned.

Another seismic survey was carried out in 1952 and more holes were drilled north and

south of Lacq No. 3 to explore the deep structure. Meanwhile, metallurgic research developed a special steel capable of resisting sour-gas corrosion at high temperatures. The steel was tested at Lacq No. 102 and No. 104 in 1955 and 1956, opening the way to commercial exploitation of the gas field 4 years after its discovery.

STRATIGRAPHY

The deepest well on the Lacq anticline was the last one drilled in 1964–1965. It reached lower Liassic at 15,452 ft (4,701 m) and was still in Liassic at 16,312 ft (4,973 m; Fig. 6). No reservoir was found in either the Liassic (Lower Jurassic) or the Dogger (Middle Jurassic).

The Malm (Upper Jurassic) is the oldest

Table 1. Productive Formations in Aquitaine Basin

Age	Lithology	Production
Oxfordian	Dolomitic strata	Gas at Meillon-Rousse, St. Marcet
Portlandian	Dolomitic strata	Gas at Lacq, Meillon-Rousse; oil at Parentis
Purbeckian-Wealdian	Sandstone	Gas at Lacq; oil at Parentis, Cazaux, Lavergne
Neocomian	Limestone and dolomite	Gas at Lacq; oil at Parentis, Lugos, Mothes, Lucats
Early Aptian	Limestone	Oil at Parentis, Lugos, Mothes, Lucats
Albian	Sandstone	Oil at Cazaux
Albian	Limestone	Oil at Mimizan

unit penetrated by most of the Lacq wells, and the order of stratigraphic description is as follows from the Malm upward.

Late Jurassic.—Overlying the Dogger (*s.l.*) is 650 ft (200 m) of hard black shale including a few gray-beige cryptocrystalline limestone beds. An abundant fauna of ammonites was found in this formation, which therefore was called *marnes à ammonites* and has been assigned to the late Oxfordian. The Upper Jurassic continues with massive gray-beige cryptocrystalline limestone 525 ft (160 m) thick, alternating in the lower part with compact microcrystalline dolomite. The presence of *Pseudocyclammina jaccardi*, alone in lower strata and associated with *Pseudocyclammina virguliana* in higher strata, indicates early Kimeridgian age.

Overlying this limestone body is 560 ft (170 m) of gray-beige cryptocrystalline limestone, the lower 230 ft (70 m) being interbedded with gray shale and anhydrite, and the upper 330 ft (100 m) alternating with light-gray calcareous shale. The assemblage of *Pseudocyclammina jaccardi* with *P. virguliana* continues into the basal part of these beds, whereas *P. virguliana* is alone in the upper part. It indicates an early Kimeridgian age.

The late Kimeridgian and part of the Portlandian consist of a largely dolomitic unit called

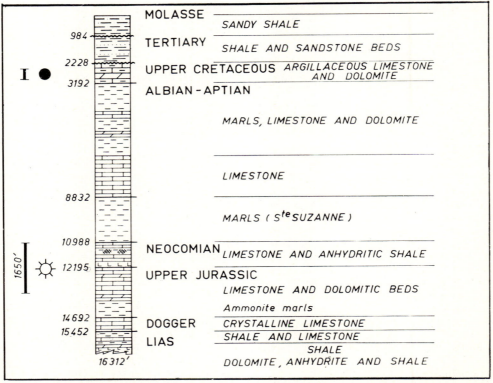

FIG. 6.—Stratigraphic column for Lacq anticline, Aquitaine basin, France, from Lacq No. 301 well. Depths in feet.

dolomie de Mano. The lower part of this formation is 230–300 ft (70–90 m) of dark-beige cryptocrystalline limestone, locally dolomitic and grading into pellet limestone toward the top. *Pseudocyclammina virguliana* is present. The upper part is homogeneous and massive, gray-beige crypto- to microcrystalline dolomite with *Iberina;* thickness ranges from 400 to 500 ft (120–150 m), with part eroded.

Wells others than Lacq No. 301 terminated at the base of the *dolomie de Mano* which forms the lower part of the gas reservoir.

Purbeckian-Wealdian (latest Jurassic–earliest Cretaceous).—The marine Upper Jurassic is overlain unconformably by a continental-lagoonal (Wealdian) sequences of alternating sandstone, dolomite, sandy limestone, and shale. The sequence ranges in thickness from 65 to 400 ft (20–120 m), and its transgressive character has been demonstrated at several places in the basin. This stratigraphic unit was called Purbeckian, although in the Lacq area it is probably equivalent to the basal Early Cretaceous.

Valanginian-Neocomian.—In all southwestern Aquitaine, as well as at Lacq, Valanginian and late Neocomian rocks are easily differentiated. The Valanginian is a homogeneous body of beige cryptocrystalline zoögene limestone locally fragmental or dolomitic. Thickness ranges from 165 to 720 ft (50–215 m). This sequence is marine and carries an abundant microfauna (*Dictyoconus, Choffatella, Miliola*), as well as numerous Mollusca.

The late Neocomian is more characteristic of brackish-water deposition and consists of the following very constant threefold succession, from base to top.

1. "Lower limestones" (*calcaires inférieurs*): dark-brown shaly fossiliferous limestone with anhydrite interbeds; contains annelids, ostracods, gastropods. Thickness is 180–620 ft (55–185 m).
2. "Laterolog shales" (*argiles du Latérolog*): so called because they are very well shown by the Laterolog. Black, anhydritic, unfossiliferous shale. A local unconformity is at the base. Thickness is 65–200 ft (20–60 m).
3. "Upper limestones" (*calcaires supérieurs*): more shaly than the "Lower limestones." Interbedded with numerous soft anhydrite layers locally a few feet thick. Fauna consists of *Choffatella*, ostracods, and annelids. Thickness is 95–650 ft (28–200 m).

Early Aptian.—The early Aptian includes the so-called *marnes de Ste. Suzanne* (Ste. Suzanne Shale), a thick shaly sequence ranging in thickness from 650 to 4,000 ft (100–1,210 m). This dark-gray, finely sandy, pyritic shale con-

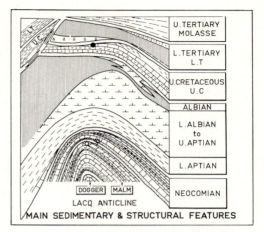

FIG. 7.—Main sedimentary and structural features of Lacq anticline. Relations between presence of oil (solid black circle) and of gas (starred circles) to major unconformities are clear. Oil of Upper Cretaceous probably came from Albian reef.

tains a very few fragments of mollusks and echinoderms. Laterally constant limestone interbeds are good and well-known electric-log markers; they indicate that this sequence transgresses the Lacq anticline. The boundary between the early and late Aptian can be either sharp or transitional. The so-called "transition zone" (*zone intermédiaire*) is present only on the western flank of the structure, where it consists of alternating limestone and shale. The limestone contains *Orbitolina* and miliolids, and resembles the late Aptian limestone. The shale is similar to the Ste. Suzanne Shale, but does not contain *Choffatella*. The total thickness ranges from a few feet to 650 ft (1–200 m).

Late Aptian-Albian.—These two units have been grouped together because, even though both have been penetrated in numerous boreholes, the boundary is uncertain and there is no definite stratigraphic break between them. They form a sedimentary body 5,000–6,500 ft (1,500–2,000 m) thick, in which both lateral and vertical facies changes are numerous and abrupt and microfaunas are not sufficiently characteristic to show facies relations.

The lower part of the sequence throughout the structures is reefoid crystalline limestone or dolomite with *Orbitolina*, miliolids, and corals.

The upper part consists of two facies.

1. Near the crest of the structure the limestone facies continues from the lower part upward to the Cenomanian boundary and forms the so-called (but improperly) "lacq reef" (*récif de Lacq*).
2. On the flanks is the argillaceous limestone and sandy spicule-shale facies.

Between these two facies, all intermediate types may be present.

Upper Cretaceous.—The lower part of the Upper Cretaceous consists of 427–750 ft (125–225 m) of gray-beige fragmental Cenomanian limestone with *Orbitolina, Miliola,* and *Prealveolina.* Overlying the Cenomanian is fragmental or chalky Turonian limestone with fragments of mollusks and echinoderms.

The lower Senonian ranges in thickness from 330 to 825 ft (100–230 m) and consists of crystalline or chalky limestone interbedded with hard gray-green shale. Secondary dolomitization affects the section from Cenomanian to early Senonian, inclusive—it is very important on the top of the anticline, but decreases and disappears downward on the flanks. The upper Senonian, known as "Aturian" facies, is argillaceous limestone grading downward into glauconitic limestone. Microfauna consists of *Lagena, Globotruncana,* and *Fissurina;* thickness ranges from 330 to 825 ft (100–230 m).

The Upper Cretaceous is eroded to the Cenomanian in the vicinity of Lacq No. 119 and No. 120. The area probably was the crest of the structure just before the Paleocene transgression (*see* Figs. 8, 9).

Paleocene-Ypresian.—The Paleocene is extremely variable in thickness (600–1,650 ft; 200–500 m), and consists mainly of shale with minor interbeds of argillaceous sandstone, locally with limestone a few feet thick at the base. Where the Upper Cretaceous has been eroded, the boreholes penetrated breccia (the so-called "Paleocene breccia"). The clasts range in age from Cenomanian to Aturian (late Senonian of Late Cretaceous). The *Globigerina* and *Globorotalia* of the argillaceous matrix may range in age from Paleocene into Ypresian (late early Eocene).

The Eocene transgression was at a maximum during Ypresian time. Ypresian strata consists of *Globigerina-* and *Globorotalia*-bearing shale. Within the shale, sandstone interbeds are more numerous than in the Paleocene.

Lutetian (middle Eocene).—The late Lutetian is present only on the north flank of the structure, where it ranges in thickness from 0 to 500 ft (0–150 m). It consists of sandy limestone grading downward into argillaceous sandy limestone. Fossils are *Nummulites,* discocyclinids, and *Fabiania.*

Molasse.—This name was given to fluviolacustrine beds ranging in age from late Eocene to Miocene. They are ocherous-colored unfossiliferous clay with sandy argillaceous limestone

pebbles; thickness increases regularly northward from 330 to 2,300 ft (100–700 m).

Structure and Structural Evolution

Structural-contour maps drawn on successive stratigraphic layers in the Lacq structure differ notably. At the base of the Tertiary, the WNW–ESE elongate anticline shows two culminations separated by a saddle. The saddle coincides in position with the top of a paleostructure where the Upper Cretaceous is partly eroded. It should be noted that the so-called "Upper Lacq" oil pool or pay zone is in the limestone and dolomite of the eastern culmination (Fig. 5).

The structure on the Neocomian is regular and elliptic, 10 mi (16 km) long and 6 mi (10 km) wide; its area is about 25,000 acres (10,000 ha.) (Fig. 8). The north flank is smooth, and the steep south flank dips sharply under the tectonically complex area of Lagor, where the Upper Cretaceous flysch facies crops out (Fig. 9). At the top of the structure, the Neocomian is 10,000 ft (3,000 m) deep; structural closure, as proved by drilling, is more than 4,600 ft (1,400 m).

With adequate well control, it has been possible to prepare a set of isopach maps, and thus to determine the successive deformations of the base of the Neocomian that have been caused by differential subsidence alone. This approach necessitates the assumption that the sea floor was practically horizontal at the beginning and end of each of the periods considered—an assumption which is fairly reasonable considering the sedimentation processes which prevailed during the Cretaceous and Tertiary Periods. The assumption is hardly to be questioned with regard to the Neocomian—the laguna-lacustrine facies of that stage would hardly be consistent with a rugged sea-floor topography on the underlying Jurassic surface.

For the end of the Neocomian (Fig. 10B), a late Portlandian structure is shown that culminates in the same place as the present oil field. The structural axis is directed northeast-southwest. The structure shows two culminations separated by a small saddle, the one in the area of the Lacq No. 120 and No. 118 wells and the other in the position of the Lacq No. 126 well and northeast of that well. Closure is negligible toward the south and east, whereas the west dip of the Jurassic strata is important.

For the end of early Aptian time (Ste. Suzanne Shale and "intermediate zone"), the foregoing reconstruction shows little change except

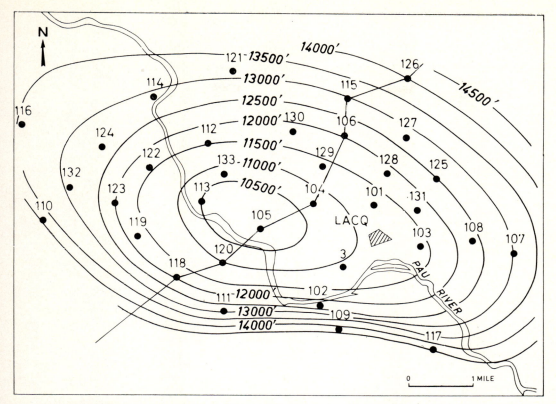

FIG. 8.—Lacq gas field, structure-contour map; datum is top of Neocomian. Structure is very regular, with south flank steeper than north flank. Contours in feet subsea. CI = 500 ft. Line of section is for Figure 9.

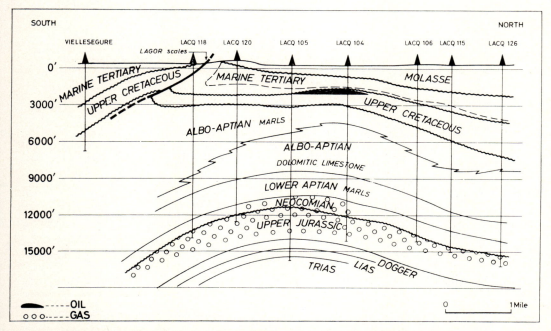

FIG. 9.—Southwest-northeast cross section of Lacq anticline. Lower Aptian marl provides seal for gas pool and Ypresian marl seals small oil pool. Vertical scale in feet. Line of section is shown on Figure 8.

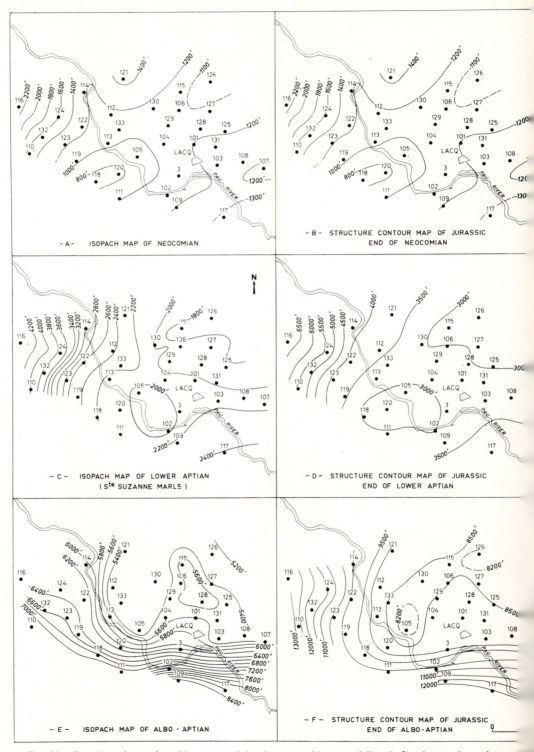

FIG. 10.—Post-Jurassic stratigraphic structural development of Lacq anticline. Left columns: isopach map Jurassic at end of Neocomian, early Aptian, Albian, early Senonian, Cretaceous, and middle Eocene times (B Contours in feet.

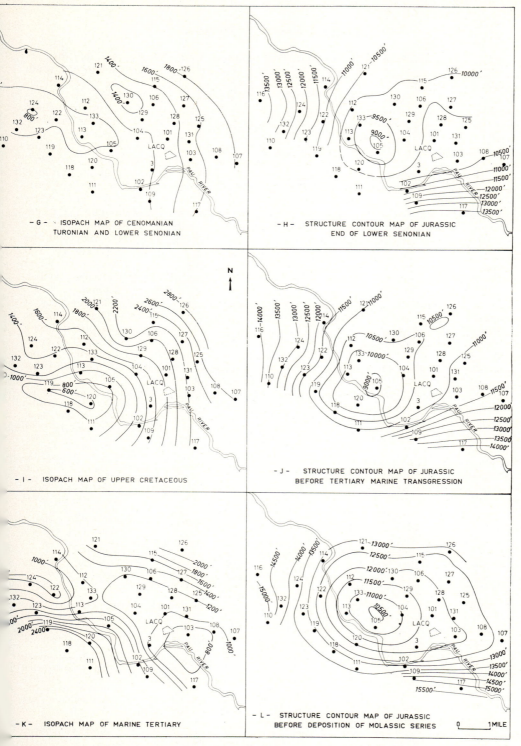

-G- ISOPACH MAP OF CENOMANIAN TURONIAN AND LOWER SENONIAN

-H- STRUCTURE CONTOUR MAP OF JURASSIC END OF LOWER SENONIAN

N

-I- ISOPACH MAP OF UPPER CRETACEOUS

-J- STRUCTURE CONTOUR MAP OF JURASSIC BEFORE TERTIARY MARINE TRANSGRESSION

-K- ISOPACH MAP OF MARINE TERTIARY

-L- STRUCTURE CONTOUR MAP OF JURASSIC BEFORE DEPOSITION OF MOLASSIC SERIES

0   1 MILE

post-Jurassic formation. (A, C, E, G, I, K). Right columns: structural reconstructions of top of Upper , J, L). For present structure, see Figure 8. First closure of anticline appears during Late Cretaceous.

for the more evident southern closure of the southern culmination, and the steeper dip of the west flank (Fig. 10D).

An important change occurred during the Albo-Aptian (Fig. 10E). Whereas during the previous epochs subsidence was most important in the western area (Lacq No. 110-No. 116), during the late Aptian and Albian the thickest deposits formed in the southern area. The corresponding isopach contours trend more definitely east-west, and southward tilting resulted in the northward shift of the area with thinner deposits (Lacq No. 130-No. 112). However, because the strata thin toward the northeast, a high area probably was present in that direction (Lacq No. 125, No. 127, No. 126).

At the end of Albian and the beginning of Cenomanian time (Fig. 10F), and under the effects of the process described heretofore, the Upper Jurassic strata took the form of a northeast-southwest-trending nose which dips regularly from Lacq No. 126 to No. 120, with a possible small culmination in the area north of Lacq No. 105.

Because of the presence of a pre-Tertiary eroded zone in the Lacq No. 119-No. 120 area, the specific influence of Late Cretaceous subsidence is more difficult to interpret. Considering only the wells in which the late Senonian is of "Aturian" facies, one may see (Fig. 10G) that the main paleogeologic features changed again during Cenomanian and early Senonian times. The area with minor subsidence trends approximately east-west with two minima, separated by a saddle, in the positions of Lacq No. 113 and No. 124. The strata thicken uniformly toward the northeast; it is most likely that they thicken also toward the flysch trough, i.e., toward the southeast. The structural reconstruction of the Upper Jurassic strata thus is very different from that for previous times. In fact, the first stage of the formation of the present structure had been the development of a dome, culminating in the Lacq No. 105 area. As far as it can be drawn, this dome seems to be fairly well closed. The previous culmination, in Lacq No. 126 area, remains as a residual anomaly on the north flank (Fig. 10H).

This interpretation is very much the same as the one which is obtained by considering the thickness of the entire Upper Cretaceous, whether it is eroded or not (Fig. 10J), and by assuming that the Paleocene-Ypresian sea transgressed an almost peneplaned surface. This method seems to prove that the pre-Tertiary movements had only a negligible effect on

the beginning Lacq structure, and that they affected only locally the structure of the south flank, which was the nearest the flysch trough.

The evolution of the structure during Paleocene and Ypresian time cannot be understood completely, because of the lack of accurate stratigraphic data and the influence of pre-"molassic" (pre-late Eocene) erosion. At the time corresponding to the base of the "molasse" (Fig. 10L), the structure had attained its final geometry. It extended along the present east-west axis; the north flank was regular, and less steep than it became under the influence of deposition of the "molasse." The top of the structure was still near Lacq No. 105, and the slopes between Lacq No. 105 and No. 116 (west flank) and between Lacq No. 103 and No. 117 (southeast) were the same as they were at the end of the early Senonian.

According to the foregoing observations, it may be concluded that the Lacq anticline reached its present stage of development during late Eocene-Pliocene time as a result of compaction and epeirogenic movements. Three conclusions can be drawn from the historic reconstruction outlined.

1. Lacq anticline is not a classic anticlinal fold, because subsidence alone seems sufficient to have created both the deep (Neocomian-Jurassic and the upper (early Senonian) traps. Thus it is unnecessary to postulate tectonic compression during the Pyrenean orogeny as a cause of the anticline. If the influence of such tectonic movements on Lacq itself may be considered to be small and limited to faults of moderate throw in proximity to the structure, it should be observed that the Pyrenean orogeny did produce important overthrusts (Lagor thrust slices) that did affect the south flank of the structure (Fig. 4). The overthrust at the limit between flysch and shelf facies of the Upper Cretaceous is part of the "North Pyrenean frontal fault zone" which follows the north border of the strongly folded Pyrenean foothills. The little information available relating to other structures on the borders of Arzacq basin, such as Audignon and Garlin, suggests that they are of the Lacq type.

2. During much of Early Cretaceous time, the main structural and stratigraphic strike was northeast-southwest. The same direction may well have dominated both the structure and the sedimentation throughout the Jurassic.

The trend of isopach lines did not shift to east-west until Early Cretaceous time, and the Lacq anticline originated on a "subsidence hinge" made by the intersection of two ancient high axes—one of Early Cretaceous age and trending northeast-southwest, and the other of Late Cretaceous and Tertiary ages and trending east-west. Such a subsidence hinge, having been active so long, is especially suitable for trapping hydrocarbons because of drainage to it and the opportunities for the improvement of the initial porosity and permeability of the rocks in such areas.

3. "Deep Lacq" structure was not closed toward the

northeast, at least until near the end of the Albian. Provided that the reservoir was not a permeability lens, this would prove that gas trapping occurred only after Albian time.

On the basis of the reconstruction (Fig. 10) of the structural evolution of the top of the Jurassic strata, it may be deduced that the "deep Lacq" reservoir was open toward the northeast at least until the end of the Albian. This suggests that the gas was accumulated and trapped after Albian time unless the reservoir was a permeability lens.

Reservoir Characteristics

The gas-producing reservoir includes the Neocomian, Valanginian, Purbeckian, Portlandian, and part of the Kimeridgian; it is nearly 1,650 ft (500 m) thick. Petrophysically this reservoir may be divided into two parts.

1. A calcareous upper part 650–1,000 ft (200–300 m) thick comprises the Neocomian and Valanginian. Porosity is very low (about 0.1 percent) except for a zone at the top of the Valanginian where it may be 5–6 percent. The permeability also is very low, less than 0.1 md, but an important network of small fissures and crevices and the high pressure of the field explain why gas could accumulate in this formation.

2. A dolomitic lower part with clayey sandstone interbeds and low-permeability limestone, 500–650 ft (150–200 m) thick, extends from the top of Purbeckian through Portlandian and forms the reservoir. Average value of porosity is 5–6 percent in the upper 350 ft, but matrix rock permeability remains low (0.1 to a few millidarcys and, exceptionally, 12 md). However, permeability is enhanced by the presence of many fractures. Toward the base, porosity decreases gradually to as little as 1 percent.

Laterally, between the holes at the top and those on the flank, there is a reduction of both the net pay thickness and fissures. Because of these factors, the field has the form of an enormous lens of fracture permeability, developed mainly on the culmination of the carbonate mass and decreasing progressively toward the flanks (Fig. 11).

Examination of the evolution of the structure in geologic time may provide an explanation for this phenomenon. On one hand, differences in the amount of net pay depend largely on the amount of clay in the Purbeckian-Wealdian sandstone. The cleanest sandstone is at the position of Lacq No. 105. It may be thought that this area, positive through the Cretaceous and Tertiary, already was high at the end of the Jurassic (marine stage) and had some influence on Purbeckian-Wealdian sedimentation.

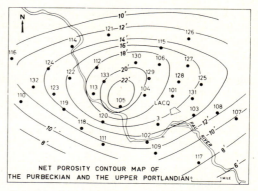

FIG. 11.—Net porosity (in feet) contour map of Purbeckian and Upper Portlandian, Lacq gas field, Aquitaine basin. CI = 2 ft.

On the other hand, the fractures of the reservoir seem to be related to the presence of the same paleogeologic high zone, rather than to faults which affected the structure during the Pyrenean orogenesis, because (1) the characteristics of the reservoir do not improve toward the south flank near the Lagor fault, and (2) the faults penetrated in the wells are plugged and do not constitute drains.

It is also conceivable that the gas was emplaced during the Late Cretaceous into a porous and fractured reservoir, and that for a reason specific to the Lacq structure the continuous subsidence brought about a progessive impairment of the reservoir on the flanks, whereas the top was unaffected because of a lower geostatic pressure and an increase of gas pressure.

Fluid

General.—Under reservoir conditions (initial pressure 9,760 psi at 13,120 ft or 4,000 m, temperature 266°F or 130°C), the Lacq gas is in a single-phase state. When the field first was developed, an expansion with separation was made at each wellhead (995 psi and about 86°F or 30°C), and yielded (1) a gaseous phase and (2) a liquid phase including water of condensation and condensate. For each 36 ft³ of dry gas, 0.057 lb of gasoline and 0.027 lb of water were recovered. The gasoline is light, yellow, limpid, aromatic, and sulfurized. The specific gravity is 0.825 at 59°F (15°C).

Presently the gas is expanded to about 1,420 psi at the wellhead and then sent to the gas plant for separation and treatment. The volumetric percentage of H_2S is 1.25 lb sulfur per 100 ft³ gas.

Reserves.—The description of the reservoir emphasizes the difficulties in evaluating *in situ* reserves by estimation of the volume of impregnated rocks.

Interstitial water saturations were approximated by measures of resistivity (Laterolog) and statistical data provided by the salt contents measured on the drilling cores. It is almost impossible to determine water saturation in the upper compact limestone, but the effect on the evaluation of the total reserves is small, for that zone contains only 15–20 percent of the reserves.

It appears reasonable to assume that water saturation is related to porosity, and it was assumed that the upper part of the reservoir where porosity is less than 1 percent was as much as 50 percent water saturated.

In the lower dolomitic part, the average porosity is more than 5 percent and average saturation is estimated with accuracy at 20 percent. The volume of reserves was calculated from the chart of "reduced isoheights"[4] of gas determined from values for each hole which eventually was completed, before the well had drilled through the entire thickness of the porous dolomitic facies of the Portlandian (Fig. 11). The vertical extension of the field was taken to be 16,400 ft (5,000 m), which is the depth of the deepest hole (Lacq No. 116)—a flank well which was still in the reservoir at total depth.

Proved in-place reserves thus calculated are 7.75 trillion ft³ standard—6 trillion for the lower reservoir (82 percent) and 1.75 trillion for the upper (18 percent). If the possible error in evaluation of water saturation is not taken into account, this value is pessimistic, because nothing is known of the deep flanks of the structure. However, analysis of pressure decline in the field after production of nearly 1 trillion ft³ shows that the reservoir acts as a closed space of constant volume. The estimate of original in-place reserves was revised, with consideration of the pressure decline and the thermodynamic behavior of the gas. The new figure thus found is 8.8 trillion ft³ standard.

Corresponding sulfur reserves amount to 50 million metric tons.

Estimation of reserves.—First is written the following law of gas:

$$PV = ZNRT \text{ at the initial time } t_o.$$

The instant t corresponds to a cumulative pro-

[4] Reduced height: thickness of pay zone times porosity.

duction C. Temperature does not change. Assuming that the volume saturated with gas is constant, the following relation is established:

$$\frac{N_o - N}{N_o} = 1 - \frac{P}{Z} \times \frac{Z_o}{P_o},$$

which establishes the value of initial reserves N_o, if C ($C = N_o\text{-}N$), P_o, P, Z, and Z_o are known. The accuracy of estimate of initial reserves depends on the accuracy of the pressure measurements and the knowledge of the compressibility factor of gas.

Taking into account the depth of the reservoir, the sour nature of the gas, and the cost of treatment, it is assumed that 80 percent of the reserves, or 7 trillion ft³, will be recovered. This figure agrees with an abandonment pressure at wellhead compatible with that of the present gathering system (about 1,400 psi). This figure is not final because it would be modified by technical and economic conditions.

FIELD DEVELOPMENT (GAS)

General.—Production is from 31 wells, about 1 mi (1.6 km) apart. Field production gradually was stabilized at 700 million ft³/day which is about 80 percent of the maximum capacity of the equipment. Most wells are capable of producing normally more than 35 million ft³/day each. The best producing well is Lacq No. 112; it attains a 35-million-ft³ output with a pressure drop of 30 psi. If a 4-in. tubing is used, the output generally is limited by the capacity of surface equipment—wellheads and gathering system. To maintain normal productivity in spite of the expected pressure decline of the field, it is planned to replace 4-in. tubing with 5-in. tubing.

Experience shows that the shut-in pressure in any well is independent of the well's history, location, and structural position; rather it depends only on when the measurement is taken. This permits the assumption that interconnecting crevices and fractures form a very efficient collecting system which feeds the gas to the wells from all parts of the reservoir.

Casing.—The drilling program included the setting of the following casing.

13⅜ in. about 2,200–2,600 ft (670–790 m) at top of Aturian.
9⅝ in. between 8,500 and 10,800 ft (2,570–3,280 m), generally at top of Ste. Suzanne Shale.
7 in. between 10,500 and 14,700 ft (3,200–4,450 m) at top of "Laterolog clays" reservoir.

Production tubing.—At the beginning of the

field development, wells were completed with double tubing:

A production string of 2⅜-in. OD, made of special steel resisting corrosion by hydrogen sulfide.

An outer string of 4-in. OD, intended to insure well security in case of damage to the inner string.

Favorable results—good performance of steel, absence of corrosion under different production rates, security of gaskets—obtained after a year's testing of a producing well at the beginning of development, made possible the use of a single 4-in. tubing. This new type of completion permitted larger flow per well, and limited the number of wells necessary to develop the field fully. The total number of drilled wells was 33.

Recently, six wells have been equipped with a liner cemented in the pay zone and perforated opposite fractures located by a flow meter. This equipment does not hamper the productivity of the well, but facilitates measurement and fishing operations with the Halliburton line. Also, a better performance of the well may be expected when the field pressure declines below hydrostatic pressure, and a better check of water inflow is provided.

Surface equipment.—The christmas tree includes a series of valves built for a service pressure of 10,000 psi. One of them is remote-controlled by radio from the central control room of the plant; another closes automatically if anything goes wrong (breaking or plugging) in the surface equipment.

The gas flows from the well at 7,000 psi pressure, which is reduced in two stages to the gathering-line pressure of 1,400 psi. The first stage is operated by the adjustable choke in the flow head where the pressure is lowered to 4,300–5,000 psi. In the second stage, the main expansion is operated through a choke after the gas is preheated to maintain a temperature above the critical temperature of formation of hydrates (75°F or 24°C) when the wells start flowing, or have a low production flow rate.

The gathering system is protected against a well's high pressure by a device including a tared valve which closes automatically when pressure rises above 1,650 psi, and a burst plate with a line to a flare. This plate is tared to the maximum allowable service pressure of the gathering system (1,850 psi).

After expansion, the gas goes through an automatic separator (WKO) which eliminates all condensation water; after corrosion and hydrate inhibitors have been added, the gas flows through a meter system, and goes to the plant.

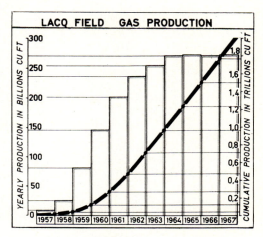

Fig. 12.—Lacq field gas production, yearly and cumulative from year of discovery.

Production

General.—For maximum efficiency in handling the wells and the production installations, the flow rate of the field is maintained at a level equal to the sum of maximal production of each of the 31 wells minus 20 percent. To insure the current average production as long as possible, and to produce the maximum volume of gas, several studies have been made.

Improvement of well potentials.—Wells have very different productivity characteristics, which are essentially dependent on fractures. They may be classified in three categories.

1. The majority of the wells are excellent wells, in which the pressure drop is less than friction losses and which generally are able to produce more than 35 million ft³/day. Their flow rate generally is limited by surface equipment and the collecting system.

2. Medium-quality wells (about 10) with production between 14 and 28 million ft³/day and rather sharp pressure drop.

3. Three wells originally thought to be nonproductive (Lacq No. 111, No. 117, and No. 124) because their open flow was not more than 3.5 million ft³/day. Stimulation tests were made on the poorest wells to open natural fractures which had been plugged by drilling mud, or to enlarge the fracture system. Hydrochloric acid, followed by acetic acid, was injected; in a further treatment, a mixture of both acids was introduced under high pressure (4,200–5,200 psi) and flow rate (70 ft³/minute). The result was a spectacular increase in the well flow rate; thus the potential flows of Lacq No. 124 and No. 111 were increased, respectively, to 14 and 17.5 million ft³/day.

Well re-equipment.—If well productivity is assumed to remain constant, the flow rate may be maintained despite the pressure drop of the field if the equipment is modified to decrease friction losses; this was done by replacement of 4-in. tubing with 5-in. in 28 wells. This replacement also permits the use of valves of larger diameter. The replacements are made progressively, according to bottom-hole pressure. This pressure must not be too high, because high pressure would prevent the introduction of fuel inside the annular space. Nor can the pressure be too low, because the well would have circulation losses.

Adjustment of gathering system.—Capacity of the gathering system is stabilized, as far as lines are concerned, by installation of a 6-in. line from single wells and 8-in. lines that are common to several wells.

Gas repressuring.—The input pressure of the plant is 1,140 psi and necessitates a minimum pressure at the wellhead. This pressure changes according to the distance to the plant, but is estimated at 1,400 psi for 700 million ft³/day. When the field pressure becomes insufficient to provide this pressure at the surface, compressors will be installed at wellheads and should insure this high production level for another 3 years.

Repressuring probably will create problems because of the nature of Lacq gas, and will necessitate important investments. However, instead of the drilling of new wells, this program is well adapted to the period between the present level of 700 million ft³/day and abandonment of the field, for as the flow rate decreases, friction losses in the gathering system also will diminish, and it will become possible to lower the service pressure in the gathering system without the friction losses being very important in terms of absolute value. Compressors would then be installed at the entrance to the plant, which would lower considerably the operating costs.

Plant pressure decrease.—As soon as the volume of gas to be treated decreases, it may be possible to lower slightly the input pressure; this will lead to an increase in the flow rate at the wells.

Pressure Decline

Since the beginning of exploitation, pressure has declined steadily (by 3,370 psi for 1.76 trillion ft³ produced), and the pressure/gas-compressibility ratio under average conditions of the field (P/Z) decreases linearly with the cumulative production of the field. This linear relation indicates that the reservoir is one of constant volume without water drive. Nevertheless, the last points measured in tests are above

the average line of the chart P/Z plotted against production figures.

This may be explained by (1) progressive contributions of gas at lower pressure which had accumulated in marginal areas slightly or not at all fractured; this would give a slight increase in volume of active gas, or (2) slow decrease of the pore volume, either by water from a space of small permeability or by shrinkage of the reservoir rocks.

Conclusions

Several features of the Lacq gas field are found in other fields of the Aquitaine basin.

1. Lacq field is on a structurally high paleogeologic zone separating two strongly subsident basins—Arzacq basin on the north, which sank continuously from Cretaceous time to the end of the Tertiary, and the Late Cretaceous–early Eocene flysch trough on the south. The recently discovered Meillon field is in a similar position about 10 mi (16 km) southeast.

2. The anticline was formed progressively during Cretaceous and Tertiary sedimentation, as were many other structures in the Aquitaine. Nevertheless, it is not a simple paleostructure but rather a subsidence hinge; this could explain why it constitutes an isolated feature in the southern part of the basin, and why attempts to find a Lacq trend have been unsuccessful. Structurally, the Meillon fold is completely independent of Lacq.

3. Hydrocarbons accumulated beneath two unconformities—pre-Tertiary, for oil of the shallow Lacq reservoir, and pre-Cretaceous, for the gas of the "deep Lacq" reservoir (Fig. 7). Similar modes of occurrence are common in most fields of the Aquitaine, but the origin of the gas (Lower Cretaceous or Jurassic) and its date of accumulation are problems which have not been answered satisfactorily.

4. The gas contains a high proportion of hydrogen sulfide, as does the gas in certain parts of the Meillon field. The currently favored hypothesis for explaining the presence of this sour gas is the reduction of sulfate from anhydrite, which is abundant in the Neocomian. The sulfur reserves make Lacq the most important sulfur field in the world and rank France third among the western sulfur producers.

5. The original reservoir pressure was abnormally high and the pool reacts as a closed reservoir of constant volume without obvious water drive. These features, specific to Lacq, may be related to the evolution of the structure during Cretaceous and Tertiary times.

SELECTED REFERENCES

Berger, Y., 1955, Gisement de gaz à pression anomalement élevée. Problèmes de forage: Inst. Français Pétrole Rev., v. 10, no. 11, p. 1453–1466.

—— 1963, L'évolution des techniques d'exploitation du gisement de gaz de Lacq: Inst. Français Pétrole Rev., no. Mors série, p. 605–616.

Bonnard, E. G., 1950, Le développement des connaissances géologiques du bassin pré-pyrénéen et la découverte du champ pétrolifère de Lacq: Inst. Français Pétrole Rev., v. 5, no. 7, p. 203–211.

—— 1950, Découverte de la structure pétrolière de Lacq et ses enseignements sur la géologie de l'avant-pays pyrénéen: Ver. Schweizer. Petroleum-Geologen u. -Igénieure Bull., v. 17, no. 53, p. 15–18.

—— and J. Schoeffler, 1958, Le développement du gisement de Lacq profond: Ver. Schweizer. Petroleum-Geologen u. -Ingénieure Bull., v. 25, no. 68, p. 36–46.

Chaillous, A., 1963, Stimulation of the deep Lacq gas wells: Inst. Petroleum Jour., v. 49, no. 472, p. 93–98.

Chaloupy, P., 1963, La stimulation des puits de Lacq profond: Gaz Europe Inf., v. 11, no. 2, p. 13–16.

Cuvillier, J., 1955, Coupe stratigraphique dans le Néocomien et la Jurassique supérieur de Lacq (forage 104: Inst. Français Pétrole Rev., v. 10, no. 5, p. 316–318.

Dieumegard, M., and M. Paillassa, 1960, Le contrôle radio-électrique des puits haute-pression du gisement de Lacq: Assoc. Française Techniciens Pétrole Bull., no. 139, p. 63–76.

Enous, R., 1958, La production du gaz de Lacq: Chimie Industrie (Juin), p. 720–742.

Feger, J., 1963, Forages à grande profondeur à Lacq: Congrés Assoc. Française Techniciens Pétrole, Toulouse (Juin), p. 11–28.

—— and L. Cauchois, 1956, Les problèmes posés par le gisement de gaz de Lacq et leurs solutions: Assoc. Française Techniciens Pétrole Bull., no. 119, p. 445–496.

—— R. Enous, and J. P. Vacher, 1958, Aspect actuel des problèmes de production: Rev. Française l'Energie, no. 94, p. 183–193.

Gayral, R., 1951, La roche magasin de Lacq: Assoc. Française Techniciens Pétrole Bull., no. 88, p. 20–29.

Jenner, P., and J. Dienesch, 1965, Histoire géophysique du champ de Lacq: Geophys. Prosp., v. 12, p. 518–540.

Ruez, G., 1964, Gisement de Lacq profond. Evolution de la pression du gisement—Réserves en place: Assoc. Technique l'Industrie du Gaz en France, Compte Rendu 81st Cong. l'Industrie du gaz, Paris, p. 358–368.

Schoeffler, J., 1956, Les gisements de Lacq, in Symposium sobre yacimientos de petroleo y gas: 20th Internat. Geol. Cong., Mexico, v. 5, p. 257–265.

Vacher, J. P., and G. Ruez, 1960, Reconnaissance des gisements de Lacq: Inst. Français Pétrole Rev., v. 15, no. 12, p. 1751–1758.

—— and ——, 1963, Calcul des réserves d'un gisement. Application au champ de Lacq: Gaz Europe Inf., v. 11, no. 1, p. 3–10.

Geology and Exploration of Sicily and Adjacent Areas[1]

JOSEPH VERCELLINO and FABRIZIO RIGO[2]
Rome, Italy

Abstract Surface seeps first attracted industrial exploration for hydrocarbons in Sicily in 1901. The first major success occurred after passage of the Sicilian petroleum law of 1950. Three major structural accumulations have been discovered—Ragusa (1953), Gela (1956), and Gagliano (1960). Recoverable reserves are estimated to be 110 million bbl at Ragusa and 90 million bbl at Gela. Although oil in place at Gela is calculated to be 1.3 billion bbl, less than 10 percent is expected to be recovered. At Gagliano, proved reserves are not defined fully, although about 700 billion ft³ (20 billion m³) of gas and 20 million bbl of condensate are estimated. Development drilling is still under way at Gagliano and Gela.

Commercial production is limited to the Tertiary Central basin and the Mesozoic Ibleo platform. The Central basin is characterized by a thick sequence of normally deposited Pliocene and Miocene terrigenous clastic rocks interspersed with chaotic gravitational slides. The Gagliano field produces from multipay Miocene-Oligocene sandstone beds. The Ibleo platform is represented primarily by carbonate rocks, and the Ragusa and Gela fields produce from thick dolomite of Triassic age.

Volcanic activity which began in the Jurassic has continued to the present, and intrusive rocks commonly are associated with the producing reservoirs.

Introduction

The Italian Mediterranean island of Sicily is separated from the mainland by the narrow Strait of Messina, and from North Africa by the Tunisian-Sicilian Channel. The area of Sicily is 9,925 mi² (25,800 km²). Approximately 5 million people inhabit the island, and the primary sources of income are agricultural products—grain, citrus fruits, olives, grapes, and vegetables. The climate is dry subtropical, with annual rainfall of 20 in. (50 cm), limited to the winter season. Topography ranges from coastal plains to mountains. Mount Etna, Europe's largest active volcano and the dominant physical feature of the island, rises to an elevation of 10,646 ft (3,245 m).

[1] Read before the 53d Annual Meeting of the Association, Oklahoma City, Oklahoma, April 23, 1968. Manuscript received, May 7, 1968; accepted, October 1, 1968.

[2] Exploration Consultants, Lungotevere Mellini 44.

The writers are grateful for the generous assistance given by Tiziano Rocco, recently retired General Manager of Exploration for AGIP, and Felice Scelsi, Chief of the Sicilian Mining Board, both of whom furnished valuable statistical data, comments, and suggestions. Drafted illustrations were prepared by Fausto Petitta and Valerio Valeri.

Tectonics and Stratigraphy

The tectonic history of Italy and Sicily is extremely complicated and nearly all types of structural deformation are found (Fig. 1). Three main orogenic events took place during the Ordovician, Permo-Carboniferous, and Tertiary. The Tertiary Alpine orogenic wave was the greatest in intensity and advanced in a generally arcuate pattern from west to east. The direction of movement was northeast in the Northern Apennines, east-northeast in the Central Apennines, almost due east in the South Apennines, and southeast in Sicily.

The effect of the Alpine orogenic wave is evident in the stratigraphic and structural configuration of Italy and Sicily—the Apennine overthrust of Mesozoic strata, the Miocene foothill exposures, the Pliocene marginal basins, and the stable, outer-margin Mesozoic shelf.

Oil and gas exploration, which began in the Northern Apennine Miocene foothill belt in the last century, has been expanded successfully into the Pliocene basin and Mesozoic shelf segments. The latter segment still must be considered virgin territory for oil and gas exploration, because most of the area is offshore and because political divisions of sea boundaries and subsequent petroleum legislation were not resolved until late 1967 and early 1968 by the countries involved—Italy, Yugoslavia, Malta, and Tunisia. At this date, there is widespread and concentrated interest in the oil and gas potential of the Italian offshore regions.

Geophysical and geologic investigations indicate that major deep-seated faulting has sculptured Sicily, and most fault alignments are NE-SW and NW-SE. The volcanic and earthquake disasters which have affected the island since the beginning of recorded history are related to the younger NE-SW alignments.

Tectonically the island can be subdivided into six geologic provinces—the Ibleo platform, the Central basin, the Trapani basin, the Miocene foothills, the Mesozoic overthrust, and the metamorphic basement segment (Fig. 2).

All commercial oil and gas production in Sicily is in the Miocene foothills–Central basin segment and the Ibleo platform.

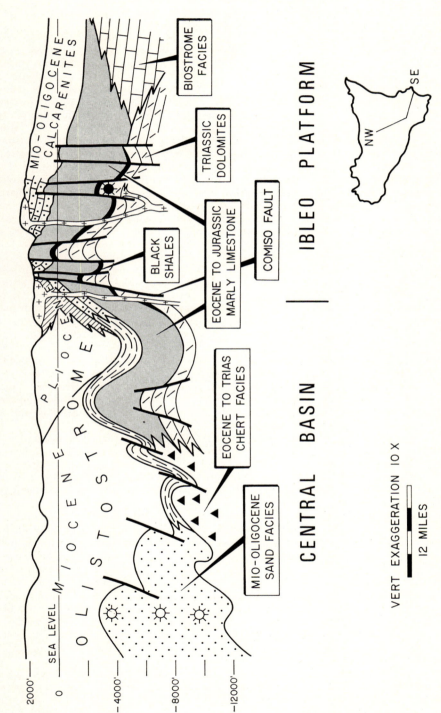

Fig. 3.—NW-SE structural cross section of Sicily. Location shown on inset map and on Figure 2.

the Ibleo platform. It is therefore essential to determine the geologic age of each structure before drilling, because structures which originated after the Early Cretaceous have proved to be barren.

Central Basin

The name "Central basin" was introduced by geophysicists as a result of a strong Bouguer gravity depression beneath the central part of the island. This negative gravity feature includes the Miocene foothill segment on the north and the Pliocene-Quaternary basin on the south. The deepest part of the Central basin corresponding to minimum gravity value is about 30 mi (48 km) southwest of the Gagliano gas field. The thickness of the Tertiary section at the minimum gravity area is unknown, but a reasonable estimate is 20,000 ft (6,000 m). The basin is characterized chiefly by a thick plastic sequence of interbedded Tertiary shale, sandstone, and "olistostroma" overlying a rigid substratum of Mesozoic carbonate rocks.

The olistostroma (olistostromes; Flores, 1955) are complex gravity-slide formations, widely distributed from the Northern Apennines to Sicily. They have an important influence on Tertiary exploration throughout much of Italy and Sicily. The olistostroma "formations" are large masses of crumpled, brecciated, and chaotic plastic shales in which rootless and rigid older blocks are "floating" (olistoliths). They are a result of large continental and submarine gravity-slide mud flows which formed during peaks of Tertiary tectonic activity. Generally, the olistostroma overlie the normally deposited Miocene and Pliocene strata, although in places they are interbedded and interfingered.

Because the olistostroma "formations" greatly hamper seismic and surface geologic studies, past exploration was limited primarily to the more easily surveyed marginal and shallower olistostroma-free areas.

The Central basin (Fig. 2) is bounded on the north by the Madonie Mountains, in which are exposed overthrusts of Mesozoic rocks; on the east by the Etna volcano and the Ibleo platform; and on the west by Mesozoic exposures which divide the Central basin from the Trapani basin. The southern flank of the Central basin is offshore. The basin originated as a deep trough which began to subside in early Miocene time and continued until the Quaternary. It was uplifted strongly during the late Quaternary, and marine Quaternary strata now are found as high as 2,300 ft (700 m) above sea level.

Mesozoic and Eocene rocks, exposed along the northern flank of the basin, consist of several hundred feet of siliceous shale, calcarenite, and cherty limestone overlying Triassic siliceous limestone. These beds grade southward into a pelagic marly calcareous sequence similar to that found on the Ibleo platform.

The Oligocene and early Miocene rocks, characterized by quartzitic sandstone and shale, range in combined thickness to 10,000 ft (3,000 m). A large percentage of the thickness is olistostroma with floating olistoliths of Mesozoic limestone. Along the southern and southeastern boundaries of the basin, the lower Miocene grades into a thinner shaly sequence with fine-grained nonporous sandstone lenses which represent the transition to the Ibleo platform calcarenite facies.

The middle Miocene is represented on the north by thick sandstone and shale with lenses of coarse-grained, terrigenous clastic rocks and conglomerate and on the south by fine-grained sandstone and shale. Porosity and permeability in the middle Miocene clastic rocks are poor along the southern coast.

The upper Miocene is evaporitic throughout the Central basin, and sedimentation was limited to localized depressions. Thick anhydrite and salt beds are common, and lack of porous reservoir beds eliminates this unit as an oil and gas objective.

The same negative evaluation can be extended to the Pliocene, which is a marly plastic shale sequence capped by porous coquina limestone. The porous unit lacks a suitable caprock because it grades into a similar Quaternary facies.

The entire area of the Central basin south of the Gagliano parallel is affected by thick olistostroma interbedded with middle Miocene, upper Miocene, and Pliocene strata. Thin lenses of olistostroma are also present in the Quaternary section.

Exploration in the Central basin is hampered greatly by lack of seismic results because of the presence of chaotic gravity slides and accompanying large olistoliths. The best exploration tool in the northern producing area of the basin is the gravity meter.

Numerous surface asphalt and gas seeps in Sicily led to shallow drilling operations in the early 1900s, but serious and orthodox exploration did not begin until 1935. Although some noncommercial gas production was found, only three wells were drilled below 4,000 ft (1,200 m).

Ragusa Oil Field

In March 1950, the Sicilian Regional Government passed the Sicilian Hydrocarbon Law which immediately led to the entry of several oil companies.

The surface geologic setting of the Ragusa anomaly and the associated prominent oil seep in Miocene strata first were observed by J. Elmer Thomas in 1922 and 1923. Although he persisted in his efforts to find financial backing to drill on the feature, it was not until after the passage of the Petroleum Law that his efforts met with success, when Gulf Oil Corporation became interested in the prospect.

After extensive surface geologic, photogeologic, magnetometer, gravimeter, reflection- and refraction-seismic studies, and shallow core drilling, the Gulf Oil Corporation began drilling Ragusa No. 1 on May 21, 1953. After finding numerous noncommercial oil shows in Tertiary, Cretaceous, and Jurassic carbonate rocks, the test reached a 1,200-ft (365 m) Lower Jurassic–Upper Triassic black shale facies with gabbroic igneous intrusives.

At 6,204 ft (1,891 m) geologists logged a coarse-grained, fractured Triassic dolomite with vugular porosity, saturated with low-gravity oil. The discovery well was drilled and cored to a total depth of 6,971 ft (2,125 m) after penetrating 767 ft (234 m) of oil-saturated dolomite. Drill-stem tests established an oil-water contact at 6,900 ft (2,103 m) for an overall pay section of 696 ft (212 m). Ragusa No. 1 was completed in open hole with production casing cemented to 6,227 ft (1,898 m). Fifty-five development wells drilled subsequently on a spacing pattern of 75 acres (30 ha.) on the flank and 37½ acres (15 ha.) on the crestal axis showed the discovery well to be 0.5 mi (800 m) from the culmination of the anomaly, and 1 mi (1,600 m) west of the major fault zone which defines the eastern flank (Fig. 4).

The gas-oil contact (Fig. 5) was well defined at 3,740 ft (1,140 m) subsea. The oil-water contact ranged in depth from 4,773 ft (1,455 m) subsea on the southern edge of the field to 5,000 ft (1,524 m) subsea in the northern part of the field.

Structurally, Ragusa No. 1 is 1,000 ft (305 m) lower than the crestal culmination. Wells drilled in Ragusa field include 2 dry holes, 4 noncommercial completions, and 49 pumpers. At the present time approximately 85 million

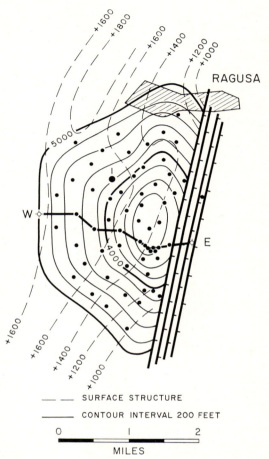

Fig. 4.—Structure map of Ragusa field, Sicily. Datum is top of Taormina Dolomite. CI = 200 ft. Line of section is location of Figure 5. Modified from Kafka and Kirkbride (1959). Location of field shown on Figure 2.

bbl of 19.4° API gravity oil has been produced. Remaining recoverable reserves are estimated to be 30–40 million bbl.

Carbon dioxide content of gas in the gas cap ranges from 15 percent at the top of the cap to 32 percent near the gas-oil contact. The carbon dioxide content of solution gas produced with the oil ranges from 63 to 83 percent. Analyses of oil samples show that Ragusa crude is very undersaturated.

Production studies of reservoir behavior confirm that the Ragusa accumulation has a strong water drive. Although the matrix porosity of the producing dolomite is low, the presence of ample fractures and large vugs creates high permeability in the reservoir.

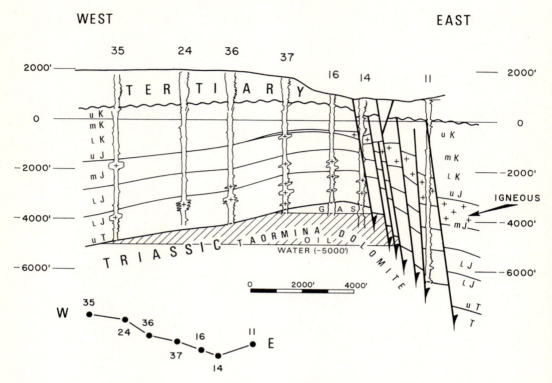

FIG. 5.—West-east structural cross section, Ragusa field, Sicily. Location of section is on Figure 4. Modified from Kafka and Kirkbride (1959).

Maximum oil and gas column indicated from subsurface contouring totals almost 2,000 ft (610 m). Productive area is about 3,000 acres (1,200 ha). Recent recoverable-reserve estimates range from 110 million to 125 million bbl—down from the 1959 estimate of 150 million bbl. Current daily production is about 10,400 bbl and 37 wells are still producing.

GELA OIL FIELD

In 1956 a second major oil field, Gela, was found 29 mi (47 km) west of the Ragusa field, when AGIP found a low-gravity crude in the Triassic Streppenosa Limestone and the underlying Taormina Dolomite (Fig. 6). Selection of the drill site was based on surface geologic, gravity-meter, and seismic-reflection and seismic-refraction studies. The Taormina Dolomite was reached at 10,919 ft (3,328 m) or 10,839 ft (3,304 m) subsea. Because the Gela anomaly is west of the Comiso regional fault (Figs. 2, 3), the producing dolomite reservoir is more than 6,500 ft (1,981 m) deeper than in Ragusa No. 1.

The API gravity of the oil found ranges from 7.3 to 17°. Laboratory studies indicate that this range in gravity values is due to fluid mobility characteristics and reservoir filtration and separation.

Subsequent development drilling showed the Gela structure to be a NE-SW-trending anticline of Mesozoic strata, with minor normal faulting and secondary closures on the crest of the main anticline. Development drilling now has extended into the offshore area and the productive limits have been defined almost completely. Eighty-three wells have been drilled. The field is nearly 7 mi (12 km) long and 2 mi (3¼ km) wide.

The oil-water contact has been established at a subsea datum of 11,539 ft (3,517 m), and gross oil column reaches 1,250 ft (381 m). As at Ragusa, the dolomite reservoir is fractured and has vugular porosity. In some wells, the dolomite has the appearance of an oil-saturated sponge. The effective permeability is low, however, and is attributed primarily to fractures.

The oil-bearing reservoirs in the Streppenosa

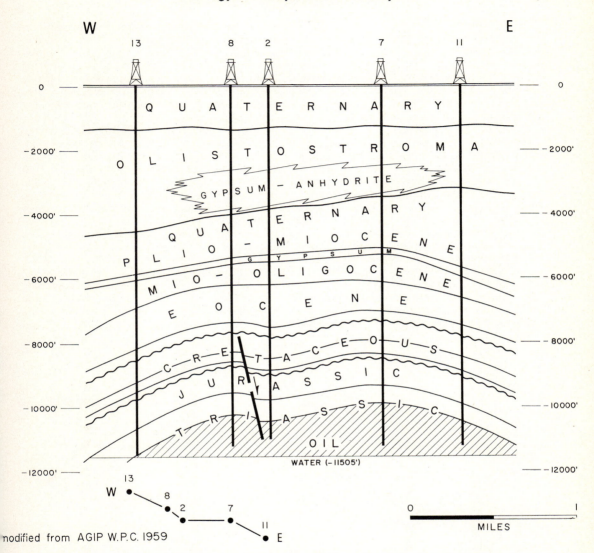

FIG. 6.—West-east structural cross section of Gela field, Sicily. Field location shown on Figure 2.

dolomite-limestone facies, overlying the Taormina Dolomite, are erratic, although in some wells the gross pay is as thick as 500 ft (152 m).

Carbon dioxide content of the solution gas averages 75 percent and the crude normally contains a large amount of dissolved gas. Most of the wells are capable of flowing. Although the flow rate is low because of the low gravity of the oil and temperature drop at surface conditions, some wells flowed more than 3,000 bbl/day of oil. Nearly all wells are pumped. Injection of diesel oil is used at times to decrease the viscosity. Average production per well during the early development phase was about 1,100 bbl/day.

Cumulative oil production to December 31, 1967, totalled 45 million bbl and daily average oil production during December was 14,500 bbl from 69 wells. Two billion ft³ (57 million m³) of gas also has been produced.

Because of the low gravity of the oil, the recoverable-reserve estimate has been reduced very recently to 90 million bbl, which is less than 10 percent of oil in place.

Gagliano Gas Field

Exploratory activity slowly shifted toward the deep Tertiary Central basin and Miocene foothills in 1958. The first important discovery was in 1960, when the AGIP No. 1 Gagliano exploratory wildcat was completed as a gas-condensate discovery from a Mio-Oligocene quartz sandstone between 8,835 and 8,869 ft (2,693 and 2,703 m) (Fig. 7).

The drill site for the Gagliano wildcat was selected partly on the basis of surface geologic studies supported by gravity, but primarily on the basis of a seismic-refraction study. One of the reasons for the drilling of the well was to obtain stratigraphic knowledge of the Miocene, Oligocene, and possibly Mesozoic facies. The test reached total depth at 13,140 ft (4,005 m) in quartz sandstone and shale indicated to be early Miocene–late Oligocene. The olistostroma gravity-slide blanket ranges from the surface to a depth of about 6,000 ft (1,829 m).

Subsequent development drilling showed that the Gagliano gravity structure is a large and well-defined flat-topped anomaly (Fig. 8), with some changes in facies and general porosity reduction toward the southeast. A down-to-the-west fault is indicated on the west flank. Twenty-seven wells have been drilled and the producing trend is now fairly well established as a gentle NE-SW-trending structure. The productive area is not known because the field limits have not been defined fully.

The average total depth of the Gagliano wells is about 10,000 ft (3,048 m). One blanket sandstone and several lenticular gas-condensate pays lie between 5,000 and 9,000 ft (1,524 and 2,743 m). Some wells are capable of commercial production from six sandstone reservoirs, and the gross pay per well averages almost 500 ft (153 m).

Average productive capacity of a well is in the range of 4–7 million ft³/day (113,200–198,100 m³) of gas for each zone completed, although some wells produce as much as 12 million ft³ (339,800 m³). The gas-condensate ratio differs, but is about 30 bbl per 1 million ft³ (28,000 m³). Daily field production during December 1967 was 115 million ft³ (3.3 million

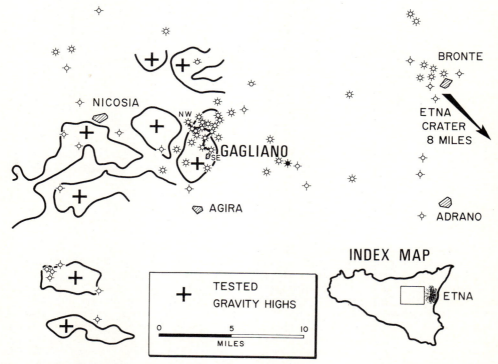

Fig. 7.—Map of Gagliano area, Sicily. Principal gravity maxima shown. Line of section (NW-SE) through Gagliano gas field is location of Figure 8.

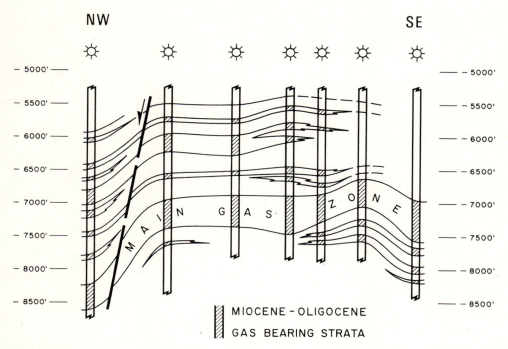

FIG. 8.—NW-SE schematic cross section through Gagliano gas field, Sicily. Location shown on Figures 2 and 7.

m³) of gas and 2,675 bbl of condensate from 24 wells. However, production still is curtailed pending completion of pipeline and market facilities. Accumulated production to December 31, 1967, totaled 61 billion ft³ (1.7 billion m³) of gas plus 1,650,000 bbl of condensate.

Formation pressures are uniform in the gas-bearing sandstone and range from 3,625 to 3,700 psi at 7,700 ft (2,347 m) subsea datum. Proved recoverable reserves are estimated to be 700 billion ft³ (20 billion m³) of gas and 20 million bbl of condensate.

FUTURE POSSIBILITIES

As indicated by Figure 9, Sicilian crude-oil production is declining, whereas gas production is increasing greatly. Recent renewed exploratory interest in the Ibleo platform and Central basin segments, plus forthcoming offshore exploration in the Sicilian-North African channel, should find additional oil or gas fields of the Ragusa-Gela-Gagliano type. Several large offshore gravity anomalies between Tunisia and Sicily should be viewed with a certain degree of optimism. In addition, the offshore area be-

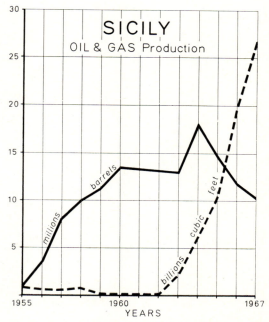

FIG. 9.—Oil and gas production, Sicily, 1955–1967.

tween the Ibleo platform and the island of
Malta is worthy of investigation. These offshore
areas soon will be opened to exploration under
the provisions of Petroleum Law 613 of July
21, 1967.

SELECTED REFERENCES

Beneo, E., 1955, Les resultats des etudes pour la
    recherche pétrolifére en Sicile: 4th World Petroleum
    Cong., Rome, sec. 1, p. 109–120.
Flores, G., 1955, *Discussion of* Beneo, E., Les resultats
    des etudes pour la recherche pétrolifére en Sicile:
    4th World Petroleum Cong., Rome, sec. 1, p. 121–
    123.
———— 1959, Evidence of slump phenomena (olisto-
    stromes) in areas of hydrocarbons exploration in
Sicily: 5th World Petroleum Cong., New York, sec.
    1, p. 259–275.
Kafka, F. T., and R. K. Kirkbride, 1959, The Ragusa
    oil field, Sicily: 5th World Petroleum Cong. Proc.,
    New York, sec. 1, p. 233–257.
Marchetti, M. P., 1956, The occurrence of slide and
    flowage materials (olistostromes) in the Tertiary
    series of Sicily: 20th Internat. Geol. Cong., Mexico,
    sec. 5, p. 209–225.
Rigo, F., 1956, Olistostromi neogenici in Sicilia: Geol.
    Italiana Boll., v. 75, p. 185–215.
Rocco, T., 1959, Gela in Sicily, an unusual oil field:
    5th World Petroleum Cong. Proc., New York, sec. 1,
    p. 207–232.
Schmidt di Friedberg, P., 1962, Introduction a la
    géologie pétroliére de la Sicile: Inst. Français
    Pétrole Rev., v. 17, p. 635–638.
Vercellino, J., 1965, Sicily due drilling boost: Oil and
    Gas Jour., August 9, p. 124–126.

Oil Fields in Mio-Pliocene Zone of
Eastern Carpathians (District of Ploieşti)[1]

D. PARASCHIV[2] and GH. OLTEANU[3]
Bucharest and Ploieşti, Rumania

Abstract The Mio-Pliocene zone in the district of Ploieşti is on the border of the western plunge of the Eastern Carpathians. During more than 100 years of geologic and exploitation activity, 35 structures productive of oil and gas in the Pliocene, Miocene, and Oligocene have been found. Oil production to January 1, 1968, was 1,725 million bbl, of which 1,528 million was from the Pliocene. The most important producing structure within the region—Moreni-Gura Ocniţei—yielded about 724 million bbl.

Hydrocarbon accumulation in the district of Ploieşti may be explained by (1) the complete development of the Pliocene rocks in the form of repeated pelite and psammite sequences, which permitted the generation, accumulation, and preservation of pools, and (2) the presence of favorable structural conditions, particularly produced by diapirism of Miocene salt which was concurrent with Pliocene sedimentation.

INTRODUCTION

The Mio-Pliocene zone and its oil fields are in the newest and most external unit of the Alpine-Carpathian area that crosses Rumania in the form of a gigantic mountainous arc (Figs. 1, 2).

In Rumania the Carpathian Mountains have been divided into three branches, of which the eastern one, known as the "Eastern Carpathians," is of interest because of hydrocarbon possibilities. They are made up of metamorphosed Precambrian as well as Mesozoic and Tertiary strata, disposed in three parallel strips —the inner crystalline-Mesozoic zone, the Cretaceous and Paleogene flysch in the middle, and the outer Neogene or pre-Carpathian zone.

Within the zones numerous units and subunits are distinguished which were tilted progressively and overlap each other from the inside to the outside.

The Neogene zone, present along the entire Carpathian Range in Rumania, shows great variation in both stratigraphy and lithofacies. In the northern part of the country only Miocene rocks are included, with very few Paleogene rocks; it is therefore known as the "Miocene zone." Near the Carpathian bend the older folded Miocene strata are partly overlain by Sarmatian strata, and farther south and west, as the flysch and Miocene units were depressed, Pliocene strata were deposited and later folded (Fig. 3). The last area has been named the "Mio-Pliocene zone."

DELIMITATION OF MIO-PLIOCENE ZONE

Herein, the name "Mio-Pliocene zone" within the district of Ploieşti is given to the whole of the folded Mio-Pliocene and Pliocene structures and strata between the flysch units (external and mediomarginal) on the north, the Moesic platform on the south, the Slănic-Buzău Valley on the east, and the Dîmboviţa Valley on the west (Fig. 3). Salt diapirism is the outstanding structural feature of the zone. The eastern and western limits are not only geomorphologic elements, but also have geologic significance. Thus, westward from the Slănic-Buzău Valley—as Pliocene strata are added to the top of the Neogene section—the zone widens from about 2 to about 16 mi in the central sector. From the Dîmboviţa Valley westward, the Mio-Pliocene zone widens still more in front of the Southern Carpathians; there the zone has broader folds and very few salt diapirs.

Considered as a whole, the Neogene zone is the transition from the Paleogene flysch to the Moesic platform. On the north, Paleogene flysch units are thrust onto the Neogene as a result of orogenic movements at the end of the Aquitanian and beginning of the Burdigalian. The most impressive thrusting is in the eastern sector where the external and mediomarginal units advanced the farthest, reducing to about 2 mi the width of the Neogene zone uncovered by the sheet; westward the amount of overthrusting decreases. Moreover, west of the Buzău River, as a result of gradual sinking of the mountainous area, the upper Miocene and Pliocene strata overlap the flysch area, forming two troughs known as the "Slănic trough" and the "Drajna trough,"

[1] Read before the 53d Annual Meeting of the Association, Oklahoma City, Oklahoma, April 25, 1968. Manuscript received, May 7, 1968; accepted, May 31, 1968.
[2] Deputy manager of Geological Department, Ministry of Petroleum.
[3] Chief geologist, District of Ploieşti.

D. Paraschiv and Gh. Olteanu

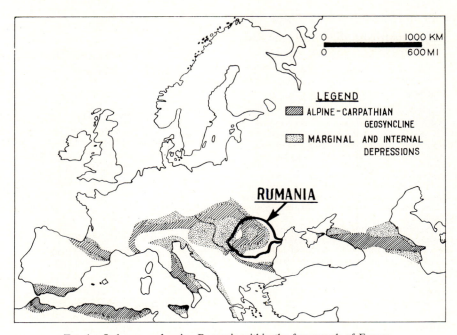

FIG. 1.—Index map showing Rumania within the framework of Europe.

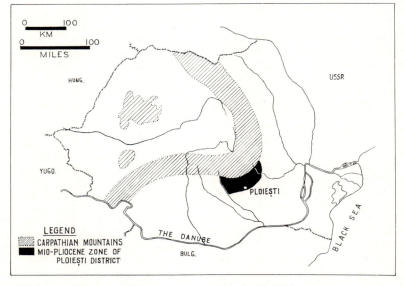

FIG. 2.—Position of Mio-Pliocene zone in Rumania.

concealing the nappe structure (Fig. 3). Drilling and seismic surveys in recent years have shown that southeastward the folded Neogene zone overlaps nearly horizontal beds of the platform (Fig. 4). The overlap occurred during the middle Sarmatian, because the lower Sarmatian of the Moesic platform is overlapped by the folded Miocene of the depression, whereas upper Sarmatian strata cover the overthrust plane.

In addition to hydrocarbon accumulations, the Mio-Pliocene zone contains coal measures and salt bodies. Because of the presence of these major natural resources, the region has been the object of numerous geologic and geophysical studies since the first half of the last century. A complete list of references to published and unpublished papers would be very voluminous. The selected references at the end of this paper are the main papers and syntheses of the last 20–30 years, and these in turn contain references to the older papers.

Investigation and exploitation of hydrocarbon accumulations in the Ploieşti region started more than 100 years ago. As technology advanced, all standard oil exploration methods were applied. Most structures were mapped by surface geologic surveys. Mapping first was on a scale of 1:100,000, but later detailed surveys were on a scale of 1:25,000 and in places on a scale of 1:10,000.

To solve regional and in places local problems, reconnaissance and semidetailed gravity, magnetic, and electric surveys have been conducted. Very detailed gravity surveys also have been made to determine the water-hydrocarbon boundary within some structures. To investigate the areas covered by flat-lying Quaternary deposits, obtain information from depth in the zone of exposed structures, and detail certain unconformities and stratigraphic traps, seismic surveys have been performed in accordance with geologic and geomorphologic conditions. On surfaces characterized by landslides or covered by thin Quaternary deposits, structural problems have been solved by structural drilling.

Stratigraphy

Most of the Mio-Pliocene zone is overlain by Pliocene and Quaternary strata. The Miocene crops out in limited areas, generally corresponding to the northern more elevated border and to the axial zones of some anticlines. The Mio-Pliocene strata overlie Paleogene, Cretaceous, and even older beds (Gri-

goraş, 1957a). As this paper concerns mainly the Tertiary strata, only brief reference is made to the section underlying the prospective units.

West of the Buzău Valley, the Eocene is developed in a marginal marly sandstone facies corresponding to the Văleni spur, and in an internal facies, the Tarcău Sandstone, corresponding to the Homorîciu spur. Transition from one facies to the other seems to be gradual (Popescu, 1952). The Eocene marginal facies, supposed to extend south below the Mio-Pliocene zone, is made up of a lower unit of green clay containing hieroglyphic sandstone and an upper unit of green and red clay (Pătruţ, 1959).

The Oligocene conformably overlies the Eocene, and also is separated into two facies, an external one, the Kliwa Sandstone, in the zone of the Văleni-Buştenari spur, and an internal one, the Pucioasa beds with Fusaru Sandstone, in the Homorîciu spur.

Just west of the Buzău Valley, the Kliwa Sandstone facies is composed from top to bottom of the following units, according to studies by Popescu (1952).

Unit	Thickness (ft)
Upper menilites[4] and diatoms	165–200
Upper Kliwa Sandstone	1,650
Podul Morii beds (predominantly marly with sandstone interbeds)	650–820
Lower Kliwa Sandstone	1,500
Dysodilic[b] shale	390–650
Lower menilites	65–130

The westernmost occurrence of the Kliwa Sandstone is near Cîmpina. West of the Cîmpina-Drăgăneasa structure and south of the Ocniţa, Runcu, and Apostolache structural alignment, the Oligocene in Kliwa facies has not been found. A well drilled recently at Moreni found only 1,500 ft of Oligocene and it is therefore difficult to indicate its facies. The Kliwa Sandstone facies generally is expected to be more pelitic in the southern part of the region, although the presence of Oligocene sandstone in the salt breccia at Băicoi may indicate that the external (Kliwa) facies is present also in that sector.

The Aquitanian overlies the Oligocene in normal succession and represents the youngest part of the Paleogene sedimentary cycle. Near

---

[4] Menilites are thin-bedded, cherty shale or black flint, of blackish or reddish color, in places associated with bituminous calcareous marl.

[5] Dysodilic refers to fine-textured, bituminous sediment (paper shales) deposited in deep water under anaerobic conditions.

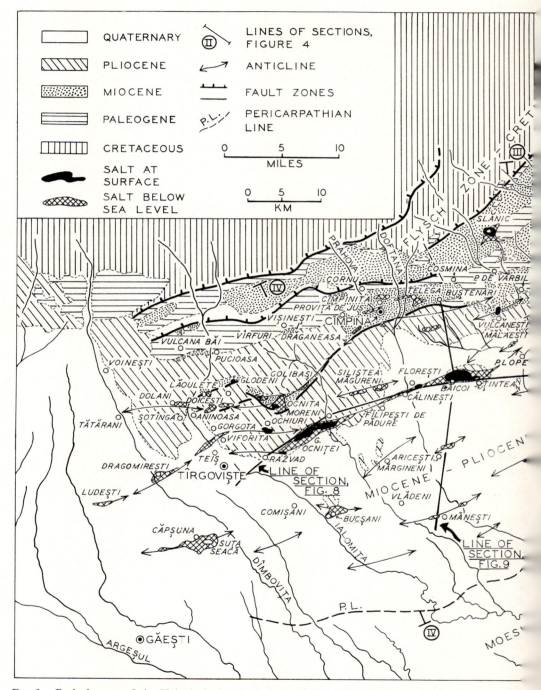

FIG. 3.—Geologic map of the Ploieşti district, Rumania. Lines of sections in Figures 4 (I-IV), 8, and 9 shown.

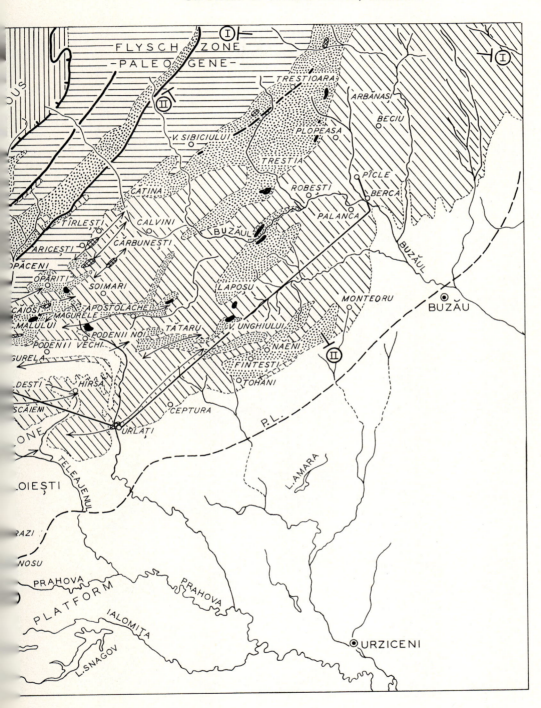

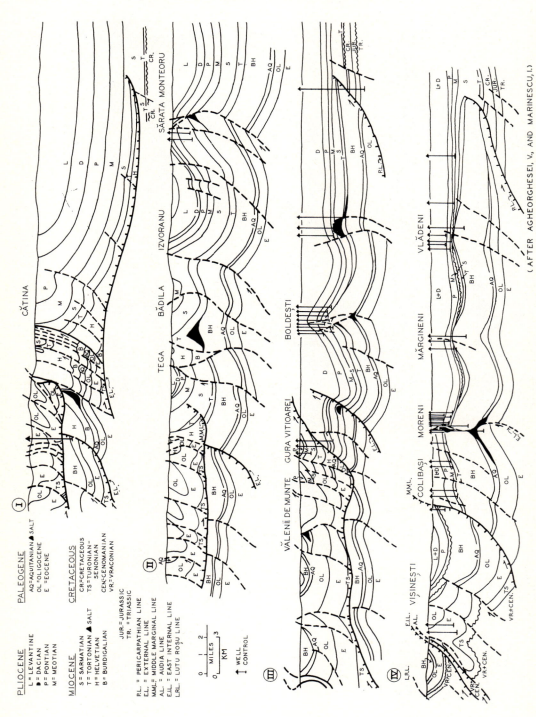

PLIOCENE
L = LEVANTINE
D = DACIAN
P = PONTIAN
M = MEOTIAN

MIOCENE
S = SARMATIAN
T = TORTONIAN ▲ SALT
H = HELVETIAN
B = BURDIGALIAN

PALEOGENE
AQ = AQUITANIAN ▲ SALT
OL = OLIGOCENE
E = EOCENE

CRETACEOUS
CR = CRETACEOUS
TS = TURONIAN–
      SENONIAN
CEN = CENOMANIAN
VR = VRACONIAN

JUR. = JURASSIC
TR. = TRIASSIC

P.L. = PERICARPATHIAN LINE
E.L. = EXTERNAL LINE
M.M.L. = MIDDLE MARGINAL LINE
A.L. = AUDIA LINE
E.I.L. = EAST INTERNAL LINE
L.R.L. = LUTU ROȘU LINE

0   1   2
|————————|
      MILES
0   3
|—————————|
      KM

⊢ WELL
  CONTROL

FIG. 4.—Cross sections through Mio-Pliocene zone. Northeasternmost at top, westernmost at bottom. Locations shown on Figure 3.

(AFTER AGHEORGHESEI, V., AND MARINESCU, I.)

the end of that cycle, because of the emergence of the Carpathian areas, the Paleogene sea receded. However, it persisted in places in the form of gulfs and lagoons. Lagoonal strata (Cornu beds) deposited there are made up of dysodilic shale (similar to that of the Oligocene), gypsum, sand, sandstone, conglomerate, and salt.

The Paleogene-Miocene boundary and the age assignment of the Cornu beds are controversial in Rumania. Latest research seems to indicate that the Cornu beds are younger than Aquitanian. The writers adopt the viewpoint expressed by many other geologists that the basal stage of the Miocene is the Burdigalian.

The Burdigalian marks the advent of a new sedimentary cycle. On the northern border of the Neogene zone it is represented by polygenic conglomerate (Brebu Conglomerate) and coarse sandstone transgressively overlying the Cornu beds.

It is to be expected that these coarse-grained rocks grade southward into finer grained beds (sandstone and marly sandstone). Therefore, the basal sandy series found in drilling (Teiş, Moreni) and assigned to the Helvetian may be partly of Burdigalian age. The thickness may be 1,800 ft and even 3,000 ft.

The Helvetian is made up of a basal, predominantly sandy complex with red and bluish intercalations, and an upper, predominantly marly, bluish unit with gypsum, dacitic tuff, and sandstone intercalations. Present thickness of the Helvetian may exceed 6,000 ft. Collected data suggest that the thickness of Helvetian deposits decreases toward the external margin of the Neogene zone. The decrease is accompanied by a partial gradation into pelitic beds.

The Tortonian is the youngest marine series of the Miocene zone and, according to the most recent investigations (Fl. Olteanu, 1951; Popescu, 1951), it is composed from top to bottom of the following members.

1. *Spirialis* marl member, a predominantly marly unit, 300–600 ft thick.
2. Radiolarian shale member, composed of lamellar clay more than 300 ft thick.
3. "Salt-breccia" member (a sedimentary breccia) which generally is found in depression zones (*e.g.,* the Meliceşti, Trestioara, Slănic, Drajna, Predeal Sărari troughs, *etc.*) It is formed of angular fragments of Eocene, Oligocene, and Helvetian sandstone as well as Jurassic limestone, green shale (both being Dobrogean elements), and schist. In some areas breccia is interbedded within the *Globigerina* marl (Cosminele Valley) or the gypsum beds. The breccia contains

intercalated *Leitha* limestone (Slănic trough) and salt bodies such as those at Slănic, Predeal-Sărari, Bădila, and elsewhere. Thickness of the "breccia" member may be as much as 1,800 ft (Grigoraş, 1961); the salt is lenticular.

4. *Globigerina* tuffaceous marly member, the thickness of which can exceed 300 ft.

The *Globigerina* tuffaceous member is assigned to the lower Tortonian, conformably overlying the Helvetian. The other three members comprise the upper Tortonian; breccia unconformably overlies the lower Tortonian.

The Sarmatian, including the Buglovian, a transition from the Tortonian to the Sarmatian, is represented by all its substages (Fig. 5). Regionally the lower Sarmatian in uplifted areas unconformably overlies older strata. In the Buglovian, and to a lesser extent in the Volhynian, pelitic deposits are predominant. The Bessarabian is a sandy facies in the region north of the group of Salcia-Apostolache structures (Atanasiu, 1948), and is a calcareous facies (*lumachelle*[6]) in the southern part. In places the Kersonian is likewise lumachellic, but also contains significant marly deposits.

The beginning of the Pliocene corresponded to a subsiding phase in the Neogene zone, expressed by a marked transgression of the Meotian and by a comparative uniformity of the facies of the first two Pliocene stages. The Pliocene generally is made up of a sequence of marl and sand (Fig. 6) in a very favorable proportion for hydrocarbon accumulations. The sequence, deposited uninterruptedly from the Meotian to the Levantine, steadily thickens southeast, and east of Buzău is approximately 30,000 ft thick. Generally the Meotian transgressively and unconformably overlies the older members. The unconformity becomes more pronounced westward and northward, as well as in the zone of the major anticlinal folds.

Because of the subsidence movements recorded in the zone of the Carpathian bend, the Meotian strata thicken from about 350 ft in the western sector, at Razvad, to 1,800 ft and even 2,000 ft in the eastern part, at Berca (Fig. 7). There is a slight north-south thickening of those beds in the central area of the Neogene zone, and a pinching out toward the borders. South of the Pericarpathian fault, the Meotian again thickens, the isopachs marking a maximum thickness in the Moesic platform. The variation of Meotian thickness can hardly be explained by the nondeposition of members, and

---

[6] *Lumachelle* is a shell conglomerate of mollusk debris.

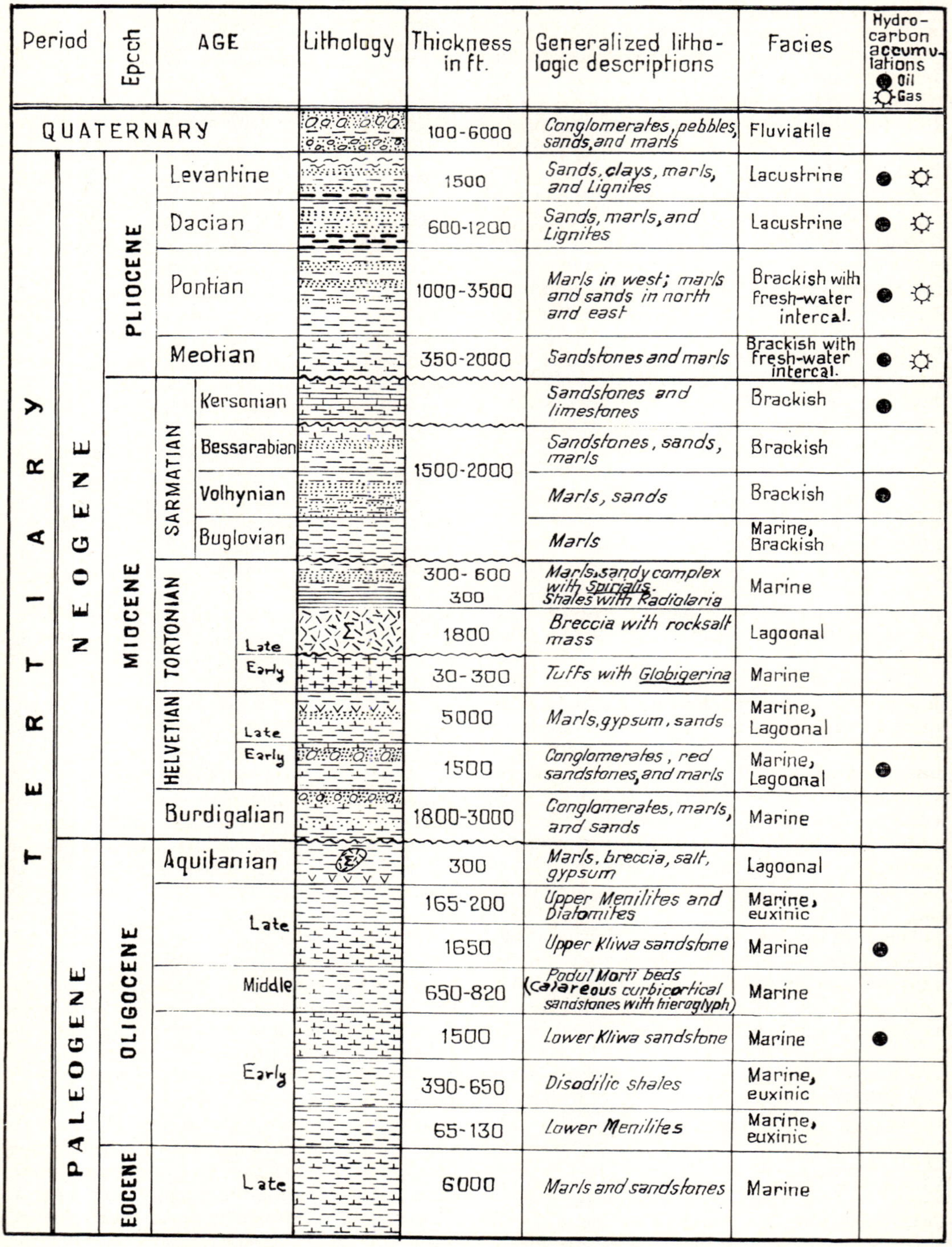

Period	Epoch	AGE		Lithology	Thickness in ft.	Generalized lithologic descriptions	Facies	Hydrocarbon accumulations ● Oil ☼ Gas
QUATERNARY					100-6000	Conglomerates, pebbles, sands, and marls	Fluviatile	
T E R T I A R Y	N E O G E N E	PLIOCENE	Levantine		1500	Sands, clays, marls, and Lignites	Lacustrine	● ☼
			Dacian		600-1200	Sands, marls, and Lignites	Lacustrine	● ☼
			Pontian		1000-3500	Marls in west; marls and sands in north and east	Brackish with fresh-water intercal.	● ☼
			Meotian		350-2000	Sandstones and marls	Brackish with fresh-water intercal.	● ☼
		MIOCENE	SARMATIAN	Kersonian		Sandstones and limestones	Brackish	●
				Bessarabian	1500-2000	Sandstones, sands, marls	Brackish	
				Volhynian		Marls, sands	Brackish	●
				Buglovian		Marls	Marine, Brackish	
			TORTONIAN		300-600 / 300	Marls, sandy complex with Spiralis. Shales with Radiolaria	Marine	
				Late	1800	Breccia with rocksalt mass	Lagoonal	
				Early	30-300	Tuffs with Globigerina	Marine	
			HELVETIAN	Late	5000	Marls, gypsum, sands	Marine, Lagoonal	
				Early	1500	Conglomerates, red sandstones, and marls	Marine, Lagoonal	●
			Burdigalian		1800-3000	Conglomerates, marls, and sands	Marine	
	P A L E O G E N E	OLIGOCENE	Aquitanian		300	Marls, breccia, salt, gypsum	Lagoonal	
			Late		165-200	Upper Menilites and Diatomites	Marine, euxinic	
					1650	Upper Kliwa sandstone	Marine	●
			Middle		650-820	Podul Morii beds (calcareous curbicortical sandstones with hieroglyph)	Marine	
			Early		1500	Lower Kliwa sandstone	Marine	●
					390-650	Disodilic shales	Marine, euxinic	
					65-130	Lower Menilites	Marine, euxinic	
		EOCENE	Late		6000	Marls and sandstones	Marine	

FIG. 5.—Generalized stratigraphic column in the Mio-Pliocene zone.

therefore it should be concluded that the sedimentation rate was lower northward and westward in this region where the basement of the basin was more stable.

The Meotian is of a fairly homogeneous facies—a repeated alternation of sand and marl. Sand and sandstone prevail in the upper and lower parts of the Meotian, whereas the marl is better developed in the middle part. Nevertheless, the homogeneity of facies is comparative, because there are slight changes both north-south and east-west (*i.e.*, the predominance of sand increases eastward and northward; Figs. 8, 9). Pelite becomes much more significant southward, in the most depressed part of the Neogene zone, where sandstone and sand have disappeared almost completely (Fig. 9).

The Pontian is made up almost exclusively of marl, which suggests a sinking of the Neogene zone. Sand is present in the Pontian near the old shoreline and the eastern border of the region. Like the Meotian, the Pontian thickens from the west (Tîrgovişte), where it is about 1,000 ft, eastward to Berca, where it is 3,500 ft.

The Dacian represents a slight transgression in comparison with the Pontian and is made up of alternating, not very dense, gray sand, sandstone, and marl, which in part are dark and carbonaceous; locally the Dacian contains a few pebbles and workable coal (lignite) intercalations. In the western part of the Mio-Pliocene zone, the Dacian consists of five lithologic units. The upper two are predominantly pelitic. Thickness of the Dacian ranges from 600 to 1,200 ft; like the older Pliocene units, it thickens eastward and in the synclinal zones, and pinches out near the uplifts.

The Levantine, exclusively of freshwater deposition, is composed of *Helix*- and *Planorbis*-bearing argillaceous sand, and fine pebble beds.

The Quaternary, also including the Cîndeşti (Villafranchian) conglomerate, terminates the sedimentary cycle started in the Meotian and in places in the upper Miocene. Quaternary strata are composed of polygenic pebbles, sand, and clay, and terrace deposits.

The foregoing stratigraphic summary shows the following main features.

The Neogene zone outside the mountain range represents the last stage of the evolution of the Carpathian geosyncline. A sequence of eastward-thickening deposits accumulated there. The sequence is approximately 90,000 ft thick in the Carpathian bend; the Pliocene alone is estimated to be about 30,000 ft thick.

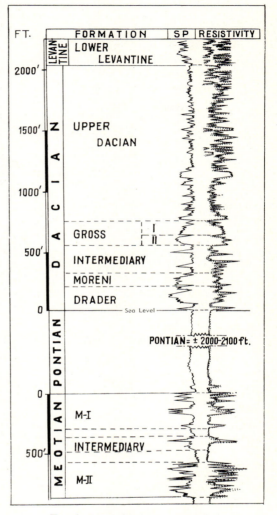

Fig. 6.—Moreni-Băicoi typical log.

This enormous mass is made up entirely of marl, clay, sand, sandstone, limestone, and evaporite.

The Paleogene and Pliocene strata were deposited continuously, whereas during the Miocene there were breaks in sedimentation at the end of the Aquitanian, the end of the early Tortonian, in places between the Tortonian and the Sarmatian, in the middle Sarmatian, and at the end of the Sarmatian. The unconformities are more pronounced near the northern border of the basin and in the zones corresponding to the major uplifts. Because of the latter fact and the thickening of the upper Miocene and Pliocene strata in the synclinal zones, it can be

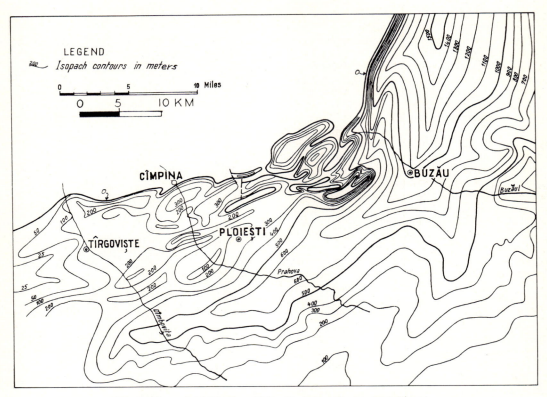

FIG. 7.—Isopach of Meotian. (After V. Agheorghesi and M. Pirvu.)

concluded that sedimentation occurred simultaneously with the tectonic movements. Sedimentation probably was continuous in the depressed areas.

The water in which the sediments were deposited became fresher continuously from the Paleogene through the Levantine. Thus, from the Paleogene through the Tortonian the sediments were marine; during the Sarmatian they were deposited in a brackish environment; and during the Pliocene they were laid down in a freshwater basin where few incursions of brackish water occurred.

Tectonics

The Mio-Pliocene zone is the outermost unit of the Carpathian area, between the flysch units and the foreland. As a whole, the Mio-Pliocene zone rises from south to north and from southeast to northwest. On this major zone of uplift, near the mountain border and westward, many fairly intricate folds spread along four main alignments (Fig. 3) can be distinguished.

The character of the folds depends on (1) the age of the strata composing them and (2) the distance from the mountainous zone. The folds are generally much sharper with increasing depth and more attenuated near the surface (Stefănescu, 1938). According to recent synthesis, it appears that at the level of the lower Helvetian the Moreni-Gura Ocniţei anticline has a reversal of 10,500 ft, which decreases to 4,500 ft at the base of the Meotian and to 3,600 ft at the base of the Dacian. This example shows that the structural elements were not formed during a single folding phase at the end of the Levantine, but that the closure and shape are the result of repeated orogenic phases. Consequently, each stage or group of stages has different tectonic characteristics. The lower Miocene has a very intricate and generally compressed structure, the Sarmato-Pliocene strata have a unique structural style, and the Quaternary conglomerate has, with some exceptions, an almost homoclinal structure.

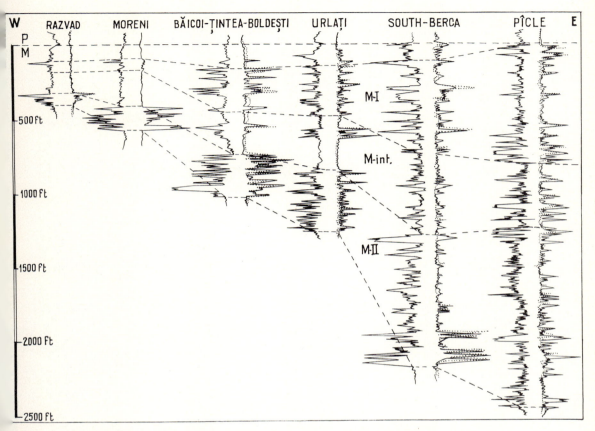

FIG. 8.—West-east electric-log correlation of Meotian in Ploieşti district. No horizontal scale. Location shown on Figure 3.

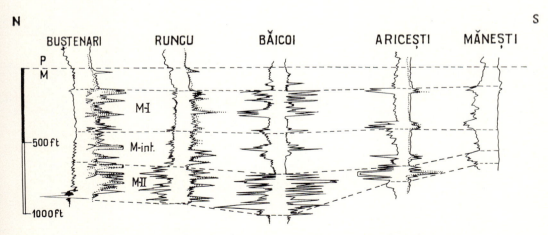

FIG. 9.—North-south electric-log correlation of Meotian in Ploieşti district. No horizontal scale. Location on Figure 3.

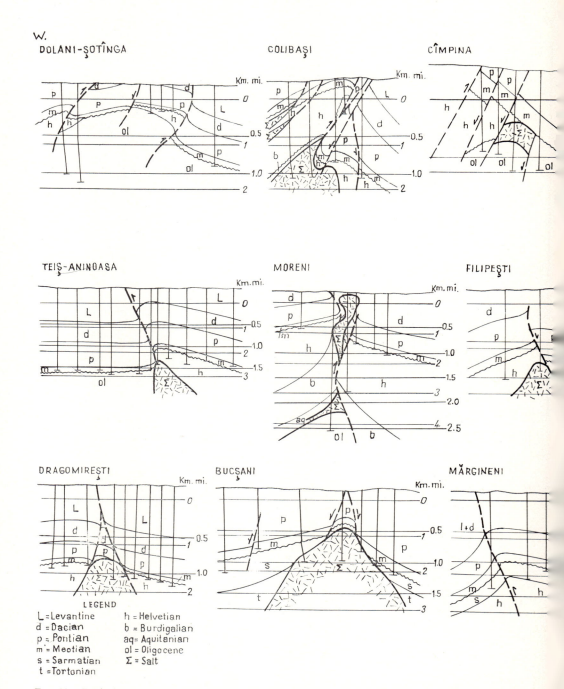

LEGEND

L = Levantine   h = Helvetian
d = Dacian   b = Burdigalian
p = Pontian   aq = Aquitanian
m = Meotian   ol = Oligocene
s = Sarmatian   Σ = Salt
t = Tortonian

Fig. 10.—Typical sections through Miocene-Pliocene structures of Ploieşti district, arranged in rows, west

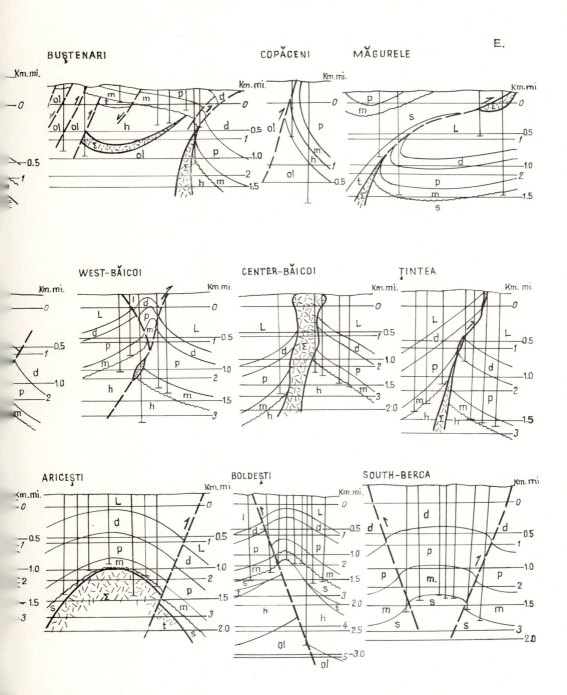

t. Northernmost at top, southernmost at bottom. Locations of structures shown on Figure 3.

The influence of the proximity of the folds to the flysch border, where the intensity of tangential movements was greatest, is obvious. The structures near the external edge of the Mio-Pliocene zone generally are slightly uplifted domes, e.g., the Mărgineni, Aricești, Bucșani, and other anticlines (Fig. 10). In the median sector of the Mio-Pliocene zone is a structural alignment composed of the Razvad, Moreni-Gura Ocniței, Florești, and Băicoi-Țintea anticlines with axially faulted large reversals, outcropping salt, and characterized as a rule by a southern flank steeper than the northern one. Outcropping salt shows slight overflowing toward the flanks, particularly toward the southern one.

Near the northern border of the Mio-Pliocene zone, even sharper, markedly faulted anticlines generally are overturned toward the south. That group of structures, which are a kind of "nascent nappes" (Atanasiu, 1948), includes the following structural units: Sotînga, Colibași, Ocnița, Cîmpina, Buștenari, Măgurele, Podenii Noi, Matița, Apostolache, and others (Fig. 10). The general thrusting of the northern flank over the southern one shows that the tangential movements proceeding from the flysch zone were stronger near the contact with the Miocene zone, and suggests that uplift was greatest along the border of the Carpathian foredeep, which continuously shifted toward the foreland.

Salt was significant in the formation of folds. Under the influence of tangential movements, temperature, and lithostatic pressure caused by the cover of more recent deposits, salt became plastic and migrated toward the less resistant zones within the axial parts of anticlines which were accentuated and made more intricate. Thus the salt shifted, in places even uprooting itself, and piled up in exotic masses. In the southern part of the region, salt accumulation was a minor phenomenon and affected the more recent beds only slightly (Fig. 10). In the median zone, however, salt partly or totally pierced the overlying sedimentary beds. On its way to the surface, salt carried along elements of its breccia as well as rock fragments from the strata penetrated during its ascent. The structures pierced by salt in the form of cores, or blades, were given the name "diapir folds" by the Rumanian geologist Mrazec (1907).

The piercement phenomena began in Miocene time and continued to the end of the Pliocene and during the Quaternary. Diapirism was favored by the occurrence of regional faults. Thus, according to recent structural syntheses, the main diapir faults with outcropping salt plugs (Moreni-Gura Ocniței, Florești, Băicoi-Țintea) are aligned along an important zone of tectonic disturbance which may represent the extension, below the Mio-Pliocene strata, of an external fault, beyond which, farther east and in Moldavia, the flysch is thrust over the Neogene zone. Another more northern alignment, along which are the Ocnița, Colibași, and Cîmpina structures, may be thrust over the mediomarginal fault, and farther east the median flysch is thrust over the external unit.

According to older reports (Macovei, 1938), Miocene zone folds have been classified into three categories—normal folds, diapir folds, and faulted folds—which characterize successive zones from south to north. The arrangement of the categories of folds also shows the trend of evolution of the Mio-Pliocene structures. In addition to the three categories of folds, there are some of an intermediate type.

Dips on the structures range from 2 to 90°, the steepest dips being near the piercing plug or along the fault lines. Generally dips decrease from north to south and increase downward from the surface. As was discussed, on the northern structural line there are overturned beds. The entire region and particularly the anticlinal folds are affected by numerous faults that have been produced to a great extent by salt piercement.

At the end of the Pliocene, tectonic movements in the region as a whole led to dislocation of as much as 9,000 ft of strata. The Cîndești conglomerate (Villafranchian) at Măgura Odobești northeast of Buzău is at an altitude of about 3,000 ft, and in some wells within the region the same beds have been found at depths of 5,500–6,000 ft.

The following are the more important tectonic elements.

1. Mio-Pliocene strata as a whole plunge from the north and northwest toward the south, and particularly toward the southeast.

2. Strata in the region have been folded during several stages. They form anticlines and synclines arranged on four main structural alignments. Characteristic of Mio-Pliocene folds is the phenomenon of salt piercement (diapirism).

3. The folds are of moderate relief in the southern part and become more pronounced and intricate toward the flysch margin, where they are overturned and even thrust. Of the four major alignments, the Gura Ocniței-Moreni, Florești, Băicoi-Țintea median structural line is the most pronounced, being surrounded by well-developed synclinal zones. Anticlines on that alignment show simpler tectonic effects than those of the northern structural line.

4. Mio-Pliocene structures are affected by numerous

Table 1. Comparison of Results of Geochemical Analysis of Pliocene and Oligocene Rocks
(After Sever Anton)

Formations	Lithologic character	Organic carbon *)	Bitumen A *)	Bitumen C *)	Asphaltene *)	Hydrocarbons *)	Non-hydro carbons *)	Total S / S from sulphates	Total Fe / FeO *)	CRo +)	CRm +)	Trask	pH
DACIAN	argillaceous marls	0.09-0.30	0.01-0.06	0.01-0.05	6-10	30-42	45-54	0.01-0.06 / 0.01-0.06	4-5.5 / 1-4.5	0.49-6.49	2.50-2.60	0.25-0.70	8.6
PONTIAN	marls	0.50-0.98	0.02-0.14	0.01-0.05	8-27	20-65	26-73	0.04-0.12 / 0.04-0.18	4.65-6.3 / 1.2-2.3	0.28-1.77	1.19-3.61	0.30-0.92	8.4-8.5
PONTIAN	marly clays	0.45-0.74	0.03-0.46	0.01-0.02	16-26	36-60	33-59	0.01-0.08 / 0.01-0.08	4.3-5.6 / 1.9-2.3	0.39-3.18	1.69-3.18	0.69-0.97	8.5
PONTIAN	calcareous microsandstones	0.29-0.39	0.02-0.03	0.02-0.03	10-15	25-32	46-59	0.04-0.06 / 0.02-0.04	4.1-4.9 / 2-2.5	1.7-2.9	3-3.4	0.40-0.50	8.6
MEOTIAN	marls	0.35-0.50	0.05-0.08	0.02-0.04	10-15	45-61	25-50	0.07-0.10 / 0.06-0.08	2.5-3.2 / 1.3-1.5	1.30-4.5	2.8-5.3	0.30-1.40	8.6
MEOTIAN	argillaceous marls	0.31-0.95	0.02-0.06	0.01-0.04	4-30	30-37	53-67	0.02-0.07 / 0.01-0.06	6.34-8.56 / 2.5-3.7	1.72-2.48	0.72-1.26	0.18-0.56	8.7-8.8
MEOTIAN	marls and marly microsandstones	0.34-1.04	0.03-0.05	0.01-0.02	25-26	28-74	23-65	0.07-0.14 / 0.01-0.09	4.4-7.4 / 1.8-2.6	1.86-3.60	1.35-2.68	0.75-1.21	8.7-8.9
MEOTIAN	marly microsandstones	0.25-14	0.06-0.08	0.01-0.02	6-27	53-68	37-45	0.08-0.12 / 0.04-0.06	1.68-3.79 / 1.2-1.6	1.12-3.12	2.40-4.08	0.40-0.67	8.3-8.6
OLIGOCENE	argillites	0.36-2.50	0.03-1.98	0.01-0.50	12.3-48.0	32.2-62.8	33.8-64.9	0.06-5.73 / 0.01-0.74	2.91-8.96 / 0.2-3	1.15-18.62	0.65-11.80	0.39-4	2.7-9.7
OLIGOCENE	calcareous and dolomitic argillites	0.15-7.65	0.02-0.75	0.02-0.37	10.9-42.5	31.9-60.08	37.9-63.8	0.08-4.60 / 0.00-0.85	1.6-5.8 / 0.03-0.7	0.38-12.88	0.25-14.70	1.5-3.5	7.1-9.7
OLIGOCENE	siliceous microsandstones	0.10-2.25	0.01-0.23	0.01-0.06	3.5-38.6	30.0-64.0	35.0-67.5	0.02-2.18 / 0-0.19	0.91-6.10 / 0.03-0.8	0.52-11.20	0.11-12.96	1-4.8	4.5-9.2

*) Percent
+) In milligrams oxide/100 gm. rock

Column headed "Trask" shows minimum and maximum of the Trask (Trask and Patnode, 1942) reducing ability in rocks analyzed from the formations shown. Indexes refer to the organic reducing ability (after removal of sulfur).

faults, partly produced by the salt diapirism that was important in controlling oil and gas accumulations.

SOURCE, RESERVOIR, AND CAPROCKS

Data on stratigraphy and tectonics of the region show that the Paleogene and Neogene sequence is made up of an alternation of clay, marl, argillaceous and marly shale, sand, sandstone, conglomerate, evaporite, and some limestone. That sequence of rocks, with a variable vertical and horizontal distribution, provided for generation, accumulation, and preservation of oil and gas.

The deposits of greatest obvious bituminous aspect are in the Oligocene Kliwa facies. Pelitic intercalations (dysodil, white marl), sandy

marl, and marly limestone in that unit have characteristics which warrant their classification as source rocks (Table 1). Aquitanian dysodilic shale also may be considered as source rock.

The Helvetian, particularly the upper part, abounds in greenish-blackish, shaly, slightly bituminous marl and bituminous foliated shale.

In the Tortonian, the hydrocarbon-generating rocks could be either the *Globigerina* marl in the lowermost part of the sequence or the radiolarian shale above the salt, which is very similar to the Oligocene dysodilic shale.

The Sarmatian contains bituminous marl, particularly in the lower part, and the Meotian contains shaly marl that also is bituminous.

The Pontian, composed exclusively of marl in the western part, may be considered the source of accumulations in its own reservoir rocks (in the eastern part of the region) as well as of part of the hydrocarbons in the lower Dacian. Moreover, the Dacian pelitic intercalations may be partly responsible.

In the Rumanian geologic literature, lively discussions have been carried on about the origin of oil accumulated in Meotian and Dacian strata, which are by far the most productive units of the Mio-Pliocene zone. A group of geologists represented by Macovei (1932, 1938), Macovei and Ştefănescu (1935), Mrazec (1931), to some extent by Atanasiu (1948), and others, were of the opinion that oil which accumulated in the Pliocene migrated along faults (particularly along those produced by the salt piercement) of Oligocene and Aquitanian age. Another group of geologists, among whom are Popescu-Voiteşti (1935a, b), Walters (1940), Preda (1957), and others, considered that most of the Pliocene deposits are autochthonous.[7] The writers provisionally share the opinions of the last-named group of geologists, on the following grounds.

If the oil had migrated from depth (Oligocene or Aquitanian), it should have impregnated not only the Pliocene reservoir rocks but also the Miocene reservoirs, such as Helvetian and Sarmatian sand and sandstone, during migration between the Oligocene and Pliocene rocks. However, oil is not present in those beds, as studies of great oil fields such as Moreni-Gura Ocniţei, Băicoi, and others demonstrate.

Analysis of the distribution of oil quality in Mio-Pliocene oil fields shows that producing zones of more recent age (Dacian) contain, with very few exceptions, asphaltic, heavy oils, whereas the older reservoirs (of Meotian and Miocene age) are saturated with lighter, paraffinic oils (Cazimir, 1934; Creangă et al., 1962; Hlauschek, 1950). This general distribution is in accordance with facts observed also in other oil basins of the world.

Studies carried out in other countries (Barton, 1934; Silverman, 1965) have led to the conclusion that youngest oils (including the very recent ones) show high specific (low API) gravity, a low natural gasoline content, and high sulfur, nitrogen, and oxygen contents.

[7] To the effect that they accumulated in reservoir rocks of the same age as the source rocks.

On the contrary, old oils show a lower specific gravity and a greater paraffin content. Hence, the evolution ("maturation") of oil starts with the heavier (younger) oils and proceeds toward paraffinic oils. The migration process itself leads to a certain evolution of oil. The results of these researches support autochthonous origin for oils within the Meotian and Dacian strata in the Mio-Pliocene zone.

Researches still being carried out show that Meotian and Pontian rocks deposited in brackish or freshwater basins have indices comparable with those of the Oligocene dysodilic shale (Table I). Investigations concerning the Dacian and Miocene rocks have not been concluded.

Reservoir rocks are the sand and sandstone bodies which are found in all Paleogene, Miocene, and Pliocene sections except the Tortonian and Pontian, where psammite is of minor importance.

Some reservoir possibilities also are offered by the Sarmatian limestone, and to a lesser extent by limestone of the Tortonian and Pontian, but they have not been found to be impregnated.

Cover rocks are present in all prospective sections. Thus, the upper dysodilic rocks cover the Oligocene Kliwa Sandstone; the Helvetian marly series, with marl, shale, and Tortonian salt, covers the reservoir rocks of the lowermost Helvetian and the Burdigalian; the Sarmatian as well as the Meotian contain covering pelitic units. For the Meotian accumulations, the most effective cover rock is the Pontian with its massive marl. In the Dacian and Levantine, marly intercalations are present and have proved to be capable of covering oil and gas accumulations in many zones.

MAIN PRODUCTIVE STRATA AND STRUCTURES

Prospecting and exploration in the Mio-Pliocene zone of the district of Ploieşti have led to the discovery of 35 oil- and gas-productive structures.

To January 1, 1968, the region had yielded about 1,725 million bbl originating from the following geologic units.

Unit	Bbl Produced
1. Dacian and Levantine	500,000,000
2. Pontian	750
3. Meotian	1,028,000,000
4. Sarmatian	89,000,000
5. Helvetian and Burdigalian	53,500,000
6. Oligocene	55,400,000

The most important oil-bearing unit is the Meotian, which has yielded almost 60 percent of the production of the region. The Dacian has yielded nearly 29 percent and the Sarmatian more than 5 percent. The least oil production has been obtained from the Pontian, mainly because of lack of reservoir rocks.

Except for the Dacian strata, the value and productivity of reservoirs decrease with the increase of depth and age of beds. Though the Sarmatian, and particularly the Helvetian and Oligocene, have not been explored completely, the results available are rather significant in that respect.

The Dacian oil and gas accumulations are concentrated in the western part of the region (Fig. 11), except for those in the eastern and northern sectors of the Mio-Pliocene zone. This is because in the northern and northeastern parts the Dacian, as well as the Levantine, was affected by erosion. The strata are preserved only in the more depressed zones, or in the zone of faulted folds where the Dacian strata are protected below the overthrust of the northern flank. In the western part of the region, sedimentary conditions were more uniform throughout the Pliocene, and therefore the Dacian pelitic intercalations, acting as cover rocks, are present across greater areas, whereas southeast and west of the Dîmbovița River the intercalations become more lenticular.

The diapiric folds with their piercement plugs offer favorable entrapment conditions, either through the walls of salt or through the faults, which have put the Dacian reservoirs in contact with the thick marl units in the Pontian or the Meotian.

Dacian oil and gas reservoirs generally are discontinuous, even within one structure. Discontinuity of accumulations might be ascribed to the dispersion of migration paths of hydrocarbons in the Meotian. Also, the Dacian petroleum accumulated after that of the Meotian when structural complexity increased as a result of both continuous salt piercement and the influence of the Rhodanian folding phase (end of Dacian). Another contributing factor was the conspicuous change of facies during the second half of the Pliocene. All those causes led to the accumulation and preservation of hydrocarbons only in certain tectonic blocks.

The most important and typical Levantine and Dacian oil field is Moreni, where there are seven productive zones composed of fairly well-compacted sandstone, named from top to bottom the Lower Levantine, Upper Dacian, Gross I, Gross II, Intermediary, Moreni, and Dräder (Fig. 6). At Ochiuri the section is similar. In the other structures, correlation with this section is difficult because of the sedimentary variation, and only the basal member, the Dräder, can be recognized. Except for the Băicoi structure and to some extent the Boldești anticline (gas), productivity of the Dacian is insignificant and generally is limited to the Dräder complex.

Because of the optimum geologic conditions of the Moreni and Băicoi structures, production from the Dacian sandstone has exceeded, on an average, 350,000 bbl/acre, and in certain blocks of the southern flank in Moreni, production has reached 860,000 bbl/acre. On the other structures, productivity is much lower.

Daily production in the Dacian is variable. The greatest daily output has been in the following wells: No. 3 Moreni-Băicoi, 32,900 bbl from the Dräder layer; No. 1 Col. Moreni, 16,150 bbl from the Moreni layer; No. 38 Ruc. Gura Ocniței, 13,200 bbl from Gross I layer; No. 27 St. R. Băicoi, 4,550 bbl from the Dräder; and No. 400 R. A. Gura Ocniței, 3,730 bbl from the Gross and Intermediary layers.

The Pontian has only a few small hydrocarbon accumulations. They are in the eastern sector of the region (Fig. 12) where increasing numbers of sandstone intercalations appear within this marly unit.

The Meotian is the most productive oil- and gas-bearing unit of the Mio-Pliocene zone in the district of Ploiești. It has proved to be oil- and gas-bearing wherever it contains reservoir rocks and has conditions favorable for preservation. The structures with oil and gas in the Meotian shown in Figure 13 are concentrated, like those in the Dacian, in the western half of the region. The scarcity of reservoirs in the eastern part of the region is the result of discontinuity or absence of the Pliocene in more uplifted zones where Miocene strata crop out in many places (Figs. 3, 13). The representative section of the Meotian in the zone of the Moreni and Băicoi structures consists of two predominantly sandy and gritty complexes at the bottom and top parts of the unit and named in abbreviation "M-II" and "M-I." The units are separated by a predominantly marly complex which also contains sandy intercalations grouped under the name of "Intermediary M" (Fig. 6).

As was noted, the thickness of the Meotian and its reservoirs decreases from east to west.

D. Paraschiv and Gh. Olteanu

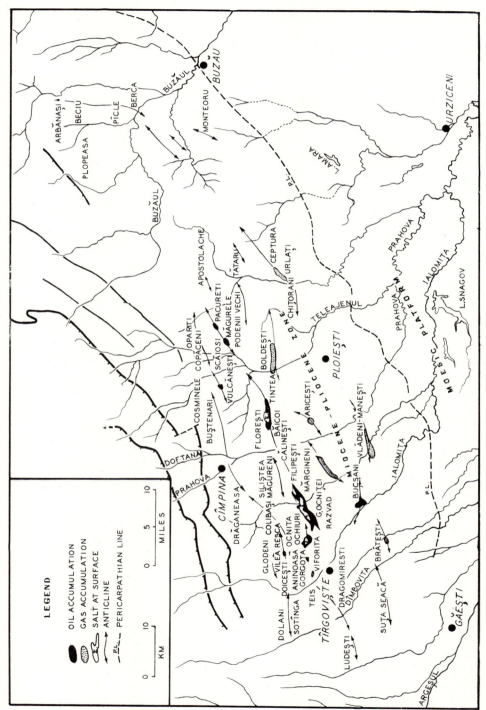

FIG. 11.—Sketch of Dacian oil and gas fields of Mio-Pliocene zone.

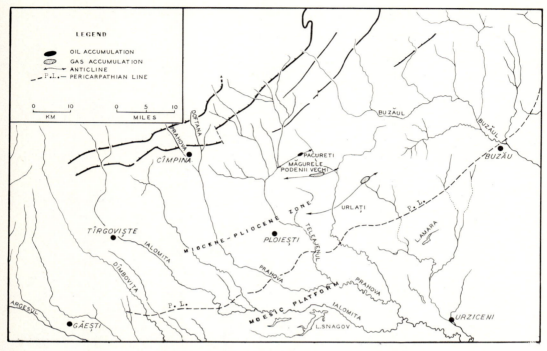

Fig. 12.—Sketch of Pontian oil and gas fields of Mio-Pliocene zone.

Thus, at Razvad-Teiş (western part), only two rather thin sandstone complexes (M-I and M-II), (46 and 25 ft, respectively) are recognized. Farther east (Moreni-Boldeşti), in addition to the sandstone beds of complexes M-I and M-II, psammite is developed in the Intermediary M and above the M-I complex ("G" sandstones). At Berca-Arbănaşi, as the Meotian thickness increases, there is also an increase in the number of sandstone complexes—to 27 units. Consequently, in contrast to two productive Meotian complexes in the western part, there are 27 in the eastern, and the massive sandstone and sand in the north pass gradually into an almost exclusively marly facies in the south. The increase in the number of sandstone beds is accompanied by a substantial decrease in (1) marl, (2) the possibilities of hydrocarbon generation, and (3) the possibilities of preservation. Thus, the productivity of the Meotian does not increase in relation to the number of sand or sandstone bodies.

The size of the productive zones differs among the structures and even among the blocks within a single structure. Sandstone beds of the bottom complex (M-II) also have both better productivity and more extensive productive areas. As a rule, the productive zones of the northern or western flanks of the structures appear to be narrower (Moreni, Berca, etc.), ranging in width from 450 to 2,100 ft. Productive zones on the southern or eastern flanks are better developed, reaching 7,300 ft at Boldeşti and 6,300 ft at Moreni.

On the Moreni-Băicoi line, average Meotian production exceeds 30,000 bbl/acre. At Boldeşti production is nearing 21,000 bbl/acre, and at Bucşani is about 8,900 bbl/acre. On the other structures, the accumulations are of less value. Compared with the Dacian strata, the Meotian shows a smaller production per surface unit area within the Moreni and Băicoi structures, because the reservoir-rock thickness of the last member is 3–10 times smaller. A compensating factor is that the productive areas of the Meotian are much more extensive than those of the Dacian. Daily average production is of a rather wide range. The most noteworthy flow rates are reported for the following wells: No. 25 S. I. Gura Ocniţei, 14,650 bbl/day; No. 42 A. R. Moreni, 10,400 bbl/day; No. 28 Un. Moreni, 9,500 bbl/day; and No 54 A. R. Boldeşti, 4,250 bbl/day.

The Sarmatian is next in productivity after

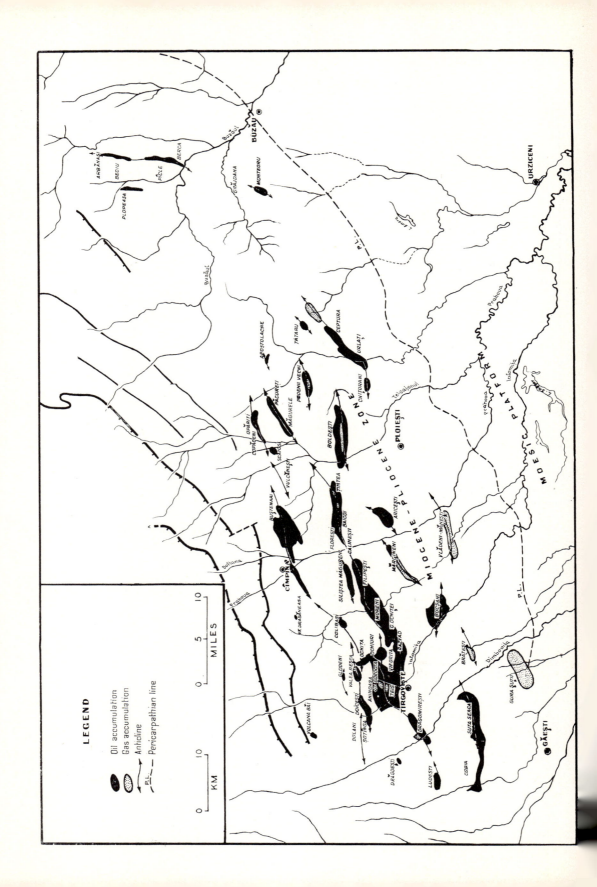

LEGEND

- ⬭ Oil accumulation
- ⬭ Gas accumulation
- ⤴ Anticline
- P.L. — Pericarpathian line

MILES
0   5   10

KM
10   0   10

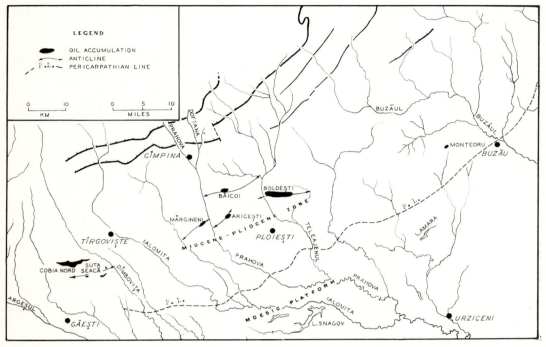

FIG. 14.—Sketch of Sarmatian oil fields of Mio-Pliocene zone.

the Meotian and Dacian. The unit is exclusively oil-bearing on the Mărgineni, Tintea, Aricești, Boldești, and Monteoru structures (Fig. 14). At Boldești, the Sarmatian is sandstone with well-rounded grains overlying a basal pelitic unit. Meotian and Sarmatian producing sandstone beds are separated by a unit of marl and gritty marl. In the Sarmatian, within the sandy unit, there is a slight unconformity which is more conspicuous in the western part of the southern flank. Sarmatian producing zones have been grouped into two complexes, Sa-I and Sa-III. In contrast with the Meotian, which is productive on the entire structure, the Sarmatian produces only on the southern flank and in a very small area just north of the axial fault (Fig. 10).

The area of distribution of Sarmatian accumulations is limited because of the absence of reservoirs or discontinuity of beds in the north-

ern, western, and eastern sectors of the region.

Helvetian oil accumulations, as well as those of the Oligocene, are localized on the northern border of the Mio-Pliocene zone and correspond generally to the alignment of the faulted structures (Figs. 15, 16). In those zones, both units contain good reservoirs and their depth is comparatively shallow. Farther south the depth may exceed 18,000 ft. On the basis of regional data it appears that southward the Helvetian and the Oligocene facies are more pelitic.

On almost all of the structures which have a northern alignment, the Helvetian and Oligocene are in direct contact with the productive Meotian (Fig. 10) and form stratigraphic traps. Where those units are in contact, part of the Helvetian and Oligocene accumulations can be attributed to the migration of Meotian oil. Examples are the Oligocene at Copăceni-Gura Vitioarei, the Oligocene and Helvetian

FIG. 13.—Sketch of Meotian oil and gas fields of Mio-Pliocene zone.

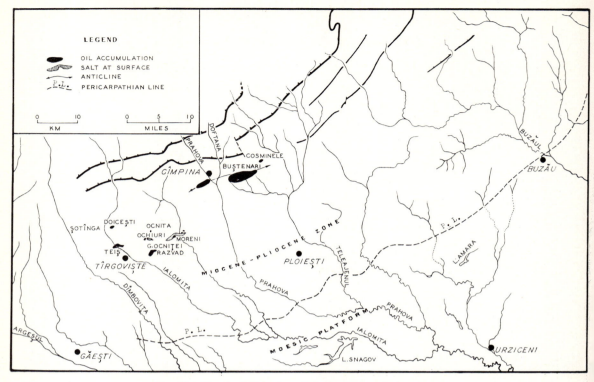

Fig. 15.—Sketch of Helvetian oil fields of Mio-Pliocene zone.

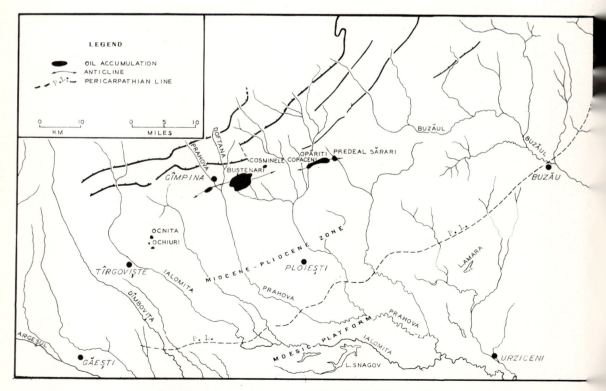

Fig. 16.—Sketch of Oligocene oil fields of Mio-Pliocene zone.

at Ocniţa and Ochiuri, and part of the Helvetian at Moreni and Teiş.

On the Cîmpina, Moreni, and Opăriţi structures productive Helvetian and Oligocene reservoirs are separated from the oil-bearing Meotian strata by marl or flushed reservoir rocks. Thus the Meotian hydrocarbons came from pre-Meotian strata, despite the fact that the quality of the Helvetian-Oligocene and Meotian oils commonly is similar.

The analysis of the amount of production from each structure is instructive. Of the total amount of oil produced in the region, the Moreni-Gura Ocniţei structure yielded about 42 percent, followed in order by the structures of Boldeşti (13.7 percent), Băicoi-Tintea (11.2 percent), and Runcu-Buştenari (7.2 percent). The combined production of the Moreni-Gura Ocniţei and Băicoi structures (both are on the same major diapiric alignment) is more than 53 percent of the oil produced in the region. If the contribution of the Boldeşti field is added, about 67 percent of the Mio-Pliocene zone oil is from three structures, the remaining 33 percent being divided among the other 32 anticlines. The Mio-Pliocene zone in the district of Ploieşti is noteworthy because of the presence of the three big structures—particularly the Moreni-Gura Ocniţei giant (Fig. 17). The other oil fields, except the Cîmpina-Buştenari and Ceptura-Urlaţi fields, are smaller and have yielded cumulative outputs ranging from 67,200,000 to 73,000 bbl. According to official statistical data, the Moreni-Gura Ocniţei giant had produced more than 724 million bbl of oil as of January 1, 1968. The output of the field reached a maximum annual level of about 37 million bbl in 1937. Oil accumulations are localized in the Dacian and the Meotian and only to a very small extent (3 percent) in the Helvetian. The factors which favored the genesis of such a giant follow.

1. Uninterrupted sedimentation during the entire Pliocene resulted in the formation of alternating pelite and psammite that created ideal conditions for the genesis, accumulation, and preservation of hydrocarbons.

2. Variation in facies of the Meotian, which becomes increasingly pelitic southward, corresponds to greater depression of the depositional basin. The southern sector in the western half of the Mio-Pliocene zone thus constituted a source of hydrocarbons that migrated toward the northern, more uplifted sector (Moreni) which has more reservoir rocks.

3. More uniform Dacian sedimentation with thicker pelitic units contributed not only to the generation of hydrocarbons but also to their preservation. East and west of the Moreni anticline the Dacian deposits are more lenticular.

4. The presence of an anticline of appreciable size (about 16 mi long) and great amplitude also was important. The structure is more uplifted and has greater amplitude than the Mărgineni, Ariceşti, and Măneşti structures on the south. The other Mio-Pliocene zone structures are either smaller or tectonically more intricate (north of the Moreni-Băicoi alignment), or despite their large size have been affected by erosion to the Miocene level (e.g., Istriţa-Monteoru, Lapoş, etc.).

5. Very well-developed adjacent synclines were present to feed both flanks of the Moreni anticline. That the southern flank was fed by a zone more extensive than the adjacent syncline is proved by the fact that about 90 percent of the Moreni oil was accumulated there, versus only 10 percent in the northern flank.

6. Diapirism played a double role. First, because salt migrated and pierced the overburden continuously, the diapiric anticlines maintained a more uplifted position during the period of sedimentation (proved by the lesser thickness of strata over the anticlines) and drew to them hydrocarbons from adjacent zones. Second, hydrocarbon accumulations were trapped against impermeable salt, as shown by the concentration of the most productive reservoirs (particularly those of the Dacian) around the diapir plug.

7. Favorable geomorphologic evolution took place during post-Levantine time; erosion did not affect the main Dacian productive and cover strata.

Other noteworthy conclusions can be derived from data on the hydrocarbon pools in the district of Ploieşti. The reservoir beds consist almost wholly of sandstone and sand. Their physical properties change laterally, partly in relation to reservoir depth and age. Porosity is 25–35 percent in the Dacian and 20–25 percent in the Meotian. Miocene and Oligocene strata indicate lower porosity values. Permeability is likewise variable, from 10 to 3,000 md, with higher values in the Pliocene. Oil accumulations are dominant over gas.

Analysis of the completely investigated zones (Dacian and Meotian) shows that in the northern sector, where the facies is coarser, the tectonic effects more intricate, and preservation conditions precarious, oil accumulations contain dissolved gas. In the middle part (zone of

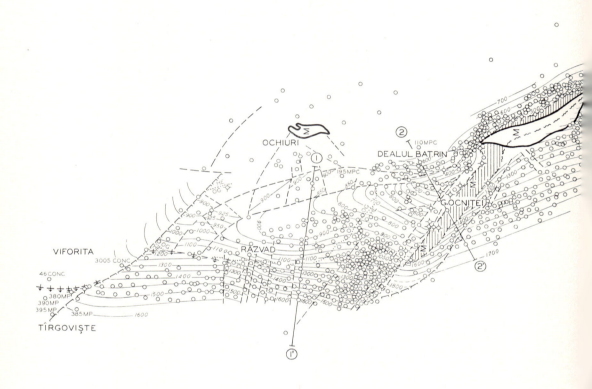

FIG. 17.—Structure map of Moreni-Gura Ocniţei Meot

main diapir folds), oil accumulations occur with dissolved gas and with gas caps. Farther south, where the strata sank deeper and the structures are of smaller amplitude, gas accumulations are predominant; thus the principles of Gussow (1954) are confirmed. As marl becomes increasingly predominant in the composition of prospective section, the amount of gas increases.

Oil quality can differ from one formation to another, from one structure to another, and even within one productive zone in a structure.

Notwithstanding the variation, Oligocene, Miocene, and Meotian oils are generally paraffinic, whereas the Dacian oil is predominantly naphthenic. In connection with this distribution, Hlauschek (1950) found that no oil intermediary between those two types is present, a fact which led him to favor an autochthonous character for the Dacian oil. More recently, Creangă et al. (1962) confirmed the results obtained by Hlauschek that naphtheno-aromatic oils are present only in the Dacian.

Dacian oils in several sectors also contain

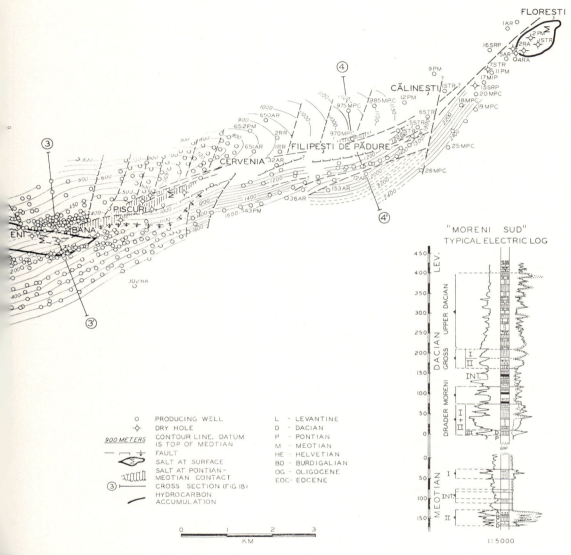

Id. Datum is top of Meotian. CI = 50 m. 1-1', 2-2', *etc.* are lines of sections in Figure 18.

some elements that are specific to the Meotian, such as in the most uplifted zone in Băicoi, near the salt fault; the Gura Ocniței zone, along a SW-NE-trending fault where the Dacian comes into close contact with the Meotian; and in the Bana sector, where it is near a salt plug. Those examples prove that oil migration from the Meotian into the Dacian occurred only in a few very limited areas. For example, the Dacian at Gura Ocniței, situated along the fault, contains only paraffin-base oil (originating in the Meotian) near the faults;

farther south (at a greater distance from the faults) oil from the same strata is of mixed character; still farther away, the oil is asphaltic, as is characteristic of Dacian oil.

In the above-mentioned localities, the specific gravity of the oil differs appreciably, ranging from 0.805 to 0.875 in the Meotian and from 0.850 to 0.926 in the Dacian. Oil of the lowest density, 0.735, has been found in well No. 21 R. A. Bucșani, which was productive in the M-I (Meotian) complex. Meotian oils have a paraffin content of 3–10 percent, whereas the

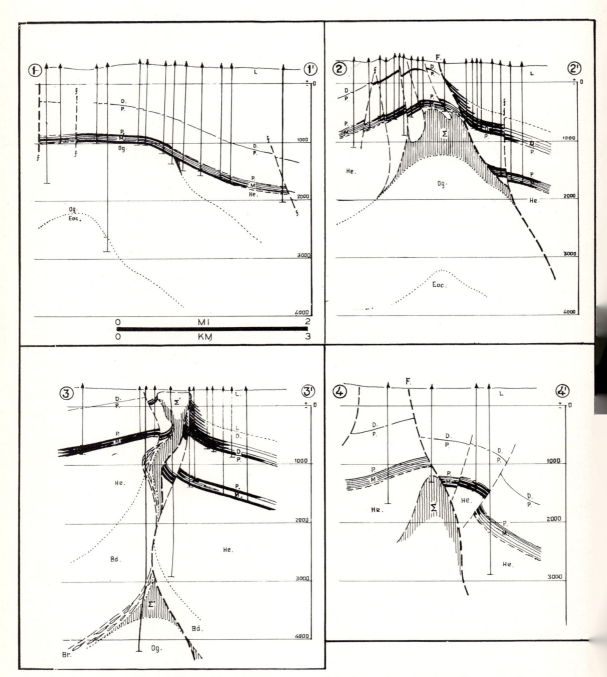

Fig. 18.—Cross sections of Moreni-Gura Ocniței Meotian oil field. Locations shown on Figure 17. Abbreviations same as for Figures 10 and 17. Vertical scale in meters.

Dacian oils contain about 1 percent (E. Cazimir, 1934). The sulfur content of the Mio-Pliocene oils does not exceed 0.3 percent. The characteristic features of Miocene and Oligocene oils are generally similar to those of the Meotian oils. With regard to dissolved gas, it was found that the Dacian gas has a CO_2 content of 10–30 percent, whereas in the Meotian CO_2 generally does not exceed 1 percent.

Most beds are stratiform. The Helvetian at Teiş and the Oligocene at Runcu-Buştenari are examples of massive deposits.

Production in most pools is associated with dissolved-gas drive or a mixed system. In some sectors at Bucşani, Moreni-Gura Ocniţei, Tintea Nord, and Boldeşti, the expanding gas cap provides the mechanism, whereas water drive is obvious at Bucşani.

Most oil pools in the Mio-Pliocene zone are in an advanced stage of exploitation. To increase the final recovery factor, either secondary recovery or pressure maintenance processes have been initiated in some fields, with water or gas being used as flooding agents.

CONCLUSIONS

The Mio-Pliocene zone in the district of Ploieşti belongs to the most external unit of the Eastern Carpathians, and is on their southwestern end, where the range sank and was loaded with a thick sequence of Neogene deposits.

The Paleogene-Neogene strata of the Mio-Pliocene zone are composed of a repeated alternation of clay, marl, shale, sand, sandstone, conglomerate, salt, and some limestone. Slight breaks of sedimentation occurred at the end of the Aquitanian, in the Tortonian, and in the Sarmatian. During the Pliocene, sedimentation was continuous, and the strata transgressively overlie increasingly older beds northward and westward.

Mio-Pliocene strata have been affected repeatedly by tangential movements, and by vertical movements as well, from which folded and faulted structures resulted. Anticlines are on four main alignments parallel with the Carpathian Range. Salt was important in the formation of folds; because of its plastic properties, it pierced the more recent deposits and reached the surface on some anticlines. Folds are more closely spaced and intricate near the flysch zone border.

Favorable conditions for hydrocarbon generation, accumulation, and preservation are offered by nearly all Oligocene, Miocene, and Pliocene units. About 35 accumulations have been found in the Dacian, Pontian, Meotian, Sarmatian, Helvetian, and Oligocene in traps of structural and stratigraphic type. On nearly all structures hydrocarbon pools are overlapping. More than 85 percent of the hydrocarbons are in the Pliocene strata, particularly in the Meotian where, because of its transgressive character and the presence of the overlying Pontian marl, conditions of accumulation and preservation were optimum. The most significant structures for hydrocarbon accumulations are Moreni-Gura Ocniţei, Băicoi-Ţintea, and Boldeşti, which correspond to the median diapir alignment.

The gigantic accumulations in the Moreni-Gura Ocniţei area were favored by (1) conditions of sedimentation—uninterrupted deposition throughout the Pliocene of alternating pelitic and psammitic beds, and presence of a more marly facies in the southern depressed part of the Mio-Pliocene zone which generated a large amount of hydrocarbons, and more uniform deposition during the Dacian; (2) tectonic conditions—the presence of a large anticline surrounded by well-developed synclines which, because of salt diapirism, was being uplifted continuously through the Pliocene, and gathered hydrocarbons from adjacent zones, diapiric salt undeniably providing a seal; and (3) geomorphologic conditions during the Levantine stage when the productive strata were protected from erosion.

SELECTED REFERENCES

Anton, Sever, unpublished works.
———— and V. Carchez, 1967, Recherches sur l'identification des roches mère d'hydrocarbures: Rev. de l'Inst. Français du Petrole, v. 22, no. 12, p. 1818–1828.
Atanasiu, I., 1932, Quelques observations sur le diapirism: Soc. România Geologie Buletinul, v. 1, p. 96–105.
———— 1948, Zăcămintele de ţiţei din România: Revistele technice AGIR, nr. 3, p. 111–125.
———— 1958, Orogénèse et sédimentation dans les Carpates Orientales: România Com. Geol. Anuarul, t. 24–25, p. 13–27.
Băncilă, I., 1958, Geologia Carpaţilor Orientali: Bucureşti, Editura Stiinţifică, 367 p.
Barton, D. C., 1934, Natural history of the Gulf Coast crude oil, in Problems of petroleum geology: Am. Assoc. Petroleum Geologists, p. 109–155.
Cantuniari, Şt., 1941, Sur le sable blanc oligocène de Vălenii de Munte (dept. de Prahova) (abs.), Rumania Inst. Géol. Comptes Rendus, t. 26 (1937–1938), p. 86–88.
Casimir, E., 1934, Composition au point de vue industriel et propriétés générales des pétroles bruts

de Roumanie: Rumania Inst. Géol. Anuarul, v. 16, p. 879–933.

Ciocirdel, R., 1952, Geologia regiunii Gornetul Cuib-Matița-Apostolache și considerațiuni generale asupra rocelor mame ale petrolului din Subcarpați: Dări de seamă ale ședințelor, Rumania, Inst. Géol., v. 36, p. 87–146.

——— and C. Stoica, 1957, Geologia zăcămintelor de țiței și gaze: Manualul Inginerului Petrolist, v. 41, cap. 3, p. 33–80.

Creangă, I., F. Dumitrescu, V. Negrescu, V. Caraiani, P. Neacșu, and S. Rădulescu, 1962, Les pétroles bruts Roumains dans la classification "Carpatica": Rev. Chimie, t. 7, no. 1, p. 111–125.

Dragoș, V., 1964, Tectonica regiunii dintre Valea Teleajenu'ui și Valea Lopatnei: Dări de seamă a'e ședințelor, Rumania Inst. Géol., v. 50, no. 1 (1962–1963), p. 349–373.

Erdman, G. J., 1965, Petroleum—its origin in the earth, in Fluids in subsurface environments: Am. Assoc. Petroleum Geologists Mem. 4, p. 20–52.

Filipescu, D., 1925, Contributions à l'étude des gisements de pétrole de Roumanie—region de Buștenari: Annales des Mines de Roumanie, no. 14, p. 531–584.

Filipescu, M., 1955, Vederi noi asupra tectonicii Flișului Carpaților Orientali: Revista Univ. C. I. Parhon și a Politechnicii București (Seria St. Naturii), nos. 6–7, p. 241–261.

Gavăt, I., 1964, Geologia petrolului și a gazelor naturale: București, Editura Didactică și Pedagogică. 303 p.

Grigoraș, N., 1956, Roci bituminoase în formatiunile geologice din R.P.R.: Analele Univ. C. I. Parhon, no. 9, p. 195–203.

——— 1957a, Géologie des gisements de pétrole et de gaz en Roumanie, in Géologie de Pétrole: 20th Internat. Geol. Cong., Mexico, sec. 3, p. 167–204.

——— 1957b, Rolul depresiunilor premontane în formarea rocelor bituminoase din R.P.R.: Analele Univ. C. I. Parhon, no. 13, p. 181–188.

——— 1961, Geologia zăcămintelor de petrol și gaze din R.P.R.: București, Editura Tehnică, 233 p.

——— and I. Petrișor, 1963, Considerațiuni privind legile de răspîndire ale zăcămintelor de petrol și gaze din R.P.R.: Comunicări științifice Secția 4, Geollogie economică, v. 5, p. 124–138.

——— E. Hristescu, and U. Suluțiu, 1963, Contributions to the knowledge about distribution laws regarding oil and gas fields in the Precarpathian Depression . . .: Oil, gas and petrochemistry in the Romanian People's Republic, Bucharest, p. 53–61.

Gussow, Wm. C., 1954, Differential entrapment of oil and gas: a fundamental principle: Am. Assoc. Petroleum Geologists Bull., v. 38, no. 5, p. 816–853.

Hlauschek, H., 1950, Romanian crude oil: Am. Assoc. Petroleum Geologists Bull., v. 34, no. 4, p. 755–781.

Krejci-Graft, K., 1929, Die rumänische Erdöllagerstätten, Schriften aus dem Gebiet der Brennstoff-Geologie, Heft, 1: Stuttgart, Verlag V. Ferd. Enke, 140 p.

Macovei, G., 1932, L'origine du pétrole des Carpates Orientales et ses roches-mères: Soc. Rom. de Geologie Buletinul, v. 1, p. 270–285.

——— 1938, Les gisements de pétrole: Paris, Masson et Cie. Editeurs, 502 p.

——— and D. Ștefănescu, 1935, Les gisements de pétrole de Roumanie, in Les Karpates et l'avant pays, v. 3: Karpacki Institut Geologiczko Naftowy, p. 31–90.

Motaș, I. C., 1952, Asupra stratigrafiei Mio-Pliocenului dintre Valea Ialomiței și Valea Dîmboviței, la nord de Tîrgoviște: Dări de seamă ale ședințelor, Rumania Inst. Géol. v. 36, p. 140–145.

Mrazec, L., 1907, Despre cute cu sîmbure de străpungere: Soc. Stiinte Bull., v. 16, p. 6–8.

——— 1926, Les plis diapirs et le diapirisme en général: Rumania Inst. Géol. Comptes Rendus t. 6 (1914–1915), p. 215–255.

——— 1931, Aperçu sur le caractère des gisements de pétrole de la Roumanie: Prague, Publications de la Faculté des Sciences de l'Université Charles, 118 p. (Résumé en Ref. Bul. Lab. Min. gen. Univ. București, v. 1, p. 136–139).

——— and I. Atanasiu, 1927, L'anticlinal diapir Moreni-Gura Ocniței: Guide des excursions, p. 171–193.

——— and W. Teisseyre, 1907, Esquisse tectonique de la Roumanie: Congrès International du Pétrole, p. 32–59.

Murgeanu, G., and O. Protescu, 1927, Géologie de la Vallée de la Prahova entre Cîmpina et Comarnic: Guide des excursions, p. 195–238.

Olteanu, Fl., 1951, Observații asupra "Brecciei sării" cu masive de sare din regiunea mio-pliocenă dintre R. Teleajen și P. Bălăneasa (cu privire specială pentru regiunea Pietraru-Buzău): Dări de seamă ale sedințelor, Rumania Inst. Géol. v. 32, p. 12–18.

——— 1952, Structura geologică a regiunii Ursei-Cimpina: Dăru de seamă ale ședințelor, Rumania Inst. Géol., v. 36, p. 125–138.

——— M. Popescu, and T. Iorgulescu, 1957, Contribuții la cunoașterea stratigrafiei Neogenului din Oltenia și din Muntenia; in Studii și ceretări de explorare, exploatare și prelucrarea țițeiului: București, Editura Tehnica, p. 11–49.

Olteanu, Gh., 1965, Salt rising mechanism in the Pre-Carpathian area of the Ploiești region: Carpatho-Balkan Geol. Assoc., 7th Cong. Repts., pt. 1, Section of Geotectonics, p. 157–163.

Oncescu, N., 1947, Structura geologică a regiunii dintre Mizil și Tîrgoviște, cu privire specială asupra cărbunilor din Dacian: Studii tehnice și economice, seria A, no. 3, p. 62.

——— 1965, Geologia României: București Editura Tehnica, 534 p.

Pătruț, I., 1959, Géologie et tectonique de la région Văleni de Munte-Cosminele-Buștenari: România Com. Geol. Anuarul, t. 26–28, p. 203–228.

Paucă, M., 1952, Le miocène presarmatien de la région de courbure des Carpates Orientales: România Com. Geol. Anuarul, t. 24–25, p. 163–176.

Popescu, Gr., 1951, Observațiuni asupra "brecciei sării" și a unor masive de sare din zona paleogenă-miocenă a jud. Prahova: Dări de seamă ale ședințelor, Rumania Inst. Géol. v. 32, p. 3–13.

——— 1952, Zona flișului paleogen în Valea Buzăului și Valea Vărbilăului: Dările de seamă ale ședințelor, Rumania Inst. Géol., v. 36, p. 145–161.

Popescu-Voltești, I., 1935a, Evoluția geologico-paleogeografică a pămîntului românesc: Univ. Cluj. Muz. Geol. Min. Rev., v. 5, no. 2, p. 211.

——— 1935b, L'état actuel des connaissances géologiques sur le problème de la génèse du pétrole des régions karpatiques roumaines, in Les Carpathes et l'avant pays, v. 3: Karpacki Institut Geologiczko Naftowy, p. 91–116.

——— 1943, Sarea regiunilor carpatice românești: București, Fund. Reg. Lit. Arta, p. 74.

Preda, D. M., 1957, Dezvoltarea geologiei petrolului in România: Petrol şi gaze, no. 9–10, p. 435–447.

Raaf, J. F. M. de, 1953, The world oilfields—the Eastern Hemisphere, *in* The science of petroleum: London, Oxford Univ. Press, v. 6, pt. 1, p. 9–17.

Saulea, E., 1956, Contribuţiuni la stratigrafia Miocenului din Subcarpaţii Munteniei: Romania Inst. Géol. Anuarul, v. 29, p. 69–83.

Silverman, S. R., 1965, Migration and segregation of oil and gas, *in* Fluids in subsurface environments: Am. Assoc. Petroleum Geologists Mem. 4, p. 53–65.

Small, W. M., 1959, Thrust faults and ruptured folds in Romanian oil fields: Am. Assoc. Petroleum Geologists Bull., v. 43, p. 455–471.

Ştefănescu, D., 1938, Le gisement pétrolifère "Bucşani": Moniteur du Pétrole Roumain, no. 7, p. 517–522.

Trask, P. D. and H. W. Patnode, 1942, Source Beds of Petroleum: Am. Assoc. Petroleum Geologists, p. 41–45.

Visoţki, I. V., 1959, Istoria dezvoltării părţii sudice a regiunii cutate a Carpaţilor Orientali: Analele Univ. C. I. Parhon Bucureşti, Seria Stiinţelor Naturii, no. 21, p. 125–131.

Walters, R. P., 1940, La présence du pétrole dans le Dacien de Gura Ocniţei-Est et la question de l'origine du pétrole en Roumanie: Rumania Inst. Géol. Comptes Rendus, t. 24 (1935–1936), p. 10–17.

——— 1946, Oil fields of Carpathian region: Am. Assoc. Petroleum Geologists Bull., v. 30, p. 319–336.

——— 1948, New approach in study of origin of oil: Am. Assoc. Petroleum Geologists Bull., v. 32, p. 1821–1822.

——— 1960a, Surani, Rumania, anticline with two erosion depleted noncontemporaneous oil reservoirs: Am. Assoc. Petroleum Geologists Bull., v. 44, p. 1638–1650.

——— 1960b, Relation of oil occurrence at Surani, Romania, to origin and migration of oil: Am. Assoc. Petroleum Geologists Bull., v. 44, p. 1704–1705.

Weeks, L. G., 1952, Factors of sedimentary basin development that control oil occurrence: Am. Assoc. Petroleum Geologists Bull., v. 36, p. 2071–2124.

Asmari Oil Fields of Iran[1]

CEDRIC E. HULL[2] and HARRY R. WARMAN[2]

London, England

Abstract The Oligo-Miocene Asmari oil fields of Iran are truly giants, most of them having recoverable reserves greater than 1 billion bbl each, and many having much more. The fields are close together in a region of relatively constant stratigraphy and structure, and have a common genetic history. The individual accumulations occupy very large rock volumes in large-amplitude folds and, although the Asmari reservoir is poor in porosity and matrix permeability, very high production rates are possible because of extensive reservoir fracturing. These rates can be maintained for very long periods because of the great vertical extent of the oil columns. The Asmari fields are prime examples of anticlinal traps and of the effect of fracturing on reservoir performance.

The Asmari reservoir is limestone of Oligo-Miocene age and consists mostly of shallow-water but nonreefal carbonate rocks, with a significant sandstone member in the northwest part of the area. The Asmari is the uppermost wholly marine unit in a shelf-carbonate sequence interspersed with shale which was deposited, with only minor interruptions, from Carboniferous through Oligo-Miocene time. At the end of the time of Asmari deposition, increasing tectonic instability caused more varied sedimentation; this phase of instability terminated in a strong orogeny (Zagros orogeny) which formed the enormous anticlinal traps in this thick sedimentary sequence.

Introduction

The Oligo-Miocene Asmari oil fields of Iran are within the sedimentary basin which is between the Arabian shield and the Central Iranian plateau and which contains a significant proportion of the world's oil reserves. The general distribution pattern of fields in this area is varied (Fig. 1). Production is from reservoirs ranging in age from Middle Jurassic to Miocene, both carbonate rocks and sandstone. All of the fields discovered are primarily structurally controlled, but the structural style changes from the Arabian side of the Persian Gulf, where gentle folds have been growing since Mesozoic time, to Iran, where the folding of the Asmari fields is very pronounced. The Zagros orogeny which formed the Asmari structures took place during very late Tertiary time.

The size and prolific production of the Asmari fields are the results of a favorable combination of stratigraphic and structural condi-

[1] Read before the 53d Annual Meeting of the Association, Oklahoma City, Oklahoma, April 25, 1968. Manuscript received, August 26, 1968; accepted, November 1, 1968.

[2] British Petroleum Co. Ltd.

tions. Within the basin very large quantities of oil have been generated in sediments of different ages at least from the Middle Jurassic to the Miocene. The age of the Asmari oil is not firmly established. There is good reason to believe that some of it originated in the uppermost Cretaceous and Eocene shaly beds underlying the Asmari. However, some of it may be older.

The scale of oil accumulation in the Asmari can be gauged by the concentration within an area less than 200 mi long and 60 mi wide of 17 fields, a quarter of which have more than 5 billion bbl of recoverable reserves each. Many of the rest have recoverable reserves in excess of 1 billion bbl.

Stratigraphy

A generalized depiction of the total sedimentary fill of the basin is shown on the right side of Figure 2.

The sedimentary section of principal interest to the petroleum industry begins with transgressive sandstone of late Carboniferous age. This unit spread over a basement complex about which little is known in Iran, except that it includes lower Paleozoic sedimentary rocks. This sandstone grades upward into Permian shallow-water limestone up to 3,000 ft thick. After the Permian, a succession of depositional cycles was laid down. Among them are some regional unconformities and regressive phases, marked by minor angular unconformities and by the deposition of terrigenous clastics and evaporites. Folding movements were small, however, and relatively quiet deposition of mainly marine sediments continued until early Miocene time. Carbonate deposition predominated in much of the area throughout this time.

The part of the sedimentary section most relevant to consideration of the Asmari oil fields is from the base of the Bangestan Group up to the Gachsaran Formation (Fig. 2, left).

For the purpose of this paper, the Bangestan is considered to be one carbonate unit 2,500–3,000 ft thick. It is developed in two interleaved facies, one neritic detrital packstone to wackestone and the other dense, fine-grained,

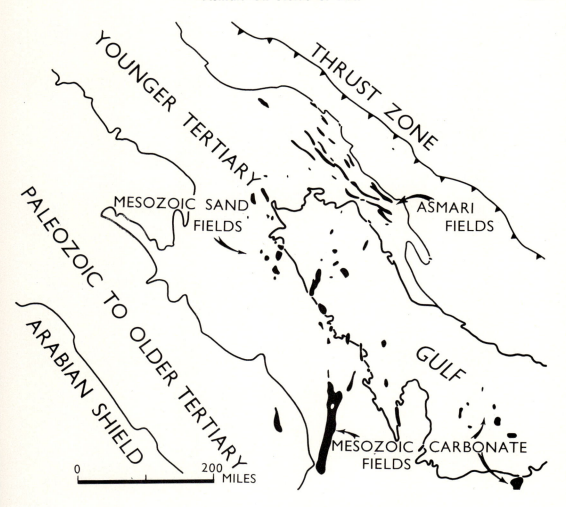

Fig. 1.—General distribution pattern of oil fields in Persian Gulf area.

calcite mudstone containing abundant *Oligoste-gina* (indicating a Cretaceous age).

Above the Bangestan is an argillaceous sequence about 2,500–3,000 ft thick, ranging in age from Santonian (Late Cretaceous) to Oligocene. The detailed stratigraphy of this sequence is complex, but the dominant rock type is dark-colored variably calcareous marine shale which has a fairly high organic content. The upper part of this unit grades into limestone, particularly in the southwestern part of the area of the Asmari oil fields.

The Asmari Limestone, which unconformably overlies shaly beds and their carbonate equivalents, of varied ages, is a major independent depositional cycle. Considering the magnitude of later tectonic events, there is strikingly little angularity between the Asmari and underlying beds. In places, both at outcrop and in wells, the basal contact is not easy to identify.

The Asmari Limestone ranges in age from Oligocene (with a rich microfauna including *Nummulites intermedius* d'Archiac, and *Nummulites vascus* Joly and Leymerie) to early Miocene. The upper Asmari has a rich microfauna including *Miogypsina* at the base and *Neoalveolina melo curdica* at the top. Contrary to some earlier reports, the Asmari shows no features related to reef development and any terms relating it to reef deposition are misleading. It consists of varied packstone and wackestone. The ratio of micrite to grain develop-

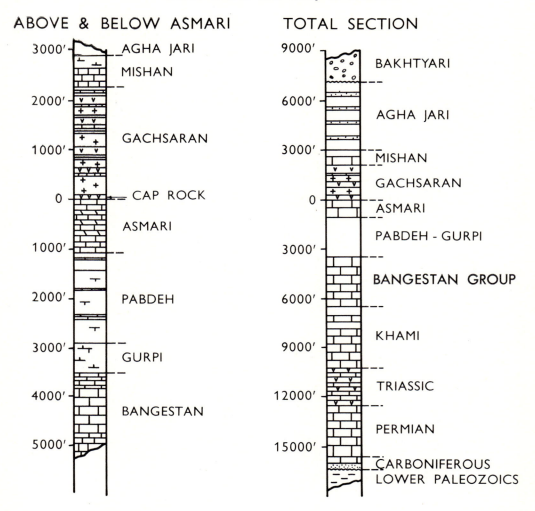

FIG. 2.—Stratigraphic section for southwest Iran. Column on right is a general depiction of total sedimentary fill of the basin. Left column shows part of section most relevant to Asmari oil fields.

ment differs in broad belts related to the overall basin shape rather than to reef developments. Diagenesis similarly affected broad belts, producing dolomitization and anhydritization. The formation of dolomite is particularly important with regard to the reservoir characteristics and in some of the fields provides the only significant porosity.

In the more northwesterly fields, such as Ahwaz, sandstone forms an appreciable part of the Asmari reservoir. The sandstone, like some Mesozoic sandstone on the west side of the Persian Gulf, is an extensive sheet that spread out from the Arabian shield. At Ahwaz the sandstone forms 50 percent of the total thick-

ness of the Asmari but it thins markedly northeast and southeast. In most of the known Asmari fields, sandstone does not constitute a significant part of the reservoir.

In the study area the Asmari ranges in thickness from 1,000 to 1,500 ft.

The usual convention in Iran is to divide the carbonate rocks of the Asmari into net pay and nonproductive rock. The division is made somewhat arbitrarily, depending on porosity, permeability, and water saturation.

The rock classified as nonproductive forms a high percentage of the total Asmari thickness, ranging throughout the study area from 25 to 75 percent. Its porosity is less than 5 percent and permeability everywhere is less than 1 md, generally much less.

The net pay carbonate rocks have average porosity normally ranging in any one field from 9 to 14 percent and permeability averaging 10 md, but in a few places exceeding 20 md.

Where significant quantities of sandstone are present, the reservoir properties are much better; porosity ranges from 20 to 30 percent and permeability values commonly are large.

An interesting feature is the apparent absence of fracturing in the sandstone, and the reservoirs with mixed sandstone and carbonate rocks have different production-behavior patterns. Some of the sandstone is uncemented and the high production rates obtained by the wells producing wholly from carbonate rocks are not possible because of the entry of sand into the borehole, and associated sand-cutting of well fittings.

Above the Asmari is a fairly thick succession composed of four formations. The lowest is the Gachsaran which consists of alternate anhydrite, salt, gray marl, and a few thin limestone beds. This unit shows considerable mobility and its thickness is difficult to estimate, though generally it ranges from 2,000 to 4,000 ft. The most important member of the Gachsaran is the "Cap Rock," a unit of constant overall character in all the Asmari fields. It consists of very compact, nodular anhydrite interbedded with thin shale beds which in one or two places are bituminous. There are also thin limestone beds with a marine microfauna of few species but abundant individuals. The "Cap Rock" forms a permeability and pressure barrier of great importance.

Below the "Cap Rock," 7,000 ft of limestone and shale (of Asmari and pre-Asmari strata) are fractured so intensively that in many places there is complete fluid connection. Above the "Cap Rock," the pressure regime in the Gachsaran, particularly in the limestone beds, has a wide range. Local pressure anomalies presumably are due largely to compaction and squeezing of the plastic mobile evaporites. These pressures are in many places nearly equal to the gross rock overburden pressure, and the mud problems associated with this part of the section are one of the main concerns in drilling in Iran.

Above the Gachsaran are marine shale and limestone of the Mishan Formation of middle Miocene age. The Mishan is well developed southeast of the Asmari oil-field belt where it is more than 2,000 ft thick, but wedges out near the northeast margin of the belt. The marine Mishan Formation passes gradationally upward into a thick redbed sequence named the Agha Jari Formation, which consists of 5,000–10,000 ft of interbedded red shale and sandstone devoid of marine fauna. The Agha Jari is overlain with marked unconformity by the Bakhtyari Formation, which consists of sandstone and well-cemented, coarse conglomerate, the pebbles and boulders of which consist mainly of Asmari and older limestone derived from the rising fold mountains of the Zagros.

STRUCTURE

The overall structure of the basin is shown in Figure 3, which is a very diagrammatic cross section from the Arabian shield to the zone of major thrusting along the northeast margin of the Zagros Mountains. The tectonic details of the whole Zagros system are very complicated. In general, the fold belt can be considered to consist of two main elements. The more northeasterly element, which forms the main chains of the Zagros Mountains, consists of relatively simple, large-amplitude folds exposing the main reservoir groups of the Asmari, Bangestan, and Khami. Southeast of this group of folds is a foothill zone in which folds of similar type and amplitude underlie a cover of younger Tertiary strata. The Asmari oil fields are in this foothill zone.

The amplitude of the folding is varied, but in several places the elevation difference between the crest of a fold and the bottom of the adjacent syncline only 5-10 mi away is more than 20,000 ft. It can be seen in Figure 4 that on Kuh-e-Khami the extrapolated elevation of the Asmari top is nearly 20,000 ft above sea level, whereas in the syncline on the southwest —only 12 mi away—it is 10,000 ft below sea level.

Critical closures on the anticlines are on the plunging ends of the structures. Because of a northwest regional tilt in the area of the main producing fields, the critical closure is on the southeast ends of the structures. The vertical difference between this spill point and the crest maximum, on most of the larger structures, is between 4,000 and 7,000 ft.

A concept of the scale of folding can be seen on the generalized cross section through the Gachsaran, Garangan, and Bibi Hakimeh structures (Fig. 4).

Two interpretations of the structural data are shown. Both seismic and drilling control

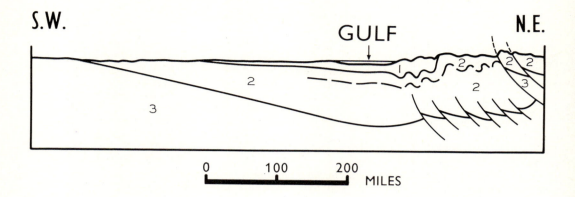

S.W.   GULF   N.E.

0   100   200
                              MILES

☐① YOUNGER TERTIARY

☐② MESOZOIC AND OLDER TERTIARY
      (INCL. SOME PALEOZOICS)

☐③ BASEMENT

FIG. 3.—Diagrammatic section from Arabian shield to zone of major thrusting along southwest margin of Zargros Mountains.

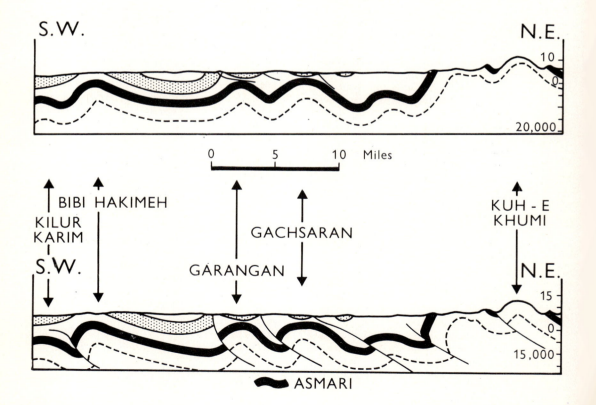

S.W.   N.E.
                                                        10
                                                         0

                                                    20,000

0   5   10   Miles

↑   ↑   ↑   ↑   ↑
BIBI HAKIMEH   KUH - E
KILUR   KHUMI
KARIM   GACHSARAN
↓   ↓   ↓   ↓   ↓
S.W.   GARANGAN   N.E.
                                                        15
                                                         0
                                                    15,000

ASMARI

FIG. 5.—Kabir Kuh, one of the larger typical Zagros folds. Shows 100 mi of its total 130-mi length. Core is formed by limestones of Bangestan Group, overlain by soft Gurpi and Pabdeh Formations. Flanking flatirons are Asmari limestone.

are restricted mainly to the gentler parts (the tops) of the folds. The structure can be drawn either with or without thrusting in the deeper beds, according to one's preference. Thrusting is known to affect the contorted Gachsaran Formation and also the more competent overlying beds. Because of the deformation of the Gachsaran, it is not possible to draw simple sections showing the trace of faults affecting both the upper and lower sets of competent formations.

Gachsaran field has a single 6,000-ft vertical column of hydrocarbons. Bibi Hakimeh is another major field, with a 3,000-ft column of hydrocarbons.

Figure 4 also shows the disharmonic relation between the simpler folds of the Asmari and older beds and the surface structures. The two are separated by the mobile evaporite complex of the Gachsaran Formation.

A good example of the problems of defining the details of the Asmari fields is the relation of two of the largest fields—Marun and Agha Jari. These fields overlap *en échelon* and, in the region of the tight syncline between their plunging ends, the two oil accumulations are horizontally 3,000 ft apart but the original oil-water contacts differ vertically by about 5,000 ft.

Figures 5–7 are oblique aerial photographs

FIG. 4.—Sections through Asmari oil fields. **Upper** is drawn without thrust faults; **lower** is drawn with thrust faults.

FIG. 6.—Kuh-e-Kialan, an Asmari fold adjacent to Kabir Kuh. Fold plunges away for 16 mi to pass beneath the Saidmarreh landslide. Scale is shown by 1,000-ft thickness of Asmari and Eocene limestone beds.

that illustrate typical Zagros folds. The first (Fig. 5) is of Kabir Kuh, which is one of the larger folds; of its total length of 130 mi, more than 100 mi can be seen on the photograph. The core of the fold, formed by limestones of the Bangestan Group, is overlain by the soft, easily eroded Gurpi and Pabdeh Formations. The Asmari forms the prominent flanking flat-irons.

Figure 6 shows an adjacent Asmari fold called Kuh-e-Kialan. The scale is shown by the 1,000-ft thickness of the Asmari and Eocene limestone beds. The Asmari fold can be seen to plunge away for 16 mi to the place where it passes beneath the Saidmarreh landslide (Harrison and Falcon, 1937). This remarkable landslide consists of approximately 65 mi³ of the Asmari which once flanked Kabir Kuh, and which slid off and travelled 15 mi across

country as one mass of ground and jumbled blocks of limestone.

Figure 7 shows folds somewhat farther back in the fold-belt system, where the folds are crowded together. The syncline is isoclinal; upsidedown Asmari strata overlie Asmari strata in normal position. The more easily eroded Pabdeh and Gurpi Formations are clearly differentiated and the visible core is limestone of the Bangestan Group.

Figure 8 shows Asmari Mountain, a typical but somewhat smaller Asmari anticline. The carapace of the mountain consists of Asmari Limestone. As can be seen from the hollow areas or holes on the southwest flank, once the more resistant limestone is breached, erosion soon cuts downward into the underlying Pabdeh shale. Surrounding the Asmari Limestone is the moderately eroded outcrop of contorted Gachsaran Formation.

OIL ACCUMULATIONS

The crude oil in typical Asmari fields generally fills the large structures to the spill point. The oil is asphaltic in general type, and gravity values in the belt of main fields range from about 30 to 38° API. The more northerly fields have the higher gravity values. On the southwest the smaller accumulations contain heavier oils, with gravity values as low as 20°; in Kuh-e-Mund, at the southeast extremity of the known Asmari fields, there is a large accumulation of 11° API oil. Sulfur content ranges from slightly less than 1 to about 3.5 percent.

Associated gas is rather unsystematically varied in composition. Some of the fields have original free gas caps, ranging from very large to small; others nearby have undersaturated oil. Some crudes are undersaturated originally by as much as 4,000 psi.

Generally the oil in any particular accumulation is of very constant composition. This homogeneity is interpreted to be a result of the extremely good fracture connection throughout the reservoir that allows mixing by convection currents which prevent gravity segregation. there is no general evidence in the Asmari of the formation of tar mats or similar heavy basal segregations.

Pressure connection *via* the fissure systems is good, and on sustained production the movement of the gas-oil and oil-water interfaces is kept relatively constant. Pressure responses to production are essentially immediate across distances of tens of miles.

There is a regional pressure gradient through the Asmari water system, with an excess above hydrostatic pressure near the mountains and approximate hydrostatic equilibrium near the

FIG. 7.—Folds farther back in fold-belt system. Syncline is isoclinal; upsidedown Asmari overlies Asmari in normal position. Softer Padbeh and Gurpi Formations are clearly differentiated, and core is limestone of Bangestan Group.

Fig. 8.—Asmari Mountain. Carapace of mountain consists of Asmari limestone, and contorted outcrops of the Gachsaran Formation surround the Asmari.

Persian Gulf. This gradient causes differences in the levels of the oil-water interfaces on the two flanks of oil fields—as much as 500 ft in some of the structures.

An interesting feature of some of the oil accumulations is the good connection through fractures across the 2,000-3,000-ft interval of shale and marl separating the Asmari from the Bangestan. Where there is such connection the oils are essentially identical in composition and type, and fluid interfaces in the two separate formations appear to move in unison as a result of production from the Asmari. In places the Asmari and Bangestan have separate oil accumulations which differ in gravity values and other properties. In such places, the two units act as separate reservoirs, and there is no pressure response or fluid exchange as a result of production from either.

Despite the generally poor porosity and matrix permeability of the limestone, production rates from the Asmari are high because of the

fractures. The Agha Jari field consistently produces approximately 1 million bbl/day from about 40 producing wells. This average production rate is common in many fields and exceptional wells are capable of producing more than 80,000 bbl/day, each. Because of the great vertical extent of the oil columns it is possible for wells to sustain these rates for many years. For example, more than 15 Asmari wells have each produced more than 100 million bbl.

Productivity indices (PI) differ in all fields, from a minimum of 2 bbl/day/psi at Lali to a maximum of 7,800 bbl/day/psi at Gachsaran. Many wells in different fields have a PI in excess of 1,000 bbl/day/psi.

Cumulative production to date (April 1968) was slightly more than 8.25 billion bbl.

Conclusion

Much has become known from the many

years of work on the detailed stratigraphy and structure of the region, but there is much to be learned, particularly about the origin, migration, and accumulation of the oil. The writers believe that, eventually, this prolific oil-producing region will provide a valuable contribution to the general understanding of oil-field formation.

In conclusion the writers wish to pay tribute to the foresight and gambler's instinct that led D'Arcy and the other early industry pioneers to prospect for oil in Iran, and to persist despite years of setbacks. The writers also pay homage to the early workers in the field who endured, a long way from home, hardships from a terrain and summer climate that have few equals for severity.

SELECTED REFERENCES

British Petroleum Co. Ltd., 1956, Geological maps and sections of southwest Persia: 20th Internat. Geol. Cong. Proc., Mexico.

Falcon, N. L., 1958, Position of oilfields of southwest Iran with respect to relevant sedimentary basins, *in* Habitat of oil: Am. Assoc. Petroleum Geologists, p. 1279–1293.

Harrison, J. V., and N. L. Falcon, 1936, Gravity collapse structures and mountain ranges as exemplified in southwestern Iran: Geol. Soc. London Quart. Jour., no. 365, v. 92, pt. 1, p. 91–102.

———— and ———— 1937, The Saidmarreh landslip, southwest Iran: Geog. Jour., v. 89, no. 1, p. 42–47.

James, G. A., and J. G. Wynd, 1965, Stratigraphic nomenclature of Iranian Oil Consortium Agreement Area: Am. Assoc. Petroleum Geologists Bull., v. 49, no. 12, p. 2182–2245.

Kent, P. E., F. G. P. Slinger, and A. N. Thomas, 1951, Stratigraphical exploration surveys in southwest Persia: 3d World Petroleum Cong. Proc., The Hague, sec. 1, p. 141–161.

Lees, G. M., 1938, The geology of the oilfield belt of Iran and Iraq, *in* Science of petroleum, ed. 1: Oxford Univ. Press, v. 1, p. 140–148.

———— 1953, Persia, *in* Science of petroleum: Oxford Univ. Press, v. 6, pt. 1, p. 73–83.

O'Brien, C. A. E., 1948, Tectonic problems of the oilfield belt of southwest Iran: 18th Internat. Geol. Cong., London, pt. 6, p. 45–58.

Slinger, F. C. P., and J. G. Crichton, 1959, The geology and development of the Gachsaran field, southwest Iran: 5th World Petroleum Cong. Proc., New York, sec. 1, p. 349–375.

Amal Field, Libya[1]

J. M. ROBERTS[2]

New York City, New York 10017

Abstract Oil was discovered in November 1959 in well B1 concession 12, Amal field, Libya, in the eastern part of the Sirte basin. The reflection seismic exploration technique was used because there is very little surface expression in this area. The field now includes more than 100,000 acres and extends north-south for 30 mi and east-west for approximately 10 mi. To date 81 wells have been drilled, 76 of which are producers. The outer configuration of the gross reservoir has been defined on the north and on the southeast. The gross oil column is as thick as 600 ft in the most favorable areas.

The Amal field accumulation is on a plunging nose extending northwest from the old ancestral basement knob, the Rakb high. The field is interpreted from seismic and gravity data to be bounded on the flanks by large fault systems.

The principal reserves in the field are contained in the Amal Formation, a fractured, quartzose sandstone of Cambro-Ordovician age, and in the Early Cretaceous Maragh Formation, a sandstone interbedded with siltstone and shale. The Amal Formation in most of the field unconformably underlies the Cretaceous Rakb Formation at a depth of 10,000 ft. The Maragh Formation is on the north and east flanks of the crest of the plunging nose; it is a transgressive marine Lower Cretaceous unit.

INTRODUCTION

Libya is on the north coast of Africa between Egypt on the east and Algeria and Tunisia on the west (Fig. 1). The United Kingdom of Libya covers about 680,000 mi^2 (1,800,000 km^2), an area 2¼ times the size of Texas. Of the 1.5 million inhabitants, about 95 percent live along the coast, mostly near the cities of Tripoli and Benghazi and along a 15-mi stretch of arable land on the Mediterranean Coast. Approximately 85 percent of the country is barren desert. Road transportation is only fair along the coast, and roads into the interior are few. The desert operations of the petroleum industry are serviced by air from Tripoli and Benghazi.

[1] Modified from a paper read before the 53rd Annual Meeting of the Association, Oklahoma City, Oklahoma, April 25, 1968. Manuscript received, May 7, 1968; accepted, December 5, 1968.

[2] Mobil Oil Libya Ltd.

I am grateful to the Mobil Oil Corporation for permission to present this information. Numerous reports restricted to company distribution served as the basis for this presentation. Special thanks are expressed to J. R. Huth, R. H. Kirk, E. Kling, M. C. Parsons, and E. Barili for their criticisms and assistance in the preparation of the paper.

The petroleum industry has been very active in Libya since 1955, when concessions first were granted. Today, 34 operators or consortia conduct petroleum operations within 150 million acres under concession.

Mobil Oil Libya Ltd. and Gelsenkirchener Bergwerks A.G.'s Amal field is in concession 12, 175 mi southeast of Mobil-Gelsenberg's pipeline terminal at Ras Lanuf on the Gulf of Sirte (Fig. 2).

REGIONAL SETTING

The Amal field is in the Sirte basin (Fig. 2), which contains the prolific oil fields of Libya. Reservoirs range in depth from about 13,000 ft to less than 500 ft, in lithologic character from carbonate rocks and sandstone to fractured granite, and in age from Early Cambrian to Oligocene.

The basement-uplift areas (Fig. 3) in the south have been relatively positive elements since Cambrian time. Fluctuation was continuous between the huge landmass on the south and the open sea on the north. At the end of Carboniferous time, the sea withdrew northward as regional uplift and erosion took place during the late Paleozoic Hercynian orogeny. In eastern Libya, marine Paleozoic strata of Ordovician to Permian ages are found. In the Sirte basin, marine Paleozoic strata found to date are limited to the northwestern and northeastern parts.

After the uplift during the Hercynian orogeny, sedimentation continued from Permian to recent time. The general distribution of sea and land remained fairly constant until middle Cretaceous time, with marine environments in western and northern Libya and mostly continental sedimentation in the east and south.

After the Hercynian orogeny and until Cretaceous time, severe tectonic deformation affected central and eastern Libya. Basement block faulting caused the formation of an intricate horst-and-graben province; fault trends were aligned either NE-SW or NW-SE. Some paralleled the Paleozoic NE-SW trends, and younger movement paralleled the Suez-Red Sea-Rift Valley direction. Extensive igneous activity accompanied these movements.

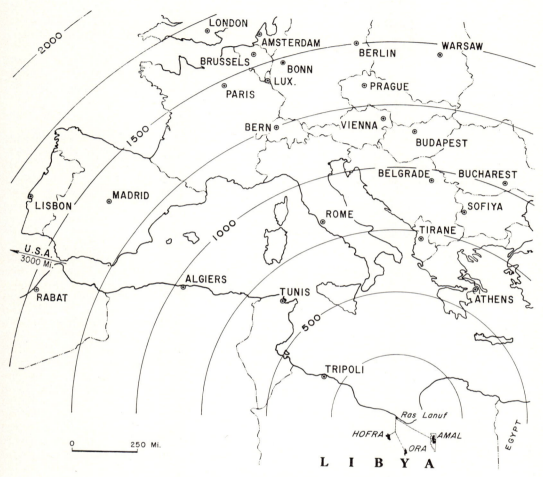

FIG. 1.—Location of Amal field, Libya.

In the Sirte basin, beginning with Albian-Cenomanian time, the sea transgressed from the north into the areas of previous fault-block development. Topographic eminences were buried slowly by marine sediments; however, syn-sedimentary tectonic movements continued along many faults and fault zones, thus contributing to differences in thickness and facies of Upper Cretaceous strata. The last fault block was covered by the sea at the very end of Late Cretaceous time (Fig. 4). In general, the Sirte basin began to be a unique basin in middle Paleocene time. Facies trends and centers of deposition shifted progressively northward. As a result, the Paleocene and younger basin extends offshore on the north.

The easternmost part of Libya remained relatively stable during Late Cretaceous and Ter-

tiary times, and is called the Cyrenaica platform (Fig. 2).

GEOLOGIC HISTORY

The Amal field on concession 12 is interpreted as being on a northwest-trending nose related to the regional Rakb high (Fig. 5). The Amal nose plunges northwest and passes through the north-central part of concession 12. Granitic rocks of the Rakb high are considered to be the source of the Amal Formation, which was deposited on the flanks of the igneous mass under predominantly continental conditions. The Amal area remained an active structural feature during the Cretacious marine transgression. Sediments deposited during the Cretaceous contributed the source beds and caprock for oil accumulation. From early Paleocene to middle Eocene times, tectonic movements within the central Sirte basin were very slight. By early Eocene time, the Amal nose again was a positive structural element (Fig. 5).

STRUCTURE

Structurally, the Amal field is on the eastern flank of the Sirte basin. Figure 5 outlines the major tectonic elements of the Amal area. The Rakb high, with the associated Rakb nose and the Amal nose, is the predominant structural feature. On the south is the Gialo high. These two structural features have had an important influence on the deposition and environment of

the entire sedimentary column across most of the area. The Rakb high was a prominent regional feature at least from Mesozoic to Oligocene time.

On the south, separating the Rakb nose from the Gialo high, is the Farigh trough. North of the Rakb high is a large low area called the Maragh low.

EXPLORATION HISTORY

Concession 12 was granted by the Libyan government to Mobil Oil Libya on December 31, 1955. The original size of the grant was 1,977,788 acres (8,004 km²) but, in accordance with the requirements of the Petroleum

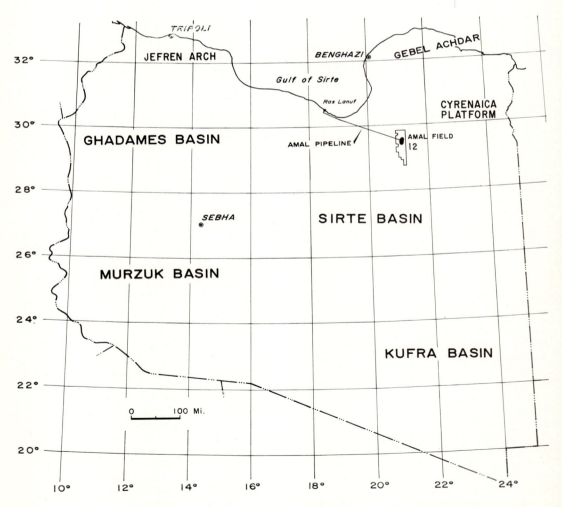

FIG. 2.—Major geologic provinces of Libya. Map shows location of Amal field and its pipeline in Libya.

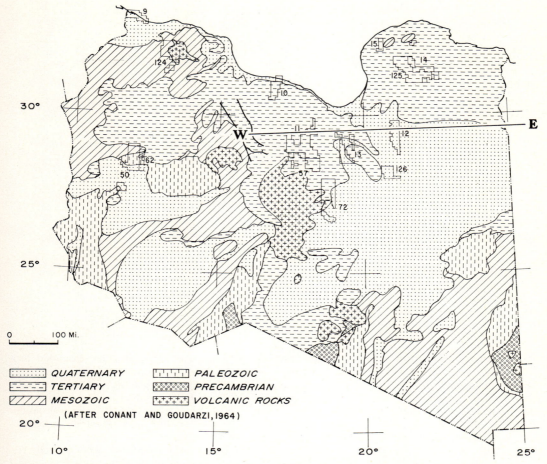

Fig. 3.—Generalized geologic map of Libya (after Conant and Gourdarzi, 1964). W-E is line of cross section of Figure 4.

Law of 1955, it has undergone final mandatory reduction to its present size of 740,559 acres (2,997 km²). In 1958 Mobil signed an agreement with Gelsenkirchener Bergwerks A.G. which gave Gelsenberg an interest in all of Mobil's Libyan holdings. Mobil was designated the operator of the combine, and wherever reference is made herein to "Mobil" it is done with recognition of this relation.

Mobil was one of the first companies to obtain concessions in Libya and initiated surface geologic surveys at the outset. Surface investigations of concession 12 demonstrated that outcrops are scarce and the area is covered largely by recent gravels and small Miocene limestone outcrops. For this reason geophysical methods were used extensively to determine the subsurface structure.

Numerous gravity and magnetic surveys were conducted in the early exploration of the Sirte basin. The first survey results were available before the 1955 acreage applications, and strongly influenced the selection of the concession 12 area.

Mobil initiated an active reflection seismic program on all of its concessions before 1960. The results of the surveys varied with the geologic and topographic conditions. Early in 1959 the reflection reconnaissance seismic program in the eastern part of concession 12 mapped several time anomalies at a shallow depth. Conventional seismic methods gave usable data from rocks of approximately middle Eocene age but failed to yield accurate deep data because of interference from strong multiple reflections originating from good shallow reflec-

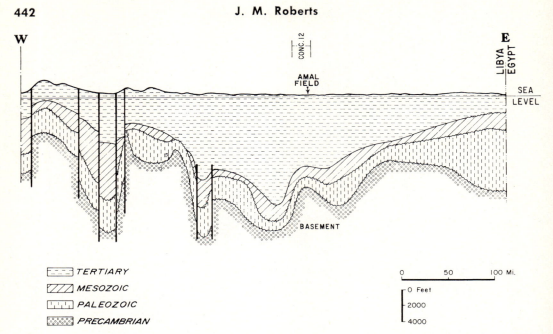

W

CONC. 12

AMAL
FIELD

LIBYA
EGYPT

E

SEA
LEVEL

BASEMENT

TERTIARY
MESOZOIC
PALEOZOIC
PRECAMBRIAN

0   50   100 Mi.

0 Feet
2000
4000

Fig. 4.—Generalized west-east structural section through northern Sirte basin, Libya. Location is shown in Figure 3.

tors. As these results were the only structural information available and regional geologic studies indicated a favorable environment, two deep tests, A1 and B1–12 (Fig. 6), were drilled on the most prominent of the shallow anomalies. The first, A1–12, in the southern part of the concession, had oil shows in basement rocks, but was dry and abandoned.

The wildcat well B1–12 (Fig. 6), in what is now the southern part of the Amal field, penetrated a Tertiary-Cretaceous section similar to that found in the first well. No oil accumulations were discovered in the shallow strata. However, the well penetrated about 600 ft of quartzitic sandstone containing oil shows, but did not penetrate the complete section. B1–12 was completed in November 1959 for 990 bbl/day through an 1⅛-in. choke from the Amal Formation. The completion of evaluation well B2–12 (Fig. 6) in November 1959 for 342 bbl/day through a ¼-in. choke from the Amal Formation extended productivity 7 mi southeast.

A wildcat drilling program was formulated to test the Amal-type deep accumulation discovered by the first wells. The test wells were located on the known shallow reflection anomalies. The E1–12 (Fig. 6) wildcat drilled in 1962 found the Amal sandstone present, and

its production of 1,224 bbl/day extended productivity 6.5 mi northwest from B1–12. In mid-1962, extensive production tests on wells B2–12 and E1–12, which are 14 mi apart, supported the concept that the area was potentially productive with good possibilities of successful development.

In 1963 a very significant wildcat well, I1–12 (Fig. 6), west of the field, found the Amal Formation below the oil-water contact. This test served to define the western limits of the reservoir at an early date.

To gain a better definition of the Amal Formation reservoir, it was decided to cover the entire remaining part of concession 12 with multiple-coverage thumper-seismic techniques. The objective was to obtain better quality reflections in the deeper strata to determine the potential size of the field.

Two thumper crews were engaged and field work was undertaken and completed in 1964. An experimental program, utilizing varied multiple seismic coverage and offset shooting incorporated with the latest playback techniques, greatly improved the deep data. These data proved to be of great value, although the productive Amal Formation was not mapped directly. An intra-Cretaceous horizon was mapped in the section just above the Amal

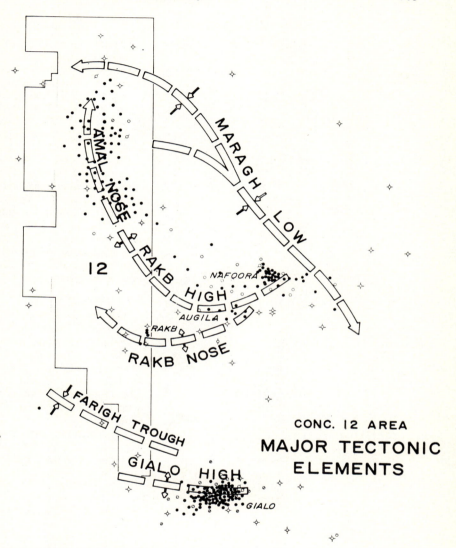

Fig. 5.—Major tectonic elements in Amal field.

Formation with fair reliability. The improved reflection seismic data and the valuable information gained from the drilling already accomplished made apparent the large size of the Amal field.

Early in 1964 the evaluation well B3–12, 4.5 mi north of B1–12 (Fig. 6), was completed in the Late Cretaceous Maragh Formation for 1,316 bbl/day through a 20/64-in. choke. The Maragh is a sandstone unconformably overlying the Amal Formation. The B3–12 well, on the eastern flank of the Amal nose, yielded both extremely important stratigraphic infor-

mation and a new producing reservoir.

Because of the distance of the field from a port and the length of pipeline required to put the field on production, the next primary goal was to prove sufficient reserves to justifiy the capital expenditure necessary to construct the pipeline and initiate production.

By the middle of 1964, 13 wells had been drilled in the Amal area, 10 of which were completed as oil wells in the Amal Formation. Two tests, B10–12 and B12–12 (Fig. 6), both in the central part of the field, discovered new reservoirs—B10–12 was completed in the

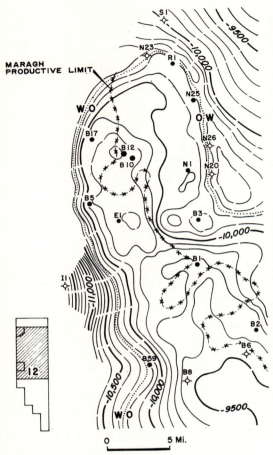

FIG. 6.—Structure contour map, Amal field, Libya. Structural datum is base of Rakb Formation (Cretaneous). Contours in feet subsea.

lower Eocene and B12–12 was completed in the upper Paleocene.

At this stage in the history of the field it was established that the proved productive area covered approximately 60,000 acres, with considerable room for extension toward the north. On this basis the next objective in the program was to develop as much production as possible on 1,000-acre spacing from the proved areas by the time the pipeline was completed so that production could begin immediately. This approach slowed the appraisal activity on the outer limits of the field for a short period while all four rigs were used in the development of production.

The construction of the 30-in. pipeline from

the Ras Lanuf terminal to the Amal field was begun in June 1965 and completed in December 1965. Construction of tanks and tie-ins to the pipeline in the field and at the terminal continued into 1966.

Two wildcats were drilled during 1966; well R1–12 (Fig. 6) extended proved productive reservoirs about 8 mi farther north. After the completion of R1–12, well S1–12 (Fig. 6) was drilled 3.5 mi north and was dry and abandoned.

Production from the Amal field with the use of temporary facilities began on February 22, 1966. The first permanent production station went on stream on April 3, 1966, and stations 2 and 3 followed on April 30 and June 11, 1966. Production began at an approximate rate of 40,000 bbl/day from each station, and addition of stations rapidly increased the total output from the field. Six production stations are now in operation.

STRATIGRAPHY

The detailed stratigraphy of the Amal field is considerably varied within the large productive area and the regional setting (Fig. 7).

Miocene-Oligocene.—The upper part of the Miocene section is composed of interbedded fossiliferous limestone, claystone, and shale. The Oligocene is made up of thick beds of sandstone interbedded with shale and a few limestone beds.

Eocene.—The top of the Eocene is a shale unit which is a potential source bed for the underlying carbonate rocks. In the middle Eocene, several prominent nummulite banks are present and are widespread throughout the area. These nummulite banks have excellent porosity values and suggest deposition in shallow water.

In the lower Eocene there is a deeper water facies in the south, accompanied by a lithologic change to porous nummulitic carbonates with increasing amounts of dolomite and anhydrite on the eastern side of the field.

Paleocene.—The Paleocene is divided into upper and lower units. The upper Paleocene is of shelf-lagoonal origin and consists of a basal dolomite-anhydrite section overlain by a fossiliferous carbonate which, in turn, is overlain by a shale unit. The shale, which can be correlated for a long distance, illustrates the stable conditions of sedimentation in late Paleocene time.

The lower Paleocene, essentially of shelf-lagoonal character, thickens west of the Amal nose into the Sirte basin. Both the upper and lower shale units of the Paleocene thin appreciably on the east, and a zone is reached where differentiation of the upper and lower Paleocene carbonate units is impossible.

Cretaceous.—The Tertiary and Upper Cretaceous section is uninterrupted and exclusively marine. It overlies a major erosion surface of Early to Late Cretaceous age, designated the Sirte unconformity. This surface has considerable erosional relief; the amount is difficult to estimate, but reaches several hundred feet. Because the Cretaceous Rakb Formation was deposited in a transgressive marine environment, the lower contact with underlying older units is unconformable, and the Rakb directly overlies the Maragh Formation, Amal Formation, or basement. The onlap relation is most pronounced on the flank of the major structural trend.

The Rakb Formation is believed to have been the source of the oil in the Amal field. The direct contact of the source beds with reservoir beds presented an ideal situation for the short-range migration and accumulation of large volumes of oil.

The Late Cretaceous (Campanian?) Maragh Formation is a transgressive marine clastic unit below the Rakb Formation in the northeastern part of the Amal field. The Maragh commonly is referred to as a sandstone, although it includes several lithologic types. At the top there is generally a vugular and crystalline sandy dolomite, grading downward into dolomitic sandstone and, lower, into glauconitic sandstone generally with varied amounts of clay cement. The sandstone has a wide range of induration and cementation. The Maragh Formation, in its normal development in the "N" (Figs. 6, 8, 9) area of the Amal field, has a basal quartzite cobble conglomerate. The cobble-size fragments were derived from the underlying Amal Formation.

The Maragh Formation onlaps irregularly the Amal nose and the southernmost occurrences appear to be associated with topographic lows on the old Sirte erosion surface. The thickness of the Maragh varies markedly in short distances. This formation commonly has a high interstitial clay content in the sandstone and in many places contains numerous shaly and silty intervals. The distribution of the Maragh in the northern part of the Amal field is shown in Figure 6.

In the low areas around the Amal nose, the Maragh Formation thickens and the Hercynian unconformity diverges from the Sirte unconfor-

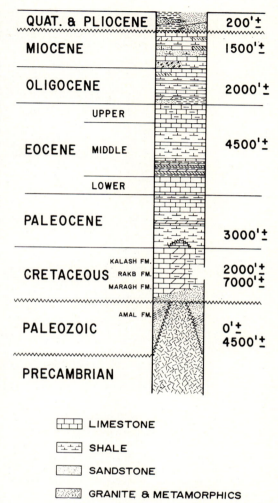

FIG. 7.—Generalized stratigraphic column, Sirte basin, Libya.

mity. A sandstone wedge is developed between the unconformities. It is extremely difficult to distinguish these sandstone units where they come together because they form a continuous clastic section.

Paleozoic.—The Amal Formation generally is described as a tight, hard, quartzose, irregularly feldspathic sandstone. Radiometric age determinations date the formation as Permian to as old as Cambro-Ordovician. The Amal unconformably overlies a complex of igneous intrusive and extrusive rocks. The oldest of the intrusives has been dated radiometrically as Cambro-Ordovician. Some igneous Permian

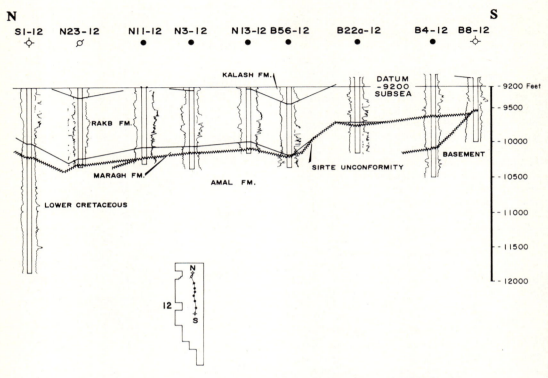

Fig. 8.—North-south electric-log cross section of Amal field. Datum is 9,200 ft subsea.

rocks form dikes, sills, or flows on the surface or within the Amal. The granitic-volcanic basement which underlies the Amal is interpreted to have formed the core of an igneous-sedimentary landmass from which the formation was derived. Deposition was moderately rapid but irregular, and was affected by alternations of marine conditions with episodic uplift in the source area, or by slight changes in the amount of igneous surface exposed to erosion. The Amal Formation is present everywhere except where truncated and eroded in the south. Northward, down the plunge of the Amal nose, the formation thickens with succeedingly younger units subcropping below the Sirte unconformity. The maximum thickness of the Amal Formation is estimated to be more than 3,000 ft.

The Amal Formation is divided into six units on the basis of texture, degree and type of cementation, and lithologic sequence within the gross characteristics of the reservoir. Significant mineralogic distinctions depend mainly on petrographic analysis. The most important distinctive petrographic characteristic is believed to be the feldspar content. Commonly, where orthoclase feldspar is present, it is altered to kaolinite and sericite-illite. These clay minerals inhibit the precipitation of silica, which may come out of solution as a result of a chemical change in the fluid environment and be deposited as filling in the voids between the quartz grains.

Basement.—Volcanic rocks of acidic to intermediate composition dated as Permian to Cambro-Ordovician are considered to be basement in the Amal field. Basement rocks are known from several wells, and are classified as granites and rhyolitic volcanic rocks.

HYDROCARBON OCCURRENCES

The Amal field accumulation is controlled stratigraphically and structurally. The accumulation extends for 30 mi north-south and the average width of the field is 10 mi east-west. The proved productive area is more than 100,000 acres. As of March 1, 1968, 85 wells had been drilled, including five classified as dry holes. Two rigs are engaged in development drilling and in the definition of the flanks of the productive area.

The Amal field reservoirs are undersaturated and contain a paraffin-base crude of 35° API gravity with a pour point of approximately 65°.

The principal reserves of the field are in the Amal and Maragh Formations. The Amal Formation does not have good reservoir characteristics, but contains a major accumulation of oil because of its huge areal extent and the thickness of the oil column. Much of the Amal Formation produces irregularly, both horizontally and vertically. The producing intervals could not be determined accurately from electric logs, and much of the production is from fractures within the formation. The average strike of the fractures is about N 20° W, approximately parallel with the plunging axis of the Amal nose. The fractures are assumed to be tensional features that developed during formation of the major Rakb high and the associated Amal nose. There is a common oil-water contact throughout the field, but in some areas it is extremely difficult to locate within the Amal Formation because of thick transition zones or nonproductive intervals within a generally tight section. In those areas where the Maragh overlaps the Amal Formation, the Maragh, with greater permeability and better reservoir properties than

the Amal, is the main productive zone (Figs. 8, 9).

On the top of the structure the Amal Formation thins by truncation toward the southeast and wedges out north of the dry holes, B6–12 and B8–12 (Fig. 8). Neither well contains the Maragh or Amal Formation, and the basal volcanic rocks are tight and nonproductive. On the basis of those wells which have penetrated the complete Amal Formation, it is possible to construct a dip on the basement surface of 5½° toward N 10° W.

The northern extent of the field is defined by the S1–12 dry wildcat well (Figs. 6, 8), which is 3.5 mi north of the R1–12 producer. The northwest edge of the field is limited by N23–12, which found the Amal Formation structurally low and tight. The west boundary of the field coincides generally with the geophysical interpretation of the west dip into a trough (Fig. 9). This area is now being investigated by drilling and the productive limits soon will be defined. On the east side of the field the Maragh and Amal Formations are low structurally, as found in wells N20–12 and N26–12 (Figs. 6, 9), which define the eastern flank of the accumulation. The Eocene and Pa-

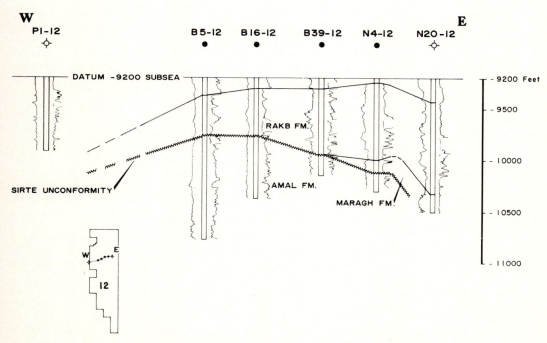

Fig. 9.—West-east electric-log cross section of Amal field. Datum is 9,200 ft subsea.

leocene have not yielded commercial production, although oil has been recovered from tests on the top of the structure in the northern part of the field in wells B10–12 and B12–12 (Fig. 6). The Amal field on March 1, 1968, had an accumulative production of 98 million bbl of oil. The daily production at that time was approximately 190,000 bbl.

SELECTED REFERENCES

Burollet, P. F., *et al.,* 1960, [Libya] *in* Lexique Stratigraphique International, v. IV, Afrique: Paris, Centre Natl. Recherche Sci., v. 4a, Libya, p. 1–62. (Compiled by Petroleum Explor. Soc. Libya Names and Nomenclature Comm., P. F. Burollet, chm.)

Colley, B. B., 1963, Libya—Petroleum geology and development: 6th World Petroleum Cong., Frankfurt, sec. 1, paper 43, 10 p. [Printed in Frankfurt, 1964.]

Conant, L. C., and G. H. Goudarzi, 1964, Geologic map of Kingdom of Libya: U.S. Geol. Survey Misc. Inv. Map I-350-A.

——— and ——— 1967, Stratigraphic and tectonic framework of Libya: Am. Assoc. Petroleum Geologists Bull., v. 51, no. 5, p. 719–730.

Desio, A., 1950, Bibliografia geologica Italiana dell'Africa sino al 1948 incluso: Rome, Ufficio Studi Ministero dell'Africa Italiana, v. 15, p. 1–34.

——— 1951, Cenno riassuntivo sulla costituzione geologica della Libia: 18th Internat. Geol. Cong. Rept., Great Britain, pt. 14, p. 47–53.

——— *et al.,* 1963, Stratigraphic studies in the Tripolitanian Jebel (Libya): Rev. Paleontologia e Stratigrafia Mem. 9.

Gillespie, I., and R. M. Sanford, 1967, The geology of the Sarir oilfield, Sirte basin, Libya: Elsevier Pub. Co., 7th World Petroleum Cong. Proc., Mexico, v. 2, p. 181–193.

Gregory, J. W., 1911, The geology of Cyrenaica: Geol. Soc. London Quart. Jour., v. 67, no. 268, p. 572–615.

Kubban, Abdul Amir, 1964, Libya, its oil industry and economic system: Beirut, Rihani Press, 274 p.

Lelubre, M., 1949, Geologie du Fezzan oriental: Soc. Géol. France Bull., 5th ser., v. 19, nos. 1–3, p. 251–261.

Sarir Oil Field, Libya—Desert Surprise[1]

ROBERT M. SANFORD[2]

Dallas, Texas 75202

Abstract The Bunker Hunt–British Petroleum Sarir oil field of Libya appears to be one of the 10 to 12 super-giants of the world. Credited with 11–13 billion bbl of oil in place, it is a water-drive field that could, and probably will, recover nearly 50 percent of its total oil. The maximum oil column is 300 ft and the area of surface closure is 155 mi[2]. The field was discovered in November 1961 on a seismically defined structure. Development drilling was continuous through the next 4 years; a pipeline and loading terminal were completed and production was begun in late 1966. The oil reservoir is Cretaceous sandstone on basement, the probable oil source being the several hundred feet of overlying Cretaceous marine shale. Structurally, the field is a combination anticline and high fault-block complex within a broad structural low.

There appears to be good fluid communication throughout the reservoir. Average porosity values are 18–19 percent, and permeability values average several hundred millidarcys, with a few 2–3-darcy streaks. All production is water free. The oil is sweet and sulfur free, though of high paraffin content.

More than 100 wells have been drilled, of which about 60 are on production and 12–14 are awaiting gathering lines; most of the others are observation wells for pressure or fluid control. There was a decline of reservoir pressure during the first year of production; however, in most of the field a sustained water drive is developing. Producing capacities of individual wells range from a few thousand barrels daily to maximum estimated open flow of 28,000–30,000 bbl/day. The field went on production at 100,000 bbl/day, which rose to 300,000 bbl within the first year. Additional field facilities, when installed, will permit even greater increases.

PROLOGUE

The Sarir oil field, owned jointly by Bunker Hunt and British Petroleum Company, is on concession 65 in the Sirte basin of east-central Libya (Fig. 1)—more specifically, in the sand-sea area of the central Cyrenaica district. The exploration of concession 65 and the development of Sarir field complex involved seemingly insurmountable problems and hardships, and many surprises, both good and bad.

Problems included (a) access, transport, and living conditions; (b) the sand-sea cover, which ruled out surface geologic mapping and necessitated the use of geophysical methods; (c) a several-hundred-foot lost-circulation zone (d)

[1] Read before the 53rd Annual Meeting of the Association, Oklahoma City, Oklahoma, April 25, 1968. Manuscript received, May 7, 1968; accepted, September 7, 1968.

[2] Hunt International Petroleum Company.

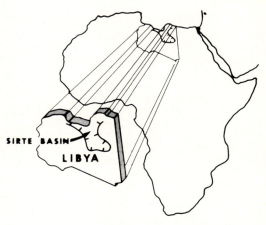

FIG. 1.—Location of Sirte basin, Libya.

pipeline-transport problems caused by the high-pour-point, waxy crude; and (e) the construction of a marine loading terminal off a limestone hillside, rather than off a flat sand beach.

During development of the field there were such unpleasant surprises as (a) the dynamiting sabotage of five field wells, (b) a long string of consecutive dry wildcats on the concession after the field discovery, and (c) several weeks of complete production shutdown after the Middle East war of June 1967.

These difficulties were offset somewhat by at least two pleasant surprises. First, the rapid increase in the number of square miles of developing field area and the mounting of reserves into the billions of barrels were certainly continually pleasing. Second, the prolonged closure of Suez Canal after the Middle East war increased demand for crude oil west of Suez, thus establishing an unexpectedly favorable market which has continued.

INTRODUCTION

The Sarir field is a water-drive reservoir of low gas-oil ratio producing average 37°, paraffin base, sulfur-free oil. Oil in place is estimated to be between 11 and 13 billion bbl. A 25-percent primary-recovery factor will give 3 billion bbl of oil. Secondary-recovery methods

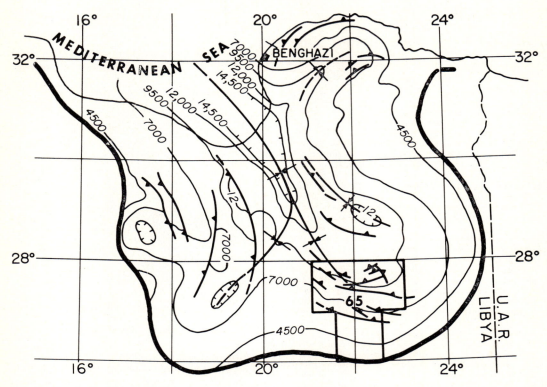

Fig. 2.—Sirte basin, Libya and original concession 65. Heavy black line shows basin outline. Datum for structural contours is unconformity surface at about mid-Cretaceous. CI = 2,500 ft.

in parts of the field, together with improved primary-recovery methods in other parts, could provide another 18–25 percent, or 2–3 billion bbl. Thus the field could have an ultimate oil recovery of 5–6 billion bbl.

The Sirte basin (Fig. 2) is mainly an upper Mesozoic and Tertiary feature developed on an old basement and eroded Paleozoic surface (Conant and Goudarzi, 1967). The main NW-SE synclinal trough underwent repeated subsidence, with accompanying fault adjustments. Several regional horst-and-graben trends which began to develop in Cenomanian (mid-Cretaceous) time remained active during Tertiary time as the basin continued to subside.

In many parts of Africa a great hiatus in mid-Cretaceous rocks is characterized mostly by sandstone of Early to middle Cretaceous age called "Continentale Intercalaire" (Intercalary Continental) deposits (for discussion of "Continentale Intercalaire" and general "Nubian Sandstone" problem, *see* Pomeyrol, 1968). Probably part of the more than 2,000 ft of sandstone in the Sarir field is of this age. Ma-

rine Late Cretaceous seas reworked these sands as they transgressed the old surface. Consequently, the drill penetrates a "sandstone-on-sandstone" sequence whose age and local unconformities are difficult to decipher. The reworked sands are at least earliest Late Cretaceous, and more likely are late Early Cretaceous. A basal sandstone of Early Cretaceous age is recognized by some geologists working in the area.

Oil is from this near-middle Cretaceous sandstone, in contact with basement and overlain by thick Cretaceous and Tertiary marine shale and limestone. The oil is at a depth of about 9,000 ft in a broad structure and in combined anticlinal and fault-trap accumulations. The drilled field area is about 160 mi² and has a maximum oil column of 300 ft; the area within the oil-water contact ring of the field is probably more than 330 mi².

Almost 100 wells have been drilled. About 90 are connected with flow lines, but generally fewer than 55 produce during a particular month. Many of the other wells are being utilized during this early stage of production as fluid-pressure and fluid-level observation holes. Day-to-day operating well potentials range up to 10,000 bbl/day; individual well potentials would range as high as 28,000–30,000 bbl/day, as observed during wild open flow at the time of the 1965 sabotage.

There are four main gathering centers in the field, and individual wells are connected by 6- and 8-in. gathering lines. Twelve-inch trunk lines connect the centers to the main 34-in. pipeline which runs north to the Tobruk marine terminal.

The field went on production in December 1966 at 100,000 bbl/day. Production gradually was increased to its present rate of about 330,000 bbl/day. The field's production capabilities are unknown; however, with ultimate pressure stabilizations, it could possibly sustain more than 700,000 bbl/day.

Concession 65, which originally contained 20,459 mi², has been reduced by three relinquishments (Fig. 3) and is now at its final size of 5,113 mi².

Sirte basin contains all of the major oil fields

of Libya (Fig. 4). The fields are on the horst ridges or high fault edges of the regional tectonic features of the basin. Local oil accumulation generally is associated with sedimentary drape and cross-faulting of these main trends. The original 20,000+ mi² of concession 65, in the southern part of the basin, previously was thought to be outside the principal oil-prospect area, which was considered to be mainly the west flank of the basin. Sarir is now the southernmost oil field in the basin.

EXPLORATION HISTORY

The original concession was about 120 mi north-south and 120 mi east-west at its widest part (Fig. 3). It was granted to Nelson Bunker Hunt in December 1957 as his second concession in Libya; he had taken concession 2 in the north in 1955. No oil had been discovered in the Sirte basin, although work was well underway on many concessions. Exploration was carried on by Hunt between 1957 and 1960, and the discoveries of 1958 and 1959 indicated major oil fields in the basin. In September 1960

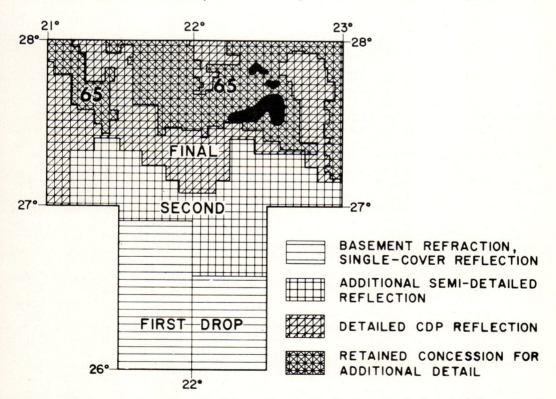

FIG. 3.—Original concession 65, Sirte basin, Libya, and parts relinquished as a result of geophysical surveys and test drilling.

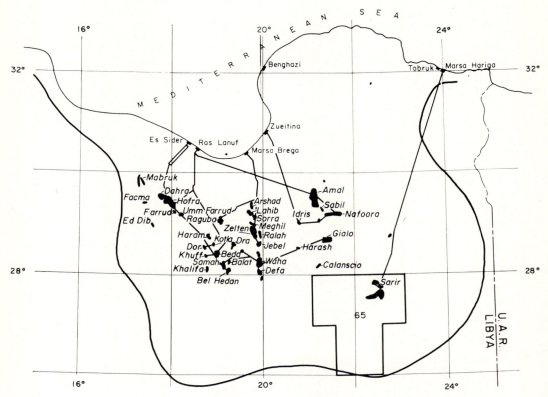

Fig. 4.—Oil fields in Sirte basin, and pipelines to shipping terminals.

Hunt turned over a half interest to British Petroleum (BP Exploration Co., Libya), which became the operator.

Air-magnetometer and gravity surveys were made over most of the original concession, and a refraction and reflection seismic reconnaissance grid was started. These surveys and a few rather discouraging wildcat holes led to the decision to relinquish the southern area (Fig. 3) in December 1962. Sizable structures were noted mostly in the northern part of the concession (Fig. 5).

Additional semidetailed reflection work and a few more wildcat holes led to the second relinquishment in December 1965. From 1965 to date seismic coverage has been by the CDP or six-fold stack technique. This enhancement of seismic data produced positive information which greatly assisted in the selection of acreage for the final land drop in December 1967. Stacked data also led to the discovery of the "L" producing area north of the main field. Figure 5 shows the principal faults and tectonic trends in concession 65, as well as all wildcats

drilled to date. Figures 6–9 and 11–13 show the present concession area after the three drops and semidetailed information on the structure and stratigraphy.

Sarir unquestionably is a discovery based entirely on geophysical methods (Fig. 6). The major highs (stippled) and lows of the concession and their relief at different stratigraphic levels were determined originally by various methods of geophysics. The stratigraphic-marker tops from wildcat wells were introduced later into the mapping. The air magnetometer revealed the presence of the large southwestern area **A** structure and its southeastern prolongation, which subsequently was found to be a buried major fault line. Sarir field, superimposed on this map, was not detectable as an anomaly on the basis of this magnetic survey.

Gravity coverage (Fig. 7) followed the magnetic survey; it also outlined the major features. The Sarir feature shows only as a nose and is definitely overshadowed by the two features on the west. Hence, the latter became the **A** struc-

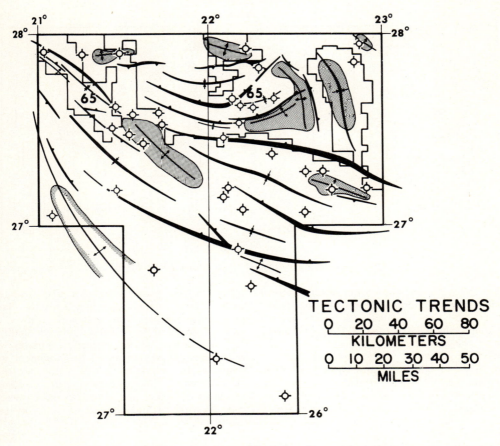

FIG. 5.—Tectonic trends in concession 65, found by geophysical surveys.

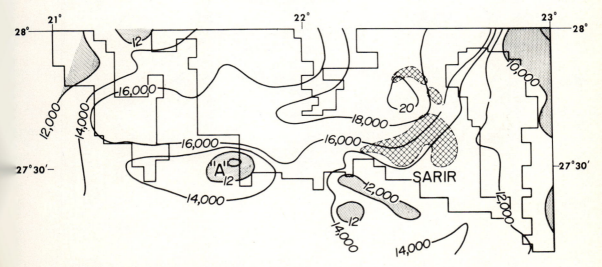

FIG. 6.—Aeromagnetometer basement-depth map of retained area. Depth in feet. CI = 2,000 ft.

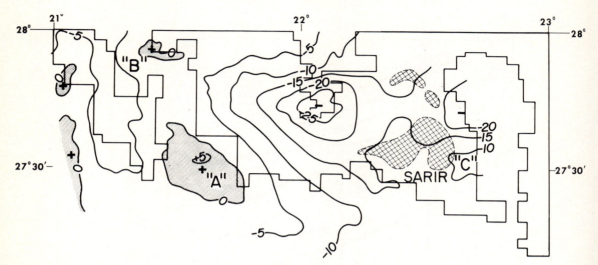

FIG. 7.—Bouguer gravity map of retained area. CI = 5 mgal.

ture at the +5-mgal contour line and **B** structure at the zero-miligal contour line. The then questionable and unnamed, eastern nosing feature was called the **C** structure; it later proved to contain the Sarir field.

Seismic reflection and refraction surveys were made in late 1960 and early 1961. The basement-depth map in Figure 8, constructed in depth from the two-way refraction time from granite and/or Precambrian metamorphic basement, also shows the major structural features. The **C** structure, comprising the Sarir producing complex (crosshatched), shows up as a pronounced basement feature. A shallow, 3,000-ft Eocene seismic-reflection horizon was mapped at the same time as the basement was mapped by refraction. The Eocene map showed only a few hundred feet of relief over the Sarir complex, in contrast to the basement refraction map which showed up to 2,000 ft of

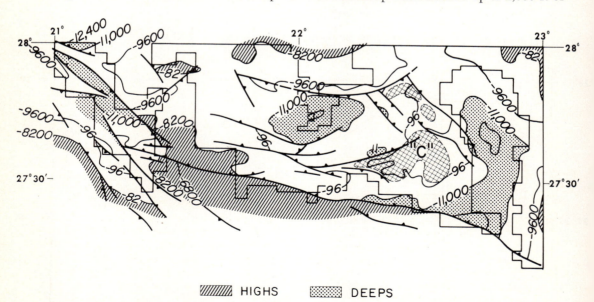

HIGHS   DEEPS

FIG. 8.—Seismic basement refraction map of retained area. CI = 1,400 ft. Cross-hatched area is **C** structure, comprising Sarir producing area.

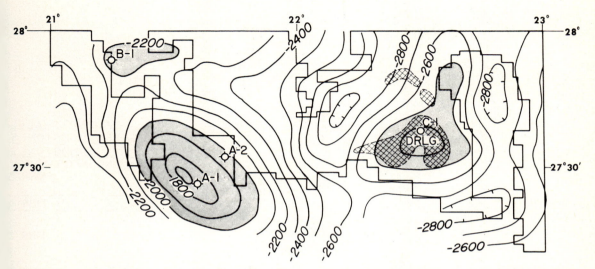

Fig. 9.—Geologic subsurface, top Eocene, November 1961 (discovery date December 1961). A-1, B-1, and C-1 are test wells drilled on **A, B,** and **C** structures.

relief. Thus, the need for one or two deeper re-flecting horizons on which to map became ap-parent and they were developed later.

By early 1962 mapping of the shallow Ter-tiary reflector and the deep basement refractor was completed. Reflection and refraction highs generally coincide. The first wildcat (A-1-65), completed in March 1961, was high and dry at 7,437 ft in a section composed mainly of car-bonate. No one knew yet of the "basal sand." The second wildcat (B-1-2) was low and dry at 10,193 ft. By combination of the seismic data with the tops penetrated in these two wells, a "top-of-Eocene" map was constructed (Fig. 9) which showed **C** as a very large structure with 300 ft of closure. It was a deeper structure than either **A** or **B;** if neither **A** nor **B** had shows or sandstone reservoirs, what chance could **C** have? The answer was provided by the C-1 well, which came in as a several thousand barrel per day discovery, producing from the upper 250 ft of a 500-ft or thicker sandstone reservoir previ-ously unknown on the concession.

The great increase of closure with depth led to the belief that the structures had been grow-ing with geologic time as a result of both reju-venated fault uplift and continued sinking and compaction of the lows. With two dry wildcats and a discovery well completed, a better pic-ture of field stratigraphy and lithology began to emerge (Fig. 10).

STRATIGRAPHY

The stratigraphic successions penetrated by the wells are generally similar, the difference being only the relatively slight thickness varia-tions of individual units (Fig. 10).

Holocene to Oligocene.—This interval con-sists of an upper zone (about 800 ft thick) of unconsolidated, slightly felspathic sand; a mid-dle zone (about 1,000 ft thick) of gray-green and red-brown shale and claystone with dolo-mite and sandstone partings; and a lower zone (about 1,000 ft thick) of fine- to coarse-grained sandstone with some claystone partings and dolomitic beds. These beds are almost un-fossiliferous but overlie well-dated late Eocene beds. In general the interval is thinnest over the crest of the Sarir structure and thickens away from it.

Upper Eocene.—The upper Eocene consists of 250–330 ft of interbedded limestone, dolo-mite, marl, and shale, which show marked lat-eral change in both lithologic character and thickness. There is a general change across the field from thinner, predominantly calcareous shale beds in the west to thicker, predominant limestone beds in the east. This variable se-quence is somewhat transitional between the middle Eocene limestone below and the epicon-tinental, detrital, post-Eocene beds above. Dat-ing of the sequence is based on the presence of

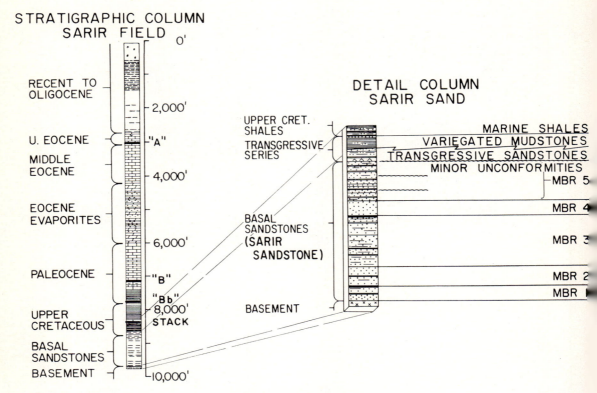

FIG. 10.—Stratigraphy of Sarir field, Libya.

the late Eocene *Nummulites fabianii* and *N. striatus.*

Middle Eocene.—This subdivision is approximately 1,200 ft of nummulitic limestone, argillaceous limestone, and marl with local beds of calcareous sandstone. The beds are very fossiliferous, containing such middle Eocene forms as *Nummulites subbeaumonti, N. curvispira, N. perforatus, N. bactchisariensis, Dictyoconus aegyptiensis,* and *Lockhartia tipperi.* Thickness of the unit is relatively constant across the large area of the field.

Eocene evaporites.—This sequence of evaporites consists of about 1,600–1,850 ft of interbedded dolomite and anhydrite. It is unfossiliferous and probably ranges in age from late Paleocene to early Eocene. It also is fairly uniform in thickness.

Paleocene.—The Paleocene is 1,600–1,900 ft thick and consists of an upper carbonate unit and a lower shale unit. The carbonate is mainly very porous dolomite with some limestone, shale, and a few stringers of anhydrite. The dolomite is mostly secondary. During the early

drilling circulation losses in the dolomite caused delays.

The limestone intercalations carry a Paleocene fauna, including such forms as *Miscellanea miscella, Alveolina ovoidea,* and *Operculina patelensis.* The lower shale is dark gray to black, marine, and carries a fauna which includes *Globorotalia pseudobulloides* and *G. trinidadensis.*

The Paleocene-Upper Cretaceous contact, in the middle of a shale sequence, was cored in one well. It coincided with a gamma-ray log peak. There is believed to be a disconformity between the Upper Cretaceous and Paleocene, apparently marked by a high level of gamma radiation; thus the boundary now is picked on this readily discernible gamma-ray log peak. The thickness variation in the Paleocene is complex. There is a general thickening from the southeast to the north and west, and also local thickening on the flanks of the structure.

Upper Cretaceous.—The Upper Cretaceous consists principally of an upper marine shale unit which unconformably overlies a lower sandstone unit; this lower unit—the Sarir Sandstone—forms the reservoir of the field and lies directly on basement. Between the two units is a third unit consisting of several hundred feet of transitional beds (Fig. 10).

The upper unit consists mainly of 750–1,000 ft of dark-gray to black marine shale with some limestone and sandstone mainly in the lower part. The interbedded limestone generally is buff, glauconitic, finely crystalline, and dense, and the sandstone generally is fine grained, glauconitic, and impervious.

The associated fauna contains such Late Cretaceous forms as *Rugoglobigerina macrocephala, Pseudogümbelina costulata, Reussella aegyptica,* and *Globotruncana* spp. The youngest forms identified are of Maestrichtian age.

The middle or transitional unit includes several hundred feet of red, green, purple, and variegated anhydritic shales with some sandstone. The shales are very different from the gray-black shale of the upper unit. The sandstone is poorly sorted, fine to coarse grained, and generally is anhydritic and impermeable. Locally some porous beds are present which yield hydrocarbons.

The transitional unit overlies the lower unit —the Sarir Sandstone—unconformably. Evidence for the unconformable relation with the Sarir consists of shale colors, presence of numerous transgressive-regressive sequences, evaporite beds, and gradational sandstone-siltstone-shale sequences.

The lower unit—the Sarir Sandstone—is the field reservoir and is divisible into five members.

SEISMIC AND SUBSURFACE CONTROL

Once the stratigraphy and lithology of the shallow section and the general geologic history of the Sarir area were known, a deeper mapping horizon or control level was needed and attainable. The geophysicists reworked all seismic data. They established what is called the "B" horizon and a phantomed "Bb" horizon, both at about 7,000–7,500 ft (*see* Fig. 10), in the Paleocene and Upper Cretaceous, respectively.

Figure 11 shows why the operators were very optimistic concerning the field during its early development stages. By March 1962, wells 1, 2, and 3 had been completed as good producers, rated at 7,500, 7,500, and 8,500 bbl/day, respectively. Number 4 was being drilled, and 5–8 were extension locations. The disappointment was great when four of the five outposts came in as marginal to nonproducing wells. Each had 100–200 ft of oil column, but the sandstone was silty to shaly and tight. The general lack of knowledge concerning the Sarir Sandstone thus became apparent.

Wildcat drilling continued in 1963 and 1964 while field development progressed. Figure 12 shows several hundred feet of structural relief at the top of the Sarir Sandstone. The bald highs surrounding the deep Sarir field area are confirmed by dry wildcats. Continued wildcat drilling also proved that the reservoir properties

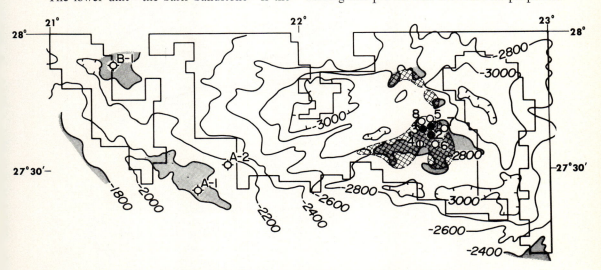

FIG. 11.—Structure as known in March 1962 with 3 wells completed and No. 4 being drilled; 5–8 are extension locations. Contour datum is seismic "A" horizon (top mid-Eocene; *see* Fig. 10).

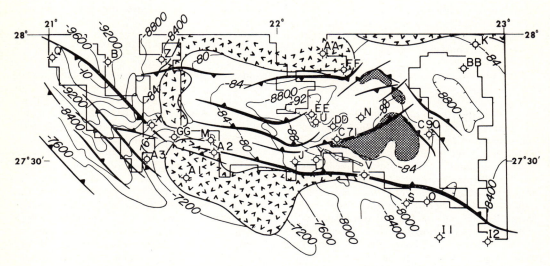

FIG. 12.—Structure contours with top of Sarir Sandstone as datum. Checked areas are basement subcrop. CI = 400 ft.

of the sandstone diminish off the west flank of the central (A-1 well) bald area. On the south the sandstone is mainly in the transition zone, and is of a poorer quality than that in the field.

Figure 13 shows the thickness of the reservoir-sandstone distribution over the present concession area. The greatest preservation of sand was in the deep trough where Sarir was discovered.

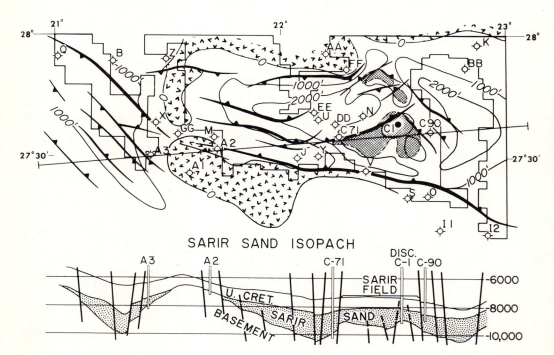

FIG. 13.—Isopach of reservoir sandstone and west-east cross section through concession area. C-1 is Sarir discovery well. Checked areas are basement subcrop. CI = 1,000 ft.

The post-Sarir uplift of the structure and subsequent erosion and reworking of the sandstone caused it to thin over the top of the structure. Sandstone is absent over the crest of the bald **A** structure, probably by erosion of Sarir Sandstone and nondeposition of younger transitional sandstone.

Normal or block faulting is apparent in the Sarir Sandstone (Fig. 13, cross section; Fig. 14). All high blocks seem to have been eroded to some degree.

As shown on Figure 10, Sarir Sandstone is divided into members. Although these divisions may be subject to later revision, they have proved to be adequate in development of the field.

Figure 14 shows the probable depositional sequence and transition from the basal (Sarir) sandstone into the overlying regressive and transgressive transitional sandstones. The main Sarir Sandstone ("Nubian," "Continentale Intercalaire," *etc.*) probably is a remnant sandstone from earlier sedimentation in the Sirte basin. It has been identified only in deep basin lows which remained below base level during the Late Cretaceous unconformable transgression. On this erosion surface, whether basement, Paleozoic rocks, or basal sandstone, the Late Cretaceous transitional sand, silt, and mud that graded upward into the Late Cretaceous marine mud were laid down.

Local faulting accompanied deposition; thus, the upthrown high edge of the Sarir field block was subjected to recurrent erosion. The eroded material was deposited downflank as a sequence of interbedded sand and mud (Fig. 14, middle panel). In time, these deposits lapped against and over the structural high and were themselves buried by the earliest deposits of the Late Cretaceous marine transgression (Fig. 14,

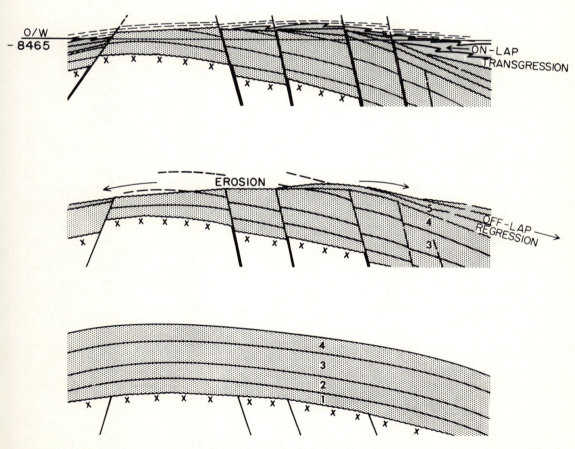

Fig. 14.—Sarir Sandstone sequence; probable transition of deposition from basal sandstone into overlying regressive and transitional sandstones.

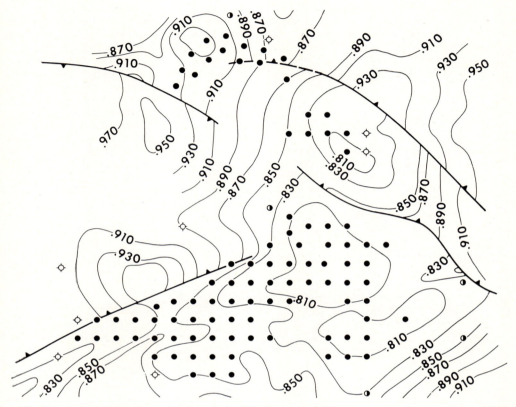

Fig. 15.—Basement refractor mapped before Sarir field discovery; 0.810 sec one-way time = approx. 9,000 ft. CI = 0.02 sec. The well pattern has been superimposed.

top panel). The oil-water contact is shown in Figure 14 to illustrate that the oil column cuts across the sandstone body and is partly in members 2–5 of the Sarir Sandstone, and in the transitional sandstone above.

In Figure 8, basement depths in the area surrounding Sarir are seen to be greater than 11,000 ft, whereas the perimeter of the structure is outlined within the 9,600-ft contour. In Figure 15 the 9,600-ft level is seen as about 0.850 sec; the crest of the closure on basement is 0.810 sec or about 9,000 ft; the deeps of 0.950 sec in the northwest part of the map are in excess of 11,500 ft.

The superimposed field wells are on a spacing of 2 km or 1.2 mi, which makes the oil-field area about 35 mi long by 25 mi wide.

Figure 16 shows the seismic "A" horizon (Fig. 10), top of the middle Eocene, as interpreted across the oil field. It is more detailed than Figure 11, which is a map of the entire concession. The basic grid was shot and the

"A" horizon interpretation was made over the field area before the wells were drilled; very few changes were needed as the field was developed.

Early in the field development, the most critical problem to resolve was the fact that 100 ft of closure at the 3,000-ft "A" horizon increases to several hundred feet at the top of the reservoir. This accentuation of structure with depth became apparent as the first 15–20 wells were drilled. The dry wildcat at the far north (L-1, Fig. 16) was drilled in 1963 on the basis of structure seen on the "A" horizon and the basement refractor; in 1966, as a result of more precise CDP methods, the south offset (L-2) became the discovery well in this north extension.

Mapping of the "B" horizon, near the base of Paleocene, generally accentuated and shifted slightly the three high areas which later proved to be the three main lobes of the oil field. However, the "B" horizon map still was not suffi-

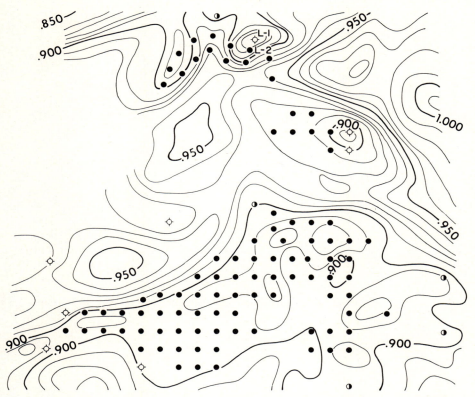

FIG. 16.—Reflection seismic "A" horizon, about top middle Eocene; 0.900 sec two-way time = approx. 3,000 ft.
CI = 0.01 sec.

ciently accurate to prevent the drilling of many dry holes, and it was little used.

DETAILED FIELD STRUCTURE

It was imperative that a control horizon be mapped on or near the top-of-sandstone level. Toward this end an experimental sixfold, multi-cover, reflection seismic survey was conducted over a part of the field where reliable, known subsurface tops had been determined. After considerable playback-center experimentation, a reliable reflector was found to be mappable just a few hundred feet above the sandstone-top unconformity; Figure 17 is a reconnais-sance view. The interval from this so-called "stack" horizon down to the top of sandstone can vary by a hundred feet or more, but it was the most accurate deep horizon yet mapped. Most of the true structure of the reservoir began to be revealed, although pre-unconfor-mity faults in the Sarir Sandstone that were not rejuvenated could not be mapped at the stack-horizon level.

CDP lines were shot from the unknown parts of the concession into the known field area where the reflecting horizons could be identi-fied. Development drilling, extension drilling, and concession wildcat drilling continued with the combined help of edge CDP coverage and normal subsurface well tops. Figure 17 shows closures of 250 ft or more, compared with only 100 ft on the "A" horizon map (Fig. 16).

Figure 18 is a south-north seismic CDP line through the west end of the field—the first line of wells east of the south-central dry hole. Here the faults in the deeper beds show up well on the 600-percent stack section. Such results as this proved beyond doubt that the actual reser-voir is faulted considerably. The large fault just north, or on the right, of well 23 is the border-ing field fault. It has about 300 ft of throw at the top of the Sarir and nearly 1,000 ft at the depth of the basement. On Figure 17 the two southwestern-edge dry holes are in segments that are downfaulted from the main field. Fig-ure 19 shows the manner in which one of these

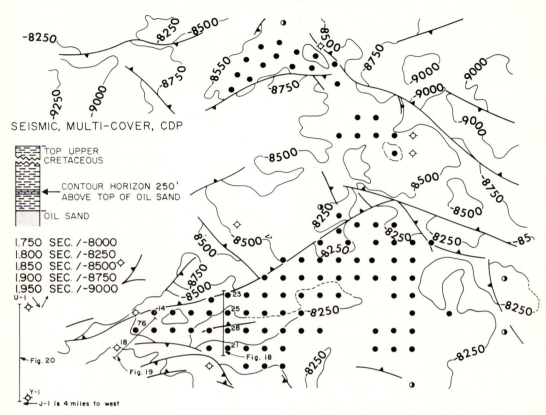

FIG. 17.—Reconnaissance map based on reflection seismic survey. Structural datum is 250 ft above top of Sarir Sandstone. CI = 250 ft. Locations of Figures 18-20 are shown.

faults grew by periodic rejuvenation, and the present measured throw at various levels. This fault has the characteristics of synsedimentary (Barakat, 1960), contemporaneous (Hardin and Hardin, 1961), or growth faults (Ocamb, 1961).

Figure 20 is a south-north multicover seismic line just west of the main field, extending from Y-1 to U-1, both basement wildcats. The stacking technique brings out basement and top-of-sandstone structure very well. What appears to be depositional thinning of the sandstone from right to left actually is caused by erosion of the sandstone at the unconformity at the top of the Sarir, as well as by truncation at the approximate stack-horizon level by the transgressive marine shales. It is not possible, seismically, to identify the individual sandstone members, but in Figure 20, member 1 probably would be subcropping beneath Cretaceous shale on the left, and members 2–5 would subcrop progressively northward.

Figure 21, a south-north CDP section in the western part of the concession, may be compared with the section between wildcats Y-1 and U-1 of Figure 20. The several down-to-the-north faults appear to have combined into two large faults. Erosion and truncation are extreme at the southern or left end of Figure 21. The sandstone thickness increases from a few hundred feet to more than 2,000 ft in about 2 mi. The sandstone is preserved in the low areas and generally is eroded from the high areas. Exceptions would be in places where the transitional sandstones above the Sarir and at the base of the overlapping transgressive sequence might lap directly onto the denuded basement. Only drilling would determine whether the thin sandstone over the high is part of the post-Sarir transitional unit or the uppermost unit (member 1) of the Sarir. Wildcats J-1 and Y-1, 4 mi west of the field area (not shown in the illustrations), penetrated several hundred feet of shaly, silty sandstone. It is still unknown

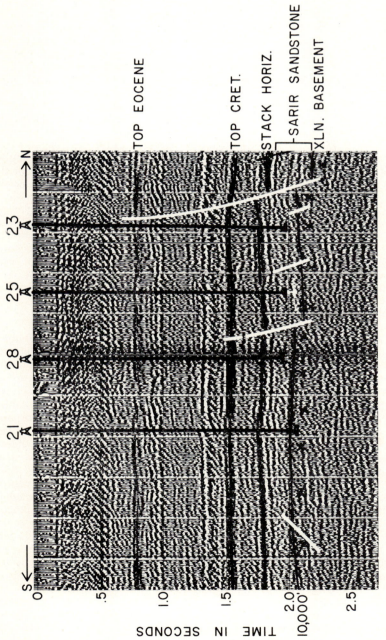

FIG. 18.—South-north seismic CPD line through west end of Sarir field. Location is shown on Figure 17. Faults are shown in white.

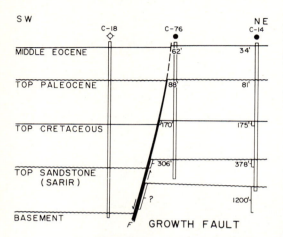

Fig. 19.—SW-NE section, southwest end of Sarir field. Fault growth is by periodic rejuvenation. Increments of growth are shown in feet. C-18 is westernmost dry hole and C-76 and C-14 are adjacent west-end field wells. Location is shown on Figure 17.

whether this section belongs to the onlapping transitional sandstones, or to member 1 of the Sarir Sandstone, or to both.

A sandstone-member subcrop (Fig. 22) shows the truncation from east to west across Sarir field. An interpretation of the structure at the top of pay sandstone, considerably simplified and without many minor faults, is given in two electric-log correlation sections (Figs. 23, 24). Figure 23, together with Figure 22, shows that the main Sarir structure is the flat crest of an upthrown block, bounded on the northwest and northeast by faults generally downthrown toward the north. The area of closure of the main field is roughly triangular, with an east-west base about 25 mi long and a north-south perpendicular distance of about 12 mi. Within the closed area dips are south and generally less than 1°. However, as an indication of the deep-seated nature of this structure, the depth to the basement between the crest of Sarir structure and the deep low just south of well 20 changes by 270 msec, or about 3,300 ft. This depth change is within a distance of 14 mi and is equivalent to an overall dip of 2.5°. The steepest dip recorded is 4.5°.

North of the main triangular structure, on the ridge separating the northwestern and northeastern basement lows, there are additional areas of closure—the Sarir North and L-area—which also are fault-bounded on the northwest and northeast. The full sequence of sandstone units is present in all three producing

areas, as seen in Figure 23. This section also illustrates the three distinct water levels of the three producing lobes. Field production has shown that faults apparently do not act as fluid barriers. Thus, the different water levels must be explained by fault displacements of impermeable against permeable beds. Between the main field and the Sarir North area of well C-19, a low block brings shale down to water level. The same may or may not be true between Sarir North and the L-area.

On Figure 24, the angularity of the unconformity from east to west is most evident. It also is seen in the subcrops in Figure 22. With the depth of basement increasing from wells 1 to 11 to 14, a true depositional thickening of the sandstone members is demonstrated, even though subsequent truncation has occurred at the sandstone-shale contact. The main part of Sarir field has a uniform oil-water level of about −8,460 ft. The sandstone is greatly thickened in dry hole 18 because it is in a downthrown, noneroded block in an off-structure position. Figure 19 shows the 1,200 ft of basement throw and 378 ft of sandstone-top throw between wells 18 and 14.

SARIR SANDSTONE (BASAL SANDSTONES)

A study of the sandstone members in Figures 23 and 24 shows that thickness ranges from about 450 ft to more than 2,000 ft throughout the field. The sandstone beds are overlain unconformably by the transgressive sequence and directly overlie a basement complex of granite and metamorphic rocks. Angiosperm pollen has been recorded at a low level in the sandstones. Thus they are dated as no older than latest Early Cretaceous (Albian), and probably are of early Late Cretaceous age.

The basal sandstones are very nonhomogeneous and several attempts at subdivision were made during the field's developmental stages. A most obvious division of the entire sandstone interval is into three parts—a lower shaly sandstone, a middle clean sandstone, and an upper shaly, silty sandstone. This division is apparent on all lithologic sections and electric logs of the area. In a further breakdown the lowest unit is called member 1, the middle unit is subdivided into members 2, 3, and 4, and the upper unit is member 5 and transition beds. The members are described in descending order.

Member 5.—Member 5 is absent in much of the field and ranges in thickness from 0 to about 450 ft. It forms an outward-thickening fringe around the margins of the main structure.

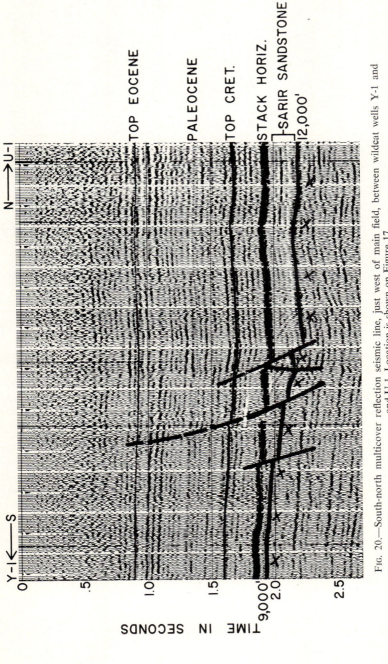

FIG. 20.—South-north multicover reflection seismic line, just west of main field, between wildcat wells Y-1 and U-1. Location is shown on Figure 17.

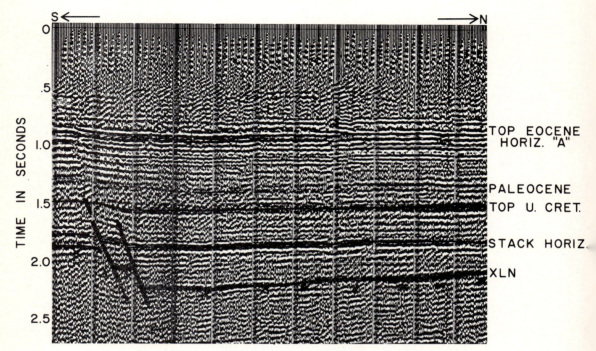

Fig. 21.—CDP section in western part of concession. Compare with Figure 20. Location is 4 mi west of Figure 20 (*see* Fig. 17).

The partial absence of the member is considered to be due to postdepositional erosion and/or nondeposition over high areas of the field. Detailed studies are still underway involving stratigraphy, lithology, and geologic history of member 5 and the overlying transitional sandstone beds.

Lithologically, member 5 is extremely variable, consisting of white to red-brown, poorly cemented, fine- to coarse-grained sandstone and shaly sandstone, with interbedded red and green shale and sandy mudstone. Although it is generally tight, several porous and permeable beds within it have almost the quality of the sandstone of member 4. Where porous and permeable beds are above the oil-water contact, they yield prolific production.

Member 4.—The 100–200-ft member 4 is the cleanest and most permeable sandstone unit. It separates the tighter sandstone of member 3 and the shalier sandstone of member 5. It is a white to buff, clean, medium- to coarse-grained, poorly sorted, poorly cemented, friable, blanket sandstone with good porosity (17–20 percent) and good permeability (several hundred millidarcys to 2- or 3-darcy

streaks). Member 4 is the main reservoir of the basal sandstone sequence.

Member 3.—Member 3, 230–820 ft thick, consists of alternating porous and tight sandstones, with some shale. The nonporous sandstone beds are cemented tightly with a clay matrix.

The porous intervals in member 3 form good reservoirs. They are generally 20–40 ft thick, with porosity and permeability characteristics similar to those of the excellent member 4 sandstones.

Member 3 is present in the entire field and was eroded only partly at the unconformity over the crest of the structure.

Member 2.—This is a fairly uniform, clean sandstone, also similar in character to member 4; however, it becomes somewhat shaly in the lower part. It ranges in thickness from 100 to 270 ft.

The member is present in its entirety across the whole field and thickens from east to west, as did member 3 before it was eroded at the top. Only in the westernmost part of the field is member 2 found above the oil-water level.

Member 1.—This basal unit of the sedimentary succession consists of an alternation of sandstone, shaly sandstone, mudstone, and siltstone. Generally it is much more shaly than the upper members. It is present on the entire structure but is greatly varied in thickness. Like members 3 and 2, it thickens from about 50 ft on the east to more than 1,000 ft on the west. The great thickness variation of this member, as well as of members 2 and 3, probably is due partly to the infilling of an irregular basement topography, and partly to proximity to a source of abundant sediment in a nearby high area on the southwest. The faulting and structural uplifts of concession 65 might not have started during the time when member 1 was deposited. Nowhere in the field is this member above the oil-water level.

Oil Reservoir

Although seismic basement control proved to be fairly accurate throughout the concession, the need for structural, stratigraphic, and reservoir data necessitated the drilling of certain selected wells to basement. Figure 25 is a structure-contour map of basement. Just within this field area the basement surface has 1,000 ft or more of relief. Subsurface control of block faulting within basement is only fair, and is somewhat less clear within the sandstone members. There are very few, if any, reliable correlation markers within the individual sandstone bodies. Locally there may be unconformable or disconformable relations between members, or within a member, as there is thought to be between members 4 and 5 and between member 5 and the transitional sandstone beds.

Figure 26 shows that, at the 2-contour line, the water-bearing sandstone is only about twice as thick as the oil-bearing sandstone around the crest of the structure (roughly at the 9,000-ft contour on Fig. 25). Discovery well C-1 on the crest had 250 ft of oil-bearing and 250 ft of water-bearing sandstone—the thinnest total sandstone sequence on the entire structure. Outward from the crest, the water-bearing sandstone thickens markedly beneath the oil column. This great increase is a most significant feature of the reservoir and is part of the key to its water-drive energy source. There is no gas cap in the reservoir, and Sarir crude has a low gas-oil ratio of between 60 and 225 scf/bbl. Except on the crestal area of thinner sandstone, the entire field has developed into a water-drive reservoir.

Figure 27 is a thickness map of the gross oil-bearing sandstone above water level. During development drilling of the field certain wells were designated to be cored, some through the entire pay section and some through only selected parts. These wells are shown by the triangle symbols in Figure 27.

In the western part of the main field, the five wells marked by squares were blown up by unknown saboteurs on May 14, 1965. The christmas trees were dynamited, and wells 48, 24, 28, and 21 went out of control and caught fire. Well 23 blew wild but did not catch fire.

Engineers, production men, and the professional fire fighters at the time estimated for well 24 a free flow of 25,000–30,000 bbl/day. The wild flow of well 48 was estimated at 30,000–40,000 bbl/day through casing and tubing during the last 4 days before it was controlled. All wild wells finally were capped, recompleted, and found to have had no actual reservoir damage. The wild flow lowered the measured bottom-hole pressures from about 3,904 psi to about 3,895 psi, or a drop in that area of less than 10 psi for the more than 1 million bbl produced by the five wells in the 6–15-day period. Nineteen months later, when the field was ready to go on pipeline production, reservoir pressure recovery had been complete. Well 24 has been producing at an average of 2,900 bbl/day and well 48 at an average of 7,000 bbl/day for 1½ years. Wellhead flowing pressures that originally were near 1,000 psi are now down about 100 psi in this part of the field, but sustained water drive is halting the decline. Flowing pressures in most of the field are still well above the 650-psi level used for potential estimates, but some sections, specifically the central, thin-sandstone area, show somewhat greater pressure drops of about 400 psi, or to about 600 psi, and pressure maintenance by water injection is being undertaken.

A structure map with the top of the pay sandstone (Sarir) as datum, contoured to accommodate the main faults known or believed to be present at this top-of-sandstone level, is shown in Figure 22. Somewhat fewer faults are shown here than are known on basement. The different oil-water contacts are marked around the three producing lobes. Drilling eventually may be done on the eastern and southern sides of the main field, as well as in the areas between the field segments; the decision to drill probably will depend on performance of the southeast-area wells when they are tied into the system.

In the center of the main field and just under the 8,200-contour elevation is well 54, one of

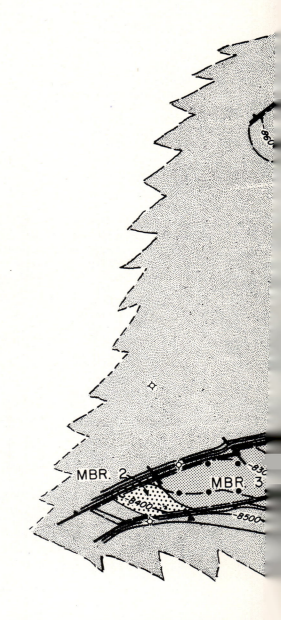

Fig. 22.—Structure on base of shale or top of pay sandstone, with subcrop of sandstone members of Sarir Sandstone superimposed. CI = 50 ft. Very fine dot pattern is member 5; fine dot pattern is member 3; coarse dots are member 2; circles are member 4.

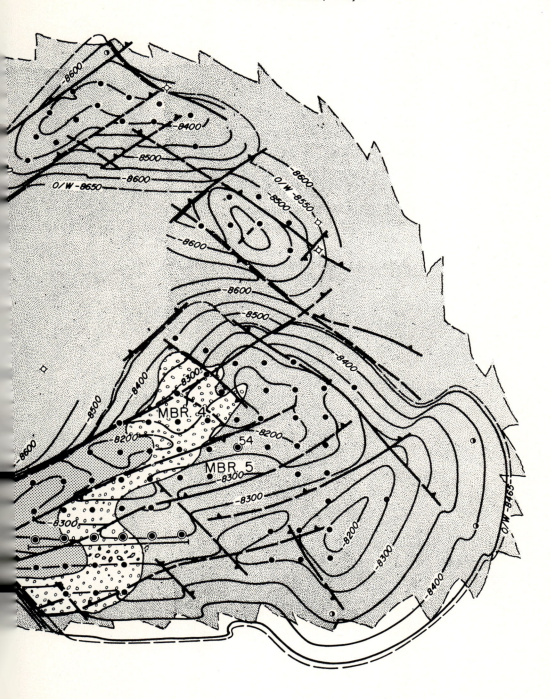

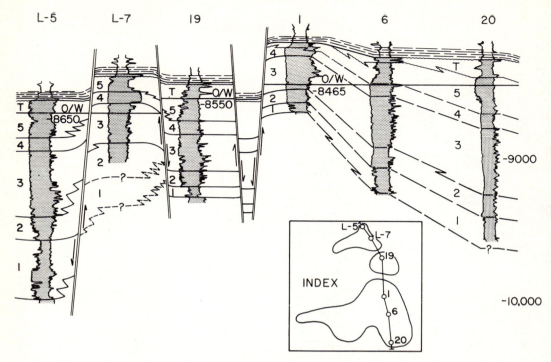

FIG. 23.—North-south section of Sarir pay.

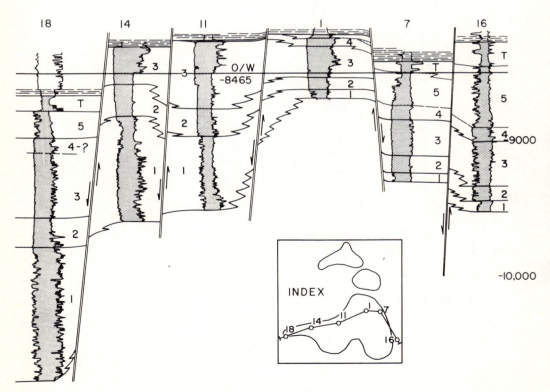

FIG. 24.—West-east section of Sarir pay.

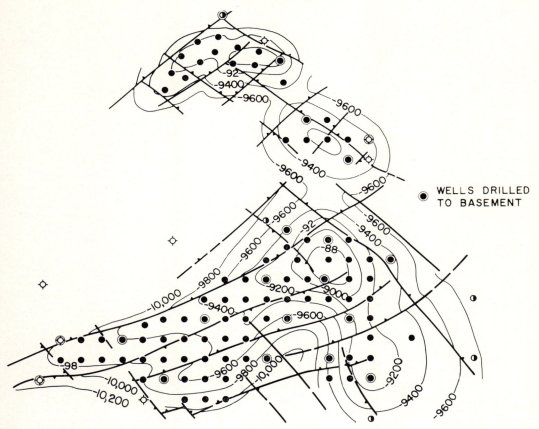

WELLS DRILLED
TO BASEMENT

FIG. 25.—Basement contours, based on information from wells drilled to basement, incorporated with seismic refraction data. Wells with double circles were drilled to basement. CI = 200 ft.

the holes in which the entire pay section was cored. Figure 28 is a representative segment of about 85 ft of the coregraph of this well. The average reservoir porosity is about 19 percent, the permeability several hundred millidarcys, and the oil saturation 18–20 percent.

This 85-ft coregraph section was selected out of the total 240-ft pay section to show basal member 5 sandstone and upper member 4 sandstone; member 4 is the most prolific producer. Porosity and permeability values are somewhat reduced in members 3 and 5, although the clean sandstone streaks in the upper part of member 5 have all the best properties of the member 4 sandstone.

PRODUCTION AND RESERVES

During field development no location was drilled intentionally in which less than 100 ft of gross oil column was expected. No wells, except fluid- or pressure-observation wells, were

perforated and opened to production any deeper than 50 ft above the oil-water contact. However, deeper perforations in some wells were being considered in early 1968. The −8,460-ft oil-water contact is common throughout the main segment of the field.

Figure 29 shows gathering lines, trunk lines, and main pipeline. The southeastern area wells, with the dashed gathering-line symbols, were not tied into a gathering center during the first 1½ years of production, but now are being connected. The oil column in this southeast section is mainly in member 5 sandstone and therefore is not of the high production quality of the rest of the field. These wells and undrilled locations are not uneconomic, but during the early days better wells were supplying the necessary oil volume. Some of the southeast area wells yield 3,000–7,000 bbl/day, and others produce less. They are expected to be in the good water-drive area.

Fig. 27.—Isopach of gross oil-bearing sandstone above water. CI = 50 ft.

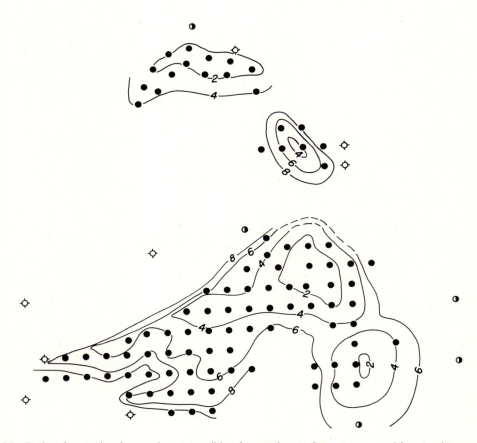

Fig. 26.—Ratio of water-bearing sandstone to oil-bearing sandstone. Strongest water drive, *i.e.* least pressure decline is in areas of highest ratio. CI = 2.

Fig. 28.—Coregraph of 85 ft of well C-54 (*see* Fig. 22), Sarir field. Shows properties of member 4, Sarir Sandstone.

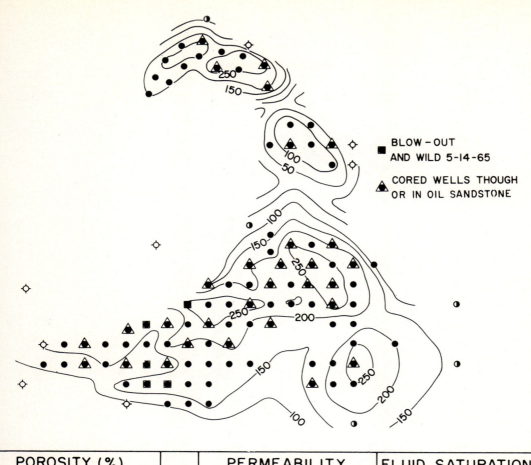

BLOW-OUT
AND WILD 5-14-65

CORED WELLS THOUGH
OR IN OIL SANDSTONE

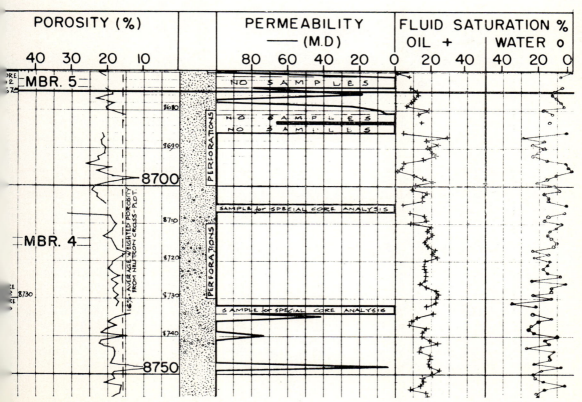

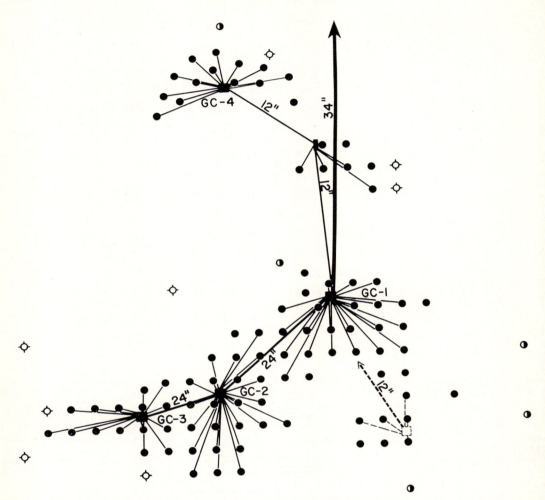

Fig. 29.—Gathering lines, center, and pipelines, Sarir field, Libya. Dashed lines are systems under construction.

For comparison of individual well potentials across the field, all Sarir wells were gauged on chokes of various sizes and thus registered various wellhead flowing pressures and volumes. Individual graphs were plotted for each well to show its producing capacity pulled down to a steady 650-psi wellhead flowing pressure. Average wellhead shut-in pressures were 970–1,000 psi before the field went on production. A total of 36 wells in the field produces 7,000–15,000 bbl/day at 650 psi, 33 produce 1,000–7,000 bbl/day, and 31 have less than the 1,000 bbl/day capability. Only 5 of 105 wells drilled—less than 5 percent—were dry holes without any oil column. Overall, the oil is a fairly light, waxy crude with a mean API

gravity of 37°, wax content of 19 percent, and a uniformly low sulfur content of less than 0.25 percent. The pour point is high and ranges from 55 to 75°F.

Possibly because of two-stage migration of oils into the structure, there are also two different gravity oils present within the formation. A low-gravity zone of tar mat is present at the base of the reservoir near the oil-water level. The tar mat is variable in thickness, but averages about 10–30 ft where present on the east side. The properties of the tar mat have not been investigated fully but samples of very viscous oil obtained from drill-stem tests at the top of the zone show it to have a mean gravity of 24.5° API, a pour point of about 160°F, a

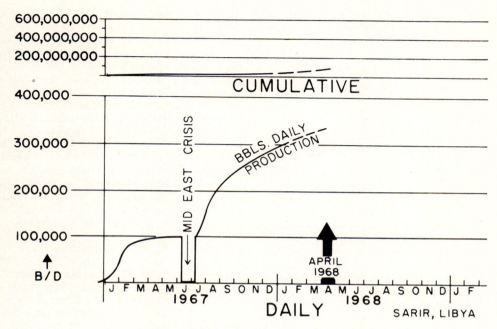

FIG. 30.—Daily production rate of Sarir field, 1967 and 1968.

wax content of about 15 percent, and an as-
phaltene content of 14–22 percent.

Most of the marginally economic wells have
been completed as reservoir-observation wells.
The field production engineers are observing

daily, weekly, or monthly, water-level rise or
fall in off-structure nonproducers, bottom-hole
shut-in and flowing pressures, wellhead shut-in
and flowing pressures, and other factors.

After testing of the gathering lines, centers,

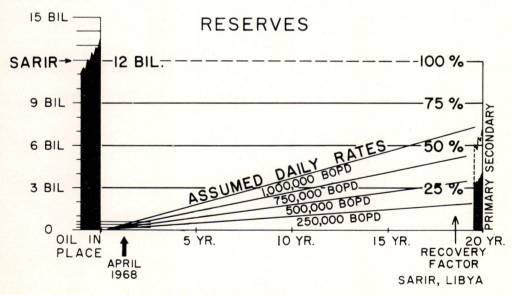

FIG. 31.—Reserves of Sarir field estimated at different percentages of recovery

Table 1. Estimated Total Potential of Completed Wells, Sarir Field

Part of Field	No. Wells	Potential BOPD	
Main Sarir			
GC-1	21	132,750	
GC-2	22	148,450	
GC-3	15	55,950	
Subtotal			337,150
Sarir North	6		9,550
"L-area"	12		83,900
Total			430,600

and main pipelines, the system went on production in December 1966. At that time, under calculated 650-psi flowing pressures, the total possible potential of the completed wells was estimated to be 430,600 bbl/day, divided as shown in Table 1.

The field installations, centers, and 34-in. pipeline actually went on production at 100,000 bbl/day; however the system had a capability at that time of about 350,000 bbl/day. The installation of additional booster stations along the line, plus separator capacity in the field, could increase the system capacity to nearly 1 million bbl daily.

Field production climbed steadily during 1967 to the 320,000–bbl/day level. Figure 30 shows the daily production rate from initiation through February 1968, with March and April estimated at about 325,000 bbl/day.[3] The June and early July 1967 Middle East war period is reflected in total shutdown. The upper graph shows cumulative production to April 1968 of 100 million bbl. This amount is the ultimate re-

[3] Field production in January 1970 is 410,000 bbl/day.

covery of what is called a giant field in the United States.

Oil reserves and oil recoveries as they now are estimated are shown in Figure 31. The four small, closely spaced lines in the lower left corner are the 0, 2, 4, and 6 hundred million increments of Figure 30.

Oil in place at Sarir is estimated to be 11–13 billion bbl. Primary-recovery methods should produce an estimated 25–27 percent of 12 billion, or slightly more than 3 billion bbl. Improved primary-recovery methods, coupled with secondary recovery, could push the total recovery to about 54 percent, or more than 6 billion bbl. This amount is about comparable with, or perhaps a few hundred million barrels more than, the estimated ultimate recovery of the East Texas field.

In Figure 31 some assumed recovery rates are traced. A production rate of 500,000 bbl/day would recover the estimated 27 percent or 3 billion bbl of oil in about 18½ years. Increased production rates would recover the oil sooner, and decreased production rates would extend the time.

SELECTED REFERENCES

Barakat, M. G., 1960, Major tectonic elements of Matzen oil field, Vienna basin: Jour. Geology United Arab Republic, v. 4, p. 19–25.

British Petroleum Co., and Hunt International Petroleum Co., unpublished reports.

Conant, L. C., and G. H. Goudarzi, 1967, Stratigraphic and tectonic framework of Libya: Am. Assoc. Petroleum Geologists Bull., v. 51, no. 5, p. 719–730.

Hardin, F. R., and G. C. Hardin, Jr., 1961, Contemporaneous normal faults of Gulf Coast and their relation to flexures: Am. Assoc. Petroleum Geologists Bull., v. 45, no. 2, p. 238–248.

Ocamb, R. D., 1961, Growth faults of South Louisiana: Gulf Coast Assoc. Geol. Socs. Trans., v. 11, p. 139–176.

Pomeyrol, R., 1968, "Nubian Sandstone": Am. Assoc. Petroleum Geologists Bull., v. 52, no. 4, p. 589–600.

Cambrian Oil Field of Hassi Messaoud, Algeria[1]

A. BALDUCCHI[2] and G. POMMIER[3]

Algiers, Algeria, and Neuilly-sur-Seine, France

Abstract Hassi Messaoud field, in the eastern part of the Algerian Sahara Desert and 350 mi (560 km) from the Mediterranean, produces 375,000 bbl/day, or about 40 percent of Algeria's production. The field is operated by SN REPAL and CFP(A).

Oil is produced from a Cambrian sandstone reservoir in a large dome. Producing depth is about 11,000 ft (3,300 m), the oil column is about 900 ft (270 m), and the productive area is 500 mi² (1,300 km²).

A total of 178 wells had been drilled to the end of 1967, 39 of which were dry or produced water, and 30 of which were producing at a rate greater than 6,000 bbl/day. In-place reserves throughout the Hassi Messaoud field are approximately 25 billion bbl; cumulative production to the end of 1967 reached 720 million bbl.

Reservoir porosity ranges from 2 to 12 percent, with an average of 8 percent; permeability ranges from 0 to 1,000 md. Accumulation is in a combination trap. On the flank, closure is structural; on the crest, it is associated with an unconformity at the base of the Triassic. The structure was formed by a late phase of Hercynian (late Paleozoic) orogeny followed by erosion that cut to the Cambrian on the crest of the field.

INTRODUCTION

The first oil in Algeria was discovered in 1956 at Edjeleh by Compagnie de Recherche et d'Exploitation des Pétroles (CREPS). By 1959 Algeria had become the tenth largest oil-producing country in the world, with an average daily production in 1967 of approximately 830,000 bbl of oil. Hassi Messaoud field contributes about 40 percent of this total.

Hassi Messaoud field is in the eastern part of the Algerian Sahara Desert (Fig. 1). It is divided into two concessions—Hassi Messaoud North and Hassi Messaoud South—which are owned jointly by Société Nationale de Recherche et d'Exploitation de Pétrole en Algérie (SN

[1] Read before the 53d Annual Meeting of the Association, Oklahoma City, Oklahoma, April 25, 1968. Manuscript received, April 27, 1968; accepted, June 17, 1968.

[2] Société Nationale de Recherche et d'Exploitation des Pétroles en Algérie (SN REPAL).

[3] Compagnie Française des Pétroles (Algérie) (CFP [A])

This paper has been compiled from published and unpublished reports written by the exploration staffs of SN REPAL and CFP(A) and their consultants. The writers thank SN REPAL and CFP(A) for permission to publish this paper.

All figures given in this paper are those available in 1967.

REPAL) and Compagnie Française des Pétroles (Algérie) (CFP[A]). The companies have approximately an equal interest in each concession. The south concession is operated by SN REPAL, discoverer of the field. SN REPAL is owned half by the Algerian government and half by the French government. The north concession is operated by CFP(A), a subsidiary of Compagnie Française des Pétroles (CFP).

Hassi Messaoud is about 350 mi (560 km) from the nearest point on the Mediterranean Coast (Fig. 1). Oil is at a depth of about 11,000 ft (3,300 m) in a sandstone reservoir of Cambrian age. The accumulation is trapped in a large dome partly eroded on the top. The field has an area of 500 mi² (1,300 km²), and an oil column of 900 ft (270 m).

GEOLOGIC OUTLINE

A vast sedimentary zone crosses northern Africa between parallels 24 and 36° N lat., from the Atlantic to the Red Sea. The Algerian Sahara forms most of the western section of this region, bounded on the north by the mountains of the High Atlas and the Saharan Atlas, and on the south by the Eglab Ridge and the Hoggar Range (Fig. 1).

Before oil exploration, geologic knowledge of the Algerian Sahara indicated that the Paleozoic outcrops of the northern Hoggar continued northward in the subsurface into a large Paleozoic basin. However, the main features of Saharan geology, mapped in the course of 15 years of exploration work for petroleum, show little relation to what was anticipated before drilling began.

The present model of the northern Sahara craton shows a series of resistant ridges oriented approximately north-south and separating the Sahara into several basins, each fairly complex. The northeastern Algerian Sahara (Fig. 2), containing the Hassi Messaoud field, is characterized by a vast platform which acquired its individual character very early and retained it during nearly the entire Paleozoic Era. Numerous unconformities in the Cambro-Ordovician are evidence that the initial uplift of this platform took place during the Sardic phase of the Caledonian orogeny. However, be-

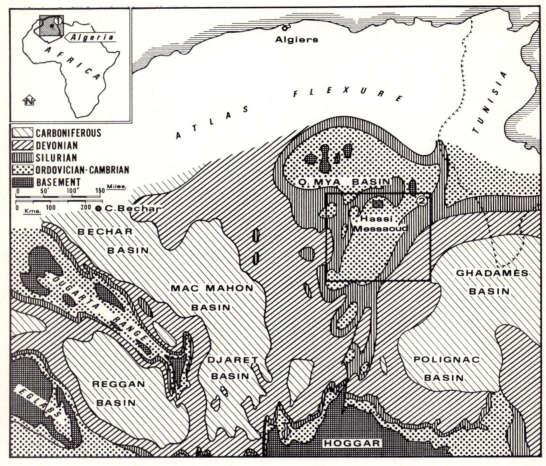

Fig. 1.—Outcrop and subcrop map of Paleozoic in northern Sahara (Algeria and Tunisia). Shows location of Hassi Messaoud. Outlined rectangle is area of Figure 2.

cause the youngest Paleozoic rocks drilled are Late Silurian, the principal uplift of the platform must have taken place during the Devonian (final phase of Caledonian movements) and ended with late Paleozoic (Hercynian) orogeny. This uplift was followed by intense erosion. A late phase of the Hercynian orogeny, at the beginning of the Mesozoic, was characterized by fracturing, extrusion of andesitic lava across large areas, and the formation of structural lows in which Permo-Triassic sands were deposited. The Messaoud-El Agreb high, already a distinct feature, persisted and was covered by a thin layer of Permo-Triassic deposits.

Subsidence of the northern Algerian Sahara

began early in the Mesozoic and continued throughout the era. This subsidence resulted in burial of the Messaoud-El Agreb high beneath the 10,000 ft (3,000 m) of Mesozoic sediments which now overlie the Hassi Messaoud structure. Figure 3 shows the present structural configuration of the top of the Paleozoic.

HISTORY OF EXPLORATION AND DEVELOPMENT

Even though no oil shows had been reported, the great Saharan sedimentary basin, by its very nature, looked sufficiently promising and attractive to launch an oil exploration program. In 1953, the Algerian government granted oil-prospecting licenses on an area of about

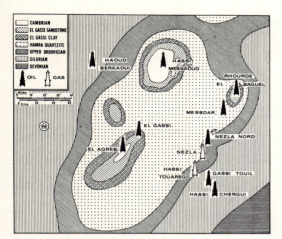

FIG. 2.—Subcrop map of Paleozoic of Oued Mya basin, eastern Algeria.

100,000 mi² (250,000 km²) in the northern Sahara to SN REPAL and CFP(A); the companies then entered into a partnership agreement.

The surface of the concession area is a plain covered by gravel and sand, and in part by sand dunes; there are no significant outcrops. Exploration began with a general gravimeter survey, which gave poor results. After the failure of many seismic-reflection tests, it was decided to carry out an extensive reconnaissance refraction survey and to drill deep wells to evaluate the sedimentary section. The first wells

were drilled near the oases of Berriane, El Golea, and Touggourt, to depths of about 10,000 ft (3,000 m). The refraction survey showed the main structural trends of the basement and Figure 4 is the refraction map from which the location of the discovery well, Md No. 1, was chosen by SN REPAL. The line OE (Fig. 4) was about 100 mi (160 km) long. About 1,000 lb (500 kg) of dynamite was exploded at each shot point; the distance between shot points and the geophones was about 15 mi (23 km). The well, on an anticlinal trend, discovered oil in May 1956 in a sandstone reservoir at a depth of 10,900 ft (3,270 m).

New refraction lines were shot concurrently with the drilling of the Md No. 1 well, and showed a large gently dipping anticline. The boundary between the CFP(A) and SN REPAL prospecting licenses is in the middle of this structure.

The next two wells, Om No. 1, 5 mi (8 km) north, and Md No. 2, 7 mi (11 km) southwest, were productive.

After a new reflection survey, it was realized that there is a major unconformity between the Mesozoic refraction horizon and the basement refraction horizon (Fig. 5); the refraction map showed the shape of the Hassi Messaoud field more clearly.

Since its discovery, Hassi Messaoud field has been under continuous development with as many as 10 rigs active. A permanent pipeline from Hassi Messaoud to the terminal of Bougie

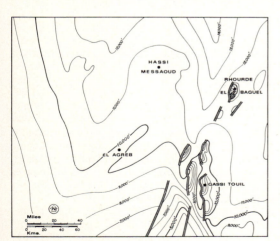

FIG. 3.—Structural contour map. Datum is top of Paleozoic. Contour elevations below sea level. CI = 1,000 ft.

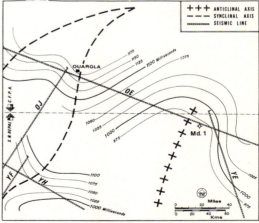

FIG. 4.—First seismic refraction map of Hassi-Messaoud area. Datum is 1,500 ft above sea level. CI = 25 msec. OE, OJ, YE, YF, and YH are seismic lines.

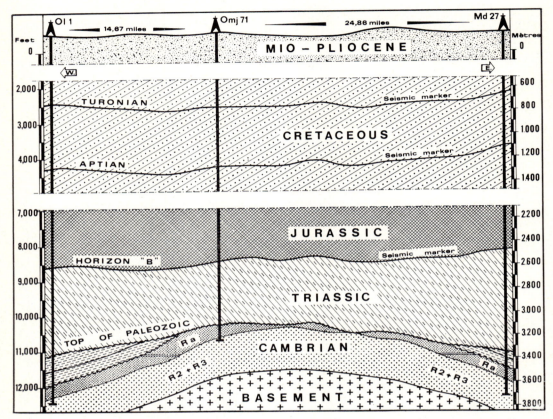

FIG. 5.—West-east cross section of Hassi Messaoud. Location of cross section is shown on Figure 11.

on the Mediterranean Coast was completed in November 1959.

STRATIGRAPHY (FIG. 6)

Basement.—Wells Md No. 2 and Om No. 81 were drilled to basement. The basement rock is pink granite porphyry, dated as Early Cambrian by Mobil Field Research Laboratory, which determined its age to be 560 ± 25 million years by the Rb/Sr method.

Cambrian.—Sandstones of Cambrian age have been divided into three zones from base to top: R3, R2, and Ra. Their petrographic characteristics are listed in Figure 7.

The R3 is 900 ft (270 m) thick near the top of the structure. It is shaly sandstone, poorly bedded, with some cross-bedding. It is composed of about 60 percent quartz and 30 percent clay, mostly illite. The median grain size is 800 μ, and the grains are angular and very poorly sorted.

The R2 is of rather uniform thickness throughout the field and consists of well-bedded, clay-cemented sandstone and siltstone. The upper R2 is 110 ft (34 m) thick and is productive locally.

The Ra is the major reservoir of the Hassi Messaoud field. It is more than 400 ft (120 m) thick on the flanks of the structure, but has been eroded deeply on the crest. It is composed of about 70 percent quartz, about 14 percent secondary silica, and 8 percent clay; in some wells the clay is mostly kaolinite. The grain size is variable, but the median is about 350 μ. Grains are rounded to subrounded.

Figure 8 is a typical thin section of Ra sandstone. Figures 9 and 10 show large outcrops of Cambrian sandstone about 500 mi (800 km) south of Hassi Messaoud field in Tassili des Ajjers. The sandstone is very similar to that in the cores from Ra sandstone.

These sandstones have large-scale cross-stratified units, separated by numerous discontinuous layers of siltstone. The cross-bedding has a unimodal orientation between north-northeast and north-northwest.

Ordovician.—The Ra is overlain unconformably by the El Gassi Sandstone. This formation is completely eroded across most of the field but reaches a thickness of 200 ft (60 m) on the flanks. The grains are small, rounded, and well sorted. The El Gassi Clay is present only in a few wells on the flanks of the field. It is dated by graptolites of Tremadocian (earliest Ordovician) age. Younger Paleozoic rocks are not present in the Hassi Messaoud field. However, Late Silurian black shale and associated rocks are present about 25 mi (40 km) northwest of the field. Devonian strata are even farther distant (Fig. 2).

Mesozoic (Figs. 5, 6).—Paleozoic rocks are overlain unconformably by a transgressive series of Triassic age, divided into two units. The lower unit consists of alternating beds of sandstone, brown or red shale, and salt. The thickness is variable because the Triassic fills in the eroded surface of the Paleozoic subcrop. The Triassic transgression came from the north, and only the upper beds of the Triassic are present across all the field. The upper or salt unit is about 1,500 ft (450 m) thick. At the base is a massive salt bed, overlain by interbedded dolomite, anhydrite, and salt.

The Jurassic is 2,800 ft (820 m) thick. The lower part is a massive bed of anhydrite, whereas the middle consists of 600 ft of evaporites at the base overlain by shale, and the upper part is interbedded limestone, dolomite,

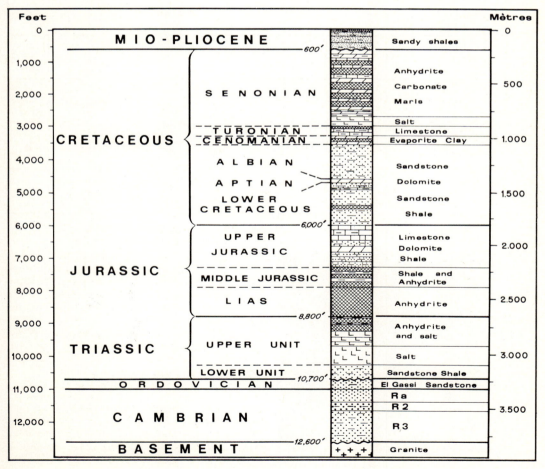

Fig. 6.—Stratigraphic columnar section, Hassi Messaoud, Algeria.

AGE	ZONE	THICKNESS IN FEET	QUARTZ %	SECONDARY SILICA %	CLAY %	TYPE OF CLAY	OTHER COMPONENTS	MEDIAN GRAIN SIZE	MORPHOSCOPY	MEAN POROSITY %	PERMEABILITY Md
CAMBRIAN	Purple Sandstone		50		40 ↘ 30		Carbonate Iron oxide	(Micron) 60 ↘ 170	Angular		0
CAMBRIAN	R₃	900 ↘ 1200	55 ↘ 60	2 ↘ 5	35 ↘ 25	Illite	Siderite Pyrite	800 ↘ 500	Angular		0
CAMBRIAN	R₂ Lower	170'	60	5	30	Illite		450	Sub Angular		0
CAMBRIAN	R₂ Upper	110'	65	9	15	Illite Kaolinite		400	Sub Rounded	11	0 to 100
CAMBRIAN	Rₐ	400'	70	14	8	Kaolinite		350	Rounded	8	0 to 100
ORDOVICIAN	EL GASSI Sandstone	200'	70	20	6		Anhydrite Dolomite	300	Rounded		0
ORDOVICIAN	EL GASSI Clay	300'	50	15	20	Illite	Carbonate Glauconite	300 400	Angular		

Fig. 7.—Properties and parameters of Paleozoic sandstone, Hassi Messaoud field.

Fig. 8.—Thin section of Ra sandstone. Polarized light, 17.5 ×.

marl, and shale. The Cretaceous is 1,500 ft (450 m) thick and changes from sandstone and shale at the base to limestone and evaporite at the top. The Albian sandstone is the main reservoir of fresh artesian water in the north Sahara.

Mio-Pliocene sandstone and shale cover the partly eroded Senonian.

GEOMETRY OF BOUNDARIES OF Rᴀ RESERVOIR

The Hassi Messaoud oil accumulation is contained in a combination trap. On the flank the closure is structural; on the crest it is associated with the Triassic unconformity. The accumulation has an area of 500 mi² (1,300 km²). The geometry of the reservoir boundaries can be

FIG. 9.—Outcrops of Cambrian sandstone, Tassili des Ajjers, NE of Hoggar on Figure 1.

FIG. 10.—Outcrops of Cambrian sandstone, Tassili des Ajjers, NE of Hoggar on Figure 1.

described by defining the lower and upper surfaces of the Ra sandstone and the oil-water contact.

The lower surface of the Ra sandstone or the top of R2 (Fig. 11) is a structurally deformed sedimentary surface. It is a gently dipping, broad, elliptical dome, 27 mi (43 km) long and 20 mi (30 km) wide. The principal axis

trends north-northeast. Average dips range from 40 ft/mi (7.5 m/km) near the center of the dome to 165 ft/mi (30 m/km) on its flanks. The domal structure is modified locally by elongate structural lows (Fig. 11) believed to be associated with faulting.

In contrast to the lower surface, which is relatively simple, the upper surface of the reser-

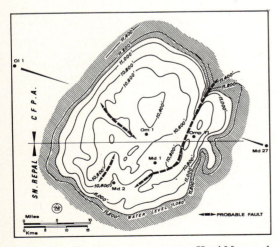

FIG. 11.—Structural contour map, Hassi-Messaoud field. Structural datum is top of R2. Elevations below sea level. CI = 200 ft. Location of cross section (Fig. 5) is shown by line Ol 1–Md 27.

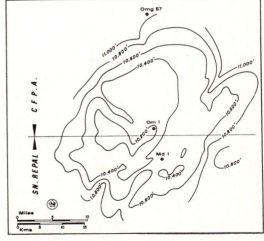

FIG. 12.—Structural contour map. Datum is top of Ra. Elevations below sea level. CI = 200 ft.

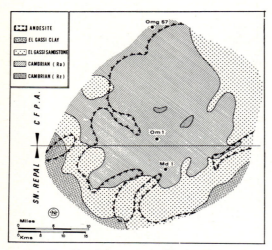

FIG. 13.—Subcrop map of Paleozoic at
Hassi Messaoud field, Algeria.

voir (Fig. 12) is much more complicated and
is the result of the following succession of
events.

At the end of Carboniferous time erosion removed most of the Paleozoic section from the
Hassi Messaoud structure (Fig. 13). The Ra
sandstone was eroded completely in two places
in the center of the structure and the El Gassi
Sandstone appears only on the flanks and in
some structural lows.

Late in the erosion cycle, the Paleozoic surface was rejuvenated with entrenchment of
streams. Andesite flows, which are widespread
regionally, spread over the flanks of the structure and into its deep valleys (Fig. 13). The
good correlations between structural lows on
the top of R2 and stream valleys on the erosion
surface justify the conclusion that the position
of the streams was determined by structural
lows on the dome.

Figure 14 is an isopach map of the Ra sandstone. Its complexities reflect the results of
three factors—the original thickness of Ra, the
truncation of the dome by erosion, and the entrenchment of the valleys. The two last factors
are the most important.

The oil-water contact was found in two wells
at a depth of 11,350 ft (3,440 m). Thickness
of the transition zone within the Ra is about
150 ft (45 m).

RESERVOIRS

The Ra reservoir is probably a single com-

municating unit through most of the field. Reservoir porosity ranges from 2 to 12 percent
(Fig. 15) with an average of 8 percent, and
permeability ranges from zero to 1,000 md.
Horizontal variations of permeability between
wells in the Ra sandstone are much greater
than vertical variations within a well. The present distribution of porosity and permeability is
irregular and appears to depend largely on
lithologic characteristics. The best zones have
large median grain size, small clay content, and
less than 15 percent secondary silica. These
zones are called "drains." Figure 16, which is
much simplified, shows the permeability distribution along a west-east section through the
field.

SOURCE ROCKS AND MIGRATION (FIG. 17)

Possible source rock for the Hassi Messaoud
crude is the overlying Triassic shale or any of
the Paleozoic black shale formations. The organic content of the Triassic red shale is too
low to be measured by the analytic techniques
available to the writers. However, two zones in
the Late Silurian black shale have a high organic carbon content. Silurian shale zones in
the area northwest of the field, tested by pyrol-
ysis, proved to be rich in kerogen and therefore
a likely source for oil and gas. These beds are
also rich in well-preserved fossil spores and organic debris. Furthermore, chemical analysis of
the Hassi Messaoud oil shows a high degree of

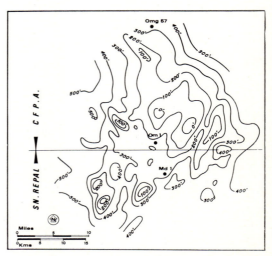

FIG. 14.—Isopach map of interval top Ra–top R2,
Hassi Messaoud field. CI = 100 ft.

maturation, which is characteristic of Paleozoic crudes.

These facts lead to the conclusion that the major source rocks for Hassi Messaoud crude are black kerogenous shale of Silurian age. If so, much of the Silurian-generated oil must have been lost during pre-Triassic erosion. It must be inferred therefore that, after the deposition of impermeable Triassic shale and evaporites, compaction during and after the deposition of Mesozoic sediments caused the Silurian shale to yield a second generation of hydrocarbons. These migrated beneath the unconformity surface until they reached the Cambrian reservoir subcrop, which they filled by downward migration.

HYDROCARBONS

There is no gas cap in the Hassi Messaoud field and the oil is highly undersaturated. Some geographic differentiations of the crude are evident. Overall it is an intermediate-base crude with a mean gravity of 43° API, low viscosity, and low sulfur content. It is very rich in gasoline. The initial reservoir pressure and temperature were 6,850 psi and 270°F (132°C), at a subsea depth of 10,500 ft (3,180 m).

OIL IN PLACE AND RECOVERABLE RESERVES

Probable reserves in place throughout the Hassi Messaoud field are approximately 25 billion bbl. Reserves recoverable by liquid expansion above the bubble-point pressure are ex-

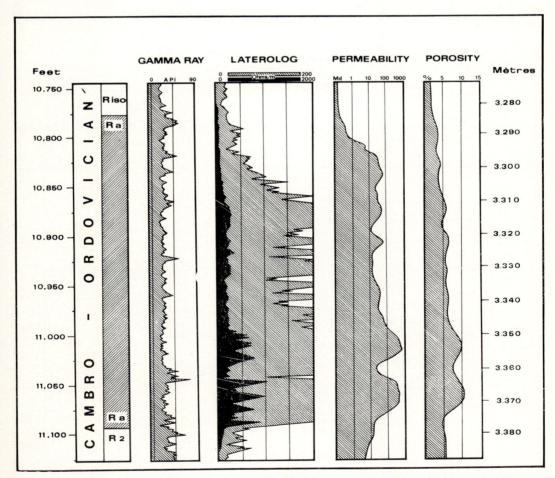

FIG. 15.—Composite log, Hassi Messaoud, showing gamma-ray, laterolog, permeability, and porosity characteristics in Cambro-Ordovician section.

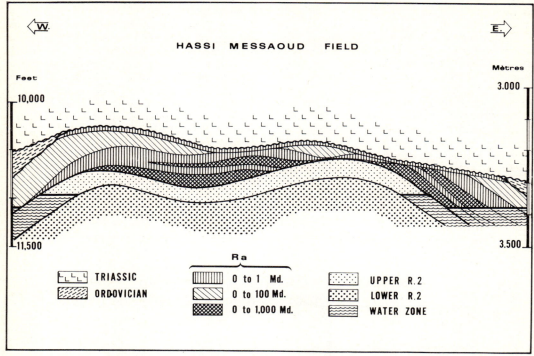

FIG. 16.—Distribution of permeability, west to east across Hassi Messaoud field.

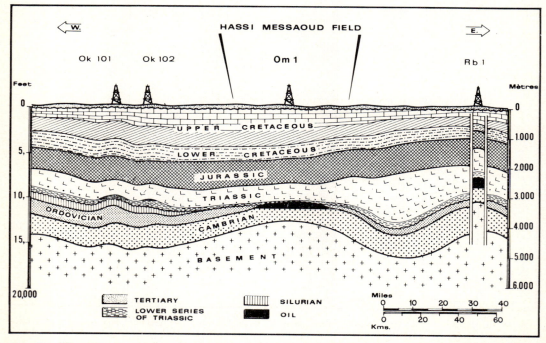

FIG. 17.—West-east cross section of Oued Mya basin, from Haoud Berkaoui to Rhourde el Baguel (Fig. 2).

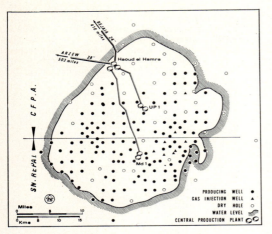

FIG. 18.—Hassi Messaoud field, Algeria.

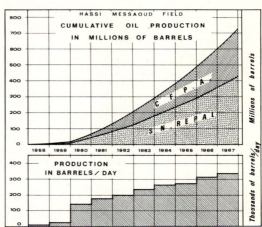

FIG. 19.—Production of Hassi Messaoud field, Algeria.

pected to be approximately 10 percent of the in-place reserves.

To test the possibility of increasing the recoverable reserves, two pilot compression plants have been built which reinject 175 million ft³ of gas a day through eight wells in selected parts of the field. Alternate injection of gas and water also is being tried.

PRODUCTION

By the end of 1967, 178 wells had been drilled in the field (Fig. 18); of these, 39 were practically dry or produced water, and 30 were producing at a rate greater than 6,000 bbl/day. Daily production is 375,000 bbl.

Cumulative production (Fig. 19) to the end of 1967 reached 720 million bbl, of which 60 percent has been produced in the Hassi Messaoud South concession.

SELECTED REFERENCES

Assoc. Française Tec. Pétrole, 1963, Aperçu sur les progrès de la géologie saharienne: Spec. vol. in memory of J. Follot, Paris, 72 p.

Bouchon, R., 1954, L'application de la sismique réfraction au Sahara: 2d Natl. Pétrole Cong. Compte Rendu, Assoc. Français Tec. Pétrole, p. 61–69.

——— C. de Lapparent, H. I. Ortynski, and G. Pommier, 1959, Le développement de la sismique réfraction dans l'interprétation géologique du Sahara nord; son rôle dans la découverte et l'étude du champ d'Hassi Messaoud: 5th World Petroleum Cong. Proc., sec. 1, p. 729–746.

Byramjee, R., and L. Vasse, 1969, Geological interpretation of Libyan and north-Saharan crude oil analyses: Amsterdam, 4th Internat. Mtg. on Organic Geochemistry Proc., in Advances in Organic Geochemistry 1968, P. A. Schenck and I. Havenaar, eds.: Oxford, Pergamon Press.

Delclaud, Ch., H. I. Ortynski, and A. Perrodon, 1957, Les découvertes de Hassi-Messaoud et de Hassi-R'Mel dans le Sahara septentrional: 3d Natl. Pétrole Cong. Compte Rendu, Assoc. Français Tec. Pétrole, p. 75–85.

——— and J. Leduc, 1965, Premiers résultats de l'injection de gaz à Hassi-Messaoud: 2d Colloque Assoc. Recherche Tech. Forage Prod., Éd. Technip, p. 427–434.

Fondeur, C., 1964, Etude pétrographique détaillée d'un grès à structure en feuillets: Rev. Inst. Française Pétrole, v. 19, p. 901–920.

Guillemot, A., P. Michel, and D. Trumpy, 1955, La prospection du paléozoïque au nord du Hoggar (Sahara): 4th World Petroleum Cong. Proc., sec. 1, p. 243–249.

Jocquel, E., 1961, Engineering the Hassi-Messaoud field: Petroleum Engineering, Sept., p. 39.

Kulbicki, G., and G. Millot, 1960, L'évolution de la fraction argileuse des grès pétroliers cambro-ordoviciens du Sahara Central: Alsace-Lorraine Service Carte Géol. Bull., t. 13, p. 147–156.

Lapparent, C. de, and P. Albert, 1960, Hassi-Messaoud: Le géant du Sahara: World Petroleum, v. 31, no. 6, p. 43–46.

——— H. I. Ortynski, and A. Perrodon, 1959, Esquisse paléogéographique et structurale des bassins du Sahara septentrional: 5th World Petroleum Cong. Compte Rendu, sec. 1, p. 705–727.

Legrand, Ph., and G. Nabos, 1962, Contribution à la stratigraphie du Cambro-Ordovicien dans le bassin saharien oriental: Soc. Géol. France Bull., 7th sér., t. 4, p. 123–131.

Malenfer, J., and A. Tillous, 1963, Etude du champ de Hassi-Messaoud: stratigraphie, aspect structural, étude de détail du réservoir: Rev. Inst. Française Pétrole, v. 18, no. 6, p. 851–867.

Migaux, L., 1960, Sismique réfraction au Sahara: Rev. Inst. Française Pétrole, v. 15, no. 10, p. 1371–1383.

Montadert, L., 1963, Le sédimentologie et l'étude détaillée des hétérogénéités d'un réservoir. Application au gisement de Hassi-Messaoud, in Colloquium of Assoc. Recherche Tech. Forage Prod., June, no. 13: Paris, Éd. Technip, 16 p.

Ortynski, H. I., 1958, La découverte du Champ de

Hassi-Messaoud: Rev. Inst. Française Pétrole, v. 13, no. 6, p. 944–949.

Perrodon, M. A., 1958, Communication sur Hassi-Messaoud et sur Hassi-R'Mel: Assoc. Française Tec. Pétrole Bull., no. 129.

———— and C. Tempere, 1960, Oil exploration in the Sahara: Petroleum (Gr. Brit.), Nov., p. 419–424.

Pommier, G., 1957, La sismique réflexion dans le Sa-hara Nord: 3d Natl. Pétrole Cong. Compte Rendu, Assoc. Français Tec. Pétrole, p. 53–61.

SN REPAL Direction Production, 1962, Lucky well is Hassi-Messaoud: World Petroleum, Feb., p. 26–28.

Tenaille, M., 1954, Problemes et évolution de la recherche du pétrole en Algérie: 2d Natl. Pétrole Cong. Compte Rendu Assoc. Française Tec. Pétrole, p. 29–38.

Triassic Gas Field of Hassi er R'Mel, Algeria[1]

PHILIPPE R. MAGLOIRE[2]

Algiers, Algeria

Abstract After the first important Saharan stratigraphic test (Berriane) had shown encouraging results, SN REPAL intensified its reconnaissance studies. Seismic refraction and reflection, combined with field geology, located the structure of Hassi er R'Mel, about 60 mi (100 km) southwest of Laghouat oasis. In 1956 the HR-1 well was spudded, and led to the discovery of the gas field of Hassi er R'Mel.

Located on the Cretaceous high zone of the M'zab area, the structure of Hassi er R'Mel is part of a zone which has been stable tectonically since the Cambrian. Above the granitic basement are Cambrian and Ordovician strata, which are covered by Silurian strata where pre-Triassic erosion was not deep.

The oldest Mesozoic rocks unconformably overlying the Paleozoic are Triassic sandstone. A lower sequence (with andesite flows) fills the erosional topography of the pre-Triassic erosion surface. The top of this sequence is the C reservoir, which shows important lateral facies variations. It is overlain by two separate sandstone reservoir zones, B (noncontinuous) and A (continuous). Above the reservoirs are the salt-bearing Triassic (1,300 ft thick), Jurassic (3,000–3,300 ft), and Cretaceous.

The structure at the top of the Triassic reservoir is anticlinal and has a NNE-SSW axis; its areal extent is about 1,000 mi^2 (2,600 km^2).

An oil-water contact is at 4,936 ft (1,500 m) below sea level. Oil shows and minor production have been found in the Ordovician quartzite and Triassic sandstone. Possibly there is a very narrow oil ring but this is unproved. Though the structure is old, evidence shows that the gas was trapped definitely during the Early Cretaceous.

Operated by SEHR (a subsidiary of SN REPAL and CFP [A]), the gas field of Hassi er R'Mel has produced 430 billion ft^3 of condensate gas since 1958. The in-place reserves now are estimated to be 70 trillion ft^3. Production could be increased considerably should the market requirements be increased; because of the limited use for the gas at present, only five wells now are producing.

Introduction

The Algerian Sahara covers part of the sedimentary fringe of the northern part of Africa (Figs. 1, 2). The Saharan sedimentary sequence is contained within it and is bounded on the south by the outcrops of the Saharan shield (Eglabs or Reguibat axis, Ahaggar, and Tibesti); on the north by the South Atlas Range on the southern border of an intercratonic geosynclinal domain later deformed by the Alpine orogeny; on the west by Morocco; and on the east by Tunisia.

[1] From a paper read before the 53rd Annual Meeting of the Association, Oklahoma City, Oklahoma, April 23, 1968. Manuscript received, July 10, 1968; accepted, October 7, 1968.

[2] SN REPAL.

The first stratigraphic reconnaissance drilling in the Sahara was carried out in 1952–1953 by SN REPAL in association with CFP(A). The first well, Berriane No. 1, revealed the presence of a sedimentary sequence about 10,000 ft (3,000 m) thick. Hydrocarbon shows, found beneath a thick salt cover, encouraged the operating company to undertake full reconnaissance work throughout its Saharan mineral rights, which at that time covered a vast area.

Much of the work was concentrated in the region known as Chebka du M'zab, where surface geology and refraction shooting indicated the culmination of the Hassi er R'Mel structure to be south of Tilremt, about 300 mi (480 km) south of Algiers. In 1956, after complementary reflection shooting, Hassi er R'Mel No. 1 was drilled, leading to the discovery of the Hassi er R'Mel gas field.

M'ZAB DORSAL STRUCTURE

The Saharan platform is characterized by several generally north-south-trending positive zones which form the boundaries of basins of different ages and evolutionary histories. The long-stable M'zab dorsal structure, first recog-

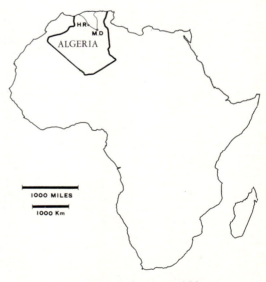

FIG. 1.—Location map, Africa.

nized from, and named after, the great uplift of Cretaceous outcrops composing the Chebka du M'zab, forms the axis separating the two principal basins of the platform. One of these basins, the Oued Mya Mesozoic basin on the east, is underlain principally by Cambro-Ordovician rocks. On the west is the Western Erg Paleozoic basin, in the central part of which, a complete Paleozoic section is present beneath a thin Mesozoic cover. In evolutionary history, the two basins are very different. West of the axis, the Paleozoic rocks plunge beneath the Western Erg basin, and east of it is the boundary of the westerly extension of the Mesozoic sequence of the Oued Mya basin. The dorsal structure also

is connected with the Talemzane Hercynian block, one of the ancient blocks which form a lineament along the northern boundary of the Sahara platform.

Pre-Hercynian orogenic features (*e.g.*, Caledonian) are difficult to identify. Figure 3 is a paleogeologic map showing Hercynian features. It is known only that there are important stratigraphic gaps in the Silurian and Ordovician toward the M'zab and Talemzane, which suggest that the M'zab dorsal structure may have formed during the Caledonian orogeny.

The structure of the Paleozoic of the Hassi er R'Mel region is mainly the result of the Hercynian orogeny. Southwest from the axis of the

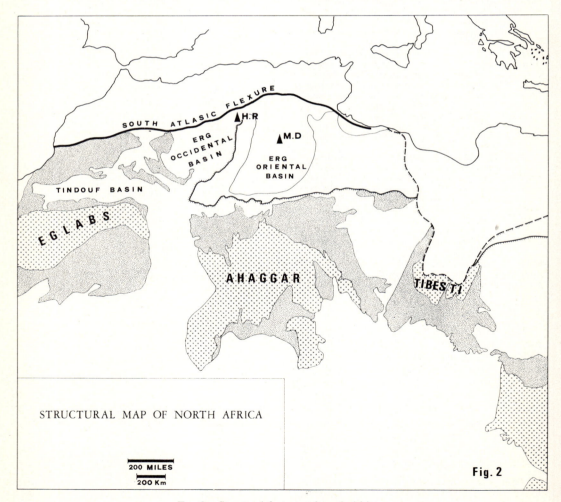

Fig. 2.—Structural features of north Africa.

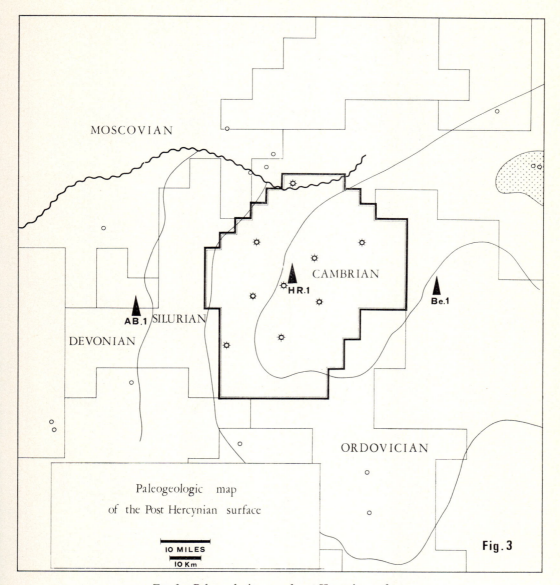

FIG. 3.—Paleogeologic map of post-Hercynian surface.

Talemzane block, the Cambro-Ordovician, Silurian, and Devonian units appear progressively below the transgressive Triassic.

During the Mesozoic the M'zab dorsal structure occupied a marginal position relative to the Oued Mya basin and to the Atlas zone on the north. The section is shown in Figure 4.

STRATIGRAPHY

Paleozoic

The basement consists of Cambrian rhyolite which has been dated radiometrically. Above the basement are several units.

The *internal Tassili Cambro-Ordovician sandstone group* includes, from base to top,

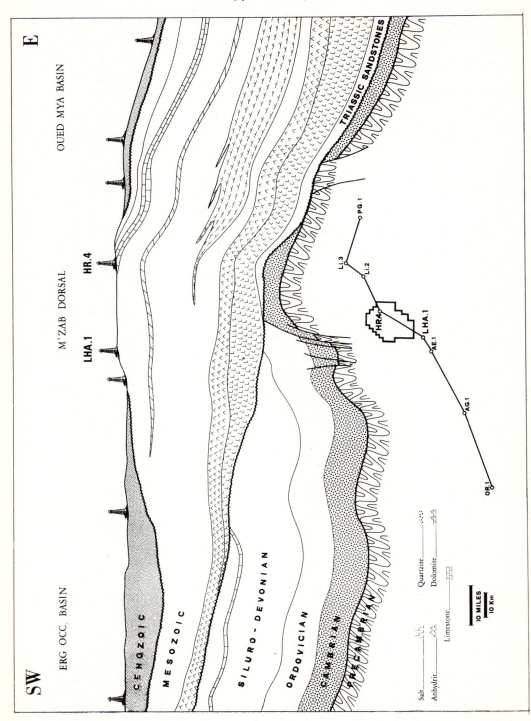

Fig. 4. West-east subsurface cross section from Erg Occidental basin to Oued Mya basin.

(1) the Hassi Merhimine Sandstone, equivalent to the Hassi Messaoud Sandstone (Balducchi and Pommier, this volume), (2) the Harich el Megta Sandstone, (3) the Bordj Nili Sandstone, (4) the El Gassi Shales and Erg el Anngueur Sandstone, (5) a probable unconformity, (6) the Ouargla Sandstone and Harich el Megta Shale, (7) a probable unconformity, and (8) the Hassi el Hadjar periglacial *ensemble* or formation, consisting of microconglomeratic shale overlain by quartzite.

The *Siluro-Devonian* consists of black, commonly carbonaceous shale. Llandoverian strata probably are absent. This unit is eroded deeply at the top.

Carboniferous strata appear in the northern part of the region. They are coastal deposits of Moscovian age, and represent a marine transgression on an unconformity surface developed on the Silurian.

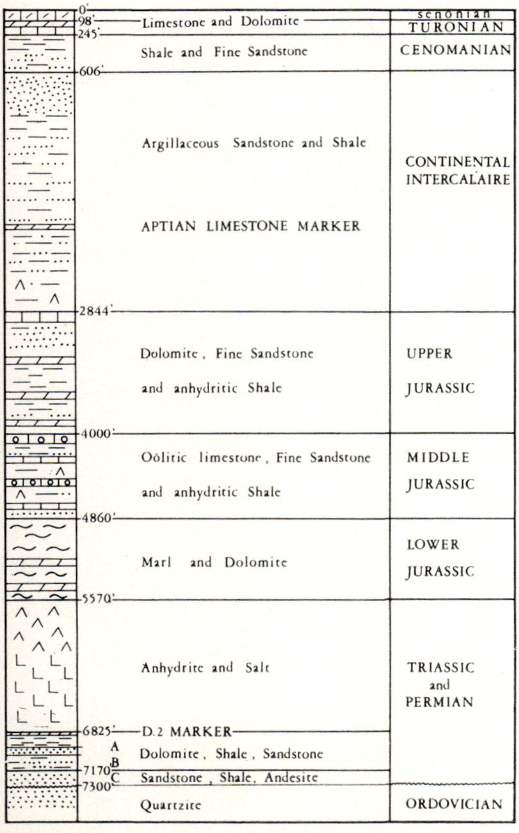

Fig. 4-A.—Stratigraphic section shown by log of H. R. No. 1 well. Depths in feet.

Mesozoic

Mesozoic formations unconformably overlie the Paleozoic. At the base is sandstone intercalated with andesitic flows, filling relief on the post-Hercynian surface. The Hassi er R'Mel reservoirs are in the sandstones of these formations.

Permo-Triassic.—The Permo-Triassic consists of three units. The lowest unit fills the relief developed on the surface after the Hercynian orogeny. Its thickness is greatly varied and in places the unit includes volcanic rocks. The lower part is probably Permian, and ranges in thickness from 0 to 260 ft (0–80 m). Its petroleum potential is nil. The upper part is the "sandstone C stratum" or Hassi er R'Mel lower reservoir. The thickness ranges from 0 to 200 ft (0–60 m).

The middle unit consists of alternating sandstone and shale subdivided into two parts: the "sandstone B stratum" or Hassi er R'Mel intermediate reservoir (thickness 0–70 ft or 0–20 m) and the "sandstone A stratum" or Hassi er R'Mel upper reservoir (thickness 50–120 ft or 15–36 m).

The upper unit includes salty shale and anhydrite having a thickness of 1,000–1,400 ft (300–420 m).

Jurassic.—From base to top, the Jurassic includes Lower Jurassic shale and dolomite (790–1,500 ft or 240–450 m thick); Middle Jurassic limestone and interbedded shale (820 ft or 250 m thick); and Upper Jurassic shale and sandstone with dolomite interbeds (1,200–1,300 ft or 360–400 m thick).

Cretaceous.—The basal Cretaceous is made up of the "Continental Intercalary" (Continental Intercalaire), a 2,300-ft (700 m) unit of multicolored shale and sandstone. Above it is Cenomanian shale and limestone (400 ft or 130 m) and Turonian dolomite (140 ft or 43 m).

STRUCTURE

The Hassi er R'Mel structure, as shown by refraction shooting (Figs. 5, 6), is along the extension of an anticlinal axis which can be seen in the Turonian and Senonian outcrops on the southeast part of the structure. A seismic reflection study carried out over the refraction high to determine the precise location for Hassi er R'Mel No. 1 could not provide, using techniques available at that time, reliable structural information below the Upper Jurassic.

As defined by wells already drilled, the anti-

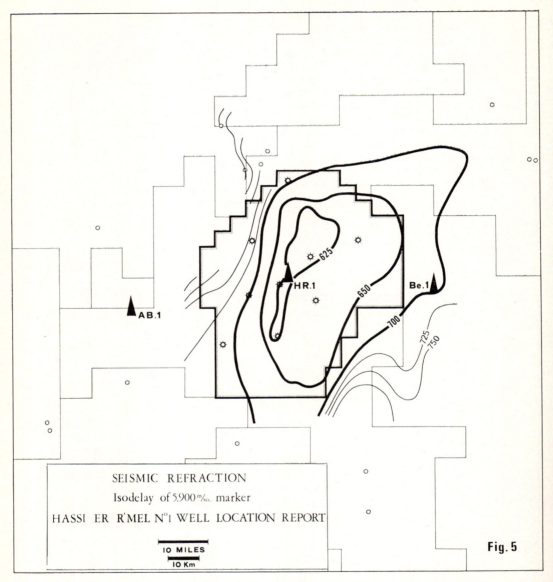

SEISMIC REFRACTION

Isodelay of 5.900 m/sec marker

HASSI ER R'MEL N°1 WELL LOCATION REPORT

10 MILES
10 Km

Fig. 5

FIG. 5.—Seismic refraction map of Hassi er R'Mel structure. Contours in milliseconds; datum plane is + 700 milliseconds.

cline covers a large area, and the zone contained within the −4,936 ft (−1,500 m) contour (gas-water contact) extends 50 mi (80 km) north-south and 28 mi (45 km) east-west. The structure includes approximately 1,000 mi² (2,600 km²). The dip is very slight and does not exceed 1–2°, on the flanks (Fig. 7).

RESERVOIR

The Hassi er R'Mel Sandstone is in the middle and lower units of the Permo-Triassic, though the reservoirs probably are Triassic. The A, B, and C reservoirs consist of fine- to medium-grained sandstone which is feldspathic or chloritic. The cement is either bituminous shale or anhydrite, and in a very few places is dolomitic. Where anhydrite is widely developed, the quality of the reservoir is considerably reduced. The shale beds separating the different sandstone strata include red-brown, silty, chloritic, and micaceous types.

Extent of reservoirs.—Reservoir C is 0–200 ft (0–60 m) thick. Its depositional pattern is related closely to the relief on the post-Hercy-

nian erosion surface. Reservoir C is restricted to the northeastern part of the structure and is absent in the center.

Reservoir B is more widespread than reservoir C, but its economic value is much reduced because of facies changes. The sandstone grades laterally into marl and silty shale, reflecting marine invasion from the east. The reservoir is 0–70 ft (0–20 m) thick.

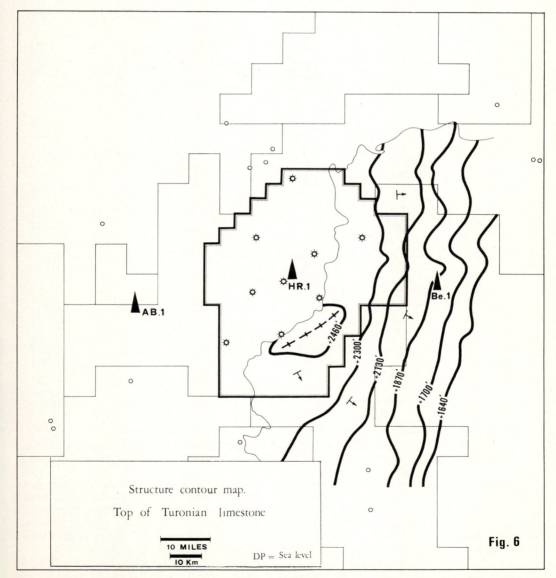

Structure contour map.

Top of Turonian limestone

10 MILES

10 Km

DP = Sea level

Fig. 6

FIG. 6.—Structure contour map. Datum is top of Turonian limestone. Contour elevations in feet above sea level.

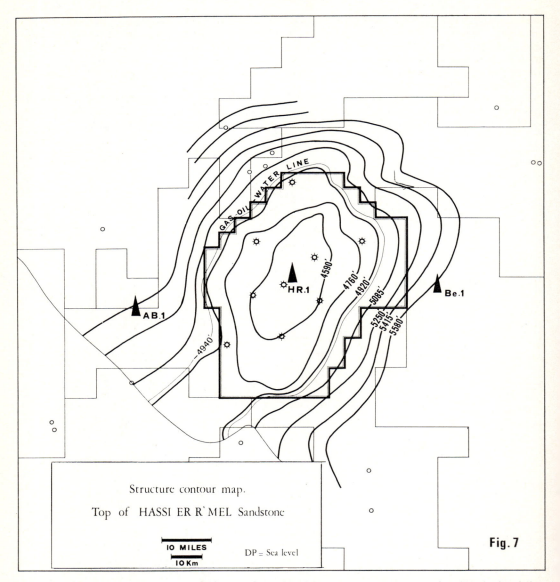

Structure contour map.

Top of HASSI ER R'MEL Sandstone

10 MILES
10 Km   DP = Sea level

Fig. 7

FIG. 7.—Structure contour map. Datum is top of Hassi er R'Mel sandstone. Contour elevations in feet below sea level.

Reservoir A, 50–120 ft (15–36 m) thick, is the most extensive reservoir, and is most consistent lithologically. It transgresses southward, and covers the entire region which was buried by older beds. (Fig. 8.)

Reservoir characteristics.—Approximately 80 percent of the total thickness of the A, B, and C reservoirs is of economic value. Porosity values change markedly, from 5 percent in the northwest part of the region to 20–22 percent in the central part of the field. Permeability values are as high as several darcys. Water saturation ranges from 5 to 25 percent. The water-gas transition zone is a few feet thick.

Migration and accumulation (Figs. 9–12). —The principal seal is Late Triassic salt and shale. The formation of traps therefore can have taken place only after the salt was deposited.

Paleotectonic maps (Figs. 9–12) demonstrate that, if there had been an early closure of essentially topographic and stratigraphic origin —formed immediately after the deposition of the sandstone—its effect would have been cancelled quickly by the formation of a homoclinal

dip upward toward the northern edge of the Sahara platform. This homoclinal dip was present through most of the Jurassic Period. (Fig. 10.)

The effect of subsidence, both toward the Atlas Mountains and the Oued Mya basin, was pronounced only during the Late Cretaceous. This subsidence caused the northern and east-

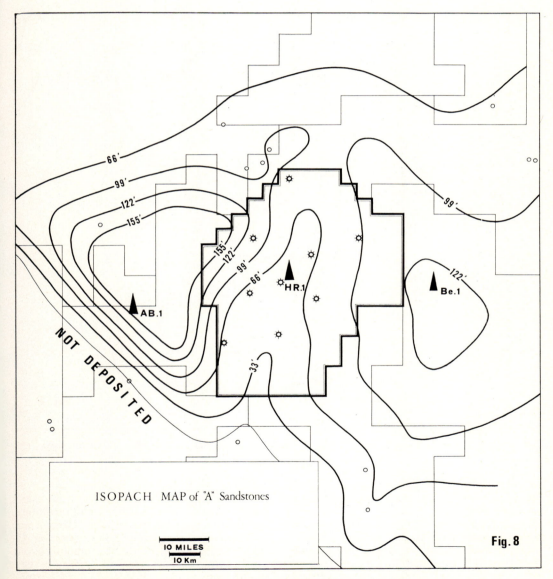

FIG. 8.—Isopach map of A sandstones. Contours in feet.

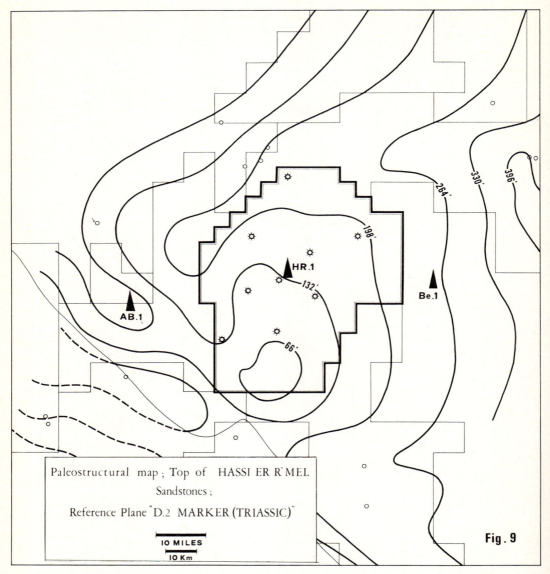

Paleostructural map; Top of HASSI ER R'MEL
Sandstones;

Reference Plane "D.2 MARKER (TRIASSIC)"

10 MILES
10 Km

Fig. 9

FIG. 9.—Paleostructure of top of Hassi er R'Mel sandstones; reference plane is D.2 marker (Triassic—*see* Fig. 4-A.)

ern closures of the structure which, during Turonian time, must have been very much as it is today. (Fig. 12.)

Water level and possibilities of oil ring.—The three reservoirs produce gas, but the gas-water contact has been located precisely only in reservoir B at −4,936 (−1,500 m). In reservoirs A and C, as much as 180 ft (55 m) of error is possible in locating the gas-water contact, espe-

cially in reservoir C.

The presumption that oil is present in reservoir C is based on the fact that small quantities of oil have been produced from Ordovician quartzite very near a zone where reservoir C and the quartzite are in direct communication and constitute a single reservoir. It is not possible, however, to assert that it would be technically feasible to exploit the oil commercially.

Reserve estimation.—In-place reserves of gas for the three reservoirs are estimated to total about 70 trillion ft³ (60° F at 1 atm = 15.5° C at 1 atm). In-place reserves proved by 10 wells amount to 55 trillion ft.³

As of March 25, 1968, the Hassi er R'Mel gas field had produced a total of 430 billion ft³ of gas and 19.5 million bbl of condensate. The field is operated by SEHR (Société d'Exploitation des Hydrocarbures de Hassi er R'Mel, a subsidiary of SN REPAL and CFP[A]).

Composition.—The molecular composition of the gas produced is as follows.

Component	Percent
Methane	78.5
Ethane	7.5
Propane	3.0
Butane and heavier gases	5.0
CO_2 and N_2	6.0
	100.0

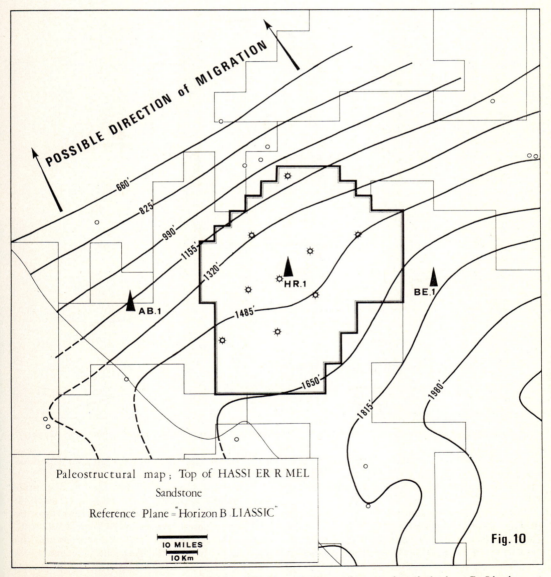

Fig. 10

Paleostructural map; Top of HASSI ER R MEL Sandstone

Reference Plane = "Horizon B LIASSIC"

10 MILES
10 Km

Fig. 10.—Paleostructure of top of Hassi er R'Mel sandstones; reference plane is horizon B, Liassic.

500 Philippe R. Magloire

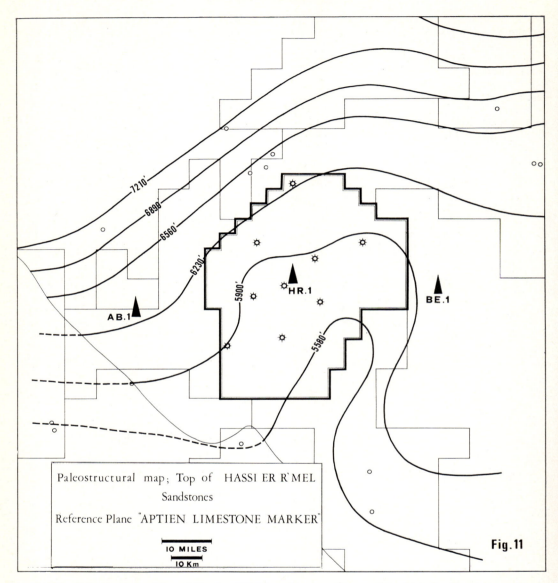

FIG. 11.—Paleostructure of top of Hassi er R'Mel sandstones; reference plane is Aptian limestone marker.

Recovery.—It is estimated that recoverable reserves may amount to 35 trillion ft³ of gas, or the equivalent in energy of 6 billion bbl of oil.

Processing and evacuation.—Because of the present limited commercial utilization of the gas, only five wells are connected with the processing plant, where the "cold frac" process is applied. Gas pressure decreases from 1,990 psi (140 kg/cm²) at separator entry to 1,000 psi (70 kg/cm²) at separator outlet. The expanded gas reaches a temperature of 3° F (−16° C).

The plant consists at present of five units with a daily capacity of 70 million ft³ each. It is connected to Arzew on the Mediterranean by a 24-in. pipeline which is 315 mi (500 km) long. The condensate is transported to the Haoud el Hamra oil-gathering system by a 186-mi (300-km) 8-in. pipeline. As prospects of gas sales increase, a second pipeline will be

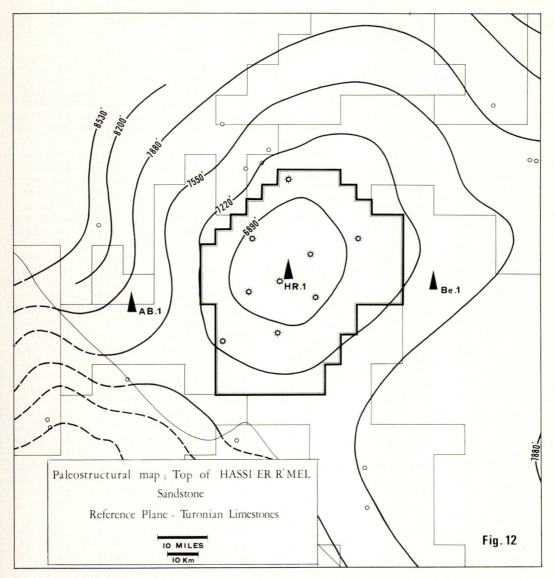

FIG. 12.—Paleostructure of top of Hassi er R'Mel sandstones; reference plane is Turonian limestone.

constructed from Hassi er R'Mel to Skikda on the Mediterranean.

REFERENCES CITED

Balducchi, A. M., and G. Pommier, 1970, Cambrian oil field of Hassi Messaoud, Algeria: this volume.

Delclaud, C., and A. Martel, 1959, Résultat de travaux de reconnaissance sur le champ de gaz de Hassi er R'Mel: Inst. Français Pétrole Rev., v. 14, nos. 4–5, p. 457–465.

—— H. I. Ortynski, and A. Perrodon, 1957, Les découverte de Hassi Messaoud et de Hassi er R'Mel dans le Sahara septentrional: Assoc. Française Techniciens Pétrole, 3d Cong. Natl. Français, Pau, Mai 1957, p. 75–85.

Ortynski, H. I., A. Perrodon, and Cl. de Lapparent, 1959, Esquisse paléogéographique et structurale des bassins du Sahara septentrional: 5th World Petroleum Cong. Proc., New York, sec. 1, p. 705–727.

Perrodon, A., 1958, Communication sur Hassi Messaoud et sur Hassi er R'Mel: Assoc. Française Techniciens Pétrole Bull. 129.

World's Giant Oil and Gas Fields, Geologic Factors Affecting Their Formation, and Basin Classification[1]

Part I

Giant Oil and Gas Fields

MICHEL T. HALBOUTY[2], A. A. MEYERHOFF[3], ROBERT E. KING[4],
ROBERT H. DOTT, SR.[3], H. DOUGLAS KLEMME[5], and THEODORE SHABAD[6]
Houston, Texas 77027; Tulsa, Oklahoma 74101; Mamaroneck, New York 10543;
Westport, Connecticut 06880; and New York City, New York 10032

Abstract In this paper a giant oil field is defined as one containing 500 million barrels or more of recoverable oil; a giant gas field has a minimum of 3.5 Tcf of recoverable gas. At least 187 giant oil fields and 79 giant gas fields are known. Together they contain an estimated, ultimate, minimum recoverable reserve of 638.77 billion bbl of oil and 1,180 Tcf of gas. Reserves of oil remaining to be produced from these giant fields amount to 521 billion bbl.

Of all giant-oil-field reserves, 58 percent is in sandstone and 42 percent is in carbonate reservoirs; 75 percent of giant-gas-field reserves is in sandstone and only 25 percent in carbonate reservoirs. A total of 29 percent of the oil and 10 percent of the gas is in strata of Tertiary age; 63 percent of the oil and 65 percent of the gas is in beds of Mesozoic age; and 8 percent of the oil and 25 percent of the gas is in Paleozoic reservoirs.

The largest number—190 (71 percent)—of the giant fields is in the Eastern Hemisphere; only 76 (29 percent) are in the Western Hemisphere; of those in the Eastern Hemisphere, 154 (58 percent of the world's total) are in a U-shaped belt 6,000 mi (10,000 km) long and 750–1,300 mi (1,250–2,000 km) wide that extends from Algeria to the Arctic Ocean at the longitude of the Polar Urals.

Introduction

At the 1968 Annual Meeting of the Association in Oklahoma City, where most of the field papers in this volume were presented, lack

[1] Manuscript received, March 31, 1970; revised June 6, 1970.

[2] Consulting geologist and petroleum engineer; independent producer and operator.

[3] The American Association of Petroleum Geologists.

[4] 580 Shore Acres Drive.

[5] Lewis Weeks Associates, Ltd.

[6] Editor, *Soviet Geography: Review and Translation,* American Geographical Society.

The writers are extremely grateful to the many individuals and petroleum companies that contributed to this paper; without the aid of many geologists in the petroleum industry, successful completion of this work would have been very difficult. Special thanks are due Kozo Kawai, Mrs. Nebo Ikonu Farrill, J. Donald Weir, Francisco Viniegra O., Stanley S. L. Chang, Y. C. Tung, K. M. Khudoley, and the All-Union Institute of Geological Researches, Leningrad, for their very substantial contributions to Tables 1 and 2. We also thank Mrs. Kathryn L. Meyerhoff for drafting, and Mrs. Carol Thompson, Miss Amy Lee Brown, and Miss Ernestine Voyles for typing the final manuscript.

of authors and limitation of time prevented full coverage of the world's fields, although the papers presented did provide a comprehensive treatment of the basic factors that produced the large hydrocarbon accumulations which we call giants. Therefore, while this symposium volume was being edited, a decision was made to add a final paper to list and characterize briefly all identifiable giant oil and gas fields of the world, as herein defined, including those of the Soviet Union and Mainland China. The tabulation of data from the Soviet Union and Mainland China is the first comprehensive presentation in the English language of ultimately recoverable oil and gas from the giant fields of these two nations.

This summary is divided into two parts. Part I lists 187 oil fields and 79 gas fields of the world identified as giants, each with estimated ultimate recoveries of 500 million bbl or more of oil, and 3.5 Tcf or more of gas, respectively (Tables 1, 2). Part I comprises the basic data on these 266 giants and their locations (Figs. 1–8) and a brief explanatory text. Part II is an interpretation of the data by Michel T. Halbouty, Robert E. King, H. Douglas Klemme, Robert H. Dott, Sr. and A. A. Meyerhoff. In this interpretation, factors believed to control the occurrence of giant fields and a classification of basins are discussed.

Although the treatment in this summary paper is more superficial than that of the individual papers which precede it, such a treatment is unavoidable because of the great number and geologic diversity of the giant fields, and the limitations of time and space. We have found certain broad similarities among many of the giant fields, as well as some unusual types of major hydrocarbon accumulations that deviate significantly from the "norm." Therefore, we believe that the summary, even though brief and generalized, will prove to be useful to petroleum geologists. If nothing else, we hope that it will provide food for a higher degree of imaginative thinking to enable us to find the giants which are yet to be discovered.

PREVIOUS SUMMARIES OF LARGE PETROLEUM FIELDS

Several studies of large fields have been made in the past. One of the most noteworthy of these—a paper which served for years as a basic reference on the subject of major oil fields—was that by Knebel and Rodríguez-Eraso (1956). The then-known 236 major oil fields of the noncommunist countries of the world, with an ultimate recoverable amount of oil from each field of more than 100 million bbl, were analyzed statistically. The study did not include gas fields, whereas the present study does, though incompletely.

A similar statistical analysis by Moody *et al.,* involving 45 oil fields of North America with ultimate recoveries of 500 million or more bbl is included in this volume (p. 8–17). Factors examined are reservoir age, lithology, depositional environment, and trap type. These authors concluded that, for these North American giants, Paleozoic age, sandstone lithology, shallow-marine environment, and structural traps are the preferred modes.

Fields of the United States with reserves of more than 100 million bbl of oil and more than 1 Tcf of gas were analyzed statistically by Halbouty (1968a; this volume, p. 91–127) to show discovery trends, geologic ages of the reservoirs, and trap types. Burke and Gardner (1969) published a brief summary of the 71 oil fields of the noncommunist world with reserves greater than 1 billion bbl each. Gas fields were not included. This, Moody *et al.'s,* and Halbouty's papers have been helpful in preparing the present summary.

TABLES 1, 2, 3, AND 4

Tables 1 and 2 list giant oil and gas fields, respectively, of the world, selected geologic and reservoir parameters, estimated ultimate recoverable reserves, and cumulative production through 1968 (earlier, in some cases), the last full year through which most data are available. The order of the fields is from those of largest reserves to those of smallest reserves.

The North American oil fields included in Table 1 are essentially the same as those listed by Moody *et al.* (this volume, p. 17). The few discrepancies are due mainly to differences in judgment concerning what individual fields to include in complexes and similar combinations.

Tables 3 and 4 are alphabetical lists corresponding to Tables 1 and 2, with literature references added. In addition to the references already cited, numerous other publications have been consulted, such as the International Oil Scouts Association's *Yearbooks,* World Petroleum Congress *Proceedings,* and several score USSR publications. The All-Union Institute of Geological Researches, Leningrad, provided a great amount of information on fields of the Soviet Union and Mainland China; additional information on Mainland China was obtained through Japanese sources. The tables have been reviewed by various petroleum geologists in several countries; these geologists have offered valuable comments, corrections, and criticisms. A selected bibliography appears at the end of the paper.

DEFINITIONS

Size of fields.—In this paper, discussion is limited to fields capable of ultimate commercial recovery of 500 million bbl or more of oil, and 3.5 Tcf or more of gas (the approximate calorific equivalent if the gas contains more than 95 percent gaseous and liquid hydrocarbons). Allowance is made for oil produced by secondary recovery or pressure maintenance if these are currently in operation or being initiated. The lower limits of field size are five times greater than the fields listed by Knebel and Rodríguez-Eraso (1956) and Halbouty (1968a). Both of these papers used the 100-million-bbl lower limit as their definitions of an oil "giant." Halbouty included gas, using 1 Tcf as the lower limit of a "giant." In this paper, we agreed that, to give coverage of giant fields on a worldwide basis, the lower reserve limit should be much higher. Hence, the figures of 500 million bbl (oil) and 3.5 Tcf (gas) were selected arbitrarily; the term "supergiants" may well apply to those fields which we include. Moreover, had we included fields with ultimate reserves between 500 and 100 million bbl, and between 3.5 and 1.0 Tcf, the total number would have been increased vastly without adding significantly to the proportion of the world's reserves being discussed, or to an understanding of the geologic factors that seem to control the accumulation of hydrocarbons into deposits of major importance.

Another possible way of defining giant fields, which was considered but not adopted, is to consider those fields giants which are dominant within the geologic or geopolitical province in which they are located. This approach has much merit, as it would result in the inclusion of such interesting fields as Matzen in the Vienna basin (Austria), Schoonebeck and Rühle in the Lower Saxony basin (Netherlands and West Germany), and Red Wash in the Uinta basin (Utah), none of which appears in our tabulation.

Field definition.—For the present purpose a field is a producing area which may contain (1) a single reservoir uninterrupted by permeability barriers that significantly affect fluid levels or pressures, (2) an area with multiple reservoirs trapped by a common structural or stratigraphic feature, or (3) an area with composite reservoirs, of different ages and lithologic types, with interconnected permeability related to the same trap. If one reservoir beneath adjacent but separate structural closures has a single, or common oil/water contact, that complex of structures is treated as a single field. Where reservoirs of different ages overlap, with the result that, in plan view, the productive area is continuous, the productive area is considered to comprise a single field. There are other factors which also may be used to define an area as a single field.

Some giant fields listed in Tables 1 and 2 once were classified as a cluster of smaller fields but, as they coalesced by development and extension drilling, there has been official recognition that subdivision is artificial. Examples are Ghawar (Saudi Arabia), Burgan (Kuwait), and Bolivar Coastal (Venezuela)—the three largest fields (*see* Table 1). There are other clusters which may be combined, especially those still under active development, but official recognition has not been given. For our purpose, several such fields are united. An example is the Amal-Nafoora-Augila field in Libya, where a different field name was adopted by each of three concession holders; production is mainly from a composite reservoir on a single major uplift, not fully connected at the present time. Bay Marchand, Timbalier Bay, and Caillou Island (Louisiana) are here classified as one giant field because they are productive above a single salt ridge, with three separate culminations, each covered by leases of different operators. Sosnino, Sovetskoye, and Medvedev in the West Siberian basin are placed together because several of the productive beds in the three culminations have common oil-water contacts (Sorokina, 1966; Vasil'yev, 1968b).

Some fields included herein may or may not qualify individually as giants, but are combined as *complexes* because of similarity of stratigraphy, trapping mechanism, reservoir content and performance, depth, *etc.*, and because of reasonable geographic proximity, even though the individual productive areas are not directly connected. These are indicated in the tables by multiple, hyphenated field names, and by such words as "complex," "greater," or "group."

One of the examples is the uniting of Infantas and La Cira fields of the Magdalena Valley, Colombia, where production is on a pair of anticlines separated by a narrow syncline; production from the oil-bearing sandstones of the two anticlines is partly connected by some wells in the syncline. The same is true of Leng-hu 2, 3, and 4 in the Tsaidam basin of Mainland China. Another example is Greater Seminole in Oklahoma, where the Bowlegs, Earlsboro, Little River, Seminole City, and St. Louis fields are located on the Seminole plateau; all produce from the Ordovician "Wilcox" sandstone. Still another is the Old Illinois complex, Crawford and Lawrence Counties, eastern Illinois, which includes as many as 15 fields along the southeast plunge of the La Salle anticline; oil is produced from Lower Pennsylvanian and Upper Mississippian sandstones and limestones. Rainbow (western Canada), composed of 54 individual pools reservoired in the Keg River Formation (Middle Devonian) reefs, is listed as one field (Barss *et al.*, this volume, Fig. 24, p. 46; M. E. Hriskevich, oral commun.).

Recoverable reserves.—Ion (1967) has discussed the problem of attempting to place the reserves of different regions of the world on a comparable basis. The figures given here for estimated ultimate recoverable reserves differ greatly in reliability. Therefore what we offer is an indication of order of magnitude rather than a result based on the uniform application of engineering calculations to known reservoir conditions; such information is not available for many fields. The most complete and consistent data available, interestingly, are from the fields of the USSR. The USSR reserve estimates, particularly those of oil, have been carefully checked by several petroleum specialists. If anything, we have tried to be conservative with USSR estimates.

Some fields listed are more than 40 years old and in an advanced stage of depletion; for these, calculation of the ultimate recoverable oil and/or gas is quite reliable. For most of the United States fields, American Petroleum Institute reserve estimates are followed closely. The figures are taken mainly from Halbouty (1968a), with some modifications based on the

(Text continued on page 509)

Table 3. Giant Oil Fields

Field	Location	Number by size (see Table 1)	Special References*
"A-100"	Libya	112	*See* footnote
Abqaiq	Saudi Arabia	10	*See* footnote
Abu Hadriya	Saudi Arabia	113	*See* footnote
Abu Jidu	Abu Dhabi	110	Hajash (1967)
Abu Sa'fah	Saudi Arabia-Bahrain	21	*See* footnote
Agha Jari	Iran	14	Ion *et al.* (1951), Mostofi (1967), Paran and Critchton (1967)
Ahwaz	Iran	23	*See* footnote
Amal	Egypt (Sinai)	51	*See* footnote
Amal-Nafoora-Augila	Libya	27	*See* footnote
Ampa, Southwest	Brunei	111	*See* footnote
Arenque	Mexico	118	Bryant *et al.* (1969), Franco (1969)
Arlan complex	USSR	33	Belov (1963), Ovanesov *et al.* (1963), Chunosov (1965), Sattarov *et al.* (1969), Viktorov and Teterev (1970)
Bahrain	Bahrain	96	Jones (1968)
Bai Hassan	Iraq	92	*See* footnote
Balakhany-Sabunchi-Ramany	USSR	53	*See* footnote
Barqan	Saudi Arabia	182	Oil and Gas Jour. (1969d), Wall Street Jour. (1969)
Barracouta	Australia	184	Hafenbrach (1969), Stewart (1969)
Bay Marchand complex	Louisiana	40	Louisiana Dept. Conserv. (1969a), Frey and Grimes (1970)
Belayim and Offshore	Egypt (Sinai)	80	*See* footnote
Berri	Saudi Arabia	25	*See* footnote
Bibi Eybat	USSR	66	*See* footnote
Bibi-Hakimeh	Iran	29	*See* footnote
Binak	Iran	105	*See* footnote
Bolívar Coastal	Venezuela	3	Young *et al.* (1956), Bryant (1961), Martínez (1970)
Bomu	Nigeria	164	Cordry (1967), Fränkl and Cordry (1967), Short and Stäuble (1967)
Boscán	Venezuela	98	Young *et al.* (1956), Martínez (1970)
Bradford	Pennsylvania	143	Newby *et al.* (1929)
Buena Vista Hills	California	149	Cunningham and Kleinpell (1934)
Burbank	Oklahoma	176	Sands (1929)
Burgan complex	Kuwait	2	May (1951), Fox (1956, 1959)
Bushgan	Iran	116	*See* footnote
Cabinda "B"	Angola	89	*See* footnote
Cheleken	USSR	145	Mirchink *et al.* (1969), Niz'yev (1969)
Coalinga	California	140	*See* footnote
Coalinga Nose	California	170	*See* footnote
Comodoro Rivadavia	Argentina	65	Fossa-Mancini (1932), AAPG-CSD (1968, 1969)
Conroe	Texas	155	Michaux and Buck (1936)
Cowden complex	Texas	163	Herald (1957)
Dahra-Hofra	Libya	137	*See* footnote
Dammam	Saudi Arabia	26	*See* footnote
Darius-Kharg	Iran	87	*See* footnote
Defa	Libya	62	*See* footnote
Dukhan	Qatar	52	Qatar Petroleum Co. Ltd. (1956)
East Texas	Texas	24	Minor and Hanna (1941), Hudnall and Eaton (1968)
Ébano-Pánuco	Mexico	120	Muir (1936), Benavides (1956)
Ekofisk	Norway	19	Gardner (1970b)
Elk Hills	California	83	Taff (1934)
El Morgan	Egypt	50	Anonymous (1970c)
Emeraude Marin	Congo (Brazzaville)	180	*See* footnote
Fadhili	Saudi Arabia	114	*See* footnote
Fahud	Oman	102	Tschopp (1967)
Fateh	Dubai	103	*See* footnote
Fereidoon-Marjan	Iran-Saudi Arabia	13	AAPG-CSD (1967, 1968, 1969)
Fyzabad Group	Trinidad	174	Ablewhite and Higgins (1968), Bower (1968)
Gach Saran	Iran	16	Slinger and Crichton (1959), Paran and Crichton (1967)
Ghawar	Saudi Arabia	1	Thralls (1955)
Gialo	Libya	60	*See* footnote
Glynsko-Rozbyshev	USSR	187	Blank *et al.* (1964), Pavlenko and Cherpak (1965)
Golden Trend	Oklahoma	186	Swesnick (1950)
Goldsmith-Andector complex	Texas	138	Herald (1957), West Texas Geol. Soc. (1966)

* General references used for nearly all of the fields above include the following: *Worldwide:* AAPG-CSD (1937–1969), Brod *et al.* (1965), Burke and Gardner (1969), Gardner *et al.* (1969), International Oil Scouts (1969), Petroleum Pub. Co. (1970); *North America:* National Oil Scouts and Landmen (1958), Beebe and Curtis (1968), Moody *et al.* (1970); *United States:* Halbouty (1968a, 1970), Oil and Gas Journal (Jan. 26, 1970); *California:* Calif. Div. Oil and Gas (1960); *Eastern Hemisphere:* Illing, ed. (1953); *Middle East:* Lees (1951), Fox (1964), Dunnington (1967); *USSR:* L'vov (1968), Shashin (1968), Vasil'yev (1968a, 1968b).

Table 3. (continued)

Field	Location	Number by size (see Table 1)	Special References*
Haft Kel	Iran	58	*See* footnote
Halibut	Australia	162	Hafenbrack (1969), Stewart (1969)
Hassi Messaoud	Algeria	38	Balducchi and Pommier (1970)
Hastings	Texas	146	Gardner (1952)
Hawkins	Texas	166	Wendlandt *et al.* (1946)
Hout	Neutral Zone	183	*See* footnote
Huntington Beach	California	121	Gale (1934)
Idd-el-Shargi	Qatar	57	Qatar Petroleum Co. Ltd. (1956)
Illinois, Old Fields	Illinois	139	Blatchley (1913)
Imo River	Nigeria	160	*See* footnote
Infantas-La Cira	Colombia	141	*See* footnote
Intisar (Idris) "A"	Libya	76	Terry and Williams (1969)
Intisar (Idris) "D"	Libya	91	Terry and Williams (1969)
Jones Creek	Nigeria	119	Short and Stäuble (1967)
Karachukhur-Zykh	USSR	157	*See* footnote
Karamai	China	134	Anonymous (1967, 1968), Ho K'o-jen (1968), Kravchenko (1968), Meyerhoff (1970b)
Karanj	Iran	84	*See* footnote
Kelly Snyder-Diamond M	Texas	72	Herald (1957), Vest (1970)
Kern River	California	151	Cunningham and Kleinpell (1934)
Khurais	Saudi Arabia	22	*See* footnote
Khursaniya	Saudi Arabia	20	Kiersznowski (1968)
Kingfish	Australia	95	Hafenbrack (1969), Stewart (1969)
Kirkuk	Iraq	7	Freeman (1951), Freeman and Natanson (1959)
Kotur-Tepe (Leninskoye)	USSR	36	*See* footnote
Kuleshovka	USSR	131	Raaben (1963), Belyayeva *et al.* (1966), Mel'nikov (1968)
La Brea-Pariñas-Talara	Peru	101	Travis (1953), Youngquist (1958)
Lama	Venezuela	61	AAPG-CSD (1958, 1965, 1966)
Lamar	Venezuela	85	Martínez (1970)
La Paz	Venezuela	99	Smith (1956), Young *et al.* (1956), Dikkers (1964), Martínez (1970)
Leduc-Woodbend	Alberta	177	Alta. Soc. Petroleum Geologists (1960)
Leng-hu complex	China	165	Ho K'o-jen (1968), Kravchenko (1968), Meyerhoff (1970b)
Lima-Indiana	Ohio-Indiana	172	*See* footnote
Long Beach	California	125	Brown (1968)
Lung-nü-ssu	China	154	Ho K'o-jen (1968), Kravchenko (1968), Meyerhoff (1970b)
Malgobek-Voznesenka-Aliyurt	USSR	45	Galyamov and Vil'tsin (1966), Plotnikov and Boltyshev (1966)
Mamontovo	USSR	46	Khafizov (1969)
Manifa	Saudi Arabia	11	*See* footnote
Mansuri	Iran	63	*See* footnote
Marun	Iran	12	*See* footnote
Masjid-i-Suleiman	Iran	69	Baniriah *et al.* (1967), Paran and Crichton (1967)
Maydan-Mahzam	Qatar	94	Jonkers (1968)
Megion	USSR	126	Beresnev and Rovnin (1964), Polushkin and Sanin (1965), Khafizov (1969)
Mene Grande	Venezuela	159	Young *et al.* (1956), Martínez (1970)
Meren	Nigeria	178	*See* footnote
Midway-Sunset	California	90	Cunningham and Kleinpell (1934)
Minagish	Kuwait	64	*See* footnote
Minas	Indonesia	34	*See* footnote
Moreni-Gura Ocnitei	Rumania	129	Paraschiv and Olteanu (1970)
Mukhanovo	USSR	74	Belov (1963), Markovskiy (1963), Yerofeyev (1969)
Murban Bab	Abu Dhabi	59	Hajash (1967), Harris *et al.* (1968)
Murban Bu Hasa	Abu Dhabi	44	Hajash (1967), Harris *et al.* (1968)
Naft-i-Safid	Iran	106	Paran and Crichton (1967)
Naranjos-Cerro Azul	Mexico	79	Benavides (1956), Viniegra and Castillo-Tejero (1970)
Natih	Oman	161	*See* footnote
Neftyanyye Kamni	USSR	49	*See* footnote
Novoyelkhov-Aktash	USSR	41	Krashin (1963), Raaben (1963), Dolgopolov and Vodovozov (1968)
Oficina (proper)	Venezuela	153	Hedberg *et al.* (1947), Martínez (1970)
Okan	Nigeria	179	Cordry (1967); Fränkl and Cordry (1967), Short and Stäuble (1967)
Oklahoma City	Oklahoma	130	McGee and Clawson (1932), Gatewood (1970)
Oktyabr'skoye	USSR	169	*See* footnote
Panhandle	Texas	73	Pippin (1970)
Paris (Faris)	Iran	43	*See* footnote
Pazanan	Iran	39	Paran and Crichton (1967)
Pembina	Alberta	70	Mills (1968)
Poza Rica	Mexico	47	Salas (1949), Colomo and Barnetche (1951), Barnetche and Illing (1956), Benavides (1956)
Pravdinsk	USSR	77	*See* footnote

Table 3. (continued)

Field	Location	Number by size (see Table 1)	Special References*
Priluki	USSR	142	Blank et al. (1964), Klitochenko and Paliy (1964)
Prudhoe Bay	Alaska	5	Oil and Gas Jour. (1970a)
Qatif	Saudi Arabia	15	See footnote
Quiriquire	Venezuela	100	Young et al. (1956), Martínez (1970)
Rag-e-Safid	Iran	31	See footnote
Raguba	Libya	107	See footnote
Rainbow	Alberta	135	Barss et al. (1970)
Rangely	Colorado	158	Pickering and Dorn (1948), Rocky Mtn. Assoc. Geologists (1954)
Raudhatain	Kuwait	18	Milton and Davies (1965)
Redwater	Alberta	136	Alta. Soc. Petroleum Geologists (1960)
Rhourde el Baguel	Algeria	181	See footnote
Romashkino	USSR	8	Begishev et al. (1963), Komarov (1969), Borisov et al. (1970)
Rostam	Iran	104	See footnote
Rumaila	Iraq	9	See footnote
Rumaila, North	Iraq	28	See footnote
Sabriya	Kuwait	35	See footnote
Safaniya-Khafji	Neutral Zone	4	See footnote
Salt Creek	Wyoming	175	Beck (1929), Barlow and Haun (1970)
Salym	USSR	117	Golovachev et al. (1966)
Samah	Libya	81	See footnote
Samotlor	USSR	6	Khafizov (1969), Chursin (1970)
Sangachaly-Duvannyy	USSR	124	Niz'yev (1969)
Santa Fe Springs	California	150	Brown (1968)
Sarir	Libya	17	Gillespie and Sanford (1967), Sanford (1970)
Sassan	Iran	75	See footnote
Seeligson complex	Texas	132	Nanz (1954)
Seminole, Greater	Oklahoma	127	Levorsen (1929)
Seria	Brunei	71	Wilford (1961)
Shaybah	Saudi Arabia	109	See footnote
Shkapovo	USSR	88	Shayevskiy and Yuferov (1963)
Sho-Vel-Tum	Oklahoma	123	AAPG (1956)
Slaughter	Texas	171	See footnote
Smackover	Arkansas	167	Shreveport Geol. Soc.(1947)
Sosnino-Sovetskoye-Medvedev	USSR	32	Sorokina (1966), Khafisov (1968)
South Pass Block 24	Louisiana	133	Louisiana Dept. Conserv. (1969a)
Starogroznyy	USSR	144	See footnote
Swan Hills	Alberta	122	Hemphill et al. (1970)
Ta Ch'ing	China	148	Anonymous (1968), Ho K'o-jen (1968), Kravchenko (1968), Meyerhoff (1970b)
Tom O'Connor	Texas	185	Mills (1970)
Tuymazy	USSR	54	Raaben (1962), Belov (1963), Zaydan (1963), Pelevin (1967)
Umm Gudair	Neutral Zone	115	See footnote
Umm Shaif	Abu Dhabi	56	Elder (1963), Banner and Wood (1964), Elder and Grieves (1965)
Usa	USSR	152	See footnote
Ust'-Balyk	USSR	42	Nikonov (1966)
Uzen'	USSR	37	Aronsov and Makhonin (1966), Trebin et al. (1966), Kozmodem'yanskiy et al. (1969), Ibrayev (1970)
Ventura Avenue	California	128	Taff (1934), Dryden et al. (1968)
Wafra	Neutral Zone	30	Nelson (1968)
Waha	Libya	97	See footnote
Wasson	Texas	156	Schneider (1943), West Texas Geol. Soc. (1966)
West Delta Block 73	Louisiana	147	See footnote
Wilmington	California	48	Mayuga (1970)
Ya-erh-hsia	China	168	Anonymous (1967, 1968), Ho K'o-jen (1968), Kravchenko (1968), Meyerhoff (1970b)
Yarino-Kamennyy Log	USSR	173	Abrikosov and Vinnikovskiy (1963)
Yates	Texas	86	Rettger et al. (1935), Herald (1957)
Yuzhno-Cheremshanka	USSR	78	See footnote
Zakum	Abu Dhabi	108	Fox and Brown (1968)
Zapadno-Surgut	USSR	67	See footnote
Zarzaïtine	Algeria	82	Compagnie Recherches Explor. Pétrole Sahara (1967)
Zelten	Libya	55	Armstrong (1967), Fraser (1967)
Zhetybay	USSR	93	Aronsov and Makhonin (1966), Byzer and Erinchek (1966), Kozmodem'yanskiy et al. (1969)
Zubair	Iraq	68	See footnote

Table 4. Giant Gas Fields

Field	Location	Number by size (See Table 2)	Special References*
Achak	USSR	41	*See* footnote
Anastasiyevsko-Troitskoye	USSR	78	Konoplev et al. (1965), Popov and Karmazina (1966)
Arkticheskoye	USSR	4	*See* footnote
Bahrain	Bahrain	14	Greig (1958)
Barqan	Saudi Arabia	61	Wall Street Jour. (1969)
Barracouta	Australia	79	Reynolds (1967), Trail (1968), Stewart (1969)
Bastian Bay	Louisiana	74	Meyerhoff, ed. (1968), Louisiana Dept. Conserv. (1969a)
Bayou Sale	Louisiana	71	Louisiana Dept. Conserv. (1969a)
Bergen	Netherlands	60	Gardner (1970)
Blanco-Basin	New Mexico	24	*See* footnote
Buyerdeshik	USSR	55	Anonymous (1970g)
Carthage	Texas	36	Rogers (1968)
Gazli	USSR	17	Berman et al. (1966)
Gidgealpa	Australia	42	Greer (1965), Smale and Trueman (1965), Williams et al. (1966)
Gomez	Texas	51	Hills (1968)
Groningen	Netherlands-West Germany	3	de Ruiter et al. (1967), Stäuble and Milius (1970)
Gubkin (Gubkinskoye)	USSR	21	Bazanov et al. (1967)
Gugurtli	USSR	72	Anonymous (1970j, k)
Hassi er R'Mel	Algeria	10	Magloire (1970)
Hateiba	Libya	23	Bowerman (1967)
Hewett	United Kingdom	52	Kent and Walmsley (1970)
Hugoton	Kansas-Oklahoma-Texas	8	Mason (1968), Pippin (1970)
Indefatigable	United Kingdom	28	Kent and Walmsley (1970)
Jalmat	New Mexico	27	Roswell Geol. Soc. (1956)
Kandym (Khadzhiy-Kandym)	USSR	63	*See* footnote
Katy	Texas	37	Allison et al. (1946)
Kazanskoye	USSR	64	*See* footnote
Kenai	Alaska	44	Kelly (1968)
Kettleman Hills North dome	California	58	Gester and Galloway (1933), Hoots and Herold (1935), Calif. Div. Oil and Gas (1960), Rudkin (1968)
Komsomol' (Komsomol'skoye)	USSR	65	*See* footnote
Korobki	USSR	77	Dolitskiy and Kucheruk (1963), Pestrikov (1963)
Krasnyy Kholm (Orenburg)	USSR	13	Anonymous (1970f)
Lacq	France	33	Winnock and Pontalier (1970)
Layavozh	USSR	16	Kortunov (1967), Anonymous (1970h)
Leman	United Kingdom	22	Kent and Walmsley (1970)
Luginets	USSR	67	*See* footnote
Maastakh	USSR	35	Afanas'yev et al. (1966), Mikulenko and Gaydeburova (1966)
Mari	West Pakistan	45	Tainsh et al. (1959)
Marlin	Australia	57	Hafenbrack (1966), Reynolds (1967), Sprigg (1968), Trail (1968), Stewart (1969)
Maykop	USSR	75	*See* footnote
Medvezh'ye	USSR	9	Anonymous (1970b)
Meillon-Rousse	France	62	Winnock and Pontalier (1970)
Messoyakha	USSR	20	*See* footnote
Mocane-Laverne	Oklahoma	53	Pippin and Poling (1961), Beebe and Curtis (1968), Pate (1968)
Monroe	Louisiana	32	Murray (1957), Berryhill (1968)
Moomba	Australia	43	C. A. Martin (1967)
Myl'dzhino (Myl'dzhinskoye)	USSR	49	*See* footnote
Novvy Port (Novoportovskoye)	USSR	40	*See* footnote
Nyda (Nydinskoye)	USSR	68	Girshgorn (1969)
Old Ocean	Texas	46	Murray (1957), Porter (1964), Halbouty (1968b)
Panhandle	Texas	11	Mason (1968), Pippin (1970)
Pazanan	Iran	6	Falcon (1958)
Pelyata	USSR	69	Anonymous (1970b)
Puckett	Texas	34	West Texas Geol. Soc. (1966), Hills (1968)
Punga	USSR	73	Rovnin (1964)
Rhourde Nouss	Algeria	48	Megateli et al. (1969)
Rio Vista	California	59	Burroughs et al. (1968)
Russkoye	USSR	26	Remeyev and Ostrowskaja (1969)
Samantepe	USSR	54	*See* footnote
Saradjeh	Iran	47	*See* footnote

* General references used for nearly all of the fields above include the following:
Worldwide: AAPG-CSD (1937–1969), Brod et al. (1965), Burke and Gardner (1969), Gardner et al. (1969), International Oil Scouts (1969), Petroleum Pub. Co. (1970); *North America:* National Oil Scouts and Landmen (1958), Beebe and Curtis (1968); *United States:* Halbouty (1968a, 1970), Oil and Gas Journal (Jan. 26, 1970); *California:* Calif. Div. Oil and Gas (1960); *Eastern Hemisphere:* Illing, ed. (1953); *Middle East:* Lees (1951), Fox (1964), Dunnington (1967); *USSR:* L'vov (1968), Shashin (1968), Vasil'yev (1968a, 1968b).

Table 4. (*continued*)

Field	Location	Number by size (See Table 2)	Special References*
Severo-Stavropol'-Pelagiada	USSR	31	*See* footnote
Shebelinka	USSR	18	*See* footnote
Shekhitli	USSR	30	Amanov and Sokolov (1969)
Shibarghan	Afghanistan	56	Grdzelov *et al.* (1969)
Shih-you-kou-Tung-hsi	China	29	Kravchenko (1968), Meyerhoff (1970b)
Solenaya	USSR	70	Anonymous (1970d)
Sredne-Vilyuy	USSR	19	Afanas'yev *et al.* (1966)
Sui	West Pakistan	38	Tainsh *et al.* (1959)
Taz (Tazovskoye)	USSR	7	Girshgorn (1969)
Tul'skiy	USSR	76	Anonymous (1970d)
Urengoy (Urengoyskoye)	USSR	1	Remeyev and Ostrowskaja (1969)
Urtabulak	USSR	39	*See* footnote
Vuktyl'	USSR	15	*See* footnote
Vyngapur (Vyngapurskoye)	USSR	25	*See* footnote
Yamburg	USSR	12	Anonymous (1970b)
Yefremovka	USSR	50	*See* footnote
Yubileynyy	USSR	2	Podobin and Sergeyev (1969), Remeyev and Ostrowskaja (1969), Anonymous (1970b)
Zapadno-Krestishche	USSR	66	Vitrik *et al.* (1969)
Zapolyarnoye	USSR	5	*See* footnote

tabulation in the *Oil and Gas Journal* of January 26, 1970 (p. 128–139), and other sources. We recognize, however, that, because of the methods employed, reserve estimates are subject to considerable upward revision, particularly in the example of newer fields that are not fully developed, and thus lack adequate production history.

Russian petroleum geologists have six categories of reserves, ranging from certain to highly speculative (Campbell, 1968[7]). These cannot be

[7] Although Russian geologists and engineers have differing opinions (*e.g.,* Mkrtchyan, 1966; Mirchink and Feygin, 1967; Brenner, 1968; Gal'person *et al.,* 1970) on how to define each category of reserves, their categories are as follows (Campbell, 1968). **A** = hydrocarbons in areas outlined by wells with proved production; reservoir lithology, character of the gas or oil, drive mechanism, pressure, permeability, hydrocarbon/water contacts, and other properties are known. **B** = hydrocarbons in areas where commercial production is established from at least two wells; the general lithologic character, core characteristics, permeability, porosity, pressure, oil or gas characteristics, and hydrocarbon/water contacts must be known from the two wells. **C**$_1$ = hydrocarbons in areas where commercial production has been proved in at least one well, geologic and geophysical data have outlined the presence of a favorable trap, the trap boundaries are reasonably well known by subsurface or seismic mapping, and study of reservoir parameters and hydrocarbon characteristics from the one well is complete and the results favorable. **C**$_2$ = reserves estimated for new structures in established oil- and gas-producing regions, or in untested parts of an already established field (*i.e.,* on an untested fault block within a field where all other fault blocks tested produce hydrocarbons commercially). An important restriction is that the structures must be sufficiently well defined

equated precisely with API standards. API "proved reserves" are both drilled and undrilled. The proved drilled reserves include hydrocarbons estimated to be recoverable by the production systems in operation, with or without fluid injection. The proved undrilled reserves include those in undrilled spacing units which are close to and related to the drilled units —so close, in fact, that there is every reason to believe that they will be found present when drilled. The Russian category **A** is somewhat narrower than the API definition, whereas Russian **A** plus **B** is much broader because it can include estimates based on the drilling of only two commercial wells on a single structure. Thus the closest approximation is that API "proved reserves" are equivalent to **A** plus special cases of **B**. **C**$_1$ cannot be included (Campbell, 1968, p. 61–62, except in special cases.

(*Text continued on page 525*)

to justify deep drilling programs. **D**$_1$ = predicted reserves in areas where oil- and gas-bearing structures have been established in a few localities, and the reservoir studies from those few localities indicate that favorable facies are present. **D**$_2$ = predicted reserves which can be assumed to exist in basins favorable for the accumulation of oil and gas deposits, but which have not been studied sufficiently for outlining the details of geologic structure and facies variation. In the present paper, estimates for Russian and Chinese fields, insofar as we could determine, are "proved plus probable" and are based solely on **A, B**, and (in some gas fields) **C**$_1$ categories. Russian and Chinese oil reserve figures are **A** plus **B** only, and are *minimum* figures, with a few possible exceptions. **C**$_2$ was used in only a few gas fields. **D** estimates were ignored.

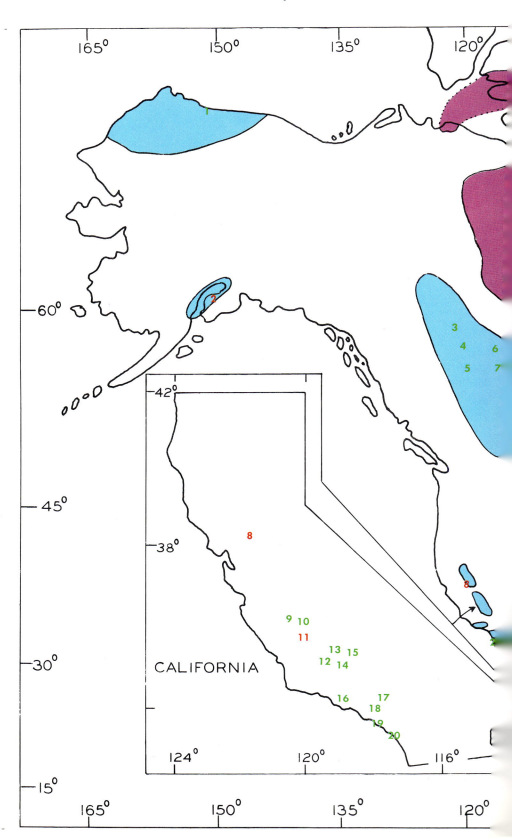

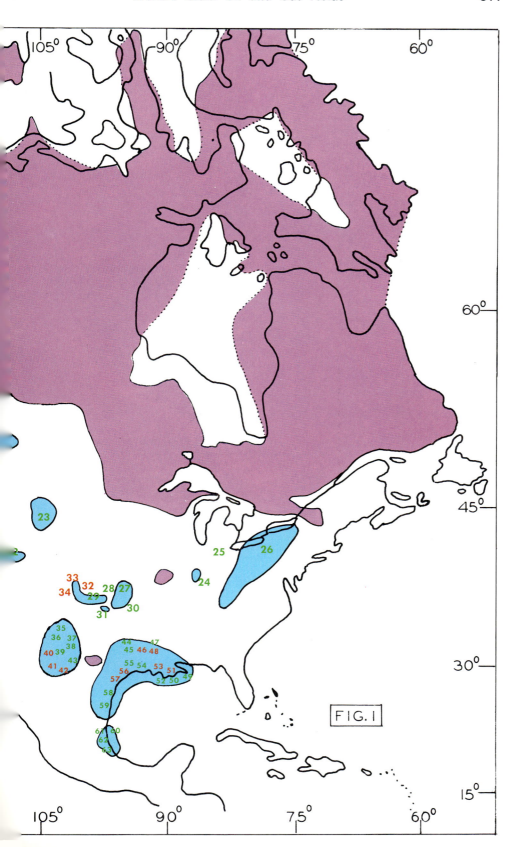

FIG. 1

← ⪻

Fig. 1. Giant Oil and Gas Fields of North America.

Map No.	Field	Type	Basin[a]	State or Province
1.	Prudhoe Bay	Oil	Colville	Alaska
2.	Kenai	Gas	Cook Inlet	do.
3.	Rainbow	Oil	Alberta	Alberta, Canada
4.	Swan Hills	Oil	do.	do.
5.	Pembina	Oil	do.	do.
6.	Redwater	Oil	do.	do.
7.	Leduc-Woodbend	Oil	do.	do.
8.	Rio Vista	Gas	Sacramento	California
9.	Coalinga	Oil	San Joaquin	do.
10.	Coalinga Nose	Oil	do.	do.
11.	Kettleman Hills, North dome	Gas	do.	do.
12.	Midway-Sunset	Oil	do.	do.
13.	Buena Vista Hills	Oil	do.	do.
14.	Elk Hills	Oil	do.	do.
15.	Kern River	Oil	do.	do.
16.	Ventura Avenue	Oil	Ventura	do.
17.	Santa Fe Springs	Oil	Los Angeles	do.
18.	Long Beach	Oil	do.	do.
19.	Wilmington	Oil	do.	do.
20.	Huntington Beach	Oil	do.	do.
21.	Blanco-Basin	Gas	San Juan	New Mexico
22.	Rangely	Oil	Uinta	Colorado
23.	Salt Creek	Oil	Powder River	Wyoming
24.	Illinois, Old Fields	Oil	Illinois	Illinois
25.	Lima-Indiana	Oil	Cincinnati arch	Ohio-Indiana
26.	Bradford	Oil	Appalachian	Pennsylvania
27.	Burbank	Oil	Cherokee	Oklahoma
28.	Oklahoma City	Oil	Nemaha Ridge	do.
29.	Golden Trend	Oil	Anadarko	do.
30.	Seminole, Greater	Oil	Seminole plateau	do.
31.	Sho-Vel-Tum	Oil	Ardmore	do.
32.	Mocane-Laverne	Gas	Anadarko	do.
33.	Hugoton	Gas	do.	Kansas-Oklahoma-Texas
34.	Panhandle	Gas and Oil	Amarillo uplift	Texas
35.	Slaughter	Oil	Permian	do.
36.	Wasson	Oil	do.	do.
37.	Kelly Snyder-Diamond M	Oil	do.	do.
38.	Cowden complex	Oil	do.	do.
39.	Goldsmith-Andector complex	Oil	do.	do.

[a] Shield areas are in violet, basins in blue. Only those basins with giant fields are colored. Green numbers are oil fields; red numbers are gas fields.

Fig. 1. (continued)

Map No.	Field	Type	Basin	State or Province
40.	Jalmat	Gas	do.	New Mexico
41.	Gomez	Gas	do.	Texas
42.	Puckett	Gas	do.	do.
43.	Yates	Oil	do.	do.
44.	Hawkins	Oil	Gulf Coast geosyncline	do.
45.	East Texas	Oil	do.	do.
46.	Carthage	Gas	do.	do.
47.	Smackover	Oil	do.	Arkansas
48.	Monroe	Gas	do.	Louisiana
49.	South Pass Block 24	Oil	do.	do.
50.	West Delta Block 73	Oil	do.	do.
51.	Bastian Bay	Gas	do.	do.
52.	Bay Marchand complex	Oil	do.	do.
53.	Bayou Sale	Gas	do.	do.
54.	Hastings	Oil	do.	Texas
55.	Conroe	Oil	do.	do.
56.	Katy	Gas	do.	do.
57.	Old Ocean	Gas	do.	do.
58.	Tom O'Connor	Oil	do.	do.
59.	Seeligson complex	Oil	do.	do.
60.	Arenque	Oil	Tampico embayment	Tamaulipas, Mexico
61.	Ébano-Pánuco	Oil	do.	do.
62.	Naranjos-Cerro Azul	Oil	do.	Vera Cruz
63.	Poza Rica	Oil	do.	do.

⪼ →

Fig. 2. Giant Oil Fields of South America.

Map No.	Field	Basin[a]	Country
1.	Fyzabad	Maturín	Trinidad
2.	Quiriquire	do.	Venezuela
3.	Oficina (proper)	do.	do.
4.	La Paz	Maracaibo	do.
5.	Boscán	do.	do.
6.	Lama	do.	do.
7.	Lamar	do.	do.
8.	Bolívar Coastal	do.	do.
9.	Mene Grande	do.	do.
10.	Comodoro Rivadavia	San Jorge	Argentina
11.	La Brea-Pariñas-Talara	Piura	Perú
12.	Infantas-La Cira	Middle Magdalena	Colombia

[a] Shield areas are in violet, basins in blue; only those basins with giant fields are shown. Green numbers are oil fields; red numbers are gas fields.

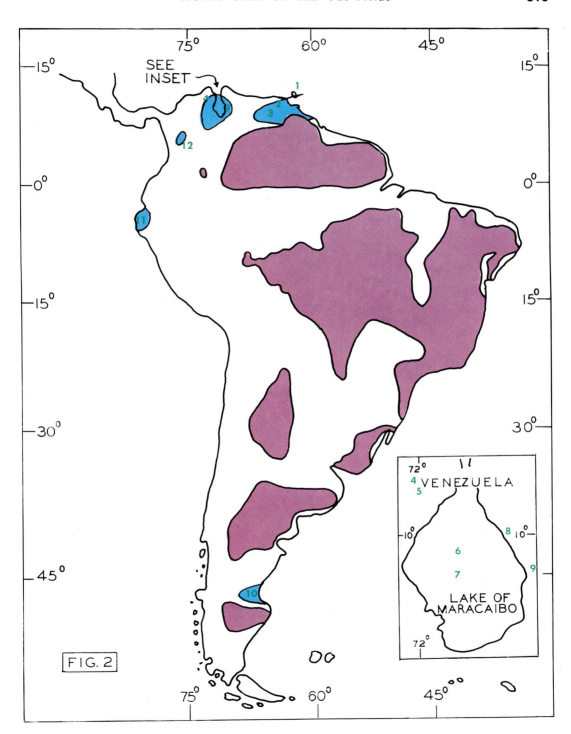

FIG. 2

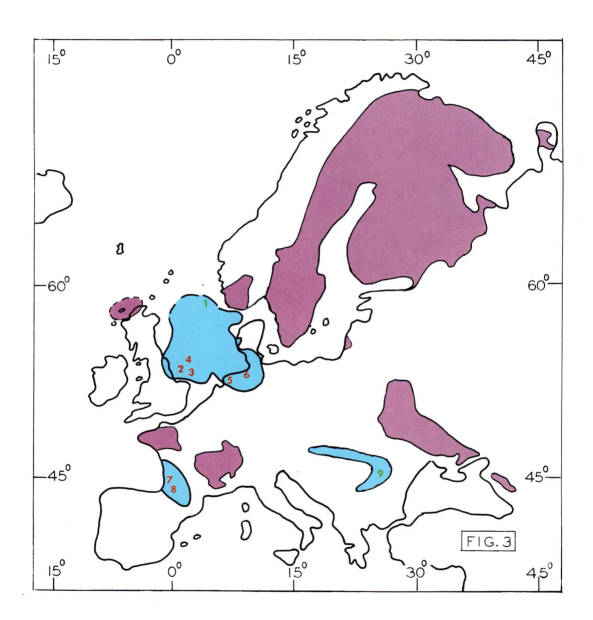

FIG. 3

← ⫷⫷⫷

Fig. 3. Giant Oil and Gas Fields of Europe.

Map No.	Field	Type	Basin[a]	Country	Map No.	Field	Type	Basin[a]	Country
1.	Ekofisk	Oil	North Sea	Norway	6.	Groningen	Gas	do.	do.—West Germany
2.	Hewett	Gas	do.	United Kingdom	7.	Lacq	Gas	Aquitaine	France
3.	Leman	Gas	do.	do.	8.	Meillon-Rousse	Gas	do.	do.
4.	Indefatigable	Gas	do.	do.	9.	Moreni-Gura Ocnitei	Oil	Sub-Carpathian depression	Rumania
5.	Bergen	Gas	do.	Netherlands					

[a] Shield areas in in violet, basins in blue; only those basins with giant fields are shown. Green numbers are oil fields; red numbers are gas fields.

⫸⫸⫸→

Fig. 4. Giant Oil and Gas Fields of Asia. Fig. 4 (Continued)

Map No.	Field	Type	Basin[a]	Country
1.	Zapadno-Krestishche	Gas	Dnepr-Donets	USSR Ukraine SSR
2.	Priluki	Oil	do.	do.
3.	Glynsko-Rozbyshev	Oil	do.	do.
4.	Yefremovka	Gas	do.	do.
5.	Shebelinka	Gas	do.	do.
6.	Korobki	Gas	Volga-Ural	Volgograd Oblast
7.	Krasnyy Kholm (Orenburg)	Gas	do.	Orenburg Oblast
8.	Mukhanovo	Oil	do.	Kuybyshev Oblast
9.	Kuleshovka	Oil	do.	do.
10.	Tuymazy	Oil	do.	Bashkir ASSR
11.	Shkapovo	Oil	do.	do.
12.	Novoyelkhov-Aktash	Oil	do.	Tatar ASSR
13.	Romashkino	Oil	do.	do.
14.	Arlan complex	Oil	do.	Bashkir ASSR
15.	Yarino-Kamennyy Log	Oil	do.	Perm Oblast
16.	Kazanskoye	Gas	West Siberian	Tomsk Oblast
17.	Luginets	Gas	do.	do.
18.	Myl'dzhino	Gas	do.	do.
19.	Yuzhno-Cheremshanka	Oil	do.	do.
20.	Samotlor	Oil	do.	Tyumen' Oblast
21.	Megion	Oil	do.	do.
22.	Sosnino–Sovetskoye-Medvedev	Oil	do.	do.
23.	Zapadno-Surgut	Oil	do.	do.
24.	Mamontovo	Oil	do.	do.
25.	Pravdinsk	Oil	do.	do.
26.	Salym	Oil	do.	do.
27.	Ust'-Balyk	Oil	do.	do.
28.	Punga	Gas	do.	do.
29.	Vuktyl'	Gas	Pechora	Komi ASSR
30.	Layavozh	Gas	do.	Nenets Nat. Okrug
31.	Usa	Oil	do.	Komi ASSR
32.	Arkticheskoye	Gas	West Siberian	Tyumen' Oblast
33.	Novyy Port	Gas	do.	do.
34.	Yamburg	Gas	do.	do.
35.	Taz	Gas	do.	do.
36.	Zapolyarnoye	Gas	do.	do.
37.	Russkoye	Gas	do.	do.
38.	Urengoy	Gas	do.	do.
39.	Gubkin	Gas	do.	do.
40.	Komsomol'	Gas	do.	do.
41.	Vyngapur	Gas	do.	do.
42.	Yubileynyy	Gas	do.	do.
43.	Nyda	Gas	do.	do.
44.	Medvezh'ye	Gas	do.	do.
45.	Solenaya	Gas	do.	Krasnodar Kray
46.	Pelyata	Gas	do.	do.
47.	Messoyakha	Gas	do.	Tyumen' Oblast
48.	Sredne-Vilyuy	Gas	Vilyuy	Yakutsk ASSR
49.	Maastakh	Gas	do.	do.
50.	Ta-Ch'ing (Daqing)	Oil	Sungliao	China Manchuria
51.	Lung-nü-ssu	Oil	Szechwan	Szechwan
52.	Shih-you-kou-Tung-hsi	Gas	do.	do.
53.	Ya-erh-hsia	Oil	Pre-Nan Shan trough	Kansu
54.	Leng-hu complex	Oil	Tsaidam	Tsinghai
55.	Karamai	Oil	Dzungaria	Sinkiang
56.	Sui	Gas	Indus	West Pakistan
57.	Mari	Gas	do.	do.

[a] Shield areas are in violet, basins in blue; only those basins with giant fields are shown. Green numbers are oil fields; red numbers are gas fields. Nos. 1-49 are in USSR; nos. 50-55 are in China; nos. 56-57 are in West Pakistan.

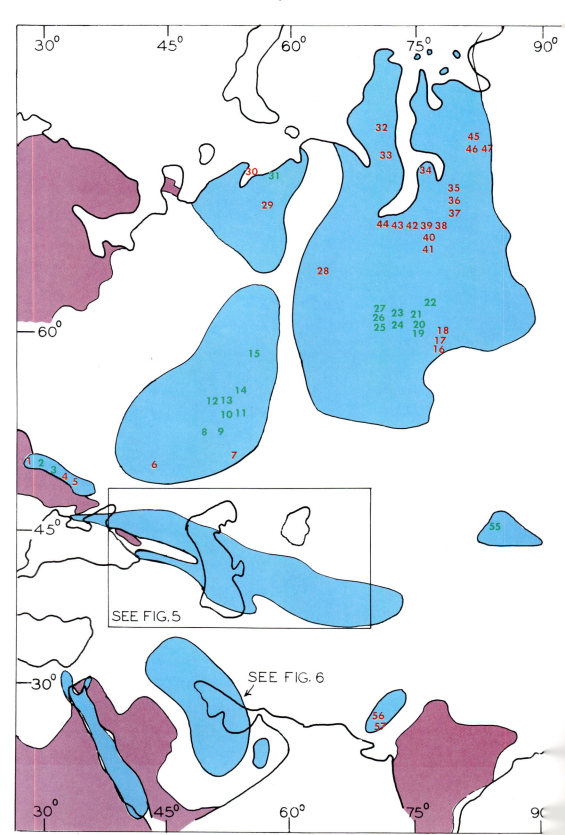

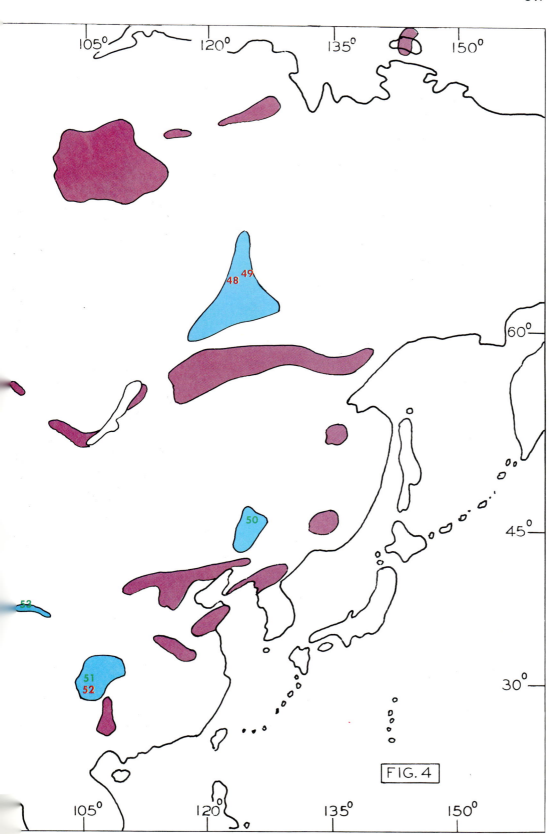

FIG. 4

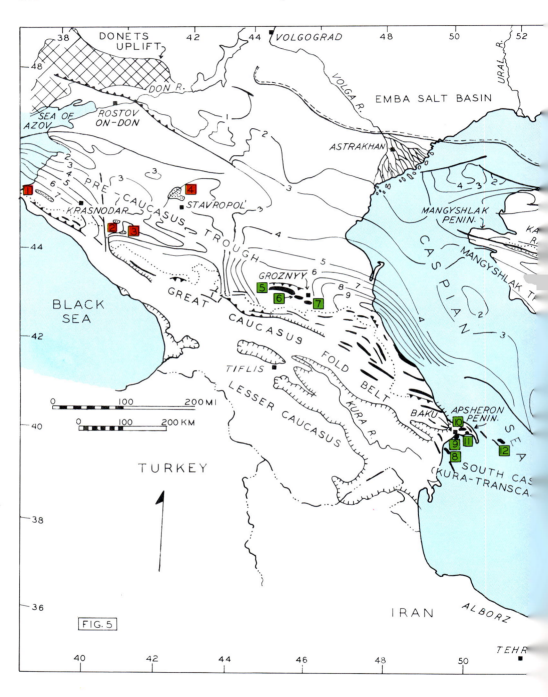

FIG. 5

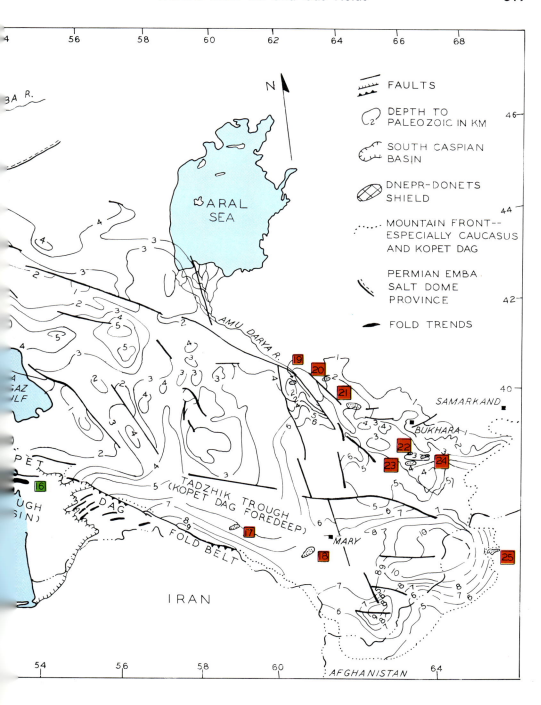

Fig. 5. Giant Oil and Gas Fields of Caucasus-Caspian-U.S.S.R. Central Asia Region.

Map No.	Field	Type	Trough[a]	Political Unit
1.	Anastasiyevsko-Troitskoye	Gas	Pre-Caucasus	Krasnodar Kray
2.	Maykop	Gas	do.	do.
3.	Tul'skiy	Gas	do.	do.
4.	Severo-Stavropol'-Pelagiada	Gas	do.	Stavropol' Kray
5.	Malgobek-Voznesenka-Aliyurt	Oil	do.	Checheno-Ingush ASSR
6.	Starogroznyy	Oil	do.	do.
7.	Oktyabr'skoye	Oil	do.	do.
8.	Sangachaly-Duvannyy	Oil	do. (Baku)	Azerbaydzhan SSR
9.	Bibi Eybat	Oil	do. do.	do.
10.	Balakhany-Sabunchi-Ramany	Oil	do. do.	do.
11.	Karachukhur-Zykh	Oil	do. do.	do.
12.	Neftyanyye Kamni	Oil	do. do.	do.
13.	Uzen'	Oil	Mangyshlak	Kazakhstan SSR
14.	Zhetybay	Oil	do.	do.
15.	Cheleken	Oil	South Caspian	Turkmen SSR
16.	Kotur-Tepe (Leninskoye)	Oil	do.	do.
17.	Buyerdeshik	Gas	Tadzhik	do.
18.	Shekhitli	Gas	do.	do.
19.	Achak	Gas	Bukhara-Khiva	do.
20.	Gugurtli	Gas	do.	do.
21.	Gazli	Gas	do.	Uzbekistan SSR
22.	Kandym	Gas	do.	do.
23.	Samantepe	Gas	do.	do.
24.	Urtabulak	Gas	do.	do.
25.	Shibarghan	Gas	Tadzhik	Afghanistan

[a] Green squares are numbers of oil fields; red squares give numbers of gas fields.

Fig. 6. Giant Oil and Gas Fields of Persian Gulf Region.

Map No.	Field	Type	Basin[a]	Country
1.	Kirkuk	Oil	Zagros fold belt	Iraq
2.	Bai Hassan	Oil	do.	do.
3.	Rumaila and Rumaila, North	Oil	Arabian platform	do.
4.	Zubair	Oil	do.	do.
5.	Raudhatain	Oil	do.	Kuwait
6.	Sabriya	Oil	do.	do.
7.	Burgan complex	Oil	do.	do.
8.	Minagish	Oil	do.	do.
9.	Umm Gudair	Oil	do.	Kuwait-Neutral Zone
10.	Wafra	Oil	do.	Neutral Zone
11.	Hout	Oil	do.	Saudi Arabia-Neutral Zone
12.	Safaniya-Khafji	Oil	do.	do.
13.	Manifa	Oil	do.	Saudi Arabia
14.	Abu Hadriya	Oil	do.	do.
15.	Khursaniyah	Oil	do.	do.
16.	Fadhili	Oil	do.	do.
17.	Berri	Oil	do.	do.
18.	Abu Sa'fah	Oil	do.	Saudi Arabia-Bahrain
19.	Qatif	Oil	do.	Saudi Arabia
20.	Dammam	Oil	do.	do.
21.	Abqaiq	Oil	do.	do.
22.	Khurais	Oil	do.	do.
23.	Ghawar	Oil	do.	do.
24.	Bahrain	Oil and Gas	do.	do.
25.	Dukhan	Oil	do.	Qatar
26.	Idd-ed-Shargi	Oil	do.	do.
27.	Maydan-Mahzam	Oil	do.	do.
28.	Murban bu Hasa	Oil	do.	Abu Dhabi
29.	Murban Bab	Oil	do.	do.
30.	Abu Jidu	Oil	do.	do.
31.	Zakum	Oil	do.	do.
32.	Umm Shaif	Oil	do.	do.
33.	Fateh	Oil	do.	Dubai
34.	Natih	Oil	Oman platform	Oman
35.	Fahud	Oil	do.	do.
36.	Shaybah	Oil	Arabian platform	Saudi Arabia
37.	Sassan	Oil	do.	Iran
38.	Rostam	Oil	do.	Iran
39.	Fereidoon-Marjan	Oil	do.	Iran-Saudi Arabia
40.	Darius-Kharg	Oil	do.	Iran
41.	Binak	Oil	Arabian-Iranian basin	Iran
42.	Bushgan	Oil	do.	Iran
43.	Bibi-Hakimeh	Oil	do.	Iran
44.	Gach Saran	Oil	do.	Iran
45.	Pazanan	Oil and Gas	do.	Iran
46.	Rag-e-Safid	Oil	do.	Iran
47.	Mansuri	Oil	do.	Iran
48.	Agha Jari	Oil	do.	Iran
49.	Marun	Oil	do.	Iran
50.	Karanj	Oil	do.	Iran
51.	Paris (Faris)	Oil	do.	Iran
52.	Haft Kel	Oil	do.	Iran
53.	Ahwaz	Oil	do.	Iran
54.	Naft-i-Safid	Oil	do.	Iran
55.	Masjid-i-Suleiman	Oil	do.	Iran
56.	Saradjeh	Gas	do.	Iran

[a] Green numbers are oil fields; red numbers are gas fields.

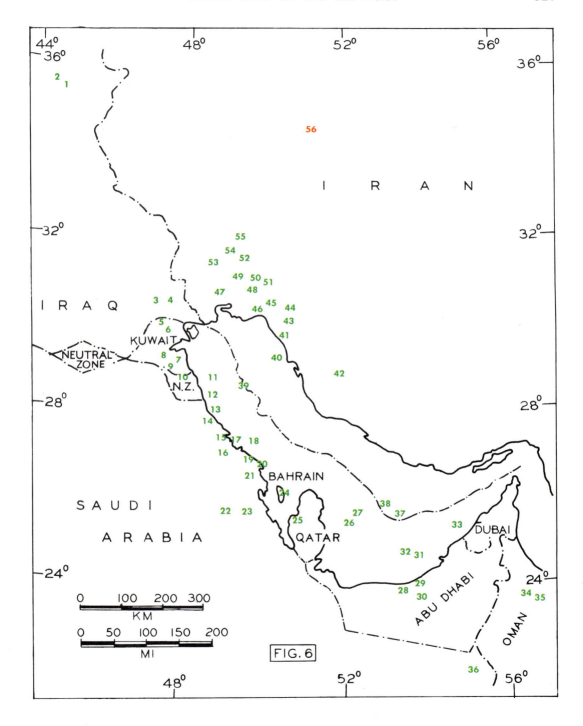

FIG. 6

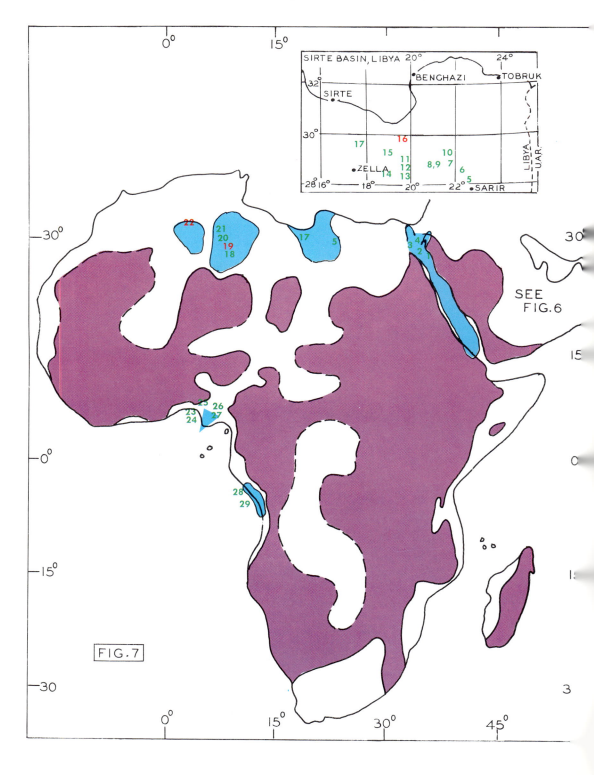

FIG.7

←—‹‹‹‹‹

Fig. 7. Giant Oil and Gas Fields of Africa and Suez.

Map No.	Field	Type	Basin[a]	Country
1.	Barqan	Oil and Gas	Red Sea graben	Saudi Arabia
2.	Amal	Oil	do.	Egypt (Sinai)
3.	El Morgan	Oil	do.	do.
4.	Belayim and Offshore	Oil	do.	do.
5.	Sarir	Oil	Sirte	Libya
6.	"A-100"	Oil	do.	do.
7.	Giola	Oil	do.	do.
8.	Intisar (Idris) "A"	Oil	do.	do.
9.	Intisar (Idris) "D"	Oil	do.	do.
10.	Amal-Nafoora-Augila	Oil	do.	do.
11.	Zelten	Oil	do.	do.
12.	Waha	Oil	do.	do.
13.	Defa	Oil	do.	do.
14.	Samah	Oil	do.	do.
15.	Raguba	Oil	do.	do.
16.	Habeita	Gas	do.	do.
17.	Dahra-Hofra	Oil	do.	do.
18.	Zarzaïtine	Oil	Polignac	Algeria
19.	Rhourde Nouss	Gas	Erg Oriental	do.
20.	Rhourde el Baguel	Oil	do.	do.
21.	Hassi Messaoud	Oil	do.	do.
22.	Hassi-er-R'Mel	Gas	Erg Occidental	do.
23.	Meren	Oil	Niger Delta	Nigeria
24.	Okan	Oil	do.	do.
25.	Jones Creek	Oil	do.	do.
26.	Imo River	Oil	do.	do.
27.	Bomu	Oil	do.	do.
28.	Emeraude Marin	Oil	Cabinda embayment	Congo (Brazzaville)
29.	Cabinda "B"	Oil	do.	Angola

[a] Shield areas are in violet, basins in blue; only those basins with giant fields are shown. Green numbers are oil fields; red numbers are gas fields.

››››—→

Fig. 8. Giant Oil and Gas Fields of Australia, New Zealand, and Malay Archipelago.

Map No.	Field	Type	Basin[a]	Country
1.	Maui	Gas	Maui	New Zealand
2.	Marlin	Gas	Gippsland	Australia
3.	Halibut	Oil	do.	do.
4.	Kingfish	Oil	do.	do.
5.	Barracouta	Oil	do.	do.
6.	Moomba	Gas	Great Artesian	do.
7.	Gidgealpa	Gas	do.	do.
8.	Minas	Oil	Central Sumatra	Indonesia
9.	Seria	Oil	North Borneo Coastal	Brunei
10.	Southwest Ampa	Oil	do.	do.

[a] Shield areas are in violet, basins in blue; only those basins with giant fields are shown. Green numbers are oil fields; red numbers are gas fields.

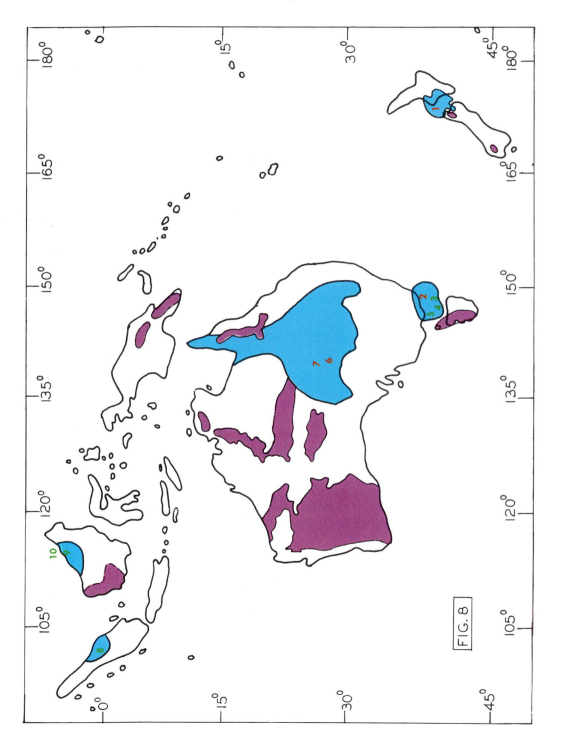

FIG. 8

Officially reported reserves in some Middle East fields have tended to be conservative. Sharp upward adjustments have been announced from time to time. For example, at the end of 1968 the reserves for 10 Saudi Arabian fields were placed by many sources at 84.4 billion bbl. In mid-1969, another calculation was published—141.8 billion bbl—though this estimate has not been confirmed by other sources. Differences of this magnitude suggest wide differences in the methods used for reserve estimation. Regardless, since World War II, significant and substantial upward revisions of reserves in Middle East fields have been made repeatedly.

Lacking other information, we have accepted some rather imprecise reports on the reserves of new fields. Some are based on an extrapolation of the operating company's projected annual rate of production from a newly discovered field. A major problem in reserve calculation in many new fields is the recovery factor to be assumed, this being the least accurately determinable of all the parameters entering into reserve calculations. Some giant oil fields, for example, have recoveries as low as 100 to 119 bbl/acre-ft, whereas others range from 600 to 1,000 bbl/acre-ft.

Acceptance of reserve estimates of such widely varying reliability, although subject to legitimate criticism, can be justified on the grounds that they provide a general measure of the capabilities and relative rank of the giant oil and gas fields of the world, and as such can serve as the basis for discussion of the geologic significance of the data in Tables 1 and 2, and their analysis in Part II of this article.

Geologic province and tectonic environment. —The characterizations of geologic province and tectonic environment shown on Tables 1 and 2 were selected as brief and expressive terms that seem appropriate for this purpose. The importance of tectonic factors is reviewed in Part II of this paper.

The *provinces* named in Tables 1 and 2 are inconsistent in size, because they range from very large basins (*e.g.*, Gulf Coast geosyncline of North America) to subdivisions of major features (*e.g.*, East Texas basin, a part of the Gulf Coast geosyncline). This does not imply, however, that areally small provinces are necessarily subdivisions of larger ones; some may be inherently small (*e.g.*, Ventura basin, California).

The *tectonic environment* generally relates to the general structural habitat of a field, rather than to the basin as a whole, although the two may be so closely related that they are inseparable.

Gas fields.—Giant gas fields are tabulated separately (Table 2) rather than inserting them in a single table with the giant oil fields on the basis of calorific equivalence of the gaseous and liquid hydrocarbons. This is justified because of basic differences (Ion, 1967, p. 30–31) and difficulties in analyzing the two types of reserves; some of the difficulties are:

1. Many giant gas fields have not been identified. In many areas, especially areas where markets are presently nonexistent, operators have not divulged data which can serve even as an "order-of-magnitude" classification of certain discovered but undeveloped reserves. Examples are found in the Persian Gulf region, the Niger delta, and offshore Brunei, and include other undeveloped gas reserves reportedly found in recent years in various parts of the world (*e.g.*, Mozambique, offshore Java, the Gulf of Guayaquil [Ecuador], offshore New Zealand, offshore Republic of South Africa, *etc.*). As a result, the tabulation of gas fields is necessarily incomplete. It will be noted that about half of the fields listed in Table 2 are in the Soviet Union. This fact may reflect the availability of data rather than the existence of an overwhelming number of giant gas deposits in the USSR, but it may also reflect the fact that the Dnepr-Donets (northern Ukraine), Pechora, West Siberian, and Vilyuy basins were exceptionally favorable habitats for gas generation and accumulation.

2. Data on volumes of gas-cap and dissolved gas are rarely available. Some fields, such as Bahrain (Persian Gulf) and Pazanan (Iran), with large gas-cap reserves from the oil ring, are listed in both tables, but undoubtedly there are many others with giant gas as well as oil reserves. This is particularly true in Russia where dissolved gas rarely is included in reserve estimates.

3. In any analysis of total recoverable hydrocarbons in a basin, the gas reserves cannot be equated with recoverable oil reserves because the recovery factor for gas is much higher. To be fully equitable, the only proper comparison would be between oil and gas in place.

4. Many gas fields contain large amounts of condensate. The amount of condensate (NGL) is shown on the tables wherever information was available. However, in the USSR, condensate reserves were not considered at all in some fields because NGL generally has not been used commercially by Russian industry. Wherever possible, recoverable condensate has been included in estimates for Russian fields, but for many fields—within and outside of the Soviet Union—an estimate was not possible.

5. We were undecided about how to treat nonhydrocarbon gases, and our opinions differ considerably. For example, 40 percent of the recoverable 5 Tcf from Mari field, West Pakistan, is CO_2 and N_2. This means that the amount of hydrocarbon gas is less than our lower limit of 3.5 Tcf for a giant. Nevertheless, several fields with large amounts of nonhydrocarbon gas are shown on Table 2 on the following grounds: (1) the

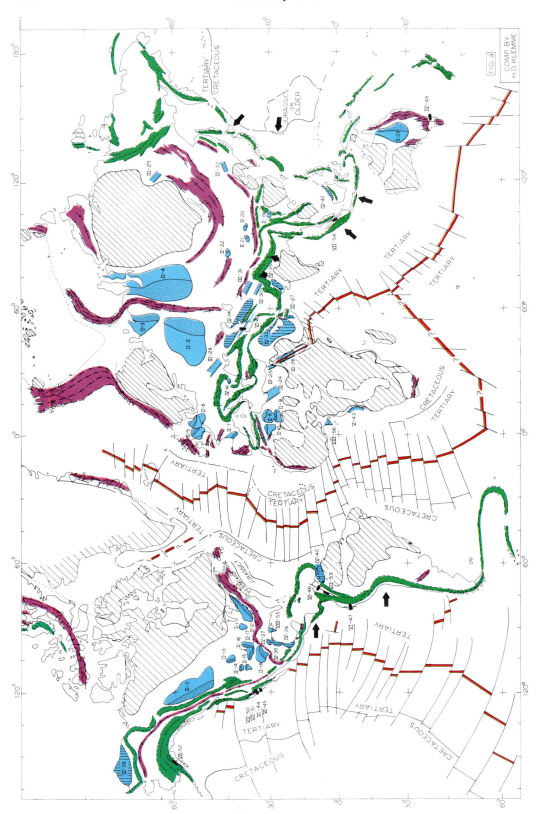

Fig. 9. Basins containing world's giant oil and gas fields.[1]

Map Explanation

LAND GEOLOGY (MODIFIED FROM UMBGROVE (1942), BROWN ET AL. (1968) CHURKIN (1969)

PRECAMBRIAN AREAS - FORMING CRATONIC SHIELD AREAS

PALEOZOIC (HERCYNIAN OR CALEDONIAN) FOLDING

POST-PALEOZOIC FOLDING

OCEAN GEOLOGY (MODIFIED FROM BULLARD (1969), DEWEY AND HORSFIELD (1970), EMERY ET AL. (1970)

CREST OF WORLDWIDE SYSTEM OF RIDGES ALONG WHICH NEW OCEANIC CRUSTAL MATERIAL IS BEING EXTRUDED ("ACCRETING PLATE MARGIN")

FRACTURE ZONE OF "TRANSFORM" MOVEMENT

PRESENT POSITION OF EXTRUDED MATERIAL AS DETERMINED BY MAGNETIC STUDIES INCLUDING AGE DATING

DIRECTION OF UNDERTHRUSTING "PLATE" MOVEMENT

BASIN GEOLOGY

CRATONIC - SIMPLE SAUCER SHAPED BASINS

CRATONIC - COMPOSITE OR MULTICYCLE BASINS

CRATONIC - RIFT BASINS

INTERMEDIATE: DOWNWARP TO OCEANIC AREA

INTERMEDIATE: COASTAL GRABEN BASIN

INTERMEDIATE: INTERMONTANE, SECOND STAGE, TRANSVERSE BASIN

INTERMEDIATE: INTERMONTANE STRIKE BASIN

INTERMEDIATE: UPPER TERTIARY DELTA

[1] Editor's note: This map, compiled by H. D. Klemme, is used as a basis for illustrating the basin classification presented herein. It should be noted that the ages shown in the ocean basins have been greatly generalized.

List of Basins

Map Number	Basin	Location	Number of Giant Fields
I- 1.	Illinois basin, U.S.A.		1
II- 2.	Volga-Ural basin, U.S.S.R.		10
II- 3.	Pechora basin, U.S.S.R.		3
II- 4.	West Siberian basin, U.S.S.R.		29
II- 5.	North Sea basin, Western Europe		4
II- 6.	Netherlands-West German basin		2
II- 7.	Aquitaine basin, France		2
II- 8.	Erg Oriental basin, Algeria		3
II- 9.	Fort Polignac basin, Algeria		1
II-10.	Erg Occidental basin, Algeria		1
II-11.	Alberta basin, Canada		5
II-12.	Appalachian basin, U.S.A.		2
II-13.	Powder River basin, U.S.A.		1
II-14.	Uinta basin, U.S.A.		1
II-15.	San Juan basin, U.S.A.		1
II-16.	Anadarko-Ardmore basin, U.S.A.		5
II-17.	Oklahoma platform, U.S.A.		3
II-18.	Permian basin, U.S.A.		9
II-19.	Szechwan basin, China		2
II-20.	Pre-Nan-Shan basin, China		1
II-21.	Tsaidam basin, China		1
II-22.	Dzungaria basin, China		1
II-23.	Great Artesian basin, Australia		2
III-24.	Sirte basin, Libya		13
III-25.	Suez graben, Egypt		2
III-26.	Red Sea graben, Egypt-Saudi Arabia		2
III-27.	Oman rift, Oman		2
III-28.	Dnepr-Donets basin, U.S.S.R.		5
III-29.	Vilyuy basin, U.S.S.R.		2
III-30.	Tadzhik basin, U.S.S.R.		3
III-31.	Bukhara basin, U.S.S.R.		6
III-32.	Sungliao basin, China		1
IV-33.	Arabian platform and Iranian basin		56
IV-34.	Pre-Caucasus-Mangyshlak-Turkmen basin, U.S.S.R.		9
IV-35.	Indus basin, West Pakistan		2
IV-36.	Colville basin, Alaska		1
IV-37.	East Texas basin, U.S.A.		2
IV-38.	Gulf Coast basin, U.S.A.		10
IV-39.	Tampico embayment, Mexico		4
IV-40.	Maturín basin, Venezuela-Trinidad		3
IV-41.	Pre-Carpathian depression, Rumania		1
IV-42.	North Borneo basin, Brunei		2
V-43.	Cabinda embayment, Angola-Congo		2
VI-44.	Ventura basin, U.S.A.		1
VI-45.	Los Angeles basin, U.S.A.		4
VI-46.	Maracaibo basin, Venezuela		6
VI-47.	Piura basin, Perú		1
VI-48.	Gippsland basin, Australia		4
VI-49.	Baku basin, U.S.S.R.		7
VII-50.	Sacramento basin, U.S.A.		1
VII-51.	San Joaquin basin, U.S.A.		7
VII-52.	Cook Inlet basin, Alaska		1
VII-53.	Middle Magdalena basin, Colombia		1
VII-54.	Central Sumatra basin, Indonesia		1
VIII-55.	Mississippi Delta, U.S.A.		4
VIII-56.	Niger Delta, Nigeria		5

ultimately recoverable amount of hydrocarbon gases is not yet known and may exceed present estimates; (2) condensate production will offset the volume of nonhydrocarbon gas, and bring the field in question into the giant category; and (3) use may be found for CO_2, N_2, and other nonhydrocarbon gases—a use which may make the CO_2, N_2, and other such gases a positive element in the ultimate "economic balance sheet" of the field.

Other factors.—The possible significances of evaporites and unconformities, as well as other factors that bear on the occurrence of giant oil and gas fields, are discussed in Part II.

Conclusions

Two hundred and sixty-six giant fields as herein defined and shown on Tables 1 and 2, contain a high percentage of the world's total produced and proved oil reserves. Addition of the gas reserves computed on a calorific oil-equivalent basis (6,000 cu ft of gas = 1 bbl of oil) reveals that 62 percent of all the giant-field reserves are in sandstone and 38 percent in carbonate reservoirs. The amount of oil and gas reserves in Tertiary rocks is 24 percent, Mesozoic 63 percent, and Paleozoic 13 percent. Excluding the Persian Gulf region, Tertiary rocks contain 40 percent, Mesozoic 39 percent, and Paleozoic 21 percent. The fact is noteworthy that 58 percent of all giant fields are in an arcuate, U-shaped band, 750 to 1,300 mi (1,250–2,000 km) wide which extends 6,000 mi (10,000 km) from central Algeria to the Arctic Ocean. *The significance of this latter fact is not known, but it is a fact and, as such, must have some explanation.*

Part II

Factors Affecting Formation of Giant Oil and Gas Fields, and Basin Classification[8]

MICHEL T. HALBOUTY, ROBERT E. KING, H. DOUGLAS KLEMME,

ROBERT H. DOTT, SR., and A. A. MEYERHOFF

Abstract Oil fields of the world with recoverable reserves more than 500 million bbl and gas fields with reserves more than 3.5 Tcf are analyzed to determine what characteristics they have in common and, on the other hand, reasons why some of these fields have unusual characteristics.

Giant hydrocarbon accumulations require that there be a giant trap, formed more or less concurrently with generation of the hydrocarbons from the organic source materials, and abundant source beds rich in organic matter. Marine sediments are dominant as source beds, but nonmarine beds also preserved the organic materials which supplied the hydrocarbons to many giant fields. Although argillaceous sediments generally trapped more organic matter, carbonates also can be sources.

The reservoir rocks of most giant fields are very porous and permeable, but there are notable exceptions; in the exceptions, total reservoir volume compensates for deficiency in reservoir quality. The reservoir rocks must be interconnected with channels of migration, or carrier beds, from the source beds. An effective seal must be present to prevent escape of hydrocarbons from the reservoir; the most efficient seal is provided by evaporites. Evaporites, in addition to sealing many important reservoirs, also may be the primary agent responsible for the development of the structures of many large fields—generally through either diapirism or *décollement*. Unconformities have had an important role in some fields in aiding trap development and in bringing carrier beds into juxtaposition with others through which hydrocarbons may migrate. Giant traps caused by lateral facies changes, changes in reservoir matrix, postdepositional diagenetic changes, and paleogeomorphic factors are few, but their absence does not mean that they are scarce; most exploration has been directed toward the structural types of trap, and searches for other types of traps have been few.

Though most giant-field reserves are in Mesozoic and Tertiary rocks, there is no preferred age of source beds or reservoir rocks. The important factor is the time when, during the sedimentational cycle of a basin, the largest amounts of organic matter were buried and preserved. It makes little difference if this time was Cambrian or Pleistocene.

Higher than normal geothermal gradients probably resulted in greater efficiency of hydrocarbon generation in certain basins.

The sources of gas include the same types of materials that generated liquid oil, but they also include other materials of vegetal origin that do not contribute significantly to the formation of oil. Hence gas can be derived from a greater variety of source materials. Volumes of gas and oil generally are inversely related to increasing depth. Downward increase of temperature results in the "phasing out" of oil and the dominance of gas. Liquids disappear and are replaced by gas in the range 5,640–8,380 m (18,500–27,500 ft); the actual depth depends on the geothermal gradient present in the field under study.

[8] Manuscript received, May 4, 1970; revised, June 6, 1970.

The depths given are not necessarily the present overburden thicknesses; many reservoirs are affected by weight of overburden of an earlier geologic time, which today may no long exist because of subsequent uplift and erosion.

We present for consideration a basin classification having different spatial relations with cratons and continental borders, and having different evolutionary development. The type of basin having the largest number of giant fields is the basin which is, or was, a downwarp toward an oceanic area. Intermontane basins, developed either as geosynclinal-type basins between rising geanticlines or in transverse downwarps, generally are smaller in size, but many such basins contain clusters of very large fields. Interior cratonic areas exposed to long periods of erosion and tectonic and/or epeirogenic activity also contain numerous large fields, but a smaller percentage of giant fields.

During the past 15 years, deeper drilling and exploration of new basins have led to the discovery of an increasing percentage of Mesozoic vs. Tertiary giant fields. The greatest future prospects for discovery of giant fields are (1) in the continental shelves of the world; (2) onshore in Asia; (3) unexplored basins containing mainly continental strata; and (4) all areas of the world where deliberate exploration for obscure traps has not been carried out—i.e., stratigraphic, unconformity-associated, and paleogeomorphic traps.

SECTION 1: FACTORS AFFECTING FORMATION OF GIANT OIL AND GAS FIELDS

In this section, we seek to find parameters of giant oil and gas fields which may explain their unusual occurrence in most productive provinces, and their ubiquity in a few of the world's sedimentary basins.

General[9]

Trap size.—A most important requisite for a giant field is a large trap with a great reservoir volume, either in a single reservoir or in multiple reservoirs, into which the oil and/or gas of a basin can migrate. Some reservoirs have great vertical thickness and others wide lateral extent (which compensates for a lesser thickness).

There is a striking contrast between some producing regions, such as north-central Texas which has a multiplicity of small traps—anticlines, porosity wedgeouts, truncations by unconformities, and reefs—with a great petrologic and textural diversity and a variety of geologic ages, and other regions in which a large part of the petroleum is stored in a single giant reservoir (*e.g.,* Burgan [Kuwait] and

[9] *Editor's note:* The various factors discussed are not necessarily in the order of importance. The fact should be borne in mind that any one of these factors may be the primary requisite in a particular region because of conditions peculiar to that region; the same order of importance of factors might not apply to another region.

Ghawar [Saudi Arabia]). Had they had a different tectonic history, some of the small field basins would have produced giant fields from the same volume of source materials.

Conversely, giant traps are present in some basins, but only small volumes of oil or gas have accumulated. An example is the Porcupine dome of central Montana. The essential barrenness of such large structures must be attributable to lack of a suitable source-reservoir relation, or to the time of formation of the structure (Kranzler, 1966). Several large structures in the Cook Inlet-Kenai basin of Alaska are barren because they formed after migration of hydrocarbons had taken place in the basin (Halbouty, 1966, p. 17).

A review of the types of trap among the giant fields listed in Tables 1 and 2 shows that, although most are anticlinal, migrating oil or gas will accumulate in a wide variety of reservoirs, some of them of rather exotic types. Examples of the latter are Quiriquire (Venezuela), Kern River (California), Mene Grande (Venezuela), Fyzabad (Trinidad), Amal-Nafoora-Augila (Libya), and La Paz (Venezuela). Many large reservoirs are filled nearly to, or to, the spill point of the trap if large quantities of migrating hydrocarbons were available during the primary time of petroleum generation.

Timing of trap development.—In terms of the geologic time scale, hydrocarbon generation from source beds takes place in a short span of time. The existence of large Pleistocene accumulations along the Gulf Coast geosyncline (Andrews, 1961) indicates that hydrocarbon generation, migration, and entrapment can, and do, occur within time spans of less than 1.0–1.5 million years. However, not all Pleistocene accumulations appear to be as large as underlying Pliocene accumulations (Meyerhoff, 1968). Therefore, it appears probable that time spans up to 10 m.y. are required for maximum accumulation to take place. This time span is corroborated by the existence of giant Pliocene accumulations in the Gulf Coast geosyncline, the Southern California basins, the Caspian Sea area, and the Tsaidam basin of northwestern China.

Relative tectonic quiescence (platform and semiplatform tectonic environments of Beloussov, 1962) is a prerequisite for giant field formation. The time span required for a giant accumulation to form is meaningless if tectonic activity disrupts communication between the source beds and the trap.

Because hydrocarbons are generated and migrate so soon within the newly deposited sedimentary sequence, the formation of large traps, contemporaneously with generation and migration, is a significant factor in giant-field formation. The existence of large traps at the time when subsurface temperature, overburden weight, and degree of compaction were sufficient to mobilize hydrocarbons in the source beds and to initiate migration into the reservoirs is important to entrapment of large volumes of petroleum. Growing anticlines, growing fault blocks, active growth faulting, and syndepositional development of porosity wedgeout and truncation traps provide these optimum conditions. Progressive growth of intrabasinal structures during the time of major deposition especially characterizes cratonic rift, second-stage transverse, and continental border geosynclines. Late-stage structures, formed in a second cycle of deformation, are more likely to entrap gas than oil. Gas can migrate greater distances and, during readjustment of structural patterns during late-stage structural evolution, gas is susceptible to remobilization (Silverman, 1965).

Hydrocarbon sources.—An obviously fundamental requirement for giant field accumulation is a favorable environment for abundant life, and for the accumulation, preservation, and conversion of organic matter to petroleum (Dott and Reynolds, 1969). Thus, organic-rich source sediments are a prerequisite.

In most basins the quantity of organic matter deposited in the rocks greatly exceeds the amount that is converted into hydrocarbons which, in turn, migrate and are trapped in exploitable reservoirs. The most significant factors in formation of giant fields are proximity of rich hydrocarbon sources to giant traps, with carrier beds or migration channels approximately parallel with bedding, by which the hydrocarbons could move into the reservoir. This is particularly true of oil; migration of gas is possible for greater distances and through rocks having lower permeability (*see* Silverman, 1965).

Argillaceous source beds generally are more easily related to petroleum in reservoirs than are carbonate sources. Hydrocarbons in the giant fields of Libya have been related by some workers to the nearby Upper Cretaceous and Paleocene dark shales; those of Algeria possibly to Silurian graptolite shale; and those of the Los Angeles and Ventura basins to shale interbedded with the reservoirs. Gulf Coast petroleum can be related clearly to sources within the beds present in each fault block (Meyerhoff, 1968). Oil generated from carbonate source rocks may have migrated shorter distances from source to trap, aided by fracturing of the rock and development of solution permeability after lithification (Hunt, 1967). Although the relative richness of argillaceous to carbonate organic sources generally favors the former, the efficiency of primary migration probably is greater in the carbonates, because much of the petroleum in some thick mud and clay sections (now shale) remained locked *in situ* because of the lack of interbedded porous zones.

Large quantities of liquid hydrocarbons in many basins have been converted to tar sands (*e.g.,* Vigrass, 1968). If one attempts to calculate the total amount of hydrocarbons generated in a basin, the quantity now locked up in asphaltic deposits must be taken into account. These undoubtedly represent a much larger original volume of liquid oil.

Marine content of total sedimentary section.—Most giant oil fields are in basins with a predominantly marine section (Hedberg, 1964, 1968; Welte, 1965; Dott and Reynolds, 1969). The Persian Gulf region is a good example of a thick, continuous sequence of marine strata from the pre-Permian to the lower Miocene, except for a Triassic continental regressive phase.

Though marine sections provide the preferred habitat for giant fields, there are notable exceptions. Giant gas fields are more likely to occur in brackish-water to continental environments, because the organic sources for some of these fields evidently were humic and lignitic. Some oil fields, such as Tertiary fields of the Gippsland basin off Australia, a few of the giant Lower Carboniferous accumulations in the Volga-Ural district of Russia (Vasil'yev, 1968 a, b; L'vov, 1968), and five of the six Chinese giant fields, are associated with paralic to nonmarine sections (Meyerhoff, 1970b). High wax crude oils such as at Sarir (Libya), Minas (Indonesia), and Uzen' and Zhetybay (Mangyshlak Peninsula, USSR) may have a nonmarine source (Aronsov and Makhonin, 1966; Hedberg, 1968).

Extension of exploration in recent years to a wide diversity of basins throughout the world has resulted in the discovery of several giant fields in continental depositional environments, once generally regarded as unfavorable for the source of hydrocarbons in large quantities. In addition, advances in geochemical and analytical techniques now permit recognition of a nonmarine source for many crude oils. Major hydrocarbon accumulations in continental environments are now well established. However, the intensity of tectonic activity, accompanied by episodes of uplift and widespread erosion, in many intermontane basins has led to destruction of numerous petroleum reservoirs, as shown by the common occurrence of tar sands in such areas.

In the United States the best-known examples of oil fields in nonmarine sequences are in the Eocene of the Uinta basin of Utah. The oil and gas of the Dzungaria, Tsaidam, Tarim, Turfan, Ordos, Pre-Nan Shan, and Sungliao basins of Mainland China were generated in continental sediments (Meyerhoff, 1970b). Other examples of oil and gas generation in continental environments are the Comodoro Rivadavia field complex of Argentina (Upper Cretaceous, Table 1), The Upper Triassic lacustrine sandstone reservoirs of Mendoza and San Juan Provinces, Argentina, and the early to late Tertiary continental strata of the Cook Inlet-Kenai basin of coastal Alaska. It is now generally accepted that large volumes of gas in the Eastern Venezuela basin, in the Permian and Triassic of the North Sea region, and other areas were generated from coal measures (Hubbard, 1950; Banks, 1959, 1966; Patijn, 1964; see also Asquith, 1966, and Krueger, 1968).

Reservoir rock.—The majority of giant field reservoirs, whether carbonate or terrigenous clastic, are very porous and permeable. None of the giant fields is productive from graywacke sandstones derived from a basic igneous source. Provenance of sand from silicic terranes is believed to be a prerequisite for large fields in sandstone reservoirs.

Poor reservoir development is unusual in giant fields. However, a very porous and permeable reservoir is not a *requirement* for the formation of giant accumulations. Notable examples of relatively poor quality reservoirs are the Oligocene-lower Miocene Asmari Limestone (the main producing zone of the Iranian fields), the Early Carboniferous Bobriki suite of the Volga-Ural district (USSR), the Devonian Bradford Sandstone of Pennsylvania, and the Late Cretaceous Cardium Sandstone of the Pembina field of Alberta. Deficiency in reservoir properties can be compensated amply by great thickness of hydrocarbon column or by great areal extent of the productive area.

The reservoir rocks in the trap must be connected with the source beds by carrier beds. Generally these are the same lithologic units as the reservoir beds themselves. They may, however, be older or younger than the reservoir rock. Carrier beds may be porous and permeable zones related to unconformities, sequences of fractured strata, porous and permeable gouge in fault zones, or zones of secondary porosity.

Evaporites.—There is a wide involvement of evaporites in source rocks, reservoir, and traps in many giant fields (Buzzalini *et al.*, 1969). Most evaporites were deposited in physically silled basins, though the sills may be chemical gradients (Scruton, 1953; Busson, 1968). Silled basins also provide optimum conditions for preservation and accumulation of organic material—hence the common association of evaporites with hydrocarbon-rich source beds.

Evaporites may be interbedded in the reservoir, form the caprock above a porous reservoir, seal a reservoir updip in a stratigraphic trap (*e.g.*, Slaughter [Texas]), or produce the structure by diapirism or formation of a sole of *décollement* (*e.g.*, Zagros Range, Iran, and Szechwan basin, China). In the Persian Gulf region evaporites play a triple role—forming the structure by *décollement* on a sole of Lower Jurassic evaporites, diapirism of late Proterozoic to Cambrian salt, and providing the caprock of many of the producing reservoirs (Jurassic). Examples of oil and gas accumulations where the association of evaporates is significant are noted in Tables 1 and 2.

Still another role of evaporites is the creation in some areas of solution-collapse pseudo-anticlines which provide important traps. Such structures are especially important in the High Plains area of the United States and parts of Canada (Parker, 1967; Swenson, 1967).

Caprock.—A satisfactory seal over a reservoir is essential to arrest upward migration of hydrocarbons. In some basins with a dominance of terrigenous clastic rocks, much hydrocarbon content has been lost because the cap above a porous stratum did not prevent escape of hydrocarbons to the surface or contamination with meteoric waters (which may result in oxidation of the hydrocarbons and conversion to a tar mat). The most efficient caprocks are evaporites. The common occurrence of evaporites in carbonate sequences enhances the efficiency of hydrocarbon accumulation in carbonate reservoirs.

Unconformities.—As shown on Tables 1 and 2, the hydrocarbon reservoirs of most known giant fields are not significantly associated with unconformities. This is especially notable in the giant fields of the Persian Gulf region, where sedimentation was essentially continuous from the late Paleozoic to the Tertiary. Nevertheless, unconformities have a major role in aiding trap development and, with the associated basal sandstones of the overlying transgressive sequence or the regressive strata below, have aided in migration of oil and gas from the source to the reservoir.

The Hassi Messaoud field of Algeria is of particular interest, because the reservoir is a Cambrian sandstone capped by a nonmarine Triassic siltstone. An organically rich Silurian graptolite shale, truncated far down the flanks of the uplift by the pre-Triassic unconformity, is believed by many to have been the original source for the oil.[10]

Geologic age.—Although most of the reserves of giant fields are in Mesozoic and Tertiary rocks, a sufficient volume is in Paleozoic rocks to indicate that the hydrocarbon-generating potential of sedimentary strata has not changed significantly through Phanerozoic time. The minority position of Paleozoic reservoirs among the giant fields may be explained as the result of destruction of all or parts of some of the ancient sedimentary basins by orogenies and by deep erosion.

As more of the world's sedimentary basins are explored, some geologic systems once regarded empirically as unlikely sources of major oil supplies have been proved to be highly important. The Triassic, for example, has produced on a small scale at Tupungato, Mendoza Province, Argentina, since 1934; more importantly at Karamai, Dzungaria basin, Sinkiang Province, China, since 1955; and most recently, the discovery at Prudhoe Bay, Alaska, confirms that great amounts of oil were generated in that system. Important Triassic reservoirs also are present in the Vilyuy basin, north-central Siberia, and the Szechwan basin, China. In the 1940s it was similarly demonstrated that the Jurassic is very important, as giant fields were found and developed in Saudi Arabia.

Statistically, approximately 50 percent of the reserves in giant fields reservoired in Paleozoic rocks are gas; and 87 percent of the total oil

and gas reserves come from Mesozoic and Tertiary rocks. More investigation and analysis of these statistical variations and of the character of Paleozoic and post-Paleozoic sediments are required.

There is no magic in geologic age *per se.* Each basin went through its own cycle or cycles of maximum marine sedimentation, and the part(s) of the cycle(s) in which the greatest number of organisms lived, and the greatest amount of organic matter was buried became the principal hydrocarbon source(s).

Truly striking examples of the statement that there is no magic in geologic age *per se* are the discovery of indigenous Proterozoic gas in the Amadeus basin of Australia, and the discovery of commercial (175 million bbl plus) Early Cambrian and Vendian (late Proterozoic) oil in the Markovo field of the Irkutsk amphitheater, Irkutsk Oblast', central Siberia (for various views on hydrocarbon potential and test results in these areas, *see* Glaessner and Daily, 1959; Glaessner, 1962 and other publications; Trofimuk *et al.*, 1964; Murray, 1965; Vasil'yev, 1968a, b).

Geothermal gradient.—Insufficient data are available to tabulate for all the giant fields the geothermal or paleogeothermal (where large-scale uplift and erosion have occurred) gradients associated with them. However, geothermal conditions may have a considerable effect on the degree to which petroleum hydrocarbons have been generated from the source materials and on their mobility during primary migration.

Philippi (1965) observed that, in the petroleum-generation process, temperature rather than pressure is the more important factor. The rate of most chemical reactions is increased by 2 or 3 times for each 10° C (18° F) rise in temperature. The amount of oil generated increases only linearly with time but exponentially with temperature. Philippi gave examples of the effect of temperature on maturity of shale hydrocarbons from the Los Angeles and Ventura basins of California; in the Los Angeles basin the average temperature gradient is 4.77° C/100 m (2.62° F/100 ft), and in the Ventura basin 2.60° C/100 m (1.46° F/100 ft).

The temperature of petroleum migration has been related to clay dehydration by Burst (1969). As compaction progresses, about 70 percent of the water is expelled during the initial phase, mostly from pore spaces and by increasing pressure. In the succeeding phase pressure is relatively ineffective, but with increased temperature part of the water within the clay

---

[10] *Editor's note:* Some workers have considered the possibility that the Algerian hydrocarbons originated in the Triassic continental beds, some of which are lacustrine. We express no opinion regarding this possibility, but believe that this viewpoint should be mentioned.

lattices is expelled, and causes a reduction of 10–15 percent of the compacted bulk volume. This volume corresponds to a change from montmorillonite to illite. The optimum period in the process for fluid redistribution and hydrocarbon migration is the second stage. Burst related this expulsion of water from clay lattices to the depths of principal oil production in the Gulf Coast geosyncline; the depth to production changes with the geothermal gradient.

Higher than normal geothermal gradients probably account for the efficiency of hydrocarbon generation from source rocks in certain basins, such as the Pre-Caucasus (including Baku), Los Angeles, and Central Sumatra basins. Where a sufficiently high temperature is reached at a shallow depth, the hydrocarbons are enabled to migrate into sandstone reservoirs before greater depth of burial has reduced their porosity by secondary quartz overgrowth.

Stheeman (1963) has shown in an interesting way the theoretical effect of the changes in temperature and pressure, in the source and reservoir rocks, on the migration and accumulation of hydrocarbons in the southwest Netherlands basin.

Factors Favoring Gas Versus Oil

The source of gas commonly includes the same source beds that generate liquid oil. However, gas sources also include other materials (chiefly coals), of vegetal origin, which do not contribute significantly to the generation of oil. The fact that gas can be generated from more than one type of source affects any reserve calculations made on the basis of the volume of source sediments originally available.

Generally, volumes of gas and oil reserves have an inverse relation with increasing depth. Exceptions to this statement include the Dnepr-Donets (northern Ukraine) basin, where the productive area of the gas- and oil-bearing strata increases with depth, and produces the "fir-tree" cross-section effect (Blank *et al.*, 1964). Such exceptions seem to be related to the depositional environment of the source rocks at different times in the basinal history.

Where depositional environments of source materials were similar through time, downward increase in temperature caused "phasing out" of oil and dominance of gas. Increased pressure resulting from weight of overburden decreases the pore volume, and results in (1) diminished recovery of liquid hydrocarbons, and (2) compression of greater volume of gas in the pore space available. Most giant gas fields are productive from depths shallower than 3,000 m (10,000 ft), but the depth relations outlined above are important with regard to such deep gas accumulations as Gomez field, Texas.

Concerning the phasing out of liquid hydrocarbons, Landes (1967) observed that hydrocarbon density decreases with increasing depth. At reservoir-rock temperatures greater than 350° F (177° C) it is unlikely that liquid hydrocarbons can exist. Liquids would not exist below 5,640 m (18,500 ft) with a thermal gradient of 2° F/100 ft, or below 8,380 m (27,500 ft) with a thermal gradient of 1° F/100 ft. This is applicable to depths below the present surface only for sedimentary rock sequences (such as on the Tertiary part of the U.S. Gulf Coast) which have not been uplifted and eroded at some time during the geologic past. Where the reservoir rocks have had an overburden that has later been partly removed by erosion, the density of hydrocarbons present is that which was imposed during the greatest earlier depth of burial, as the chemical processes that occur at higher temperatures generally are not reversible.

In the foreland zone of folded miogeosynclines the same thermal effect (as from depth of burial) has resulted in conversion of hydrocarbons to dry gas. Landes (1967) referred to the induration of sedimentary rocks in such areas as "eometamorphism." The accompanying processes have long been a part of the carbon-ratio theory, under which respective oil- and gas-bearing zones may be differentiated.

Some large gas fields, notably those of the North Sea basin, probably originated from second-stage coalification of the underlying Carboniferous coals, as a result of the weight of overburden imposed by late Mesozoic and Tertiary sedimentation. Volatile hydrocarbons were expelled and moved upward to accumulate in Permian and Triassic reservoir rocks.

The giant gas fields of the northern West Siberian basin are associated with a regressive, nonmarine-to-brackish facies of the Cenomanian (middle Cretaceous). Terrestrial vegetal debris probably formed lignins and humus which were the sources of the gas.

Although few delta gas fields are listed in Table 2, the small number is chiefly a result of lack of information. Large gas reserves are present in the Mississippi, Niger, and Nile deltas (*see* Beebe and Curtis, 1968). The sediments include a high proportion deposited in brackish water, with a great amount of terrigenous organic matter—a potential source of large quantities of methane.

Clastic sediments in general contain humic and lignitic terrestrial organic matter. Organic matter associated with carbonates, on the other hand, is chiefly proteinaceous, and is generally of marine origin. Shallow gas accumulations in many sandstone reservoirs probably originated from the terrigenous organic matter in the associated argillaceous rocks.

The foregoing points probably account fairly satisfactorily for most of the giant gas fields listed. However, others such as Panhandle-Hugoton and Jalmat are anomalous. Because gas can migrate much more freely than oil, some of the major accumulations in carbonate reservoirs may have been derived from deeper or more distant sources. Another factor to be considered is that the gaseous hydrocarbons in some fields of this type originally may have been liquids and were converted to gas following deep burial. They remained in the gaseous phase after uplift and erosion transferred the reservoir to an environment of low temperature and pressure.

SECTION 2: BASIN CLASSIFICATION

In the preceding pages the factors conducive to formation of giant fields within a sedimentary basin have been examined, as well as the factors likely to have produced giant reserves of liquid versus gaseous hydrocarbons. We now seek to relate the incidence of giant fields to sedimentary basins of different types. For the purposes of this discussion, eight major basin types are recognized. Three of these types are developed on continental-type crust of the cratons; the other five are developed on intermediate-type crust at the borders of present or former continental plates. In addition to the basin types discussed herein, there are also sedimentary basins beyond the continental margins, such as the Blake Plateau, floored with oceanic crust (Hedberg, 1970). Though many of these oceanic basins have favorable oil and gas prospects they are not discussed here because they are not presently productive and are beyond the reach of presently economic operations.

Relation of Field Size to Basin Tectonics

The classification given here generally follows the tectonic types of basins defined by Weeks (1952) and Uspenskaya (1967). Although cratonic basins (*i.e.*, basins underlain by continental crust) predominate in continental areas, some cratonic basins may extend offshore for considerable distances (North Sea, United Kingdom; Pechora, USSR; Aquitaine,

France, basins), whereas intermediate basins (*i.e.*, basins underlain by crust intermediate in type between that beneath continents and that beneath oceans), though generally associated with the continental margins, may be found in areas now considered continental (San Joaquin, California; South Caspian and Baku areas, USSR, basins). Therefore, the present continental margin does not everywhere correspond to the division between oceanic and continental crust.

This latter fact may result from the emplacement of oceanic crustal material along mid-ocean ridges according to an hypothesis of sea-floor spreading (Dewey and Horsfield, 1970; Fig. 9). According to this hypothesis, fracturing (rifting) has separated, or is separating, some continental areas and cratonic basins (North Sea, Pechora, Aquitaine), whereas other continental margins are being moved together (South Caspian). Many of the cratonic basins are associated with Precambrian and Paleozoic fold belts which seem to form a concentric pattern around Precambrian shield areas (Fig. 9). In contrast, the intermediate-crust basins appear to be associated more closely with the post-Paleozoic fold trends. This observation suggests the possibility that many cratonic basins might be related to pre-Mesozoic continental accretion, whereas the younger intermediate basins might be related to post-Paleozoic sea-floor spreading[11].

The generally younger ages of the reserves in intermediate basins (Table 5) indicate a post-Paleozoic origin of almost all of these basins. Toward the continental margin, or from the zone of cratonic basins toward the zone of intermediate basins, the basin sediments not only become younger (Weeks, 1952), but many basins are smaller and relatively richer in hydrocarbons (see estimates of hydrocarbon recovery per cubic mile of sediments, Table 5). Approximately 63 percent of the total reserves of all giant fields (526 billion bbl or gas equivalent) are found in the intermediate basins and 87 percent of all reserves in giant fields are in

[11] *Editor's note:* It should be clearly emphasized that the authors are fully aware that there are geologists who do not agree with the sea-floor spreading or continental drift concept, and, consequently may not agree with the basin classification outlined here. In fact, among the authors of this paper there are differences of opinion on the concept (*see* Meyerhoff, 1970a), as well as on the classification. Nevertheless, the authors unanimously believe that the discussion presented here may generate ideas which could lead to other basin concepts and/or classifications, or modification of that given in this paper.

Table 5. Basin Types and Characteristics[1]

CRUSTAL TYPE	CRATONIC			INTERMEDIATE				
BASIN TYPE →	TYPE I INTERIOR SIMPLE SAUCER SHAPED	TYPE II INTRA-CONTINENTAL-COMPOSITE FORELAND SHELF, REMOTE INTERIOR AND INTERMONTANE	TYPE III GRABEN OR HALF GRABEN (RIFT)	TYPE IV EXTRA-CONTINENTAL-DOWNWARP TO SMALL OCEAN BASIN	TYPE V STABLE COASTAL GRABEN OR "PULL APART"	TYPE VI INTERMONTANE SECOND STAGE TRANSVERSE	TYPE VII INTERMONTANE SECOND STAGE STRIKE	TYPE VIII UPPER TERTIARY DELTA
EXAMPLE (FIELD NUMBER)	I-1	II-10	III-24	IV-37	V-43	VI-44	VII-53	VIII-55
Total reserves (in billions of bbl)[2]	0.67	240.0	50.0	450.0	1.7	54.0	12.0	8.0
Number of basins[3] with giant fields	1	22	9	10	1	6	5	2
Number of giant fields	1	89	36	90	2	23	11	9
Number of basins with giant fields found after World War II	0	5	5	2	0	1	1	1
Estimated recovery of oil or gas (bbl[2] per cu mi of sediments in basin)								
a. Range from poorest to richest of basins	10,000 to 70,000	40,000 to 170,000	25,000 to 250,000	50,000 to 450,000	5,000 to 40,000	100,000 to 2,500,000	50,000 to 1,000,000	175,000 to 220,000
b. Average	35,000	100,000	140,000	200,000	18,000	250,000	150,000	200,000
Reserves by reservoir type (billions of bbl)[2]	Sandstone 0.34 Carbonate 0.33	195.0 45.0	31.5 18.5	200.0 250.0	1.7? ?	54.0 —	12.0 —	8.0 Ss. 61% — Car. 39%
Age of reserves (billions of bbl)[2]								
Tertiary	—	7.0	12.0	80.0	1.1	54.0	12.0	8.0→174=23%
Mesozoic	—	43.0	30.0	365.0	0.6	—	—	— →538=65%
Paleozoic	0.67	90.0	8.0	5.0	—	—	—	— →104=12%
Structural trap types	Regional arch	Regional arches, reefs	Tilted blocks, horsts, reefs	Large, growing anti-clines and fault blocks	Fault block with strati-graphic change?	Faulted anti-clines and fault blocks	Faulted folds with strati-graphic change	Flowage structures, "rollover" anticlines
Relation to evaporites	None	Common	Common	Common	Possible?	Absent	Absent	Either salt or uncom-pacted clay
Percent of basin's reserves in the largest fields of basin								
a. Range of largest field in all basins	8–18	6–70	15–45	5–50	All producing basins average ↓	11–90	14–50	3–7
b. Average of first four largest fields	13, 10, 8, 5	32, 9, 6, 5	27, 11, 8, 8	22, 14, 7, 3	55, 15, 10, 6	36, 9, 5, 4	27, 11, 7, 5	5, 3, 3, 3
Hydrocarbon types— percent gas in giant fields	?	65	38	4	?	1	15	7
Gas content in basins this type	Average	Much above average	Average	Average	Low	Below average	Below average	Much above average
Gravity of oil	High	High	High	Intermediate to high	Intermediate	Low to inter-mediate	Variable	Intermediate to high
Sulfur content High =0.5%+ Low =0.5%−	Carb.—high Ss.—mixed to low	Carb.—high Ss.—mixed to low	Low	Intermediate to high	Low	High to variable	Variable	Low

Compiled by H. D. Klemme.
Based on oil (or gas equivalent at 6,000 cu ft = bbl).

[3] There are 56 basins with giant fields (*see* Figures 1-8).

post-Paleozoic sedimentary rocks. Although the volume of unmetamorphosed sedimentary rocks has not been calculated according to different geologic ages, the age of present reserves is related disproportionately to geologic time and also is apparently disproportionate to the available volume of sedimentary rocks of the Cenozoic, Mesozoic, and Paleozoic.

The following is a brief characterization of the different types of basins recognized herein.

Cratonic-crust basins.—I. *Interior: simple, saucer-shaped basins* are in continental interior areas and have a single cycle of sediments, generally of Paleozoic age and in the form of shelf or embayment facies (*e.g.,* Illinois, Michigan, and Williston basins, USA). Basement uplifts or major arches within the basin over which stratigraphic variations occur provide the principal trap types. These basins have produced only one giant field (Old Illinois) and have generally low hydrocarbon recovery rates per unit volume of sediments. A possible reason for the small number of giant fields in such basins is the vulnerability of the craton to repeated uplift and erosion.

II. *Intracontinental: composite or multicycle basins* are in the more exterior parts of continental cratonic areas, and commonly are associated with a series of first-cycle Paleozoic carbonate shelf or platform sediments overlain by second-cycle terrigenous clastics derived from a nearby orogenic uplift or orogen (Fig. 9 and Table 5). Such basins differ considerably in size—ranging from subcontinental and miogeosynclinal (*e.g.,* West Texas-New Mexico Permian [USA], Volga-Ural [USSR], Alberta [Canada], Erg Oriental and Occidental [Algeria]) to small cratonic intermontane basins (Rocky Mountain basins [USA]). They also differ in shape and general tectonic character. Further subdivision of this type of basin obviously is required.

Within this broad group, certain characteristics appear to stand out. Depositionally these basins are generally embayments where wide, flat-bottomed seaways developed on the cratonic platform. Evaporites are common in these basins, either as local caprock or basinwide caprock, and recovery of hydrocarbons per unit volume of sediment is greater in the basins which have evaporites than in similar basins without evaporites. In many places, unconformities which developed over major cratonic, basement uplifts localize the accumulations. In some basins the largest field holds 70 per cent of the basin's recoverable reserves, whereas in

others the largest field has only 6–7 percent of the recoverable reserves. In the latter, small stratigraphic and structural traps play a dominant role in accumulation, whereas in the former, with a large percentage of basin reserves in one field, a major uplift or arch which dominates the basin has been the locus of accumulation (*e.g.,* Romashkino and Novoyelkhov-Aktash fields on the Tatar arch, Volga-Urals basin, USSR). In the smaller and more deformed basins (such as intermontane basins of the Rocky Mountains which display considerable uplift at their margins), post-uplift erosion may have truncated and breached fields on these margins. Truncation and loss of hydrocarbons result in low recovery per unit volume of sediments, whereas the larger, less deformed basins with major basinwide uplifts and hinge lines have greater recovery factors.

Class II basins have a disproportionate share of gas. This can be related to sedimentary facies and basin history. The clastic sequences which are so abundant in these basins have large volumes of coaly material—an important source of methane—whereas the carbonates in these basins contain much proteinaceous material instead. In many basins the overburden pressure and higher temperature caused by a great thickness of sediments deposited above the hydrocarbon source beds may have caused a conversion of liquid hydrocarbons to gas. Although the reservoirs may now be at shallower depth because of uplift and erosion, the hydrocarbons remain in a gaseous state.

III. *Graben or half-graben rift basins* are in the more exterior parts of cratons. Many had open access to oceanic basins during late Mesozoic and Tertiary deposition. They range in size from single narrow rifts (*e.g.,* Suez graben, Egypt) to broad basins subdivided into a series of large horst-and-graben structures (Sirte embayment, Libya). In Africa many of these rifts are related to the continental East African rift system which may be incipient or dormant fracture zones similar to, and related to, the mid-ocean ridges. A semi-riftlike character is evident in many cratonic basins, although most of their characteristics require that they be classified as class II cratonic-composite or multicycle basins (*e.g.,* Hassi Touareg horst-graben axis in the Algerian Erg basins; the alternating horst-and-graben movements in the Noord Holland part of the Netherlands-German basin; and the Tertiary graben-like trough in the North Sea basin). Graben or half-graben rift basins include some of the richest hydrocarbon yields within the cratonic basin category, particularly where extensive evaporite caprocks (1) cover the source beds and reservoir-rock sequence in a basin, and (2) tend to (a) limit leakage of hydrocarbons, and possibly (b) disseminate the heat from moderately high heat flow centers (below the incipient rifts). Under such conditions, the evaporite seal acts as a "lid on a kettle."

Structural types associated with rifts include horst-and-graben development, block uplifts, and draped anticlines. Local barriers to the adjacent oceanic areas during sedimentation and basin fill in many areas are associated with reef structures along regional horst blocks or platforms. In some basins deep barred troughs within the rift zones were the sites of deposition of evaporite and euxinic dark mud during early stages of marine deposition. On the average the largest field contains nearly a quarter of the basin's reserves.

Four of the eight cratonic-rift basins which contain giant fields have been discovered since World War II; these are the Sirte embayment (Libya), the Vilyuy basin (USSR), the Red Sea graben, and the Oman rift (southeastern Arabian Peninsula).

Intermediate-crust basins.—IV. *Extracontinental: downwarps toward oceanic areas* may include one or more troughs. They are along the margins of continents and open into "small ocean basins" (Menard, 1967), such as the Tethyan zone, Gulf of Mexico, Caribbean Sea, Arctic Canadian basin, and the China Sea. These basins and the next type (Class V) commonly have uplifts or ridges developed on their continental margins, which act as barriers and traps for land-derived sediments (Werenskiold, 1953; Emery, 1969). The general outline of this type of basin parallels the trend of nearby post-Paleozoic fold belts, though in many places the subsequent troughs and fold belts have a trend which is oblique or normal to the Paleozoic and early Mesozoic depositional and structural trends.

Where the younger basins have a strike paralleling the older platform areas, the platform-like areas may be considered to be cratonic in behavior because their margins parallel the late Mesozoic-Tertiary fold belts and/or the present continental margins. Many of these basins are composite, with first-cycle Paleozoic and early Mesozoic carbonates and terrigenous clastics of shelf facies, and thick basinal second-cycle late Mesozoic and Tertiary sediments deposited in foredeep troughs along the margins of stable platforms (Arabia-Iran, Mangyshlak [southern

USSR], eastern Venezuela) or along truncated fold belts (Alaska North Slope, North Borneo, Tampico [Mexico]). Some of these basins have been referred to as "successor" basins (P. B. King, 1969b) or epieugeosynclines (Kay, 1951).

Class IV basins are among the richest hydrocarbon-bearing basins of the world. The outstanding example is the Persian Gulf region which is unique in its number and magnitude of giant and supergiant fields. Although the Arabian-Iranian basin contains 86 percent of the total reserves of the 11 basins of this class, the others are also relatively rich in hydrocarbons per unit volume of sediment. Large anticlines or uplifts with stratigraphic thinning (showing structural growth concomitant with sedimentation) account for the large accumulations of hydrocarbons. Where extensive evaporites are present, the richness of the basin is greater. The presence of gas seems to be related to the sedimentary facies of a basin; larger gas accumulations occur in terrigenous clastic facies than in carbonates. Intermediate- to high-API gravity values, and high- to intermediate-sulfur contents are common in Class IV basins. The number of giant fields depends on the size of the traps and the relative richness of the basin. A large trap in a low-yield basin (*i.e.*, bbl per unit volume of sediment) may contain the largest field in the basin, and hold up to 70 percent of the basin's total reserves; in contrast, high-yield basins, even with large traps, have so much oil that the largest fields contain only 6–7 percent of the basin reserves.

These basins, like the cratonic composite basins (Class II) can be subdivided further, because they include several subtypes.

The classification of basin position of major oil fields by Knebel and Rodríguez-Eraso (1956)—mobile rim, deep basin, hinge line, and shelf—differs from that used here, partly because of change in emphasis and partly because of difference in interpretation. The most notable example of a difference in interpretation is the reclassification of the fields of the Arabian mainland and the offshore fields in the Persian Gulf. We classify them as Arabian platform fields. We have done this on the basis of regional tectonic and sedimentational analyses made since the mid-1950s. Before the mid-1950s, most geologists believed that the Persian Gulf was a miogeosynclinal axis, and that the major fields on the southwest occupied a hinge-line position. Various studies, notably that of Stöcklin (1968), have shown that the Arabian platform has a gentle regional northeastward dip toward the Zagros folded belt, and that the present Persian Gulf is a late Cenozoic feature superimposed on this. The Arabian platform structures have a submeridional trend, which impinges against the northwest trend of the Zagros folds near the Iranian shoreline. The Zagros fold belt was a miogeosyncline whereas, on the Arabian platform, the uplifts resulted from block faulting and/or deep-seated salt diapirism. The Zagros folds may in large part be the result of *décollement* of the Upper Jurassic and younger strata over Lower Jurassic or older evaporite gliding planes.

V. Stable coastal, or coastal graben and fault basins are present along the continental margins of the Atlantic and parts of the Indian Ocean. They may represent the end phase of cratonic-rift basins, in that sea-floor spreading may have separated or pulled apart initial rift basins by oceanic distances (*e.g.,* coastal basins of central West Africa, eastern South America, and eastern North America). The basins include down-to-the-sea faults and downwarps with Tertiary and Mesozoic brackish to marine sediments draped over fault-block structures. Most of these basins have an offshore ridge which appears to be a basement uplift that acts as a dam to contain the shore-derived sediments (Emery, 1969; Hedberg, 1970). Evaporites form in many of these basins. However, evaporite deposition depends on the tectonic development of the basin—whether it is restricted by a sill at one end of the initial rift, or later by fault block or ridge uplifts acting as sills, which are developed parallel with the coast or rift trend. The evaporites not only may form a basinwide caprock series but also develop flowage features ranging from anticlinal swells to piercement diapirs. The presence of evaporites appears to enhance the hydrocarbon potential of these coastal basins. Such basins to date have yielded relatively poor hydrocarbon recovery per unit volume of sediments. Only two giant fields (Cabinda "B" and Emeraude Marin in West Africa) are associated with such basins. However, such basins may be more prospective than they have appeared, because technology has only recently made the relatively more prolific downdip, seaward side of the basins available for exploratory drilling. The two known giant fields (Cabinda "B" and Emeraude Marin) were found in 1966 and 1969, respectively.

VI. Intermontane: second-stage transverse basins also are present along continental mar-

gins. They have been called epieugeosynclines (Kay, 1951), backdeeps (van Bemmelen, 1954), median-zone basins (Weeks, 1952), and successor basins (P. B. King, 1969b). They contain Upper Cretaceous to Tertiary clastic sequences deposited in depressions which trend at right angles to the coast. They are transverse to Mesozoic or older fold belts, which generally parallel the coast (Fig. 9). The Tertiary terrigenous clastic strata form a second cycle of sediments above the deformed fold belts of eugeosynclinal deposits which parallel the continental margin. Examples are the Ventura and Los Angeles basins, on strike with the Transverse Ranges, which are normal to the Tertiary fold belt and Mesozoic eugeosyncline of California. The east-west Gippsland basin of southeastern Australia is transverse to the deformed north-south Tasman eugeosyncline of eastern Australia, the north-south South Caspian basin of the USSR is transverse to the east-west Caucasus, Alborz, and Kopet Dag trends, and the Maracaibo basin of Venezuela is oblique to the Mérida Andes trends (although the basin does parallel the Perijá Andes on the west). The cause, or causes, of the tranverse trend of these basins is not clear. Many are opposite major zones of strike-slip (transcurrent or transform) faults in either the oceanic or cratonic crusts. Transverse basins are unusual in that they are small and generally very rich in hydrocarbon content. Their sediments consist of thick series of alternating sandstone and shale, and production comes from multiple-pay sandstones, generally through a large depth range. The basic petrology of many of the eugeosynclinal metamorphic and intrusive rocks that surround these basins tends to result in the deposition of "tight" or "dirty" reservoir sandstones. However, if the eugeosynclinal fold belt contains silicic plutonic bodies that were emplaced before basin formation, cleaner reservoir rocks are deposited. The terrigenous clastic series generally are of marine to paralic origin (turbidite sandstones are an exception; *e.g.,* Ventura basin [Barbat, 1958]), and contain rich source beds, regardless of the character of the reservoirs. The basins have been deformed by block movements, with the result that the rapid subsidence of the basin is accompanied by, or associated with, faulted or tilt-block anticlines over which the clastic sequence wedges out, or is draped.

Where turbidity-current deposits are present in such basins, they greatly enhance the migration of hydrocarbons from source beds to reservoir beds, or act as reservoirs themselves. Commonly the transverse basins seem to be associated with zones of relatively high geothermal gradient or "hot spots" which may have influenced the efficiency of hydrocarbon formation and migration from source beds into reservoir rocks. The exact relation between high heat flux and hydrocarbon formation in these basins is not known.

In general the richest hydrocarbon-bearing transverse basins appear to possess a combination of the following factors:

1. Tertiary terrigenous clastics beds derived from granitic terranes with paralic to marine depositional environments;
2. silled basin, opening toward the sea;
3. growing, block structures; and, in some basins,
4. turbidity deposits or blanket sandstones, which can serve as "carrier beds" and/or reservoirs.

In addition, either paleo- or present geothermal gradient may enhance the relative richness of a basin, provided that the foregoing conditions are present. Further analysis of the heat flow around continental margins and oceanic crustal areas is needed to determine if a significant relation exists.

The factors discussed above appear to be negated if the reservoir sandstones are poorly developed or if the basin has been so deformed that truncation of many traps has resulted, as was described in the case of the smaller cratonic intermontane basins (Class II).

It appears that the giant fields in these basins are on major uplifts. Where the uplift has an unconformity developed over it caused by "two-way basin tilt," a "supergiant" forms (e.g., Bolivar Coastal field of Venezuela).

In the transverse basins there is both a wide variability of hydrocarbon types and a wide range of hydrocarbon recovery per unit volume of sediment, caused in large part by the large difference in reservoir-rock types formed in such basins.

VII. *Intermontane strike basins* are found mainly in the Tertiary and Mesozoic fold belts which occupy the continental margins around the Pacific Ocean, and along the Tethyan trend extending from the Mediterranean to the Far East. These basins are small, commonly have the form of grabens or half-grabens, and contain Tertiary terrigenous clastic strata deposited as second-cycle paralic to marine sediments on the deformed, metamorphosed, and intruded Mesozoic eugeosynclinal troughs that developed around the Pacific, Caribbean, and Tethys seaways. In general they are similar to the transverse basins, except that they generally have a different regional tectonic orientation and a greater length-to-breadth ratio. These basins have been referred by other authors to the various classifications given the transverse basins. The same geologic parameters which cause transverse basins to develop important reserves also are operative in strike basins. The hydrocarbon types found are quite variable, and these differences generally are related to variations in sedimentary facies from basin to basin.

Granitic plutons were intruded during the Jurassic and Cretaceous around the Pacific and along the Tethys during the growth of these basins. From these plutons in the resultant uplifted ranges, thick molasse deposits were shed from the highlands into the nearby basins. Van Houten (1969) suggested that these processes were related to the "new global tectonics," involving sea-floor spreading and movement of crustal plates.

VIII. *Tertiary deltas* in various parts of the world have unusual characteristics that require them to be regarded as a separate type of basin. These "sediment piles," which fan out in "birdfoot" to semicircular form, are built across continental margins where major river systems enter the sea. Many have had their greatest buildup since Miocene time (*e.g.*, Nile, Niger, Ganges, *etc.*), but others began to form in early Tertiary time (*e.g.*, Rio Grande, Mississippi). The ages of some of these deltaic buildups are discussed by numerous authors, among them Hardin (1962), Short and Stäuble (1967), and Meyerhoff (1968).

The delta sediments are nonmarine to marine terrigenous clastic strata with a general topset, foreset, and bottomset sequence of deltaic sandstone and shale (Short and Stäuble, 1967; Meyerhoff, 1968). Within the deltas, rollover anticlines develop on the downthrown sides of down-to-the-coast growth faults (Hardin and Hardin, 1961; Ocamb, 1961). Development of these structures is facilitated by gliding on listric surfaces of salt or "undercompacted" shale (Dickey *et al.*, 1968). The recovery of hydrocarbons from deltas indicates that they are as hydrocarbon-rich as any of the basin types; however, the size of reserves in their largest fields generally is less than 5 percent of the total of the delta. Deltas appear to contain a disproportionately large volume of gas. This may reflect in part the paralic facies of many of the source beds which contained a large amount of the terrigenous organic matter.

Conclusions and Results

Two hundred and sixty-six giant fields, as defined herein, contain a very high percentage of the world's total proved and produced oil reserves. Addition of the gas fields reveals that 62 percent of all reserves in giant fields are in sandstone reservoirs and 38 percent in carbonate reservoirs. Exclusion of the Persian Gulf giants leaves 59 percent of the reserves in sandstone reservoirs.

The amount of oil and gas in Tertiary rocks is 24 percent, in Mesozoic rocks 63 percent, and in Paleozoic rocks 13 percent. Exclusion of the Persian Gulf fields reveals that the Tertiary contains 40 percent of the giant proved and produced reserves, the Mesozoic 39 percent, and the Paleozoic 21 percent.

A striking fact which stands out clearly in Tables 1 and 2 is that 237 of the 266 giant fields are in anticlines. Only 25 fields are classified as stratigraphic, or partly stratigraphic, and only 78 fields have important production related to an unconformity. We do not believe that this overwhelming preponderance of anticlinal fields is necessarily an accurate reflection of "things as they are." Instead, we believe that many giant stratigraphic, unconformity-associated, and paleogeomorphic fields remain to be found.

Historically, petroleum exploration efforts have been directed toward the discovery and exploitation of the anticlinal trap. Geological and geophysical methods are designed and used deliberately to find the anticline, the rollover, the fold, and the dome. Giant stratigraphic, unconformity-associated, and paleogeomorphic traps undoubtedly exist in greater numbers than Tables 1 and 2 would indicate. Hence, future exploration, to be successful and to keep the world supplied with ample reserves for a greater length of time, *must* be oriented toward searching for the more obscure and difficult-to-find accumulations (Halbouty, 1966, 1968a, 1969, this volume; Martin, 1966).

Another striking result of this study is that so many fields are listed which never have been listed before (Knebel and Rodríguez-Eraso, 1956; Halbouty, 1968a and this volume; Moody *et al.*, this volume). The many new listings are the direct result of new discoveries during the past 15 years, and the listing of fields discovered in Russia and Mainland China. This is the first time, to the best of our knowledge, that a comprehensive list of Russian and Chinese giants has been published in the western, or noncommunist, literature.

Still another result of this study is the revelation of the fact that many giant fields in sandstone reservoirs were formed under nonmarine conditions. Traditional geologic thinking is that the source of oil and gas (coal gas excepted) must be marine. The existence of so many nonmarine giant accumulations hopefully will inspire explorationists to search for oil and gas in rocks that formed in nonmarine environments.

Finally, this study makes it clear that, in the future, increasing numbers of giants will be found, most of them offshore, but many such fields are awaiting discovery in the less-explored areas of every continent. The majority of future onshore discoveries will be in Asia; offshore giants will be found on all of the world's shelves, including the broad shelf of the Russian Arctic; giants in nonmarine strata will be found in more places; and giant stratigraphic, unconformity-associated, and paleogeomorphic traps should be found everywhere when the explorationist learns to reorient his thinking habits away from traditional, increasingly unrewarding approaches.

Selected References

Introduction

The following bibliography is designed primarily to list key references on the stratigraphy, structure and, in some cases, production characteristics of the giant fields listed on Tables 1–4 of this paper. *No attempt was made to make the bibliography complete.* To have made such an attempt would have required months of additional research and, in the final analysis, would not have added a great deal. This last statement is true because most of the references listed will lead the interested reader to additional publications which are on the subject of his interest but which do not apppear here. With only a few exceptions, the references given provide some important information on each field, field complex, or group of fields mentioned in this paper. A few references are included to support and document key points and assertions made by the writers; most of these are cited in the text.

The bibliography for Russian and Chinese fields is very incomplete, but is more complete than most published in the noncommunist world. Many are classics, or "firsts" of their type. Some are listed for the first time in the American literature.

Selected Bibliography

AAPG, 1956, Petroleum geology of southern Oklahoma, v. 1: Am. Assoc. Petroleum Geologists, 402 p. (various authors).

AAPG-CSD, 1937–1969, Development papers, domestic and foreign: Am. Assoc. Petroleum Geologists Bull., v. 21–53.

AAPG-SEPM-SEG, 1952, Los Angeles meeting guidebook—field trip routes, geology, oil fields: 290 p.

Abaie, J., H. J. Ansari, A. Badakshan, and A. Jaafarie, 1964, History and development of the Alborz and Sarajeh fields of central Iran: 6th World Petroleum Cong., Frankfurt am Main, Proc., sec. 1, p. 697–713.

Ablewhite, K., and G. E. Higgins, 1968 (1969), A review of Trinidad, West Indies oil development and the accumulations at Soldado, Brighton Marine, Grande Ravine, Barrackpore-Penal and Guayaguayare: 4th Caribbean Geol. Conf., Trinidad and Tobago, Trans., p. 41–73.

Abrikosov, I. Kh., and S. A. Vinnikovskiy, 1963, Working the Yarino-Kamennyy-Log field: Geologiya Nefti i Gaza, v. 7, no. 10 (Engl. trans., in Petroleum Geology, McLean, Va., December 1968, p. 546–549.

———— V. V. Semenovich, I. A. Blinnikov, and L. I. Moreva, 1969, Itogi geologorazvedochnykh rabot na nefti i gaz v SSSR za 1968 g. i zadachi na 1969 g.: Geologiya Nefti i Gaza, no. 6, p. 1–8.

Afanas'yev, Yu. T., Ye. A. Gaydeburova, and K. I. Mikulenko, 1966, Perspektivy neftegazonosnosti triasovykh i yurskykh otlozheniy Viluyuskoy gemisineklizy i Predverkhoyanskogo progiba: Geologiya Nefti i Gaza, no. 10, p. 21–25.

Alberta Society of Petroleum Geologists, 1960, Oil fields of Alberta: Calgary, Alberta Soc. Petroleum Geologists, 272 p.

———— 1967, Oil fields of Alberta supplement 1966: Calgary, Alberta Soc. Petroleum Geologists, 136 p.

Aleksin, A. G., G. G. Yudin, V. F. Markov, I. V. Shabatin, Ye. M. Borisenko, B. A. Polosin, and Ye. P. Sayenko, 1969, Perspektivy neftegazonosnosti yugo-voctochnoy chasti Stavropol'skogo Kraya i smezhnykh territoriy: Geologiya Nefti i Gaza, no. 6, p. 31–33.

Alikhanov, E. N., and V. F. Solovyov, 1967, Oil and gas bearing capacity prospect of the USSR offshore shelves: Amsterdam, Elsevier Pub. Co., 7th World Petroleum Cong., México, Proc., v. 2, p. 863–869.

Aliyev, T. U., M. S. Arabadzhi, Yu. M. Vasil'yev, V. S. Mil'nichuk, Yu. I. Pilipenko, Ye. I. Safonov, and M. N. Charygin, 1966, Novyye dannyye po geologii i gazonosnosti shakhpakhtinskogo podnyatiya na Ust-yurte: Geologiya Nefti i Gaza, no. 5, p. 29–34.

Allison, A. P., R. L. Beckelhymer, D. G. Benson, R. M. Hutchins, Jr., C. L. Lake, R. C. Lewis, P. H. O'Bannon, S. R. Self, and C. A. Warner, 1946, Geology of Katy field, Waller, Harris, and Fort Bend Counties, Texas: Am. Assoc. Petroleum Geologists Bull., v. 30, no. 2, p. 157–180.

Álvarez, M., Jr., 1951, Geological significance of the distribution of the Mexican oil fields: 3d World Petroleum Cong., The Hague, Proc., sec. 1, p. 73–85.

Amanov, S. A., and V. Ya. Sokolov, 1969, Krupneysheye gazovoye mestorozhdeniye Turkmenistana: Ashkhabad, Problemy Osvoyeniya Pustyn', no. 6, p. 72–75.

American Association of Petroleum Geologists and United States Geological Survey, 1967, Basement map of North America: scale, 1:5,000,000.

Andrews, D. I., 1961, Indigenous Pleistocene production in offshore Louisiana: Gulf Coast Assoc. Geol. Socs. Trans., v. 11, p. 109–119.

Anonymous, 1964, New centers of oil and gas production in West Siberia and South Mangyshlak: Geologiya Nefti i Gaza, v. 8, no. 1 (Engl. trans., in Petroleum Geology, McLean, Va., May 1969, p. 1–4).

———— 1966, Uskorit' podgotovku neftegazonosnoy bazy na Yuzhnom Mangyshlake: Geologiya Nefti i Gaza, no. 4, p. 1–4.

———— 1967, Petroleum industry in Communist China: Taipei, Fei-ch'ing Yen-chiu, v. 1, no. 9, p. 52–57 (in Chinese) (in trans., Washington, D.C., U.S. Dept. Commerce, Joint Pub. Research Service Document 43561, Trans. on Communist China, Dec. 5, 1967, Economic Document no. 36).

———— 1968, China's Taching oil field: eclipse of an industrial model: U.S. Information Service, Current Scene, Developments in Mainland China, v. 6, no. 16 (Sept. 17), 10 p.

———— 1969a, New forward leap in China: London, Petroleum Press Service, v. 36, no. 11, p. 405–407.

———— 1969b, The *Challenger* will sail on!: Ocean Industry, v. 4, no. 11, p. 58–60, 84.

———— 1969c: Pétrole Informations, no. 1141 (Nov. 14), p. 74.

———— 1969d, Russia uncovers Arctic reserves: World Oil, v. 169, no. 7 (Dec.), p. 66–67.

———— 1970a, Production mondiale en 1969: Pétrole Informations, no. 1145 (Jan. 23), p. 38.

———— 1970b: Turkmenskaya Iskra, February 21.

———— 1970c, El Morgan leads the way in Egypt: World Petroleum, v. 41, no. 2 (February 1970), p. 46.

———— 1970d: Sovetskaya Rossiya, March 12.

———— 1970e: Sovetskaya Latviya, March 14.

———— 1970f: Moscow, Pravda, March 17.

———— 1970g: Moscow, Pravda, March 28.

———— 1970h: Turkmenskaya Iskra, April 11.

———— 1970i: Moscow, Pravda, April 20.

———— 1970j: Turkmenskaya Iskra, May 1.

———— 1970k: Moscow, Pravda, May 4.

Antropov, P. Ya., 1958, On some achievements of geological surveying in the Chinese People's Republic: Washington, U.S. Dept. Commerce, Joint Pub. Research Service Document L-1008-N (reprinted by Am. Geol. Inst., 1960, Internat. Geol. Review, v. 2, no. 12, p. 1071–1077).

Armstrong, T., 1967, Giant Libyan oil field predicted: Oil and Gas Jour., v. 65, no. 8 (May 1), p. 118–120.

Aronsov, V. Ye., and A. K. Makhonin, 1966, O nekotorykh osobennostyakh stroyeniya neftyanykh i gazovykh mestorozhdeniy Yuzhno-Mangyshlaka: Geologiya Nefti i Gaza, no. 9, p. 21–28.

Arshinova, N. K., 1969, Osobennosti razrabotki i rezul'taty izucheniya neodnorodnosti svity NKP Surakhanskogo neftyanogo mestorozhdeniya: Geologiya Nefti i Gaza, no. 5, p. 45–49.

Asquith, D. O., 1966, Geology of Late Cretaceous Mesaverde and Paleocene Fort Union oil production, Birch Creek Unit, Sublette County, Wyoming: Am. Assoc. Petroleum Geologists Bull., v. 50, no. 10, p. 2176–2184.

Babalan, G. A., G. P. Ovanesov, L. A. Pelevin, A. B. Tumasyan, I. I. Kravchenko, Sh. I. Valeyev, E. M. Khalimov, and N. E. Lantev, 1969, Pervyye rezul'taty opytno-promyshlennykh rabot po primeneniyu PAV pri zavodnenii: Neftyanoye Khozyaystvo, no. 6, p. 41–45.

Baker, N. E., and F. R. S. Henson, 1952, Geological conditions of oil occurrence in Middle East fields: Am. Assoc. Petreolum Geologists Bull., v. 36, no. 7, p. 1885–1901.

Bakirov, A. A., and A. M. Pronina, 1962, Neftegazonosnyye oblasti Blizhnego Vostoka i Yugo-Vostochnoy Azii; geologicheskiye usloviya regionalnogo neftegazonakopleniya: Moscow, Gosudar. Nauch.-Tekh. Izd. Lit. Geol. i Okhrane Nedr., 208 p.

Balducchi, A., and G. Pommier, 1970, Cambrian oil field of Hassi Massaoud, Algeria: this volume, p. 477–488.

Balestrini, C. C., 1969, El petróleo en Latinoamérica: Soc. Venezolana de Geólogos, Bol. 4, no. 1, p. 3–24.

Baniriah, N., G. C. Beckman, and J. Birks, 1967, Repressuring of the Masjid-i-Sulaiman oilfield from a deep underground Jurassic gas leak and remedial killing operations: Amsterdam, Elsevier Pub. Co., 7th World Petroleum Cong. Proc., México, v. 3, p. 765–771.

Banks, L. M., 1959, Oil-coal association in central Anzoátegui, Venezuela: Am. Assoc. Petroleum Geologists Bull., v. 43, no. 8, p. 1998–2003.

———— 1960, Densidad de perforación y recuperación de petróleo en Venezuela Oriental: 3d Venezuelan Geol. Cong. Mem., v. 3, p. 1070–1118.

———— 1966, Geologic aspects of origin of petroleum: discussion: Am. Assoc. Petroleum Geologists Bull., v. 50, no. 2, pp. 397–400.

Banner, F. T., and G. V. Wood, 1964, Lower Cretaceous-Upper Jurassic stratigraphy of Umm Shaif field, Abu Dhabi marine areas, Trucial Coast, Arabia: Am. Assoc. Petroleum Geologists Bull., v. 48, no. 2, p. 191–206.

Barbat, W. F., 1958, The Los Angeles basin area, California, *in* Habitat of oil: Am. Assoc. Petroleum Geologists, p. 62–77.

Barlow, J. A., Jr., and J. D. Haun, 1970, Regional stratigraphy of Frontier Formation and relation to Salt Creek field, Wyoming: this volume, p. 147–157.

Barnetche, A., and L. V. Illing, 1956, The Tamabra Limestone of the Poza Rica oilfield: 20th Internat. Geol. Cong., México, 38 p.

Barr, K. W., F. Morton, A. R. Richards, and R. O. Young, 1951, Relationships between crude oil completion and stratigraphy in the Forest Reserve field of southwest Trinidad: 3d World Petroleum Cong., The Hague, Proc., sec. 1, p. 345–358.

———— S. T. Waite, and C. C. Wilson, 1958, The mode of oil occurrence in the Miocene of southern Trinidad, B.W.I., *in* Habitat of oil: Am. Association Petroleum Geologists, p. 533–550.

Barss, D. L., A. B. Copland, and W. D. Ritchie, 1970, Geology of Middle Devonian reefs, Rainbow area, Alberta, Canada: this volume, p. 18–49.

Bazanov, E. A., F. G. Gurari, V. D. Kozyrev, V. D. Nalivkin, Yu. A. Pritula, L. I. Rovnin, M. Ya. Rudkevich, A. A. Trofimuk, and S. G. Sarkisyan, 1967, Gas- and oil-bearing provinces of Siberia: Amsterdam, Elsevier Pub. Co., 7th World Petroleum Cong., México, Proc., v. 2, p. 109–120.

Beck, E., 1929, Salt Creek field, Natrona County, Wyoming, *in* Structure of typical American oil fields: Am. Assoc. Petroleum Geologists, p. 589–603.

Becker, L. E., and J. B. Patton, 1968, World occurrence of petroleum in pre-Silurian rocks: Am. Assoc. Petroleum Geologists Bull., v. 52, no. 2, p. 224–245.

Beebe, B. W., and B. F. Curtis, 1968, Natural gases of North America—a summary: Am. Assoc. Petroleum Geologists Mem. 9, v. 2, p. 2245–2355.

Begishev, F. A., G. S. Vakhitov, S. A. Sultanov, and I. P. Cholovskiy, 1963, Regulation of processes of working Horizon D_1 of the Romashkino field: Geologiya Nefti i Gaza, v. 7, no. 10 (Engl. trans., *in* Petroleum Geology, McLean, Va., December 1968, p. 556–561).

Beloussov, V. V., 1962, Basic problems in geotectonics: New York, McGraw-Hill Book Co. Inc., 816 p.

Belov, A. V., 1963, Distribution of permeability in the main productive horizons of the oil fields of the Ural-Volga: Geologiya Nefti i Gaza, v. 7, no. 12 (Engl. trans., *in* Petroleum Geology, McLean, Va., December 1968, p. 680–685).

Belov, C. A., I. O. Brod, M. S. Burshtar, S. T. Korotkov, D. V. Nesmeyanov, and A. I. Tasaturov, 1960, Oil and gas productivity of the Fore-Caucasus in connection with the occurrence of oil and gas accumulations in piedmont oil and gas basins, *in* Regional and structural problems in oil geology: 21st Internat. Geol. Cong., Copenhagen, Proc., pt. 11, p. 75–86.

Belyayeva, A. I., O. I. Rogoza, and S. A. Semenova, 1966, Nekotoryye resul'taty dinamicheskogo analiza otrazhennykh voln na ploshchadi Kuleshovskogo mestorozhdeniya nefti: Geologiya Nefti i Gaza, no. no. 3, p. 42–46.

Bemmelen, R. W. van, 1954, Mountain building: The Hague, Martinus Nijhoff, 177 p.

Benavides G., Luis, 1956, Notas sobre la geología petrolera de México, *in* Symposium sobre yacimientos de petróleo y gas: 20th Internat. Geol. Cong., México, v. 3, p. 351–562.

Benbow, D. D., 1968, Case history—Mereenie field: Australian Petroleum Exploration Assoc. Jour., v. 8, pt. II, p. 114–119.

Beresnev, N. F., and L. I. Rovnin, 1964, Megion oil field and oil-gas prospects of the Nizhne-Vartovskoye dome: Geologiya Nefti i Gaza, v. 8, no. 1 (Engl. trans., *in* Petroleum Geology, McLean, Va., May 1969, p. 18–22).

Berman, L. B., G. D. Margulov, I. B. Rozenberg, Ya. N. Basin, V. S. Neyman, and Yu. V. Uchastkin, 1966, O potentsial'nykh debitakh osnovnykh produktivnykh gorizontov gazovogo mestorozhdeniya Gazli: Geologiya Nefti i Gaza, no. 12, p. 17–21.

Berryhill, R. A., 1968, Monro field, Union, Morehouse, and Ouachita Parishes, Louisiana, *in* Natural gases of North America: Am. Assoc. Petroleum Geologists Mem. 9, v. 1, p. 1161–1167.

Blank, M. I., P. T. Pavlenko, L. S. Palets, A. M. Sinichka, and S. Ye. Cherpak, 1964, Some regularities in the distribution of oil and gas pools in the Dnieper-Donets depression: Geologiya Nefti i Gaza, v. 8, no. 4 (Engl. trans., *in* Petroleum Geology, McLean, Va., October 1969, p. 192–199).

Blatchley, R. S., 1913, The oil fields of Crawford and Lawrence Counties, Illinois: Illinois Geol. Survey Bull. 22, 442 p.

Bliznichenko, S. I., 1966, Neftegazonosnost' yursko-neokomskikh otlozheniy Privasyugan'ya: Geologiya Nefti i Gaza, no. 2, p. 47–53.

Bonnard, E. G., 1950, Discovery of oil at Lacq dome, Basses-Pyrénées, southwestern France: Am. Assoc. Petroleum Geologists Bull., v. 34, no. 7, p. 1584.

———— A. Debourle, H. Hlauschek, P. Michel, V. Perebaskine, J. Schoeffler, R. Seronie-Vivien, and M. Vigneaux, 1958, The Aquitainian basin, southwest France, *in* Habitat of oil: Am. Assoc. Petroleum Geologists, p. 1091–1122.

Borger, H. D., 1952, Case history of Quiriquire field, Venezuela: Am. Assoc. Petroleum Geologists Bull., v. 36, no. 12, p. 2291–2330.

—— and E. F. Lenert, 1959. The geology and development of the Bolivar Coastal field at Maracaibo, Venezuela: 4th World Petroleum Cong., New York, Proc., sec. 1, p. 481–498.

Borisov, Yu. P., O. I. Dorokhov, M. M. Ivanova, and R. Sh. Mingareyev, 1970, Vnedreniye novoy sistemy razrabotki na Romashkinskom neftyanom mestorozhdenii s vnutrikonturnym zavodneniyem: Neftyanoye Khozyaystvo, no. 1, p. 7–11.

Bouchon, R., H. I. Ortynski, C. de Lapparent, and G. Pommier, 1959, Le développement de la sismique réfraction dans l'interprétation géologique de Sahara-Nord—son rôle dans la découverte et l'étude du champ d'Hassi Messaoud: 4th World Petroleum Cong., New York, Proc., sec. 1, p. 729–746.

Bower, T. H., 1968 (1969), Geology of Texaco Forest Reserve field, Trinidad, W. I.: 4th Caribbean Geol. Conf., Trinidad and Tobago, Trans., p. 75–86.

Bowerman, J. N., 1966, Petroleum developments in North Africa in 1965: Am. Assoc. Petroleum Geologists Bull., v. 50, no. 8, p. 1681–1703.

—— 1967, Petroleum developments in North Africa in 1966: Am. Assoc. Petroleum Geologists Bull., v. 51, no. 8, p. 1564–1586.

Boyarova, Ye. D., 1966, O geologicheskom stroyenii Yuzhno-Kuybyshevskogo gazoneftenosnogo rayona: Geologiya Nefti i Gaza, no. 3, p. 28–33.

Brenneman, M. C., 1960, Estudio químico de los petróleos crudos de la cuenca de Maracaibo: 3d Venezuelan Geol. Cong. Mem., v. 3, p. 1025–1069.

Brenner, M. M., 1968, O proportsiyakh v neftedobyvayushchey promyshlennosti i metodologii ikh opredeleniya: Geologiya Nefti i Gaza, no. 11, p. 48–53.

British Petroleum Co. Ltd., 1956, Oil and gas in southwest Iran, in Symposium sobre yacimientos de petróleo y gas: 20th Internat. Geol. Cong., México, v. 2, p. 33–72.

Brod, I. O., V. G. Vasil'yev, I. V. Vysotskiy, K. N. Kravchenko, V. G. Levinsov, M. S. L'vov, V. B. Olenin, and B. A. Sokolov, 1965, Neftegazonosnyye basseyny zemnogo shara: Moscow, Izd. "Nedra," 598 p.

Brown, D. A., K. S. W. Campbell, and K. A. W. Crook, 1968, The geological evolution of Australia and New Zealand: London, Pergamon Press, 409 p.

Brown, J. R., 1968, Gas in Los Angeles basin, California: Am. Association Petroleum Geologists Mem. 9, v. 1, p. 149–163.

Bryant, H., 1961, Utilización de registros de neutrones de pozos en producción para determinar contactos gas/petróleo en las formaciones del Campo Costanero de Bolívar: 3d Venezuelan Geol. Cong. Mem,. v. 4, p. 1555–1568.

Bryant, W. R., A. A. Meyerhoff, N. K. Brown, Jr., M. A. Furrer, T. E. Pyle, and J. W. Antoine, 1969, Escarpments, reef trends, and diapiric structures, eastern Gulf of Mexico: Am. Assoc. Petroleum Geologists Bull., v. 53, no. 12, p. 2506–2542.

Bullard, E., 1969, The origin of the oceans: Sci. America, v. 221, no. 2 (Sept.), p. 66–75.

Burk, C. A., M. Ewing, J. L. Worzel, A. O. Beall, Jr., W. A. Berggren, D. Bukry, A. G. Fisher, and E. A. Pessagno, Jr., 1969, Deep-sea drilling into the Challenger Knoll, central Gulf of Mexico: Am. Assoc. Petroleum Geologists Bull., v. 53, no. 7, p. 1338–1347.

Burke, R. J., and F. J. Gardner, 1969, The world's monster oil fields and how they rank: Oil and Gas Jour., v. 67, no. 2, p. 43–49.

Burroughs, E., G. W. Beecroft, and R. M. Barger, 1968, Rio Vista gas field, Solano, Sacramento, and Contra Costa Counties, California: Am. Assoc. Petroleum Geologists Mem. 9, v. 1, p. 93–103.

Burst, J. F., 1969, Diagenesis of Gulf Coast clayey sediments and its possible relation to petroleum migration: Am. Assoc. Petroleum Geologists Bull., v. 53, no. 1, p. 73–93.

Busson, G., 1968, La sédimentation des évaporites: Paris, Muséum Nat. d'Histoire Naturelle, n.s., sér C, t. 19, fasc. 3, p. 125–169.

Buttin, R., 1959, Le pétrole en Afrique: 4th World Petroleum Cong., New York, Proc., General Volume, p. 143–152.

Buzzalini, A. D., F. J. Adler, and R. L. Jodry, eds., 1969, Evaporites and petroleum: Am. Assoc. Petroleum Geologists Bull., v. 53, no. 4, p. 775–1011.

Byzer, B. I., and P. T. Erinchek, 1966, K metodike oprobovaniya skvazhin mnogoplastovogo Zhetybayskogo mestorozhdeniya Yuzhnogo Mangyshlaka: Geologiya Nefti i Gaza, no. 10, p. 57–59.

California Div. of Oil and Gas, 1960, California oil and gas fields—maps and data sheets, Pt. 1, San Joaquin-Sacramento Valleys and northern coastal regions: State of California Dept. Nat. Resources, p. 1–495.

—— 1961, California oil and gas fields—maps and data sheets, Pt. 2, Los Angeles-Ventura basins and central coastal regions: State of California Dept. Nat. Resources, p. 496–913.

Cameron, J. C., 1967, Australia's oil and gas potential—a review of the main sedimentary basins: Amsterdam, Elsevier Pub. Co., 7th World Petroleum Cong., México, Proc., v. 2, p. 151–160.

Campbell, R. W., 1968, The economics of Soviet oil and gas: Baltimore, The Johns Hopkins Univ. Press, 279 p.

Caribbean Petroleum Co. Staff, 1948, Oil fields of the Royal Dutch-Shell Group in Western Venezuela: Am. Assoc. Petroleum Geologists Bull., v. 32, no. 4, p. 517–628.

Casey, J. N., and M. C. Konecki, 1967, Natural gas—a review of its occurrence and potential in Australia and Papua: Amsterdam, Elsevier Pub. Co., 7th World Petroleum Cong., México, Proc., v. 2, p. 265–288.

Chin Jih Ta Lu (Mainland China Today), 1967a: Taipei, no. 282 (June 16), p. 29.

—— 1967b: Taipei, no. 283 (July 1), p. 16.

Choubert, G., and A. Faure-Muret, 1968, Tectonic map of Africa: Paris, UNESCO, Nature Resources, v. 4, no. 2, p. 18–19, scale: 1:5,000,000.

Christian, H. E., Jr., 1969, Some observations on the initiation of salt structures of the southern British North Sea, in The exploration for petroleum in Europe and North Africa: London, Inst. Petroleum, p. 231–250.

Chu, T. O., 1924, The oil fields of China: Am. Assoc. Petroleum Geologists Bull., v. 8, no. 2, p. 169–177.

Chunosov, P. I., 1965, Vliyaniye plotnosti setki skvazhin na poteri nefti na Novo-Khazinskoy ploshchadi Arlanskogo mestorozhdeniya: Geologiya Nefti i Gaza, no. 4, p. 22–24.

Churkin, M., 1969, Paleozoic tectonic history of the Arctic basin north of Alaska: Science, v. 165, no. 3893, p. 549–555.

Chursin, F., 1970, The future of Samotlor: an interview with M. Svishchev, Deputy Director of Giprotyumen'neftegaz Institute: Moscow, Pravda, January 14.

Claracq, R., G. Couraud, and J. Aymon, 1964, Le dévonien inférieur du bassin de Fort Polignac; un champ dévonien typique de la C.R.E.P.S., Zarzaïtine: 6th World Petroleum Cong., Frankfurt am Main, Proc., sec. 1, p. 773–795.

Claret, J., and C. Tempère, 1967, Une nouvelle région productrice au Sahara Algérien: l' anticlinorium d'Hassi Touareg: Amsterdam, Elsevier Pub. Co., 7th World Petroleum Cong., México, Proc., v. 2, p. 81–100.

Colomo, J., and A. Barnetche, 1951, The Poza Rica field: 3d World Petroleum Cong., The Hague, Proc., sec. 2, p. 647–664.

Commission for the Geological Map of the World, 1964, Carte géologique de l'Amérique du Sud: Rio de Janeiro, Ministerio Minas e Energia, Dpto. Nac. Prod. Mineral, 2 sheets, scale, 1:5,000,000 (Available at Geol. Soc. America, Boulder, Colo.).

Compagnie de Recherches et d'Exploitation de Pétrole au Sahara, 1967, Maintenance of reservoir pressure in Zarzaïtine oilfield, Algerian Sahara, *in* 3d symposium on the development of petroleum resources of Asia and the Far East: New York, U. N. ECAFE Mineral Resources Devel. Ser., no. 26, v. 2, p. 160–177.

Conant, L. C., and G. M. Goudarzi, 1967, Stratigraphic and tectonic framework of Libya: Am. Assoc. Petroleum Geologists Bull., v. 51, no. 4, p. 719–730.

Cordry, E. A., 1967, Petroleum developments in central and southern Africa in 1966: Am. Assoc. Petroleum Geologists Bull., v. 51, no. 8, p. 1587–1625.

Coustau, H., J. Gauthier, G. Kulbicki, and E. Winnock, 1969, Hydrocarbon distribution in the Aquitaine basin, *in* The exploration for petroleum in Europe and North Africa: London, Inst. Petroleum, p. 73–85.

Criado Roque, P., C. de Ferrariis, A. Mingramm, E. Rolleri, I. Simonato, and T. Suero, 1959, The sedimentary basins of Argentina: 4th World Petroleum Cong., New York, Proc., sec. 1, p. 883–900.

Cronen, A. D., 1969, Well spacing in North Sea gas fields: London, Inst. Petroleum Jour., v. 55, no. 543, p. 141–152.

Cserna, Z. de., 1961, Tectonic map of Mexico: Geol. Soc. America, scale, 1:2,500,000.

Cunningham, G. M., and W. D. Kleinpell, 1934, Importance of unconformities to oil production in the San Joaquin Valley, California, *in* Problems of petroleum geology: Am. Assoc. Petroleum Geologists, p. 783–805.

Dakhnov, V. N., and E. M. Galimov, 1966, Ustanovleniye karstovogo tipa kollektora po otnositel'nomu soderzhaniyu isotopov ugleroda C^{13}/C^{12} v izvestnyake i vo vtorichnom kal'tsite: Geologiya Nefti i Gaza, no. 6, p. 59–63.

Daniel, E. J., 1954, Fractured reservoirs of Middle East: Am. Assoc. Petroleum Geologists Bull., v. 38, no. 5, p. 774–815.

Demaison, G. J., 1965, The Triassic salt in the Algerian Sahara, *in* Salt basins around Africa: London, Inst. Petroleum, p. 91–100.

Dewey, J. F., and B. Horsfield, 1970, Plate tectonics, orogeny and continental growth: Nature, v. 225, no. 5232, p. 521–525.

Dickey, P. A., C. R. Shriram, and W. R. Paine, 1968, Abnormal pressures in deep wells of southwestern Louisiana: Science, v. 160, no. 3828, p. 609–615.

Dickie, R. K., 1966, Nigeria, the Federal government's control of the oil industry and the development of crude oil production in the Niger delta: London, Inst. Petroleum Jour., v. 52, no. 506, p. 38–45.

Dietz, R. S., 1961, Continent and ocean basin evolution by spreading of the sea floor: Nature, v. 190, no. 4779, p. 854–857.

Dikkers, A. J., 1964, Development history of the La Paz field, Venezuela: London, Inst. Petroleum Jour., v. 50, no. 492, 330–333.

Dolgopolov, K. V., and S. A. Vodovozov, ed., 1968, Yevropeyskiy Yugo-Vostok: Moscow, Izd. "Mysl'," 795 p.

Dolitskiy, V. A., and Y. V. Kucheruk, 1963, Use of detailed paleogeological maps in exploration for oil and gas pools: Geologiya Nefti i Gaza, v. 7, no. 11 (Engl. trans., *in* Petroleum Geology, McLean, Va., February 1969, p. 611–616).

Dott, R. H., Sr., and M. J. Reynolds, 1969, Sourcebook for petroleum geology: Am. Assoc. Petroleum Geologists Mem. 5, 471 p.

Dryden, J. E., R. C. Erickson, T. Off, and S. W. Yost, 1968, Gas in Cenozoic rocks in Ventura-Santa Maria basins, California: Am. Assoc. Petroleum Geologists Mem. 9, v. 1, p. 135–148.

Dunnington, H. V., 1958, Generation, migration, accumulation, and dissipation of oil in northern Iraq, *in* Habitat of oil: Am. Assoc. Petroleum Geologists, p. 1194–1251.

—— 1967, Stratigraphical distribution of oilfields in the Iraq-Iran-Arabia basin: London, Inst. Petroleum Jour., v. 53, no. 520, p. 129–161.

Dupouy-Camet, J., 1953, Triassic salt structures, southwestern Aquitaine basin, France: Am. Assoc. Petroleum Geologists Bull., v. 37, no. 10, p. 2348–2388.

Elder, S., 1963, Umm Shaif oilfield, history of exploration and development: London, Inst. Petroleum Jour., v. 49, no. 478, p. 308–315.

—— and K. F. C. Grieves, 1965, Abu Dhabi marine areas geology, *in* Le pétrole et la mer: 1st Internat. Cong., Monaco, sec. 1, paper no. 127, 8 p.

Elias, M. M., K. Y. Lee, and R. J. Sun, 1966, Atlas of Asia and eastern Europe to support detection of underground nuclear testing, volume IV. Features affecting nuclear underground testing: U.S. Geol. Survey, 7 sheets, scale, 1:5,000,000.

Elmore, W. Z., 1958, Sante Fe Springs [California] oilfield: Pacific Section AAPG Guidebook—Los Angeles and Ventura regions, p. 100–104.

Emery, K. O., 1969, Continental rises and oil potential: Oil and Gas Jour., v. 67, no. 19 (May 12), p. 231–236, 238, 240.

—— E. Uchupi, J. D. Phillips, C. O. Bowin, E. T. Bunce, and S. T. Knott, 1970, Continental rise off eastern North America: Am. Assoc. Petroleum Geologists Bull., v. 54, no. 1, p. 44–108.

Falcon, N. L., 1958, Position of oil fields of southwest Iran with respect to relevant sedimentary basins, *in* Habitat of oil: Am. Assoc. Petroleum Geologists, p. 1279–1293.

Fedorov, S. F., 1962, Stupenchataya migratsiya nefti i gaza (Multistage migration of oil and gas): Sovetskaya Geologiya, no. 7, p. 8–25 (*in* trans., Am. Geol. Inst., 1964, v. 6, no. 2, p. 263–276).

—— 1963, Usloviya formirovaniya zalezhey nefti i gaza v Bashkirii (Genesis of oil and gas pools in Bashkiria): Akad. Nauk SSSR Doklady, v. 150, no. 6, p. 1340–1343 (*in* trans., Am. Geol. Inst., 1965, Acad. Sci. USSR Doklady, v. 150, p. 89–92).

Fernandes, O. O., and E. Bertrand, 1969, Trinidad and

Tobago, *in* C. H. Neff, 1969, Review of 1968 petroleum developments in South America, Central America, and Caribbean area: Am. Assoc. Petroleum Geologists Bull., v. 53, no. 8, p. 1629–1642.

Filippov, B. V., 1967, Tipy prirodnykh reservuarov nefti i gaza: Leningrad, Izd. "Nedra," 123 p.

Fohs, F. J., 1947, Oil reserve provinces of Middle East and southern Soviet Russia: Am. Assoc. Petroleum Geologists Bull., v. 31, no. 8, p. 1372–1383.

———— 1948, Petroliferous provinces of Union of Soviet Socialist Republics: Am. Assoc. Petroleum Geologists Bull., v. 32, no. 3, p. 317–350.

———— 1962, Petroliferous provinces of U.S.S.R.—a revision: Am. Assoc. Petroleum Geologists Bull., v. 46, no. 11, p. 1973–1989.

Fossa-Mancini, E., 1932, Faults in Comodoro Rivadavia oil field, Argentina: Am. Assoc. Petroleum Geologists Bull., v. 16, no. 6, p. 556–576.

Fox, A. F., 1956, Oil occurrences in Kuwait, *in* Symposium sobre yacimientos de petróleo y gas: 20th Internat. Geol. Cong., México, v. 2, p. 131–160.

———— 1959, Some problems of petroleum geology in Kuwait: London, Inst. Petroleum Jour., v. 45, no. 424, p. 95–110.

———— 1964, The world of oil: London, Pergamon Press, 221 p.

———— and R. C. C. Brown, 1968, The geology and reservoir characteristics of the Zakum oil field, Abu Dhabi, *in* 2d AIME regional technical symposium: Dhahran, Soc. Petroleum Engineers of AIME, Proc., p. 39–66.

———— and C. V. Rollinson, 1966, Technical co-ordination of oilfield development, *in* Proceedings—petroleum: Melbourne, 8th Commonwealth Mining Metal. Cong. Pub., v. 5, p. 255–269 (History of Burgan and Umm Shaif fields).

Franco, A., 1969, Mexico's big-oil dreams materializing in the Gulf: Oil and Gas Jour., v. 67, no. 23 (June 9), p. 33–36.

Fränkl, E. J., and E. A. Cordry, 1967, The Niger delta oil province: recent developments offshore and onshore: Amsterdam, Elsevier Pub. Co., 7th World Petroleum Cong., México, Proc., v. 2, p. 196–209.

Fraser, W. W., 1967, Geology of the Zelten field, Libya, North Africa: Amsterdam, Elsevier Pub. Co., 7th World Petroleum Cong., México, Proc., v. 2, p. 259–264.

Freeman, H. A., 1951, Techniques in the Kirkuk oilfield: 3d World Petroleum Cong., The Hague, Proc., sec. 2, p. 683–695.

———— and S. G. Natanson, 1959, Recovery problems in a fracture-pore system: Kirkuk field: 4th World Petroleum Cong., New York, Proc., sec. 2, p. 297–317.

Frey, M. G., and W. H. Grimes, 1970, Bay Marchand-Timbalier Bay-Caillou Islands salt complex, Louisiana: this volume, p. 277–291.

Froelich, A. J., E. A. Krieg, and R. M. Hopkins, 1968, Recent geophysical results, northern Amadeus basin: Australian Petroleum Exploration Jour., v. 8, pt. II, p. 104–113.

Gale, H. S., 1934, Geology of Huntington Beach oil field, California: Am. Assoc. Petroleum Geologists Bull., v. 18, no. 3, p. 327–342.

Gal'person, Ye., F. Dunayev, R. Mingareyev, M. Timofeyev, and N. Titkov, 1970, Retsenziya na knigu M. M. Brenner, "Ekonomika neftyanoy i gazovoy promyshlennosti": Neftyanoye Khozyaystvo, no. 1, p. 65–68.

Galyamov, G. Kh., and P. K. Vil'tsin, 1966, O formiro-

vanii neftyanoy v verkhnemelovykh otlozheniyakh Malgobekskogo rayona Checheno-Ingushskoy ASSR: Geologiya Nefti i Gaza, no. 6, p. 17–19.

Gamero, G. A., 1968, Importancia de la sedimentología en el estudio de los yacimientos petrolíferos de la cuenca del Lago de Maracaibo: Soc. Venezolana de Géologos, Bol. 3, no. 1, p. 9–15.

García Rojas, A., 1949, Mexican oil fields: Am. Assoc. Petroleum Geologists Bull., v. 33, no. 8, p. 1336–1350.

Gardner, F. J., 1952, Reference report on oil and gas fields of Texas upper Gulf Coast, R. R. Comm. Dist. Three: Houston, Five Star Oil Report, Petroleum News Corporation, 484 p.

———— 1969, Siberia answering Russia's oil SOS: Oil and Gas Jour., v. 67, no. 26 (June 30), p. 70–72.

———— 1970a, NAM's Dutch gas monopoly is cracked: Oil and Gas Jour., v. 68, no. 4 (Jan. 26), p. 96–97.

———— 1970b, Huge North Sea find has world vibrating: Oil and Gas Jour., v. 68, no. 21 (May 25), p. 33–36.

———— L. Van Dyke, and D. Stormont, 1969, Oil looks for growth to continue: Oil and Gas Jour., v. 67, no. 52 (Dec. 29), p. 89–152.

Gatewood, L. E., 1970, Oklahoma City field—anatomy of a giant: this volume, p. 223–254.

Geological Society of Australia, Tectonic Map Committee, 1960, Tectonic map of Australia: Australia Bur. Mineral Resources Geology and Geophysics, scale, 1:2,534,400.

Gere, W. C., G. W. Horton, J. N. Harstead, and D. F. Russell, 1964, Oil and natural gas, *in* Mineral and water resources of Utah: Utah Geol. Mineralog. Survey Bull. 73, p. 51–60, Table 3.

Gerth, H., 1955, Der geologisch Bau der südamerikanischen Kordillere: Berlin, Gebrüder Borntraeger, 264 p.

Gester, G. C., and J. Galloway, 1933, Geology of Kettleman Hills oil field: Am. Assoc. Petroleum Geologists Bull., v. 17, no. 10, p. 1161–1193.

Gill, W. D., 1967, The North Sea basin: Amsterdam, Elsevier Pub. Co., 7th World Petroleum Cong., México, Proc., v. 2, p. 211–219.

Gillespie, J., and R. M. Sanford, 1967, The geology of the Sarir oilfield. Sirte basin, Libya: Amsterdam, Elsevier Pub. Co., 7th World Petroleum Cong., México, Proc., v. 2, p. 181–193.

Girshgorn, L. Sh., 1969, Osnovnyye cherty tektonicheskogo stroyeniya platformennogo kompleksa otlozheniy severa Zapadnoy Sibiri: Geologiya Nefti i Gaza, no. 2, p. 38–42.

Glaessner, M. F., 1958, The oldest fossil faunas of South Australia: Geol. Rundschau, Bd. 47, Hft. 2, p. 522–531.

———— 1960, Precambrian fossils from South Australia: 21st Internat. Geol. Cong., Copenhagen, Rept., pt. 22, p. 59–64.

———— 1962, Pre-Cambrian fossils: Biol. Review, v. 37, p. 467–494.

———— and B. Daily, 1959, The geology and late Precambrian fauna of the Ediacara Fossil Reserve: South Australia Museum Rec., v. 8, p. 369–401.

Goding, Y. U., G. Dickenstein, V. Danisevich, L. Zhukovsky, and C. Semenovich, 1960, Geological structure and oil and gas possibilities of the Middle Asia Epi-Hercynian platform, *in* Regional and structural problems in oil geology: 21st Internat. Geol. Cong., Copenhagen, Proc., pt. 11, p. 67–74.

Golovachev, V. S., F. K. Salmanov, and A. V. Tyan,

1966, Geologicheskoye stroyeniye i perspektivy nefte-gazonosnosti Salymskogo neftegazonosnogo rayona Zapadno-Sibirskoy nizmennosti: Geologiya Nefti i Gaza, no. 10, p. 10–14.

González de Juana, C., and L. Ponte Rodríguez, 1951, Fundamental geological characteristics of the Venezuelan oil basins: 3d World Petroleum Cong., The Hague, Proc., sec. 1, p. 41–55.

Grdzelov, L. I., S. D. Ivanov, Yu. M. Malinovskiy, V. A. Rutskov, B. Sharafi, and A. Vatan'yar, 1969, Novyye dannyye o geologicheskom stroyenii i neftegazonosnosti Severnogo Afganistana: Sovetskaya Geologiya, no. 6, p. 79–87.

Greer, W. J., 1965, The Gidgealpa gas field: Australian Petroleum Exploration Assoc. Jour. for 1965, p. 65–68.

Gregg, J. W., and J. Wahl, 1964, Performance history of the Pembina Cardium pool, Alberta, Canada: 6th World Petroleum Cong., Frankfurt am Main, Proc., sec. 2, p. 415–449.

Grieg, D. A., 1958, Oil horizons in the Middle East, *in* Habitat of oil: Am. Assoc. Petroleum Geologists, p. 1182–1193.

Grigoras, N., and N. Constantinescu, 1969, Geological activity for oil and gas in the Socialist Republic of Rumania, *in* The exploration for petroleum in Europe and North Africa: London, Inst. Petroleum, p. 147–159.

Guariguata, R. C., and J. A. Richardson, 1960, Producción petrolífera del basamento en el oeste del Lago de Maracaibo: 3d Venezuelan Geol. Mem., v. 3, p. 985–1008.

Gurari, F. G., A. E. Kantorovich, and G. B. Ostryy, 1966, O roli diz'yunktivnykh narusheniy v protsesse formirovaniya zalezhey nefti i gaza v yurskikh i melovykh otlozheniyakh Zapadno-Sibirskoy nizmennosti: Geologiya Nefti i Gaza, no. 2, p. 5–11.

Guzmán, E. J., 1967, Reef type stratigraphic traps in Mexico: Amsterdam, Elsevier Pub. Co., 7th World Petroleum Cong., México, Proc., v. 2, p. 461–470.

—— R. Suárez, and E. López-Ramos, 1955, Outline of the petroleum geology of Mexico, *in* Proceedings of the Conference on Latin-American Geology: Austin, Univ. Texas, p. 1–30.

Guzmán Reyes, A., K. R. Tripp, and P. H. Klootwijk, 1967, The use of mathematical models in analyzing Venezuelan reservoirs: Amsterdam, Elsevier Pub. Co., 7th World Petroleum Cong., México, Proc., v. 2, p. 653–667.

Hafenbrack, J. H., 1966, Recent developments in the Bass and Gippsland basins: Australian Petroleum Exploration Assoc. Jour. for 1966, p. 47–49.

—— 1969, Drilling results from several south-eastern Australia offshore areas: Australian Petroleum Exploration Assoc. Jour., v. 9, pt. 2, p. 146–148.

Hajash, G. M., 1967, The Abu Sheikdom—the onshore oilfields history of exploration and development: Amsterdam, Elsevier Pub. Co., 7th World Petroleum Cong., México, Proc., v. 2, p. 129–139.

Halbouty, M. T., 1966, Stratigraphic-trap possibilities in Upper Jurassic rocks, San Marcos arch, Texas: Am. Assoc. Petroleum Geologists Bull., v. 50, no. 1, p. 3–24.

—— 1967, Salt domes—Gulf region, United States and Mexico: Houston, Gulf Pub. Co., 425 p.

—— 1968a, Giant oil and gas fields in United States: Am. Assoc. Petroleum Geologists Bull., v. 52, no. 7, p. 1115–1151; this volume, p. 91–127.

—— 1968b, Old Ocean field, Brazoria and Matagorda Countries, Texas: Am. Assoc. Petroleum Ge-

ologists Mem. 9, v. 1, p. 295–305.

—— 1969, Hidden trends and subtle traps in Gulf Coast: Am. Assoc. Petroleum Geologists Bull., v. 53, no. 1, p. 3–29.

Hamilton, S. V., 1968, Wilmington field, Los Angeles County, California: Am. Assoc. Petroleum Geologists Mem. 9, v. 1, p. 164–168.

Hardin, F. R., and G. C. Hardin, Jr., 1961, Contemporaneous normal faults of Gulf Coast and their relation to flexures: Am. Assoc. Petroleum Geologists Bull., v. 45, no. 2, p. 238–248.

Hardin, G. C., Jr., 1962, Notes on sedimentation in the Gulf Coast geosyncline, U.S.A., *in* Geology of the Gulf Coast and central Texas and guidebook of excursions: Houston Geol. Soc., p. 1–15.

Harris, T. J., J. T. C. Hay, and B. N. Twombley, 1968, Contrasting limestone reservoirs in the Murban field, Abu Dhabi, *in* 2d AIME regional technical symposium: Dhahran, Soc. Petroleum Engineers of AIME, Proc., p. 149–187.

Heatzig, G., and R. Michel, 1968, Petroleum developments in North Africa in 1967: Am. Assoc. Petroleum Geologists Bull., v. 52, no. 8, p. 1489–1511.

—— and —— 1969, Petroleum developments in North Africa in 1968: Am. Assoc. Petroleum Geologists Bull., v. 53, no. 8, p. 1700–1727.

Hedberg, H. D., 1956, Petroleum developments in Africa in 1955: Am. Assoc. Petroleum Geologists Bull., v. 40, no. 7, p. 1582–1632.

—— 1964, Geologic aspects of origin of petroleum: Am. Assoc. Petroleum Geologists Bull., v. 48, no. 11, p. 1755–1803.

—— 1968, Significance of high-wax oils with respect to genesis of petroleum: Am. Assoc. Petroleum Geologists Bull., v. 52, no. 5, p. 736–750.

—— 1970, Continental margins from the viewpoint of the petroleum geologist: Am. Assoc. Petroleum Geologists Bull., v. 54, no. 1, p. 3–43.

—— L. C. Sass, and H. J. Funkhouser, 1947, Oil fields of Greater Oficina area, central Anzoátegui, Venezuela: Am. Assoc. Petroleum Geologists Bull., v. 31, no. 12, p. 2089–2169.

Hemphill, C. R., R. I. Smith, and F. Szabo, 1970, Geology of Beaverhill Lake reefs, Swan Hills area, Alberta: this volume, p. 50–90.

Henson, F. R. S., 1951, Observations on the geology and petroleum occurrences of the Middle East: 3d World Petroleum Cong., The Hague, Proc., sec. 1, p. 118–140.

Herald, F. A., ed., 1957, Occurrence of oil and gas in West Texas: Texas Univ. Bur. Econ. Geology Bull. 5716, 442 p.

Hess, H. H., 1960, Evolution of ocean basins: Princeton Univ., preprint, 38 p.

—— 1962, History of ocean basins, *in* Petrologic studies: a volume to honor A. F. Buddington: Geol. Soc. America, p. 599–620.

Heybroek, P., U. Haanstra, and D. A. Erdman, 1967, Observations on the geology of the North Sea area: Amsterdam, Elsevier Pub. Co., 7th World Petroleum Cong., México, Proc., v. 2, p. 905–916.

Hills, J. M., 1968, Gas in Delaware and Val Verde basins, West Texas and southeastern New Mexico: Am. Assoc. Petroleum Geologists Mem. 9, v. 2, p. 1394–1432.

Ho K'o-jen, 1968, The developments in Red China's petroleum industry: Taipei, Fei-ch'ing Yueh-pao (Mar. 1), p. 52–65 (in Chinese) (*in trans.,* Washington, D. C., U.S. Dept. Commerce, Joint Pub. Research Service Document 45194, Trans. on Communist China, Apr. 29, 1968, Economic Document no. 11).

Holmes, A., 1931, Radioactivity and earth movements: Geol. Soc. Glasgow Trans., v. 18, pt. 3, p. 559–606.

Hoots, H. W., and S. C. Herold, 1935, Natural gas resources of California, *in* Geology of natural gas: Am. Assoc. Petroleum Geologists, p. 113–220.

Hornabrook, J. T., 1967, Seismic interpretation problems in the North Sea with special reference to the discovery well 48/6–1: Amsterdam, Elsevier Pub. Co., 7th World Petroleum Cong., México, Proc., v. 2, p. 837–856.

Hossin, A., 1964, Calcul de la porosité utile dans les grès argileux: Paris, Compagnie Française Pétroles Notes et Mém., no. 7, 94 p.

Hotchkiss, H., 1962, Petroleum developments in Middle East and adjacent countries in 1961: Am. Assoc. Petroleum Geologists Bull., v. 46, no. 7, p. 1241–1280.

Hourcq, V., 1966, Le bassin côtier congolais, *in* Sedimentary basins of the African coasts, pt. 1, Atlantic Coast: Paris, Assoc. African Geol. Surveys, p. 197–206.

Hriskevich, M. E., 1967, Middle Devonian reefs of the Rainbow region of northwestern Canada, exploration and exploitation: Amsterdam, Elsevier Pub. Co., 7th World Petroleum Cong., México, Proc., v. 3, p. 733–763.

Huang, T. K., C. C. Young, Y. C. Cheng, T. C. Chow, M. N. Bien, and W. P. Weng, 1947, Report on geological investigation of some oil-fields in Sinkiang: Nanking, Natl. Geol. Survey China (Geol. Survey China Mem.), Ser. A, no. 21, 118 p.

Hubbard, B., 1950, Coal as a possible petroleum source rock: Am. Assoc. Petroleum Geologists Bull., v. 34, no. 12, p. 2437–2351.

Hudnall, J. S., and R. W. Eaton, 1968. Geology of Woodbine Formation in East Texas oil field and related areas: Am. Assoc. Petroleum Geologists Bull., v. 52, no. 3, p. 532–533.

Hull, C. E., and H. R. Warman, 1970, Asmari oil fields of Iran: this volume, p. 428–437.

Hunt, J. M., 1967, The origin of petroleum in carbonate rocks, *in* Developments in sedimentology—9B, Carbonate rocks—physical and chemical aspects: Amsterdam, Elsevier Pub. Co., p. 225–251.

Ibrayev, B., 1970, Narusheniya skhemy razrabotki Uzenskogo mestorozhdeniya vedut k poteryam nefti: Moscow, Sotsialisticheskaya Industriya, March 21, p. 2.

Illing, V. C., ed., 1953, The world's oilfields; the Eastern Hemisphere, *in* The science of petroleum, v. 6, pt. 1: Oxford Univ. Press, 174 p.

Ingram, W. L., 1968, Long Beach [California] oil field: California Div. Oil and Gas, Summ. Operations, v. 54, no. 1, p. 5–15.

Internat. Oil Scouts Assoc., 1969, International oil and gas development, review of 1968: v. 39, pt. 2 (Production), 911 p.

Ion, D. C., 1967, The significance of world petroleum reserves: Amsterdam, Elsevier Pub. Co., 7th World Petroleum Cong., Mexico, Proc., v. 1B, p. 25–36.

———— S. Elder, and A. E. Pedder, 1951, The Agha Jari oilfield, south-west Persia: 3d World Petroleum Cong., The Hague, Proc., sec. 1, p. 162–186.

Iraq Petroleum Co. Ltd., 1956, Geological occurrence of oil and gas in Iraq, *in* Symposium sobre yacimientos de petróleo y gas: 20th Internat. Geol. Cong., México, v. 2, p. 73–101.

Irving, E., and W. A. Robertson, 1969. Test for polar wandering and some possible implications: Jour. Geophys. Research, v. 74, no. 4, p. 1026–1036.

Isacks, B., J. Oliver, and L. R. Sykes, 1968, Seismology and the new global tectonics: Jour. Geophys. Research, v. 73, no. 18, p. 5855–5900.

James, G. A., and J. G. Wynd, 1965, Stratigraphic nomenclature of Iranian oil consortium agreement area: Am. Assoc. Petroleum Geologists Bull., v. 49, no. 12, p. 2182–2245.

Jenkins, O. P., comp., 1943, Geologic formations and economic development of the oil and gas fields of California: California Bur. Mines Bull. 118, 773 pp.

Jones, A. G., ed., 1961, Reconnaissance geology of part of West Pakistan; a Colombo Plan cooperative project: Toronto, Hunting Surveys Corp., 550 p.

Jones, R. A., 1968, Successful workover technique and recompletion in the Bahrain field, *in* 2d AIME regional technical symposium: Dhahran, Soc. Petroleum Engineers, of AIME, Proc., p. 461–480.

Jonkers, E. W., 1968, Water injection, Maydan Mahzam field, offshore Qatar, *in* 2d AIME regional technical symposium: Dhahran, Soc. Petroleum Engineers of AIME, Proc., p. 415–425.

Kalinin, N. A., 1964, Results of exploration for oil and gas in the USSR for 1963 and main directions for 1964–1965: Geologiya Nefti i Gaza, v. 8, no. 2 (Engl. trans., *in* Petroleum Geology, McLean, Va., June 1969, p. 69–73).

Karogodin, Yu. N., 1968. Ob odnoy osobennosti razmeshcheniya neftyanykh i gazovykh gigantov v razreze osadochnogo chekhla Zapadnoy Sibiri (One feature of the vertical distribution of giant oil and gas pools in the sedimentary cover of western Siberia): Akad. Nauk SSSR Doklady, v. 183, no. 6, p. 1393–1395. (*in trans.*, Am. Geol. Inst., 1969, Acad. Sci. USSR Doklady, v. 183, p. 119–120).

Kas'yanov, M. V., 1966, Nomenklatura, korrelyatsiya i neftegazonosnost' peschanykh plastov v Srednem Priob'ye Zapadno-Sibirskoy nizmennosti: Geologiya Nefti i Gaza, no. 10, p. 25–32.

Kaufman, G. F., 1958, Petroleum developments in Far East in 1957: Am. Assoc. Petroleum Geologists Bull., v. 42, no. 7, p. 1709–1726.

Kay, M., 1951, North American geosynclines: Geol. Soc. America Mem. 48, 143 p.

Kazarinov, V. P., M. V. Kassyanov, U. K. Mironov, D. V. Nalivkin, L. I. Rovnin, N. N. Rostovtsev, N. G. Chochia, and A. A. Trofimuk, 1960, Geology and oil and gas content of the western Sibirian [*sic*] lowland, *in* Regional and structural problems in oil geology: 21st Internat. Geol. Cong., Copenhagen, Proc., pt. 11, p. 57–66.

Kelly, T. E., 1968, Gas accumulations in nonmarine strata, Cook Inlet basin, Alaska: Am. Assoc. Petroleum Geologists Mem. 9, v. 1, p. 49–64.

Kent, P. E., 1967, Progress of exploration in North Sea: Am. Assoc. Petroleum Geologists Bull., v. 51, no. 5, p. 731–741.

———— 1969, The geological framework of petroleum exploration in Europe and North Africa and the implications of the continental drift hypothesis, *in* The exploration for petroleum in Europe and North Africa: London, Inst. Petroleum, p. 3–17.

———— and P. J. Walmsley, 1970, North sea progress: Am. Assoc. Petroleum Geologists Bull., v. 54, no. 1, p. 168–181.

Kersting, C. C., 1964, Petroleum developments in Egypt, *in* Guidebook to the geology and archaeology of Egypt: Petroleum Exploration Soc. Libya, 6th Ann. Fld. Conf., p. 179–184.

Khafizov, F. Z., 1969, Nekotoryye osobennosti stroyeniya i neftenosnost' produktivnogo plasta B$_{VS}$ Nizhne-Vartovskogo svoda: Geologiya Nefti i Gaza, no. 2, p. 42–45.

Khalimov, E. M., 1965, O forme setki skvazhin dlya mnogoplastovykh neftyanykh mestorozhdeniy: Geologiya Nefti i Gaza, no. 4, p. 19–22.

Khanin, I. L., V. Ye. Gavura, and Yu. I. Dubov, 1969, Razrabotka produktivnykh plastov A$_3$ i A$_4$ Yakushkinskogo neftyanogo mestorozhdeniya: Neftyanoye Khozyaystvo, no. 11, p. 28–33.

Kiersznowski, S. F., 1968, The performance of the Khursaniyah Arab reservoir system, in 2d AIME regional technical symposium: Dhahran, Soc. Petroleum Engineers of AIME, Proc., p. 557–571.

King, P. B., comp., 1969a, Tectonic map of North America: U.S. Geol. Survey, 2 sheets, scale, 1: 5,000,000.

———— 1969b, The tectonics of North America—a discussion to accompany the tectonic map of North America: U.S. Geol. Survey Prof. Paper 628, 94 p.

King, R. E., 1965, Petroleum exploration and production in Europe in 1964: Am. Assoc. Petroleum Geologists Bull., v. 49, no. 8, p. 1176–1231.

———— 1966, Petroleum exploration and production in Europe in 1965: Am. Assoc. Petroleum Geologists Bull., v. 50, no. 8, p. 1625–1680.

———— 1967, Petroleum exploration and production in Europe in 1966: Am. Assoc. Petroleum Geologists Bull., v. 51, no. 8, p. 1512–1563.

———— 1968, Petroleum exploration and production in Europe in 1967: Am. Assoc. Petroleum Geologists Bull., v. 52, no. 8, p. 1439–1488.

———— 1969a, Petroleum exploration and production in Europe in 1968: Am. Assoc. Petroleum Geologists Bull., v. 53, no. 8, p. 1649–1699.

———— 1969b, Oil and gas exploration and production in the Soviet Union, in The exploration for petroleum in Europe and North Africa: London, Inst. Petroleum, p. 181–190.

Kinzekeyev, A. R., 1964, Effectiveness of using structural-exploration (core) drilling in Tataria: Geologiya Nefti i Gaza, v. 8, no. 2 (Engl. trans., in Petroleum Geology, McLean, Va., June 1969, p. 118–122).

Klemme, H. D., 1957, Regional geology of circum-Mediterranean region: Am. Assoc. Petroleum Geologists Bull., v. 42, no. 3, p. 477–512.

Klitochenko, I. F., and A. M. Paliy, 1964, Results of oil-gas exploration in the Ukrainian SSR for 1963: Geologiya Nefti i Gaza, v. 8, no. 4 (Engl. trans., in Petroleum Geology, McLean, Va., October 1969, p. 188–192).

Klubev, V. A., 1963, Morphogenetic classifications of oil-gas-bearing tectonic structures of the Volga-Ural region: Geologiya Nefti i Gaza, v. 7, no. 9 (Engl. trans., in Petroleum Geology, McLean, Va., October 1968, p. 511–517).

Knebel, G. M., and G. Rodríguez-Eraso, 1956, Habitat of some oil: Am. Assoc. Petroleum Geologists Bull., v. 40, no. 4, p. 547–561.

Koesoemadinata, R. P., 1969, Outline of geologic occurrence of oil in Tertiary basins of west Indonesia: Am. Assoc. Petroleum Geologists Bull., v. 53, no. 11, p. 2368–2376.

Kolganov, V. I., and L. G. Yugin, 1966, K metodike opredeleniya nizhnego predela poristosti i pronitsayemosti plastov: Geologiya Nefti i Gaza, no. 11, p. 50–53.

Komarov, A. I., 1969, Nekotoryye prichiny izmeneniya vodo-neftyanogo kontakta i smeshcheniya zalezhi nefti v gorizonte D$_1$ Romashkinskogo mestorozhdeniya: Geologiya Nefti i Gaza, no. 3, p. 37–42.

Konoplev, Yu. V., A. Kitsenko, and B. V. Kalichenko, 1965, Issledovaniye vozmozhnosti opredeleniya VNK v usloviyakh IV gorizonta Anastasiyevsko-Troitskogo mestorozhdenniya nefti metodom impur'skogo neytron-neytronnogo karotazha: Geologiya Nefti i Gaza, no. 4, p. 44–48.

Korshunov, E., 1968, Development of Tatar petroleum industry reviewed: Moscow, Ekonomicheskaya Gazeta, no. 34, p. 7–8 (in trans., Washington, D.C., U.S. Dept. Commerce, Joint Publ. Research Service, JPRS-46,559, p. 26–33).

Kortunov, A. K., 1967, Gazovaya promyshlennost' USSR: Moscow, Izd. "Nedra," 323 p.

Koshlyak, V. A., Yu. S. Vinitskiy, I. L. Zubik, M. A. Yunusov, and Yu. V. Golubev, 1969, O roli biogermnykh massivov v formirovanii strukturnogo plana podstilayushchikh otlozheny (Role of bioherms in the development of the structural plan of underlying deposits): Akad. Nauk SSSR Doklady, v. 185, no. 6, p. 1331–1334 (in trans., Am. Geol. Inst., 1969, Acad. Sci. USSR Doklady, v. 185, p. 66–69).

Kosov, B. M., 1962, Glavneyshiye itogi geologorazvedochnykh rabot v Sibiri i perspektivy ikh dal'neyskogo razvitiya (Principal results of geological exploration in Siberia and the prospects of its further development): Sovetskaya Geologiya, no. 4, p. 3–14 (in trans., Am. Geol. Inst., 1963, Internat. Geol. Review, v. 5, no. 10, p. 1290–1296).

Kozlov, A. L., and O. N. Aliyeva, 1966, Opredeleniye ob'yema produktivnoy chasti plasta pri malom kolichestve proburennykh skvazhin: Geologiya Nefti i Gaza, no. 12, p. 10–17.

Kozmodem'yanskiy, V. V., K. V. Kruchinin, I. G. Mikhaylenko, and I. A. Khafizov, 1969, O razmerakh i amplitude lokal'nykh struktur Yuzhnogo Mangyshlaka: Geologiya Nefti i Gaza, no. 5, p. 31–32.

Kranzler, I., 1966, Origin of oil in lower member of Tyler Formation of central Montana: Am. Assoc. Petroleum Geologists Bull., v. 50, no. 10, p. 2245–2249.

Krashin, P. A., 1963, Geology and oil productivity of the clastic Devonian sediments of the Novo-Yelkhov field in Tataria: Geologiya Nefti i Gaza, v. 7, no. 12 (Engl. trans., in Petroleum Geology, McLean, Va., April 1969, p. 686–689).

Kravchenko, K. N., 1965, Sec. 31, Gruppa zapadnokitaiskikh basseynov; sec. 32, Prednan'shanskiy basseyn; sec. 33, Gruppa basseynov vostoka i yugovostoka Kitaiskoy Narodnoy Respubliki, p. 286–326, in Brod, I. O., and I. V. Vysotskiy, eds., Neftegazonosnyye basseyny zemnogo shara: Moscow, Izd. "Nedra," 598 p.

———— 1968, Kitai, p. 321–349, in Vysotskiy, I. V., ed., Geologiya nefti, spravochnik, tom 2, kniga 2. Neftyanye mestorozhdeniya zarubezhnikh stran: Moscow, Izd. "Nedra," 804 p.

———— E. B. Movshovich, and B. A. Sokolov, 1969, Neftegazonosnyye basseyny i mestorozhdeniya Pakistana: Geologiya Nefti i Gaza, no. 2, p. 53–57.

Krivosheva, V. A., A. M. Sinichka, and S. Ye. Cherpak, 1966, Nekotoryye vzglyady na usloviya i vremya formirovaniya zalezhey nefti i gaza v Dneprovsko-Donetskoy vpadine: Geologiya Nefti i Gaza, no. 1, p. 40–44.

Krueger, M. L., 1968, Occurrence of natural gas in Green River basin, Wyoming, in Natural gases of North America: Am. Assoc. Petroleum Geologists

Mem. 9, v. 1, p. 780–797.

Krylov, A. P., 1969, New methods of working petroleum deposits: Akad. Nauk SSSR Vestnik, v. 39, no. 6, p. 78–89 (*in trans.*, Washington, D.C., U.S. Dept. Commerce, Joint Pub. Research Service, JPRS 48,823, p. 105–118).

Kudo, T., 1966, Chūgoku tairiku no sekiyu-shigen (Petroleum resources of continental China): Tokyo, Ajia Keizai Kenkyūjo (Inst. Asian Economic Affairs), Kaigaitōshi-sankōshiryō (Research materials on overseas investments), v. 9, 61 p. (in Japanese).

Kudryavtsev, N. A., 1963, Quantitative relationships between hydrocarbons of sedimentary rocks and oils: Geologiya Nefti i Gaza, v. 7, no. 9 (Engl. trans., *in* Petroleum Geology, McLean, Va., October 1968, p. 518–523).

Kulakhmetov, N. Kh., I. I. Nesterov, N. N. Rostovtsev, and M. Ya. Rudkevich, 1966, Perspektivy neftegazonosnosti melovykh otlozheniy Nadym-Purskogo mezhdurech'ya (Zapadno-Sibirskaya nizmennost'): Geologiya Nefti i Gaza, no. 10, p. 1–6.

Landes, K. K., 1967, Eometamorphism, and oil and gas in time and space: Am. Assoc. Petroleum Geologists Bull., v. 51, no. 6, p. 828–841.

Langton, J. R., and G. E. Chin, 1968, Rainbow Member facies and related reservoir properties, Rainbow Lake, Alberta: Am. Assoc. Petroleum Geologists Bull., v. 52, no. 10, p. 1925–1955.

Law, J., 1957, Reasons for Persian Gulf oil abundance: Am. Assoc. Petroleum Geologists Bull., v. 41, no. 1, p. 51–69.

Lees, G. M., 1951, The oilfields of the Middle East: 3d World Petroleum Cong., The Hague, Proc., General Volume, p. 94–101.

Lehner, P., 1969, Salt tectonics and Pleistocene stratigraphy on continental slope of northern Gulf of Mexico: Am. Assoc. Petroleum Geologists Bull., v. 53, no. 12, p. 2431–2489.

Leicester, P., 1966, North Sea geology and geophysics, *in* Proceedings—petroleum: Melbourne, 8th Commonwealth Mining Metal. Cong. Pub., v. 5, p. 65–75.

Levchenko, I. V., and Yu. K. Mironov, 1965, Novyye gazokondensatnyye mestorozhdeniya Tomskoy oblasti, svyazannyye s yurskimi otlozheniyami: Geologiya Nefti i Gaza, no. 7, p. 8–13.

Levorsen, A. I., 1929, Greater Seminole district, Seminole and Pottawatomie Counties, Oklahoma, *in* Structure of typical American oil fields, v. 2: Am. Assoc. Petroleum Geologists, p. 315–361.

Louisiana Dept. Conservation, 1969a, Annual oil and gas report for 1968: State of Louisiana, 189 p.

—— 1969b, Secondary recovery and pressure maintenance operations in Louisiana, 1968 report: State of Louisiana, 187 p.

L'vov, M. S., 1968, Resursy prirodnogo gaza SSSR: Moskova, Izd. "Nedra," 224 p.

L'vovskiy, Yu. M., 1963, Principles of tectonic regionalization in connection with exploration: Geologiya Nefti i Gaza, v. 7, no. 11 (Engl. trans., *in* Petroleum Geology, McLean, Va., February 1969, p. 617–623).

Magloire, P. R., 1970, Triassic gas field of Hass er R'Mel, Algeria: this volume, p. 489–501.

Markovskiy, N. I., 1963, Zakonomernost' razprostraneniya nefti i gaza v nizhnem karbone na vostoke Russkoy platformy (Trends in occurrence of oil and gas in Lower Carboniferous in eastern Russian platform): Sovetskaya Geologiya, no. 6, p. 75–93 (*in trans.*, Am. Geol. Inst., 1965, Internat. Geol. Review, v. 7, no. 5, pp. 803–815).

Martin, C. A., 1957, Moomba—a South Australian gas field: Australian Petroleum Exploration Assoc. Jour., v. 7, pt. II, p. 124–129.

Martin, R., 1966, Paleogeomorphology and its application to exploration for oil and gas (with examples from western Canada): Am. Assoc. Petroleum Geologists Bull., v. 50, no. 10, p. 2277–2311.

Martínez, A., 1970, Giant fields of Venezuela: this volume, p. 326–336.

Mason, J. W., 1968, Hugoton-Panhandle field, Kansas, Oklahoma and Texas: Am. Assoc. Petroleum Geologists Mem. 9, v. 2, p. 1539–1547.

Matveyev, V. G., and Yu. A. Volozh, 1965, Novyye perspektivnyye struktury na Yuzhnom Mangyshlake v rayone neftyanykh mestorozhdeniy Zhetybay i Uzen': Geologiya Nefti i Gaza, no. 2, p. 28–30.

Maxwell, J. C., 1966, The Mediterranean, ophiolites and continental drift: Princeton Univ., preprint, 23 p.

—— 1968, Continental drift and a dynamic earth: Am. Scientist, v. 56, no. 1, p. 35–51.

May, L. G., 1951, A discussion of petroleum engineering problems in the Burgan field, Kuwait: 3d World Petroleum Cong., The Hague, Proc., sec. 2, p. 697–700.

Mayuga, M. N., 1970, Geology and development of California's giant—Wilmington oil field: this volume, p. 158–184.

McCamis, J. G., and L. S. Griffith, 1968, Middle Devonian facies relations, Zama area, Alberta: Am. Assoc. Petroleum Geologists Bull., v. 52, no. 10, p. 1899–1924.

McGee, D. A., and W. W. Clawson, Jr., 1932, Geology and development of Oklahoma City field, Oklahoma County, Oklahoma: Am. Assoc. Petroleum Geologists Bull., v. 15, no. 10, p. 957–1020.

McMillen, R. E., 1963, The Maracaibo-Falcón basin, *in* Symposium on the petroleum geology of South America: Tulsa (Okla.) Geol. Soc. Digest, v. 31, p. 129–145.

Megateli, A., A. Said, and D. G. Sarber, 1969, Exploration in Algeria: past, present, future, *in* The exploration for petroleum in Europe and North Africa: London, Inst. Petroleum, p. 271–278.

Mel'nikov, N. V., ed., 1968, Toplivno-energeticheskiye resursy: Moscow, Izd. "Nauka," 631 p.

Menard, H. W., 1967, Transitional types of crusts under small ocean basins: Jour. Geophys. Research, v. 72, no. 12, p. 3061–3073.

Mencher, E., H. J. Fichter, H. H. Renz, W. E. Wallis, H. H. Renz, J. M. Patterson, and R. H. Robie, 1953, Geology of Venezuela and its oil fields: Am. Assoc. Petroleum Geologists Bull., v. 37, no. 4, p. 690–777.

Meyerhoff, A. A., ed., 1968, Geology of natural gas in south Louisiana, *in* Natural gases of North America: Am. Assoc. Petroleum Geologists Mem. 9, v. 1, p. 376–581.

—— 1970a, Continental drift: implications of paleomagnetic studies, meteorology, physical oceanography, and climatology: Jour. Geology, v. 78, no. 1, p. 1–51.

—— 1970b, Developments in Mainland China, 1949–1968: Am. Assoc. Petroleum Geologists Bull., v. 54, no. 8, p. 1567–1580.

Michaux, F. W., Jr., and E. O. Buck, 1936, Conroe oil field, Montgomery County, Texas, *in* Gulf Coast oil fields: Am. Assoc. Petroleum Geologists, p. 789–832.

Mikulenko, K. I., and Ye. A. Gaydeburova, 1966, K voprosu o sopostavlenii gazoneftenosnykh gorizontov nizhneyurskogo vozrasta Vilyuyskoy gemisineklizy i Predverkhoyanskogo progiba: Geologiya Nefti i Gaza, no. 2, p. 43–47.

Mills, B. A., 1968, Solution-gas reserves in Pembina Cardium oilfield, Alberta, Canada: Am. Assoc. Petroleum Geologists Mem. 9, v. 1, p. 698–704.

Mills, H. G., 1970, Geology of Tom O'Connor field, Refugio County, Texas: this volume, p. 292–300.

Milton, D. I., and C. C. S. Davies, 1965, Exploration and development of the Raudhatain field [Kuwait]: London, Inst. Petroleum Jour., v. 51, no. 493, p. 17–28.

Mina, P., M. T. Razaghnia, and Y. Paran, 1967, Geological and geophysical studies and exploratory drilling of the Iranian continental shelf—Persian Gulf: Amsterdam, Elsevier Pub. Co., 7th World Petroleum Cong., México, Proc., v. 2, p. 871–903.

Minato, M., S. Honjo, and M. Ishii, 1967, Electron microscopic study of oil bearing calcilutite, Khafji oil field, Neutral Zone: Amsterdam, Elsevier Pub. Co,. 7th World Petroleum Cong., México, Proc., v. 2, p. 447–456.

Minor, H. E., and M. A. Hanna, 1941, East Texas oil field, Rusk, Cherokee, Smith, Gregg, and Upshur Counties, Texas, *in* Stratigraphic type oil fields: Am. Assoc. Petroleum Geologists, p. 600–640.

Mirchink, M. F., 1964, Status of theory and practice of working oil fields: Geologiya Nefti i Gaza, v. 8, no. 6 (Engl. trans., *in* Petroleum Geology, McLean, Va., December 1969, p. 307–316).

———— V. A. Benenson, L. P. Dmitriyev, N. Ya. Kunin, and Yu. K. Yuferov, 1969, O zadachakh dal'neyshego razvitiya geologopoiskovykh rabot na nefti i gaz na Mangyshlake: Geologiya Nefti i Gaza, no. 4, p. 6–12.

———— and M. V. Feygin, 1966, Otnositel'no obespechennosti razvitiya dobychi nefti i gaza zapasami: Geologiya Nefti i Gaza, no. 8, p. 1-7.

———— and ———— 1967, Otsenka sovremennogo sostoyaniya zapasov nefti vysshikh kategoriy: Geologiya Nefti i Gaza, no. 11, p. 34–38.

———— N. A. Krylov, A. I. Letavin, L. I. Rovnin, V. P. Tokarev, M. V. Feygin, and Y. G. Ervie, 1967, New major oil- and gas-bearing areas of young platform of the USSR (western Siberia, Mangyshlak, etc.): Amsterdam, Elsevier Pub. Co., 7th World Petroleum Cong., México, Proc., v. 2, p. 101–108.

Mkrtchyan, M., 1966, Normirovaniye rentnykh platezhey v dobyvayushchey promyshlennosti: Planovoye Khozyaystvo, no. 10, p. 50–60.

Moody, J. D., J. W. Mooney, and J. Spivak, 1970, Giant oil fields of North America: this volume, p. 8–17.

———— and M. C. Parsons, 1963, Petroleum developments in Africa in 1962: Am. Assoc. Petroleum Geologists Bull., v. 47, no. 7, p. 1348–1396.

Moore, E. L., and J. A. Shields, 1952, Chimire field, Anzoátegui, Venezuela: Am. Assoc. Petroleum Geologists Bull., v. 36, no. 5, p. 857–877.

Moore, G. T., 1969, Interaction of rivers and oceans—Pleistocene petroleum potential: Am. Assoc. Petroleum Geologists Bull., v. 53, no. 12, p. 2421–2430.

Morgan, W. J., 1968, Rises, trenches, great faults, and crustal blocks: Jour. Geophys. Research, v. 73, no. 6, p. 1959–1982.

Mostofi, B., 1967, Progress in the Iranian oil industry during the past three years, *in* 3d Symposium on the development of petroleum resources of Asia and the Far East: U.N. ECAFE Mineral Resources Devel. Ser., no. 26, v. 1, p. 71–76.

———— and E. Frei, 1959, The main sedimentary basins of Iran and their oil prospects: 4th World Petroleum Cong., New York, Proc., sec. 1, p. 337–348.

Muir, J. M., 1936, Geology of the Tampico region, Mexico: Am. Assoc. Petroleum Geologists, 280 p.

Murray, G. E., 1957, Geological occurrence of hydrocarbons in Gulf coastal province of the United States: Gulf Coast Assoc. Geol. Socs. Trans., v. 7, p. 253–299.

———— 1965, Indigenous Precambrian petroleum?: Am. Assoc. Petroleum Geologists Bull., v. 49, no. 1, p. 3–21.

Nalivkin, D. V., ed., 1966, Geologicheskaya karta SSSR: Minist. Geol. SSSR, 2 sheets, scale, 1:7,500,000.

Nanz, R. H., Jr., 1954, Genesis of Oligocene sandstone reservoir, Seeligson field, Jim Wells and Kleburg Counties, Texas: Am. Assoc. Petroleum Geologists Bull., v. 38, no. 1, p. 96-117.

National Oil Scouts and Landmen's Assoc., 1958, Oil and gas field development in U.S. and Canada, review of 1957: Yearbook, v. 28, 1133 p.

National Science Foundation, 1969, Initial reports of the Deep Sea Drilling Project, volume I: Natl. Sci. Foundation, National Ocean Sediment Coring Program, NSFSP-1, 672 p.

Nelson, P. H., 1968, Wafra field, Kuwait-Saudi Arabia Neutral Zone, *in* 2d AIME regional technical symposium: Dhahran, Soc. Petroleum Engineers of AIME, Proc., p. 101–120.

Newby, J. B., P. D. Torrey, C. R. Fettke, and L. S. Panyity, 1929, Bradford oil field, McKean County, Pennsylvania, and Cattaraugus County, New York, *in* Structure of typical American oil fields, v. 2: Am. Assoc. Petroleum Geologists, p. 407–442.

Nikonov, V. F., 1964, Composition of the oils of the area of east-west trend of the River Ob: Geologiya Nefti i Gaza, v. 8, no. 1 (Engl. trans., *in* Petroleum Geology, McLean, Va., May 1969, p. 18–22).

———— 1966, Osnovnyye cherty geologicheskogo stroyeniya i neftenosnost' Ust'-Balykskogo mestorozhdeniya: Geologiya Nefti i Gaza, no. 2, p. 21–28.

Niz'yev, V. A., 1969, Novyye dannyye o glubinnom stroyenii i neftegazonosnosti shel'fa vostochnoy chasti Yuzhnogo Kaspiya: Geologiya Nefti i Gaza, no. 4, p. 15–19.

Ocamb, R. D., 1961, Growth faults of south Louisiana: Gulf Coast Assoc. Geol. Socs. Trans., v. 11, p. 139–175.

Oil and Gas Journal, 1969a, Russian gas reserves continue to soar: Oil and Gas Jour., v. 67, no. 26 (June 30), p. 70–72).

———— 1969b, Soviets building Siberia's first 48-in. pipeline: Oil and Gas Jour., v. 67, no. 38 (September 22), p. 125.

———— 1969c, Soviets expect oil output to pass 6,530,000-b/d target: Oil and Gas Jour., v. 67, no. 42 (October 20), p. 58.

———— 1969d, Two strikes set Red Sea to boiling: Oil and Gas Jour., v. 67, no. 50 (December 15), p. 31.

———— 1970a, Spacing rules set for big Prudhoe Bay: Oil and Gas Jour., v. 68, no. 3 (January 19), p. 36.

———— 1970b, Modest gains forecast for 1970: Oil and Gas Jour., v. 68, no. 4 (January 26), p. 113–139.

———— 1970c, Where are the reserves around the United States?: Oil and Gas Jour., v. 68, no. 4 (January 26), p. 128–139.

———— 1970d, Russia sees bigger natural gas market: Oil and Gas Jour., v. 68, no. 9 (March 2), p. 34–35, 38.

Ortynski, H. I., A. Perrodon, and C. de Lapparent,

1959, Esquisse paléographique et structurale des bassins du Sahara septentrional: 4th World Petroleum Cong., New York, Proc., sec. 1, p. 705–728.

Osborne, N., 1966, Petroleum geology of Australian New Guinea, *in* Proceedings—petroleum: Melbourne, 8th Commonwealth Mining Metal. Cong. Pub., v. 5, p. 99–121.

Osment, F. C., R. M. Morrow, and R. W. Craig, 1967, Petroleum geology and development of the Cook Inlet basin of Alaska: Amsterdam, Elsevier Pub. Co., 7th World Petroleum Cong., México, Proc., v. 2, p. 141–150.

Ovanesov, G. P., E. M. Khalimov, and M. S. Sayfullin, 1963, Status and ways of improving the working of the Arlan oil field: Geologiya Nefti i Gaza, v. 7, no. 10 (Engl. trans., *in* Petroleum Geology, McLean, Va., December 1968, p. 539–545).

———— I. A. Yakupov, and M. A. Kamaletdinov, 1963, Oil-gas prospects of the Zilair synclinorium: Geologiya Nefti i Gaza, v. 7, no. 12 (Engl. trans., *in* Petroleum Geology, McLean, Va., April 1969, p. 663–667).

Owen, R. M. S., and S. N. Nasr, 1958, Stratigraphy of the Kuwait-Bastra area, *in* Habitat of oil: Am. Assoc. Petroleum Geologists, p. 1252–1278.

Pakiser, L. C., and R. Robinson, 1966, Composition of the continental crust as estimated from seismic observations: Am. Geophys. Union Monogr. 10, p. 620–626.

P'ang, Chu-hsiang, and G. E. Ryabukhin, 1961, Geologicheskoye stroyeniye, vnutrennykh vpadin Tsentral-'noy Azii i ikh neftegazonosnost' (The geologic structure of the intermontane basins of Central Asia and the presence of oil and gas therein): Vysshikh Uchebnykh Zavedeniy Izv., Geologiya i Razvedka, no. 7, p. 3–21 (*in trans.*, Am. Geol. Inst., 1963, Internat. Geol. Review, v. 5, no. 8, p. 985–998).

Paran, Y., and J. G. Crichton, 1967, Highlights of exploration for petroleum in Iran 1956–1965, *in* 3d Symposium on the development of petroleum resources of Asia and the Far East: U. N. ECAFE Mineral Resources Devel. Ser., no. 26, v. 1, p. 215–224.

Paraschiv, D., and Gh. Olteanu, 1970, Oil fields in Mio-Pliocene zone of eastern Carpathians (District of Ploiești [Rumania]): this volume, p. 399–427.

Parker, J. M., 1967, Salt solution and subsidence structures, Wyoming, North Dakota, and Montana: Am. Assoc. Petroleum Geologists Bull., v. 51, no. 10, p. 1929–1947.

Pate, J. D., 1968, Laverne area, Beaver and Harper Counties, Oklahoma: Am. Assoc. Petroleum Geologists Mem. 9, v. 2, p. 1509–1524.

Patijn, R. J. H., 1964, Die Entstehung von Erdgas infolge der Nachinkohlung im Nordosten der Niederlande: Erdöl und Kohle-Erdgas-Petrochemie, Jahrg. 17, Hft. 1, p. 2–9.

Patterson, A. M., and A. A. Arneson, 1957, Geology of Pembina field, Alberta: Am. Assoc. Petroleum Geologists Bull., v. 41, no. 1, p. 937–949.

Pavlenko, P. T., and S. Ye. Cherpak, 1965, Osobennosti razvitiya strukturnykh elementov Glinsko-Rozbyshevskogo vala i zakonomernosti prostranstvennogo raspredeleniya v nikh zalezhey nefti i gaza: Geologiya Nefti i Gaza, no. 6, p. 8–12.

Pelevin, L. A., 1967, Bashkiriya—odin iz krupneyshikh neftedobyvayushchikh rayonov strany: Neftyanoye Khozyaystvo, no. 11, p. 27–32.

Perrodon, A., 1960, Aperçu des principaux résultats obtenus en Libye: Assoc. Français Tech. Pétrole, B,

no. 142, 16 p.

Pestrikov, A. S., 1963, Use of large potential probes for evaluating the productivity of carbonate sediments: Geologiya Nefti i Gaza, v. 7, no. 10 (Engl. trans., *in* Petroleum Geology, McLean, Va., December 1968, p. 595–597).

Peterson, J. A., 1966, Stratigraphic *vs.* structural controls on carbonate-mound hydrocarbon accumulation, Aneth area, Paradox basin: Am. Assoc. Petroleum Geologists Bull., v. 50, no. 10, p. 2068–2081.

Petroleum and Economic Digest, 1969, U.A.R.: a new producing area unveiled: Dublin, Internat. Economic Pub. Ltd., no. 1, Nov. 15–30, p. 18–19.

Petroleum Publishing Co., 1970, International Petroleum Encyclopedia: Tulsa, 400 p.

Philippi, H. W., 1965, On the depth, time and mechanism of petroleum migration: Geochim. et Cosmochim. Acta, v. 29, no. 9, p. 1021–1049.

Picard, M. D., 1957, Red Wash-Walker Hollow field, stratigraphic trap, eastern Uinta basin, Utah: Am. Assoc. Petroleum Geologists Bull., v. 41, no. 5, p. 923–936.

———— 1968, Outline of occurrence of Pennsylvanian gas in Four-Corners region: Am. Assoc. Petroleum Geologists Mem. 9, v. 2, p. 1327–1356.

Pickering, W. Y., and C. L. Dorn, 1948, Rangely oil field, Rio Blanco County, Colorado, *in* Structure of typical American oil fields, v. 3: Am. Assoc. Petroleum Geologists, 516 p.

Pippin, L., 1970, Panhandle-Hugoton field, Texas-Oklahoma-Kansas—the first fifty years: this volume, p. 204–222.

———— and L. Poling, 1961, Mocane-Laverne field, *in* Oil and gas fields of Texas and Oklahoma Panhandles: Amarillo, Panhandle Geol. Soc., p. 221–228.

Plotnikov, N. A., and N. N. Boltyshev, 1966, O metodike promyshlennoy razvedki neftyanykh zalezhey v verkhnemelovykh karbonatnykh treshchinnykh kollektorakh na territorii Checheno-Ingushskoy ASSR: Geologiya Nefti i Gaza, no. 5, p. 39–42.

Podobin, A. G., and V. B. Sergeyev, 1969, Metodika i rezul'taty primeneniya ispytateley plastov na razvedochnykh ploshchadyakh Tomskoy oblasti: Neftyanoye Khozyaystvo, no. 12, p. 55–58.

Polushkin, A. S., and V. P. Sanin, 1965, K voprosu o metodike promyshlennoy razvedki Megionskogo mestorozhdeniya: Geologiya Nefti i Gaza, no. 12, p. 16–21.

Popov, V. K., and T. S. Karmazina, 1966, Geofizicheskiye metody kontrolya za razrabotkoy Anastasiyevsko-Troitskogo gasoneftyanogo mestorozhdeniya: Geologiya Nefti i Gaza, no. 7, p. 7–12.

Porter, P., 1964, The Old Ocean field: Brazoria and Matagorda Counties, Texas: London, Inst. Petroleum Jour., v. 50, no. 492, p. 321–329.

Powers, R. W., 1962, Arabian Upper Jurassic carbonate reservoir rocks: Am. Assoc. Petroleum Geologists Mem. 1, p. 122–192.

Qatar Petroleum Co. Ltd., 1956, Symposium on the geological occurrence of oil and gas, *in* Symposium sobre yacimientos de petróleo y gas: 20th Internat. Geol. Cong., México, v. 2, p. 161–169.

Raaben, V. F., 1962, Nekotoryye dannyye v pol'zu vertikal'noy migratsii neftey v kamennougol'nyye i permskiye otlozheniya v predelakh Volgo-Ural'skoy oblasti (Vertical migration of petroleum into Carboniferous and Permian deposits of the Volga-Ural region): Akad. Nauk SSSR, ser. geol., Izv., no. 10, p. 90–96 (*in trans.*, Am. Geol. Inst., 1964, Internat. Geol. Review, v. 6, no. 8, p. 1416–1421).

——— 1963, Zakonomernosti izmeneniya svoystv neftey po stratigraficheskomu razrezu paleozoya v predelakh Volgo-Ural'skoy oblasti i metodika ikh izucheniya (Change in properties of crude oil throughout Paleozoic section of Volga-Urals region, and
method of their study): Sovetskaya Geologiya, no. 5,
p. 63–75 (in trans., Am. Geol. Inst., 1965, Internat.
Geol. Review, v. 7, no. 4, p. 681–688).

——— M. V. Dakhnova, and O. L. Nechayeva, 1967,
Zakonomernosti izmeneniya kachestvennoy kharak-
teristiki neftey i poputnykh gazov v predelakh
Volgo-Ural'skoy oblasti, in Buyalova, N. I., ed.,
Geologo-ekonomicheskiye issledovaniya poiskovo-
razvedochnykh rabot po nefti i gaza v SSSR: Moscow, Izd. "Nedra," p. 47–69.

Radchenko, O. A., 1965, Geokhimicheskiye zakonom-
ernosti razmeshcheniya neftenosnykh oblastei mira
(Geochemical regularities in the distribution of the
oil-bearing regions of the world): Leningrad, Izd.
"Nedra" (in trans., Israel Program for Sci. Trans.,
Jerusalem, 1968, in cooperation with U.S. Dept. In-
terior-NSF, Washington, D.C.: Springfield, Va., U.S.
Dept. Commerce, Clearinghouse for Federal Sci. and
Tech. Inf., 312 p.).

Rahman, H., 1964, Geology of petroleum in Pakistan:
6th World Petroleum Cong., Frankfurt am Main,
Proc., sec. 1, p. 659–683.

Remeyev, O. A., 1965, O zalezhakh nefti v yurskom
produktivnom plaste Shaimskogo neftenosnogo rayona: Geologiya Nefti i Gaza, no. 7, p. 13–16.

——— and K. W. Ostrowskaja, 1969, Urengoj—die
grösste Erdgaslagerstätte der Welt: Berlin, Zeitschr.
für angewandte Geologie, Bd. 15, Hft. 6, p. 297–300.

Rettger, R. E., J. B. Carsey, and J. E. Morero, 1935,
Natural gas in West Texas and southeast New Mexico, in Geology of natural gas: Am. Assoc. Petroleum Geologists, p. 417–457.

Reynolds, M. A., 1967, A comparison of the Otway
and Gippsland basins: Australian Petroleum Exploration Assoc. Jour., v. 7, pt. II, p. 50–58.

Richtofen, F. Freiherr von, 1912, Atlas von China,
zweite Abt., Das südliche China: Berlin, Dietrich
Reimer (Ernst Vohsen) Verlag, 54 pl.

Roberts, J. M., 1970, Amal field, Libya: this volume, p.
438–448.

Rocky Mountain Association Geologists, 1954, Oil and
gas fields of Colorado: 302 p.

Rodríguez-Eraso, G., 1955, Some geological problems
of the northern Maturín basin, Eastern Venezuela,
in Proceedings of the Conference on Latin-American geology: Austin, Univ. Texas, p. 64–82.

Rogers, R. E., 1968, Carthage field, Panola County,
Texas: Am. Assoc. Petroleum Geologists Mem. 9, v.
1, p. 1020–1059.

Rostovtsev, N. N., and I. I. Nesterov, 1965, Zapadno-
Sibirskaya nizmennost'—novaya neftyanaya baza
SSSR: Geologiya Nefti i Gaza, no. 7, p. 1–8.

Roswell Geological Society, 1956, The oil and gas
fields of southeastern New Mexico: 375 p.

Rovnin, L. I., 1964, Punga natural gas field in Tyumen'
region: Geologiya Nefti i Gaza, v. 8, no. 1 (Engl.
trans., in Petroleum Geology, McLean, Va., May
1969, p. 9–12).

Rubio, F. E., 1960, Condiciones de las acumulaciones
de petróleo en los campos costaneros del Distrito
Bolívar, Lago de Maracaibo: 3d Venezuelan Geol.
Cong. Mem., v. 3, p. 1009–1024.

Rudd, E. A., and M. F. Glaessner, 1967, Review of
Australian petroleum exploration, 1966: Australian
Petroleum Exploration Assoc. Jour., v. 7, pt. II, p.
11–15.

——— and H. R. Katz, 1968, Petroleum developments
in southwest Pacific region during 1967: Am. Assoc.
Petroleum Geologists Bull., v. 52, no. 8, p. 1592–
1603.

——— and ——— 1969, Petroleum developments in
southwest Pacific region during 1968: Am. Assoc.
Petroleum Geologists Bull., v. 53, no. 8, p. 1808–
1820.

——— and D. Kear, 1966, Petroleum developments in
southwest Pacific during 1965: Am. Assoc. Petroleum Geologists Bull., v. 50, no. 8, p. 1782–1791.

——— and ——— 1967, Petroleum developments in
southwest Pacific region during 1966: Am. Assoc.
Petroleum Geologists Bull., v. 51, no. 8, p. 1669–
1676.

Rudkevich, M. Ya., and S. I. Shishigin, 1965, O prirode neftyanykh mestorozhdeniy Shaimskogo rayona
v svyazi s izucheniyem kollektorov-rezervuarov: Geologiya Nefti i Gaza, no. 7, p. 17–21.

Rudkin, G. H., 1968, Natural gas in San Joaquin Valley, California: Am. Assoc. Petroleum Geologists
Mem. 9, v. 1, p. 113–134.

Ruiter, H. J. de, G. Van der Laan, and H. G. Udink,
1967, Development of the north Netherlands gas
discovery in Groningen: Amsterdam, Elsevier Pub.
Co., 7th World Petroleum Cong., México, Proc., v.
2, p. 221–229.

Salas, G. P., 1949, Geology and development of Poza
Rica oil field, Veracruz, Mexico: Am. Assoc. Petroleum Geologists Bull., v. 33, no. 8, p. 1385–1409.

Sands, J. M., 1929, Burbank field, Osage County, Oklahoma, in Structure of typical American oil fields, v.
1: Am. Assoc. Petroleum Geologists, p. 220–229.

Sanford, R. M., 1970, Sarir oil field, Libya—desert surprise: this volume, p. 449–476.

Sattarov, M. M., E. M. Timashev, and V. B. Sergeyev,
1969, Analiz i puti uluchsheniya sostoyaniya razrabotki Arlanskogo neftyanogo mestorozhdeniya: Nef-
tyanoe Khozyaystvo, no. 1, p. 24–29.

Savchenko, V. P., V. L. Vinogradov, and Yu. I. Ya-
kovlev, 1965, Lobovoy i tylovoy effekt i yego poisko-
voye znacheniye: Geologiya Nefti i Gaza, no. 7, p.
36–40.

Schaub, H. P., and A. Jackson, 1958, The northwestern
oil basin of Borneo, in Habitat of oil: Am. Assoc.
Petroleum Geologists, p. 1330–1336.

Schneider, W. T., 1943, Geology of Wasson field, Yoa-
kum and Gaines Counties, Texas: Am. Assoc. Petroleum Geologists Bull., v. 27, no. 4, p. 479–523.

Scruton, P. C., 1953, Deposition of evaporites: Am.
Assoc. Petroleum Geologists Bull., v. 37, no. 11, p.
2498–2512.

Shabad, T., 1969, Basic industrial resources of the
U.S.S.R.: New York, Columbia Univ. Press, 393 p.

Shashin, V. D., 1968, Neftedobyvayushchaya promyshlennost' SSSR 1917–1967: Moscow, Izd. "Nedra,"
350 p.

Shatsky, N. S., president, 1962 (1964), Carte tectonique internationale de l'Europe: Moscow, Internat.
Geol. Cong., Comm. Geol. Map of the World, Subcomm. Tectonic Map of the World, scale, 1:
2,500,000.

Shayevskiy, Yu. I., and Yu. I. Yuferov, 1963, Status of
working of Shkapovo field: Geologiya Nefti i Gaza,
v. 7, no. 10 (Engl. trans., in Petroleum Geology, Mc-
Lean, Va., December 1968, p. 550–556).

Shcherbina, B. Y., 1970: Sovetskaya Rossiya, February
28.

Shimkin, D. B., 1953, Minerals—a key to Soviet
power: Harvard Univ. Press, 452 p.

Short, K. C., and A. J. Stäuble, 1967, Outline of geology of Niger delta: Am. Assoc. Petroleum Geologists Bull., v. 51, no. 5, p. 761–779.

Shreveport Geol. Soc., 1947, Reference report for 1945, v. 1 and v. 2: 503 p.

Silverman, S. R., 1965, Migration and segregation of oil and gas: Am. Assoc. Petroleum Geologists Mem. 4, p. 53–65.

Sitter, L. U. de, 1964, Structural geology: 2d ed., New York, McGraw-Hill Book Co., 551 p.

Slinger, F. C. P., and J. G. Crichton, 1959, The geology and development of the Gachsaran field, southwest Iran: 4th World Petroleum Cong., New York, Proc., sec. 1, p. 349–376.

Smale, D., and N. A. Trueman, 1965, The mineralogy and petrology of the Permian sandstones at Gidgealpa, South Australia: Australian Petroleum Exploration Assoc. Jour. for 1965, p. 152–158.

Small, W. M., 1959, Thrust faults and ruptured folds in Romanian oil fields: Am. Assoc. Petroleum Geologists Bull., v. 43, no. 2, p. 453–471.

Smith, D. A., 1961, Geology of South Pass Block 27 field, offshore, Plaquemines Parish, Louisiana: Am. Assoc. Petroleum Geologists Bull., v. 45, no. 1, p. 51–71.

Smith, J. E., 1951, The Cretaceous limestone producing areas of the Mara and Maracaibo district, Venezuela: 3d World Petroleum Cong., The Hague, Proc., sec. 1, p. 56–72.

———— 1956, Basement reservoir of La Paz-Mara oil fields, western Venezuela: Am. Assoc. Petroleum Geologists, v. 40, no. 2, p. 380–385.

Sokolova, K. N., 1965, Effektivnost' gazokarotazhnykh rabot v razlichnykh Volgo-Ural'skoy neftegazonosnoy provintsii: Geologiya Nefti i Gaza, no. 12, no. 39–41.

Sorgenfrei, T., 1963, De franske olie- og gasfelter i Sahara: Dansk Geol. Foren. Meddel., Bd. 15, Hft. 2, p. 263–267.

———— 1969, A review of petroleum development in Scandinavia in The Exploration for petroleum in Europe and North Africa: London, Inst. Petroleum, p. 191–208.

———— and A. Buch, 1964, Deep tests in Denmark, 1935–1959: Danmarks Geol. Undersøgelse, r. III, no. 36, 146 p.

Sorokina, Ye. G., 1966, Litologiya i kollektorskiye svoystva produktivnogo plasta B-VIII Sosninsko-Sovetskogo mestorozhdeniya nefti v Zapadnoy Sibiri: Geologiya Nefti i Gaza, no. 2, p. 28–33.

———— 1969, Litologiya i kollektory produktivnogo plasta B₈ (Sredniy Valanzhin) tsentral 'noy i yugo-vostochnoy chasti Zapadno-Sibirskoy plity: Geologiya Nefti i Gaza, no. 1, p. 38–42.

Sovetskava Rossiya, 1970: January 13.

Sprigg, R. C., 1968, Review of Australian petroleum exploration, 1967: Australian Petroleum Exploration Assoc. Jour., v. 8, pt. II, p. 8–15.

Stäuble, A. J., and G. Milius, 1970, Geology of Groningen gas field, Netherlands: this volume, p. 359–369.

Stehli, F. G., 1964, Permian zoogeography and its bearing on climate, in Problems in palaeoclimatology: London, Interscience Publishers, p. 537–549.

Steineke, M., R. A. Bramkamp, and N. J. Sander, 1958, Stratigraphic relations of Arabian Jurassic oil, in Habitat of oil: Am. Assoc. Petroleum Geologists, p. 1294–1329.

Stephenson, M., 1951, The Cretaceous limestone producing areas of the Mara and Maracaibo districts, Venezuela: reservoir and production engineering: 3d World Petroleum Cong., The Hague, Proc., sec. 2, p. 665–682.

Stewart, W. J., 1969, The physical properties of Gippsland basin hydrocarbons: Australian Petroleum Exploration Assoc. Jour., v. 9, pt. 2, p. 149–152.

Stheeman, H. A., 1963, Petroleum development in the Netherlands, with special reference to the origin, subsurface migration and geological history of the country's oil and gas resources: Koninkl. Nederlandse Geol. Mijn. Gen. Verh., Geol. Ser., deel 21, pt. 1, p. 57–95.

———— and A. A. Thiadens, 1969, A history of the exploration for hydrocarbons within the territorial boundaries of the Netherlands, in The exploration for petroleum in Europe and North Africa: London, Inst. Petroleum, p. 259–269.

Stille, H., 1940, Einführung in den Bau Amerikas: Berlin, Gebrüder Borntraeger, 717 p.

Stöcklin, J., 1968, Structural history and tectonics of Iran: a review: Am. Assoc. Petroleum Geologists Bull., v. 52, no. 7, p. 1229–1258.

Stowell, C. C., 1967, West Bay-South Pass Block 24 area, Plaquemines Parish, Louisiana, in Oil and gas fields of southeast Louisiana, v. 2: New Orleans Geol. Soc., p. 179–189.

Stroganov, V. P., 1969, O vliyanii glinistoy pokryshki na raspolozheniye kontaktov gaz-voda i gaz-nefti na mnogoplastovykh mestorozhdeniyakh: Geologiya Nefti i Gaza, no. 1, p. 35–38.

Strubell, W., 1968, Über die Gewinnung und Verwendung von Erdöl im alten China: Erdöl und Kohle-Erdgas-Petrochemie, Bd. 21, Hft. 7, p. 435–436.

Swenson, R. E., 1967, Trap mechanics in Nisku Formation of northeast Montana: Am. Assoc. Petroleum Geologists Bull., v. 51, no. 10, p. 1948–1958.

Swesnick, R. M., 1950, Golden Trend of south-central Oklahoma: Am. Assoc. Petroleum Geologists Bull., v. 34, no. 3, p. 386–422.

Taff, J. A., 1934, Physical properties of petroleum in California, in Problems of petroleum geology: Am. Assoc. Petroleum Geologists, p. 177–234.

Tainsh, H. R., K. V. Stringer, and J. Azad, 1959, Major gas fields of West Pakistan: Am. Assoc. Petroleum Geologists Bull., v. 43, no. 11, p. 2675–2700.

Tal'-Virskiy, B. B., and Yu. F. Ivantsov, 1969, Chardzhouskoye podnyatiye—novyy vysokoperspektivnyy rayon Bukharo-Khivinskoy neftegazonosnoy oblasti: Geologiya Nefti i Gaza, no. 4, p. 12–14.

Tercier, J., 1948, Le flysch dans la sédimentation alpine: Eclogae Geol. Helvetiae, v. 40, no. 2, p. 163–198.

Terman, M. J., and C. C. Woo, 1967, Atlas of Asia and eastern Europe to support detection of underground nuclear testing; v. 2, tectonics: U.S. Geol. Survey, p. 1–9, 7 sheets, scale, 1:5,000,000.

Terry, C. E., and J. J. Williams, 1969, The Idris "A" bioherm and oilfield, Sirte basin, Libya—its commercial development, regional Palaeocene geologic setting and stratigraphy, in The exploration for petroleum in Europe and North Africa: London, Inst. Petroleum, p. 31–48.

Thralls, W. H., 1955, Ghawar oilfield, giant among giants: London, The Oil Forum, v. 9, no. 4, 121–123.

———— and R. C Hasson, 1956, Geology and oil resources of eastern Saudi Arabia, in Symposium sobre yacimientos de petróleo y gas: 20th Internat. Geol. Cong., México, p. 14–32.

Tkhostov, B. A., and V. S. Klyucharev, 1969, Ob effektivnosti izolyatsionnykh rabot pri obvodenii zalezhey v neodnorodnykh plastakh: Neftyanoye Khozyaystvo, no. 5, p. 29–32.

Trail, T. R., 1968, Some aspects of the Upper Cretaceous-lower Tertiary section in the Gippsland basin of Victoria: Australian Petroleum Exploration Assoc. Jour., v. 8, pt. II, p. 67–77.

Travis, R. B., 1953, La Brea-Pariñas field, northwestern Peru: Am. Assoc. Petroleum Geologists Bull., v. 37, no. 9, p. 2093–2118.

Trebin, G. F., A. V. Savinikhina, and Yu. V. Kapyrin, 1966, Plastovyye nefti mestorozhdeniya Uzen': Geologiya Nefti i Gaza, no. 5, p. 24–26.

Trofimuk, A. A., V. G. Vasil'yev, I. P. Karasev, S. P. Kosorotov, M. M. Mandel'baum, A. M. Mustafinov, and V. V. Samsonov, 1964, Main problems of prospecting the Markovo oil field in eastern Siberia: Geologiya Nefti i Gaza, v. 8, no. 1 (Engl. trans., in Petroleum Geology, McLean, Va., May 1969 p. 13–18).

—— V. S. Vyshemirskiy, A. M. Dmitriyev, V. V. Ryabov, O. P. Vyshemirskaya, I. A. Olli, and T. I. Shtatnova, 1969, O sravnitel'nom izuchenii gigantskikh mestorozhdeniy nefti s ispol'zovaniyem logiko-diskretnogo analiza: Geologiya Nefti i Gaza, no. 6, p. 17–19.

Tschopp, R. H., 1967, Development of the Fahud field: Amsterdam, Elsevier Pub. Co., 7th World Petroleum Cong., Mexico, Proc., v. 2, p. 243–250.

Umbgrove, J. H. F., 1942 (repr. 1947), The pulse of the earth: The Hague, Martinus Nijhoff, 179 p.

United Nations, ECAFE, 1963a, Proceedings of the second symposium on the development of petroleum resources of Asia and the Far East: U.N. ECAFE Mineral Resources Devel. Ser., no. 18, v. 1, 556 p.; v. 2, 511 p.

—— 1963b, Case histories of oil and gas fields in Asia and the Far East: U.N. ECAFE Mineral Resources Devel. Ser., no. 20, 161 p.

Us, Ye. M., and K. S. Kozhina, 1966, Ob opredelenii zony fil'tratsii burovogo rastvora v kollektory na mestorozhdeniyakh zapadnogo Predkavkaz'ya: Geologiya Nefti i Gaza, no. 5, p. 36–39.

Uspenskaya, N. Yu., 1967, Principles of oil and gas territories subdivisions and the classification of oil and gas accumulations: Amsterdam, Elsevier Pub. Co., 7th World Petroleum Cong., México, Proc., v. 2, p. 961–969.

Van Houten, F. B., 1969, Molasse facies: records of worldwide crustal stresses: Science, v. 166, no. 3912, p. 1506–1508.

Vasil'yev, V. G., ed., 1968a, Gazovyye mestorozhdeniya SSSR: spravochnik: Moscow, Izd. "Nedra," 687 p.

—— ed., 1968b, Geologiya nefti, tom 2, kniga 1. Neftyanyye mestorozhdeniya SSSR: Moscow, Izd. "Nedra," 687 pp.

—— 1968c, Razvitiye poiskovo-razvedochnykh rabot na gaz v SSSR: Gazovaya Promyshlennost', no. 3, p. 4–6.

Vaypoplin, Yu. V., and V. N. Kislyakov, 1966, Tektonicheskoye stroyeniye severnoy chasti Zapadno-Sibirskov nizmennosti: Geologiya Nefti i Gaza, no. 10, p. 6–10.

Vest, E. L., Jr., 1970, Oil fields of Pennsylvanian-Permian Horseshoe atoll, West Texas: this volume, p. 185–203.

Vigrass, L. W., 1968, Geology of Canadian heavy oil sands: Am. Assoc. Petroleum Geologists Bull., v. 52, no. 10, p. 1984–1999.

Viktorov, P. F., and I. G. Teterev, 1970, Osobennosti razrabotki i vozmozhnaya nefteotdacha zalezhey VI plasta Arlanskogo mestorozhdeniya: Neftyanoye Khozyaystvo, no. 1, p. 31–34.

Vine, F. J., 1966, Spreading of the ocean floor: new

evidence: Science, v. 154, no. 3755, p. 1405–1415.

—— 1968, Magnetic anomalies associated with mid-ocean ridges, in The history of the earth's crust: Princeton Univ. Press, p. 73–89.

—— and D. H. Matthews, 1963, Magnetic anomalies over oceanic ridges: Nature, v. 199, no. 4897, p. 947–949.

Viniegra, F., and C. Castillo-Tejero, 1970, Golden lane fields, Veracruz, Mexico: this volume, p. 309–325.

Vitrik, S. P., V. G. Dem'yanchuk, and P. F. Shpak, 1969, Perspektivy otkrytiya gazovykh i neftyanykh mestorozhdeniy v Vostochno-Ukrainskoy neftegazonosnoy oblasti: Gazovaya Promyshlennost', no. 5 (May), p. 8–11.

Wall Street Journal, 1969, Tenneco Inc. announces Red Sea gas discovery off Saudi Arabia coast: New York, Wall Street Jour., Dec. 10, p. 18.

Wallis, W. E., 1967, Offshore petroleum exploration, Gippsland and Bass basins—southeast Asia: Amsterdam, Elsevier Pub. Co., 7th World Petroleum Cong., México, Proc., v. 2, p. 783–791.

Walters, R. P., 1960, Surani, Rumania, anticline with two erosion-depleted, non-contemporaneous reservoirs: Am. Assoc. Petroleum Geologists Bull., v. 44, no. 10, p. 1638–1650.

Waring, W. W., and D. B. Layer, 1950, Devonian dolomitized reef, D3 reservoir, Leduc field, Alberta: Am. Assoc. Petroleum Geologists Bull., v. 34, no. 2, p. 295–312.

Weeda, J., 1958a, Oil basin, of east Borneo, in Habitat of oil: Am. Assoc. Petroleum Geologists, p. 1337–1346.

—— 1958b, Oil basin of east Java, in Habitat of oil: Am. Assoc. Petroleum Geologists, p. 1359–1364.

Weeks, L. G., 1948, Highlights on 1947 developments in foreign petroleum fields: Am. Assoc. Petroleum Geologists Bull., v. 32, no. 6, p. 1093–1160.

—— 1952, Factors of sedimentary basin development that control oil occurrence: Am. Assoc. Petroleum Geologists, v. 36, no. 11, p. 2071–2124.

—— 1958, Habitat of oil and some factors that control it, in Habitat of oil: Am. Assoc. Petroleum Geologists, p. 1–61.

—— 1965, World offshore petroleum resources: Am. Assoc. Petroleum Geologists Bull., v. 49, no. 10, p. 1680–1693.

—— and B. M. Hopkins, 1967, Geology and exploration of three Bass Straits basins, Australia: Am. Assoc. Petroleum Geologists Bull., v. 51, no. 5, p. 742–760.

Weller, J. M., 1944, Petroleum possibilities of Red basin of Szechuan Province, China: Am. Assoc. Petroleum Geologists Bull., v. 28, no. 10, p. 1430–1439.

Welte, D., 1965, Relation between petroleum and source rock: Am. Assoc. Petroleum Geologists Bull., v. 49, no. 12, p. 2246–2268.

Wendlandt, E. A., T. H. Shelby, Jr., and J. S. Bell, 1946, Hawkins field, Wood County, Texas: Am. Assoc. Petroleum Geologists Bull., v. 30, no. 11, p. 1830–1856.

Wennekers, J. H. L., 1958, South Sumatra basinal area, in Habitat of oil: Am. Assoc. Petroleum Geologists, p. 1347–1358.

Werenskiold, W., 1953, Geosynclines: Am. Geophys. Union Trans., v. 34, no. 5, p. 776.

West Texas Geological Society, 1966, Oil and gas fields of West Texas symposium: 397 p.

Wilford, G. E., 1961, The geology and mineral resources of Brunei and adjacent parts of Sarawak, with descriptions of Seria and Miri oilfields: Borneo Geol. Survey Mem. 10, 319 p.

Williams, G. K., R. M. Hopkins, Jr., and D. A. McNaughton, 1965, Pacoota reservoir rocks, Amadeus basin, N.T., Australia: Australian Petroleum Exploration Assoc. Jour. for 1965, p. 159–167.

———— and ———— 1966, Pacoota reservoir rocks, Amadeus basin, N.T., Australia, in Proceedings—petroleum: Melbourne, 8th Commonwealth Mining Metal. Cong. Pub., v. 5, p. 223–233.

Wilson, J. T., 1965, Evidence from ocean islands suggesting movements in the earth, in A symposium on continental drift: Royal Soc. London Philos. Trans., ser. A, v. 258, no. 1088, p. 145–167.

———— 1966, Did the Atlantic close and then re-open?: Nature, v. 211, no. 5050, p. 676–691.

Winnock, E., and Y. Pontalier, 1970, Lacq gas field, France: this volume, p. 370–387.

World Oil, 1969, Arctic challenge: v. 169, no. 7 (Dec.), p. 61–65.

Wyoming Geological Association, 1957, Wyoming oil and gas fields symposium: 484 p.; supplement, 1961, p. 485–579.

Yanshin, A. L., 1966, Tectonic map of Eurasia: 22d Internat. Geol. Cong., India, Comm. Geol. Map of the World, Subcomm. Tectonic Map of the World, Sci. Commun., p. 103–109, scale, 1:5,000,000.

Yaroslavov, B. R., 1969, Ob effektivnosti razlichnykh sistem zavodneniya na Nikolo-Berezovskoy ploshchadi Arlanskogo mestorozhdeniya: Neftyanoye Khozyaystvo, no. 4, p. 23–27.

Yenikeyev, P. N., and S. Kurmanov, 1966, Itogi geologorazvedochnykh rabot na nefti i gaz v Kazakhstane za 1965 i ikh zadachi na 1966 g.: Geologiya Nefti i Gaza, no. 3, p. 1–6.

Yerofeyev, V. F., 1969, O prirode teplovykh anomaliy Volgo-Ural'skogo neftegaznogo basseyna: Sovetskaya Geologiya, no. 5, p. 81–90.

Young, G. A., A. Bellizzia, H. H. Renz, F. W. Johnson, R. H. Robie, and J. Mas Vall, 1956, Geología de las cuencas sedimentarias de Venezuela y de sus campos petrolíferos, in Symposium sobre yacimientos de petróleo y gas: 20th Internat. Geol. Cong., México, v. 4, p. 161–322.

———— ———— and ———— 1956, Geología de las cuencas sedimentarias de Venezuela y de sus campos pertrolíferos: Caracas, Minist. Minas Hidrocarburos, Dir. Geología Bol., Pub. Esp. 2, 140 p.

Youngquist, W., 1958, Controls of oil occurrence in La Brea-Pariñas field, northern coastal Peru, in Habitat of oil: Am. Assoc. Petroleum Geologists, p. 696–720.

Zaydan, S., 1963, Geothermal characteristics of the Tuymazy oil field: Geologiya Nefti i Gaza, v. 7, no. 12 (Engl. trans., in Petroleum Geology, McLean, Va., April 1969, p. 663–667).

Zhelonkin, I., A. R. Kinzikeyev, and S. Kh. Aygistova, 1964, Variation in the main parameters of the oils of some fields of eastern Tataria and western Bashkiria: Geologiya Nefti i Gaza, v. 8, no. 3 (Engl. trans., in Petroleum Geology, McLean, Va., August 1969, p. 158–162).

Zubov, I. P., and V. A. Kirov, 1969, O metodike planirovaniya vosproizvodstva zapasov nefti i gaza: Geologiya Nefti i Gaza, no. 1, p. 57–59.

Title Index

Author Index

Keyword Index